W9-BRW-408

A CONCISE HISTORY OF THE WORLD

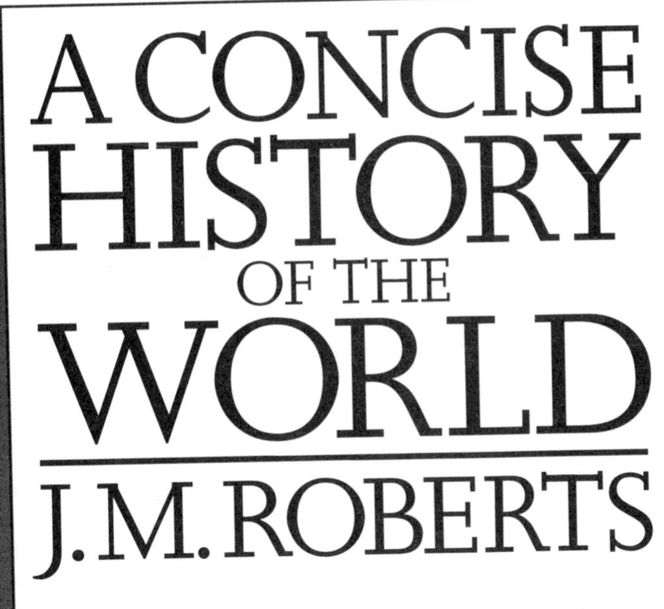

A CONCISE HISTORY OF THE WORLD

J. M. ROBERTS

NEW YORK
OXFORD UNIVERSITY PRESS
1995

First published in Great Britain 1993
by Helicon Publishing Ltd.,
42 Hythe Bridge Street, Oxford OX1 2EP

Copyright © 1993 J. M. Roberts

Maps and chronologies
copyright © 1993 Helicon Publishing Ltd.

Design by Peter Lawrence, Oxprint
Page layout and makeup by Roger Walker
Picture research by Jane Lewis

Published in the United States of America 1995 by
Oxford University Press, Inc.,
200 Madison Avenue, New York, New York 10016, U.S.A.

Oxford is a registered trademark of Oxford University Press

All rights reserved. No part of this publication may be reproduced, stored in a retrieval system, or transmitted, in any form or by any means, electronic, mechanical, photocopying, recording, or otherwise, without the prior permission of Oxford University Press.

Library of Congress Cataloging-in-Publication Data
available on request from the Publisher.

ISBN 0–19–521151–0

1 3 5 7 9 8 6 4 2

Printed in the United States of America
on acid-free paper

CONTENTS

CHAPTER 13: THE LATEST AGE: THE LONG RUN

CHAPTER 14: THE LATEST AGE: UPHEAVAL

CHAPTER 15: THE LATEST AGE – AN ERA OF UNREST

LIST OF MAPS

LIST OF CHRONOLOGIES

page

INTRODUCTION

Some years ago I wrote a *History of the World*. When my publisher suggested to me that I should go on to write a *Concise History of the World* (and one illustrated more profusely, too), I knew it would be much more challenging than writing the bigger book had been. This is not a condensation. You cannot just boil down general history, reducing pages to paragraphs, and paragraphs to sentences. Topics must be distributed differently, sometimes in different sequences if the main lines are to be brought out in a shorter space. Some things have to go altogether in any attempt to tell a story in briefer terms, and so some others will loom larger than before. That is as it should be; history is about what is important in answering the questions the historian wants to answer and however small his topic seems to be, and however long his book, he never can actually do more than say what is important and relevant from his point of view and in answering his questions.

Some readers may be surprised at what – or who – is or is not treated as important and relevant in the following pages. They may find an explanation in the prefaces I wrote to editions of this book's larger predecessor, where my approach was set out at some length. Equal time is not given here to every part of the world and every section of humanity. Though its direction and shape has often been settled by what large numbers of men and women did, there is nothing democratic about history. Big numbers are very important in deciding the ways things go when, for example, people move about in great migratory movements, set up patterns of farming or manufacture which shape the destinies of important areas of the planet, turn in large numbers to religious teachers and sects, or even to political leaders, and so on. But mere numbers do not make for historical importance. China has in historical times always contained a large proportion of the human race, but her impact on the rest of the world was for a long time virtually non-existent. Women make up approximately half the human race, but though they have sometimes exercised much power and influence in some places the overwhelming bias of most societies has been to give much more power and influence to men, so that claims can plausibly be made that history is the story of injustice and repression imposed upon the female sex by 'patriarchal' arrangements. It is not only what the numbers do which always matters, but what happens to them.

The most defensible and common sense way of identifying what is important in history is to ask how many people it affected in the end. Some ideas, some geographical facts, some discoveries, some societies, even some individuals, have changed the lives of hundreds, perhaps thousands, of millions. They determined that certain courses of action were open to humanity, others closed. This leaves room in so brief an account of world history for only a few personal names, it may be remarked: they are those of people who changed the possibilities open to their fellow-humans. This does not mean that the history of the millions of the unnamed, unknown, easily forgotten, is not worth studying. Everything in the past is worth study. A summary, though, registers the fact that those millions might have lived very different lives had the more influential humans not done things which changed the world – and all history is some kind of summary.

Because this book has evolved from a predecessor, I must once more express my thanks to all those who helped me and whom I have thanked in earlier prefaces. But two persons must now be given special recognition for their contribution to the present volume. I owe much gratitude to my UK publisher, David Attwooll, for suggesting this undertaking in the first place, and to my UK editor, Anne-Lucie Norton, for spurring an often-flagging author over a course more obstacle-ridden than either of us (I hope) had anticipated. Strangely, though I did not always feel it as work proceeded, I find myself now profoundly grateful to them both, and warmly thank them.

J.M.R.

A CONCISE HISTORY OF THE WORLD

1
PREHISTORY

BEGINNINGS

History is a word which traditionally means two different things: what happened, and a true account of what happened. In the second sense, it is always a selection from the past. Even the history of the whole world, though, is not a selection from all the past. We can ignore most of Time. We need not go back to the 'Big Bang' from which our universe emerged in order to understand where we are now. History is the story of human beings, and it is the human past which concerns us. Even when historians look at things beyond human control, such as geography, climate and, for immense tracts of time, disease, they do so only because that helps to explain why men and women lived and died in some ways rather than others. Or, to put it another way, history in the first sense – what happened – is what happened to and has been done by humans.

This cuts down the past we have to deal with by quite a lot, but still leaves an enormous amount to be tackled. Nor does it make it very clear where we are to begin. In theory, it could be with the first human being. But we do not know when or where he or she appeared, even if we can make responsible guesses within fairly broad limits. And, to make things more difficult, not everyone agrees about what sort of creatures in early times might be thought of as 'human'; and where the line between them and other animals is crossed.

A clear line is very hard to draw. People do not now talk (as they once did) about 'apemen' and 'missing links'. Physiology helps us to classify data but what we call 'human' is still a matter of a definition and disagreement is possible. What is surely and identifiably unique about humans is not just their possession of certain characteristics, but what they do with them. What human beings have always shown is a cumulative capacity such as no other species has ever shown to create change. They have made their own history – though, of course, within limits. Those limits are now very wide indeed, though once they were so narrow that we cannot pick out the first step which took human evolution away from the determination of nature. Human history began when the inheritance of genetics and behaviour which had until then provided the only means of survival was first broken through by conscious choice.

Once there was any power to break free, in however small a degree, from the determinism of nature, a huge difference had been made. From that point, human culture was progressive; it was increasingly built by deliberate selection within it as well as by accident and natural pressure, by the accumulation of a capital of experience and knowledge, and by exploiting it. That is really the place where our story should start, if we could identify it.

Evolution

Most biologists nowadays accept some form of a hypothesis which accounts for evolution by 'natural selection'. Not all do, but many would agree that evolution operates through environment, a particular environment favouring the survival of some genetic strains, and disfavouring others. Genetic messages which are favoured are carried forward into the next generation: those which are disfavoured are not, because the environment eradicates them before the genetic inheritance has been passed on. For example, certain mechanisms for conserving body heat exist in some species which enable them to go on living in cold climates – penguins, for example. Other species without those mechanisms cannot survive in those climates, and so they are only to be found elsewhere, where they can.

This may seem remote from the human story, but it is necessary to an understanding of its roots in a very distant past. Biological evolution long inched forward with incredible slowness within the possibilities offered by diverse habitats, at first to different organisms and later to different animals. The deciding factor in making these different habitats available was climate. About forty million years ago a long climatic phase began to draw to a close which had favoured great reptiles – of which the dinosaurs are the most celebrated. The world was growing colder. As new climatic conditions restricted their habitat, the great reptiles disappeared (some believe that other factors than climate were at work). But the new conditions suited other animal strains which were already about, among them some mammals whose tiny ancestors had appeared two hundred million years or so earlier. They now inherited the earth, or a considerable part of it. With many breaks in sequence and accidents of selection on the way these strains were themselves to evolve into the mammal families of today, among them our own.

This is by no means the end of the story of climate as an evolutionary selector. Although they took hundreds of millennia, even millions of years, to run their full course, huge swings of temperature still occurred. The resulting extremes, of freezing on the one hand and aridity on the other, choked off some possible lines of development. Conversely, in other times and places, the onset of appropriately benign conditions allowed some species to flourish and encouraged their spread into new habitats. In this immensely long process and even before the appearance of the creatures from which humanity was to evolve, climate was both setting the stage on which human history would take place and shaping, by selection, the eventual genetic inheritance of humanity itself.

Fifty-five million or so years ago, primitive mammals were of two main sorts. One, rodent-like, remained on the ground; the other took to the trees. The competition of the two families for resources was thus lessened. Strains of each survived to become eventually creatures we know today. The second group are now called the prosimians. We are among their descendants, for they were the ancestors of the first primates. What survived in the next phase of evolution were genetic strains among the prosimians which were best suited to the special uncertainties and accidental challenges of the forest.

The risky, sometimes sun-dappled forest environment put a premium on the capacity to learn. Strains prone to accident in such conditions were wiped out. Those whose genetic inheritance could respond and adapt to the surprising, sudden danger of deep shade, confused visual patterns, and treacherous handholds survived. Among those which prospered (genetically speaking) were some species with long digits which were to develop into fingers and the oppositional thumb which makes it easy to grasp a branch – and, later, a tool. Others evolved towards three-dimensional vision, a diminished sense of smell, and the characteristic anthropoid face. This story need not be followed further in details (it is not only highly complicated, but one still much open to debate among specialists). It is the necessary background, though, to the understanding of one of the outcomes, the main branch of the primate family to which humans belong, the hominids.

Hominids

The first apes and monkeys – then the most highly developed primates – appeared 25 million or so years ago. They had much bigger bodies and brains than the earlier primates. Among them evolved some strains particularly well-suited to the savannahs which, as a result of climatic change, were beginning to encroach on the forests. These creatures were better able to cope with new conditions than the 'pongids', as the ape/monkey families are termed. Some of them walked upright and could run in a way an ape cannot. From them evolved the evolutionary line called 'hominids'. Fossil remains suggest that between three and four million years ago we can begin to distinguish among them some who can be reckoned among our 'ancestors'. The earliest traces all come from south and east Africa. In what is now Tanzania, at a place called the Olduvai gorge, discoveries in the 1950s turned earlier ideas about the origins of mankind topsy-turvy. For the next forty years scholars went on extending by new discoveries the time-span within which to argue about human origins. There were to be many other important discoveries in Africa and the conclusions first drawn are now very much questioned. But the remains found in the Olduvai gorge are still the best starting-point for any account of what we know.

They were not at first sight very impressive. A deal of searching produced only a collection of bits and pieces which could all be got on to a large table. Over a thousand teeth were found and a dozen or so skulls, but only three hipbones and one shoulder-blade, and a scatter of other bones. Nonetheless, these were the relics of creatures of a type hitherto unknown. It was labelled 'Southern African ape' – *Australopithecus africanus* in the scientific classification system – and it looked as if these particular specimens were between two and three million years old.

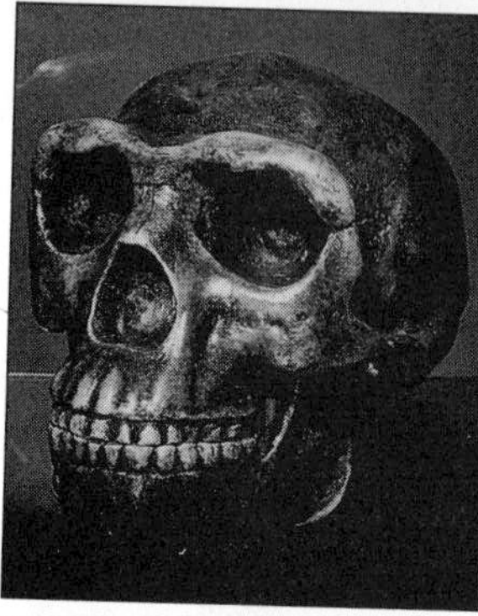

Australopithecus was about four feet tall and had hips, legs and feet more like those of humans than those of apes. The skull was fairly ape-like, though, held a brain about as big as that of a gorilla, and had a massive jaw. The forelimbs ended in true hands, with fingertips that flattened at the end like human fingers. What is more, some deductions about behaviour can be made; there was evidence of habitation in the same place for long periods and some scholars have also thought that Olduvai provides the first evidence of hominid building, in the shape of a windbreak of stones. Evidently, the creatures who lived there were meat-eaters, for there were smashed bones about which had been broken up so that their marrow and the brains from skulls could be got at. As there was no fire to cook them, they must have been eaten raw. The meat was probably not very attractive, because it is likely that it was carrion found by scavenging, but that it was eaten at all is very important. For one thing, meat is a concentrated food, so that those who eat it do not have to eat so continuously or often. For another, it presented technical problems; smashing bones and cutting off strips of skin or flesh cannot have been easy for creatures descended from vegetarians and therefore without the sharp teeth and claws of some mammals. They would need some help in solving these problems. That may explain one of the most exciting discoveries made at Olduvai: *Australopithecus* was a tool-maker.

Human qualities

Excavation revealed that different kinds of stone had been brought to Olduvai after being picked up elsewhere. So, they had been selected. By chipping flakes off them to give a jagged edge, they had then been turned into crude choppers which could be held in the hand – the first consciously made cutting tools. Here, about two million years ago, technology begins. Olduvai is the first landmark in the long, slow spreading across the world of purposefully made implements. They were to be refined, improved and enormously multiplied, but stone

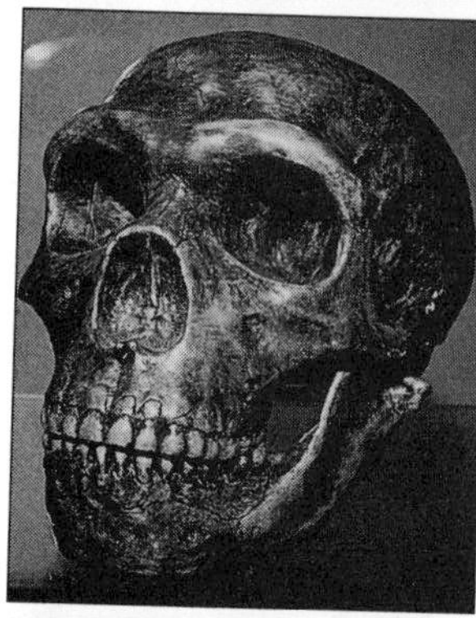

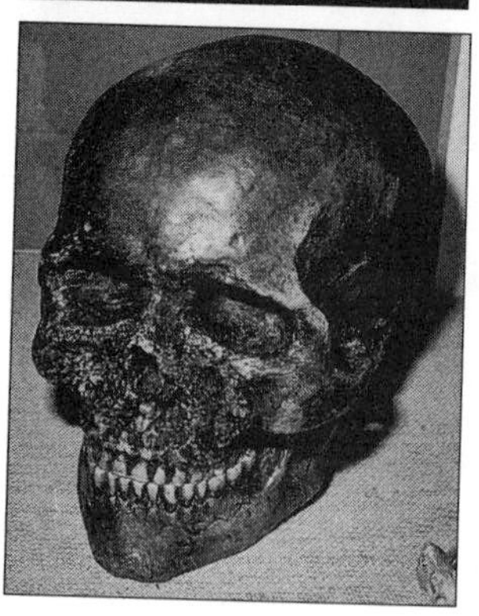

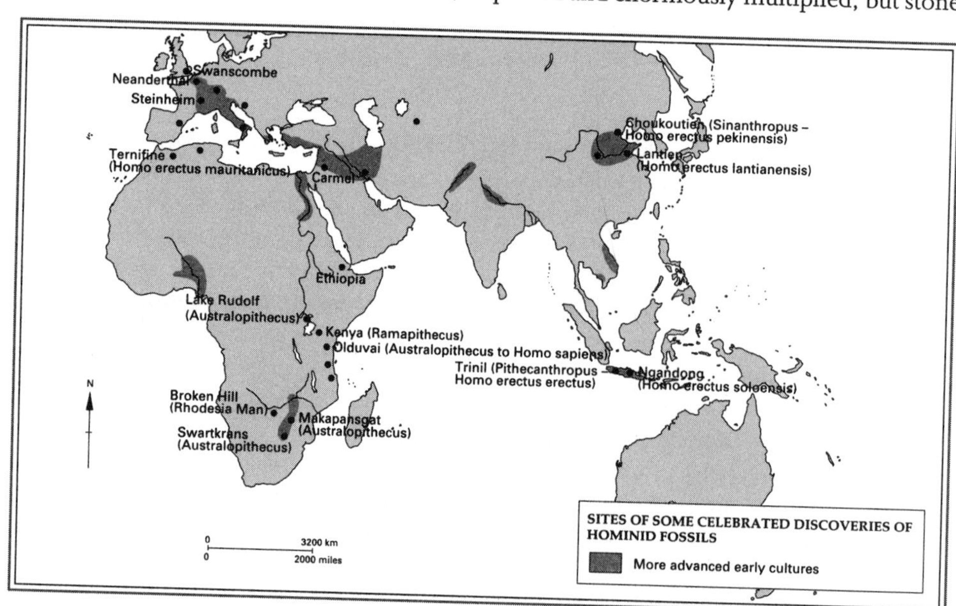

The Olduvai Gorge today. Once a great lake, it is now hard to imagine its appearance in primeval times, when there were deposited in its deepest layers relics which millions of years later would transform ideas of human origins.

tools provided for most of human prehistory and even for most of the time our own species has existed the main equipment for dealing with the environment.

Many problems about what was found at Olduvai remain unsettled; above all, that of deciding whether these creatures really are part of human ancestry at all. They are certainly hominids but they may belong to another strain than that from which we descend, only our remote cousins, so to speak, instead of our direct progenitors. Though they share some characteristics with later humans, they were also very different – in the shape of their skulls, for example, and their small stature. Some argue that there is a common 'ancestor' in some other strain, further back, from which both we and *Australopithecus* descended. But this debate is probably less important than another: where do they come on the scale of evolution? Have we really reached with them some elementary kind of human being? Another way of putting this would be to ask when prehistory begins and natural history is left behind. But such questions are largely about definitions: answering them depends on what you have chosen to regard as the essence of being human.

Some people have said it is a matter of language. But other animals make noises which are signals, or appear to gesture: when does this become language? As for building, ants, bees, birds and beavers do it very impressively. Another suggested criterion is tool-using. This, too, is not something which is absolutely confined to human beings. People have observed apes using sticks to poke at insects' nests or to help them to reach objects out of their grasp. Yet the very few higher primates other than humankind which have been seen to use tools do so only in a very simple way (as these examples suggest) and they do not seem to *make* tools with a sense of purpose and the idea that a future benefit will result; perhaps tool-making is a better test. Man is the only creature whom we know to think deliberately about possible changes in the environment and ways to produce them.

Facing page: (from top) Skull of Australopithecus Africanus; Reconstruction of skull of Homo Erectus; Skull of Neanderthal man; Reconstructed skull of Cro-Magnon man (Homo sapiens sapiens).

The Olduvai evidence suggests one reason for thinking that the early inhabitants of that site were different in kind from other primates. Since stones, and probably meat too, were brought there from some distances, and since it also seems likely that because of their physique the offspring of *Australopithecus* could not cling to their mothers for foraging expeditions as do those of other primates, we may have at Olduvai the first known example of creatures with a home base. Among the primates, only human beings have places where the females and young stay for long periods while the males search for food and bring it back to them. This has suggested something else about the mind of *Australopithecus*, too: however crudely and vaguely, these creatures seem to have been able to exercise some forethought, resisting the temptation to break up their food and eat it where it was found, and showing enough self-control to carry it home for consumption. The curbing of natural impulses with future satisfaction in view lies at the roots of everything humanity has achieved. It is the mechanism psychologists call 'inhibition'. This is the faint beginning of conscious planning.

But later discoveries have now complicated matters further. *Australopithecus* now seems to have existed in several species and elsewhere in East Africa evidence has recently been found of some other creatures living at the same time which seem even more like later humans, at least in physique. So australopithecines were not the only hominids on the scene two or three million years ago. Because they seem much more 'manlike' in physical terms, some of the other creatures about at the same time have therefore been given the family name of *Homo* (the Latin word for 'man'). One of them whose remains turned up a few years ago near Lake Rudolf in Kenya seems to have been about five feet tall and to have had a brain about twice the size of a modern chimpanzee.

It is not easy for laymen to pick their way among evidence that scholars argue fiercely about it. And we ought never to forget that we are for the most part at the mercy of what we are lucky enough to find. We do not have to believe that what was found at Olduvai was typical of the rest of the world, or even of East Africa, around 2,000,000 BC – it may just be what happens to have survived. The picture could be transformed if other evidence turned up. But at least it seems probable that the origins of mankind are to be sought in Africa, where three (and perhaps four) million years ago creatures of more than one species in some ways already like humans lived. Some of them – the australopithecines in particular – seem to have gone on to be very successful in biological terms; there is evidence that by about 1,000,000 BC they could be found all over the world outside the Americas and Australia.

Homo erectus

Though the story rapidly becomes very complicated after about 2,000,000 BC scholars have now become very good at making distinctions between different hominids. What matters for our purpose, though, is simply to note that the biological strain leading to humanity showed its superior efficiency by the fact of surviving at all. Across climatic ups and downs spanning millions of years, the survival and spread of species with human characteristics show that however feebly developed they seem to us, they adapted to changing conditions as many species could not. Though we are talking of climatic changes which took tens of thousands of years, this was rapid by comparison with the millions of years of much steadier conditions which lay in the past and they included what we call the 'Ice Ages', each lasting between fifty and a hundred thousand years. They covered big areas of the northern hemisphere (including much of Europe, and America as far south as modern New York) with great ice sheets, sometimes a mile or more thick. Scholars have now distinguished seventeen or nineteen (there is argument about the exact number) such glaciations since the onset of the first, over three million years ago. The most recent came to an end some ten thousand years ago.

They had huge impacts on life and evolution. The slow onset of the ice was decisive and sometimes disastrous for what lay in its path. Our landscapes are still shaped by its scourings and gougings thousands of centuries ago. The vast inundations which followed when the ice melted must in their turn also have been locally catastrophic, destroying the habitats of creatures which had adapted to the challenge of arctic conditions. Yet they also created new opportunities; new species could spread into the areas uncovered by the thaw. Possibly more important still for the global story of evolution, though, were the coolings and warmings which took place thousands of miles from the ice itself. New environments appeared; aridification and the spread of grassland both transformed the life-chances of existing species, some of which form part of the human evolutionary story. Africa was far from even the greatest ice-fields.

Climate can of course still be very important today, but only locally, and in a much shorter term; it no longer absolutely determines where the human species can survive. In prehistoric times it did, and could be offset only by slowly acquired physical and mental characteristics. Among the hominids, some strains were better equipped than others to grapple with nature and, in a measure, to master it. We can return to the human thumb, whose remote origins have already been mentioned. It is a much more useful and impressive instrument than we often remember. It can be brought together with the first finger to hold very small objects, and it can curl round the haft of a weapon or tool to give a good grip. It makes it much easier to go in for tool-making (as well as playing the piano, painting a picture, or hitch-hiking, of course). But the crucial human superiority to other animals was probably a bigger brain. Perhaps this was what allowed some hominids to take thought for the morrow. Together with the existence of the home base, that made survival easier. Hominid children take a long time to grow up and thoughtfulness and a home base made it more likely that they would survive to become parents in their turn. Some rest and possible recovery from sickness or accident also became possible. However it worked, the upshot is clear; somehow the species with the most 'human' characteristics were slowly screened somewhat from nature's harsh mechanism of evolutionary selection. Hitherto, nature had worked by eliminating genetic strains unable to adapt physically to environmental challenge. Once prudence, forethought and skill made it possible for some of them to avoid disaster, a new force was at work in selection – and it looks very much like what we call human intelligence. It is at this junction, as it were, that we begin to look for the station at which we get off for humanity. It provides the first signs of that positive, conscious impact upon environment which marks the earliest human achievements.

All such thoughts, though, are still very speculative. We must never forget that we have very thin evidence for much of prehistory. Europe, for example, has just one cave where tools similar to those of *Australopithecus* in Africa have been found and they date from just over a million years ago.

Not until we are well past that time, do we begin to have evidence of another important step in human evolution. After appearing about 800,000 BC, a creature of new physical type spread gradually over the whole 'Old World' (that is, the land-mass of Africa, Europe and Asia, which is now separated from the Americas and Australasia) by about 250,000 BC. Its earliest forms are called *Homo habilis* ('clever man') and later developments into another species have been called *Homo erectus* ('upright man').

The prehistoric archaeologist's material: the lower jaw of a specimen of Homo Habilis.

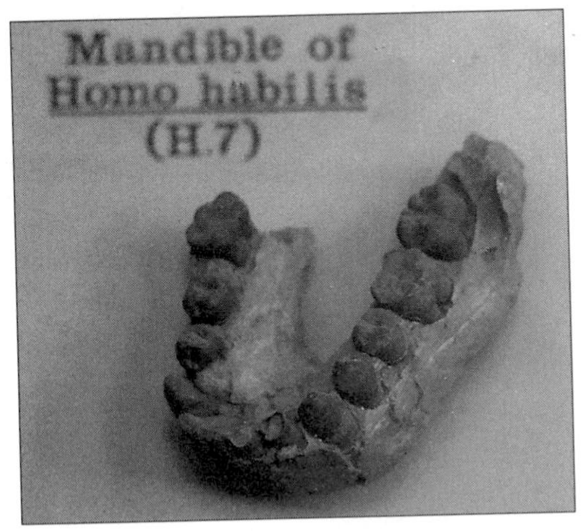

The crucial difference between *Homo erectus* and earlier manlike creatures is size – above all, size of brain. One strain among them developed a brain about twice the size of *Australopithecus* and much nearer that of modern man than the brain of any earlier hominid. There were other large-brained manlike creatures about and pieces of some of them have turned up as far away as China and Java, but it looks as if the type originated in Africa and then spread to Europe and most of Asia. Scholars have tried to measure how long this spread took. One theory is three or four thousand years, during which *Homo erectus* spread on the average at something like a mile a year northwards across what is now the Sahara and the Near East to France. Though the species has left no fossil remains in Europe, it lasted about ten times longer than has mankind as we know it. *Homo erectus* can be traced by special tools: the so-called stone 'hand axe' is the most important evidence. By plotting the places where it is found, scholars have discovered where the species flourished.

Hunting

In spite of the huge gaps in our knowledge, there can be no doubt that *Homo erectus* was a very successful species. Where did its bigger brains come from? Here are more mysteries, but it seems most likely that a change in diet explains the emergence of a new physical type. Eating more meat would have helped creatures with larger stature than the average to survive and reproduce. This may well be connected with the appearance of the first specialized skill: hunting.

The first meat-eaters seem either to have relied on small prey like reptiles and rodents or to have eaten the carrion found where bigger animals had died. Right at the beginning of the archaeological record, an elephant and perhaps giraffes and buffaloes helped provide meat consumed at Olduvai, but for long after this the bones of smaller animals vastly preponderate in the rubbish. Scavenging for them must have been unreliable, and changed into true hunting – big-game hunting, too – only very slowly. But change it did and the consequences were very great. Alongside a growth in the size of the animal remains found among the relics of primeval diet can be seen a parallel growth in the size of the skulls of the eaters. By about 300,000 BC elephants were being killed and cut up on the spot; the remains of large numbers of these (at one site, about fifty) have been found. Over the same period (and down to later times, too) the shape of the teeth and jaws of manlike creatures slowly evolves from those of predominantly vegetable-eating species. There is a chicken-and-egg aspect to this. The better diet to be got from organized hunting was only available to creatures already at least advanced enough to carry out such a complicated operation. This once more shows the speeding-up of evolutionary change. An enormous new range of capacities and skills comes into being somewhere between 1,000,000 and 100,000 BC, and they made the first human societies possible.

A NEW GRASP OF THE WORLD

Before big-game hunting could become possible someone had to know a lot about animals' habits and to be able to pass this knowledge on to others, both those engaged in the cooperative enterprise of hunting, and from generation to generation. Some kind of speech, therefore, must have existed. It has been argued that genetic selection which led to changes in the shape of the brain favoured the development of language. How *Australopithecus* communicated may never be known, but even lower primates have ways of doing so. Perhaps the first steps in the organization of language would have been the breaking up of calls like those of other animals into particular sounds capable of re-arrangement. This would make different messages possible. But other changes may have helped, too. Better vision, a growing sense of the world as lots of separate objects, and the making of new things (tools) were all going on simultaneously over hundreds of thousands of years in which language was evolving. Together they slowly made possible the coming of abstract thought (thinking about things not actually present). And this trend must have been reinforced when hunting made record and memory more important still.

A big range of techniques and skills was also needed for hunting. To trap and kill such monsters as the mammoth or woolly rhinoceros with the help only of weapons of stone or wood was enormously difficult. Numbers and discipline were needed to drive them to a killing-ground favourable because of a bog in which a weighty creature would flounder, or because it offered good vantage points or secure platforms to the hunters. Once dead, the victims presented further problems. With nothing but wood and stone tools, they had to be cut up. The meat had then to be carried home.

Bigger meat supplies were a tiny step towards a little leisure; the consumers were released temporarily from the drudgery of ceaseless rummaging about for small, but continuously available, dollops of nourishment and had time to do something else. They could add to the

Examples from our best line of evidence for a million years or so of men and manlike creatures – their technology. The scraper on the left is from Libya, the hand axe on the right from Hampshire: thousands of years separate their manufacture. They are both very different and very much the same, speaking for the same prolonged phase of human development, the hundreds of millenia during which different traditions of shaping and working stone are our best guide to the variety of what was going forward.

existing technology. Besides making more elaborate, bifacial hand-axes, *homo erectus* has also left the first certain evidence of constructed dwellings, the earliest worked wood, the first wooden spear and the earliest container, a wooden bowl. Inventing things on this scale shows a pace of development and a capacity quite different from what has gone before. Minds were forming ideas of objects before manufacture was begun. Some have even argued that simple forms (triangles, ellipses and ovals) used in huge numbers of stone tools, can be seen as the first beginnings of art – the production of objects giving pleasure because of their form, as well as being useful.

The coming of fire

The greatest of the technical and cultural advances of *homo erectus* was to learn to manage fire. The earliest evidence of its use comes from China (c.600,000 BC) but it does not show that fire could be made. Probably *homo erectus* never got so far. Still, that the species could make use of it was an enormous gain, the most important single change in technology before the coming of agriculture. It was the first chemical tapping of energy other than by the conversion of food inside the body.

Many peoples have had legends of heroic figures or magical beasts who first seize fire, often from the gods. Perhaps this reflects a dim memory that the first fire was taken from a natural source, whether from volcanic activity, an outbreak of natural gas, or a blazing forest. However it was obtained, the use of fire was revolutionary – though we must remember that it took hundreds of thousands of years to develop its full impact. Immediately, it meant warmth and light, the conquest of the cold and the dark and therefore the extension of the habitable environment into them, even if only a little way at first. Families could survive in colder regions than before, and could live in temperate zones with a little more ease. By occupying caves whose darkness had previously made them unusable, they were safer from the weather. Animals could now be driven out of their lairs and kept out (perhaps this was how the idea came of using fire to drive big game in hunting). Wooden spears could be hardened in fires. Cooking became possible. As a result, eating became easier; marrow can be sucked out of cooked bones but getting it out raw is a laborious business. Gibbons and gorillas have to spend much of their time simply chewing their raw food; cooking saved time, for food softened by it did not have to be chewed so long. Time was thus made available to do other things. More important still, substances indigestible in their raw state could become sources of food; distasteful or bitter plants could be made edible. This must have increased food supply (and therefore made population growth a little easier). It may also have stimulated attention to the variety and availability of plant life and so have launched the science of botany and the art of cookery. Finally, in the long run, eating cooked food helped to alter the shape of the face and the form of the teeth.

Cooking would have encouraged further restraint on immediate impulses, too: you put off eating and did not give way to immediate appetite by swallowing raw food. The focus of the cooking fire as a source of light and warmth would have brought people together around it after dark and helped to make a group more aware of itself as a community. They would have talked somehow: the development of language – of whose origins we know little – must have speeded up in this setting. Finally, fire slowly brought new distinctions between members of the group. At some point, fire-bearers and fire specialists appeared, beings of awesome and mysterious importance, for on them might depend life and death for the rest of the group. They carried and guarded the great liberating tool, and controlled its power to break up the iron rigidity and discipline of night and day and even that of the seasons.

During the age of *Homo erectus*, then, fire had already a little offset the pressures of the great external rhythms of the natural world on hominid life. Life was already less dominated by routine and less automatic than it had been for *Australopithecus*; it was now far removed from that of animals merely programmed by instinct and genetic endowment. *Homo erectus* could make choices. This is the best ground for saying that with this species we are already on the human side of any definition of the difference between apes and men, however cramped and miserable its life may appear.

Early society

It is easier to form a picture of the material circumstances of ancient life than of what was going on inside the bigger brains now grappling with those circumstances. But looking at the material remains is really all that we can hope to do. That can tell us something. It is worth thinking once more about big game hunting. Once *Homo erectus* became dependent on meat, he became a parasite on the herds of game – and therefore had to follow them about, or explore new territory in search of them – and was more likely to settle and multiply in some places than others. There home bases would be established. Some of them seem to have been occupied for thousands and thousands of years.

The family which lived in the home base was developing, too. The existence of that base already made it likely that the future human family would be very different from animal families. This became clearer as the predecessors of *Homo sapiens* grew bigger. The larger heads required to accommodate bigger brains, for instance, meant both that children would be bigger before birth – and this was reflected in changes in the female pelvis which permitted the birth of offspring with larger heads – and also that a longer period of growth after birth was needed for children to mature. No physiological change in the female could provide ante-natal accommodation to protect them until physical maturity. Human children, in consequence, unlike most mammals, whose offspring mature within months, need maternal care long after birth. Prolonged infancy, dependence and the support of children by the family and society during immaturity meant that human families developed in a way very unlike the families of other animals. Part of this was a result of genetic selection; large litters ceased to be the way in which the survival of the species was assured. Instead, human societies have learnt to give more and longer attention to the protection, nurture and training of their young (now often running on into their twenties). Sharper differences between the patterns of life of males and females appeared. Hominid mothers were very much more tied down than the mothers of other primates and fathers became more involved in the provision of food by hunting, which demanded arduous and prolonged activity in which the females could not easily join.

Another result of prolonged infancy was that learning and memory became more and more important. Here, too, we seem to cross a line with *Homo erectus*. The genetic programming of humanity's predecessors is somehow replaced by conscious learning about the environment and reflexion on it. Somewhere a great change has occurred in which tradition and culture – the things the members of a community learn from one another – take over from physiological inheritance as a factor in evolutionary selection, though we may never actually be able to say exactly where this change occurs.

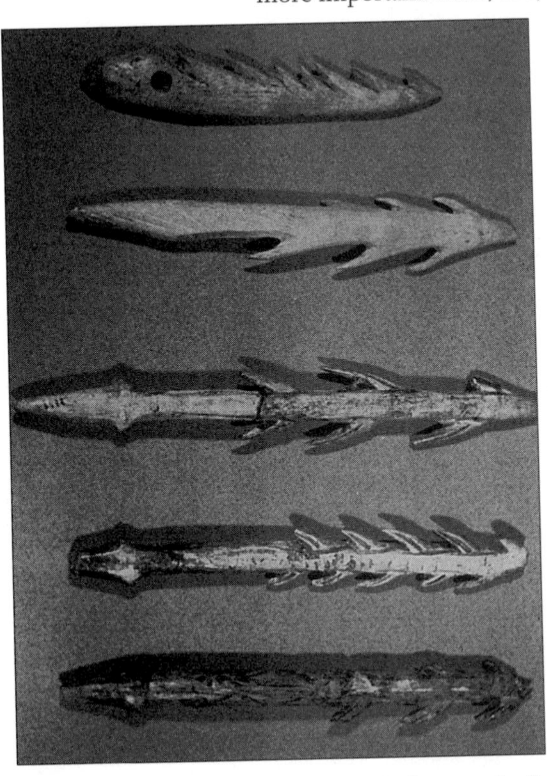

An evolving technology in the late prehistoric period (c. 3000 BC); stone and flint implements then had added to them others based on new materials and new skills, in this case the working of bone to make harpoon tips for fishing.

Of course the physiological inheritance still remains very important. Clearly it mattered greatly to the future shape of human society, for instance, that one particular genetic strain long ago gave our species a unique sexual characteristic. Among all other mammals, both the sexual attraction exercised by the female on the male and her fertility are restricted to certain periods. We speak of animals as being 'on heat' at these times, and when they are in this condition their lives are very much disrupted. If they had then to look after young, they could not possibly continue to nurture them. Human females do not work like this and this is very important. If they were like other animals, their slowly maturing offspring would have often been neglected in infancy and could hardly have survived. It may have taken a million years or so for a genetic strain with a sexual characteristic which dispensed with 'oestrus', as it is called, to emerge, but once it did so, the consequences for the future development of humanity were enormous and were to affect much else we take for granted about the way we live. The fact that human females were continuously attractive to human males (and not only so at periods when each sex was being regulated by automatic mechanisms of attraction) must have made individual choice a much more important factor in mating. This is the beginning of a very long and obscure road which leads to later notions of sexual love. Together with the longer childhood and greater dependency made possible by better food-gathering, it points ahead also to the stable and enduring human family unit – father, mother and offspring – who remain together and constitute a real community. This is an institution which only humans have, even if not every human family is like that and modern social arrangements often now tell against it.

Once more, though, we are much at the mercy of guesswork. We ought to be very cautious. We can say little that is certain about the social life of our predecessors. Yet it is difficult not to feel that (however slowly) what happened to early hominids laid down the main lines of much of human life long before humans like ourselves existed. Culture and tradition gradually took over from genetic mutation and natural selection as the primary source of change among the hominids – or, to put it another way, what was learnt was becoming as important for survival as what was biologically inherited. The groups with the best 'memories' of effective ways of doing things and increasing power to reflect upon them would have carried forward human evolution fastest. Attempts have been made to trace its progress in the actual physiological evolution of the cortex, the mass of cells making up the brain itself.

Yet when we try to think about prehistoric mental processes, we can be sure of little except that they were almost inconceivably different from our own. All we can say is that the life of *homo erectus* looks more like human, than pre-human, life. Physically, the creature's brain was of an order of magnitude comparable to our own, even if its skull was somewhat differently shaped. *Homo erectus* made tools in different styles in different places, built shelters, took over natural shelter by exploiting fire, and sallied forth from it to hunt and gather food. It did this in groups showing some discipline and able to transmit ideas by speech, and founded on the home base and a distinction between the activities of males and females. There may even have been other specializations; fire-bearers, or old creatures whose memories made them the data banks of their 'societies', could in some measure be supported by the labour of others.

There is nothing to be gained by looking, though, for some prehistoric dividing line. The evidence for one does not exist. All we can confidently say is that things had turned out in a certain way. When one or more sub-species of *Homo erectus* evolved at last into *Homo sapiens*, a new physical type, a big achievement and heritage was already pretty secure in their grasp. Individuals come naked into the world, but humanity did not. It bore with it from the past all, indeed, that made it humanity.

HOMO SAPIENS

On the scale of prehistoric time our own species, *Homo sapiens*, has not been about on earth for very long. *Homo erectus* exploited the world successfully about ten times longer than we have done. But there are no firm lines and for thousands of years many different species of hominids must have lived on earth, the end of one branch of the family overlapping by many generations the early generations of another. This is one reason why it is difficult to date the next important hominid species, beyond saying that it flourished somewhere between 250,000 and 50,000 BC. By the last date it was about to give way to human beings recognizably like ourselves.

The Neanderthals

The story begins with the discovery in Europe of two skulls about a quarter of a million years old. They were of a different shape from those of *Homo erectus* and had yet bigger brains. Still, they were not much like those of modern humans and they do not tell us much more than that something other than *Homo erectus* already existed. Then an Ice Age descends and no more relics of them or of other hominids turn up until fossils appear from about 100,000 BC, in the middle of the next warm period. These show a big step forward, for they are the first traces within the family *Homo* of the species called *sapiens*. Strictly, it is called *Homo sapiens neandertalensis*, being thus labelled after the place where the first skull announcing its existence was found – the Neanderthal valley in Germany, not far from Düsseldorf. Neanderthals would have looked somewhat unlike most modern human beings, not having much of a chin and being very thick-set in build. Indeed, the first skull found was so oddly shaped that it was thought by some scientists to be the skull of a modern idiot. Some modern people may somewhat resemble Neanderthals; we need not exaggerate the differences, and we know they were right-handed, like most modern humans, so there are similarities which go deep. The crucial difference from earlier hominids is their bigger brains.

Because the Neanderthals cannot easily be linked to earlier hominids, some people have been tempted to think they must have been in some way cut off from them, perhaps by the ice. As more evidence has turned up, though, it has become clear that they were very widely spread, from western Europe and Morocco at one extreme, to China at the other, with other settlements in the Near East, Iraq and Iran. The earliest Neanderthal in China may indeed go back to 200,000 BC. At all events, they were certainly living over much of Europe and Asia in about 80,000 BC, just before the onset of another Ice Age (and one of the worst). Though they look physically primitive to modern eyes, the Neanderthals were very successful, much

Stone tools were refined as skills improved. These, from Neanderthal sites, show (left) the stone 'core' from which 'flakes' to make individual tools have been struck off, (centre) a side-edge scraper, and (right) a pointed axe.

more capable and mentally more advanced than anything which had appeared on earth before them. As the ice returned, they did not go south but did their best to cope with the cold, moving into caves for protection and using fires kept going in hearths dug in their floors. Caves can hardly have been very comfortable places. Besides being smoky from the fire (which probably burnt animal fat and bones as wood grew scarcer in the new tundras produced by the glaciation), they would have become smelly from the rubbish on the floor, and damp at the back, where the fire's heat would not penetrate. But caves made it easier to survive. Those chosen nearly always looked south so as to assure as much sunlight as possible, and may have had skin curtains at their mouths in the winter months.

This was putting technology to use in a new way. The Neanderthals wore skins, too, and made better-shaped stone tools than earlier tool-makers. The growth in power over nature suggests that they had complex languages, though some scientists have argued that their skulls suggest that those parts of the brain which deal with speech are not so well developed as other parts. Yet they must have had some very advanced ideas – and therefore words in which to express them because they also did something else quite new: they buried their dead. Some graves show signs of careful attention. Near Samarkand the body of a Neanderthal child was buried inside a ring of animal horns; another grave, in Iraq, contained a man's body surrounded with masses of wild flowers and grasses placed there before he was buried. Such things are important. Not only are they the beginning of a long and rich line of evidence about the past obtained by digging up graves, but because they suggest an enormous important change in thinking. Why, after all, did the Neanderthals bury their dead?

The exact answer we do not and probably cannot know, but perhaps some Neanderthals were beginning to experiment with ritual. They may have sought to control nature by carrying out certain acts in order to make things happen. Perhaps, even, these burials and the faint traces of rituals involving animals which are to be found on other Neanderthal sites mark the beginnings of religion. Someone may have already begun to believe in another world, invisible but powerful, even one in which an afterlife might be possible. We do not know. However we speculate, though, it seems clear that with the Neanderthals we reach a new mental level. The new indicators, like the size of the Neanderthal skull and the big brain it held tell us these are humans, who can in some ways think abstractly – as we can. They had a new order of mental resources with which to face the challenges of the last Ice Age, to which they adapted successfully, lasting long into the cold era. Probably they lived on besides other human stocks, perhaps sometimes mating with them, perhaps contending with them, though, they were in the end vanquished, genetically speaking.

They were supplanted by the branch of the family to which we belong, *Homo sapiens sapiens*. It seems to have evolved separately from the Neanderthals and to represent a different line of hominid descent, with lighter skulls, smaller faces and straighter limbs. Its members are traceable first in the Levant, the Near East and the Balkans between 50,000 and 40,000 BC. Perhaps they advanced northwards and west as the ice retreated. By the latter date they were well-established in western Europe (where *Homo sapiens* has sometimes been called 'Cro-Magnon Man'). Arrival in the Far East seems to have been later. Somewhere about 30,000 BC, too, humans had crossed the Bering Strait after the retreat of the ice and thus entered the Americas (until then, so far as we know, without hominid inhabitants). For the next fifteen thousand years or so, their successors moved slowly southward until human beings lived all over the later Americas. Meanwhile, others had reached Australia, where the first human remains have been dated to about 25,000 BC.

This was a huge achievement. No other primates had spread so widely. Nor was any later to do so, except through human agency. Yet although human beings spread throughout the world, and at a remarkable rate by comparison with earlier species, for a long time there were not many of them. One scholar has suggested that in 40,000 BC there were probably only about 10,000,000 in all. (Others have suggested 20,000 at most for the whole of France in Neanderthal times.) It was a world thus almost unimaginably different from our own, yet one in which human beings had already achieved amazing things and were showing more clearly than ever their power to change it for themselves.

Physical type

To a biologist, human beings are all members of the same species. There may be very little precise similarity between any two individuals (except identical twins), and no animals vary individually in appearance so much as we do. Nonetheless, we are all human beings. Striking divisions of appearance among us – in colour of skin, shape of eyes and nose, straightness or crinkliness of hair and so on – do not prevent human beings from choosing partners with different physical features from their own and from having children who in varying measure share physical features with both parents. It is important to remember this, because at different times some people have thought that the main differences of appearance between human beings based on racial characteristics reflected deeper differences, too, which were also part of their physical heritage. This is no longer a favoured view, though it still has supporters. The differences in the way different groups of people behave are very real, but their origins now tend to be seen in circumstance, different traditions and ways of life – in culture, all that they learn from their community – and not in a genetic endowment like that of skin colour or shape of feature.

Such physical differences nonetheless still divide mankind into certain large groups of markedly different appearance. Most people fall into one of what were long and traditionally called the main 'races' of mankind, so far as appearance goes. Though these divisions are not genetically absolute and are now distrusted as liable to be the basis of unjust discriminatory practice, they were what human beings actually saw and recognized for the whole period of historical record. On this basis, and setting aside complex distinctions which are for most purposes more important, the bulk of the world's population falls into three major groups, Negroid, Caucasoid and Mongoloid. Negroid peoples are black or very dark skinned, usually have woolly hair and vary much in stature. They often have broad, flattish noses and thick lips. Most peoples of this physical type are to be found around the equatorial regions of Africa, but others live across the Indian Ocean as far away as New Guinea, Fiji, the Philippine islands and even Tasmania. Caucasoids, too, were scattered at an early date over a broad area, to the north of the Negroid peoples, in North Africa, Europe, western Asia. They have often been called 'white', though their skins vary in colour from very fair (so fair that it cannot tan) to deep brown. There are Caucasoids in India and Ceylon who even have black skins, but otherwise are much like Europeans in that they have relatively narrow noses and lips, straight or wavy hair and a tendency to grow heavy beards. Similarly, the skin of the Mongoloids can vary from yellow to dark brown. Their other characteristics are straight, usually black hair, scanty beards, flattish faces and slanted, almond-shaped eyes. Their main homelands are in central Asia, China and Japan, but they also live in Malaysia and Indonesia. Some scholars believe that the original American Indians, both north and south, were Mongoloid, but this is not settled; the immigrants to the Americas of 20,000 or so years ago may have belonged to groups more Caucasoid in physical character.

These three main groups are now so distributed about so much of the earth, and have interbred so much, that mankind today is very complicated in appearance, as well as genetic inheritance. There are, too, still peoples elsewhere, notably in the southern hemisphere, who do not fit into these three groups, notably the Australian aborigines, or the Bushmen of the Kalahari desert. Of the way these major physical divisions of humanity appeared, there is little than can be said. Except for its bones, the human body does not keep well. Skin, fat, and other tissues all decay rapidly. We can look at skulls from long ages ago, but bodies from which we can safely make deductions about appearance are only available from a few thousand years back. By then, the main distinctions of appearance were well-established. It is probable, though, that the three major groups appeared in different areas and spread out from them until they met the others and sometimes mingled with them. But the basic mechanisms originally at work were probably those of natural selection, by environment, of genetic strains well-adapted to some areas, climates, diets, altitudes, latitudes, and so on, once early humans had acquired the means of surviving there. In cold climates, for example, Mongoloids are efficient biological machines because they have more fat just under the skin than Caucasoids or Negroids. Caucasoids, because they lived for the most part in more temperate regions, did not need this; nor did they need the skin pigmentation protecting black peoples against sunlight; the sun was not so fierce in Europe and western Asia. This does not take us far. But we know also that changes of diet and environment can produce striking changes in physique and appearance in only a few generations. It seems likely, then, that only after about 40,000 BC, in the era during which *Homo sapiens* spread out to establish himself all over the globe and confront new conditions, did what we think of as the enduring physical differences between humans appear. Only those differences made it possible for human life to establish itself easily in some regions at all. When they did, they fairly soon led to a racial distribution which remained fairly stable until only recently – only since about AD 1500 has it changed much.

HUMANITY IN THE OLD STONE AGE

Most of the time human beings have been on earth falls in what has long been called 'the Stone Age', a fairly familiar term, but one used very loosely, and one of three invented to talk about pre-history: the Stone, Bronze and Iron ages. Scholars keep coming back to them because more precise and complicated ways of describing prehistory have their own disadvantages. They are based on the fact that mankind successively learnt how to use stone, bronze and iron. In other words, they classify human development; they get away from divisions of time based on rocks, biology or climate such as geologists and palaeontologists employ and concentrate on what human beings have done and the tools they use. But there are some awkward things about this way of looking at the human past, useful as it is. There are no clear endings and beginnings to such divisions. Even very recently, the Stone Age was still going on in some parts of the world, though only just. There are still a few people alive who live with tools not much better than those of prehistoric men, though their numbers are rapidly shrinking. Another awkwardness of the particular term 'Stone Age' is that it has to cover an enormous stretch of time, during much of which we are not really sure whether we are dealing with humans in the full sense at all, though we are certainly dealing with makers and users of stone tools (*Homo erectus*, for example). Because of this, prehistorians have divided up the Stone Age further still. We need not bother very much in this book with most of the terms they use, but one of them, the 'Palaeolithic' (or 'Old Stone Age', from Greek words meaning old stones) takes up by far the greater part of human prehistory and continues down to about 10,000 BC. This spans the last really cold spell and the period during which *Homo sapiens* established himself widely, a sub-division sometimes called the 'upper Palaeolithic' (because, archaeologically speaking, the evidence about it tends to be found in the upper levels of excavations). Over such a long period – perhaps 30,000 years – general statements about what life was like should be cautious. Climates were very different in different parts of the world, fauna and flora offered different opportunities and presented different challenges, and human beings were all the time growing more diverse in the ways they did things, in what we can call their traditional cultures.

Stone Age (Palaeolithic) Chronology and Terminology

600,000 bc	Lower Palaeolithic period begins.
600,000–540,000	Ice Age.
540,000–480,000	Interglacial period.
480,000–430,000	Ice Age.
430,000–240,000	Interglacial period
240,000–180,000	Ice Age.
180,000–120,000	Interglacial period.
120,000–10,000	Ice Age.
100,000	Middle Palaeolithic period begins.
50,000	Upper Palaeolithic period begins. Cro-Magnon man replaces Neanderthal man.
10,000	Beginnings of the Neolithic Revolution as ice recedes and human settlement begins.

Tool-making

It is certain, though, that the pace of change was very slow in the upper Palaeolithic. There would have been little variety during most of it in the way people lived except those imposed by the availability of food. Hunting and gathering must have been the rule. Cro-Magnon Europeans seem to have been especially expert as fishers and hunters, and developed new techniques, with nets and barbed arrows. The tool-kit of *Homo sapiens* was better than that of any predecessor and went on improving. Long before, in pre-human times, the first cutting tools had been made of whatever suitable material came to hand; examples of pebbles, quartzite and even fossil wood have been found. But as time passed, flint was used more and more. After hundreds of thousands of years, in some places it may even have become somewhat scarce, because early methods of shaping it were very wasteful. Sometimes nine-tenths of a lump had to be flaked away to give an edge. In upper Palaeolithic times, primitive craftsmen began to work out new ways of making flint tools. A core of flint rather like a tube or cone was shaped in such a way that there could be flaked from it 'blades' of flint whose sides were nearly parallel and with a pretty uniform thickness. Not only was this economical, but such blades could be refined further to make very thin and beautiful tools. The best of these seem to have been made in Europe and the Levant.

Among new and specialized tools thus made possible were chisels for carving wood or bone called 'burins'. They had narrow edges for cutting, gouging and engraving, backed by broad edges to give them strength. They must have helped the human advance into new regions, especially northward, because they made it possible to use antlers or mammoth tusks for spearheads and harpoons. Ivory and antler are stronger than wood and more elastic than flint, and such implements lasted longer than earlier ones. It has been suggested that burin-making (whose techniques can be traced from Europe across the northern hemisphere as far as the Bering Strait and Alaska) may have been what made possible human invasion of the Americas. Burins are not found in the southern hemisphere, which may be negative evidence of their special value in working materials which provided food and clothing in cold climates. Another new flint-working skill was the making of tiny flint 'blades' to be fitted to weapons and other instruments made of wood or bone. One of them was the sickle or reaping knife in use in several places by the end of the old Stone Age. It does not show that people were growing crops, but that they had learnt to recognize and collect certain wild grain plants. Animals were hunted more easily, too, by hunters with bows and flint-tipped arrows, spears and spear-throwers. These weapons were very important; the bow and spear-thrower greatly increased the velocity – and therefore the range, accuracy and killing-power – of missile weapons. Many of these things were made from the new materials of bone, antler and wood. Some of them could be used to tap new sources of food by making fish-hooks and harpoons from them. Bone also made needles possible and very fine examples of them survive from the last hunting cultures of the Palaeolithic.

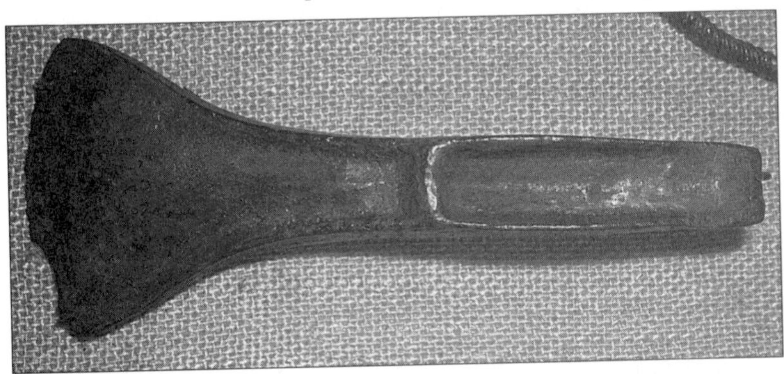

A Bronze age tool: a chisel or 'palstave' shaped to fit into a handle.

Life styles

Much of this suggests, correctly, that there was more food available. Yet in spite of this and in spite of a basic physical similarity to ourselves, people in the Stone Age were still shorter and lighter than later humans. They cannot have had a balanced diet (Neanderthals seem to have suffered from vitamin deficiencies) and their meat must often have been semi-putrid. Still, even today people like to eat game which is high. Probably few Stone Age humans survived to reach forty. Those who did would have lived on to a pretty miserable old age by our standards, racked by the pains of arthritis and rheumatism, suffering from scurvy and threatened with death by every broken bone or, even though they would not have had sugar to help rot them as we do, by rotting teeth. This would go on being true for many people in the world for a long time to come, of course.

Fire-using had greatly enlarged the choice of where to live. The first evidence of fire-making dates from about 30,000 BC. It must have made life easier. But clothes and man-made dwellings also show other ways of mastering the environment in the upper Paleolithic. Without them it would have been impossible for mankind to settle over the whole surface of the globe as it has done. There were no textiles, but skins were cut into thongs and strips which were then softened and made flexible in flint thong-shapers, to provide clothes. The oldest clothed body which has been discovered dates from about 35000 BC. It was found in Russia and was dressed in fur trousers and a decorated shirt.

In the Upper Palaeolithic we also begin to see faint signs of purpose-built dwellings. Other animals can build, but only in narrow, inherited ways programmed by instinct. Human beings can build anywhere, adapting style and technique to local climate, terrain and materials, and on whatever scale suits their purposes. Probably the main materials used at first for shelter (other than piled stones) were mud and reeds, which quickly decay. Also, until the very end of the Old Stone Age humans had to move about with the herds of game they hunted over the year, and though this seems to have brought them back time and again to certain caves which have traces of long occupancy, it did not leave permanent building behind. Nevertheless, some remains of huts from about 9000 BC have been found in the Near East. On the plains of eastern Europe (where there were few caves) there seem to have been quite big dwellings with frames made of mammoth bones or tusks covered in skins. They had floors sunk two or three feet into the ground and were sometimes grouped in settlements of some hundreds of people. This suggests the beginnings of a new scale of social organization.

The first art

Clearly, humanity had come a very long way by the late Stone Age. One of the most striking signs of it comes from a relatively small part of western Europe, where we find the finest evidence we yet possess of the beginnings of art. It begins with little stockpiles of red ochre collected by Neanderthals about fifty thousand years ago. They have been much studied, but we really do not know what they were for. The first difficulty in talking about the origins of art is that so little evidence survives of its first expressions. It is a reasonable guess that early humans scratched patterns in mud, daubed their bodies with colour, stuck flowers or feathers in their hair or danced in complicated patterns as did later peoples – but, if they did, nothing of that survives. A second difficulty is that we have no knowledge why the humans who produced the first art which survives took the trouble to do so. We do not know what they thought they were doing. In later, historical, times many peoples have painted their faces and bodies, and have done so for very different reasons. Some may have done this in ancient times, whether for religious or practical (camouflage) reasons, or as a part of their sexual culture, whether for pure fun or for its own sake. But what the ochre was for, we are unlikely ever to know for sure. If archaeologists thousands of years from now had as evidence of some twentieth-century societies nothing but a few packets of eye shadow and lipstick tubes, they would be hard put to it to say what the purpose of these substances might be.

Nevertheless, from somewhere about 35,000 BC onwards we have a pretty steady supply of data from Europe. It was produced over a long period, until about 10,000 BC. Elsewhere, caves and rocks in Africa, for example, also display lots of prehistoric painting and carving. But so far nothing as old as European palaeolithic art has been found. Much of what remains of it comes from a pretty restricted area, a number of sites in south-western France and northern Spain. The oldest things found there are little decorated and coloured objects, often of bone and ivory – carved spear-throwers, for example. Often they have engravings of animals on them. Then, in about 20,000 BC we begin a period perhaps 5,000 years long – which has left a splendid series of paintings and carvings on the walls and roof of caves. Animals provide most of the subject-matter of these decorations. The last phase of palaeolithic art follows this the stag often looming larger in the subject-matter – and finishes in a burst of decorated tools and weapons. And then, puzzlingly, this tradition seems to die out and there is no more really fine art for six thousand years.

Perhaps one of the first essays in European realism, from a cave near Lascaux in the Dordogne; the hunter falls back before the bison which charges him, his throwing-stick useless beside him on the ground, the spear having found its mark in the belly of the enraged beast whose entrails drag on the ground.

Boys discovered this cave at Lascaux in 1940, after it had been lost for thousands of years. Scholars still argue about the meaning of the great cave paintings of the Upper Palaeolithic; whether they are to be interpreted as elements in a magical ritual or as, in some dim sense, the beginnings of something conceivable as 'pure' art, are still debated. What remains unambiguous is the power of the impressions they can still produce.

This is a very impressive, but also a very mysterious survival. Many guesses have been made about how it can be explained. The great sequences of cave pictures have especially interested scholars. For one thing, these are often put in out-of-the-way and difficult corners, and can only have been painted or carved with the help of artificial light. Obviously the huge predominance of animal subjects is important, too: the palaeolithic artist did not spend his time on landscape or even very much on human beings. Interestingly, human beings are shown always very unrealistically, in an abstract, stylized way, whereas the animals are often drawn with close attention to detail. Perhaps to draw something realistically was to have power over it. Some scholars have tried to work out patterns in the way certain animals repeatedly appear, but this does not get us very far. It is a reasonable guess, though, that in societies without writing, these patterns carried messages to those who looked at them. Remembering the hints of burial practices in Neanderthal times, it is tempting to think that religious or magical rituals went on in these dark caves. Perhaps, if they did, they were connected with attempts to influence the movements and behaviour of the game animals on which early humans depended for their living. This would fit in with there being more pictures of stags as time passed – the reindeer and mammoth of the earlier sequences would have gradually died out as the ice slowly retreated.

We really know very little about the first great art. There is enough of it, nevertheless, to show that human beings in the late Palaeolithic were capable of astonishing mental achievements and of observing the world around them acutely. It may be that as they lost faith in their ability to influence the behaviour of the animals – which was, though they did not know it, determined by climate – so they lost the incentive to produce art. It is unlikely that they practised art for its own sake, far less because people would buy it, as in later times. But that it was art in the full sense of careful, controlled, imaginative creation of beautiful and moving things, appealing to us not only because of what can be done with them but because of what they are, there can really be no doubt.

THE COMING OF AGRICULTURE

Art has to be paid for even when it is not bought; the people who carried out the great cave-paintings, however simple their needs, could not go out to look for food while they were engaged in their work. So some small surpluses over immediate need must have been available even in early hunting and gathering societies. The decisive step in making them bigger was achieved when people learnt to grow and harvest crops, and to tame and exploit animals. This was the discovery – or invention – of agriculture. It must already be clear that the whole story of human beings is one of continual change, much of it man-made, but some steps in that story stand out as of quite special importance. Agriculture is one of them, like the mastery of fire, or the coming of speech. It was almost the last of the great strides made by mankind in pre-history; it changed life so much and so deeply that nothing since would have been possible without it.

Different parts of the world achieved agriculture at different times. Climate and physical setting must be the basic reason, and explain also why some peoples arrived at agriculture on their own, but much later than the Old World of Eurasia (in the Americas, for example), and why some never got there at all without help from the outside (like prehistoric western Europe). The oldest traces of cultivated plants are said to date from about 10,000 BC and were found in south-east Asia; these were early forms of millet and rice, both still important today in that region. About 8,000 years later, people in central America learnt to grow a kind of sweet potato and a primitive maize. Much more is known, though, about the early stages of agriculture in the Near East, where, between about 9000 and 6000 BC many of the cereal grains we still use became widespread.

The Fertile Crescent

People sometimes call the crucial area 'the Fertile Crescent', an arc of territory running up from the Nile delta, through Palestine and the Levant, turning eastwards along the hills of Anatolia and finishing on the highlands between Iran and the Caspian, the other side of the river valleys of Mesopotamia. Much of it now looks pretty uninviting, but ten thousand years ago good rainfall and fertile soil made it well-wooded, with its woods full of game. The growth of forests as the last Ice Age rolled back was less fierce than further north and they were easier to clear. The ancestors of later cereals grew on the hills – wild barley, emmer (a

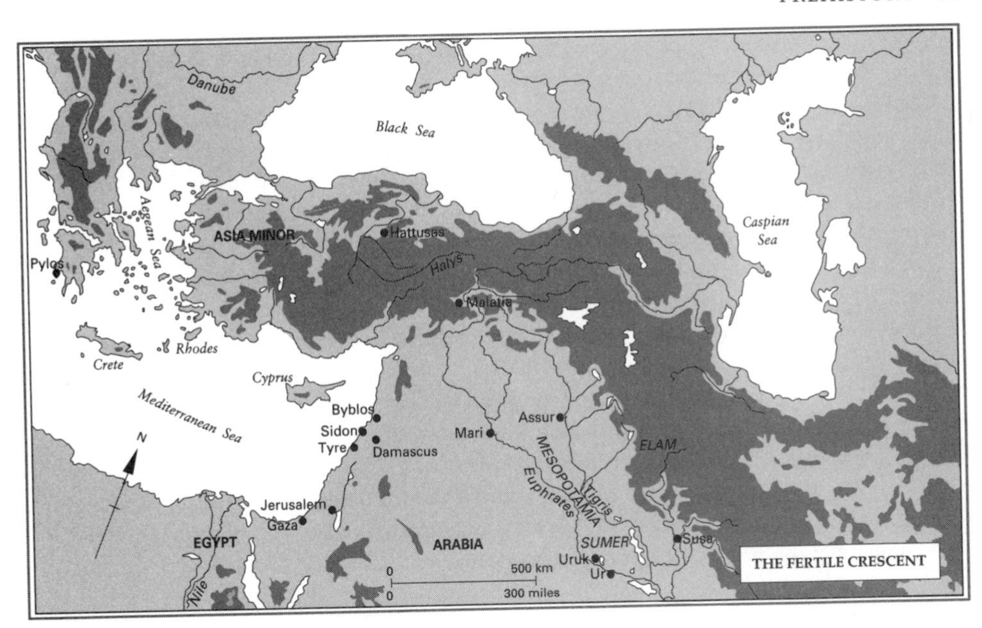

THE FERTILE CRESCENT

wild wheat) and many grasses. From this area newly discovered techniques of planting and harvesting seem to have spread, both into south-eastern Europe and into the Nile valley.

A possible time-scale is that in about 9500 BC people were harvesting wild grasses and grains in Asia Minor; by 7000 BC the first planting and cultivation had begun in the Levant and Mesopotamia; in the next three thousand years these practices moved west as far as (roughly) the Rhine; and by 3000 BC farming had reached western Europe and the British Isles. Elsewhere new ways of raising food may have been arrived at independently, but to the east and north of the Mediterranean they seem to have been passed on by the neighbours.

Learning how to keep livestock had almost as revolutionary an effect as growing crops. The dog had already been domesticated by the hunting peoples of Europe, another big step in harnessing natural energy for human use. The next was to round up wild animals and keep them in herds, culling them for food, for their skins and wool, or for their antlers and bones. There were plenty of animals about in the Fertile Crescent which turned out to be highly manageable by man. Sheep and goats (or their ancestors) were especially plentiful there, while pigs of a kind lived wild almost world-wide. Once the principle of keeping them as living resources instead of merely hunting them had been grasped, much else was to follow – milking, for example, or the taking of eggs from domestic poultry. Later would come the use of animals for riding, carrying or traction.

The domestication of four animals has provided the backbone of livestock exploitation ever since – goats, sheep, pigs and cattle. They are all members of the mammal families of the northern hemisphere and usefully complement one another. Goats are tough beasts, surviving on poor pasture, and provide meat, milk, skin and wool. Sheep will live wherever there is grass (of which there was plenty on the hills of the temperate zone) and can be used in the same ways as goats. The first traces of someone keeping sheep come from about 9000 BC in northern Iraq. Pigs provide meat, will root for their food in woods and forests, grow quickly and have large and frequent litters. Cattle produce meat, milk and hides, and can also be used to pull and carry loads. Some of the best evidence of domesticated animals comes from the bones of those which were eaten in early agricultural settlements. Nearly always, they are the remains of young animals, killed before they had reached maturity and therefore culled from managed herds (the animals killed by hunting peoples were nearly always full-grown).

Around the cultivation of grains and the herding of animals evolved agricultural systems. In some parts of the world, though, only a part of the revolution took place. When the inhabitants of Central America took to growing crops, they did not go on to domesticate animals, probably because few suitable ones were available until introduced far later by the followers of Columbus. Only in the Andes, where the llama provided an all-purpose meat, milk and wool producer, as well as a bearer of burdens, were the skills of the herdsman mastered. Isolation, therefore, soon began to differentiate life in the Americas and in the Old World land masses. But within the Old World, too, difficulties of communication may explain big differences in the way agriculture developed. It seems likely, for instance, that whereas knowledge and crops spread fairly easily from the Near East through North Africa to western Europe, and even, slightly less easily, up the Danube valley, their movement eastwards into Asia was much harder. Furthermore, big climatic differences stood in the way of doing the same thing there. One result was that when China came to be an agricultural country, crops which suited local conditions were derived from local plants and were not imported as had been grains to Europe from the Near East. The outstanding example was rice which, interestingly, does not really need animals in its cultivation, but huge inputs of human labour. This may be an explanation why the only animal the Chinese bred for eating was the pig, though various cattle were later used for work. But archaeologists are still much divided and uncertain about the origins of agriculture in China, so that it is best not to be dogmatic.

A changing world

The revolution in the conditions of human life launched by the coming of agriculture took thousands of years. Early agricultural settlements were short-lived; the first farmers were still probably pretty mobile cultivators, likely to practise what is called 'slash-and-burn' agriculture, something still seen among primitive peoples. An area of forest is chosen (the soil is likely to be good because of accumulated leaf-mould and decayed rubbish) and the trees are killed by ringing them. Then they are burnt, the stumps removed where possible, and crops are planted among the roots. After a few years undergrowth again becomes too thick or perhaps the soil is exhausted, and so another site has to be chosen. This was for a long time all the farming that could be done. People knew no better ways of tilling the soil, and it was very difficult to clear ground with stone tools and antler picks. As the centuries went by, though, fields appeared in some places as a result of continuous reoccupation and recultivation; when this happened, the early farmers began to be more tied down in one place – they became sedentary, as the anthropologists put it. Because archaeologists have uncovered many of these early places, this was one of the first effects of agriculture on human behaviour which we know about.

As for the effect on the non-human environment, ancient vegetables and crops would have looked very different from ours. Those we now use are much bigger; sometimes they differ from wild varieties in shape, colour and size because humans have intervened in the evolutionary process. By choosing some strains for cultivation and rejecting others, humans began at an early date to alter the balance of nature, which would have produced a very different selection had not human choice cut across it.

The most important of the new crops were all grains or seeds. Leaf, root and bulb crops could be cultivated and some soon were, but wheat and barley, peas and lentils all kept much better. Once dried, they could be stored. They could thus provide food for the winter and for bad years, when crops failed. This was another liberation from the rigidities of natural rhythms; though not complete – for sowing and harvesting made their own, new, demands – it was another step toward that freedom to do something else than merely hunt for enough nourishment to survive which had begun with meat-eating hominids so long ago.

The effects of human decision would have shown among animals, too. The first domesticated sheep and pigs must have been much smaller, scraggier beasts than the prosperous-looking, square-fleeced creatures nibbling English hillsides, or the pink monsters dozing in our pigsties. Nor have there been changes merely in size. In most domesticated animals (and birds) the muzzle has got shorter; as humans have assured their food-supply, strains with less-well-developed teeth and jaws than are needed in the wilds have been able to survive. Modern animals may even be less 'brainy' than their wild ancestors; because they have been protected by man from so many natural enemies and dangers, strains have been able to survive in which some parts of the brain handling messages from the outside world remain somewhat under-developed. More striking still, the colour of animals becomes less uniform as they are domesticated; this is because removal from their natural habitat enables varieties to survive which might have perished in the wild because of the hazard presented by their colouring.

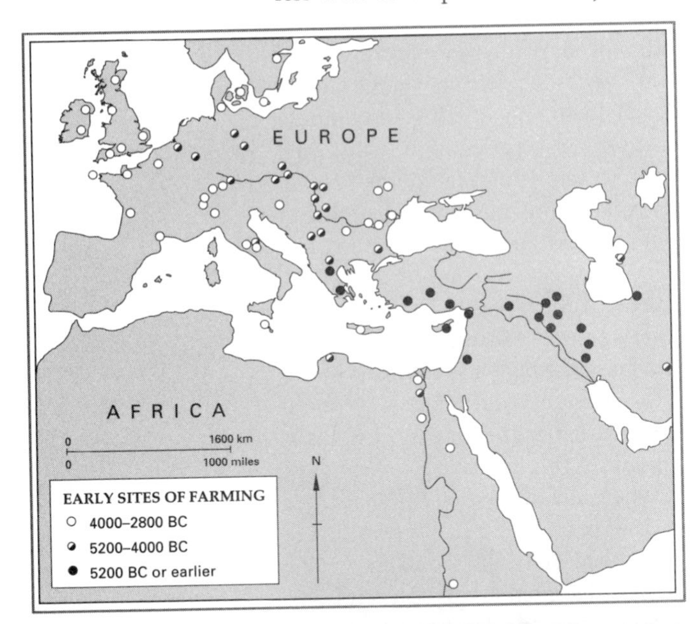

What we cannot know is how far or when humans began to try to produce such changes as these. We are now used to the idea that sheep with heavier fleeces, or cows with better milk yields, can be bred selectively, but to begin with there must have been a great deal of accident involved. The first reasons for intervention with natural

Jericho was one of the first settlements to leave a substantial archaeological record of its defences, which by 6000 BC already included this thirty-foot watchtower with an internal staircase and a defensive ditch cut into rock without the help of metal tools.

selection may also have been different from those which would now operate. An early herdsman might, for example, have chosen to breed animals because of markings which pleased him or which could easily be recognized. We simply do not know.

The most important result of the new methods and, indeed, of any agriculture at all, is clear enough, though: much more food became available. Very much depends on the kind of game, plants and soil, but a human family in a society dependent on gathering and hunting for its food has to have hundreds of acres at its disposal in order to get enough, while even with primitive agriculture about twenty-five acres will do. This was the result of the first big jump in food production. What has happened since may make it look tiny, but the coming of agriculture brought the first dramatic increase in the human food supply after hunting, and one of even greater potential.

More plentiful food brought a rise in human numbers; if there was more to eat, then more mouths would survive to eat it. We cannot make very good estimates, but all the signs of what it meant can be read by archaeologists in the remains of the bigger settlements themselves – the first villages. Their appearance makes us fairly certain that other changes in social life were going on, too. Continuous occupation of the same settlement became more normal; there was less need to follow game or seasonal plants about. As this happened, buildings of greater solidity appear. One famous example is Jericho, where already before 9,000 BC there was a village on the site of a never-failing spring. A thousand years later it had grown in size until its mud-brick houses covered eight or nine acres and had big walls round them. The inhabitants clearly felt they had something to protect and possible enemies to protect it from. They owned wealth, and human beings had already discovered that one quick way to acquire it was to take it from those who had it already.

In such places as Jericho new patterns of life gradually emerged as such communities grappled more and more successfully with the special demands and opportunities of their environment. This was not what had happened after earlier Ice Ages when there had been no animal as resourceful as *Homo sapiens* to take advantage of what they left behind. Since, though, for a long time there were not many people in the world, the communities they lived in tended long to remain isolated from one another. Then, as they evolved new skills and met new challenges in their own settings, they grew apart in their ways of living, in what we may call their cultures, which became more and more complicated. People raised in them must always have taken them for granted (as most people in the world still take their own ways for granted). The importance of routine ways of doing things must have seemed overwhelming if the community was to continue and so for a long time different cultures continued to draw apart. No doubt language took different forms as it had to meet different needs. Even today, many different languages can still be found side-by-side in areas which are still primitive. Almost every tribe has its own, meeting its own precise needs. Great worldwide languages such as modern English or Spanish only appear at a very late date and are the result of civilization.

Such differences offset the fact that all these communities still lived in what we should think a very similar way. They all rested on a very simple technology, though one much more elaborate than anything available even a few thousand years earlier. It was also of immense potential for generating further, and faster, change, as the next distinguishable phase of humanity's evolution was to show.

THE NEOLITHIC REVOLUTION

The word 'Neolithic', like 'Palaeolithic', has an ending which suggests a connexion with stone. People have sometimes wished to distinguish phases within the 'Stone Age'. They came to speak of both a 'Mesolithic' era and a 'Neolithic' era. Like all other divisions of prehistory these eras begin and end at different times in different places and are not marked by sudden changes. People did not wake up one morning and suddenly find they were in a different era; what happened was that their ways of doing things (particularly of making certain tools from stone, which is the easiest thing for archaeologists to observe) gradually changed. The end-result was clear enough, although there had never been a precise moment at which one could say the change took place. This is the way societies, which are complicated things, still change, even though they now do so rather faster. The Mesolithic really does not matter much for our purposes, interesting and important as it is to specialists. The Neolithic, on the other hand, marks a very important phase of human development indeed.

Why is this? After all, in its strict sense, archaeologists only use the word 'Neolithic' to denote a culture in which ground and polished stone implements replace those made by flaking or chipping. This does not by itself seem enormously exciting. Much more is involved. The Neolithic phase of human existence also brought about a number of other changes of huge importance going far beyond the ways of using stone which are its convenient markers. Nearly all of them, though they might have their roots in the distant past, could only take place on a large scale and over a wide area because of the earlier discovery of agriculture. It made possible the bundle of developments which have been called the 'Neolithic revolution' and which should only be separated into distinct elements for convenience in describing them. All human societies are tied together in complicated ways; each part only operates and develops in the way it does because the other parts are there too. There is no pressing reason to describe any one particular change as crucial, once we have recognized the enormous importance of the provision of more food by agriculture. Let us therefore start at an obvious, if arbitrarily chosen, point, the stone technology which gives the era its name.

Technological change

For tens of thousands of years heavy cutting was done with hand-axes, stone choppers held in the fist, their edges sharpened by chipping flakes away. The Neolithic tools were still of stone, but were smoother, with blades ground and polished on other, harder, stones. Moreover, blades came to be set in hafts. This greatly increased the energy going into a cutting-power already improved by shape and polish. Some Neolithic stone axes have been fitted with modern hafts and have been tried out on woodland; they are formidable instruments, with a long life, because when blunted they can be given a fresh edge by the grinding process which produced them in the first place. Here was a tool of great importance to agriculture. It made easier the clearing of woods, brush and scrub for crop-growing. Stone hoes for breaking the ground followed. A change in stone technology, that is to say, is closely connected with the food-producing revolution. We do not know exactly the sequence of causes and effects, but a connexion is obvious.

Many such connexions between different changes in human life during the Neolithic era justify its claim to be crucial in the story of mankind. It took five or six thousand years for them all to be worked out, and the overall effect was the greatest acceleration of economic and social development before the coming of steam-power in the eighteenth and nineteenth centuries AD. Not only could more people be fed, but more could be fed without themselves taking part in the raising and bringing in of food. There may have been a few specialists in earlier times but economic and technical specialism became both more likely and much easier with the invention of agriculture. Inside bigger and settled communities more craftsmen carried out tasks for which others gave them their livelihood – making tools, or decorative objects, for example. Some of their activities were of great importance in bringing about both new differences between the way life was lived by different individuals and new technical discoveries.

One such was pottery. It may have been first made in Japan in about 10,000 BC, but it was widespread in the Fertile Crescent a couple of thousand years later. Bowls and containers of wood and perhaps of stone had already been used in earlier times. With agriculture and foods which could be stored they became more necessary. Someone observed (perhaps by accident) that firing clay could change its nature. That meant that containers could be made much more quickly and easily. In its turn, this made it easier to cook in different ways (and therefore to enlarge still further the range of human foods) as well as to store food (and therefore to make an assured supply a little less of a problem). Moreover, pottery could be decorated easily, both by daubing it with colour and by changing its shape, and so offered a new medium for art, though for a long time it was pretty crude. All this adds up, therefore, to a wide-ranging series of effects. Woven fabric was another Neolithic invention. Humans had worn skins for thousands of years and, at their best, their clothes must have looked like those still worn in the last century by North American Indians, or by Eskimaux in this (both can until recently be regarded as Stone Age peoples). But in Neolithic times, the first woven materials – textiles – appeared in the Near East.

Farming, pottery, textiles: these alone imply vast changes within a few thousand years. It is, in fact, nothing less than the preparation of the nursery for the next phase of human development, the one we still live in, the era of civilization. Nor are these all the changes we can associate with the Neolithic. One more, overlapping into the age of civilization (and this is why some scholars do not count it as part of the Neolithic), was the discovery of how to use metal.

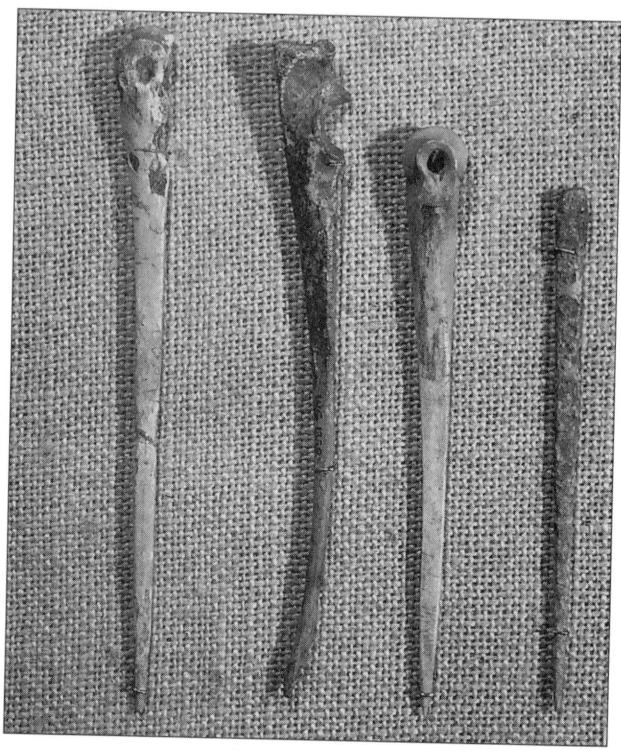

From bone, needles could be made, once clothes were worn – such as these, found at Avebury, a major British prehistoric site.

The coming of metallurgy

In the long run, metal-using changed the human world almost as much as farming. The change came more gradually, though, and is much less easy to pin down, because many other things had to happen before its full effects could be felt. There was not for a long time much ore available, even when people had found out how to treat it, and the early use of metal was only occasional and scattered, and not very influential. Nonetheless, the earliest traces fall well within the Neolithic era. At some time between 7000 and 6000 BC, copper, the first metal to be exploited, was in use at a site in Anatolia, a region where the ore was fairly easily found. Cyprus and some of the Aegean islands also provided it. The evidence shows copper-using spreading gradually westwards into the Mediterranean basin as time passes, then in Italy and Hungary on the mainland, as well as in the British Isles.

Early copper was worked by hammering. This was only possible with very pure ores. The next step was to cast it – that is to say, to heat it until it became liquid and then pour it into moulds. Finally came the discovery that impure ore could be refined (again by heat) to provide purer metal. Once these processes were known, metallurgy had its basic pattern set for thousands of years. Development went on by finding new sources of ore, by experimenting with them and discovering different metals, by specializing them for certain purposes, by providing greater heat to work and refine them and by blending metals to make artificial materials – alloys would be our word.

The first alloy seems to have been bronze, a blend of tin and copper; about one part of tin in ten of copper is plenty for good bronze. It was so important that it has given its name to the Bronze Age. For a long time it was thought that bronze was first made and used in Mesopotamia somewhere shortly before 3000 BC. It was known also that it was made in

China about a thousand years later than this. But recently, bronze objects found in north-eastern Thailand have been dated as early as 3600 BC, earlier than any others. No doubt more evidence will be needed before scholars agree which ancient society was the first to use bronze. The chief point is that the Bronze Age was the phase of human society during which its main metal requirements were met by the use of this alloy, one much superior to copper. It could be given (and would keep) a much better cutting edge (copper was not even as good as flint in this respect) and it could be cast in moulds much more easily and so more easily varied in shape. Its discovery, of course, made places where tin could be found important, too; sometimes these were in the same areas as copper deposits.

How progress was made we do not know. Possibly someone noticed that copper left in an oven used for baking pottery melted – and so could be moulded. Technology has always advanced because of 'spin-off'; the by-products of one advance are the major steps in another. Modern non-stick frying-pans make use of knowledge about heat-resisting materials developed for building space rockets. Once spin-off is noted, specialists then begin to experiment with the result, and further spin-off can follow. Some scholars have recently argued that the process of refining ore was not discovered accidentally, but that it was the outcome of what we might now call 'research' and conscious experiment. If so, it is clear evidence of how deliberately, by the Neolithic era, humans were change-makers.

Gold, too, was discovered and used at an early date. It was probably more easily available in the ancient world in deposits near the surface than in later times. Its use, though, was almost purely decorative; copper had been turned into ornaments, but had soon been put to use for weapons and tools. This was not possible with gold.

Even bronze was less important for weapons and tools than iron, whose exploitation is hardly part of prehistory at all. It made its appearance only after the first civilizations were already well established and was widely used only after 1000 BC, a very late date. All users of iron have lived since civilization began, even if they themselves were not civilized, and therefore, in the era we usually think of as historical, not prehistorical. But it is logical to discuss iron here, because its coming is really the completion of the story of metallurgy in early times. Prehistorians still talk of the 'Iron Age' and of 'Iron Age cultures' too; it is another phrase which describes not an era of time but a phase of material culture. The Iron Age can be thought of as both the culmination and the ending of the Neolithic era, though iron-using peoples long lived alongside others using only stone tools.

It is usually agreed that iron-working, like the working of copper, began in Asia Minor, and there is much disagreement about exactly where. No doubt a start in that part of the world is to be explained not only by the presence of the raw material, but also by growing experience in working other metals. For refining, iron has to be heated to a much higher temperature than copper. People in Anatolia already knew how to make ovens hot enough for pottery to bake in them by firing them with charcoal and blowing air through them. Even so, they could not reach a high enough temperature to cast iron in moulds, like copper and bronze. For a long time, iron could only be made by beating it into shape – it could be 'wrought', but not 'cast'.

Iron-making spread rapidly. Europeans later called Celts were among the finest iron-workers (they had been very good at bronze, too, though in that skill the Chinese long led the field) but only well after iron was adopted elsewhere. The first people who much used it were Anatolians called Hittites who ruled a large empire in the Near East in about 1500 BC and were to show that military victory was much easier for those who had iron weapons. An iron sword was much tougher than a bronze one – let alone a copper dagger or stone axe. But iron changed history more through its application to agriculture. Iron tools were better than anything else for tilling the soil. They meant easier and deeper digging and therefore more food still. Better crops followed, and so did the use of plants which needed deeper rooting. Tree-felling became easier. Nevertheless, the impact of iron was very slow and for a long time

A new scale of wealth: copper and bronze treasure from a Palestinian site of the fourth millenium BC.

very expensive. Only a hundred years ago wooden ploughs were the rule in Russia and millions of them are still at work all over the world.

Wood-working was transformed by metal. Even copper and bronze tools had already made possible work of a level we can reasonably term 'carpentry'; iron brought another stride forward. It gave people more objects to use and enjoy and it made likely even more specialisation of skills. The importance of ore-bearing areas steadily grew. Once into the age of civilization, Europe's landmass would be more important to the outside world than ever before because of it and the interest of outsiders would increase. The European peoples were for a long time backward by comparison with those elsewhere, but their continent was full of easily available ores and thick forests left by the retiring ice which could provide fuel. Metal-seeking prospectors from the Near East were begining to look there well before 3000 BC and a thousand years later several specialised metallurgical regions existed, notably in Spain, Greece and central Italy. Soon, Europe would be a major manufacturing area as well as a producer of ores. Metallurgy had some bad environmental effects, too. The stripping of timber reserves probably began not with the clearing of ground for agriculture, but with the felling of trees to provide charcoal for smelting metals. But this story takes us too far ahead for an account of pre-history. It is the beginning of a huge theme – the slowly unrolling story of the way in which Europe rose in the end to a world dominance owing much to the exploitation of her mineral and technological resources.

AT THE EDGE OF HISTORY

Most animals which live in groups – ants, bees, herds of deer look very well-regulated. It seems that they are better at keeping rules in their societies than are human beings. Yet that is only because they are in fact very different from humans. They are not actually obeying rules (as we understand them) at all, but are behaving almost automatically; they do things because they are programmed by their genes or by patterns of behaviour imprinted so deep that we call them 'instincts'. They could not behave otherwise if they wanted to, indeed, they cannot want to.

This is not how human societies work. They have to take account of basic human nature, its needs and drives, but they provide many different ways of doing so and these ways have often been deliberately chosen by their members. Right round the world, for example, men and women are sexually attracted to one another, live together and have children. But there are many different sets of rules about the ways in which they may do this and these rules are made by humans, not by nature. In England the law will not allow you to have more than one wife or husband at the same time, whereas in some countries it is legal to do so. Or, to take a quite different example, in much of Europe a few hundred years ago, it was not possible to follow certain trades (being a shoemaker, for instance) without belonging to a special guild of those who practised that trade and being regulated by its rules about how you did your work. For various reasons this system crumbled away and for much of the eighteenth and nineteenth centuries there were fewer legal restraints on following the trade you wanted and working at it as you pleased. But in some countries it was for a long time difficult to do certain jobs without belonging to the appropriate trade union. This does not at first sight have much to do with prehistory, but it should direct our attention to the fact that what we call 'social institutions' – ways of organizing people to do things – are to some extent matters of choice for different societies, and that they will vary a lot from one society to another. This was true even in early times.

Differences in traditional ways of making things show that some human communities had become very distinct already at the end of prehistoric times. The carving out of their own traditions had begun a long way back in the palaeolithic. But differences became much sharper with settlement and that, of course, was another result of the coming of agriculture. High up the valley of one of the tributaries of the Tigris, in the Kurdish hills, there was by 6500 BC a little village at a place called Jarmo. Its people did not yet know how to fire pottery, but they already had houses of clay, sometimes with stone foundations and more than one room. Some of their dwellings had built-in furniture, such as ovens, or basins in the floor. The inhabitants could cut and grind bowls from smooth stone and also made from it ornaments such as beads and bracelets. They practised agriculture and had domesticated sheep, goats, oxen, pigs and, of course, dogs. Already at Jarmo a little wealth marked off some – the owners of prized ornaments or weapons – from others. There was a role for specialists in such a place. Its affairs had to be regulated and the harvesting and storing of crops seen to. Yet Jarmo probably only contained a couple of hundred people. Far to the west, in Palestine, Jericho probably had about three thousand inhabitants by that time. So big a group, to say nothing of the problems of managing and retaining control of an important oasis, would have made specialised skills and ordered government much more necessary than in Jarmo. One of the things which was happening at least in the Near East during the Neolithic, then, was that bigger groups were evolving to which people gave loyalty and obedience. Human life was already well down the road away from the nomadic tribe to the organization of human social life in settled territorial units under the same laws which is the way of governing people we are familiar with today.

How men and women saw one another's roles in these early societies is still hidden from us, but must have been rooted in biological and economic facts already mentioned. Because human infants – the future of the tribe – require so much prolonged care, the division of labour between the sexes, whereby men went out to hunt and gather food while women stayed at the home base, was probably well-established before communities became more settled. On this division would grow different traditions of education, the boys going about more with the men as they grew to be able to keep up (or at least not to be a nuisance) in the hunt, while women may have learnt to observe carefully the plant life near at hand to the base, to gather in it specially useful and nutritive crops. Perhaps in many places they already provided the main labour force for agriculture – as they still do today. Before we reach the age of history, we can be sure that much of what was to be taken for granted for thousands of years, and was to endure until the present, was already in place.

5000 BC has no special merit as a date except that it is easily remembered, but it will do as an imaginary point of vantage. By then, the physical shape of the world was very nearly what it is today. The forms of continents, natural barriers and natural channels of communication have not changed much since. Climate, too (since seven thousand years are hardly significant in comparison with the upheavals of the hundreds of millenia before the last Ice Age) can be regarded as stabilized; only short-term fluctuations need concern historians in what follows. Ahead lay the age (in which we still live) in which most change was to be man-made.

Some parts of the world were by 5000 BC already well into the Neolithic era; human life was thus already very diverse and complicated. It was quite unlike the life of *Homo erectus*, which, for all its great advances, had still been very much the same everywhere. Yet *Homo erectus* had himself lived a life very different from that of the poor, vulnerable specimens of *Australopithecus* who crouched about the lake which filled what is now the Olduvai gorge 2 million years ago, armed with means of survival hardly better than the beasts prowling nearby with whom they shared their food-supply.

By Neolithic times we are in a world full of human variety and potential. This variety was to increase. Some human communities were to progress rapidly; some would not. New forces would operate in human development as different peoples came into contact and learnt from one another, or reflected further on their own experience and plunged forward into new experiments. More and more, that is to say, human diversity could stem from mankind's ability to change things consciously as well as from the brute facts of environment. The result would be more untidy still; there have hardly ever been such possibilities of differences of human experience as exist in the world today. But, it was already far from uniform in 5000 BC. There is no clear line marking off the end of one human era from another, only a blurred, ragged-edged time zone, with some people forging ahead on their way to civilization and others still stuck firmly in the Stone Age where some of them were to remain for thousands of years yet.

In the prehistoric Ukraine the bones of mammoths were assembled to provide frameworks for huts and were then presumably covered in skins. This reconstructed example has been dated to c. 20,000 BC.

Yet this was a world where there had already been an enormous acceleration of change. It had begun far back in prehistory and it is important to get the perspective right. Someone alive today, who recalls that the aeroplane did not exist in 1900, that nuclear power is only fifty years or so old, that many African nations had not been invented forty years ago, and that Aids was only identified as a disease in the 1980s, may well be excused the feeling that for centuries on end not much changed in, say, the Middle Ages. Over much of Europe people were still farming in the fifteenth century, for example, in the way they did in the ninth. Yet if we look at European art – at buildings, say – in AD 800 and then five hundred years later, it is obvious that there has been a very big change indeed. In the first art of all, though, that of the upper Palaeolithic, experts tell us that the great cave-decorations show virtually no stylistic change for perhaps five or six thousand years. And if we were to go further back, the long persistence of certain ways of making tools from stone would give us examples of even more gradual development. In even earlier times, the actual physiological evolution of human beings can be observed, but had been of glacier-like slowness by comparison even with the sometimes barely perceptible changes of palaeolithic art.

The crucial reason why change had already speeded up a lot by 5000 BC is that by then the main source of innovation had moved from natural forces to human beings themselves. By the end of prehistory the human story is increasingly one of choice. Human beings are making more and more decisions to act and adapt in certain ways to meet their problems and to develop certain ways of doing things, to utilise certain materials or skills. This is why what we can reckon the most important change of all came about somewhere right at the beginning, though we cannot exactly know when or where, when some creature perhaps hardly recognisable to us as human first began to think of the world as a collection of separate objects distinct from himself. If we knew when it was, it might serve as the best definition of the beginning of the prehistory of man; it opened the way to using the world, and that is the story of the whole change from life shaped blindly by nature to one shaped by human culture and tradition. After that, something like compound interest began to work. As more and more human beings appeared, there was not only a bigger pool of talent to draw upon but there was also more and more human achievement and experience to use. Even small communities did not have to go through the tedious business of learning everything from scratch.

As we know only too well, this has not meant that human beings have not created new problems as fast as they solved old ones, nor that the story of mankind is not full of slippings backward and tragic failures too. But it helps to emphasize a central fact about human beings as they approached the era when some of them were to create the first civilizations – that they are unique in consciously setting about using and changing the world. One of the few good descriptions of *Homo sapiens* is that he is, above all, a change-making animal. The evidence of that lies in what he has done – his history. Behind it lies all that has been only glanced at so far in this book, millions of years in which creatures were taking shape in ways which decided that among the primates, human beings alone were formed as creatures which could mould their own destinies. Even if for a long time they could only do so within narrow limits, the earliest evidence of the effect of that is very ancient.

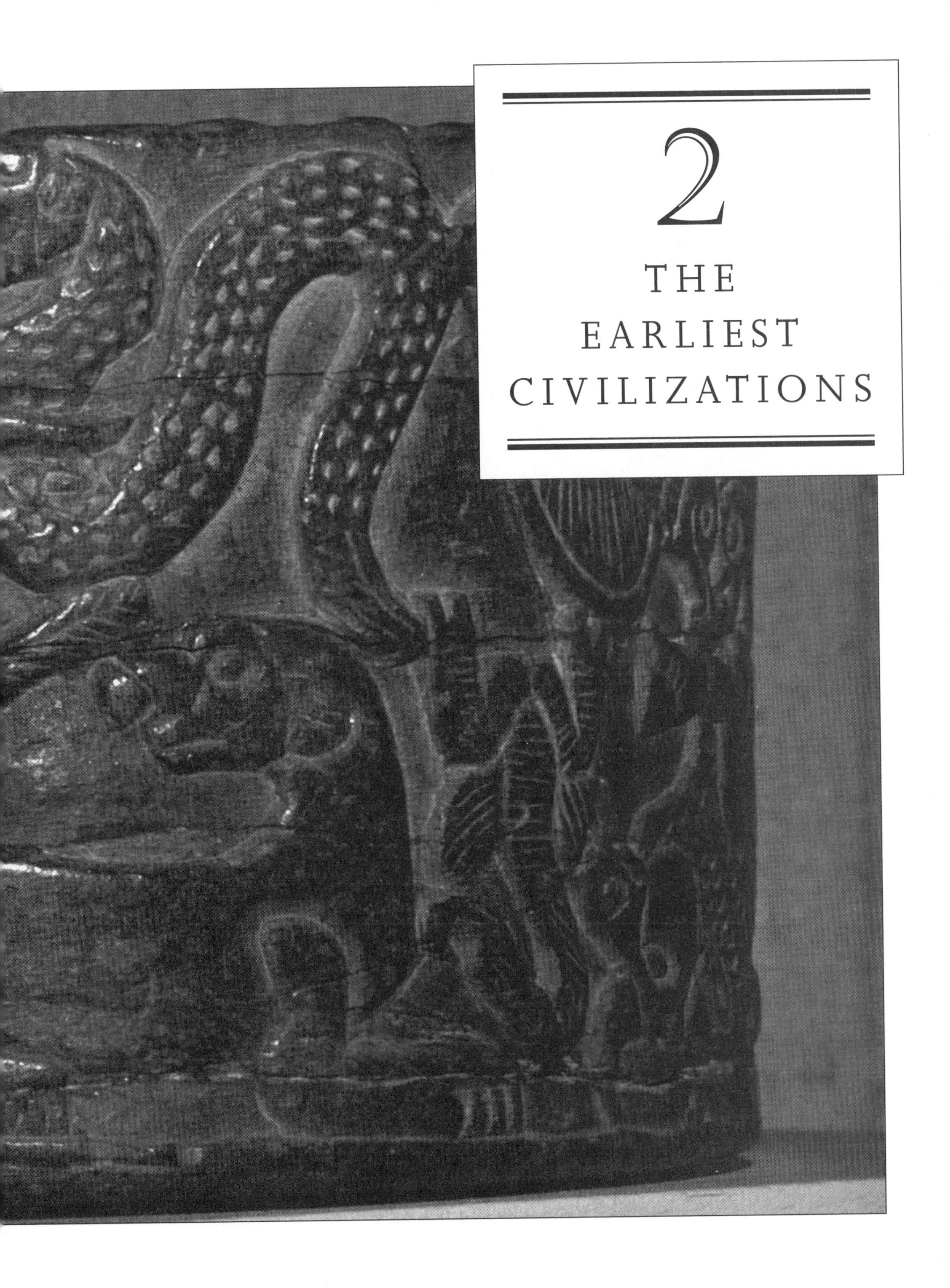

2
THE EARLIEST CIVILIZATIONS

THE ROOTS OF CIVILIZATION

Some people argue that, even if we agree about who they were, what was done by the first humans is not really history at all. They point out that we can really only begin to understand people and write real history when we have evidence which can give us a fairly good idea of what they thought. In practice, we have to have, and to understand, words. 'History' as the story of the human past cannot, therefore, go back much further than the first writing. Two Chinese characters are claimed to have been found which date from as early as 5500 BC, but the first writing we can be sure about was invented between 3500 and 3000 BC. Within a few hundred years, written records in the form of inscriptions on stone or clay tablets were being made. Swollen vastly in due course by papyrus, parchment and paper, a growing flood of written documents thereafter gives us evidence about the era during which all the most striking and rapid changes in the way people live have occurred. Human life has been transformed in the last five thousand years and that, say some scholars, is what 'history' is about. It is the era of written record.

Traditionally, what happened before writing is called 'pre-history', and historians have left it to other scholars. Non-documentary evidence was all that could be used to find out about it. Yet in prehistory much was settled to ensure that one day humans would know how to write – as well as engineer, build, organize and many other things. Though so much has happened so recently (in the last 5000 years or so), it did not start happening without any preparation. In sheer weight of years most of the story of human beings lies in prehistoric times, and that is why we have begun there. In looking at the beginnings of civilized life, we have to start with what lies behind it.

What our ancestors brought with them into the era of civilization was very important in shaping what the entities we call civilizations were going to be like. If people had not already discovered how to make clothes, they would simply have been unable to live on much of the globe's surface which is now inhabited. The adoption of clothing, with all that followed, from the introduction of sewing and weaving, down to the invention of special materials for space-travel, is one sign of the human capacity to adapt to different environments. All sorts of other things might have been different without it; signals of sexual identity for instance. But there is no need to speculate about such matters; the main point is simple enough. Without their inheritance from the past, the first human communities which produced civilization would not have been able to do so. Some, in fact, never did, and had to wait for civilization to come to them from outside.

To settle what constitutes a 'civilization' is a little like pinning down in time the first human beings. There is a shaded area in which we know a shift occurs, and we can agree about when a line has been crossed; but to look for exact dates is pointless. All over the Near East around 5000 BC farming villages could provide agricultural surpluses. Some of these little settlements had complex religious practices and could make elaborate painted pottery, already a widespread art in the Neolithic era. But by civilization we usually mean something more than the presence of ritual, art or a certain technology, and certainly something more than the mere agglomeration of human beings in the same place, though scale is important. It is best to start with what people have been able to agree about.

THE FIRST CIVILIZATIONS

The first civilizations emerged between 3500 BC and 500 BC. They begin the age in which all significant change to human life has been man-made. In them can be sought foundations of our own world; they still determine much of the cultural map of the world even today. They were the outcome of particular mixes of human skills and natural facts which came together and in each case made possible a new order of life based on the exploitation of nature. Though the earliest civilizations all appeared within a few thousand years barely a moment on the vast scale of prehistory – they were neither simultaneous, nor equally successful. Some of them raced ahead to lasting achievements while others declined or disappeared, even after spectacular flowerings. Yet all of them showed an increase in the rate and scale of change dramatic by comparison with anything achieved in earlier times.

A rough chronology begins about 3500 BC in Mesopotamia, when the first acknowledged civilization becomes visible there. In Egypt, civilized life appears at a later but still early date, perhaps about 3100 BC. In Crete, by about 2000 BC, we have another marker in what is called Minoan civilization. From that time we can disregard questions of priorities in the eastern Mediterranean and Near East: it was already by then a region where a complex of civilizations influenced and interplayed with one another. Meanwhile perhaps by 2500 BC, civilization had emerged in India; China's starts later, towards the middle of the second millennium BC, and it is a sufficiently isolated example to show that interaction does not have to be a big part of explaining what happens. From that time, except in Central and South America, there are no civilizations to be explained which appear without external stimulus, shock or inheritance from others already mature.

It is not easy to see much that the first civilizations had in common except their complete dependence on local agriculture, their achievement of writing, and their organization of society on a new scale in cities. Even if their technology was advanced by comparison with that of their uncivilized predecessors, they still all worked with little beyond muscle power, animal and human, to carry out their material purposes, and their shape and development were still determined very much by their setting. Yet they had begun to nibble at the restraints of geography and had a growing ability to exploit and overcome them. The currents of wind and water which directed early maritime travel are still there today; even in the second millennium BC men were learning to put them to use. This meant that at a very early date the possibilities of human interchange were already considerably greater than in prehistoric times.

This makes it rash to dogmatize about origins or why civilizations arose in the first place. Civilization did not appear in any standard way. No doubt it was likely always to result from the coming together of a number of factors predisposing a particular area to throw up something dense enough to be recognized later as civilization, but we do not know what catalysts or detonators were at work to speed up the process. Different environments, different influences from outside and different cultural inheritances from the past meant that humanity did not move everywhere at the same pace towards the same outcome. A favourable geographical setting was essential, but culture was important, too. Peoples had to be able to take advantage of an environment, to rise to challenges. River valleys like those of Mesopotamia, the Indus valley, China and Egypt were obviously favourable settings; their rich and easily cultivated lands could support fairly dense populations of farmers in villages which would then grow to form the first towns. But civilizations have also appeared in very different settings, in meso-America, Minoan Crete and in Greece. To the last two, we know important influences came from the outside, but Egypt and the Indus valley were at least in touch with Mesopotamia at very early dates. External contacts may be important even in

Some Landmarks in Agricultural Development in Early Civilizations

8000–5000	Cultivation of domesticated wheats and barleys, domestication of sheep, goats, cattle in SW Asia, Anatolia, Greece, Persia, and the Caspian basin. Planting and harvesting techniques transferred from Asia Minor to Europe. Millet cultivated in China.
3400–3100	Flax used for textiles in Egypt, and first evidence of ploughing, raking, and manuring there.
3000	Asses used as beasts of burden in Egypt. Sumerians growing barley, wheat, dates, flax, apples, plums, and grapes.
2900	Domestication of pigs in E. Asia.
2540	Reputed start of Chinese silk industry.
2500	Domestic elephants in the Indus valley. Potatoes grown in Peru.
2350	Wine-making in Egypt.
1600	Cultivation of vines and olives in Crete.
1500–1300	Shadoof used for irrigation in Egypt, and evidence of aqueducts and there.
1400	Iron ploughshares in use in India.
1200	Domestic camels in Arabia.

China, almost insulated from the outside world though it seems at first sight. People used to argue that we should look for one central source of civilization from which all others came. But there is not only the awkward case of isolated civilizations in the Americas to deal with; it is also very hard to get the timetable of the supposed diffusion to fit as radio-carbon dating improves our knowledge of early chronology.

It is easier, in fact, to recognize early civilization when it is in place than to know how it came to be there, and about the exact process no absolute and universal statements can plausibly be made. Civilization is the name we give to the interaction of human beings in a very creative way, when, as it were, a critical mass of cultural potential and material resources has been built up, and human capacities are released for development which becomes in large measure self-sustaining. It brings together the co-operative efforts of more people than hitherto, usually by bringing them together physically in larger masses, too. 'Civilization' is a word connected with a Latin word meaning 'city'. More than any other institution the city has provided the critical mass required and it has fostered innovation better than any other environment so far. Inside the first cities the wealth produced by agriculture was used for the upkeep of priestly classes which elaborated complex religious structures and encouraged the construction of great buildings with more than merely economic functions. Much bigger resources than in earlier times were thus allocated to something other than immediate consumption. The result, as writing became available, was a new and intensive storing of enterprise and experience. The accumulated culture gradually became more and more effective as an instrument with which to change the world.

With the coming of civilization, peoples in different parts of the world grew more rapidly unlike one another. The most obvious fact about early civilizations is their startling variety of style, but because it is so obvious we usually overlook it. The coming of civilization brings growing and faster differentiation in dress, architecture, technology, behaviour, social forms and thought. Prehistory no doubt threw up societies with different life-styles, different routines, different mentalities, as well as different physical characteristics, but the lack of wealth and specialized technology limited the range of differences. With the first civilizations differentiation becomes much more obvious, and it is the product of the creative power of civilization itself. From the first civilizations to our own day there have always been alternative and potentially stimulating and fertilizing models of human society existing at the same time, however much or little they knew of one another. Much of this new potential for variety is now very hard to recover, because we have still so little evidence in the earliest stages of civilization about the life of the mind except institutions (so far as we can understand them), the conscious symbols and unconscious statements of art, and the ideas embodied in literature.

The interplay of Culture in the Fertile Crescent

The mutually stimulating interactions of different cultures first become obvious in the Near East. A turmoil of racial comings and goings for three or four thousand years both enriched and disrupted the Fertile Crescent. It was to be for most of historic times a great crucible of cultures, a zone not only of settlement but of transit, through which poured people exchanging ideas, institutions, languages and beliefs, many of which still influence us today.

The root cause of the wanderings of peoples was probably over-population in their homelands. Yet the world's population in about 4000 BC has been estimated only at between eighty and ninety millions. In the next four thousand years it was to grow by about fifty per cent to about one hundred and thirty millions; this increase is at a rate tiny by comparison with what came later. It shows the relative slowness with which men added to their power to exploit the natural world. Yet it marks a huge demographic break with prehistoric times and it rested on a very fragile margin of resources. Drought or desiccation could dramatically and suddenly destroy an area's food supplies. The immediate results must often have been the threat of famine, one of the prime movers of early history. For thousands of years food could not easily be moved about, except on the hoof. Droughts, catastrophic storms, even a few years of marginally lower or higher temperatures, could force peoples to get on the move and so threw together different traditions. In collision and cooperation they learnt from one another and the potential of their societies thus increased.

Linguistic differences provide distinctions among the peoples of early civilized times in the Fertile Crescent. They can all be assigned either to stocks which evolved in Africa north and north-east of the Sahara (called 'Hamitic'), to the 'Semites' of the Arabian peninsula, to the 'Indo-Europeans' who, from southern Russia, had spread also by 4000 BC into Europe and Iran, or to what may be termed 'Caucasian' from Georgia and the Caucasus region. In about 4000 BC most of the Fertile Crescent seems to have been occupied by the last. They are the main protagonists of the earliest Near Eastern history. But by then Semitic peoples, too, already began to penetrate the region; in the middle of the third millennium BC they were well established in central Mesopotamia, across the middle reaches of the Tigris and Euphrates. The interplay and rivalry of the Semitic peoples with the Caucasians, who were able to hang on to the higher lands which enclosed Mesopotamia from the north-east, is a continuing theme in the early history of the area. By 2000 BC other peoples whose languages form part of what is called the 'Indo-European' group have also entered on the scene. One of them, the Hittites, pushed into Anatolia from Europe, while their advance was matched from the east by that of another, the Iranians. Between 2000 BC and 1500 BC branches of these sub-units disputed and mingled with the Semitic and Caucasian peoples in the Crescent itself, while the interplay of Hamites and Semites lies behind much of the political history of old Egypt. This confused scenario is, of course, highly impressionistic and its detail uncertain. None the less, whatever its exact nature and cause, wanderings of peoples within this general pattern are the background against which the first civilization appeared and prospered.

Sumer

It did so in the southern part of Mesopotamia, an ancient Greek name for what is now Iraq. In this seven-hundred-mile-long land formed by the river valleys of the Tigris and Euphrates human beings had long lived in some numbers, and in Neolithic times it was thickly studded with farming villages. Some of the oldest settlements of all seem to have been in the extreme south. Southern Mesopotamia then joined the sea about a hundred miles north of the point at which it does today and centuries of drainage by the two great rivers from up-country and annual floodings had built up a soil of great richness in the areas round the deltas. Crops would grow easily if water was continuously and safely available; this was usually possible, for though rain was slight and irregular, the river-bed was sometimes above the level of nearby land. Here, at an early date, was the possibility of growing more than was

This little picture of a scene at a Sumerian dairy was executed in limestone and shell set into bitumen. It probably decorated the walls of a temple dedicated to a fertility goddess often symbolized by a cow.

needed for daily consumption, the surplus allowing the appearance of town life. Furthermore, fish could be taken from the nearby sea.

This setting was both a challenge and an opportunity. The Tigris and Euphrates could suddenly and violently change course: the marshy, low-lying land of the delta had therefore to be managed by banking and ditching and canals had to be built to carry water away. Techniques like those first employed to form the platforms of reed and mud on which were built the first homesteads of Mesopotamia could still be seen in use there even in this century. Patches of cultivation would be grouped where the soil was richest and the drains and irrigation channels they needed could only be managed properly if they were managed collectively. No doubt the social organization of reclamation was a consequence, and with it went some kind of legitimized authority. However it happened, the seemingly unprecedented achievement of making land from watery marsh must have been the forcing house for a new complexity in the way Mesopotamians lived together.

As their numbers grew, more land was taken to grow food. Sooner or later men of different villages would come face to face with others intent on reclaiming marsh which had previously separated them from one another. Different irrigation needs may even have brought them into contact before this. There was likely to be a choice: fight or cooperate. Somewhere along this path it made sense for men to band together in bigger groups for self-protection as well as management of the environment.

One physical result was the town. Mud-walled at first to keep out floods and enemies, then raised above the waters on a platform, it was logical for the local deity's shrine to be the place chosen for a larger settlement: he stood behind the community's authority, which would be exercised by his chief priest, who became the manager of a little theocracy competing with others.

Something like this explains the difference between southern Mesopotamia in the third and fourth millennia BC and the other zones of Neolithic culture with which by then it was in contact. Mesopotamia and Neolithic Anatolia, Assyria and Iran all had much in common. But only in this relatively small area did a pattern of Near Eastern villages begin to grow faster and harden into something else – the first true urbanism, with many centres, and the first observable civilization, that of Sumer, the ancient name for the southernmost part of Mesopotamia. Scholars are still divided about when the Sumerians – that is, those who spoke the language later called Sumerian and probably Caucasian in root – arrived in the area: by 4000 BC they were well established. But the population of civilized Sumer came to be a mixture of races, perhaps including even earlier inhabitants of the region, and showing both foreign and local elements in its culture.

Not very differently from their neighbours, the early Sumerians lived in villages and had a few long and continuously occupied important cult centres. One, at a place called Eridu, probably originated in about 5000 BC. It grew steadily well into historic times and by the middle of the fourth millennium there was a temple there which some have thought was the original model for Mesopotamian monumental architecture, though only the platform on which it rested now remains. Such cult centres began as places of devotion and pilgrimage, with no considerable resident populations. Nonetheless, cities later crystallized around them and this helps to explain the close relationship religion and government always had in ancient Mesopotamia.

Early Sumer

c.5000	Sumerian language in use.
c.4000	Settlement on future site of Babylon.
c.3500	Sumerian language appears in written form.
c.2800	Earlier Sumerian dynasty.
c.2350–2200	Reign of Sargon I and dynasty of Akkad.
c.2150	Gutians and Amorites overthrow Akkadian dynasty.
c.2000	Akkadian rule restored as third dynasty of Ur.
c.1800–1600	Bablyonian ascendancy over Sumer.

Writing

Sumerian civilization lasted roughly from 3300 to 2000 BC. At a very early date it produced from cylinder seals little pictures which were rolled on to clay, and from these, Sumerians developed simplified pic-

Gilgamesh - ruler of Uruk - Mesopotamia Sumerian Iraq

An impression left by a Mesopotamian cylinder seal from the early centuries of third millennium. The seal was rolled over soft clay to form a continuous pattern like a frieze. This one, called the 'Gilgamesh' seal, represents a recurrent pattern of animals, men, or mysterious bearded, semi-human bulls in combat. Others show a hero holding helpless or at bay two animals, one on each side, and this was to be a pattern cropping up again and again in the art of many different traditions and peoples.

tures (pictograms) made on clay tablets with a reed stalk, a big step towards true writing. This evolved into a style called cuneiform, which used signs and groups of signs to stand for sounds and syllables. When cuneiform was available, much better communication of information than ever before was possible. It made much easier the complex operations of irrigating lands, harvesting and storing crops, and so made for more efficient exploitation of resources. It also immensely strengthened government and its links with the priestly castes who at first monopolized literacy.

Besides the records which now begin to survive in much greater quality, the invention of writing opens more of the past to us in another way, because it preserves literature. We can at last begin to deal in hard currency when talking about ideas. The oldest story in the world is the Epic of Gilgamesh, a tale which appears in Sumerian times and written down soon after 2000 BC. Gilgamesh was a real person, ruling at the city of Uruk, and he became also the first individual and hero in world literature. His is the first personal name which must appear in this book. To a modern reader the most striking part of the Epic includes the coming of a great flood which obliterated mankind except for a favoured family who survived by building an ark; from them sprang a new race to people the world after the flood subsided. This is not to be found in the story's oldest versions, but in a separate poem which turns up in many Near-Eastern forms (and in due course in the Jewish Old Testament) and its incorporation in the Sumerian epic is easily understandable. Floods were the typical disaster in lower Mesopotamia; archaeological attempts to identify a single, cataclysmic flood behind the legend of the Ark have not been convincing, but there is plentiful evidence of recurrent inundations.

It is hard to get at history through the Epic, let alone relate it to the historical Gilgamesh. From the water eventually emerges the land: perhaps, then, what we are being given is a Sumerian account of the creation of the world, of genesis. In the later Christian Bible, too, earth emerges from the waters at God's will, an account which satisfied most educated Europeans for a thousand years. We owe something of our own intellectual ancestry to a mythical reconstruction by the Sumerians of their own pre-history, when farming land had been created out of the morass of the Mesopotamian delta.

Sumerian ideas were widely diffused in the Near East long after the focus of history had moved away from Mesopotamia. Versions and parts of the Epic – to stick to that text alone for a moment – have turned up in the archives and relics of many peoples in different parts of this region in the second millennium BC. Though later to be lost to sight until rediscovery in modern times, Gilgamesh was for two thousand years or so a name to which literature in

The invention of writing was momentous. This limestone tablet is one of the earliest documents we have of the stages by which it happened. It was found in southern Mesopotamia and has been dated to about 3500 BC. The little pictures are of hands, feet and some kind of sledge (perhaps used for threshing grain); they mark a big stride towards using signs to stand for things, a process already foreshadowed in symbols on pottery. Later, the Egyptians clung to such 'pictographic' forms of writing, elaborating them into the 'hieroglyphics' which recorded their life and business for thousands of years. By then, the Mesopotamian tradition had taken another direction, away from picture-writing towards patterns of signs made of identical marks in different arrangements which could be easily formed by a reed-stalk stamp on a soft clay tablet. This process took about fifteen hundred years after these tablets were made.

many languages could knowingly refer. The Sumerian language lived on for centuries in temples and scribal schools, too, much as Latin later lived on for the learned in Europe after the collapse of the western classical world of Rome.

Sumerian religion

Literary and linguistic tradition embody ideas and images which impose, permit and limit different ways of seeing the world. They have their own historic weight. The Epic of Gilgamesh tells us something of the gods of early Mesopotamia, and probably the most important ideas kept alive by the Sumerian language were religious. By about 2250 BC a pantheon of individual gods more or less personifying the elements and natural forces had emerged. It was to be the backbone of Mesopotamian religion for thousands of years and the beginning of theology. Originally, cities had their own gods, but they formed a loose hierarchy which both reflected and shaped views of human society. To each of them was given a special activity or role; there was a god of the air, another of the water, another of the plough and a goddess of love and procreation (but also of war). At the top of the hierarchy was a trinity of three great male gods: the father of the gods, a 'Lord Air', without whom nothing could be done, and a god of wisdom and of the sweet waters that literally meant life to Sumer. This reveals a view of the supernatural much more complex and elaborate than anything elsewhere at so early a date. Significantly, temples grew bigger and more splendid (in part, because of a tradition of building new ones on mounds enclosing their predecessors) as the centuries passed. Sacrifices were offered in them to ensure good crops. We hear of one built with cedars brought from the Lebanon and copper from Anatolia. No other ancient society at that time gave religion quite so prominent a place or diverted so much of its collective resources to its support. Perhaps this was because no other ancient society left men feeling so utterly dependent on the will of the gods. Lower Mesopotamia in ancient times must have been a flat, monotonous landscape of mudflats, marsh, water. There were no mountains for the gods to dwell in on earth with men, only the empty heavens above, the remorseless summer sun, the overturning winds, the irresistible power of flood-water, the blighting attacks of drought against which there was very feeble protection. The gods lived in these elemental forces, and might be approached in the high places which alone dominated the plains, the brick-built towers and Ziggurats (faintly remembered in the later biblical Tower of Babel). The Sumerians, not surprisingly, saw themselves as a people created to serve, to labour for the gods.

The gods – though no Mesopotamian could have put it in these terms – were the conceptualization of attempts to control environment, to resist the sudden disasters of flood and dust-storm, to assure the continuation of the cycle of the seasons by the repetition of the great spring festival when the gods were again married and the drama of Creation was re-enacted. After that, the world's existence was assured for another year. Nothing else was, as it were, on offer. Men later came to ask that religion should help them to deal with the inevitable horror of death. The Sumerians and those who inherited their religious ideas seem to have seen the next world as a gloomy, sad place; in this lies the root of the later notion of Sheol, of Hell. Yet a Sumerian king and queen of the middle of the third millennium were followed to their tombs by their attendants who were then buried with them (perhaps after taking some drug). It may be that the dead were thought to be going somewhere where a great retinue and gorgeous jewellery would help. As for the political aspects of Sumerian religion, all land belonged ultimately to the gods; the king, probably a king-priest as much as a warrior-leader in origin, was their vicar. Around him stood a priestly class, the cultivators of special skills and knowledge. In this respect, too, Sumer originated a tradition, that of the seers, soothsayers, wise men of the East. They had charge of the first organized system of education, based on memorizing and copying.

Sumerian life

Among the by-products of Sumerian religion were the first true likenesses of human beings. Sometimes they are grouped in processions; thus is established one of the great themes of pictorial art. Two others are also prominent: war and the animal world. Some have also seen in Sumerian portraits the psychological qualities which made the astonishing achievements of their civilization possible, a drive for pre-eminence and success. What is certainly visible in Sumerian art is a daily life hidden from us in earlier times. Given Sumer's widespread contacts and the agriculture base it shared with neighbouring peoples, it may tell us something of life as it was lived over much of the ancient Near East. Seals, statuary and painting all reveal a people often clad in a kind of furry – goat-skin or sheep-skin? – skirt, the women sometimes throwing a fold of it over one shoulder. The men are often, but not always, clean-shaven. Soldiers wear the same costume but carry weapons and sometimes wear a pointed leather cap. Luxury seems to have consisted in leisure and possessions, especially jewellery. Its purpose often seems to be the indication of status; if so, that is another sign of growing social complexity.

The head of the family was the patriarchal husband. He married after a contract with the bride's family and presided over a household of his relatives and slaves, a pattern common until very recently in many parts of the world. Yet there are interesting nuances. Sumerian women seem less down-trodden than their sisters in many other, and later, Near-Eastern societies. Semitic and non-Semitic traditions may diverge in this. Sumerian stories of their gods suggest a society very conscious of the power of female sexuality; the Sumerians were the first people to write about passion. Their law (whose influence can be traced well past 2000 BC) into post-Sumerian times certainly gave women important rights. A woman was

Between four and five thousand years ago, little statues like these were carved by Mesopotamians for shrines where they were to pray and worship on behalf of those they represented. Perhaps they were statues of the sculptors and their families, perhaps those of wealthy men who commissioned them. The second is more likely: stone was scarce and must have been expensive in Sumer (the tallest of these figures is only about thirty inches high). These examples were discovered in what had been a pit under the floor of a temple at Tel Asmar, on a tributary of the Tigris. Besides throwing light on the way the ancient Mesopotamians worshipped, they tell us something of their daily appearance.

not a mere chattel; even the slave mother of a free man's children had rights. Women as well as men could seek separations and could hope for equitable treatment after divorce. Though a wife's adultery was punishable by death while a husband's was not, this can be understood in the light of concern over inheritance and property. Only much later did Mesopotamian law begin to emphasize the importance of virginity and to impose the veil on respectable women. Both were signs of a hardening and more cramping role for them.

By the end of their history as an independent civilization the Sumerians had learnt to live in large communities; one city alone is said to have had thirty-six thousand males. This made big demands on building skill, but even more were made by great monumental structures. Lacking stone, southern Mesopotamians first built in reeds plastered with mud, then with mud bricks dried in the sun. By the end of the Sumerian period brick technology was advanced enough to build very large buildings with columns and terraces; the greatest, the Ziggurat of Ur, had an upper stage over a hundred feet high and a base two hundred feet by a hundred and fifty. The earliest surviving potter's wheel (found at Ur) makes another technological step; this was the first known exploitation of rotary motion. On it rested the large-scale production of pottery which made it a man's trade and not, like earlier pottery, a woman's. Soon, by 3000 BC, the wheel was applied to transport. Another invention of the Sumerians was glass, and specialized craftsmen were casting in bronze early in the third millennium BC.

This raises further questions: where did the raw material come from? There is no metal in southern Mesopotamia. Even in Neolithic times, too, the region must have imported flint and obsidian for agricultural implements. A widespread network of external contacts evidently existed, both with the Persian Gulf peoples and with the distant Levant and Syria. Certainly by 2000 BC Mesopotamia was obtaining goods – though possibly indirectly – from the Indus valley. With some fragmentary documentation, this suggests a dimly emerging interregional trading system which was already creating patterns of economic interdependence.

Yet the basis of society remained agriculture for local provision. Barley, wheat, millet and sesame were grown in quantity; the first may have been the main crop, and no doubt explains the frequent evidence of the presence of alcohol in ancient Mesopotamia. In the

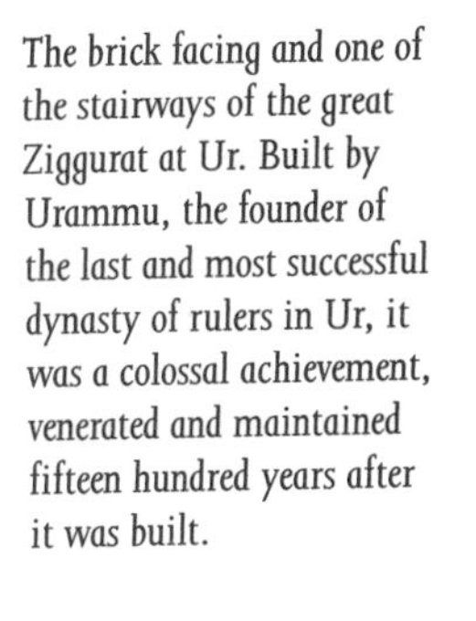

The brick facing and one of the stairways of the great Ziggurat at Ur. Built by Urammu, the founder of the last and most successful dynasty of rulers in Ur, it was a colossal achievement, venerated and maintained fifteen hundred years after it was built.

easy soil of the flood bed iron tools were not needed to achieve intensive cultivation; the great contribution of technology here was in the practice of irrigation, building and the growth of government. Such skills accumulated slowly; the evidence of Sumerian civilization is spread over fifteen hundred years of history, so far discussed almost as if nothing happened during it, as if it were an unchanging whole. Of course it was not. Whatever reservations are made about the slowness of change in the ancient world and though it may now seem to us very static, we know these were fifteen centuries of great change for the Mesopotamians even if much of it remains obscure, and even its dating is for much of the time only approximate.

Political change

Three broad phases can be marked out. The first phase of Sumer, lasting from about 3360 BC to 2400 BC, is its archaic period. The narrative story is one of wars between city-states, their waxings and wanings. Fortified cities and the application of the wheel to military technology in clumsy four-wheeled chariots are some of the evidence for it. Towards the middle of this phase, local dynasties begin to establish themselves with some success. Originally, Sumerian society seems to have had some representative, even democratic basis, but a

A royal portrait from the post-Akkadian and neo-Sumerian era. Gudea was ruler in Lagash in about 2100 BC. Mysteriously, the city seems to have escaped submission to the Gutians and to have prospered under Gudea and his son. No doubt this has something to do with the survival of about twenty different statues of this king, all of them showing the complacency and strength which literary sources suggest he drew from his life of prudent pious service to the gods. The statue is a fine example of a resumed Sumerian tradition.

growth of scale led to the distinction of kings from the early priestly rulers; they may have emerged as warlords appointed by cities to command their forces who clung to power when the emergency which called them forth had passed. From them stemmed the dynasties which fought one another.

The sudden appearance of a great individual then opens a new phase. He was Sargon I, a king of a city higher up the Euphrates called Akkad whose site is undiscovered. He conquered the Sumerian cities between 2400 and 2350 BC and inaugurated an Akkadian supremacy over them. His people came from among those Semitic tribes which had long pressed in on the civilizations of the river valleys from outside. There exists a sculpted head which is believed to be his portrait; if it is, it is one of the first royal images. He was the first of a long line of empire-builders and is alleged to have sent his troops as far afield as Egypt and Ethiopia. Sargon's rule was not based on the relative superiority of one city state to another; he set up a unified empire integrating the cities into a whole. They left behind a new style of Sumerian art marked by the theme of royal victory. Akkadian empire was not the end of Sumer, though, but an interlude and its second main phase. It expressed a new achievement in organization. By Sargon's time a true state has appeared. Lay and priestly authority had completely diverged. Palaces appeared beside the temples in the Sumerian cities; the authority of the gods lay behind the palaces' occupants, too.

The invention of professional soldiery probably played a part in this. Disciplined infantry, moving in formation with overlapping shields and levelled spears, appear on monuments from Ur. Sargon, it was boasted, had 5400 soldiers eating before him in his palace. This, no doubt, was the outcome of conquests which had provided the resources to maintain such a force. If state power had originated in the special challenges and needs of Mesopotamia, and the primary duty of the ruler to organize big works of irrigation and flood control, the power to assemble labour for this could also provide soldiers. As weapons became more complex and expensive, professionalism would be more likely. One source of Akkadian success was the use of a new weapon, the composite bow made of strips of wood and horn.

A century and a half after Sargon, under his great-grandson, the Akkadian hegemony was overthrown, apparently by mountain peoples called Gutians, and the third, last, phase of Sumer, called by scholars 'neo-Sumerian', began. For another two hundred years or so, until 2000 BC, hegemony again passed to the native Sumerians whose centre was Ur. The first king of the Third Dynasty of Ur who exercised this ascendancy called himself King of Sumer and Akkad, whatever that might mean in practice. Sumerian art in this phase showed a new tendency to exalt the power of the prince; rulers sought to embody their grandeur in bigger and better ziggurats. Administrative documents show the Akkadian legacy, too. Perhaps the aspiration to wider kingship reflects this inheritance. The tributaries of the last successful kings of Ur stretched from Susa, on the frontiers of a land called Elam on the lower Tigris, to Byblos on the coast of Lebanon.

This was the sunset era of the first civilization. It did not disappear, but its individuality was about to be lost in the general history of Mesopotamia and the Near East. A great creative era was over. From one relatively small area, the horizons of our history now need to expand. Enemies abounded on Sumer's frontiers. In about 2000 BC, the Elamites came and Ur fell to them. Why, we do not know, but after intermittent hostility for a thousand years this has been seen as the outcome of a struggle to control routes giving access to the highlands of Iran and the minerals the Mesopotamians needed. At all events, it was the end of Ur and the distinctive Sumerian tradition, now merged in the swirling currents of a world of more than one civilization. Yet for fifteen centuries or so Sumer had built up the subsoil of civilization in Mesopotamia, just as its precivilized forerunners had built up the physical land itself. The Sumerians left behind writing, a literature, a mythology, monumental buildings, an idea of justice and legalism, and the roots of a great religious tradition. It is a wonderful record and the seed of much else; the diffusion of civilized ways had already gone far when Sumer died.

MESOPOTAMIA AFTER SUMER

The Near East was by then a growing confusion of peoples. The Akkadians had pushed up originally from the great Semitic reservoir of Arabia to finish in Mesopotamia. The Gutians, who took part in the Akkadians' overthrow, were Caucasians, from among the original stocks of the area. The Amorites, a Semitic people which had spread far and wide and joined the Elamites to overthrow the armies of Ur and destroy its supremacy, had established themselves in Damascus, Assyria, or upper Mesopotamia, and Babylon in a series of kingdoms which stretched as far as the coast of Palestine. Southern Mesopotamia they continued to dispute with the Elamites. In Anatolia their neighbours were Hittites, an Indo-European people which crossed from the Balkans in the third millennium. At the edges of this chaotic picture stood other vigorous peoples.

Babylon

One landmark is a new empire in Mesopotamia: Babylon. One of its kings, Hammurabi, would have a secure place in history if we knew nothing of him except his reputation as a law-giver; his code is the oldest statement of the legal principle of an eye for an eye. He was also the first ruler to unify the whole of Mesopotamia; though the empire was short-lived, the city of Babylon was to be from his time the symbolic centre of the Semitic peoples of the south. Hammurabi may have become its ruler in 1792 BC; his successors held things together until some time after 1600 BC, when Mesopotamia was once more divided.

At its height the first Babylonian empire ran from Sumer and the Gulf north to Assyria, the upper part of Mesopotamia. Hammurabi ruled the cities of Nineveh and Nimrod on the Tigris, Mari high on the Euphrates, and controlled that river up to the point at which it is nearest to Aleppo. Seven hundred or so miles long and about a hundred miles wide, this was a great state, the greatest, indeed, to appear in the region up to this time (Ur had been a looser, tributary affair). It had an elaborate administrative structure, and Hammurabi's code of laws is justly famous, though this owes something to chance. As probably happened to earlier collections of judgements and rules which have only survived in fragments, Hammurabi's was cut in stone and set up in the courtyard of temples for the public to consult; it happened to survive. At greater length and in a more ordered way than earlier collections it assembled some 282 articles, dealing comprehensively with a wide range of questions: wages, divorce, fees for medical attention and many other matters. It was not legislation, but a declaration of existing law, and to speak of a 'code' may be misleading unless this is remembered. Hammurabi assembled rules already current; he did not create them *de novo*.

This body of 'common law' long provided one of the major continuities of Mesopotamian history. Sadly, perhaps, as time went by its penalties seem to have harshened by comparison with Sumer, but in other respects Sumerian tradition survived in Babylon in law.

The code's provisions included laws about slaves. Like every other ancient civilization and many of later times, Babylon rested on slavery. Slavery was the fate likely to await the loser of any ancient war of early history, and his women and children, too. Under the first Babylonian empire, regular slave-markets already existed and steady prices indicated a fairly regular trade, in which slaves from certain districts were especially prized. Though the master's hold on the slave was virtually absolute, some Babylonian slaves enjoyed remarkable economic independence, engaging in business and even owning slaves on their own account. They had legal rights, if narrow ones. It is hard to say what this meant in practice. Generalities dissolve in the light of evidence about the diversity of things slaves might do; if most lived hard lives, then so, probably, did most people. In ancient times civilization rested on a great exploitation of man by man; if that was not felt to be very cruel, this is only to say that no one thought any other way of running things was conceivable.

At Mari and one or two other places, shaving the head while growing a luxuriant beard seems to have been fashionable. This portrait statue comes from the first half of the third millenium bc, when the Sumerian tradition of sculpture was becoming more naturalistic and individual – this is a likeness of a steward, Ibihil, one of a few models whose name we know.

Babylonian thought

Babylonian civilization remains a legend of magnificence. The great palace of Mari, high up the Euphrates, had walls in places forty feet thick around its courtyards, and three hundred or so rooms forming a complex drained by bitumen-lined pipes running thirty feet deep. It covered an area measuring 150 by over 200 yards and in it were found great quantities of clay tablets revealing the business of Babylonian government. They give us evidence of the life of the mind, too. The Epic of Gilgamesh then took the shape in which we know it.

The Babylonians' astrology pushed forward the observation of nature. Hoping to understand their destinies by scanning the stars, they founded a science, astronomy, and established an important series of observations. By 1000 BC the prediction of lunar eclipses was possible. Within another two or three centuries the path of the sun and some of the planets had been plotted with remarkable accuracy against the positions of the apparently fixed stars. Babylonian mathematics, drew upon the intellectual achievements of the Sumerians, to whom we owe the technique of expressing number by position as well as by sign (as we, for example, can reckon the figure 1 as one, one-tenth, ten or several other values, according to its relation to the decimal point). The Sumerians had known about the decimal system, too, but did not exploit it, and arrived at a method of dividing the circle into six equal segments. From this the Babylonians progressed to the circle of 360 degrees and the hour of sixty minutes. They also worked out mathematical tables and an algebraic geometry of great practical utility.

Astronomy had begun in the temple, in the contemplation of celestial movements announcing the advent of festivals of fertility and sowing, and Babylonian religion, too, held close to Sumerian tradition. Its cosmogony began, like that of Sumer, with the creation of the world from watery waste (the name of one god meant 'silt') and the eventual fabrication of Man as the slave of the gods. In one version, gods turned men out like bricks, from clay moulds. It was a cosmic picture which suited absolute monarchy. Like the old cities, Babylon also had a civic god, Marduk; gradually he elbowed his way to the front among his Mesopotamian rivals. This took a long time, but Hammurabi said that the Sumerian gods had conferred the headship of the Mesopotamian pantheon upon Marduk, bidding him to rule over all men for their good. After the twelfth century BC Marduk's status was usually unquestioned. Meanwhile, Sumerian continued to be used in the Babylonian liturgies, in the names of the gods and the attributions they enjoyed.

Hammurabi's achievement did not long survive him. He had overthrown an Amorite kingdom established in Assyria at the end of the hegemony of Ur, but this was a temporary success. For the next thousand years Assyria was to be a battleground and prize; then it came to overshadow Babylon and the centre of gravity of Mesopotamian history moved decisively northwards. The Hittites who had established themselves in Anatolia in the last quarter of the third millennium BC pushed slowly forwards in the next few centuries; they, too, took cuneiform and adapted it to their own Indo-European language. By 1700 BC they ruled the lands between Syria and the Black Sea. Then, they turned southwards against a weakened and shrunken Babylonia. Hammurabi's dynasty and achievement finally came to an end. When the Hittites withdrew other peoples ruled and disputed Mesopotamia for a mysterious four centuries of which we know little except that during them the separation of Assyria and Babylonia which was to be so important in the next millennium was made final. By then, though, the focus of world history had shifted away from Mesopotamia.

ANCIENT EGYPT

Soon after they appear in Sumer, the first signs of civilization can be seen in Egypt. If the Egyptians learnt from Sumer (and it seems likely), we do not know how, what, or to what extent. Much more of the explanation is supplied, as in southern Mesopotamia, by the setting. Prehistoric climatic change had gradually dried up most of Egypt outside the valley of the Nile itself. Yet that narrow strip of fertile land was enough. The mud washed down from the interior highlands and deposited there made agriculture easy. On the banks of silt 1100 kilometres long and anything from six to twenty wide, the first Egyptians were able to start farming. Their land slowly turned into a long straggling oasis, surrounded by desert and

The landscape of Egyptian agriculture has been vividly preserved for us in painting on the walls of tombs. This vintage scene is from one at Thebes, for centuries Egypt's cultural and religious centre of gravity.

rock. It was importantly different from ancient Mesopotamia as a setting for a new stage of human development. The Egyptians needed no such reclamation works as the Sumerians. The Nile was friendlier than the Tigris and Euphrates. Like them it flooded each year, but it did so predictably; its floods were not sudden, surprising disasters but so regular that they set the pattern of the agricultural year. The Nile was a huge clock, regulating the life of the ancient Egyptians in a rhythm unvaried from year to year.

In about 3300 BC, substantial numbers of people already lived along some five to six hundred kilometres of the lower Nile in villages and hamlets not much separated from one another. These Egyptians at first seem to have thought of themselves as members of clans rather than settled communities. Cities did not develop along the river until thousands of years had passed, perhaps because there were no powerful neighbours to threaten the farmers and encourage them to live in towns for protection. But about these early days it is difficult to know much because the Nile destroyed the evidence each year, either washing it away to the delta, or burying it deep in the banks of soil which slowly rose above high-water mark as the centuries passed. We know that these early Egyptians knew how to make papyrus boats, to work hard stone, and to hammer copper into articles for daily use. Somewhere about the middle of the fourth millennium, too, there are signs of contact with other areas – notably Mesopotamia. There is a sudden sense of rapid crystallization.

Information rapidly becomes much more abundant about early civilized Egypt than about anywhere else at so early a date. Egyptians had from almost the start a form of writing, called 'hieroglyphic'. It was in origin pictographic, that is, consisting of little pictures which stood for the names of things; with time they came to stand for sounds. Because it was always harder to write, hieroglyph never spread as did cuneiform, though it had just as long a life; the last known example was not written until AD 394. Thereafter, as mastery of writing it was lost, so it became unintelligible. Then, at the beginning of the nineteenth century, the 'Rosetta Stone' was brought back to France from Egypt. It was inscribed in Greek and the later 'demotic' Egyptian as well as in hieroglyph, and this made translation possible. It opened the way to understanding ancient Egypt as never before, because of the large number of inscriptions on tombs, monuments and papyrus which survived.

From these hieroglyphics there emerges a narrative. Egypt was by 3000 BC already organized in two kingdoms, northern and southern, Lower and Upper, Egypt. Soon, the records

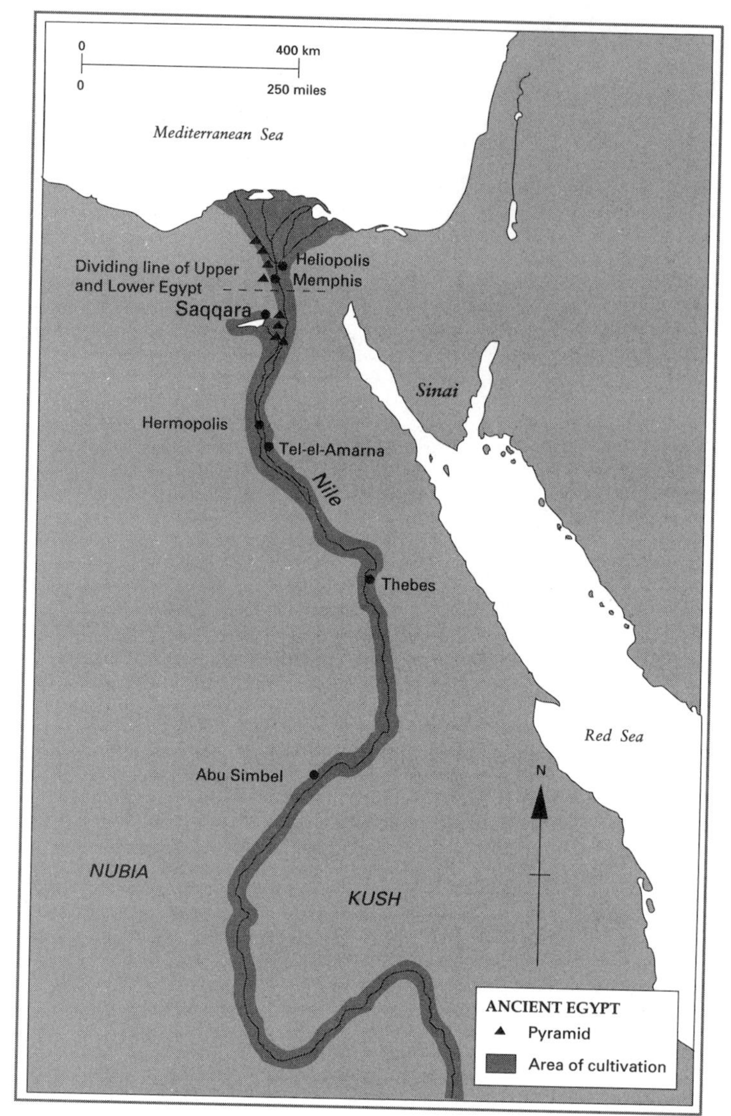

say, a king from the south called Menes conquered the north and established a dynasty which lasted until 2884 BC, ruling from Memphis in Lower Egypt. This was a realm approximately 1000 kilometres long – a much bigger affair than any other contemporary state. What its government meant is hard to say, but it is an impressive achievement to have established even a claim to rule so big an area. More striking still, this began some 2000 years during which Egypt was usually under one ruler, one religious system, and one pattern of government and society, while no important influence intruded from the outside. There were ups and downs; the state was sometimes strong and prosperous, sometimes weak and poor. Nonetheless, this is a still astonishing continuity and it made possible great achievements whose physical remains would long fascinate mankind as its greatest visible inheritance from antiquity.

The historical records enable scholars to talk about ancient Egypt as a series of dynasties in three major divisions of time. The first, termed the 'Old Kingdom' runs from 2664 to 2155 BC. There follows a century or so of upheaval before the 'Middle Kingdom' then begins in 2051. It lasted until 1786 BC when another disturbed period opened, to end in 1554 BC with the beginning of the 'New Kingdom'. Into this tripartite scheme are then fitted two 'intermediate' periods, and the dates of the dynasties. It can be conveniently closed off at the ethos

Egyptian Dynasties

I-II	Protodynastic c. 3200–2665 BC
III-VIII	Old Kingdom 2664–2155 BC
IX-XI	First Intermediate 2154–2052 BC
XII	Middle Kingdom 2052–1786 BC
XIII-XVII	Second Intermediate 1785–1554 BC
XVIII-XX	New Kingdom 1554–1075 BC

R. A. Parker's table in *The Legacy of Egypt*, 2nd edn, ed. J. R. Harris (Oxford, 1971), pp. 24-5.

beginning of the first millennium BC, when the Egyptians' greatest achievements were behind them, although there was to be an independent Egypt under its own rulers until 30 BC, when it came to an end with the suicide of the legendary Cleopatra.

Egyptian kingship

The state itself was the embodiment of Egyptian civilization. It was centred first at Memphis, the capital of the Old Kingdom. Later, under the New Kingdom, the capital was normally at Thebes. These two places were great religious centres and palace complexes; they were not really cities with lives apart from government. Partly this was because Egypt's kings had not emerged as 'big men' in a city-state community which originally deputed them to act for it. Nor were they simply men like others, subject to gods who ruled all men, great or small. They were themselves to be gods. At an early stage Egyptian monarchs already had an impressive authority, as their huge images on early monuments show; they inherited it ultimately from prehistoric kings who had a special sanctity because of their power to assure prosperity through successful agriculture. They were believed to control the annual rise and fall of the Nile: life itself, no less, to the riparian communities. The first rituals of Egyptian kingship which are known to us are concerned with fertility, irrigation and land reclamation. The earliest representations of Menes show him excavating a canal.

Under the Old Kingdom the idea appears that the king, or 'Pharaoh', was the absolute lord, and soon he was venerated as a god. Justice is 'what Pharaoh loves', evil 'what Pharaoh hates'; he is divinely omniscient and so needs no code of law to guide him. Later, under the New Kingdom, the Pharaohs were to be depicted with the heroic stature of great warriors; they are shown in their chariots, mighty men of war, trampling down their enemies and confidently slaughtering beasts of prey, but Egyptian kingship remained sacred and awesome. 'He is a god by whose dealings one lives, the father and mother of all men, alone by himself, without an equal', wrote an Egyptian civil servant of the Pharaoh as late as about 1500 BC.

By then Egypt had an elaborate and impressive hierarchy of bureaucrats. Usually, the most important came from the nobility; a few of the greatest were buried with a pomp rivalling that of the Pharaohs. Less eminent families provided thousands of scribes to staff and service an elaborate machine of government. They were trained in a special school at Thebes and their ethos can be sensed through texts which list the virtues needed to succeed as a scribe: application to study, self-control, prudence, respect for superiors, scrupulous regard for the sanctity of weights, measures, landed property and legal forms.

Most Egyptians, though, were peasants, providing labour for great public works and the surplus upon which a noble class, the bureaucracy and a great religious establishment could subsist. The land was rich enough, and was increasingly improved by irrigation techniques which were some of the earliest manifestations of the remarkable capacity to mobilize collective effort which Egyptian government long showed. Vegetables, barley, emmer were the main crops of the fields laid out along the irrigation channels; the diet they afforded was supplemented by poultry, fish and game (all of which figure plentifully in Egyptian art). Cattle were used for traction and ploughing at least as early as the Old Kingdom. With little change this agriculture sustained Egypt until modern times.

Egyptian building

It also made possible public works in stone unsurpassed in their time. Houses and farm buildings were made of mud brick. They were not meant to outface eternity. The palaces, tombs and memorials of the Pharaohs were a different matter, social and administrative as well as architectural triumphs. Under the direction of a scribe, thousands of slaves and sometimes regiments of soldiers would be deployed to cut and man-handle into position huge pieces of stone carefully dressed with first copper and then bronze tools and often elaborately incised and painted. With only such assistance as was available from levers and sleds – no winches, pulleys, blocks or tackle existed – and by the use of colossal ramps of earth, a succession of still-startling buildings was produced. The most famous are the pyramids, which dominate the great complexes of buildings housing the king after death. Among those of the Third Dynasty, at Saqqara, near Memphis, one, the 'Step Pyramid', was the masterpiece of the first identified architect, Imhotep, chancellor to the king. He was later to be deified – as the god of medicine – as well as revered as astronomer, priest and sage. The beginning of building in stone was attributed to him and it is easy to believe that something so unprecedented as the two-hundred-foot-high pyramid was seen as evidence of godlike power. During the Fourth Dynasty, though, still greater pyramids were completed at Giza. Cheops' pyramid was twenty years in the building and huge quantities of stone (between five and six million tons) were brought to it from as much as 500 miles away. This colossal construction is perfectly orientated and its sides, 750 feet long, vary by less than ten inches. Unsurprisingly, the Pyramids later figured among the Seven Wonders of the World. Nor, of course, were they the only great monuments of Egypt. At other sites there were great temples, palaces, the tombs of the Valley of the Kings.

These huge public works explain why the Egyptians were later thought to have been great scientists: people could not believe that such building did not require the most refined mathematical and scientific knowledge. Yet though Egyptian surveying was highly skilled and Egyptian civil servants were accomplished civil engineers, only elementary mathematics was needed to build as they did. Competence in mensuration and the manipulation of certain formulae for calculating volumes and weights was sufficient, and this was as far as

The step pyramid at Saqqara was conceived as a staircase to heaven and built as a monument to a Third Dynasty King in about 2650 BC by Imhotep. He was an administrator and engineer of genius who for the first time fully exploited techniques for the handling and shaping of stone which were just coming to maturity.

Egyptian mathematics went. The Egyptians did not rival Babylon in the sciences. The inscriptions in which Egyptian astronomical observations were recorded commanded centuries of respect from astrologers, but their scientific value was low and their predictive quality relatively short-term. Their one solid achievement was the calendar. The Egyptians established the solar year of 365 days, which they divided into twelve months, each of three 'weeks' of ten days, with five extra days at the end of the year (an arrangement, it may be remarked, revived in 1793 when the French revolutionaries sought to replace the Christian calendar by one more rational).

Religion

The religious life of ancient Egypt also greatly struck foreigners. Yet it remains something with which it is difficult to come to grips, an all-pervasive framework, as much taken for granted as the circulatory system of the human body, rather than an independent structure such as what later came to be understood as a church. It was not consciously seen as a growing, lively force: it was, rather, one aspect of reality, a description of an unchanging cosmos. But this, too, may be a misleading way of putting it; we have to remember that concepts and distinctions which we take for granted did not exist for the ancient Egyptians. The boundary between religion and magic, for example, hardly mattered for them.

Whatever religion in ancient Egypt meant, for almost the whole duration of their civilization, the ancient Egyptians show a remarkably consistent tendency to seek through it a way of penetrating the variety of the flow of ordinary experience so as to reach a changeless world most easily understood through the life the dead lived there. Perhaps the pulse of the Nile is to be detected here, too; each year it swept away and made new, but its cycle was ever recurring, changeless, the embodiment of a cosmic rhythm. The supreme change threatening men was death, the worst expression of the decay and flux which was their common experience. Egyptian religion seems from the start obsessed with it: its most familiar embodiments, after all, are the mummy and the grave-goods from funeral chambers preserved in our museums. Under the Middle Kingdom it came to be believed that all men, not just the king, could expect life in another world. Accordingly, through ritual and symbol, through preparation of the case he would have to put to his judges in the afterworld, a man might prepare for the afterlife with a reasonable confidence that he could achieve the changeless well-being offered by it in principle. The Egyptian view of the afterlife was, therefore, unlike the gloomy version of the Mesopotamians; men could be happy in it. The centuries-long struggle to assure that they would explains the obsessively elaborate care shown in preparing tombs and conducting the deceased to his eternal resting-place.

The Egyptian pantheon was huge. There were about two thousand gods and several important cults, some originating in prehistoric animal deities. One, Horus, the falcon god, was also god of the dynasty. These animals underwent a slow but incomplete humanization; artists stick their animal heads on to human bodies. Their relationships were rearranged in fresh patterns as the Pharaohs sought through the consolidation of their cults to achieve political ends. In this way the cult of Horus was consolidated with that of the sun-god, of whom the Pharaoh came to be regarded as the incarnation. It was not the end of the story. Horus later underwent another transformation, to appear as the offspring of Osiris, the central figure of a national cult, and his consort Isis. This goddess of creation and love was probably the most ancient of all – her origins, too, go back to the pre-dynastic era, she is one development of a ubiquitous mother-goddess of whom evidence survives from all over the Neolithic Near East. She was long to endure, her image, the infant Horus in her arms, surviving into the Christian iconography of the Virgin Mary. Among the most celebrated variations of a doctrinal and speculative kind, must be reckoned the attempt of a fourteenth-century pharaoh to establish the cult of Aton, another manifestation of the sun. It has been seen as the first monotheistic religion.

Egyptian art and technology

The gods loom large in ancient Egyptian art, but it contains much more besides. It was based on a fundamental naturalism which, however restrained by conventions of expression and gesture, gave it for two millennia at first a beautiful simplicity and later, in a more decadent style, an endearing charm and approachability. It permitted a realistic portrayal of scenes of everyday life. Rural themes of farming, fishing and hunting are deployed; craftsmen are shown at work on their products, and scribes at their duties. For some two thousand years, artists were able to work satisfyingly within the same classical tradition while borrowing from foreign influences, with a central strength and solidity which never wavers. It must have been one of the most impressive visual features of Egypt to a visitor in ancient times; what he saw was all of a piece, the longest and strongest continuous tradition in the whole history of civilized art.

Some ancient Egyptian art has survived thanks to the invention (as early as the First Dynasty) of papyrus – strips of reed-pith, laid criss-cross and pounded together into a homogeneous sheet. This was a real contribution to the progress of mankind. It did more for communication than hieroglyph. Papyrus was cheaper than skin (from which parchment was made) and more convenient (though more perishable) than clay tablets or slates of stone. It was therefore to be the most general basis of correspondence and record in the Near East until well into the Christian era, when the invention of paper arrived from the Far East

Egyptian goldsmiths at work, pictured on a Theban tomb of a Pharaoh's first minister during the Eighteenth Dynasty.

(and took its name from papyrus). Soon after the appearance of papyrus, writers began to paste sheets of it together into a long roll: thus the Egyptians invented the book, as well as the material on which it could first be written. A huge proportion of what we know of antiquity comes to us on papyrus.

The supposed prowess of her religious and magical practitioners and the spectacular embodiment of a political achievement in art and architecture explains much of Egypt's continuing prestige. Yet if her civilization is looked at comparatively, it seems neither very fertile nor very responsive. Stone architecture is the only major innovation for a long time after the coming of literacy; the Egyptians invented the column. Technological history suggests a people slow to adopt new skills, reluctant to innovate once the creative jump to civilization had been made. There is no definite evidence of the presence of the potter's wheel before the Old Kingdom; for all the skill of goldsmiths and coppersmiths, bronze-making does not appear until well into the second millennium BC and the lathe only much later still. The bow-drill was almost the only tool for the multiplication and transmission of energy available to Egyptian craftsmen. Though papyrus and the wheel were known under the First Dynasty, Egypt had been in touch with Mesopotamia for getting on for two thousand years before she adopted the well-sweep, by then long in use to irrigate land in the other river valley.

Life for the Egyptian poor was hard, but not unremittingly so. The major burden must have been conscript labour services. When these were not exacted by Pharaoh, then the peasant would have considerable leisure at those times when he waited for the flooding Nile to do its work for him. The agricultural base was rich enough, too, to sustain a complex and variegated society with a wide range of craftsmen. About their activities we know more than of those of their Mesopotamian equivalents, thanks to stone-carvings and paintings. The great division of this society was between the educated, who could enter the state service, and the rest. Slavery was important, but, it appears, less fundamental an institution than elsewhere in the ancient Near East.

Women in ancient Egypt

Tradition in later times remarked upon the seductiveness and accessibility of Egyptian women. It helps to give an impression of a society which gave them somewhat more independence and higher status than their sisters in other civilizations. Some weight must be given to an art which depicts court ladies clad in fine and revealing cottons, exquisitely coiffured and jewelled, wearing the carefully applied cosmetics to whose provision Egyptian merchants gave much attention. We should not lean too strongly on this, but even the pictorial impression of the way in which women of the Egyptian ruling class were treated is important, and it is one of dignity and independence. The Pharaohs and their consorts and other noble couples – are sometimes depicted, too, with an intimacy of mood found nowhere else in the art of the ancient Near East before the first millennium BC. Their suggestion of a real emotional equality can hardly be accidental.

The beautiful and charming women who appear in many of the paintings and sculptures may reflect a certain political potential for their sex which was lacking elsewhere. The throne often in practice descended through the female line. An heiress brought to her husband the right of succession; hence there was much anxiety about the marriage of princesses. Many royal marriages were of brother and sister, apparently without unsatisfactory genetic effects; some Pharaohs married their daughters, but perhaps to prevent anyone else marrying them rather than to ensure the continuity of the divine blood. Some female consorts exercised important power and one even occupied the throne, being willing to appear ritually bearded, in a man's clothes, and taking the title of Pharaoh.

There is also much femininity about the Egyptian pantheon, notably in the cult of Isis, which is suggestive. Literature and art stress a respect for the wife and mother which goes beyond the confines of the circle of the notabilities. Both love stories and scenes of family life

Musicians and dancers entertain guests at a feast in an Egyptian house under the New Kingdom.

emphasize a tender eroticism, relaxation and informality. Some women were literate and there is even an Egyptian word for a female scribe, but there were, of course, not many occupations open to women outside the home except those of priestess or prostitute. If they were well-off, however, they could own property and their legal rights seem in most respects to have been akin to those of Sumerian women.

The Old and Middle Kingdoms

For all the records, it is difficult to keep in perspective Egypt's relations with the world outside or the ebb and flow of authority within the Nile valley. There are huge tracts of time to account for when it is difficult to be sure exactly what was going on and what was its importance. For nearly a thousand years after Menes, Egypt's history can be considered in virtual isolation. It was to be looked back upon as a time of stability when Pharaohs were impregnable. Yet under the Old Kingdom there has been detected a decentralization of authority; provincial officers show increasing importance and independence. The Pharaoh, too, still had to wear two crowns and was twice buried, once in Upper and once in Lower Egypt; that division was still real. Relations with neighbours were not remarkable, though a series of expeditions was mounted against the peoples of Palestine towards the end of the Old Kingdom. The First Intermediate period which followed saw the position reversed and Egypt was the invaded, rather than the invader. No doubt weakness and division helped Asian invaders to establish themselves for a time in the valley of the lower Nile.

The Middle Kingdom was effectively inaugurated by a powerful king who reunified the kingdom from his capital at Thebes. For about a quarter-millennium after 2000 BC, Egypt enjoyed a period of recovery and there was a new emphasis on order and social cohesion. The divine status of the Pharaoh subtly changes: not only is he God, but it is emphasized that he is descended from gods and will be followed by gods. The eternal order will continue unshaken after bad times have made men doubt. It is certain, too, that there was expansion and material growth. Great reclamation work was achieved in the marshes of the Nile. Nubia, to the south, between the first and third cataracts, was conquered and its goldmines exploited. Egyptian settlements were founded even farther south, too, in what was later to be a mysterious kingdom called Kush. Trade leaves more elaborate traces than ever before

A huge building, the mortuary temple of Amenhotep III, once stood on the site in theTheban plain now marked only by two towering statues, of which this is one, each over sixty feet high and now known as the Colossi of Memnon.

and the copper mines of the Sinai were now worked again. Yet the Middle Kingdom ended in political upheaval and dynastic competition.

The Second Intermediate period of roughly two hundred years was marked by another and far more dangerous incursion of foreigners remembered as the Hyksos. Not much is known about them, but they were possibly a Semitic people, who used the military advantage of iron-fitted chariots to win overlordship in the Nile delta. Seemingly, they took over Egyptian conventions and methods, even maintaining the existing bureaucrats at first, but this did not lead to assimilation and under the Eighteenth Dynasty they were evicted; this was the start of the New Kingdom, which followed up victory in the years after 1570 BC by pursuing the Hyksos into their strongholds in south Canaan and in the end occupying much of Syria and Palestine.

The New Kingdom

The New Kingdom in its prime was internationally very successful and left rich physical memorials. There was under the Eighteenth Dynasty almost a renaissance of the arts, a transformation of military techniques by the adoption of Asiatic devices such as the chariot, and, above all, a huge consolidation of royal authority. Interestingly, too, this was when the throne was for a time occupied by a woman, Hatshepsut. For a century or so Egypt won further military glory, Hatshepsut's consort and successor, Thotmes III, carrying the limits of the empire to the Euphrates. Monuments recording the arrival of tribute and slaves or marriages with Asiatic princesses testify to Egyptian pre-eminence; at home the period is generally regarded as the peak of Egyptian artistic achievement, though foreign influences (from Crete) can be discovered.

Towards the end of the New Kingdom, signs of multiplied foreign contacts begin to show also that the world outside Egypt had already changed in many ways. Even Thotmes III had taken seventeen years to subdue the Levant and he had to leave unconquered a huge empire ruled by people called the Mitanni who dominated eastern Syria and northern Mesopotamia. Later, a Mitanni princess married a pharaoh and the New Kingdom came to rely on the friendship of her people to protect Egyptian interests in this area. Egypt was being forced out of the isolation which had long protected her. But the Mitanni were under growing pressure to the north from the Hittites, one of the most important of the peoples whose ambitions and movements break up the world of the Near East more and more in the second half of the second millennium BC. Things went on changing.

Egypt reached its peak of prestige and prosperity under Amenhotep III (c.1410–375 BC). It was the greatest era of Thebes and he was fittingly buried there in the largest tomb ever prepared for a king, though nothing of it now remains but the fragments of the huge statues the Greeks later called the colossi of Memnon (a legendary hero, whom they supposed to be Ethiopian). His successor, Amenhotep IV, attempted a religious revolution, the substitution of a monotheistic cult of the sun-god Aton for the ancient religion. To mark his seriousness, he changed his name to Akhnaton and founded a new city at Amarna, 300 miles north of Thebes, where a temple with a roofless sanctuary open to the sun's rays was the centre of the new creed. The opposition his religious revolution provoked helped to cripple him on other fronts. Meanwhile, Hittite pressure was telling on the Egyptian dependencies; Akhnaton could not save the Mitanni. They lost all their lands west of the Euphrates to the Hittites in 1372. A civil war then preceded their kingdom's disappearance thirty years or so later. The Egyptian imperial sphere was crumbling.

The era of decline

Amenhotep IV had changed his name because he wished to erase the reminiscence of the cult of the old god Amon; his successor and son-in-law changed his name from Tutankhaton to register its restoration and the overthrow of the attempted religious reform. It may

Akhnaton (or *Amenhotep IV*), one of the most celebrated kings of ancient Egypt, but famous also for the disorder he introduced to Egyptian life through the promotion of the cult of *Aton*, the Sun God, bitterly opposed by the priests of *Amon*.

On the walls of a temple Ramses III celebrates his victory over the 'sea-peoples' at the beginning of the twelfth century BC. *This fragment shows Egyptian vessels in action against the invaders, who may have included peoples later termed Philistines and Achaeans.*

have been gratitude for this that led to the magnificent burial in the Valley of the Kings which was given to Tutankhamon after only a short and otherwise unremarkable reign. When he died, the New Kingdom had two centuries of life ahead, but they were centuries of only occasionally interrupted and steadily accelerating decline. Later kings made efforts to recover lost ground and sometimes succeeded; the waves of conquest rolled back and forth over Palestine and at one time a pharaoh took a Hittite princess as a bride as his predecessors had taken princesses from other peoples. But there were yet more new enemies appearing; even a Hittite alliance was no longer a safeguard. The Aegean was in uproar, the islands 'poured out their people all together' and 'no land stood before them', say the Egyptian records. These 'sea peoples' were eventually beaten off, but the struggle was hard. From about 1150 BC the signs of internal disorganization, too, are plentiful. One king, Ramses III, died as a result of a conspiracy in the harem; he was the last to achieve some measure of success in offsetting the swelling tide of disaster. We hear of strikes and economic troubles under his successors and sacrilegious looting of the royal tombs at Thebes. The age of Egypt's imperial power was in fact over.

So was that of the Hittites, and of other empires of the end of the second millennium. The world which was the setting of Egypt's glories was itself passing away. Much of the explanation of Egypt's decline must be sought in that. Yet it is impossible to resist the feeling that the end of the New Kingdom exposes weaknesses present from the beginning. The creativity of Egyptian civilization seemed, in the end, strangely to miscarry. Colossal resources of labour were massed under the direction of outstanding civil servants, but only to set up the greatest tombstones the world has ever seen. Craftsmanship of exquisite quality was employed, but to make grave-goods. A highly literate elite utilizing a complex and subtle language and possessing, in papyrus, a material of unsurpassed convenience, deployed them copiously in texts and inscriptions, but left to humanity no great philosophical or religious idea. It is difficult not to sense an ultimate sterility, a nothingness, at the heart of this glittering *tour de force*. Only its sheer staying-power remains amazing. It worked for a very long time, undergoing at least two phases of considerable eclipse, but recovering from them, seemingly unchanged. Survival on such a scale is a great material and historical success; what remains obscure is why it should have stopped at survival. In the end, even Egypt's military and economic power made little permanent difference to the world and her civilization was never successfully spread abroad.

Of course, in early times all social and cultural change was slow and often imperceptible. Used as we are to change, it is difficult for us to sense the huge inertia possessed by any successful social system (one, that is, which enables men to grapple effectively with their physical and mental environment) in the ancient world. Innovation had far fewer and far more occasional sources then than now. The pace of history is rapid in ancient Egypt if we think of prehistory; it seems glacially slow if we reflect how little daily life must have changed between Menes and Thotmes III, a period of more than fifteen hundred years (and comparable to that which separates us from the end of Roman Britain). Only very slowly could technology or economic forces exert such pressures for change as we take for granted. As for intellectual stimuli, these could hardly be strong in a society dedicated to the inculcation of routine and preparation for death.

ASIA'S FIRST CIVILIZATIONS

By 1000 BC – an arbitrary date – civilization was well-established east of the Fertile Crescent and Egypt. Both India and China had by then evolved distinctive patterns of civilized life. Whatever their peripheral contacts with lands further west, they were quite unlike earlier examples. Indigenous cultures had created in those countries ways of thinking and behaving which were, over a huge area, very distinctive. They were the heartlands of the two dominating geographical (and later cultural) divisions which divide Asia east of Afghanistan and the Iranian highlands and south of Siberia. The Indian zone is bounded by the Himalayas and their neighbouring mountain ranges, the Burmese and Siamese highlands and the shores of the Indonesian archipelago. The other, east Asian, zone principally consists of the huge mass of China, but includes also Korea, Japan and Indo-China. For most of historical times, the civilizations which appeared in India and China tended to dominate these two big sub-divisions of Asia. Immensely varied in terrain, climate and geography, they contain some of the greatest rivers in the world, all of which – the Indus, Ganges, Brahmaputra, Mekong, Yangtze and Yellow rivers – drain vast quantities of water from the inner Asian highlands. Two provided valleys which were to be the sites of Asia's first civilized life.

Looked at a little more closely, the Indian sub-continent is about the size of Europe. It was for a long time virtually isolated by geography. Until the sixteenth century, India has rarely been invaded except by the passes of the north-west. The mountains there – and still more those further north – are some of the highest in the world; to the north-east lay belts of jungle. On its other two sides, the Indian peninsula looks out on the huge expanses of the Indian ocean. This has given it a very distinctive and varied, but tropical climate. The northern mountains keep away the worst of the icy winds of Central Asia (though north India can be bitterly cold in winter) but the long coasts are open to the rain-laden clouds which roll in from the oceans to water the arid plains of northern India in the annual monsoon. This sets India's climatic and therefore agricultural clock, bringing rain during the hottest months of the year, and ensures that the southern highlands of the Deccan remain heavily forested. Agriculture seems to have established itself first in the north-western alluvial plains of the Indus.

China's cultural zone is even larger. It is itself bigger than the United States; from Peking to Hong Kong, more or less due south, is 1200 miles as the crow flies. Its huge expanse contains many climates and many regions. In summer the north is scorching and arid while the south is humid and subject to floods; the north looks bare and dustblown in the winter, while the south is always green. Civilization in China has tended always to spread from north to south and to be stimulated by currents from Mongolia and Central Asia. The major topographical divisions of this vast country are set by three great river valleys which drain the interior and run across the country roughly from west to east. They are, from north to south, the Hwang-Ho, or Yellow River, the Yangtze and the Hsi. Until the coming of the European, they did not help to link China to the outside world. She was long as isolated as India. Much of China is mountainous and except in the extreme south and north-east her frontiers still sprawl across and along great ranges and plateaux. The head-waters of the Yangtze, like those of the Mekong, lie in the high Kunlun, north of Tibet. These highland frontiers are great insulators. The arc they form is broken only where the Yellow River flows south into China from inner Mongolia.

EARLY INDIAN CIVILIZATION

Civilization in India is older than in China, but has a more disjointed history. In some ways, ancient India is with us still, visible and accessible as in no other early centre of civilization. At the beginning of this century, many Indians still lived as all our primeval ancestors must once have lived, by hunting and gathering. The bullock-cart and the potter's wheel of many villages today are, as far as can be seen, identical with those used four thousand years ago. Gods and goddesses whose cults can be traced to the Stone Age are still worshipped at village shrines. Social arrangements whose main lines were set well before 1000 BC still regulate the lives of millions of Indians, Christians and Moslems as well as Hindus.

The main streets of Mohenjo-Daro, thirty to forty feet wide, ran north and south,probably because of the direction of the prevailing winds. The city was better ordered and planned than those of earlier Mesopotamia or later India. Even its carefully laid-out drains were provided with inspection holes. It was a rich well-ordered place, apparently investing its wealth in its comfort and material well-being, rather than in temples and palaces.

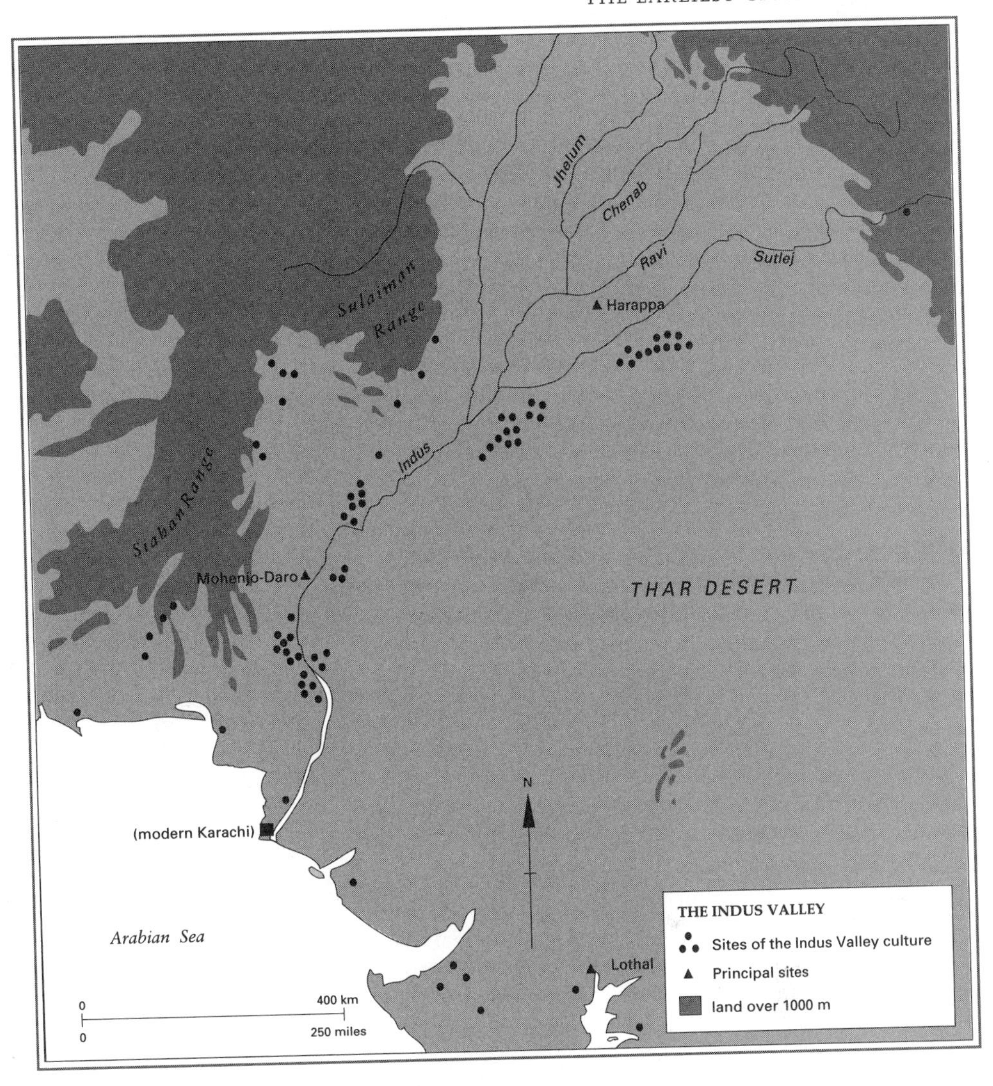

The dark-skinned peoples of the type called Dravidian, whose descendants are now mainly to be found in south India, lived also in the north about five thousand years ago. They may even have been the aboriginal Indians, though we cannot be sure. Many peoples have entered the sub-continent through the passes of the north-western mountains; perhaps some other ethnic movement may have touched off civilization in the Indus valley, where agriculture began in India and where there is the first evidence of pottery made on a wheel. But no one can say for certain. Although the idea that Indian civilization was taken from the Near East has attracted some scholars, it may simply be that Indians arrived at civilization for themselves, as the Mesopotamians had done.

The evidence that civilization had appeared in India by about 2500 BC has been provided by archaeologists working at forty or fifty sites – among them, the ruins of two whole cities, at Mohenjo-Daro and Harappa, both close to the Indus, but about four hundred miles apart. They were big and impressive places, perhaps containing thirty thousand people each at their greatest extent and between two and two and a half miles in circumference. Obviously, the Indus valley could not have been at that date the arid region it later became, because civilization on such a scale (and 'Harappan' civilization, as it is called, stretched over a very wide area) would have demanded a very productive agriculture. There must also have been ways of controlling the Indus, always prone to flood, with drainage and irrigation systems, before city life could emerge. That suggests, in turn, high levels of organizational, administrative and technological skill. One crucial invention may have been baked brick, the material of which the Harappan cities are built. Unlike the sun-dried brick of Mesopotamia, this would provide good material for flood-control in a valley lacking stone. Baked brick could be used for dams, culverts and canals.

There were docks, too, one at Lothal being connected by a mile-long canal to the sea. Such evidence of trade with the outside world (perhaps in cotton goods, of whose existence this civilization provides the earliest evidence) reinforces the impression of wealth left by the remains of the cities themselves. Each had a citadel and residential areas with houses laid out regularly on a grid pattern. The streets were not paved, but the houses were solidly built of standardized brick and were well-provided. Brick-lined sewers carried away water from wash-places and latrines, with covered man-holes for cleaning and inspection. Rubbish shoots from the houses collected other kinds of waste. Large baths or tanks in public places, like those of countless Indian villages today, have also been discovered. These may not have been only for water; together with the evidence of the attention to sanitation they suggest that the lasting emphasis of Indian religion on bathing and the ritual ablution of later Hinduism may have very ancient origins.

Few pieces of sculpture have attracted so much speculation as this soapstone figure of a priest or deity from Mohenjo-Daro. The ornamentation of the cloak perhaps reflects a theme to be found in Mesopotamia and Persia, the arrangement of the same garment anticipates the later style in which robes were worn by Buddhist monks, the modelling of the head and features suggests faint reminiscences of Sumer. Nothing about the relationship of sculpture from Mohenjo-Daro – of which this is one of only a few fragmentary examples – to the sculpture of other times and places can safely be deduced from it.

Harappan civilization was literate. Its writing is to be found on thousands of seals which seem to have been used for marking bundles of goods sold abroad, as well as a few fragments of pottery. The calligraphy – a pictographic form – has not yet yielded much to scholarly inquiry, beyond a suggestion that the Harappan language was akin to Dravidian tongues still surviving in southern India. Literacy was, nonetheless, clearly important in the working of an effective and far-ranging system of administration. Weights and measures seem to have been standardized over a wide area and large, surely public, granaries were built in the cities. Anchor-stones found in the dock at Lothal show that big vessels docked there, and traces of the seals used on bundles of goods have turned up as far away as the upper end of the Persian gulf: Harappan commerce was far-ranging. It is known, too, that ideas and techniques from the Indus spread throughout Sind and the Punjab, and down the west coast of Gujarat. The process took centuries, though, and the picture revealed by archaeology is at present too confused for a clear picture to emerge.

Aryan India

Within a couple of centuries after 2000 BC, Indus civilization came to an end. Why, we do not know. Some hold that it may have been a victim of an early manmade environmental disaster. Perhaps it arose from excessive tree-felling to feed the brick-kiln fires and so to the wrecking of the delicate balance of agriculture on the banks of the Indus. Or, clearing for cultivation or pasture may have caused erosion and dessication and a collapse of productivity. If either or both of these occurred, they would have made control of the devastating Indus floods and alterations of course – always potential hazards – much harder. We really do not know.

There is also the possibility that another force was at work, adding to the troubles of a land already devastated by over-exploitation of the environment, and it has attracted support because it may link Indian history to major disturbances elsewhere. Harappan civilization seems to come to an end in about 1750 BC. That coincides strikingly with the irruption into the sub-continent of a new group of invading peoples, the so-called Aryans. They were 'Indo-European' and, like that word, 'Aryan' is a linguistic term. But it has customarily and conveniently been specially applied to one group of the Indo-European peoples who began to enter India from the Hindu Kush in about 2000 BC, at about the time when other Indo-Europeans were flowing into Iran. This was the beginning of centuries during which waves of these migrants washed deeper and deeper into the Indus valley and the Punjab.

Eventually, they reached the upper Ganges. Scholars do not favour the idea that they alone destroyed the valley cities, nor did they obliterate the native peoples, but no doubt much violence marked their coming, for the Aryans were warriors and nomads, armed with bronze weapons, bringing horses and chariots. In the Indus cities, skeletons have been found which suggest struggle. There are plenty of signs, though, that elsewhere the native population lived beside the newcomers keeping their own beliefs and practices alive and producing some fusion of Harappan with later ways.

The Aryans had no culture so advanced as what they found. Writing disappears with their arrival, not emerging again until the middle of the first millennium BC; cities, too, have to be re-invented, and when they appear again lack the elaboration and order of their Indus valley predecessors. The Aryans seem only slowly to have given up their pastoral habits and settled to agriculture, spreading east and south from their original settlement areas in a sprawl of villages. This took centuries. Not until the coming of iron was it complete and the Ganges valley colonized. By then, Aryan culture had nevertheless made decisive contributions to Indian history.

The first was to lay the foundations of Indian religion. Their own centred on sacrificial concepts; through sacrifice the process of creation which the gods achieved at the beginning of time was to be endlessly repeated. Agni, the god of fire, was very important, because it was through his sacrificial flames that men could reach the gods. Great importance and standing was given to the brahmans, the priests who presided over these ceremonies. Among other deities, two of the most important were Varuna, god of the heavens, controller of natural order and the embodiment of justice, and Indra, the warrior god who, year after year, slew a dragon and thus released again the heavenly waters which came with the breaking of the monsoon. We learn about them and others in the Rig-Veda, a collection of more than a thousand 'Vedic' hymns performed during sacrifice, which was accumulated over centuries.

The Rig-Veda seems to reflect an Aryan culture already changed by settlement in India and not Aryan culture as it had existed at earlier times. It was at first passed on as oral tradition, and its sanctity made its exact memorization essential. It was still almost certainly largely uncorrupted from its original form (first put together in about 1000 BC) when first written down at some time after AD 1300. Together with later Vedic hymns and prose works, it is the best source of information for Aryan India, whose archaeology is cramped for a long time because the Aryans built in wood, not the brick of the Indus valley cities.

The India described in the Rig-Veda stretches from the western banks of the Indus to the Ganges. Within it Aryan peoples and dark-skinned native inhabitants lived side by side in societies whose fundamental units were families and tribes. In them certain social patterns of Aryan origin were to prove of enduring importance. Their basis is now called caste. Along with their religion, and their language, Sanskrit (which is the basis of Indian languages still spoken today), the caste system is the major legacy of Aryans to later India. Castes, as they evolved, became groups of people following the same occupation and alone entitled to do so. Membership of caste is hereditary and, ideally, caste members marry only with one another, share special ritual practices and obligatory acts and, if strict, will only eat food pre-

52808

pared by other members of the same caste. There eventually came to be hundreds of castes and sub-castes. As society became more complex, more and more marriage and eating taboos were codified. The demands of the system eventually became a primary regulator of Indian society, and in many Indians' lives the most significant. By modern times, thousands of local and occupational castes, together with the ties of tribe, family and locality, formed a structure of power in Indian society as influential as formal political institutions and centralized authority have ever been.

The system had begun with a simple division of Aryan society into three classes: brahmans, warriors and farmers. These were not at first so closely defined nor so exclusive as they later became; for some centuries, it appears, people could move from one caste to another. The only unleapable social barrier in early times seems to have been that between Aryans and non-Aryans. A fourth social class seems to have been singled out rigidly; this contained the original native population, darker-skinned than the invaders, who wanted to keep separate from them and therefore saw them as outside the three-class system altogether. The non-Aryans, members of the new fourth class, became the 'unclean'; because they were non-Aryans they could not take part in religious sacrifices, or study or hear the Vedic hymns. In the end, these 'unclean', originally identified from a wish to preserve ethnic purity, became the 'untouchables' of modern India, a class to which is left the dirty work of cleaning and scavenging, so looked down upon that some brahmans still feel that the shadow of a sweeper falling across food pollutes it.

Early Aryan tribal society already had kings and by about 600 BC there was a scatter of sixteen or so kingdoms well established in the Ganges valley. This was the outcome of centuries of steady pressure eastwards and south-eastwards. Peaceful settlement and intermarriage seem to have played as big a part in it as conquest, though exactly how cannot be confidently disentangled from mythology. Gradually, during this era, the centre of gravity of Aryan India shifted from the Punjab to the Ganges valley as Aryan culture was adopted by the peoples already there. That was already by the seventh century BC the major centre of Indian population; perhaps the cultivation of rice made this possible. A second age of Indian cities began, the earliest of them market-places and centres of manufacture, bringing together specialized craftsmen. The great plains, together with the development of larger and better-equipped armies (we hear of the use of elephants), favoured the consolidation of larger political units. The existence of coinage and the rediscovery of writing make it likely that they had governments of growing solidity and regularity. Something about them and the great names involved can be found in later documents, above all in two great Indian epics, the Ramayana and the Mahabharata, whose texts were for the first time written down as we know them in about AD 400.

Whatever the local characteristics of the Indian cultures, Aryan civilization continued to spread. After advance towards Bengal down the Ganges valley, it moved south along the west coasts towards Gujarat, and then towards the central highlands of the sub-continent. There, Aryan settlement seemed to stop. The Deccan has always been cut off from the north by jungle-clad hills, the Vindhya. Internally, too, the south is broken and hilly, and this did not favour the building of large states. Instead, south India remained fragmented, some of its peoples persisting, thanks to their inaccessibility, in the hunting and gathering cultures of a tribal age. But there is a danger of distortion. The written evidence and the classical texts make it all too easy to forget the existence of the non-Aryan half of the sub-continent. Such archaeological evidence as there is from the south shows a clear and continuing cultural lag even in the early period between the area of the Indus system and the rest of the sub-continent, to which the river was to give its name. In the south, bronze and copper only begin to appear well after the Aryan arrival in the north. Once outside the Indus system, too, there are no metal sculptures, no seals and fewer terracotta figures from pre-Aryan times. The survival of Dravidian languages in the south shows the region's persistent isolation.

Estimates of ancient populations are notoriously unreliable but India's has been put at about 25 million by 500 BC, which may have been roughly a quarter of the whole population of the world at that time. India's early history nevertheless is important because of its continuing effect in shaping hundreds of millions of lives today, not because its big population had much impact in antiquity. This is above all true of religion. Classical Hinduism crystallized in the first millennium BC. Buddhism, also, then appeared in India, a great world religion in the making. Because what humans do is so much a matter of what they believe they can do, these religions are keys to Indian history; it is the making of a culture that is its pulse, not the making of a nation or an economy, and to this culture religion was central.

Early Indian religion

Its deepest roots go very deep indeed. A seal from Mohenjo-Daro already shows a figure who looks like an early Shiva, one of the major popular cult figures of Hinduism, and stones like the lingam, the phallic cult-object which is his emblem in modern temples, have been found in the Harappan cities. Worship of Shiva may be the oldest surviving religious cult in the world, though it shows many important Aryan characteristics. Other Harappan seals seem to suggest a religious world centred about a mother-goddess and a bull. The bull survives to this day, the Nandi of countless village shrines all over Hindu India (and newly vigorous in his latest incarnation, as the electoral symbol of the Congress Party).

Vishnu, another focus of modern popular Hindu devotion, is more clearly an Aryan. Vishnu joined hundreds of local gods and goddesses still worshipped today to form the Hindu pantheon. Whatever survived from the Harappan (or even pre-Harappan) past, the major philosophical and speculative traditions of Hinduism stem from Vedic religion and the Aryan legacy. Sanskrit is the language of religious learning still used today in the Dravidian-speaking south as much as in the north; it was a great cultural adhesive and so was the religion it carried. The Vedic hymns provided the nucleus for a system of religious thought more abstract and philosophical than primitive animism. Out of Aryan notions of hell and paradise, the House of Clay and the World of the Fathers, as they were called, there gradually evolved the belief that action in life determined human destiny.

An immense, all-embracing structure of thought slowly emerged from such elements, a world view in which all things are linked in a huge web of being. Souls might in different lives pass through different forms in this vast whole; they might move up or down the scale of being, between castes, for example, or even between the human and animal worlds. The idea of transmigration from life to life, its forms determined by proper behaviour, was linked to the idea of purgation and renewal, to the trust in liberation from the transitory, accidental and apparent, and to belief in the eventual identity of soul and absolute being in Brahma, the creative principle. The duty of the believer was the observation of Dharma – a virtually untranslatable concept, but one which embodies something of the western ideas of a natural law of justice and something of the idea that men owe respect and obedience to the duties of their station.

These developments took a long time and the complicated steps by which Vedic tradition changed into classical Hinduism remain obscure. At the centre of the early evolution had been the brahmans who had a key role in the sacrificial rites of Vedic religion. They seem to have soon come to terms with the gods of an older world. Not until about 700 BC are there signs of the advent of a more philosophical approach to be found in scared texts called the Upanishads, a mixed bag of devotional utterances, hymns, aphorisms and reflexions of holy men pointing to the inner meaning of the traditional religious truths. They give much less emphasis to personal gods and goddesses than earlier texts and also include some of the earliest ascetic teachings which were to be so visible and striking a feature of Indian religion, even if only practised by a small minority. The Upanishads met the need some felt to look outside tradition for religious satisfaction. Doubt appears to have been felt about the sacrificial principle. New patterns of thought had begun to appear at the beginning of the historical period and uncertainty about old beliefs is already expressed in the later hymns of the Rig-Veda. Classical Hinduism was to embody a synthesis of ideas like those in the Upanishads (pointing to a monistic conception of the universe) with the more polytheistic and popular tradition upheld by the brahmans.

ANCIENT CHINA

For about two and a half thousand years there has been a Chinese nation using a Chinese language. This registers a continuing experience of civilization hardly disturbed from the outside rivalled only by that of ancient Egypt. China's government as a single unit has long been taken to be normal, in spite of intervals of division and confusion. This experience shaped a Chinese historical identity as much cultural as political. In China, culture made unified government easier. Somehow at a very early date she crystallized certain institutions and attitudes suited to her circumstances, and they would endure.

In northern China, ground just above the flood level of the Yellow River begins to yield evidence of agriculture on an exhaustive or semi-exhaustive basis from about 5000 BC. From this area it spread both north to Manchuria and to the south. Soon there appeared within the key area complex cultures using jade and wood for carving, domesticating silk-worms, and making ceremonial vessels in forms which were to become traditional. Perhaps prehistoric Chinese even used chopsticks. In other words, Neolithic times already display much that is characteristic of later China. One sign is the widespread use of millet, a grain well adapted to the sometimes arid farming of the north and the basic staple of Chinese diet until about a thousand years ago.

One of the first images of a human face – from pre-Shang Honan. It may represent a shaman or magician wearing some kind of professional costume (the ruff suggests something special), but this is speculation. It is at least as probable that the form was determined by the shape of the clay pot for which this was the lid. Other such pots to which human faces form lids have been found.

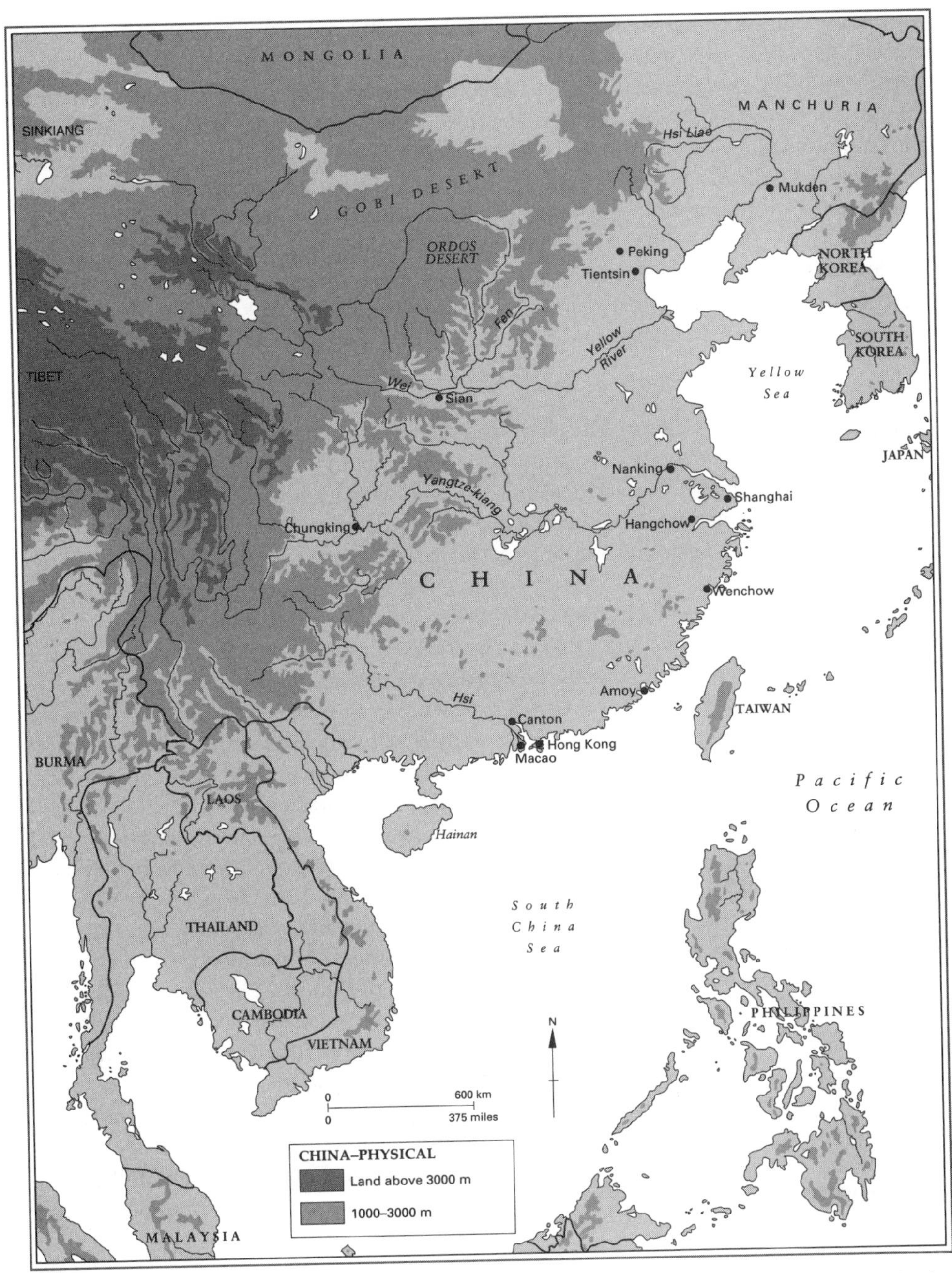

Ancient writers and legends identified for China a specific inventor of agriculture, which no doubt registers the importance of its coming. Though little can in fact be inferred confidently or clearly about early social organization, it was assumed that 'under heaven every spot is the sovereign's ground' and this may reflect ancient ideas that all land belonged to the community as a whole. In early times there can also be attributed the appearance of a clan structure and totems. Kinship is almost the first institution which can be distinguished and survived to be important in historical times. The evidence of the pottery, too, suggests some complexity in social roles; things were being made in Neolithic times which cannot have been intended for the rough and tumble of everyday use; a stratified society seems to be emerging before we reach the historical era.

The agriculture which made advanced culture possible was for a long time confined to north China. Many parts of this huge country only took up farming much later and well into the historical era. In the crucial northern area, though, tradition and scholarship agree that the story of civilization begins under rulers from a people called the Shang, the first name with independent evidence to support it in the traditional list of dynasties which was for a long time the basis of Chinese chronology. From the late eighth century BC we have better dates, but there is no chronology for early Chinese history as good as that of Egypt. Nonetheless, somewhere about 1700 BC (and a century each way is an acceptable margin of approximation) a tribe called the Shang, which enjoyed the military advantage of the chariot, imposed itself on its neighbours over a sizable stretch of the Yellow River and some 40,000 square miles of northern Honan.

Shang China

Shang Kings were major figures. They lived and died in some state; slaves and sacrificial victims were buried with them in deep and lavish tombs. Though Shang government seems to have been largely a matter for warrior landlords who were leading members of aristocratic lineages with semi-mythical origins, it achieved a standardized currency and was able to build fortifications and cities on a scale requiring mass labour. The Shang court had scribes and archivists; this was a literate monarchy, working in what seems to have been the first truly literate culture east of Mesopotamia (unless the literacy of the Indus valley can be shown to have been more elaborate than can be inferred at present). Shang civilization, too, had an influence extending beyond the area of Shang political control.

In early times, decisions of state, as well as lesser ones, were taken by consulting oracles. Turtle shells or the shoulder-blades of certain animals were engraved with written characters and a heated bronze pin was then applied to them so as to produce cracks on the reverse side. The direction and length of these cracks in relation to the characters would then be considered and the oracle read accordingly. This provides the evidence for the foundation era of Chinese language, for the characters on the oracle bones (which were kept as records) are basically those of classical Chinese. The Shang had about 5000 such characters; though not all can be read, the structure of the language is known to be like that of modern Chinese – monosyllabic and pictographic, depending on word order, not on inflections, to convey meaning. The Shang, in fact, were already using a form of Chinese. The readers of the oracles, the so-called shih, were the fore-runners of the later scholar-gentry class; they were indispensable experts, the possessors of hieratic and arcane skills. The language was thus always the possession of a relatively small élite which not only found its privileges rooted in it but also had an interest in preserving it against corruption or variation. It was of enormous importance as a unifying and stabilizing force. Written Chinese became a language of government and culture transcending divisions of dialect, religion and region, and calligraphy was to remain high on the scale of Chinese art. With it the élite could tie a huge and diverse country together.

The Chou era

The Shang succumbed in the end to another tribe from the west of the valley, the Chou, probably in 1027 BC. Under the Chou, many of the already elaborate Shang governmental and social structures were preserved and further refined. Burial rites, bronze-working techniques and decorative art survived in hardly altered forms. The Chou period saw the consolidation and further diffusion of this heritage and the hardening of the institutions of a future Imperial China. Interestingly, the Chou thought of themselves as surrounded by barbarian peoples waiting for the benevolent effects of Chou tranquillization. In fact, Chou supremacy rested on war. Government was usually a matter of a group of notables and vassals, some more dependent on the dynasty than others, offering in good times at least a formal

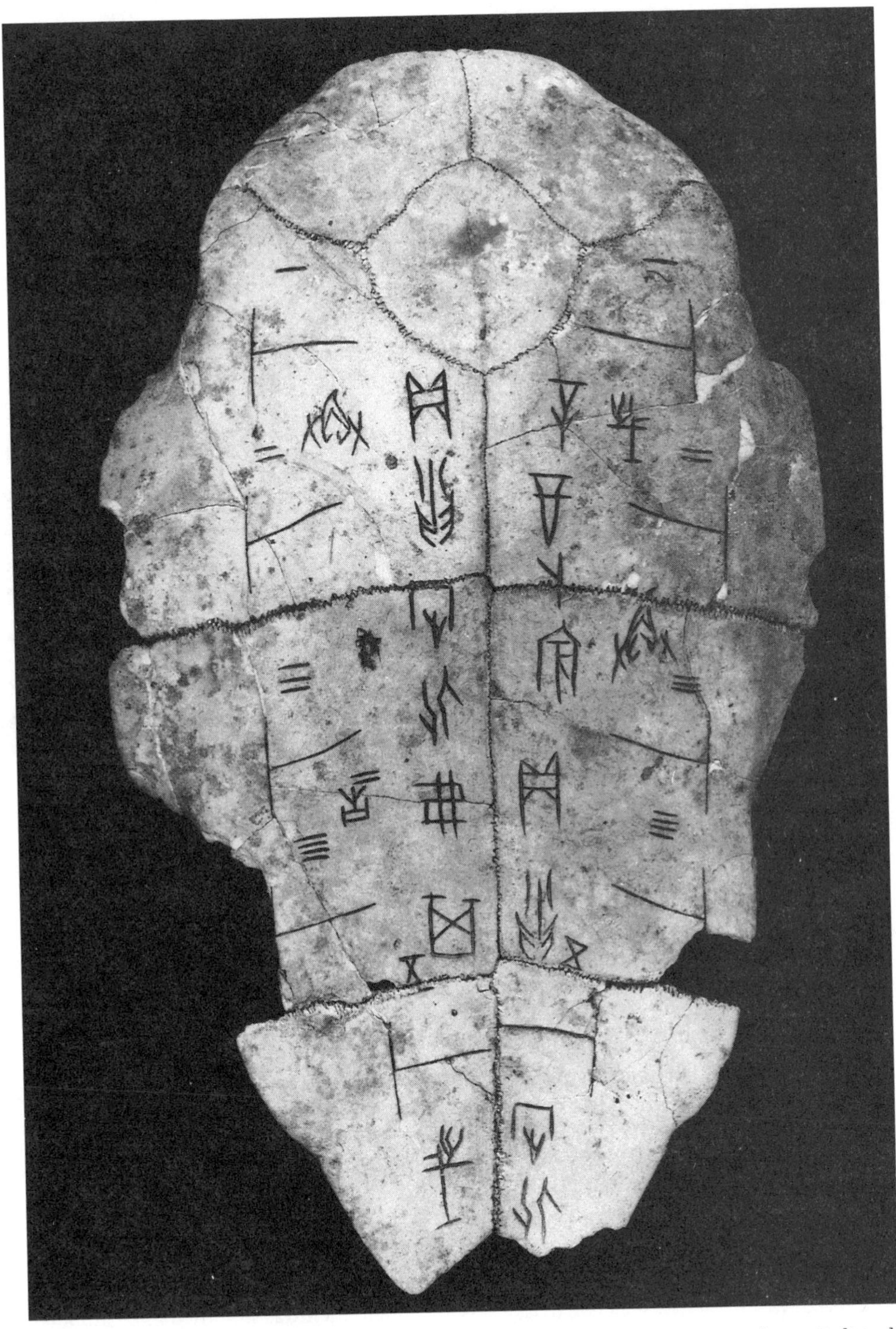

An especially fine and complete example of an oracle on a tortoise shell which has survived almost intact from Shang times. After cracks had been observed in answer to the questions previously engraved on the shell or bone, these answers were then themselves also inscribed; thus a permanent record was available for consultation. Archaeologists have now recovered more than a hundred thousand fragments of the Shang oracles which have provided examples of some five thousand of the first Chinese characters (about a third of them have been interpreted).

acknowledgement of its authority and increasingly sharing a common culture. Political China (if it is reasonable to use such a term) rested upon big estates which had sufficient cohesion to have powers of long survival and sometimes their original lords turned into rulers who could be called kings, served by elementary bureaucracies of their own.

In about 700 BC, barbarians drove the Chou from their ancestral centre to a new home further east, in Honan. The dynasty did not end until 256 BC, but the next distinguishable epoch dates from 403 to 221 BC and is significantly known as the Period of the Warring States. Historical selection by conflict now grew fierce. Big fish ate little fish until one only was left and all the lands of the Chinese were for the first time ruled as one great empire, under the Ch'in, from whom the country was to get its name – a good point at which to pause. So far, the traditional Chinese historical record accounts for fifteen hundred years or so of dimly discernible struggles of kings and over-mighty subjects. It does not provide much of a story-line, though basic processes were going on for most of this time which were very important for the future.

One was a continuing diffusion of culture outwards from the Yellow River basin. Chinese civilization began as a matter of tiny islands in a sea of barbarism. Yet by 500 BC it was the common possession of scores, perhaps hundreds, of statelets and feudatories scattered across the north, and had also been carried into the Yangtze valley. This had long been a swampy, heavily forested region very different from the north and inhabited by far more primitive peoples. At the end of the Period of Warring States the stage of Chinese history is about to be much enlarged. Chou influence – in part thanks to military expansion – irradiated this area, and helped to produce the first major culture and state in the Yangtze valley.

Chinese technology of the fourth century BC: fragments of a stone mould for casting bronze knives, in this instance for use as currency.

Early Chinese society

Under both Shang and Chou the ground plan of a future society was already present in a fundamental division between a landowning nobility and the common people, most of whom were peasants. They and their descendants, the vast majority of the population for centuries, ultimately paid for all that China produced in the way of civilization and state power. We know little of their countless lives. There is one good physical reason for this: the Chinese peasant alternated between his mud hovel in the winter and an encampment where he lived during the summer months to guard and tend his growing crops. Neither left much trace. For the rest, he appears sunk in the anonymity of his community (he does not belong to a clan), tied to the soil, occasionally taken from it to carry out other duties and to serve his lord in war or hunting.

The distinction of common people from the nobly born was enduring. In later times the gentry were exempt from the beatings which might be visited on the commoner (though, of course, they might suffer appropriate and even dire punishment for more serious crimes). The nobility long enjoyed a virtual monopoly of wealth, too, which outlasted its earlier monopoly of metal weapons. The crucial distinctions of status, though, lay in the nobleman's special religious standing through a monopoly of certain ritual practices. Only a nobleman belonged to a family – which meant that he had ancestors, reverence for whom, and propitiation of whose spirits, dated from before Shang times.

The family was a legal refinement and subdivision of the clan, of which there were about a hundred, within each of which marriage was forbidden. Each was supposed to be founded by a hero or a god. The patriarchal heads of the clan's families and houses exercised special authority over its members and were all qualified to carry out its exacting and time-consuming rituals and thus influence spirits to act as intermediaries on the clan's behalf with the powers which controlled the universe. These practices came to identify persons entitled to possess land or hold office. The clan offered a sort of democracy of opportunity at its own level: any of its members could be appointed to the highest place in it, for they were all qual-

Principal Chinese Dynasties

SHANG? 1523–?1027
CHOU? 1027–?256
CH'IN 221–206
(having annihilated CHOU in 256 and other rival states afterwards)
FORMER HAN 206 BC–AD 9
HSIN AD 9–23
LATER HAN 25–220
WEI 220–265 | SHU 221–263 | WU 222–280
WESTERN CHIN 265–316
SIXTEEN KINGDOMS 304–439 | EASTERN CHIN 317–420
LIU SUNG 420–479
SOUTHERN CH'I 479–502
NORTHERN WEI 386–581
LIANG 502–557
CH'EN 557–589
WESTERN WEI 535–557 | EASTERN WEI 534–550
NORTHERN CHOU 557–581 | NORTHERN CH'I 550–577
SUI 581–618
T'ANG 618–907
FIVE DYNASTIES 906–960 | TEN KINGDOMS 907–979
NORTHERN HAN 951–979
(reckoned as one of the Ten Kingdoms)
SUNG 960–1126
(the extreme north of China being ruled by the LIAO 947–1125)
CHIN 1126–1234 | **SOUTHERN SUNG** 1127–1279
YUAN 1279–1368
(having succeeded the CHIN in North China in 1234)
MING 1366–1644
CH'ING 1644–1912

ified by the essential virtue of a descent whose origins were godlike. In this sense, a king was only *primus inter pares*, a patrician outstanding among all patricians.

The common people found its religious outlets in nature gods. To propitiate and worship the spirits of mountains and rivers had been an important kingly duty from early times, but in China nature cults had less formal influence than in other religious systems. Yet the heart of the ruling house's claim to obedience was its religious superiority. Through the maintenance of ritual, it had access to the goodwill of unseen powers, whose intentions might be known from the oracles. When these had been interpreted, the ordering of the agricultural life of the community was possible, for they regulated such matters as the time of sowing or harvesting. Much turned, therefore, on the religious standing of the king; it was of the first importance to the state. Under the Chou appeared the idea that there existed a god superior to the ancestral god of the dynasty and from whom was derived a 'mandate of heaven' to rule. This was the introduction of another idea fundamental to the Chinese conception of government and it was to be closely linked to the notion of a cyclic history, marked by the repeated rise and fall of dynasties. Inevitably, it provoked speculation about what might be the signs by which the recipient of a new mandate should be recognized.

The earliest records of Chinese government do not leave an impression of a very busy monarchy. Apart from making extraordinary decisions of peace or war, the king seems to have had little to do except fulfil his religious duties, hunt, and initiate building projects or (as did some Chou kings) agricultural colonization. Ministers who regulated court life slowly emerge, but the king was a landowner who for the most part needed only bailiffs, overseers and a few scribes. No doubt much of his life was spent on the move about his lands. The only other activity in which he needed expert support was that concerned with the supernatural. From this was to grow an intimate connexion between government and the determination of time and the calendar, both very important in agricultural societies. The necessary techniques were based on astronomy, and though they came to have a respectable basis in observation and calculation, their origins were magical and religious.

Iron and cities

The Chou period had come to an end amid increasing signs of social disturbance. Change probably sprang from the pressure of population upon resources. This was why iron, probably in use by about 500 BC, was so important. As elsewhere a sharp rise in agricultural production (and therefore in population) followed its introduction. The first iron tools which have been found come from the fifth century BC; iron weapons came later. At an early date, they were made by casting, and iron moulds for sickle blades have been found dating from the fourth or fifth centuries. Chinese technique in handling the metal was thus well advanced in early times. Whether by evolution from bronze casting or after experiments with pottery furnaces which could produce high temperatures, China somehow arrived at the casting of iron at about the same time as knowledge of how to forge it; temperatures high enough for casting were not available elsewhere for another nineteen centuries or so.

Another important change under the later Chou was the growth of cities. The earliest probably derived their locations from landowners' temples used as administrative centres for their estates. About them and the temples of the popular nature gods communities collected, often on plains near rivers. Under the Shang we find stamped-earth ramparts, specialized aristocratic and court quarters and the remains of large buildings. At Anyang, a Shang capital in about 1300 BC, there were metal foundries and potters' kilns as well as palaces and a royal graveyard. The later Chou capital, Wang Ch'eng, was a rectangle of earth walls each nearly three kilometres long. By 500 BC there were scores of cities, often with three well-defined areas: a small enclosure where the aristocracy lived, a larger one inhabited by specialized craftsmen and merchants, and the fields outside the walls which fed the city. The merchants' craftsmen's quarters were separated from those of the nobility by walls and ramparts round the latter, but they, too, fell within the city's own walls – a sign of a growing need for defence. In the commercial streets of cities of the Warring States Period could be found shops selling jewellery, curios, food and clothing, as well as taverns, gambling houses and brothels.

The heart of Chinese society, none the less, still lay in the countryside. As the Chou period came to the end, the landowning class showed unmistakable signs of a growing independence of its kings. Its economic supremacy was rooted in customary tenure, and ownership – theoretically granted by the king – extended not only to land but to carts, livestock, implements and, above all, people. Labourers could be sold, exchanged, or left by will. The nobleman had always had a monopoly of arms, too, and, as time passed, only noblemen could afford the more expensive weapons, armour and horses which increasingly came into use. By about 600 BC, it seems clear that the Chou king had been reduced to dependence on the greatest nobles. Disorder and growing scepticism about the criteria governing the right to rule came to a head in the profound and prolonged social and political crisis of the last, decaying centuries of the Chou and the Period of Warring States and produced an important burst of thought about the foundations of government and ethics. One school of teachers, the 'Legalists', urged that law-making power, ritual observances, should be the key principle of the state; there should be one law for all, ordained and vigorously applied by one ruler. This struck some as little more than a cynical doctrine of power, but in the next few centuries kings found it attractive. The debate went on for a long

Early Chinese architecture, of which little has survived, visible to us through Han pottery, such as this model of a watchtower. Such models were made to be placed in the tombs of the rich along with other replicas of possessions, including animals and slaves, they had enjoyed in life.

time. Criticism of the Legalists was particularly undertaken by followers of the most famous of all Chinese teachers, Confucius.

Confucius and Chinese culture

It is convenient to call K'ung-fu-tzu by the latinized version of his Chinese name, Confucius, though it was given to him by Europeans in the seventeenth century, more than two thousand years after his birth in the middle of the sixth century BC. He was to be more profoundly respected in China than any other philosopher. What he said – or was said to have said – shaped his countrymen's thinking for two thousand years. He came from a shih family of the lesser nobility and had spent some time as a minister of state and an overseer of granaries. When he could not find a ruler to put into practice his recommendations for just government he turned to meditation and teaching. His aim was to present a purified and more abstract version of the truths he believed to lie at the heart of traditional practice and thus to revive personal integrity and disinterested service in the governing class. He was a reforming conservative. Somewhere in the past, he thought, lay a mythical age when each man knew his place and did his duty; to return to it was Confucius' ethical goal. He advocated a principle of order – the attribution to everything of its correct place in the great whole of experience. This came out in a strong predisposition to support institutions promoting order – the family, hierarchy, seniority – and a reverence for the many nicely graded obligations which linked them. Such teaching was likely to produce men who would respect the traditional culture, emphasize the value of good form and regular behaviour, and seek to realize their moral obligations in the scrupulous discharge of duties.

It was immediately successful in that many of Confucius' pupils won fame and worldly success (though his teaching deplored the conscious pursuit of such goals, urging, rather, a gentlemanly self-effacement). But it was also successful in a much deeper and long-lived sense, since generations of Chinese civil servants were later to be drilled in the precepts of behaviour and government which he laid down. Confucian texts (not all of them authentic) came to be treated with something like religious awe. They were used for centuries in a unified and creative way to mould generations of China's rulers in precepts of which Confucius was believed to have approved (the parallel with the later use of the Christian Bible, at least in Protestant countries, is striking). Yet Confucius had not much to say about the supernatural and was not, in the ordinary sense of the word, a 'religious' teacher (which may explain why other teachers had greater success with the masses). He was most concerned with practical duties. Later Chinese thought, too, seems less troubled than other intellectual traditions by agonized uncertainties over such matters as the reality of the actual or the possibility of personal salvation. Nor was the mapping of knowledge by systematic questioning of the mind about the nature and extent of its own powers to be of much interest to of Chinese philosophers. The lessons of the past, the wisdom of former times and the maintenance of good order came to have more importance in it than pondering theological enigmas and philosophical puzzles, far less seeking reassurance in the arms of the dark gods. The tone of Chinese intellectual tradition after Confucius, owes less to the teaching of individuals, moreover, than does the methodical and interrogatory European tradition.

Systems rivalling Confucianism nonetheless appeared. Lao-Tse, a teacher whose vast fame conceals the fact that we know virtually nothing about him, was supposed to be the author of the text which is the key document of a philosophical system later called 'Taoism'. This advocated the positive neglect of much that Confucianism upheld; respect for the established order, decorum and scrupulous observance of tradition and ceremonial, for example. Taoism urged identification with a submission to a conception already available in Chinese thought and familiar to Confucius, that of the Tao or 'way', the cosmic principle which runs through and sustains the harmoniously ordered universe. The practical results of this were likely to be political quietism, non-attachment, and an idealization of simplicity and poverty.

Another and later sage, the fourth-century Mencius (Meng-tzu), taught men to seek the welfare of mankind in a development of Confucian teaching rather than a departure from it. But all schools of Chinese philosophy had to take account of Confucian teaching, so great was its prestige and influence, and its total and ultimate effect is imponderable. Its great period of influence lay still in the remote future at the time of Confucius' death, when it was to set standards and ideals for the directing élites down to our own day. Its teachings accentuated a preoccupation with the past among them which was to give a characteristic bias to Chinese historiography, and may have had a damaging effect on scientific enquiry. Many of its precepts – filial piety, for example – also filtered down to popular culture through stories and the traditional motifs of art. It thus further solidified a civilization many of whose most striking features were well entrenched by the third century BC.

China's art is still the most immediately appealing and accessible side of what now remains of ancient China. Of the architecture of the Shang and Chou, not much survives; their building was often in wood, and the tombs do not reveal very much. Excavation, on the other hand, reveals a capacity for massive construction; the wall of one Chou capital was made of pounded earth thirty feet high and forty thick. But it is smaller and more plentiful objects which reveal a civilization capable already in Shang times of exquisite work, above all in its ceramics, unsurpassed in the ancient world, and the great bronzes which begin to be made in early Shang times and continue thereafter uninterruptedly. The art of casting sacrificial containers, pots, wine-jars, weapons, tripods was already at its peak as early as 1600 BC. Bronze casting appears so suddenly and at such a high level of achievement that people long sought to explain it by transmission of the technique from outside. But there is no evidence for this. Nor is there any to show that Chinese bronzes reached the outside world in early times; no discovery of them elsewhere can be dated before the middle of the first millennium BC. Nor are there many discoveries outside China at earlier dates of the other things to which Chinese artists turned their attention, to the carving of stone or jade, for example, into beautiful and intricate designs. Apart from what she absorbed from her barbaric nomadic neighbours China had little to do with the outside world until well into the historical era, it seems. No more than India's did China's civilization show much potential for expanding beyond its cradle, huge as that was.

3
FOUNDATIONS OF OUR WORLD

INTERPLAY AND INTERCHANGE

By 1000 BC several patterns of civilization had appeared in the Near East and eastern Mediterranean. Some had even come and gone. Great powers were already used to dealing with one another across the region by formal diplomacy. Artisans, merchants and mercenaries could move around in search of their living. Ideas, artistic style and technical skill were already being consciously exchanged and borrowed all round it. It was the first truly cosmopolitan group of societies and, for a long time, the only part of the world where such vitality and cross-fertilization were to be evident. In it can be discerned the beginnings of trends which were, by our own century, to dominate world history. The best place to begin to understand this is again with comings and goings of peoples, migrations and folk movements. They provided much of the dynamic of what was going on. Over the whole of the second millennium BC, India, Iran, Mesopotamia, the Levant, Egypt and the Aegean all show the impact of the peoples whose languages we call Indo-European. One view is that they came from southern Russia, probably well before 2000 BC. Perhaps climatic changes in Central Asia drove other peoples to the east to migrate westward, and so put pressure on the 'Indo-Europeans' to move in their turn. One route they followed was into the Balkan peninsula down into Greece and Thrace and from there across into Anatolia. Others meanwhile had come down through the Caucacus and from Turkestan into Bactria, Iran (which is actually the same word as 'Aryan') and India. Meanwhile, besides the 'Danubian' Indo-Europeans in south-east Europe, others had moved westward into Germany, northern France and even the British Isles, where we may leave them for the moment.

In the Fertile Crescent, these peoples came into conflict with the old empires which had dominated it since the overcoming of Sumer. The Semitic peoples were moving about, too. The Old Testament of the Christian Bible tells us that Abraham, the figure Jews now place at the start of their traditional history, set out from Ur across Mesopotamia and the Levant in search of pasture. This is probably folk-memory. Many tribes were on the move at that time (and the Indo-Europeans were nomadic pastoralists, too), disputing with one another the grazing and water of the Fertile Crescent, whose advanced cultures were attractive and could be borrowed from, or imitated.

Migrations must have enormously accelerated the spread of civilized skills. Literacy is an indicator. In 2000 BC, it was still largely confined to the river-valley civilizations, though cuneiform had spread throughout Mesopotamia and two or three languages were being written in it; in Egypt, while monumental inscriptions were hieroglyphic, day-to-day writing was done in a simplified form called hieratic, and cuneiform was used for diplomacy. A thousand years or so later, literate peoples can be identified by their languages all over the Near East, and in Crete and Greece, too. New scripts were being invented. One, in Crete in about 1500 BC, takes us to the edge of a new world, for it reveals a people using a form of Greek. Based on an alphabet devised for a Semitic language and borrowed from the Phoenicians, the script of the first western literature was in use by about 800 BC.

Trade and travel

Like language, long-distance trade is both a symptom and motor of change. For a long time, evidence of it only comes through archaeology which throws up such evidence as the Harappan docks, or the discovery of foreign seals. From it, we know that tin was brought from Mesopotamia and Afghanistan, as well as Anatolia, to what we should now call 'manufacturing' centres. The copper of Cyprus turns up in so many places, that it must have been widely traded, and so does that of mainland Europe; mine-shafts in what is now former Yugoslavia were sunk sixty and seventy feet below ground even before 4,000 BC. As for non-metallurgical products, the cedars of Lebanon later celebrated in the Bible were being supplied to Egypt, amber was brought from the Baltic to the Aegean, and spices from the Far East came through the Red Sea to Egypt, all before 1000 BC. Economic geography slowly changed as such links developed. Centres of special commercial skill and prosperity appeared. The Cretans and early mainland peoples of Greece did much trading; Bahrain drew merchants from India and Mesopotamia. As the millennium drew to its close, the greatest of all the ancient trading peoples, the Phoenicians of the coastal cities of the Levant, were just about to enter a golden age of prosperity.

The decoration of this incense burner stand of the twelfth century BC shows a man carrying a copper ingot shaped like an ox hide. It was found in Cyprus, a major Bronze Age producer of copper.

The carriage of bulk goods was difficult, at least on land, where it was a matter of asses and donkeys until camels were domesticated in the middle of the second millennium BC. This opened an environment hitherto almost impenetrable, the waterless desert, and created the caravan trade of Asia and Arabia. Wheeled transport for a long time had probably only local importance, given the poor quality of early roads and axles, though carts were in service in Mesopotamia about 3000 BC, in Syria around 2250 BC, in Anatolia two or three hundred years later and in mainland Greece about 1500 BC. But for goods in quantity, carriage by water was already simpler and cheaper than by land, as it would remain until the coming of steam railways. Neolithic peoples had been able to make long journeys by sea in dug-out canoes. The oar provided the first motive power for long sea crossings as well as for close handling. The Egyptians of the Third Dynasty then put a sail on a sea-going ship; the central mast and square sail were the beginning of navigation relying on anything but currents or human energy. Improvements of rigging came slowly over the next two millennia. But the ships of antiquity for the most part were square-rigged and prevailing winds therefore settled the pattern of maritime communication. Nonetheless, merchants could trade profitably and in growing quantities by sea. Long before camel caravans, ships carried the gums and resins of southern Arabia up the Red Sea. Others went back and forth around the Aegean. By the thirteenth century BC, ships which could carry more than 200 copper ingots were sailing about the eastern Mediterranean, a few centuries later some of them were being fitted with watertight decks.

As late as 1000 BC, it is true, we can still not always be very sure what such traffic meant. People appear to have exchanged goods and services before they were civilized but this was perhaps more like agreed redistribution within the community. In historic times some peoples have had chiefs presiding over a common store, 'owning', in a sense, everything the community possessed, and doling out enough to keep society working smoothly. This may be what lay behind the centralization of goods and supplies in Sumerian temples. It was a long time, too, before there was any widely recognized medium of exchange – money. The first evidence comes from Mesopotamia, where accounts were recorded in measures of grain or silver before 2000 BC. Copper ingots seem to have been sometimes treated as monetary units over much of the Mediterranean in the late Bronze Age, but the first officially sealed means of exchange which survives comes from Cappadocia in the late third millennium BC; its silver ingots were a true metal currency. Yet we have to wait until the seventh century BC for the first coins. People could get along without them. The Phoenicians, legendary for commercial skill and acumen, did not have a currency until the sixth century BC; Egypt, a centrally controlled economy, did not adopt a coinage until two centuries after that. Yet, the exchange of goods went on. Not all of it can have been what would now be termed 'trade', nor is it clear that what looks like a market is, in the ancient world, always a place where values were arrived at by bargaining. The era of historical record brings evidence of transfers of commodities seen as tribute, symbolic or diplomatic gifts between rulers, votive offerings. As late as the nineteenth century AD, the Chinese empire still thought of foreign trade in terms of tribute from the outside world and the pharaohs looked at business done with the Aegean in somewhat similar ways, to judge by tomb paintings. Such transactions might include the transfer of standard objects such as tripods or vessels of a certain weight, or rings of uniform size; they thus had some of the characteristics of currency.

War and technology

One part of the story of developing civilization is of technology transfer, another liberating and stimulating force. Unsurprisingly, some of it first shows in warfare, where the search for advantage appears to be a constant of human behaviour. The Indo-European war chariots and cavalrymen transformed operations in open country over the whole Near East. Though Sumerian fighting men are depicted trundling about in clumsy four-wheeled carts, drawn by asses, this was probably only a means of moving generals about or getting a leader into the mêlée, so that spear and axe could be brought to bear. The true war chariot is a two-wheeled fighting vehicle drawn by horses, usually with a crew of two, one man driving, the other using it as a platform for missile weapons, especially arrows shot from the bow. The Kassites, a people of Indo-European stock, were the first we know to use this form of fighting. They could draw upon the high pastures to the north and east of the Fertile crescent and their reserves of horses, animals for a long time rare in the river valleys where they were the prized possessions of kings and great leaders. Warfare was much changed by riding horses. The skill came from the Iranian highlands, where it may have been practised as early as 2000 BC, and spread through the Near East and Aegean in the next millennium. A cavalryman proper does not merely move about in the saddle but fights from horseback; it took a long time for the art of managing horse and bow or spear at the same time to evolve. The armoured horseman, charging home and dominating foot-soldiers by sheer weight and impetus, was only to appear much later still.

Eventually, cavalry and chariots appeared in the armies of all the great kingdoms of the Near East. During the second millennium BC chariot parts began to be made of iron, notably their wheels, which were hooped with it. The other military advantages of iron as a material for weapons are obvious and the use of iron spread rapidly through the Near East and beyond in spite of attempts by those who first had it in Anatolia to keep it to themselves. Iron ore, though scarce, was more plentiful than copper or tin. Already in the eleventh century BC it was used for weaponry in Cyprus (some have argued that steel was produced there, too) and it spread to the Aegean soon after 1000 BC, a date which can serve as a rough division between the Bronze and Iron Ages, though no more than a helpful prop to memory, for parts of what we may call the 'civilized world' long went on living in a Bronze Age culture. Together with the 'Neolithic' elsewhere, the Bronze Age lived on well into the first millennium BC, fading away only slowly like the smile on the face of the Cheshire cat. For a long time, after all, there was not much iron to go round.

New differences

Technology, like much else when history and prehistory run still side by side, does not provide much of a chronology, but this need not trouble us. Some dates are clear in the story of states and empires, at least, though it is best not to focus too particularly. It is the trend that matters most, and it is both clear and paradoxical: individual peoples and regions, though in more and more frequent contact and discovering more to share, yet become increasingly distinct. Tribes and peoples took on firmer identities as their governments crystallized in continuing and

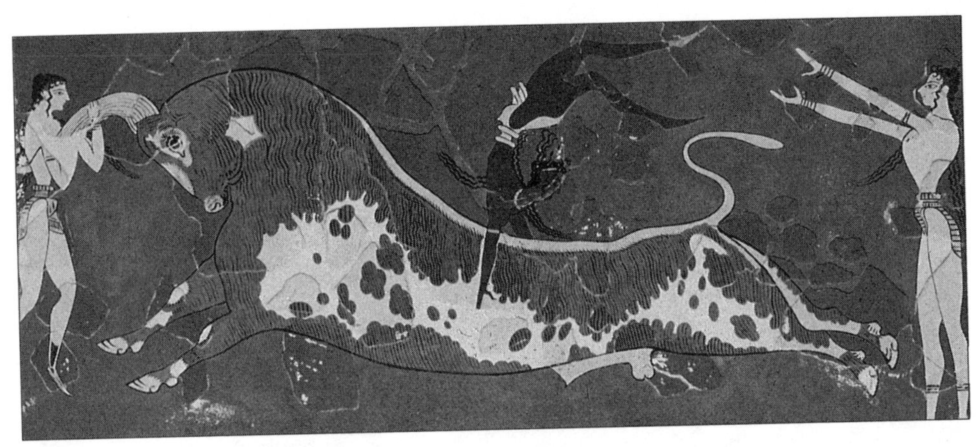

This damaged but reconstructed fresco from Knossos appears to show different stages of the acrobatic bull-leaping practised there. Its ritual content remains unknown,but the bull played a central part in Minoan mythology, perhaps as a representation of the elemental force of earthquakes not uncommon in the Aegean. More than a thousand years later, Greeks at Ephesus s till celebrated a feast at which Poseidon, the `earth-shaker' and god of the sea, was addressed as a bull and had black bulls sacrificed to him.

institutionalized forms, often linked to religion. The dissolution of empires into more viable units is a familiar story from Sumer to modern times, but some areas emerge again and again as enduring nuclei of distinctive tradition. Even in the second millennium BC, some structures are getting more solid, and show greater staying-power. When such things happen, institutions become more particular and escape from generalities covering all early civilizations. Though there is a new cosmopolitanism, societies take very different paths through it.

There is a new scope for differentiation. Literacy helped to pin down tradition and a sense of being a special group. Before civilization began, art had already established itself; perhaps it had done so as an autonomous activity not necessarily linked to religion or magic (often so linked though it continued to be). Style takes different forms in different places and special leisure activities appear. Gaming-boards appear in Mesopotamia, Egypt, Crete, and everywhere kings and noblemen hunted with passion, and were entertained in their palaces by musicians and dancers. Boxing seems to have been popular first in Bronze Age Crete, an island where a unique and probably ritualistic sport of bull-leaping was also practised.

AEGEAN CIVILIZATION

Though the life of the Near East and eastern Mediterranean of the second millennium BC was already rich in interconnexion and interplay, some of its centres of civilization were both specially successful and important. One such was in Crete, the largest of the Greek islands. An advanced people had lived there through Neolithic times who may have had contacts with Anatolia, though the evidence is indecisive. Yet something spurred them to striking achievements. By about 2500 BC there were important towns and villages of stone and brick on the Cretan coasts whose inhabitants practised metal-working and cut seals and jewels. They shared much of the culture of mainland Greece and Asia Minor and exchanged goods with other Aegean communities. There then came a change. About 2000 BC they began to build the great palaces which are the major monuments of what we call Minoan civilization. The greatest of them, Knossos, was first built about 1900 BC. Nothing quite as impressive appears anywhere else among the islands.

Minoa

'Minoan' is simply a label for the civilization of Bronze Age Crete which came to exercise a cultural hegemony over more or less the whole of the Aegean. It has no other connotation. The adjective is taken from the name of a legendary king, Minos, who may in fact never have existed. Much later, the Greeks believed – or said – that he was a great ruler who lived at Knossos, parleyed with the gods, and married Pasiphae, the daughter of the sun. Her monstrous offspring, the Minotaur, devoured sacrificial youths and maids sent as tribute from Greece at the heart of a labyrinth eventually penetrated successfully by the hero Theseus, who

The Greeks always looked back on prehistoric Crete as a mysterious land of legend. This vase of about 500 BC shows the hero Theseus, about to ward off a boulder hurled at him by the Minotaur, the monster at the heart of the labyrinth of King Minos.

slew him. This is a rich and suggestive theme but there is no evidence to support the story. Minos may even have been a titular identification of several Cretan rulers.

Minoan civilization lasted some six hundred years, but only the outlines of its history can be put together. Its towns were linked in some dependence on Knossos. For three or four centuries they prospered, trading (if that is the word) with Egypt and the Greek mainland. In late Neolithic times, agricultural advance led not only to better cereal-growing but to the cultivation of the olive and vine in Crete. The island seems then, as today, to have been even better suited climatically and geologically for the production of these two great staples of later Mediterranean agriculture than either the other Aegean islands or mainland Greece. Both could be grown where grains could not, and their discovery changed Mediterranean possibilities. Immediately they permitted a larger Cretan population and new human resources meant yet other potentials. But more people must also have meant new demands, for organization and government, for the regulation of a more complex agriculture (Crete also exported wool) and the handling of its produce.

Minoan civilization came to a peak in about 1600 BC. A century or so later, its palaces were destroyed, perhaps by earthquakes. Recent scholarship has identified a great eruption in the island of Thera at a relevant time; one sug-

Ancient Crete

2600–2000	Minoan Period: first cities in eastern Crete and circular burial chambers
2000–1570	Middle Minoan Period – building of first palaces at Knossos, Mallia and Phaistos; contacts with Egypt and Greece; adoption of pictographs
1700–1600	Arrival of Luvians; appearance of Linear A script; era of palace building influence in the Aegean
1570–1425	Middle to Late Minoan period – development of commerce and sea power; pictographs supplanted by Linear A; building of palaces at Knossos, Phaestus and Hagia Triada
c.1500–1400	Palace of Knossos twice destroyed by earthquakes and rebuilt
c.1400–1300	Achaean settlements from Mycenae begin to displace native population; use of Achaean Linear B; palace of Knossos destroyed by fire
c.1300	Phaistos and Hagia Triada reoccupied; new settlements from Knossos in western Crete
1200-1100	Dorians destroy Knossos

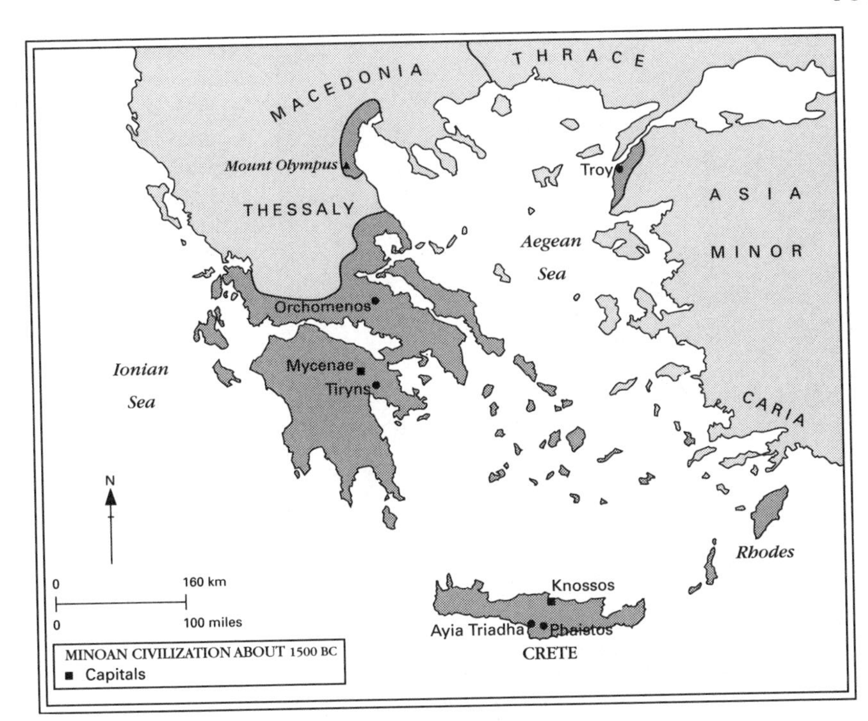

gested scenario is that tidal waves and earthquakes in Crete, seventy miles away, followed by the descent of clouds of ash which blighted Cretan fields explain the catastrophe. If so, natural cataclysm, though it broke the back of one culture, was not the end of early civilization in Crete. There were still some fairly prosperous times to come, and though the ascendancy of the indigenous civilization of Crete was, in effect, over, Knossos was occupied for another century or so by people from Greece and seems still to have prospered. Then, early in the fourteenth century BC it, too, was destroyed by fire and this time it was not rebuilt. So ends the story of early Cretan civilization.

More than a thousand years later, Greek tradition had it that Minoan Crete had dominated the Aegean through naval power. This idea has been much blown upon. The Minoans may have had a lot of ships, but they were unlikely to be specialized at this early date and there is no hope in the Bronze Age of drawing lines between trade, piracy and counter-piracy. Nevertheless, the Minoans felt sure enough of the protection the sea gave them to live in unfortified towns near to the shore on only slightly elevated ground, and they exploited the sea as other peoples exploited their natural environments. This stimulated interchange of products and ideas. Minoans had close connexions with Syria before 1550 BC. Someone was then taking their goods up the Adriatic coasts. Even more important was their penetration of Greece. The Minoans may well have been the most important single conduit through which the goods and ideas of the earliest civilizations reached Bronze Age Europe. Elsewhere, Cretan products begin to turn up in Egypt in the second millennium BC and the art of the New Kingdom shows Cretan influence. There was even, some scholars think, an Egyptian resident for a time at Knossos, presumably to watch over important interests, and some scholars argue that Minoans fought with the Egyptians against the Hyksos. Cretan vases and metal goods have been found too, at several places in Asia Minor, and a wide range of other products – timber, grapes, oil, wood, and even opium – were probably supplied by the Minoans to the mainland. In return, they took metal from Asia Minor, alabaster from Egypt, ostrich eggs from Libya. It was a busy world.

Wealth had made possible some splendour. The Minoan palaces are its finest relics but the towns were well built too, with elaborate piped drains and sewers. Less practical achievements were artistic rather than intellectual; Minoans seem to have taken their mathematics from Egypt and left it at that. Their religion went under with them, apparently leaving nothing to the future, but Minoan art influenced the style of civilization elsewhere and is still spectacular. Its genius was pictorial and reached a climax in palace frescoes of startling liveliness and movement. Here is a really original style, copied across the seas, both in Egypt and in Greece. Through other palatial arts, too, notably the working of gems and precious metals, it was to shape fashion abroad.

Minoan art also provides evidence about the Cretans' style of life. They seem to have dressed scantily, the women often being depicted bare-breasted; the men are beardless. There

For this almost spherical flask of about 1500 BC, ten inches or so in diameter, a Cretan artist found a decorative motif in the fronds of seaweed and octopus so plentiful on the shores of the Aegean islands. It is an outstanding example of the relating of a naturalistic design to the form of the artwork.

is an abundance of flowers and plants to suggest a ready appreciation of nature's gifts; Minoans do not seem to have found the world an unfriendly place. Minoan wealth is attested by the rows of huge and beautiful oil-jars found in the palaces. A concern for comfort and what cannot but be termed elegance comes clearly through the dolphins and lilies which decorate the apartments of a Minoan queen. Archaeology has also provided evidence of a singularly unterrifying religious world, though we have no texts. Nor can we penetrate its rituals, beyond registering the frequency of sacrificial altars, double-headed axes, and the apparent centring of cult in a female figure who is perhaps a Neolithic fertility goddess such as was to appear again and again all over the Near East as the embodiment of female sexuality, the later Astarte and Aphrodite. In Crete she appears elegantly skirted, bare-breasted, standing between lions and holding snakes. Whether there was also a male god is less clear.

But the appearance of bulls' horns in many places and frescoes of these noble beasts are suggestive when linked to later Greek legend (Minos' mother, Europa, had been seduced by Zeus in the shape of a bull; his wife Pasiphae's monstrous coition was with a bull from which was born the half-bull, half-man Minotaur), and to the obscure but obviously important rites of bull-leaping. Whatever it was, Cretan religion does not seem gloomy; frescoes of sports and dancing do not suggest an unhappy people.

Minoan government remains obscure. The palace was some sense an economic centre – the great store – the apex of an advanced form of a system of redistribution by the ruler. It was also a temple, but not a fortress. In its maturity it was the centre of a highly organized structure whose inspiration may have been Asian. The Minoans were literate, and kept records; we know nothing of their literature, but have a little knowledge of Minoan administration through a huge collection of thousands of tablets, which at least show what Minoan government aspired to, a supervision far closer and more elaborate than anything conceivable for a long time yet outside the Asian empires and Egypt.

Though many of the tablets still cannot be read (the earliest are in a form of pictographs, some of whose symbols are borrowed from Egypt), the latest of them are written in a script (called 'Linear B', as opposed to the 'Linear A' of the earlier ones) which is now agreed to be the earliest written version of Greek we possess. They date from 1450 or so to 1375 BC. This fits archaeological evidence of the successful arrival of newcomers from the mainland about that time who appear to have pushed aside the native Cretan rulers and to have presided over the last phase of Minoan civilization. Once again, it seems to be part of the story of the Indo-European peoples which crops up at so many places in this mysterious era. Later still, other mainlanders followed them to Crete, colonising it successfully after the final collapse of Knossos and the Minoans' disappearance from world history.

The Mycenaeans

A few centuries before that, Cretan culture had much influence in mainland Greece. Survivors of the Neolithic peoples whom later Greeks called 'Pelasgoi' were still to be found scattered about in the northern Aegean as late as 500 BC, but had by then long been displaced or conquered by those whom later Greeks saw as their forbears and called 'Achaeans'. They had arrived in Attica and the Peloponnese somewhere about 2500 BC, and spoke Indo-European languages. They were sheep-herders, warlike and knowing how to use chariots, and seem to have given much more importance to men than to women in society. Their cult objects are very different from those of religions centred on female deities such as were common in the Near East and the Aegean itself just before their arrival. By comparison with the Cretans, they were barbarians.

One of their settlements in a valley in the Peloponnese at Mycenae became the centre of a civilization. Though much more advanced than anything which had hitherto appeared in Greece, it was less developed than that of Minoan Crete, to which it owed much. It appeared in about 1600 BC and over five or six centuries spread through most of mainland Greece. In about 1300 BC Hittite kings from Anatolia were writing to the king of Mycenae as a person of importance, with whom it would be useful to do business. It is likely that the Mycenaean kings were advised by councils of headmen; they were not like the rulers of eastern empires, with numerous administrators and complex archives. Tablets from Pylos in the western Peloponnese show that some sort of officialdom existed (and are a mark of Minoan influence), but it may be realistic to think of Mycenaean society as a collection of large estates, one of which belonged to a king whose prestige or power made others accept him as an overlord. It may not mean much more than tribal or family organization somewhat stepped up in scale and under the headship of kings.

Mycenaean civilization was not, in fact, to leave much behind, except a few spectacular buildings and some fine gold objects. Although superior to the Pelasgoi, the barbarians who

became over the centuries 'Mycenaeans' remained distinguished mainly by their fighting-power. The skills represented most spectacularly at Mycenae, spread to many of the islands of the Aegean when Mycenae replaced the Minoan trading supremacy with its own around 1400 BC or so. Its pottery exports sometimes displaced Minoan and beads manufactured in Britain from Baltic amber have been found at Mycenae. Perhaps peoples enriched by trade could have disproportionate importance while great powers like Egypt and the Hittite empire were in trouble in an age of migrations.

The Phoenicians

Among the most important early trading peoples were the Phoenicians of the Levant. They had a long and troubled history, and claimed improbably that they had arrived in Tyre in about 2700 BC. They were certainly well settled on the coast of the modern Lebanon in the second millennium BC when the Egyptians bought cedar-wood from them. Behind the narrow coastal strip which was the historic channel of communication between Africa and Asia lay a shallow hinterland, poor in agricultural resources, cut up by hills running down from the mountains to the sea so that the coastal settlements found it difficult to unite. Like the Arabs of the Red Sea, their inhabitants became seafarers because geography urged them to look outwards rather than inland.

The 'Treasury of Atreus' at Mycenae was in fact a royal tomb, built in about 1325 BC with highly developed techniques, shown both in the masonry of the conical roof and the use of huge monolithic masses such as the lintel (which weighs over a hundred tons).

Left: Cretan religion has attracted much speculation; what goddess can have been represented by the little faience statuette from Knossos? More can perhaps be inferred from it about Cretan dress than about Cretan mentality: other archaeological finds suggest that she wears much the same costume as that of court ladies in Minoan times.

Weak at home – they came under the sway of several powers in turn – it cannot be entirely coincidental that the Phoenicians became prominent only after the great days of Egypt, Mycenae and the Hittite empire. They prospered in others' decline; the Phoenician cities of Byblos, Tyre and Sidon enjoyed their brief golden age after the great Minoan era of trade was long past. Ancient writers (including those of the Old Testament) stressed their reputation as traders and colonizers. Phoenician dyes were long famous and much sought after down to classical times and commercial need must have stimulated Phoenician inventiveness; their alphabet is a distant ancestor of our own (though no remarkable Phoenician literature survives). Trade was their speciality, and they came in the end to base themselves more and more on colonies or trading stations, sometimes where others had traded before them. There were in the end some twenty-five of them up and down the Mediterranean, the earliest at what is now Larnaca in Cyprus, at the end of the ninth century BC, the westernmost just beyond the straits of Gibraltar on the site of Cadiz. They even exchanged goods with the savages of Cornwall. The planting of colonies may reflect a time of troubles which overtook the Phoenician cities after that brief and profitable independence at the beginning of the first millennium. In the seventh century Sidon was razed to the ground and the daughters of the king of Tyre were carried off to the harem of the Assyrian king. With that Phoenicia was reduced to its colonies and little else.

The Phoenicians and Mycenaeans were traffickers in civilization; the Minoans had been something more. True originators, they had not only taken from the great established centres of culture, but remade what they took before diffusing it again. But all such, middlemen and creators alike, helped to shape a more rapidly changing world. The search for minerals would take explorers and prospectors further and further, even into the barbarian unknown of northern and western Europe. Trade was always slowly at work, eating away isolation, changing peoples' relations with one another, imposing new shapes on the world. But it is not always easy to relate it to the bubbling of the ethnic pot in the Aegean or the troubled history of the Asian mainland from the second millennium BC onwards.

THE FIRST GREEKS

At the end of the thirteenth century the great Mycenaean centres were destroyed, perhaps by earthquakes, though new barbarian invasions of mainland Greece had begun by then. Mycenaean civilization collapsed; life must have continued but the kingly treasures dis- appeared, the palaces were not rebuilt. In some places the established resident peoples hung on successfully for centuries; elsewhere they were exploited as serfs or driven out by new waves of migrants who had been on the move from the north since about 1200 BC. Not always settling the lands they ravaged, these invaders swept away the existing political structures and the future would emerge from their kinships, not from Mycenaean institutions. The picture is confused. What can be called the Dark Ages of the Aegean close in from about 1000 to 700 BC.

This silver bowl came originally from Cyprus, the site of important Phoenician settlements, whose art, as the sphinxes, Egyptian figures and Assyrian warriors on this example show, draws on a wide diversity of motifs.

Later times inherited from Mycenae myths and traditions handed down orally by bards and above all the language – a primitive form of Greek – in which they were sung. This was why men back hundreds of years later and saw the Achaeans as the first real Greeks. What exactly happened under their successors is uncertain. We really do not know much about it, though there was a big deterioration, certainly. The population fell and it seems that writing died out. Yet potters did not stop potting nor smiths, presumably, forging, nor farmers tilling the ground. Mankind had already built up so much cultural capital that society could by now stand a lot of strain. Civilized societies can fall backward from their peaks, and even collapse, but they usually contribute to the slowly accumulating patrimony of the whole human race, to its knowledge and collective experience, so that a total, irreversible regression to barbarism is unlikely except in small areas.

The Dorians

The new invaders, speaking versions of Greek, were broken up into many raiding parties and little groups of settlers. Language has helped scholars to trace some of their movements, in particular those of the people which came to dominate the southern Peloponnese, later called Laconia. These were the Dorians, and they went on from the mainland to be settlers of Rhodes, Cos, Crete and other islands, their dialect marking their advance. Linguistic evidence suggests that Ionia, the south-western coast of Asia Minor, was probably settled by a different group of Achaean fugitives from central Greece and the Peloponnese who had been driven out by the Dorians. Folk movements thus extended the Greek world. But in some places, notably Attica, the newcomers do not seem to have had the upper hand (to judge again by language), even though Mycenaean rule disappeared. What was going on was not so much the arrival of Greeks in the Aegean basin, but the making of Greeks from a mixture of tribal peoples by Aegean experience. They were becoming Greeks by being there.

Though the outcome is fairly clear, how it all happened is still hard to see in detail. It is very difficult to put together any chronological account until about 700 BC, or to say what caused it. The most important effect was the dispersion of Greek-speakers in communities which were the start of scores of later Greek cities around the Aegean. Except for Laconia and Attica, parts of which were ruled as regions containing several cities, each of these communities was small even tiny – and independent. They were, to use a word Greeks left to us, 'autonomous' or self-governing, not a part of anybody else's empire. During the Dark Ages some of them grew in size until they had ten thousand or more inhabitants. It was common for them to have a high place or 'Acropolis', as their centre and the home of the shrines of their gods (an idea originating in Asia centuries earlier) and to be ruled by kings (of whom the first were probably the leaders of warbands or pirate gangs) who were later replaced by councils of the most important landowners.

EMPIRES AND PEOPLES ON THE NEAR EASTERN MAINLAND

Meanwhile, the story of the Near East is for a long time one of struggles over the slowly-growing wealth of the best-defined agricultural region of the ancient world. The deserts and steppes surrounding it had no comparable resources. Invaders came and went rapidly, therefore, driven by envy and greed, sometimes leaving new communities behind them, sometimes setting up new states to replace those they overthrew. Patterns in such disorder could hardly have been grasped by those to whom these events would only have come home occasionally and suddenly, when (for instance) their homes were burned, their wives and daughters raped, their sons carried off to slavery – or, less dramatically, when they discovered that a new governor was going to levy higher taxes. Such events would be very upsetting, no doubt. Yet millions of people must also in those times have lived out their lives unaware of any change more dramatic than the arrival one day in their village of the first iron sword or sickle, never questioning ideas and institutions unchanged for many generations. Furthermore, invaders encountered well-established centres of government and population, powerful and long-lasting political structures, numerous hierarchies of specialists in administration, religion and learning. New arrivals could obliterate less of what they found in the Near East than was the case in the Aegean. The dynamism and violence of the region as it moved from the Bronze to Iron Ages was real, certainly, but there is a danger of exaggerating its impact on well-established patterns.

The Hittites

Towards the beginning of the second millennium BC, the Hittites, another Indo-European people, arrived in Asia Minor to settle in Anatolia at about the same time that Minoan civilization was rising to its greatest triumphs. Far from primitive, they had a legal system of their own and soon absorbed much of what Babylon could teach. The Hittites long enjoyed a virtual monopoly of iron in Asia; together with their skill in fortification and mastery of the chariot, it made them a military scourge to Egypt and Mesopotamia. The raid which cut down Babylon in about 1590 BC was something like the high-water mark of the first Hittite 'empire'. Eclipse and obscurity followed, and then a renaissance of Hittite power in the first half of the fourteenth century. For a while, a Hittite hegemony stretched from the shores of the Mediterranean to the Persian Gulf, dominating all of the Fertile Crescent except Egypt. But like other ancient empires it crumbled, the end coming in about 1200 BC.

By then, Hittites no longer enjoyed a monopoly of iron; around 1000 BC it is to be found in use all over the Near East, no doubt spread by the newly-arriving waves of yet more Indo-European peoples who were throwing everything into turmoil. There is a notable closeness of timing between the collapse of the last Hittite power (at the hands of a people from Thrace called the Phrygians) and the attacks of 'sea peoples' recorded in the Egyptian records. The 'sea peoples' were another symptom of upheaval. Armed with iron, from the beginning of the twelfth century BC they were raiding the East Mediterranean mainland, ravaging Syrian and Levantine cities. Some of them may have been 'refugees' from Mycenae; Achaeans were among those who took part in attacks on Egypt at the end of the century and it now seems that it was a raid by them in about 1200 BC which has been immortalized as the Siege of Troy. One group among these wanderers settled in Canaan, the land between the Dead Sea and the west, in about 1175 BC, where they are commemorated still by a modern name derived from their own: Palestine. But Egypt was the major victim of the sea peoples, who at one time even wrested the Nile delta from Pharaoh's control. Under great strain, in the early eleventh century, Egypt broke apart for a time and was disputed between two kingdoms. Nor were the sea peoples her only enemies. At one point, a Libyan fleet appears to have raided the delta, while in the south, though the Nubian frontier did not yet present a problem, an independent kingdom emerged round about 1000 BC in the Sudan which would later be troublesome. A tidal surge of barbarian peoples was wearing away the old imperial structures just as it had worn away Mycenaean Greece.

A Hittite clay tablet, inscribed in a cuneiform rendering of the Hittite language.

Hebrews

In the midst of all this, there probably took place an event we cannot date and known only through tradition written down centuries later. This was the flight from Egypt of a people Egyptians called Hebrews and the world (much later) came to call Jews. No people of such insignificant origins and, for a long time, such tiny numbers has had so disproportionate a historical impact. The Jewish legacy was to change the world. Its origins are to be sought among the Semitic, nomadic peoples of Arabia, but not much is definitely known of them. Probably in pre-historic times, various tribes who were their ancestors were already pressing into the Fertile Crescent. The start of the traditional Jewish history is the age of the patriarchs, embodied in the biblical accounts of Abraham, Isaac and Jacob. It may well be that men who were the origins of these gigantic and legendary figures actually existed. If they did, it was in about 1800 BC, amid the confusion following the end of Ur, from which the Old Testament tells us that Abraham came to Canaan, a quite plausible assertion. Though those who were to be remembered as his descendants became known in the end as 'Hebrews', this name is a rendering of a word meaning 'wanderer' which does not appear before Egyptian writings and inscriptions of the fourteenth or thirteenth centuries BC, long after their first settlement in Canaan. It is probably the best name to give the tribes with which we are concerned at that time; 'Jews' is a name better reserved for a later era.

The Bible depicts Abraham's people first as tribes of herdsmen, quarrelling with neighbours and kinsmen over wells and grazing, still vulnerable, and liable to be pushed about the Near East by the pressures of drought and hunger. For centuries they must have been hardly distinguishable from many similar wanderers. One group among them named in the Bible as the family of Jacob went down into Egypt, we are told, perhaps in the early seventeenth century BC. As its story unfolds, we learn of Joseph, the great son of Jacob, rising high in Pharaoh's service. We might hope here for help from Egyptian records and perhaps there was such a royal servant during the Hyksos ascendancy, when large-scale disturbance might lead to the improbable pre-eminence of a foreigner in the Egyptian bureaucracy. Unfortunately, there is no evidence either to confirm or to disprove it. There is only tradition, as there is only tradition for all Hebrew history until about 1200 BC. The Old Testament books which contain it only took their present form in the seventh century BC, perhaps eight hundred years after the story of Joseph, though older elements can be and have been distinguished in them. None of this would matter much, except to Jews and scholars, were it not for future events whose roots lay in the unique religious vision of this tiny, not very easily identifiable people.

The Hebrews were, so far as we know, to be the first people to arrive at an abstract notion of God – indeed, they came to forbid his representation in images. Quite plausibly, it has been pointed out that a number of more or less contemporaneous forces in the ancient Near East were likely to make monotheistic religious views attractive. The protection of local deities cannot have seemed very reliable as great upheavals and disasters swept again and again across the peoples of the region after the first Babylonian empire. The religious innovations of Akhnaton and the growing assertiveness of the Mesopotamian cult of Marduk have both been seen as attempts to respond. Yet only the Hebrews arrived at a coherent and uncompromising monotheism, to be completed in essentials by the eighth century BC. The earliest Hebrew religion was probably polytheistic, but monolatrous – like other Semitic peoples, the Hebrew tribes had believed that there were many gods, but worshipped only one, their own. The next stage of refinement was the idea that the people of Israel (as the

Through the Old Testament the Jewish tradition provided for centuries to later Christians a picture of human history and a rich body of symbolism and story which was the mainspring of Christian art and mythology: here, a fourteenth century *AD* fresco from San Gimingnano sets out for those who could not read the Book of Exodus, the story of the Israelite crossing of the Red Sea.

tribe of Jacob came to be called) owed exclusive allegiance to Yahweh, its tribal deity, later known worldwide as Jehovah. he was a jealous God who had made a covenant with his people to bring them again to the promised land, the Canaan to which he had already brought Abraham out of Ur. The covenant was a master idea. Israel was assured that if it did something, then something desirable would follow.

The first monotheistic religion

The exclusive demands of Yahweh opened the way to monotheism; the time came when the Israelites felt no respect for other gods. Nor was this all. At an early date Yahweh's nature was already distinctive. That no graven image was to be made of him was the outstanding feature of his cult. At times, he appears (like other gods) in an immanent dwelling place, such as a temple made with hands, or even in manifestations of nature, but, as the Israelite religion developed, he could be seen as transcendent and omnipresent: 'Whither shall I go from thy spirit? or whither shall I flee from thy presence?' asks the author of one Jewish hymn (Psalm cxxxix, 7). Yahweh's creative work was something else sharply distinguishing Jewish from earlier Mesopotamian tradition. Both depicted Man's origins in a watery chaos; 'the earth was without form, and void; and darkness was upon the face of the deep', says the book of Genesis. For the Mesopotamian, though, matter of some sort had always been there; the gods only rearranged it. It was different for the Hebrew; Yahweh had already created chaos itself. He was for Israel what was later described in the Christian creed, 'maker of all things, by whom all things are made', and He made Man in his own image, as a companion, not as a slave; Man was the culmination and supreme revelation of His creative power, a creature able to know good from evil, as did Yahweh Himself, and moving in a moral world set by Yahweh's own nature as the only author of right and justice.

The implications of such ideas were to take centuries to clarify and emerge. At first, they were muffled in the superstitions of a tribal society looking for its god's favour in war and distress. Later Jewish tradition placed great emphasis on the miraculous exodus from Egypt, a story dominated by the heroic and mysterious figure of Moses, and when the Hebrews came to Canaan they were probably already and consciously grouped round the cult of Yahweh. The biblical account of the wanderings in Sinai probably reflects the crucial time when this first national consciousness was forged. But, once more, the much later record of biblical tradition is all that there is to depend upon. It is certainly credible that the Hebrews should at last have fled from harsh oppression in a foreign land – an oppression, too, stemming from burdens imposed by huge building operations. Moses is an Egyptian name, and it is likely that there was a historical original of the great leader who dominates the biblical story of exodus and holds the Hebrews together in the wilderness. In the traditional account, Moses founded the Law by bringing down the Ten Commandments from his encounter with Yahweh. This was the occasion of the renewal of the covenant by Yahweh and His people at Mount Sinai; perhaps it represents a formal return to its traditions by a nomadic people whose cult had been eroded by long sojourn in the Nile delta. Yet the Commandments themselves cannot be convincingly dated until much later than the time when Moses might have lived.

The biblical account must be treated with respect, nonetheless; it contains much that can be related to other sources. Archaeology at last comes to the historians' help with the arrival of the Hebrews in Canaan. The story of conquest told in the book of Joshua fits evidence of destruction in the Canaanite cities in the thirteenth century BC. What we know of Canaanite culture and religion also fits the biblical account of Hebrew struggles against local cult practice and a pervasive polytheism. Palestine was disputed between two religious traditions throughout the twelfth century. It now seems likely that the Hebrews drew support from other nomadic tribes who accepted the cult of Yahweh. After settlement, although the tribes quarrelled with one another, they continued to worship Yahweh, the only uniting force among them, for tribal divisions formed Israel's only political institution.

Kings and prophets

The Hebrews were in many ways less advanced culturally than the Canaanites, whose script they borrowed, as they did their building practice, though without always achieving much; Jerusalem was for a long time a little place of filth and confusion, not comparable to the Minoan towns built centuries before. Military necessity provoked the next stage in the consolidation of a nation. Challenge from the Philistines (the Indo-European people which gave its name to Palestine) seems to have stimulated the emergence of the Hebrew kingship at some time about 1000 BC. With it appears another institution, that of the special distinction of men called prophets.

The prophets were not soothsayers such as the Near East already knew (though this may have been the tradition which formed the first of them) but preachers, poets, political and moral critics. Their status depended essentially on the conviction that God spoke through them. One of them, Samuel, anointed (and thus, in effect, designated) both Saul, the first king, and his successor, David. When Saul reigned, the Bible tells us, Israel had no iron weapons, for the Philistines took care not to endanger their supremacy by permitting them. None the less, the Jews learnt from their enemies; the Hebrew words for 'knife' and 'helmet' both have Philistine roots. Saul won victories, but died at last by his own hand. His work as a state-builder was completed by David who, of all the great figures of the Old Testament remains outstandingly credible, although outside its pages there is no evidence that he existed. The literary account, confused though it is, tells of a noble-hearted but flawed and all-too-human hero who ended the Philistine peril and reunited a divided kingdom after Saul's death. Jerusalem then became Israel's capital and David imposed himself upon the neighbouring peoples. Among them were the Phoenicians (though they had helped him against the Philistines); this was the end of Tyre as an important independent state. David's son and successor, Solomon, though, was the first king of Israel to win major international standing. He launched expeditions to the south against the Edomites and built a navy. Conquest and prosperity followed, but this success is perhaps only further evidence of the eclipse of the older empires. Solomon was undoubtedly a king of great energy and drive and his achievements were not only military. The legendary 'King Solomon's Mines' have been said to reflect memories of the first copper refinery of which there is evidence in the Near East, though this is disputed. The building of the Temple (after Phoenician models) was only one

A Jewish king makes obeisance to the Assyrian ruler Shalmaneser III in 841 BC. Jehu, the king shown here, may be the first Jew of whom a picture survives. Above the figures is an inscription in cuneiform.

of his public works, though perhaps the most important, for it gave the worship of Yahweh a more splendid form than ever before and an enduring focus.

Yet in the end Israel was to be remembered not for the great deeds of her kings but for the ethical standards announced by her prophets. They brought to a new height the Israelite idea of God. Few preachers have had such success. They shaped the ties of religion with morality which were to dominate for thousands of years not only Judaism but two world religions, Christianity and Islam. Through them the cult of Yahweh evolved into the worship of a universal God, just and merciful, stern to punish sin but ready to welcome the sinner who repented. This was the climax of religious development in the ancient Near East. Religion could henceforth be separated from locality and tribe. The prophets also bitterly attacked social injustice. They announced that all men were equal in the sight of God, that kings might not simply do what they would; they proclaimed a moral code which was a given fact, independent of human authority. The preaching of adherence to a god-given moral law thus became a basis for a criticism of existing political power. Since the law was not made by man the prophets could always appeal to it as well as to their divine inspiration against king or priest. The prophets made one of the great intellectual jumps of mankind; the heart of political liberalism is the belief that power must be used within a moral framework independent of it, and its tap-root lies in their teaching.

Israel had prospered in the eclipse of paramount powers, but after Solomon's death in 935 BC her history took a turn for the worse. The kingdom split again. Israel, in the north, gathered ten tribes together around a capital at Samaria; in the south the tribes of Benjamin and Judah still held Jerusalem in the kingdom of Judah. In 722 BC the Assyrians obliterated Israel and its ten tribes disappeared from history in mass deportations. Judah, more compact and somewhat less in the path of great states, survived a century longer.

Israel and Judah

c.1010	Saul proclaimed king of Israel, but later defeated by the Philistines
c.1006–960	Reign of David; Judah and Israel united; Philistines defeated; Jerusalem becomes the political and religious capital; conquest of the Canaanite lands
c.960–635	Reign of Solomon; Aramaic province gains independence from Israel and Edomite kingdom re-established; re-organisation of the kingdom into a centralised state; re-building of Jerusalem
926	Israel divided into southern (Israel) and northern (Judah) kingdoms
Israel	
878–871	Reign of Omri; order restored following period of internal strife; Samaria becomes capital of Israel
871–852	Reign of Ahab; his marriage to Phoenician Jezebel; introduction of worship of Baal; resistance led by the phrophet Elijah
845–17	Elijah has Jehu proclaimed king; suppression of Phoenician culture and religion; pressure from Assyria
722	Assyrians destroy Samaria after a three-year seige and absorb Israel; deportations
Judah	
845–839	Reign of Athaliah; descendants of Omri and David eradicated; introduction of worship of Baal
725–697	Reign of Hezekiah; unsuccesful attempts to shake off Assyrian influence
587	Nebuchadnezzar II destroys Jerusalem and deportations follow

THE LAST AGE OF MESOPOTAMIAN EMPIRE

The civilized tradition of Mesopotamia began its final flowering in the eighth century. Nineveh, the city high up the Tigris which had replaced the ancient capital of Assur, became a political focus as Babylon had once been. The turmoil which had followed the crumbling of Hittite power was to be disciplined by a new Assyrian empire. The Old Testament tells us of Assyrian armies moving again and again against the Syrian and Jewish kingdoms until they conquered. This was a hard time for losers. Assyrian empire did not rely on the vassalization of kings and the creation of tributaries but swept native rulers away and installed Assyrian governors. Often, too, it scooped up peoples (like the ten tribes of Israel) in mass deportations.

Soon after 729 BC, when they seized Babylon, the Assyrian armies not only destroyed Israel, but invaded Egypt and annexed the delta. By then Cyprus had submitted to Assyrian rule, Cilicia and Syria had been conquered. Finally, in 646 BC, came the last important conquest, part of the land of Elam, whose kings dragged the Assyrian conqueror's chariot through the streets of Nineveh. A standardized system of government and law now spanned most of the Near East. The marching of armies and the driving of the deported about within it sapped its provincialism. Aramaic, long spoken as a *lingua franca* in Syria and Mesopotamia, spread as a common language.

Monuments of undeniable impressiveness commemorate the empire's great formative power. A great palace built at the end of the eighth century at Khorsabad, near Nineveh, half a square mile in area, was embellished with more than a mile of sculpted reliefs. The profits of conquest financed a rich and splendid court. Ashurbanipal (668-626 BC) left behind him his own monuments (including obelisks carried off to Nineveh from Thebes), and a great collection of tablets made for his library. He was a man with a taste for learning and antiquities who accumulated copies of all that he could discover of the records of ancient Mesopotamia, and to him we owe much of our knowledge of Mesopotamian literature. The ideas that moved this civilization are thus more accessible than those of its predecessors. The frequent representation of Assyrian kings as hunters is traditionally part of the image of the warrior-king, but may also consciously identify him with legendary conquerors of nature in the remote Sumerian past. Yet the stone reliefs which commemorate the great deeds of Assyrian kings also repeat, monotonously, another tale, that of sacking, enslavement, impalement, torture, and the Final Solution of mass deportation. Assyrian empire had a brutal foundation. Its army was recruited by conscription and armed with iron weapons. It

In the Middle Ages, Christian kings were urged to consider David as both a model and a warning. This fifteenth-century manuscript illumination shows the start of the Biblical story (2 Samuel, II) of his adultery with the lovely Bathsheba – whom he here summons by note after seeing her bathing – and his subsequent contriving of the death in battle of her husband, Uriah the Hittite.

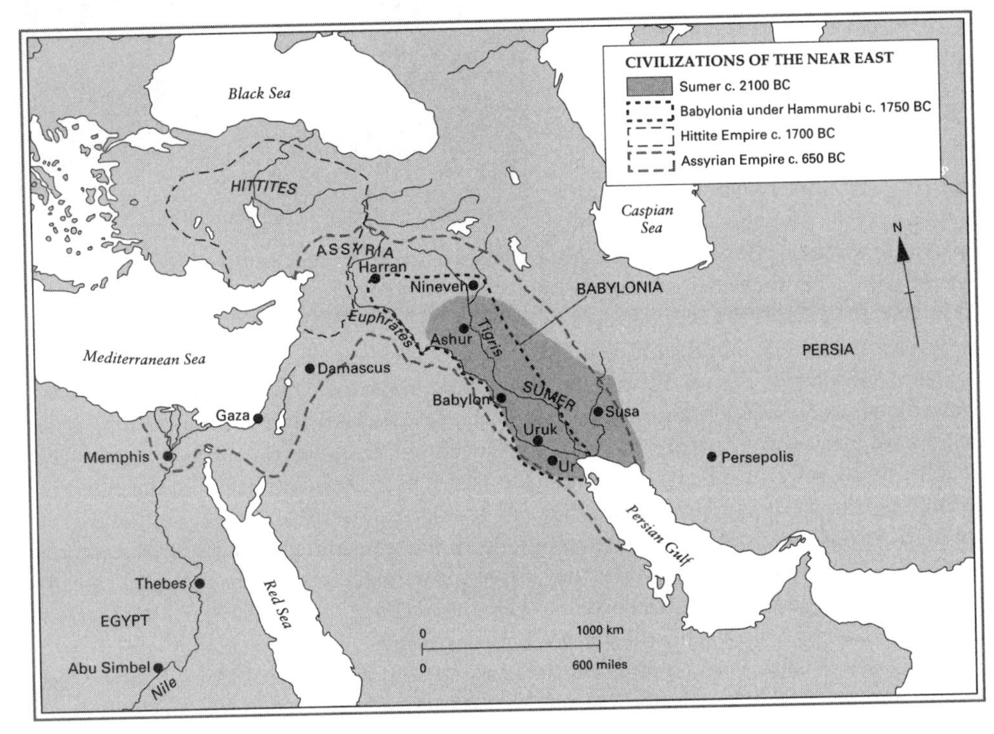

was a coordinated force of all arms, with siege artillery able to breach walls until this time impregnable, and even some mailed cavalry. Perhaps too, it had a special religious fervour. The god Assur is shown hovering over the armies as they go to battle and to him kings reported their victories over unbelievers.

Yet the last Assyrian empire quickly waned. Possibly there was too great a strain on its numbers. The year after Ashurbanipal died, it began to crumble. Babylonian rebels were supported by the Chaldeans and by a new neighbour, the kingdom of the Medes. Its entrance onto the stage of history marks an important change. The Medes had long been distracted by yet another barbarian people from the north, the Scythians, who poured down into Iran from the Caucasus (and at the same time along the Black Sea coast towards Europe). These light cavalrymen, fighting with the bow from horseback, announce the appearance of a new force in world history, the nomadic peoples of Central Asia. The Scythians pushed other peoples before them while the last of the political units of the Near East based on the original Caucasian inhabitants was being gobbled up by themselves, Medes or Assyrians. When Scyths and Medes joined forces, they pushed Assyria over the edge, too; she passes from history when the Medes sacked Nineveh in 612 BC.

Slingers of the Assyrian army, carved in relief on the walls of Nineveh.

The last Babylonian empire

This thunderbolt was still not quite the end of the Mesopotamian tradition. Assyria's collapse left the Fertile Crescent open to new masters. In the north the Medes pushed across Anatolia until halted at the borders of Lydia. They drove the Scyths back into Russia. An Egyptian pharaoh made a grab at the south and the Levant, but was defeated by a Babylonian king, Nebuchadnezzar, who gave Mesopotamian civilization an Indian summer of grandeur. His, the last, Babylonian empire ran from Suez, the Red Sea and Syria across Mesopotamia and the old kingdom of Elam (by then ruled by a minor Indo-European dynasty called the Achaemenids). If for nothing else, Nebuchadnezzar would be remembered as a great conqueror. In 587 BC, after a Jewish revolt, he destroyed Jerusalem and, again, the Temple. Three great deportations drove the tribes of Judah into captivity, thus imposing on them an Exile as experience so formative that after it we may properly speak of the 'Jews', the inheritors and transmitters of a national tradition henceforth easily traced. Meanwhile, they helped to embellish Nebuchadnezzar's capital, whose 'hanging gardens' or terraces were to be remembered as one of the Seven Wonders of the World. He was undoubtedly the greatest king of his day.

The glory of Babylonian empire came to a focus yearly in the cult of Marduk, now at its zenith. At a great New Year festival all the Mesopotamian gods – the idols and statues of

A Scythian wall-hanging, of the fifth or fourth century BC, discovered in Siberia, showing a seated goddess, dressed possibly in a style reflecting Chinese influence, and a Scythian horseman.

provincial shrines – came down the rivers and canals to take counsel with Marduk at his temple and acknowledge his supremacy. Borne down a processional way three-quarters of a mile long (which was, we are told, probably the most magnificent street of antiquity) or landed from the Euphrates nearer to the temple, they were carried into the presence of a statue of the god which, the Greek historian Herodotus reported two centuries later, was made of two and a quarter tons of gold. No doubt he exaggerated, but it was indisputably magnificent. The destinies of the whole world, whose centre was this temple, were then debated by the gods and determined for another year. Thus theology reflected political reality. The re-enacting of the drama of creation was the endorsement of Marduk's eternal authority, and of the absolute monarchy of Babylon. The king had responsibility for assuring the order of the world and therefore the authority to do so.

It was the last flowering of a long-lasting tradition. Provinces were lost under Nebuchadnezzar's successors. Then came an invasion in 539 BC by new conquerors from the east, called Persians. The passage from worldly pomp and splendour to destruction had been swift and the Old Testament book of Daniel telescopes it in a magnificent closing scene, Belshazzar's feast – an account only written three hundred years later and badly astray on important facts. Yet its emphasis has a dramatic and psychological truth. In so far as the story of antiquity has a turning-point, this is one. An independent Mesopotamian tradition going back to Sumer is over and we stand at the edge of a new world.

THE EMERGENCE OF PERSIA

More than half the story of civilization (counted in years) is over by the middle of the first millennium BC. A world of a few highly distinctive civilizations had by then already given way to one in which a larger and larger area of one part of the globe shared lasting governments, technology, organized religion, city life – and changed more and more rapidly as the interplay of different traditions increased. Small things show the change as well as great. On the legs of huge statues at Abu Simbel, seven hundred miles up the Nile, sixth-century Greek mercenaries in the Egyptian army cut inscriptions (just as twenty-five hundred years later

English regiments left their badges and names cut into the rocks of the Khyber Pass). The military and economic drive of the Mesopotamians and their successors, the movements of the Indo-Europeans, the coming of iron and the spread of literacy had thoroughly mixed up the once-clear patterns of the Near East. The greatest upheavals of the ancient *Völkerwanderung* were over as the Near East emerged from the late Bronze Age; the political structures left behind by them would be levers of the next era of world history.

The tomb of Cyrus the Great, Pasargadae.

For some, their time was past. This was Egypt's fate. After 1000 BC she underwent a decline which seems to reveal an inability to change or adapt. To survive the first attacks of the iron-using peoples and beat off the Peoples of the Sea was the New Kingdom's last triumph: thereafter the symptoms are unmistakably those of a machine running down. At home kings and priests disputed power; abroad, Egypt's suzerainty declined to a shadow. A period of rival dynasties was briefly followed by a reunification which again took an Egyptian army to Palestine, but by the end of the eighth century a dynasty of Kushite invaders had established itself. In 671 BC it was ejected from Lower Egypt by the Assyrians. Ashurbanipal sacked Thebes, but, as Assyrian power ebbed, there was again an illusory period of Egyptian 'independence'. By this time, evidence of a new world towards which Egypt had to make more than political concessions can be seen in the establishment of a school for Greek interpreters and of a Greek trading enclave with special privileges in the

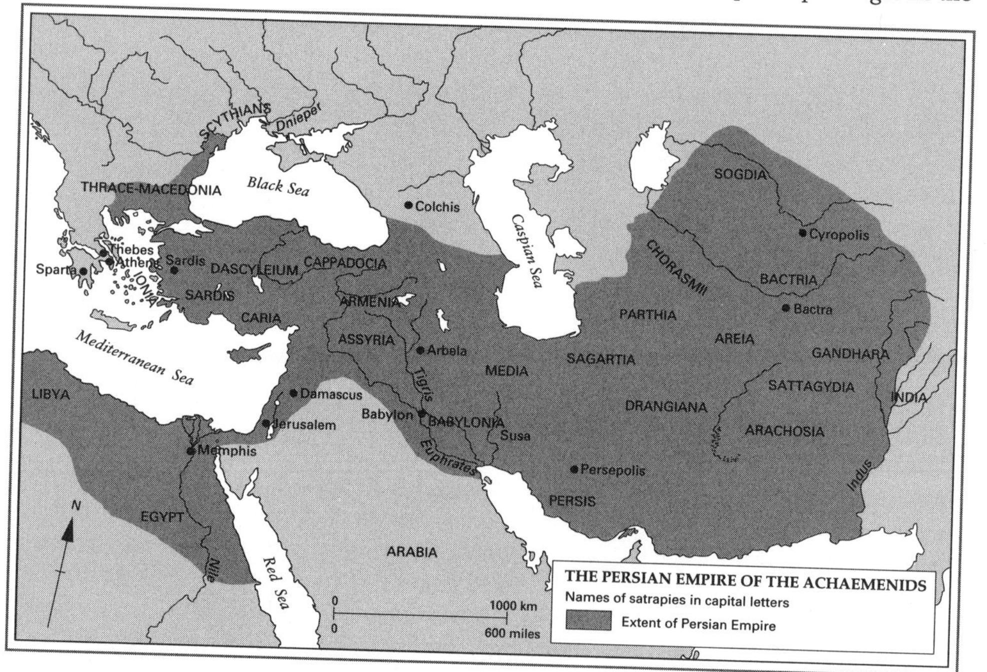

delta. Then, in the sixth century, Egypt again went down to defeat first at the hands of Nebuchadnezzar (588 BC) and sixty years later (525 BC), before the Persians. From the fourth century BC to the twentieth AD, Egypt was to be ruled by foreigners or immigrant dynasties and passes from the forefront of history.

The ruler who overthrew both Babylon and Egypt is remembered as Cyrus, King of Persia. The word 'Iran' does not appear until about AD 600 but in its oldest form means 'land of the Aryans'; and with an irruption of Aryan tribes from the north, the history of Persia begins a few centuries before that. Among the newcomers were the especially vigorous and powerful Medes and Persians. The Medes moved west and north-west to Media; the Persians had gone south towards the Gulf, establishing themselves on the edge of the Tigris valley and in the old lands of Elam. Fars was the ancient name for their new kingdom.

The Achaemenid emperors

Cyrus was descended from a family called Achaemenid. He was the first Persian to live in memory as a world-historical figure, recognized as such by other would-be conquerors who strove in the next few centuries to emulate him. In 549 BC he humbled the last independent king of the Medes and so created a united kingdom. Thenceforth the boundaries of conquest rolled outwards to enclose the largest empire yet seen. Only in the east (where he crossed the Hindu Kush and was eventually killed fighting the Scythians) did Cyrus find it difficult to stabilize his frontiers. His success owed much to his kingdom's wealth in minerals, above all in iron, and in the high pastures of the valleys lay a great reserve of horses and cavalrymen. His government was different in style from its predecessors; brutality was not celebrated in official art and Cyrus was careful to disturb his new subjects as little as possible. There are some notable hints of a wish to conciliate them; the protection of Marduk was solicited for his assumption of the Babylonian kingship and at Jerusalem he launched the third rebuilding of the Temple. Provincial governors were required to produce little beyond the tribute which replenished the treasuries of Persia. The result was a diverse but powerful empire.

Though with setbacks a-plenty, it provided for nearly two centuries a framework for the Near East. Over large areas there were longer periods of peace than for many years. Cyrus's son added Egypt to the empire; yet he died before he could deal with a pretender to the throne whose attempts encouraged Medes and Babylonians to seek to recover their independence. The restorer of Cyrus's heritage was a young man who also claimed Achaemenid descent, Darius.

Though he did not achieve all he wished, Darius (who reigned 522–486) carried still further the boundaries of the empire. He failed, as Cyrus before him, to make much headway against the Scythians, but his work rivalled that of his great predecessor. His own inscription on the monument recording his victories over rebels ran 'I am Darius the Great King, King of Kings, King in Persia', a recitation of an ancient Achaemenid title whose braggadocio he adopted. Decentralization was carried further with the division of the empire into twenty provinces, each under a satrap who was a royal prince or great nobleman. Royal inspectors surveyed their work and a royal secretariat corresponded with them. Aramaic became the administrative language. Government rested on better communications than any yet seen; roads were built along which, at their best, messages could travel at two hundred miles a day.

Darius planned a great new capital at Persepolis, where he was eventually buried. Intended as a colossal glorification of the king, Persepolis reflected the diversity and cosmopolitanism of the empire. Assyrian colossi, man-headed bulls and lions guarded its gates as they had done those of Nineveh. Its decorative columns are an Egyptian device borrowed through Ionian stone-cutters and sculptors. Greek details are to be found also in the reliefs and decoration and a similar mixture of reminiscences is to be found in the royal tombs not far away, whose conception recalls Egypt's Valley of the Kings. Persian culture was always open to influence from abroad. Vedic and Persian religion mingled in Gandhara, but both

At Persepolis, the guards depicted at the side of the great staircase leading to the Audience Hall would probably have been brightly coloured, their weapons and accoutrements gilded. The overall effect must have been splendid, though now almost impossible to imagine.

were Aryan. The core of Persian religion was sacrifice and centred on fire. By the age of Darius the most refined and official of its cults had evolved into what has been called Zoroastrianism, which spread rapidly through western Asia with Persian rule. Though probably never more than a minority cult, it was to influence both Judaism and the mysteries which form so much of the context of early Christianity; the angels of Christian tradition and the hellfire which awaited the wicked both came from Zoroaster, its founder. Of him, though, we know little, except that he taught that the world was the setting of an eternal struggle between a good god of light and an evil spirit of darkness. Zoroastrianism's sacred texts are contained in a collection called the Avesta or Zend-Avesta and still in use by the Parsee community of India (whose name derives from the same word as 'Persia').

Persia suddenly pulled even more peoples into a common experience. Indians, Medes, Babylonians, Lydians, Greeks, Jews, Phoenicians, Egyptians were for the first time all governed by one empire. Indian mercenaries fought in the Persian armies as had Greeks in those of Egypt. Men lived in cities throughout the Near East and around much of the Mediterranean, too, sharing a literacy now expressed in many scripts. Shared agricultural and metallurgical techniques stretched even further; the Achaemenids transmitted the irrigation skills of Babylon to Central Asia and brought rice from India to be planted in the Near East. When Asian Greeks came to adopt a currency it would be based on the sexagesimal numeration of Babylon. In this richness, a basis of the future world civilization is beginning to be discernible, and is time to turn to those who laid it out, the Greeks.

THE MEDITERRANEAN WORLD

For about a thousand years after 500 BC the lands round the eastern Mediterranean, including the Levant, Syria, Europe as far east as the Rhine and Danube, and the coasts of the Black Sea, are at the centre of an important story. They were never cut off from what went on outside them, and were, indeed, more aware of it than ever before. Ideas, knowledge, customs, and the peoples carrying them flowed in from the north and east and changed the region's life very much as time passed. Yet as they grappled with and sometimes welcomed the forces pressing on them from the outside, the peoples living in this area made something quite new of them and that moves the focus of world history from the Fertile Crescent to the west. The Mediterranean and Aegean peoples put a civilization of their own solidly in place. Three of them above all contributed to the character it evolved over a thousand years: Greeks, Romans, Jews. Many of the most decisive steps which founded and set that character, were taken by the Greeks between about 700 and 350 BC. Allowing for the absence of what Roman and Jewish influences were to bring to it, this was when much we still take for granted and much more that shaped the story of later Europe was defined.

Though they had the word 'Europe', it would have meant nothing like what we think of as Europe to the Greeks. Their world was the Mediterranean basin, not the land mass to its north, and in ancient times that sea was very self-contained. Almost everywhere behind its coasts there swiftly rose high hills and mountains. Only in Libya and Egypt are there broad tracts of country running inland, most of them now desert, though this was not always so. Elsewhere, the high ground is only separated from the sea by narrow coastal plains. The Black Sea is fed by the great river valleys of Russia and the Balkans and pours water into the Mediterranean; it has vast plains on its north side but mountainous eastern and southern shores. Only three really important rivers run into the Mediterranean itself and provide access to the interiors of the countries around it. The Nile is the longest, stretching inland for 1000 kilometres before coming to the First Cataract. The Rhone flows down the middle of France, draining the western Alps. The Ebro rises only 60 or 70 kilometres from Spain's northern and Atlantic coast before making its way to the sea through the broad valley of Aragon. The only other major European river valley which matters here, that of the Po in northern Italy, empties into the Adriatic.

As for climate, for the most part the Mediterranean lands are warm (but not too warm) and reasonably well watered by the rainfall on their hills and mountain ranges. The coasts would have been much greener 2500 years ago; much of the cover of trees and shrubs has since been stripped from the hills by grazing, which has led to the washing-away of topsoil. The forests of Lebanon remain legendary and in North Africa farming went on much further inland, holding at bay deserts more distant than nowadays.

All round the Mediterranean most people got their living in very similar ways. Those who lived on the coasts fished in well-stocked waters. On the narrow plains, they grew wheat and barley, and further up the slopes, vines and olives. On the hills they herded sheep and goats. They lived much in the open air, because the climate was not fiercely cold except in the mountains in winter. As well as having so much in common, the peoples of the Mediterranean were also more and more linked to one another by sea as time passed. Until modern times travel by water was almost always easier than by land, and there was plenty of long-distance maritime commerce by 500 BC. When new influences appeared in any part of the Mediterranean they were always likely to spread more easily and quickly than in the land-locked empires of the East. This was particularly true in the basin of the Aegean, the sea between Crete, Greece and Turkey. Its topography and climate are Mediterranean, but it has a number of special features of its own. The coastal plains and valley floors of its mainland tend to be smaller than further west; the coast is craggier and more indented with little harbours and refuges. There are hundreds of islands, some of them mere rocks incapable of growing anything, but others very fertile. It is not a big area; from the coast of Thrace to Crete is about 600 kilometres, and from Troy to the nearest coast of mainland Greece only about 250. The water is often rough, and strong prevailing winds made navigation awkward except by certain routes, but even in Minoan times men could sail about it from place to place with confidence once they understood its conditions. This had encouraged trade.

As early as Minoan times the ideas and skills of the lands of ancient civilization – Mesopotamia, Syria, Egypt – had been taken up by islanders and coast-dwellers of the Aegean. In an even more remote age, agriculture may well have first spread from the Levant to the islands and from there to Macedonia. Through the Levant ports and the Nile delta, the peoples of the Near East could influence the Aegean far more easily than they could India, Central Asia and China, shut off as they were by natural barriers and distance. Terrain also explains why the Aegean peoples could not have much direct contact with western Europe even if it supplied metals and other raw materials to the richer lands of the Near East through the Phoenicians.

Greek shipping: a sixth century BC bowl from Athens shows a merchantman, powered by sail only, and a warship with fifty or so oarsmen and ram for attacking enemy ships.

Greece

The western wall of the Aegean is provided by mainland Greece. Behind its straggling, rocky, indented coast lies a land falling into three natural geographical divisions. The south is a big peninsula, almost an island, called the Peloponnese. It is attached by a very narrow neck (the isthmus of Corinth) to another peninsula, larger than the Peloponnese and itself jutting out from south-eastern Europe. Its southern tip is called Attica, and it runs north to the highlands of Thessaly. Finally there is the mountain region of the extreme north, Macedonia. Its geography has always made Greece a difficult country to invade by land except by a few well-defined routes. Attacking it from the sea has usually been easier.

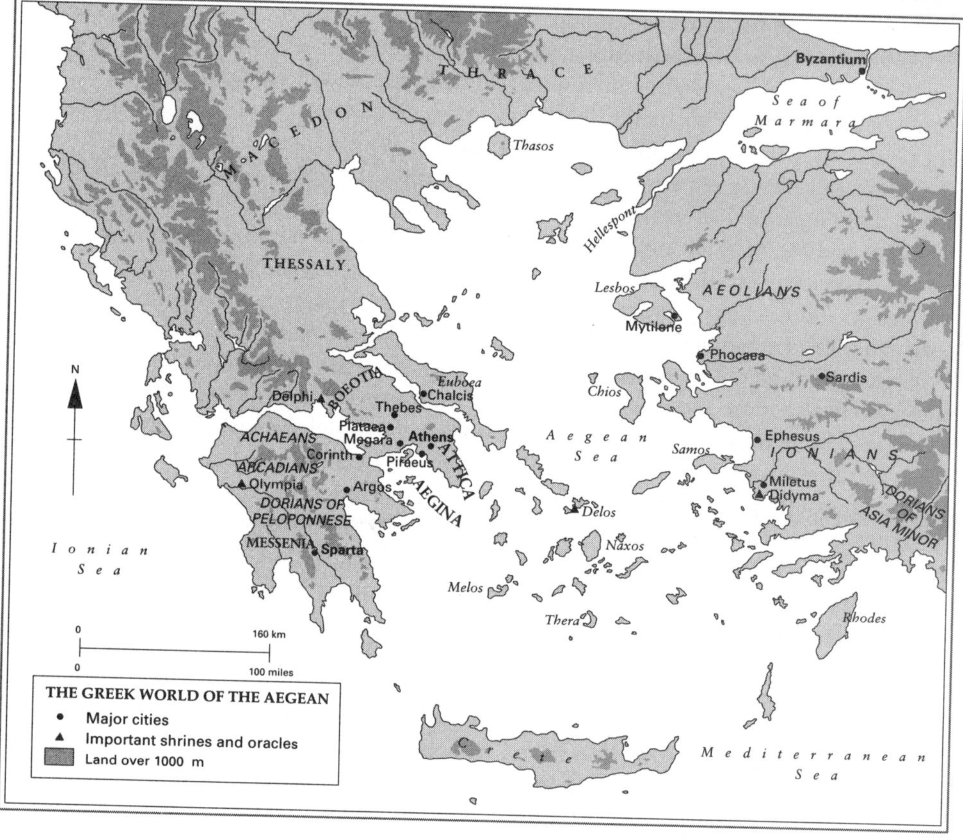

Individual artists begin to emerge from the anonymous flow of early art in the sixth century. One master of potting and the painting of pots was Exekias of Athens *whose signature appears on this cup decorated with the ship of Dionysus. The god of wine lies back in his ship holding a drinking-vessel, perhaps contemplating the vine which decorates its mast and breaks out into bunches of grapes over his head, or perhaps amused by the leaping dolphins into which (legend said) he had transformed a gang of pirates he had encountered on his voyage.*

From Dark Age Greece emerged a new sort of city life, shaped by the fragmentation of its topography. Some Greek cities were to make such a special contribution to the future that this deserves special note, even at the risk of running ahead of our story. In their early days they were often ruled by kings and some were to go on calling some of their officials by that title in later times. Once we have good historical records, though, we find that in the seventh century they were usually governed by 'aristocrats' – a Greek word meaning the 'best people'. Aristocrats were landowners and rich enough to buy the expensive arms, armour and horses which made them leaders in war. At first they ruled over other Greeks who were, in the main, farmers; in most of the Greek world this was always to be how most free men got their living.

The simple early Greek societies had already begun to grow more complicated by 600 BC. Foreigners who were craftsmen and traders lived in some of them; they were called 'metics' and were not given the same rights as the native-born inhabitants. Their presence and the services they offered reflect an upsurge in trade. Because there was more money about, too, some men grew richer. One sign was the growing use of metal currency in the Greek world. People and communities began to specialize in different kinds of business or manufactures – Athens, for example, in pottery.

As wealth grew, more men acquired land. They could also afford arms and armour, and in the seventh century BC there appeared a new sort of warrior. The 'hoplites', as the Greeks called them, were infantry, wearing bronze helmets and body-armour, carrying shields and spears. With them Greek warfare suddenly changed. Earlier fighting had been very much a matter of single combat between the few who could afford weapons and armour, which made them much more formidable than most of those who followed them to battle. Now battles began to be won by disciplined masses of hoplites. They kept careful formation, each man being protected on his right-hand side by the shield of his neighbour. On the little valley-floors which were the usual place for Greek battle (since the aim was normally to destroy or defend the crops grown there) a good hoplite formation was almost invulnerable if it charged as a mass and kept its ranks unbroken. As more men shared military experience, and as discipline and drill won battles, power began to slip from the old aristocracies. They ceased to be the only people controlling armed force. This important change gave rise to politics, one of the great Greek inventions.

The polis

We take the word 'politics' for granted and often do not bother to think very much about what we mean by it. A rough definition might be this: 'a way of running public affairs by making decisions about them after public discussion of different possible courses of action'. This may sound abstract, but there is a big difference between such public business and what follows from the arbitrary will of a ruler, as the Greeks well knew from looking at Persia or Egypt. The word politics itself (like 'political') is Greek in origin. It comes from the Greek word for a state or independent city: polis. That meant more than just a place; in fact, the Greeks did not speak of 'Athens' doing this, or 'Thebes' doing that, but of what was done by 'the Athenians' and 'the Thebans'. Polis is often translated 'city-state', for want of a better term, but the idea behind it was not just that of an agglomeration of people living in the same place and it was certainly much more than we often mean nowadays by 'state'. The polis was a community. It did not include everyone who lived in the city and its surrounding countryside, but consisted only of citizens: those who had to take their place in the hoplite ranks in war and were entitled to a say, however small, in their common affairs. Slaves, metics and women could not be citizens. It was nearer to membership of a tribe than our modern notion of citizenship. In fact, the earliest social institutions within cities were based on kinship.

The first citizens of the Greek states were those who could afford arms to take their place in the hoplite ranks and fight to defend their heritage. Many fierce struggles must have taken place about which we now know very little, but, little by little, new men everywhere won admission to citizenship in the sixth century BC. Government by aristocrats gave way to government by popular strong men – 'tyrants', the Greeks called them – but these in their turn were replaced eventually by governments with a broader base. 'Oligarchy' and 'democracy' are two other words of Greek origin we still use. Some cities were ruled by the well-off (the 'oligarchies'), others by the majority of free men (the 'democracies'). Almost everywhere, though, rule by hereditary leaders declined.

For a long time most cities were fairly small, feeding themselves from the farms of the little valleys in which they were usually sited. Even in later times, only an unusual polis had 20,000 citizens. Most citizens, therefore, could feel much more personally involved in public life than can the citizens of a modern state. They took part as a body in things which we have hived off to private organizations like clubs and churches. In the assemblies which ran public business a Greek would get to know how his friends, enemies and acquaintances felt on questions that affected them all. All these things made life in the polis an intense, demanding business, but also an exciting one. Some men emerged as the first representatives of a new trade: politicians, who sought to persuade their fellow-citizens of what they should do, and did their persuading in the assembly. It is not surprising, then, that Greeks came to think that the polis gave men the chance to be themselves – to release all the potential of their human nature – in a way no other kind of human organization did. You could, they thought, be civilized in a polis as you could not be elsewhere; man, said one Greek philosopher, was a creature made by nature to live in one. The Greeks also had a word for someone who, as it were, 'dropped out' into private preoccupations and would not take part in or concern himself with public affairs: 'idiot'.

A new civilization

When trade revived after the Dorian invasions, in the ninth century or so, enterprising men began to venture abroad and put together again old networks of commerce now decayed. Many of the things they traded were transported in pots and jars whose remains make it possible to trace the process. Design and finish had gone badly downhill at the end of the Mycenaean period, but even around 1000 BC pottery was beginning to be made in a new and very exciting style, first, perhaps, at Athens. It was very simple, but pleasing to the eye, using abstract decoration – lines, concentric circles and bands of colour – which has led to its name: 'geometric'. It is often very beautiful and as it developed became much more complicated. Not until the eighth century BC, after existing for about 250 years, did it begin to show human figures and they, too, were drawn at first in a very geometrical and abstract way. Pottery was a sign that life in the Aegean was again getting more civilized.

The ships on some of the vases of the eighth century tell the same story; there was a Greek Aegean already in existence not long after the year with which later Greeks began their chronology – the year 776 BC. That was the date of an important event, the first Olympic Games, which took their name from the place where they were held – Olympia, in western Greece. Scholars have questioned the date (and it seems, in any case, that an older festival of some sort lay behind these boxing, running, singing and dancing events) but it is a good marker for the beginning of Greek history. The Games went on being held more or less continuously every four years for about a thousand years. Though by about 200 BC they had become mainly a professional show attracting tourists from abroad, in early days teams of amateurs from all over Greece turned up at them to represent their cities; this was the only festival which regularly brought Athenians, Thebans, Spartans and the representatives of many more cities together. It helped them to think of one another as Greeks.

Language and Hellenism

The Greeks called themselves 'Hellenes' (our word 'Greek' comes from Latin, the language of the later Romans). By the end of the Dark Ages many different Aegean communities felt that however much they might quarrel among themselves, they had much in common. Above all, they shared a language. In the eighth century BC Greek was just about to develop in a new way. Its first written form, the script of the tablets found at Knossos and in the Mycenaean palaces, had been used only for accounting. Greek only acquired the form it has today when Greeks took (perhaps through trade) the Phoenician device of the alphabet and adapted it to their own needs. The earliest inscription in the new Greek characters which has been found is on a jug from about 725 BC. When Greek became a written language another step had been taken in the growth of an idea that for all their differences, Greeks had much in common. This was felt so strongly, indeed, that their word for a non-Greek was based on the idea that he was someone who could not speak Greek: he was barbaraphonoi – someone making an unintelligible noise like 'bar-bar-bar'. 'Barbarian' still means an uncivilized person.

Written language pins down thought. For centuries, bards and story-tellers must have repeated the tales, songs and legends of the peoples from whom the Greeks had sprung.

Early in the sixth century BC *an anonymous sculptor carved this herdsman, one of the earliest Attic statues to be found on the Athenian Acropolis. An inscription tells us that the subject may have been Rhombos, who to judge by his dress was probably a wealthy man. Yet he has been caught in a pose which is simple and even consciously so, one evocative of the pastoral and rural background of classical civilization. Similar figures, carrying lambs (or even rams) on their shoulders, have been found as far away as Crete and were dated a century or so earlier than this one.*

Memorized literature of this sort is often very enduring, but its details gradually undergo changes as those who are telling the stories or singing the songs try to bring out a point that seems especially apt, or to introduce an allusion which will make their performance more effective. When tales are written down, there is less scope for individuals to alter them. No doubt there were already hundreds of well-known tales about gods and heroes in ancient Greece long before any of them could be written down. One group of stories, though, became central to later Greek culture and education because the first written works of Greek literature were taken from them. These were the stories and legends of an Achaean expedition to Troy, a city in Asia Minor, the background to the long poems we call the *Iliad* and the *Odyssey*, together about 28,000 lines long and two of the greatest works of literature in any language. The first gets its title from the Greek name for Troy – Ilium – whose siege is in progress while the poem tells us, essentially, about a few days in the life of a great Achaean hero, Achilles. The second poem is called after its hero, Odysseus, who went to Troy on the Achaean expedition. It tells of his subsequent ten years' a-wandering after the end of the siege, of his marvellous adventures and resourcefulness, of his homecoming at last and triumph over those who had tried to usurp his place while he was away. Traditionally these poems were thought to have been written by one man, a blind poet called Homer, who may have lived on the isle of Chios, but this is only one of many facts about them which are still disputed.

To the Greeks the expedition to Troy and the great deeds of the heroes who fought there were history, and their own history, too. The combat between Hector and Achilles depicted on this early fifth–century vase was one of the great episodes of the story.

Probably, they were first written down in the seventh century BC. If so, the *Iliad* and *Odyssey* were by then only the latest versions of very old materials, containing ideas and scraps of fact from many centuries, jumbled together without any sense of what was historically appropriate. A very rough parallel might be to imagine a twentieth-century poem about Christopher Columbus' discovery of America (misdated by a few centuries) in which, though he travelled in sailing-ships, he was taken to have a radio, and to meet natives who often dressed, spoke, and thought like Spaniards, while into the central theme had also been woven details and dim memories, and some of the language, of *Beowulf*. This mixed-up material has given scholars plenty to argue about. Archaeology has been brought to bear on it and often it turns out that reality must have been very different from what the poem describes. The great ten-year siege of Troy, for example, was probably more like a swift Viking raid by a few hundred freebooters on a little settlement no more than three or four hectares in extent. Still, there are some details mixed up in the process which throw light, too, on the Dark Age from which some of what went into making up the poems was descended.

But this was not why Homer's great poems were important to the Greeks, to whom they came to be something like sacred books, in some ways as central as the Old Testament was to Jews. Professional bards for generations made a living out of going around reciting them, and if a Greek had formal education at all, it was grounded in them. Homer summed up what it was that made Greeks think of themselves as different; his poems are the first documents of a Greek self-consciousness. They contained, Greeks believed, their ancient history, essential information about their gods, the ways they were related to humans and were likely to behave, explanations of human destiny and the purpose of life, guidance on ethics, examples of good manners, the qualities which made for a good life and much else. In the *Iliad* and *Odyssey* they found texts to settle disputes, standards by which to judge behaviour and the most splendid example of how their language could be used. These poems helped form Greek ideas and taste for hundreds of years and therefore of later mankind for even longer. Significantly, the Greeks usually did not refer to Homer by his name, but simply as 'the Poet'.

Greek religion

There is a great deal in Homer about gods and goddesses, and therefore about the religion the Hellenes also shared. But the gods and goddesses whose names have come down to us, though they became the pantheon of a whole classical world, were far from the whole story of Greek religion. In approaching it, we have to try to rid ourselves of some of the associations which now cluster around the word 'religion'. Greeks had no clear body of doctrine, no 'clergy' who were specialists in such matters (their priests and soothsayers had narrower functions), no 'church' of believers with a special corporate organization. What they had was a jumble of myths, ideas and superstitions, none of which was a matter of enforced belief for all Greeks, but some of which tried to make sense of deep and continuing human problems – of the insecurity of good fortune, for example, or of the fate which the Greeks

Dionysus, depicted in about 500 BC, holds in his hands the vine-staff and the empty cup which symbolize his role as god of wine. To the Greeks he was also much more –the focus of a major religious cult. Probably Dionysus was in origin a nature god but his cult evolved into an interesting and psychologically fruitful complement to that ofApollo, who was the embodiment above all of the qualities of harmony, rationality, equity. The cult of Dionysus gave expression to another side of human nature. It stressed abandon, frenzy, even violence. In the winter months, when Apollo was supposed to be absent from his shrine at Delphi, troops of young women called Maenads would follow the priests of Dionysus to remote places in the mountains for the secret rites of the cult,dancing in ecstasy, impervious to cold and exposure, destroying animals which fell in their way and, perhaps, human intruders too.

contraposto

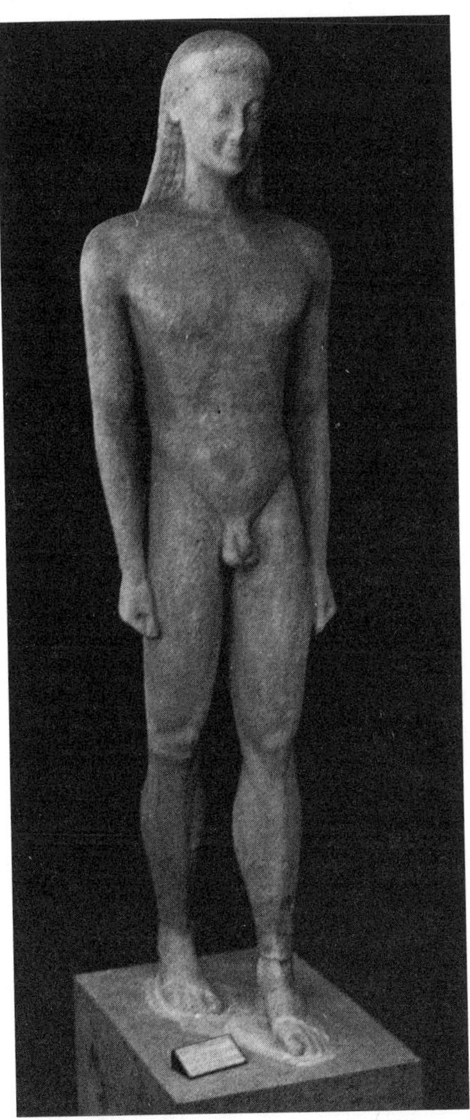

The inspiration of Greek sculpture came first from Asia, then from Egypt. The basic posture of the statues known as kouroi – youths – reflects this. The four-square stance is meant to be seen from the front; one leg is in advance of the other, as if in a walking position. The rigidity of the posture, the clenched fists and set smile, all characterized this type of statue from its first appearance in the Greek islands (this example is from Melos)until it gave way to a more relaxed, naturalistic manner at the beginning of the fifth century. Such statues stood over graves or as dedications to gods; they were not portraits,but figures only one degree removed in their nature from the anonymous cult figures of shrines.

called nemesis awaiting people who flouted the rules by which life is run and who became too big for their boots. Myths were a way of coming to grips with the puzzles of life. We have similar helpful myths, but they tend to be related to science rather than gods: we say that someone's behaviour is 'explained' by an unhappy childhood, or by pressure of work, and this makes us feel we can do something about problems, by, for example, going to a doctor or taking a holiday. When this turns out well, our belief is reinforced. Like us, the Greeks had remedies based on faith, but on faith in things other than those we trust nowadays. Theirs often seemed to work, too, and religious practice meant carrying out rituals which would keep the gods in a good mood.

There were many different things the individual Greek might do as a part of his religion. There were ritual cults practising 'mysteries' re-enacting great natural processes of germination and growth, or the passage of the seasons. Omens could be interpreted and, on important matters, oracles could be consulted. The chief oracle was that of Apollo at Delphi, but people made long pilgrimages to many others to get guidance about their destiny. Then there were the gods of each city, living in the temples usually to be found on the local Acropolis, and served respectfully in civic festivals in which religious ritual mingled with games and theatrical performances, or by sacrifices at household and wayside altars. Athena might be worshipped elsewhere, but at Athens she was the special guardian of the city. To these should be added the respect given to hundreds of very minor gods, demons and powers of nature at thousands of shrines. From the amount of attention given them and the efforts to placate them (for instance, by sacrifice), it may well be that their cults were the most popular part of Greek religion.

The gods and goddesses shared by all Hellenes might intervene in human life, it was thought, but it may be that they received more attention at the official level than from private individuals. They are the gods who appear in Homer. The greatest of them dwelt, thought the Greeks, on Mount Olympus and their names – Zeus, Ares, Aphrodite – passed later into the European heritage of myth and legend. They are the clearest examples of what Greeks thought divinities to be. The remarkable thing about them is that they are very human. Zeus, king of the gods, is certainly at times a terrible figure, hurling thunderbolts about, but he is also a well-meaning, often bungling middle-aged Greek gentleman who is somewhat overgiven to chasing girls. Aphrodite, goddess of love and fertility, is very much a woman with her own vanities, likes and dislikes. These deities do not stand aloof from human affairs, and when intervening in them, show very human emotions. Homer shows Poseidon, god of the sea and of earthquakes, dogging the path of Odysseus with misfortune because of a grudge against him, while the hero's side is taken (and he is helped) by Athena, the virgin goddess of war and wisdom, whose favourite he is.

Religion makes up a deep irrational subsoil which it is easy sometimes to overlook when we look at Greek civilization in its later maturity. It produced views of the world which were sometimes incoherent or self-contradictory. It borrowed and incorporated elements from abroad like the Asian myth of golden, silver, bronze and iron ages. Yet the outcome was a religious experience different from that of other peoples. Homer, perhaps, did as much as anyone to order the Greek supernatural in this way and he does not give much space to popular cults. A later Greek critic grumbled that he 'attributed to the gods everything that is disgraceful and blameworthy among men: theft, adultery and deceit', and he was right. As Homer showed it, the world of the gods operated much like the actual world. Much as it may have owed to Egypt and the East, Greek mythology and art usually continued to present its gods as better (or worse) men and women, a world away from the monsters of Assyria and Babylonia, or from Shiva the many-armed. This suggests a momentous change of intellectual attitude; if gods were like men, perhaps men, then, could be godlike?

The Greek sphere

Greece itself is tiny, but it was part of a much larger world. The fact that no Greek in Greece or the islands lived more than seventy kilometres from the sea must have meant that sea-travel (which is what much of the Odyssey is about) was less frightening than to more inland peoples. This helps to explain why the Greeks quickly set out to explore and put to use the wider environment. From the start, much of the Aegean itself was settled by colonists and emigrants. Mainland Greece offered little chance of agricultural plenty. Already, as early as the tenth-century population growth led to pressure on available land. This was the deep background to a great age of colonization; and the creation, by the end of it, in the sixth century, of a Greek world stretching far beyond the Aegean, from the Black Sea in the east to the Balearics, France and Sicily in the west and Libya in the south. Other forces, too, had been at work. While Thrace was colonized by agriculturalists looking for land, other Greeks settled in the Levant or south Italy in order to trade. The Phoenicians had led the way and this may have encouraged Greeks to emulate them – or may even simply have suggested sites for colonies to which Greek cities could export surplus population. The first Greek settlement in the west Mediterranean was founded by an expedition from two cities, Chalkis and Eretria, in the bay of Naples in about 750 BC. In the next century or so many other cities set out to

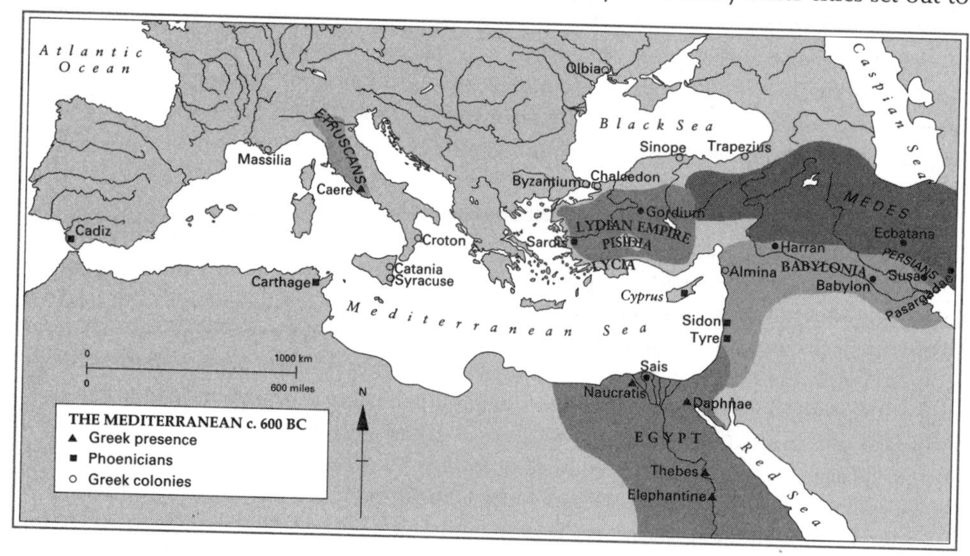

Etruscans early Romans

create daughter settlements which would become self-sufficient and autonomous, like themselves. Colonies took root on many cultivatable plains in Sicily as well as on the North African and French coasts – one was the origin of the later Marseilles. Syracuse, in Sicily, was the most successful; later, roughly between 500-350 BC, in what are called 'classical' times, she was, with Athens, the only Greek city of more than a hundred thousand inhabitants, a reminder of the relatively small human scale of Greek civilization.

After snapping up the good sites in Sicily and Italy, Greek colonists turned east. But there were political problems. No city looking for sites for new settlements could risk taking on the great land powers on its own. The direction of colonization therefore turned north from the Levant. Some Greeks settled on peninsulas in the northern Aegean, pushing out or enslaving the surviving Pelasgoi, while others sailed through the Dardanelles and into the Black Sea, the favoured area of a second wave of Greek colonizing, about a century after the first. By 500 BC there was a chain of Greek cities all round the Black Sea. Some of them obviously had farming potential, but there is among them a more marked emphasis on trade than in the western colonies, and they did much to stimulate business in the Aegean. To the north of them dwelt barbarians in the modern sense of the word. Among them, the Scythians, originally nomadic, were beginning in the sixth century to take up agriculture on the plains between the Don and Danube. Further west, too, in Thrace, the Greeks encountered more fiercely barbaric mountain peoples.

Italy and the Etruscans

No great empires faced the Greeks in the west, but Carthage, a North African city-state founded by Phoenicians somewhere around 800 BC much concerned them. It had grown over a couple of centuries to be greater in power and wealth than the old Phoenician Levant cities of Tyre and Sidon, and menaced the Greeks of southern Italy and Sicily. Yet in the long run, the real danger in the west lay elsewhere, to the north. When Greek colonization was in full swing, only a cluster of shepherds' huts on a couple of hills overlooking the Tiber marked the site of the capital of an empire still to be made, that of Rome. Its later impact on history and the eagerness of the Romans themselves to invent stirring and fabulous tales about their origin make it very hard to picture the tiny scale on which their story actually began. Traditionally Rome is the city of Seven Hills. Archaeologists disagree about whether the date should be nearer 1000 or 800 BC, but the tops of two of them, inhabited by graziers who had come there from the neighbouring highlands, were the site of the earliest settlement. The pasture would have been better there than farther inland, the two sites were easily defensible, and there was a convenient crossing of the River Tiber at that point – the lowest practicable one, in fact, before the sea.

Though its site must always have been likely to make this a key-point in communications, these tiny settlements might never have come to anything had it not been for the arrival shortly before 600 BC of people now called 'Etruscans', from the Greek name for them. They remain somewhat mysterious. It seems most likely that they first came by sea from the Balkans in the tenth century, though they may have been joined later by immigrants from Asia Minor about 700 BC. Certainly, by the time they moved onto the Roman site, the Etruscans were already a mixed people with knowledge of skills and cultures at that date foreign to Italy. Since arriving in the peninsula, they had become skilled metal-workers and they exploited the rich iron ore deposits of Elba as well as those on the mainland.

Much remains unknown about them, but they seem to have lived in a loose federation of cities ruled by kings. Etruria, the homeland they dominated, was at one time very large, running from the River Po in the north down to the coastal plain south of the Tiber. They were literate and had adopted the Greek alphabet – probably from Greek cities in the south – but many of their inscriptions have not yet been understood. They were city-dwellers (one of their cities, Caere, had about 25,000 inhabitants in 600 BC, many Greeks among them), and

These figures of a man and his wife reclining on a couch have something of the feel of archaic Greek art about them. They come, in fact, from Cerveteri in Italy: they form the lid of an Etruscan sarcophagus made in about 500 BC.

their arrival in large numbers at Rome must have transformed the place. The Romans later stressed the idea of the city and the citizen (they began their calendar *ab urbe condita* – 'from the founding of the city', which was to be wrongly dated 753 BC in the Christian style) but this was far from all that they owed to the Etruscans. Through them they came to Greek civilization. From them they inherited the Greek myth that Aeneas, a Trojan hero mentioned in the *Iliad*, had escaped the overthrow of Troy and had sailed west to found Rome. Many Roman customs, such as the holding of gladiatorial 'games', wearing the toga, and reading auguries, also came from the Etruscans. The basic Roman religious beliefs took shape in Etruscan times, too. Even the wolf which figures in one legend about the founding of Rome, the story of Romulus and Remus, is probably a relic of the Etruscans' cult of that animal. The later Roman attention to drainage may also have been something they learnt from Etruscans.

Rome was a city early in the sixth century BC. Either then or soon afterwards we can cease to think of two peoples on the same site, for, although the city stayed bilingual for a time, an amalgamation of Etruscans and earlier inhabitants (sometimes called Latins) came about. Once this was complete, it is sensible to talk about 'Romans' as a distinct people. The years from 600 to 500 BC are still very obscure (not least because the later Romans invented so many legends about them) but during this time Rome acquired several institutions which were to last a very long time. One was the compulsory mobilization of the citizens for military service; military duties went with civil rights in early Rome. Another was the restriction of effective government to 'patricians' or nobles; these took their name from their position as *patres*, or heads of families.

Traditionally, the last Etruscan king was expelled in 510 BC (though this is probably three or four years out). Rome can by then be said to have emerged from the Etruscan chrysalis. The Etruscans, meanwhile, were having a bad time. They were turned out of Campania by the Greeks, who also took Elba from them. For a time the new Roman city-state continued to have friendly relations with neighbouring Etruscan cities, since she had to face grave threats from other neighbours and the communications she controlled mattered to Etruscan trade. But the fifth century BC saw the beginning of a long period of war with the Etruscan cities too. At roughly the same time come the last traces of Etruscan presence at Rome – inscriptions and trade-goods, for instance. Rome had by then become truly independent, though nothing of the world power she was to become.

THE GREEK MIRACLE

Though the Etruscans sometimes pressed hard on the Greeks of Italy in the sixth century BC, and the Carthaginians always threatened those in Sicily, both were held off: at the beginning of the fifth century it was the eastern Greeks who were having a rough time at the hands of their neighbours. This was not new. The Greek Ionian cities had always been open to blackmail or invasion by Asian powers and at the beginning of the fifth century it looked as if mainland Greece too was threatened. Greeks and Persians came more and more closely into contact in Asia Minor as the Persians conquered the non-Greek kingdoms there, pushed across the Dardanelles and occupied cities on the Thracian coast. Then, in 499 BC, the Greeks of Ionia revolted against the demands made on them; some of them were especially annoyed when the Persians interfered in their internal affairs – for example, by supporting tyrants against their subjects. With the help of some of the mainland cities, the Ionians were for a time successful. But eventually they were overcome. The Persians then decided to punish the mainland Greeks for their support of the revolt.

One naval attack failed, but another Persian fleet set out in 490 BC. The army it carried duly landed, only to be beaten by the Athenians at Marathon, where the discipline of the hoplites showed that even the great Persian host was not invincible. But the Persians came back ten years later, this time by land down the coast. A great bridge of boats was built at the Dardanelles for their army to cross to Europe, before moving west and south, and the Persian fleet covered its flank as it moved slowly forward. The Spartans took the lead in the Greek defence, and at the pass of Thermopylae, Leonidas, the Spartan king, and 300 Spartan soldiers were overwhelmed but left an imperishable legend of heroism to the future. The Persians pressed on. Attica had to be given up to them, Athens was taken and destroyed and the Greeks fell back on Corinth, massing their fleet in the bay of Salamis. By now it was autumn. Perhaps because he feared the onset of winter – which can be harsh in Greece – the Persian king decided to make an end of it and attacked the Greek ships. But in the narrow waters of Salamis he lost the advantage of numbers. The Greeks shattered his fleet, and, without its support and supply, the Persian army had to retreat. In the next year (479 BC) it was defeated at Plataea and on the same day the Greeks won another great victory at Mycale, on the Asian coast, where a second Persian fleet was burned. Though the war dragged on for years after this, this was really the end of the Persian threat and it opened the greatest age of Greek history.

A Greek warrior arms himself: from a fifth-century Athenian dish.

hoplites

The Ionian cities were liberated under the leadership of the Athenians. The following decades were marked by growing Athenian power, especially at sea, and it led to a fear of Athens among the other states, especially Sparta. Athens and Sparta show how big the contrast might be between one polis and another. Sparta was ruled by a large aristocracy (about 5000, according to fifth-century writers) in a very austere, somewhat puritanical way. Luxury was forbidden; Spartans were not supposed to own gold or silver. They played little part in the colonizing movement and remained wholly agricultural, conquering more land from their neighbours as required. The Spartans were not rich but held down a large population of serfs – called 'helots', whose revolts they feared – and prided themselves on their military achievements; the hoplite tradition was especially vigorous among them.

The Athenian ascendancy

Persian Defeat

By contrast, fifth-century Athens, rebuilt and restored to prosperity after the disaster of 480 BC, was a commercial city. Of all the Greek states, her government was the most democratic, all decisions being taken by the general assembly of the citizens (not that they would all be able to be present, or that they were a majority of the inhabitants) and civic officials were chosen by lot. In the aftermath of the Persian wars the Athenians were tempted to something like a bid for Hellenic leadership. A league was formed to support a common fleet to fight the Persians (the Spartans did not join) to which members at first contributed ships, but then began simply to pay money to the Athenians to build and man them. When members began to think the Persian danger remote and refused to pay up, the Athenians forced them to do so. Even when peace was made (in 449 BC) the Delian League (as it was called because its headquarters was at first the island of Delos) went on. At its height, 150 states were paying tribute to Athens.

Top dogs are rarely popular, but the Athenians won support and admiration as well as hostility. They tended to interfere with the internal affairs of other cities and to back up democratic regimes where they existed. The richer citizens of other cities did not like this and it was they who paid the taxes which went into the tribute to Athens. On the other hand, the poor

The Acropolis was the site of the most sacred place in Athens, the home of the divine guardians of the city. Under Pericles' leadership there began to be built upon it (from money subscribed to resist Persia by the allies of Athens) the astonishing range of buildings which made it the greatest shrine of Greek classical architecture. The most celebrated of its buildings is the Parthenon, first a temple to Athene, later an Orthodox church, a Catholic cathedral, an Islamic mosque and finally a gunpowder magazine until it blew up in 1687. Then began the era of ruin and spoilation which lasted until very recent times.

majority, who were supported by the Athenians (often by force), did not mind taxing their rich fellow-citizens, nor, it seems, worry that the money raised was increasingly spent not on the Athenian navy but on beautifying Athens with splendid buildings and monuments. Fifth-century Greeks often recognized that Athens was in a sense their cultural leader, a model for the rest of Greece. Slowly, though, the Greek world began to look more and more divided. On the one side were many democratic states looking to Athens for leadership. This associated democracy with the anti-Persian struggle and the Athenians' naval supremacy (somewhat in the way many nineteenth-century Englishmen thought that the spread of civilization and constitutional government was inseparable from the Royal Navy which guaranteed their empire). On the other side were the more oligarchic and aristocratic states anxious to avoid trouble with the Persians and fearing the further extension of Athenian power.

Greek life

Although the Greeks of the fifth century lived under different constitutions and although they thought about the world in a hugely different way from the men and women of the first civilizations, some things had not changed very much since the end of the Dark Ages. Even today, we sometimes forget, many people in the world are peasants, and in the Greek world most people still got their living on the land, where, even with iron tools, agriculture was still laborious and pretty primitive. As for manufacturing, only a few hundred potters kept up the big export trade in pottery from Athens. The largest shield factory in the city was thought enormous because 120 people worked in it. There were relatively few smiths, stone-cutters, armourers, jewellers and other specialists; agriculture was the backbone of the economy, as it still is today in many countries. It did not produce great wealth; though olives and the vine had increased its potential, Greek soil is not usually rich and the range and quality of crops available remained poor throughout classical times. Plots were very small – even a rich man had as little as 20-30 hectares of mixed cornfields and vineyards according to one sixth-century classification of Athenian citizens by wealth. The overall tendency for a very long time was for estates to be sub-divided again and again at inheritance, and most freemen would have been smallholders by our reckoning. Reliance on a smallholding agriculture meant that life was hard and simple. When we look at the great ruins of Hellas – at the Parthenon at Athens, for example, or the many Greek temples which survive more or less complete – we are in danger of getting a false impression of Greek life. These were public buildings, paid for by collective resources. Most Greeks would have lived in pretty humble little houses, eating plain food, without slaves, or even servants.

Within the polis the great dividing line was not wealth, nor that between free and unfree, but between those who were citizens and those who were not. There were many poor among the citizen body of a democracy like Athens. In many places, the growing impoverishment of the peasants was a recurring problem. There seemed to be more and more landless freemen as time passed. Where trade flourished, too, 'metics' or resident foreigners grew in numbers. One estimate is that about forty per cent of the male population of fifth-century Athens were non-citizens of one sort and another, but this was unusual. More dependent on trade and manufactures then most cities, Athens also had proportionately more wealthy men. Old, aristocratic families provided some of them – those who lived on incomes from their estates – and this helps to explain a prejudice against earning money by entering a profession or by trade. But the ranks of the wealthy were also added to by rich merchants. What wealth meant is a different matter: probably not much, in comparison with modern patterns of consumption, but a great deal in contrast with the life of the peasant.

Slavery

The only other important group of inhabitants of the polis with a distinct legal status were the slaves. In archaic times the losers in wars were sometimes enslaved, though the men were usually killed and only the women, who could be put to work at domestic drudgery and exploited for their sexual services, were likely to be spared. Apart from such conquests, slaves were either born to slavery, condemned to it by a court, or bought in one of the great markets of Asia Minor. A feeling grew up in the fifth and fourth centuries BC that there was something wrong and unnatural about Greeks having Greeks as slaves; but there were Greek slaves all the same. That there were fewer slaves in Greece than in the great oriental empires or in later Roman times, has much to do with the small-holding character of Greek agriculture. Small farmers could just about rub along getting a living for themselves and their families. They could afford neither to buy a slave, nor to feed him, because he could not do enough to earn his keep. There were no great estates relying on slave-labour. Most slaves were to be found in the towns, where they worked at all kinds of tasks as servants and craftsmen. One who became famous was Aesop, the story-teller. In the fifth century about a quarter of the population of Athens seem to have been slaves, though no section of the economy absolutely depended on their labour, except the silver-mines owned by the state. Slaves who were not in full-time personal service were hired out in gangs and paid like free labourers; they worked alongside free men on the same tasks. The slave had to give his master a part of his wages, but his position was practically often not very different from that of a poor free labourer.

Slaves could be freed (and could also buy their freedom), though this does not seem to have happened very often. However, their lot was not likely to be much improved by being free if they were already working for wages. We do not hear of revolts by slaves (the Spartan helots, or serfs, were a different matter), but silence does not tell us much. Probably most domestic slaves were not treated too badly, but we know that in the silver-mines of Attica they had to endure very harsh conditions, although this would have been less striking to ancient Greeks than it is to us. Everyone, after all, had by modern standards a hard life in those days. What was distinctive about the slave was that another man had absolute power over him.

Women in ancient Greece

Free women, too, had no citizenship. Some evidence suggests that they led pretty cramped and sheltered lives in other ways, too, but there were differences of custom. Most Greeks seem to have thought that Spartan girls were given too much freedom (and much deplored the very brief gym-slips in which they exercised together with the boys), while the women of a wealthy household in Athens, for instance, lived in separate quarters, locked off at night from the rest. This may suggest the seclusion of the eastern harem, but its purpose was probably to stop men getting at the servant-girls: if they became pregnant or had young children, they would be less use as servants and meant more mouths to feed. Still, we also know that respectable married women were likely to be veiled when they went out, did not leave the house alone and were not expected to talk to anyone they met. The Greeks liked parties – their pottery shows that – but there seems to have been nothing of the easy atmosphere of the mixed gatherings of the ladies and gentlemen of Egyptian tomb-paintings; Greek men might never meet their friends' womenfolk. If they did meet a woman at a party, she was almost certainly a professional entertainer called a 'hetaira'. Some were famous enough for their names to have come down to us, and they were more than prostitutes, with skills in singing, conversation and dancing. They were, though, by no means respectable, since their charms were for sale.

There was virtually no activity outside the home open to a Greek lady of good family, in fact. Poor women could work for others, but a lady could not. No woman could become a nurse, actress, scribe or anything similar because such female professions did not exist. Girls, it seems, were not usually thought worth educating. At home, though, they had plenty to do. Greek women not only washed the family clothes, but made them, probably after weaving the material from thread they had themselves spun. The management of the household was complicated and time-consuming.

One reason why women in Athens (to take a particular city) had fewer legal rights than their menfolk was that Greek society (like virtually every other before our own and like much of the world even today) thought in terms of the family rather than the individual. Society was patriarchal, women could not hold property or conduct business and were always the legal wards of their husbands or nearest male relatives. If a daughter was left sole heiress to her father's estate, her nearest male kinsman was entitled and enjoined to claim her in marriage to ensure that the property remained in the family. Beyond such specifics, it is hard to say anything in general terms about exactly how Greek women were regarded. One problem is that literature in domestic life hardly makes an appearance. Yet we know that women went to the theatre in Athens; they must have watched and listened to Antigone, Electra, Jocasta, Medea – the great female characters of Greek tragedy – and many other very varied female roles. They can hardly have made sense of them if they were themselves simply empty-headed drudges. On tombstones and vases there are pictures of deceased wives and mothers taking leave of their families and they suggest deep affection; there does not seem to be evidence suggesting the lack of respect involved in, say, the veiled and confined life of an oil sheikh's wife today. Socrates' wife nagged him; she certainly did not behave subserviently, and many Greek wives must have been like her. All in all, it is best to be cautious in judging Greek attitudes to women. Homer said that 'there is nothing finer than when a man and his wife live together in true union, sharing the same thoughts' and that was something every educated Greek must have read.

Greece has left to posterity the name and reputation of one woman poet – Sappho – but few Greek women are likely to have been able to read in spite of the suggestion of this image of a girl with a scroll.

When small, Greek children were brought up by their mothers but, if the boys were going to go to school (girls never did), they left their mothers' charge at an early age. The education received by a Greek boy of a family which could afford it placed great emphasis on learning by heart – we hear of boys learning the whole of Homer in this way – and literature together with writing, music and gymnastics made up most of the curriculum. The aim was to produce the 'whole man', to provide a rounded education which would fit someone to take his place in the polis, sharing its values and tastes, rather than to train in specialist skills – something Greeks thought best left to slaves. There were no universities until something like one appeared at Athens, in the late fifth century, but the general standard of literacy – to judge by the Athenian use of public notice boards and inscriptions – seems to have been quite high.

The Greek intellect

While the grip of custom was strong, and Greeks were deeply attached to tradition, they produced quite suddenly a rush of achievements many of which were startlingly novel. Ever since, people have wondered how it happened. Some have called it 'the Greek miracle', so amazing do they find it. It happened over a period longer than the great classical age itself. In about four hundred years, Greeks invented politics, philosophy, much of arithmetic and geometry (those are all in origin Greek words) and notions of art accepted by Europeans almost until our own day. This huge step shows how different was Greek civilization from its predecessors. It was just much more creative. Central to this was the new importance the Greeks gave to rational, conscious inquiry about the world they lived in. The fact that many of them continued to be superstitious and believe in magic should not obscure this. Because of the way they used reason and argument, they gave human beings a better grip on the world they lived in than any earlier people had done. Greek ideas were not always right, but they were worked out and tested in better ways than earlier ones. The Greek miracle made an immense contribution to the development of the powers of the human mind. So intense an effort to grapple with the deepest problems of thought and life had never before been made, and there was not to be another like it for a long time.

Thales, one of the founders of Greek science: a Roman copy of a Greek original, but since Thales lived before the age of classical portraiture his appearance remains conjectural.

One outstanding example is Greek science, which was quite unlike any earlier attempt to approach the natural world. In sixth-century Ionia a number of thinkers first put forward explanations about the way the universe worked in terms of laws and regularities rather than of gods and demons. The Greek philosopher Democritus even arrived at the idea that all matter was made up of 'atoms' – a theory about two thousand years ahead of its time; it did not catch on. Instead the Greeks were to leave behind the idea that all matter was made up of four 'elements' – earth, water, air and fire – combined in different ways in different substances. This was not so near the truth as the atomic theory, but it made further investigation possible and kept science going roughly until the seventeenth century AD. In much the same way the teaching of Hippocrates (a Greek from Cos who was a pupil of Democritus) was the basis of medicine until very recent times. There is great difficulty in disentangling the truth about him, but it is clear that his teaching is the real beginning of the scientific study of health, in observing symptoms and the effects of treatments, making sensible recommendations about diet and separating knowledge from superstition. The 'Hippocratic Oath' called after him is the foundation of medical ethics to this day.

Even more important contributions to the future were made by Greek mathematics. This story, too, starts away from mainland Greece. At Crotone in southern Italy, there lived in the second half of the sixth century BC a philosopher called Pythagoras who was one of the first people to argue deductively – that is, by applying pure logical argument to first principles or axioms. This was important not only because of the advances in arithmetic and geometry which followed, but because it helped to make other people think clearly and rigorously about problems that were not mathematical. But Pythagoras is probably best known for the theorem about right-angled triangles which is named after him, though it is in fact of a later date.

There is no contemporary portrait of Socrates, but this Roman copy probably represents a tradition originating in a true likeness.

Philosophy

One of the most famous of all Greeks in insisting on the importance of rigorous thought was the late fifth-century Athenian, Socrates. He wrote no books and we know of him only through what other people tell us. Most of what he is thought to have said and taught (and he was clearly one of the finest teachers who has ever lived) is recorded in a series of 'dialogues' or conversations, set down by his greatest pupil, the philosopher Plato. People still argue about whether what Socrates is reported by Plato as saying was really what he taught or whether it is what Plato would have wished him to say, but the message is clear enough: the most important thing a man can do, says Socrates, is to try to understand how he can live a good life. And what is the good life at which man should aim? The only way of finding out about this with certainty, Socrates taught, is to examine carefully arguments about such ideas as good, justice, truth – to scrutinize, in short, the values men live by.

Socrates also said much else, but the most important thing about his teaching is its general drift and the way he did it rather than the conclusions he arrived at. He seemed to question almost everything normally taken for granted. In the end he was brought to trial at Athens in 399 BC, charged with denying the gods recognized by the state and corrupting the young with his teaching. He was said to have done this in many ways: by teaching disrespect for Athenian institutions; by mocking democracy and public morality (notably by citing passages of Homer in a mischievous way); and by teaching the young to disobey their parents. Such charges may have masked political enmity. There is no doubt, though, about the legality of his trial and condemnation. Democracy is no more sure to tolerate unconventional views than any other form of government and Socrates was ordered to commit suicide. He did so; he seems to have thought the state had a perfect right to condemn him, which perhaps tells us something about the loyalty the polis could call on from its best citizens. Ever since there have been men willing to shake us out of our complacency by questioning everyday beliefs and bringing us up short by looking at familiar ideas in a new light. Socrates has been accused of exaggerating the power of reason, and of using it only negatively, but to expose error and remove intellectual rubbish is a necessary step towards discovering the truth. His teaching, though, did not help to hold traditional structures together, and the polis rested in the end on unquestioned assumptions, as does every human institution.

Plato, inspired by Socrates, tried to go further. He thought that reason provided the certainty that such concepts as justice, beauty, goodness really existed in a world made up of ideas. He did not mean by this that they existed in the sense that they were in someone's mind (as one might say, 'I have an idea') but that somewhere there was a world of changeless reality beyond the changing material world. This reality, which could be reached by the human soul (which he, like Socrates, distinguished from the body) through the use of reason, was made up of such ideas. Plato did not think much of the way most people behaved (and very little of the Athenian democrats who had condemned his teacher). He believed that most people would never be able to live the good life which the real world of ideal 'forms' would reveal. Nevertheless, his teaching was very important. It kept people thinking about all sorts of problems right down to our own day and, in particular, founded an important tradition of thought called Idealism – the belief that in some way or another there exists a world which is more real than that of material experience, is understandable by reason and is not just a matter of incomprehensible magic.

Plato's greatest pupil was Aristotle. He came from Thrace and wrote about so many things – biology, physics, mathematics, logic, literature, psychology, ethics, politics that he left behind enough for learned men to build on for two thousand years and laid out the main ways in which people thought about these subjects until very recent times. Aristotle was a less abstract thinker than Plato; he liked to collect and classify facts and ideas so as to make clear the general laws which underlay them. He was a great observer. In all (though it is almost impossible to judge such a matter) his influence may have been even wider-ranging than Plato's. What is certain is that these two Greek philosophers long dominated the history of rational thought as no other two thinkers have done.

The first historians

The Greeks also made another great intellectual step forward in the fifth century BC by inventing scientific history. *Istorie* was a Greek word; meaning 'enquiry'. One Greek from Asia Minor, Herodotus, is often called the 'father of history'; he was the first man to enquire about events in time and he did so in the first prose work of art in a European language ('researches' would be a possible translation of his title), a huge account of the interplay of Greece and Persia which came down to the end of the Persian war. It is really a history of a world – Herodotus' world. There are tall stories in it, but it is based on a serious consideration of witnesses and accounts of events. His successor, the Athenian Thucydides, was even more scrupulous in his inquiries in the book he wrote towards the end of the century to explain a great struggle called the Peloponnesian war which had erupted inside the Greek world. He has been admired even more than Herodotus for his attempt to explain 'why' as well as 'how' things happened.

In philosophy, science, mathematics and history the growth of reason and intellectual power was carried forward by the Greeks faster than ever before. They also made great contributions to the arts, among other things, founding the European – or, if we prefer, western – theatre. Greek drama had its roots in religious festivals, notably those of Dionysus, god of wine. At these festivals choral songs were recited and in the sixth century speeches by an individual were added to this. The first actors thus came into being. From these simple beginnings more changes followed until, in the fifth century BC, a great series of tragic plays were written which (together with Homer) are the peak of Greek literary art. They were performed only on semi-religious occasions, civic festivals of importance to all the citizens, and they often retold familiar stories and legends which had religious and supernatural themes woven into them. About three hundred tragedies were performed at Athens in the fifth century;

Beside the Parthenon on the Acropolis at Athens stands a small temple called the Erechtheum, or house of Erechtheus, one of the legendary kings of Athens. It contained a small wooden image of the goddess Athene supposed to have been placed there by the gods, and was a shrine as sacred as the Parthenon itself – from some points of view it was even more sacred. The south porch probably the finest examples of the female figures called Caryatids and used by the Greeks (as they had been earlier used in the Near East) as a variant on the usual columns for the support of architectural loads. How they came to be named is not exactly known, but it may have been as the result of a joke comparing them to the (allegedly) robust ladies of Caryae, a district of Sparta.

thirty-three by three great tragedians, Aeschylus, Sophocles and Euripides, are all that survive. In them, the audience was given a fresh look at old and familiar stories, perhaps to bring out some new point not previously likely to be in their minds, though at the heart of Athenian tragedy always lay an emphasis on the mysterious workings of the laws which govern human life and the sad destinies that lie in wait even for the fortunate. There were comedies, too; Aristophanes, another Athenian was the first great comic writer for the stage. By his time comedy was turning from slapstick into a means of commenting on public life – he made fun of Socrates, for example, and wrote the first play about female emancipation (mocking it, like most authors who took up the theme for the next couple of thousand years).

Among the physical legacy of Greek culture many great works of architecture and sculpture survive and long provided models for future generations. Yet we have lost much of what the Greeks would have seen, simply because though stone and marble last well, paint, wood and fabrics do not. The beautiful ruins on the Acropolis in Athens would have looked much more garish when they were cluttered up with little shrines and when their statues and friezes were painted in bright colours; yet that is how Greeks would have seen them. In their architecture, the Greeks borrowed much from Asia – the column, for example, which probably came from Egypt, was another inspiration – though they later evolved a style which was all their own. Their first ideas for statues, too, also probably came from Egypt or further east, but again the Greeks developed them into something truly original. Their greatest achievements were in representing the human form; gradually the stiff, ritualistic stances of early statues gave way to natural, easy poses. Greeks seem to have delighted in showing just how splendid a thing the human being could be in both mind and body.

THE PELOPONNESIAN WAR

For over a quarter of a century, from 431 to 404 BC, a great struggle raged (with only brief interruptions) over the whole Greek world. It is named the Peloponnesian war because one side consisted of a league of Peloponnesian states led by Sparta against Athens. At one time or another almost every Greek state was involved in it. It was so important in scale and so unprecedented that the first historical monograph was written by Thucydides to explain why it had happened. Historians ever since have agreed that it may have been a turning-point in the history of civilization.

Given the suspicion and susceptibility of many Greeks, the origins of the war do not seem very remarkable. There had been growing irritation over the Athenian ascendancy and Sparta had led the opposition to it in what we might think of as a 'cold war' for about forty years. Fighting which broke out in 460 had lasted several years; from time to time, too, Athens and Sparta had each had to take up arms against dissident allies and satellites. Many Athenians seem to have thought another war was bound to come. It had to be settled whether Athens was to be top dog among the Greek states. The Spartans feared Athenian intentions and had sympathizers wherever there were Greeks who opposed the democratic regimes which Athens increasingly seemed to favour. There was jealousy of Athenian wealth and commercial power too, notably on the part of the people of Corinth, another great trading city. The war started, indeed, when Corcyra (today Corfu), a dependency of Corinth, rebelled and appealed to the Athenians for help. To some at Athens, this seemed an opportunity too good to be missed. Corcyra was on such an important route to the west (in the days when ships tried to keep in sight of land) that there was a strategic advantage at stake. When the Athenians helped the Corcyrans, the Corinthians complained to the Spartans. Other complaints were made by other cities against Athenian high-handedness, but Pericles, the patri-

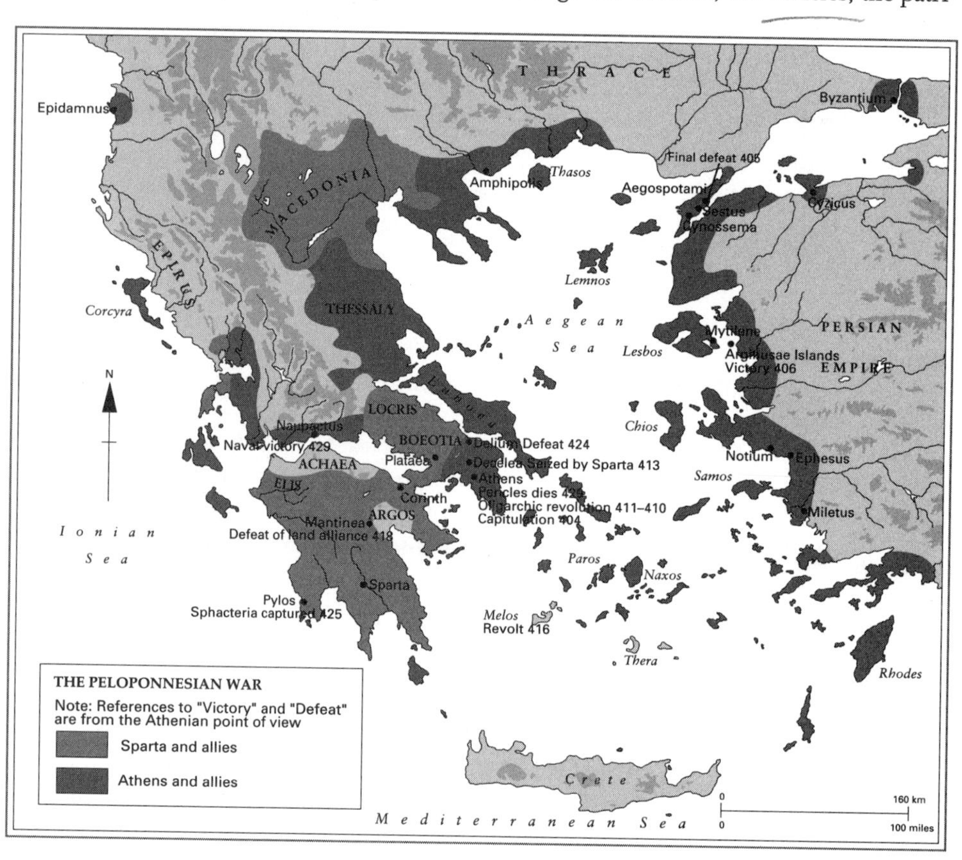

Not much pottery survives from Sparta; her citizens were debarred from manual labour and looked down on craftsmen. What has survived is for the most part from the sixth century; some was exported and includes pieces of high quality. This example has a certain commercial interest, for it represents the king of Cyrene in North Africa supervising the weighing and storing of bales of a medicinal plant which was one of his kingdom's most important exports. Workmen call out the weights, and his pets play round his chair as he watches.

otic demagogue who dominated Athens, urged his countrymen not to make concessions. Instead, he told them that Athens was a model for the rest of Greece. And so in 431 BC. the Spartans began what they called a war of liberation.

The allies of the Athenians and the Spartans were scattered about all over the Greek world, though, broadly speaking, the Athenians had Ionia and the islands on their side, and the Spartans the Peloponnese (colonies sometimes followed mother-cities, sometimes did not). The war was really one of sea against land. The Spartans invaded Athenian territory each campaigning season; the Athenians retired behind the walls of the city, allowed their lands to be ravaged and occupied, but fed themselves by sea, and carried out long-distance naval operations. The result was a long deadlock. Though the Athenians at one time suffered terribly from plague, they were not conquerable by a Greek army which lacked the siege techniques and engineering skills of the armies of Persia. Pericles died in 429 BC and peace was made for a few years in 421. But then the Athenians started the war again. A great expedition was planned against Syracuse, richest of the supporters of the Peloponnesian League. Not only was it a disaster – the Athenians lost half their army and all their fleet – but it led the Spartans to ask the Persians for help. In return for a promise to restore the old Persian domination over the Greek cities of Asia, the Persians paid for a fleet to help the cities which wanted to shake off Athenian control. This time, after years of fighting, the destruction of their fleet, blockade and starvation, the Athenians were forced to give in.

This struggle has great historic importance over and above its unique scale (turning back the earlier Persian invasions had been a matter of two sea battles and two on land; wars between Greek states had usually been affairs of a short summer campaign ending in a battle between two hoplite armies). To cite Thucydides, it was not only prolonged to an immense length but was unparalleled for the misfortunes it brought in a short time on the Hellenes.

The Strygil which this young Greek athlete holds was a body-scraper of wood or bone used to clean the skin after exercise, working or massage. The boy attendant holds a flask of oil which may be needed.

The only gainers were the Persians. The Spartan victors at first tried to bully other Greek states through their military might, as the Athenians had once dominated them through naval power, but then had to fight coalitions in their turn. When the famed Spartan army was defeated at Leuctra in 371 BC, it was clear that Sparta was no more able than had been Athens to dominate the Greek cities and impose some sort of imperial unity upon them. The war showed, in fact, that even the strongest of the Greek states could not enjoy unquestioned supremacy over the rest. The Greeks could neither keep out of one another's affairs nor accept the domination of any one polis. Perhaps Greek unity was impossible unless imposed by conquest, and no polis had enough power for that. Many of them had long since ceased to be self-sufficient. They had many trading interests and had to be involved in politics over the whole length of the Greek world in order to look after them. Athens could not live without importing corn and exporting wine, oil and manufactures.

The war also threw great strains on the internal working of the Greek communities. We have to be careful, because we know so much more about the history of some than of others. At Athens, the specialization required for prolonged warfare by both soldiers and sailors which led to the increasing use of mercenaries, the social divisions which appeared as a result of trade (merchants grew rich) and were made to seem unfair because of the way the war went (farmers saw their lands ravaged), the bitterness resulting from attacks on the politicians who had brought on disaster, all helped to produce a revolution which for a while replaced democracy with oligarchy. But something was going on which went deeper than just the replacement of one political regime by another. The old idea of the polis as essentially a unity of citizens all of whom took an equal part in all of its activities was no longer appropriate to the scale which the life of the state had come to have.

Why the polis did not survive may not be the most important question, though. Later ages were to look back on ancient Greece and see it as a gigantic success, whatever its political failures. They saw it as a whole; in part because they did not know so much about Greece as we now do, and were ignorant of distinctions scholars now draw between different times and places. This was a fruitful error, because what Greece was to be thought to be was as important to the future as what she really was. The meaning of the Greek experience was to be represented and reinterpreted, and ancient Greece was to be rediscovered and reconsidered and, in different ways, reborn and re-used, for more than two thousand years. Europeans were to go back to it again and again to ponder its meaning. This was only due recognition. For all its shadows Greek civilization was quite simply the most important extension of humanity's grasp of its own destiny down to that time. Europe has never ceased to benefit from the capital Greece laid down, and through Europe the rest of the world came in time to draw on the same account.

4
THE ROMAN WORLD

MACEDON AND HELLENISM

A new Mediterranean and Near Eastern world was in large measure defined by Greeks. Paradoxically, this came about during the decay of the city-states. As they became feebler and less able to resist outside interference after the disaster of the Peloponnesian war, a new force began to make itself felt on the northern fringe of Hellas, the kingdom of Macedon. Some people – Macedonians, for the most part – claimed it to be a Greek state and part of the Greek world. The Macedonians spoke Greek and attended Hellenic festivals; their kings claimed to be descended from Greek families – from Achilles, the great Achaean hero of the *Iliad*, no less. But many Greeks disagreed. They thought the Macedonians were a barbarous lot, barely civilized, and certainly not on a par with the cultivated peoples of the Aegean and Sicilian cities.

Undoubtedly Macedon was a rougher, tougher place than, say, Athens or Corinth, and its kings had to manage an aristocracy of mountain chiefs who might not have been much impressed by Socrates. Yet Macedon changed the course of Greek history, thanks to the coincidence of a number of favourable facts. One was the appearance there in 359 BC of an able and ambitious prince – ambitious, among other things, that Macedon should be recognized as Greek – Philip II, regent of the kingdom. Circumstances were much in his favour; the Greek states were worn out by their long struggles and Persia had undergone a series of revolts which left her weak. Macedon was rich in gold and so could pay for a strong and effective army, whose effectiveness owed much to Philip's personal efforts. He had pondered Greek military methods while at Thebes in his youth. He decided that the answer to hoplite tactics lay in a new formation, the phalanx of ten ranks of infantry armed with pikes. These were twice as long as ordinary spears. Those who carried them stood farther apart than hoplites so that the pikes of men in the rear stuck forward between those in the front ranks. The result was a hedgehog-like array of points, a formidable weapon. To back it up, there were armoured cavalry and a siege-train of heavy weapons such as catapults.

Alexander the Great

The Macedonian army was so good, indeed, that under Philip and his son it ended the independence of the mainland Greek cities and an era of human history, the age of the polis. 335 BC, when Thebes was razed to the ground and its inhabitants enslaved as a penalty for rebellion, will do as a marker. There were a few later revolts, but the great age of classical Greece was over. This might be enough in itself to ensure Macedon's kings a place in history, but there were more spectacular changes still to come in the reign of Philip's son, Alexander, one of the few men in history who has traditionally been called 'Great'. So glamorous did he seem to his successors that the legends which surrounded his name led to his being idolized for thousands of years. Though, first and foremost a soldier and conqueror, he was also much more. Unfortunately no contemporary biography of him survives and many of the facts of his life and personality remain obscure. Still, it is clear that he was a decisive force not only in Greek but in world history, from 334 BC, when he crossed Asia to attack the Persians at the head of an army drawn from many Greek states, to 323 BC, when he died in Babylon (perhaps of typhoid) only thirty-three years old.

Philip of Macedon and Alexander the Great

358–336	Reign of Philip II of Macedon.
338	Battle of Chaeronea secures Philip of Macedon control of mainland Greece.
336–323	Reign of Alexander.
334	Alexander invades Persian empire, winning battle of Granicus near the Dardanelles.
333	Alexander defeats Darius III of Persia at Battle of Issus.
332	Siege of Tyre and capture of Egypt by Alexander.
331	Final defeat of Darius III at battle of Gaugamela and his death the following yar.
330–24	Alexander campaigns in the E Persian provinces and invades India.
323	Death of Alexander

Alexander was a passionate Hellene. He revered the memory of Achilles, his supposed ancestor, and

For more than two thousand years the story of Alexander the great conqueror fascinated men who dreamed they might carry out exploits as astonishing. No doubt it added to its glamour that Alexander's physical image had to be evoked from representations of him of doubtful authenticity. This one is at least second-hand – a copy in mosaic by a Roman artist of a Hellenistic painting long since perished. The original may have been near-contemporary (it may even have been commissioned by Cassander, the king of Macedon, murderer of Alexander's mother, wife and heir); by the time it was embodied in this version it is more an idealization of a soldier, pictured in the vigour of his onslaught on the Persians, than a portrait. The huge mosaic at Pompeii (sixteen feet by eight) from which this detail is taken may have been a copy of one painting in a series which represented the whole cycle of Alexander's almost immediately legendary life.

carried with him on his campaigns a treasured copy of Homer. He had been tutored by Aristotle. He was a brave – and sometimes reckless – soldier as well as a shrewd general and a great leader of men who, once he had made conquests, could behave with sympathy to peoples whose rulers he overthrew. He was also violent; once while drunk it seems he killed a friend in a brawl. He may also have agreed to his father's murder.

Whatever his defects, they did not impede a staggering record of success. He defeated the Persians in Asia Minor at the battle of Issus, and then marched through the length of their empire, first southwards through Syria to Egypt, then back north and east to Mesopotamia, pursuing the Persian king Darius III, who died while still on the run; that was the end of the

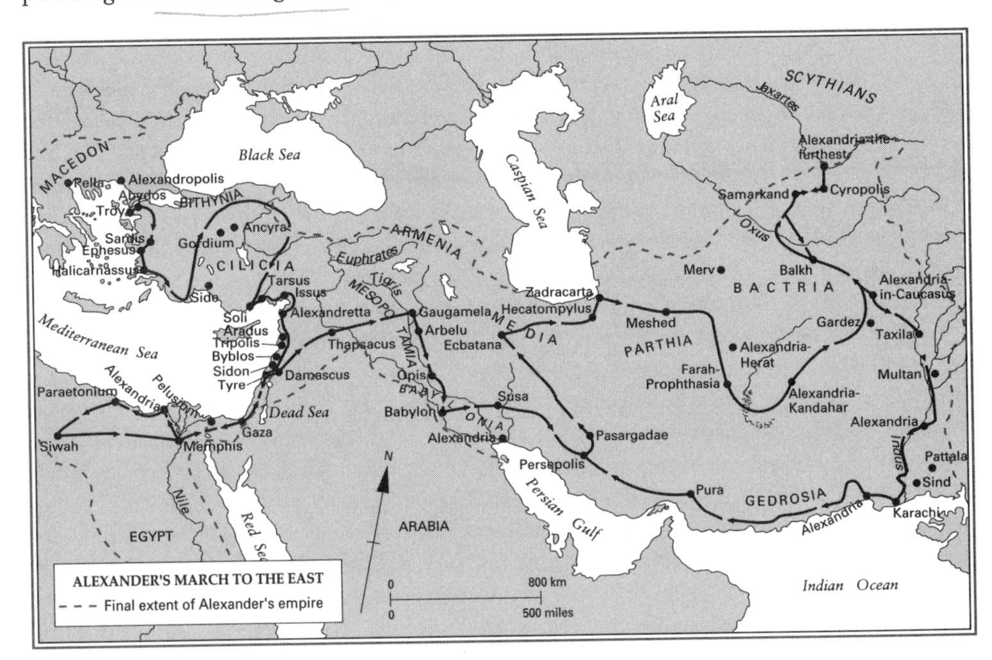

Achaemenid empire. On went Alexander across Iran and Afghanistan, the Oxus and beyond to Samarkand. He founded a city on the Jaxartes. Then he came south again, to invade India. Two hundred kilometres or so beyond the Indus, well into the Punjab, his weary generals made him turn back. A terrible march back down the Indus and along the northern coast of the Persian Gulf followed until he reached Babylon. And there Alexander died.

His short life had much more to it than simple conquest. His 'empire' soon fell apart in that it ceased to have a single centre of government, but he spread Hellenic influence where it had never been felt before. Alexander founded many cities (often named after him: there are still several Alexandrias let alone other places with names which disguise his own somewhat more) and he mixed Greeks and Asians in his army so that they learnt from one another and it became a more cosmopolitan force. He enlisted young Persian nobles and once presided at a mass wedding of 9000 of his soldiers to eastern women. The former officials of the Persian king were kept in post to administer his conquered lands. Alexander even adopted Persian dress, which did not go down well with his Greek companions; nor did they like it when he made visitors to his court kowtow to him as Persian kings had done.

To overthrow the mightiest empire of his age and end the age of the Greek cities (in Asia as well as Europe) were world-shaping deeds, though their full impact was not at once obvious. Many of the positive results were only to appear after his death; then, though, in Greek and non-Greek lands alike the effects would be felt of the Greek ideas and standards he spread far and wide. This is why the words 'Hellenism' and 'Hellenistic' have been coined and applied both to the age which followed his death and to the area formerly covered by his empire (roughly speaking, the zone between the Adriatic and Egypt in the west and the mountains of Afghanistan in the east). The empire itself did not hold together for long; Alexander left no heir who could take over from him and his generals soon began fighting over the spoils.

Alexander's successors: the Hellenistic world

It took forty years or so for the lands of the former empire to settle down into a new pattern as a group of kingdoms, each ruled by one of Alexander's men, or a descendant of one. They are sometimes referred to as the 'Successors' or 'Diadochi'. The richest of these kingdoms was in Egypt, where a Macedonian general named Ptolemy seized control. He was able to get hold of Alexander's body and had it buried in a splendid tomb at Alexandria; this gave him special prestige and pre-eminence as its guardian. Ptolemy founded the last Egyptian dynasty of antiquity. It was to rule Egypt until 30 BC (when the last of the Ptolemies, the famous Cleopatra, died) as well as Palestine, Cyprus and much of Libya. Yet Egypt was not the biggest of the Successor states. Although Alexander's Indian conquests had passed to an Indian king, the family of Seleucus (another Macedonian general) for a time ruled an area running from Afghanistan to the Mediterranean; the Seleucid kingdom did not remain as big as this, though. Early in the third century BC a new kingdom of Pergamon was set up in Asia Minor, and in Bactria yet another kingdom was founded by Greek soldiers. As for Macedon itself, after being invaded by barbarians it passed to a new dynasty, while the old Greek states, loosely organized from time to time in leagues, continued to moulder away (though some of them had hoped to recover independence at Alexander's death). All such ups and downs need not concern us here. The difference they made to history was that from them emerged what Alexander's conquests had made possible and likely – a framework within which Greek ideas and civilization took root as never before. Greek became the official language of the whole Near East and more widely used as an everyday language too, above all in its new cities.

These were founded in considerable numbers (especially in the Seleucid territories) and Greek immigrants were encouraged to settle in them, but they were very different from the old Greek cities of the Aegean. For one thing, they were much bigger. Alexandria in Egypt, Antioch in Syria and the Seleucid capital near Babylon each soon had nearly 200,000 inhab-

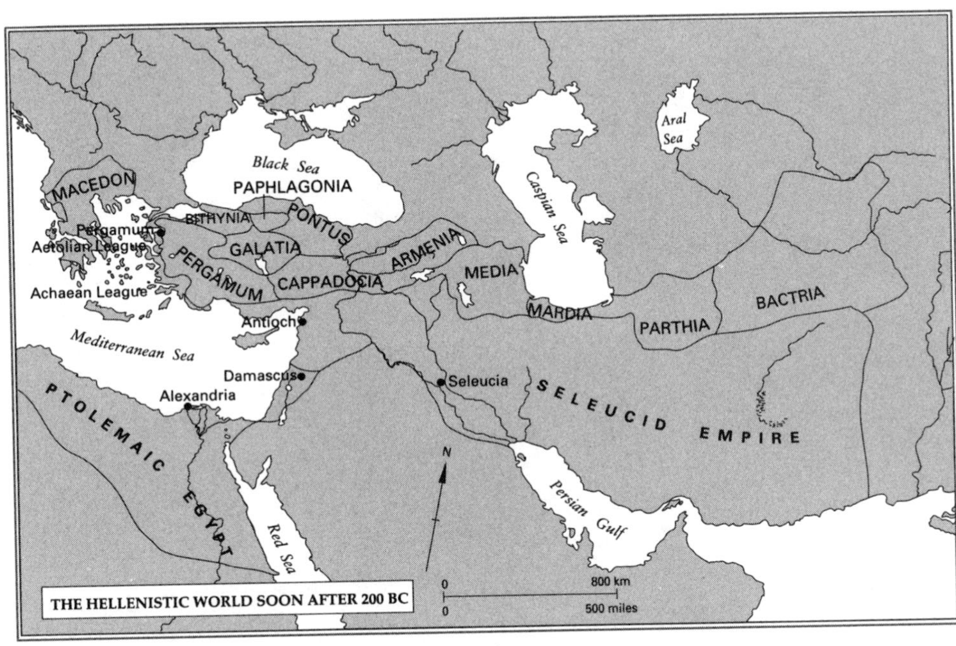

THE HELLENISTIC WORLD SOON AFTER 200 BC

itants. Nor were they in any sense autonomous. The Seleucids, for example, governed through provincial rulers and machinery which they took over from the old Persian empire – which had been a barbarian despotism in the eyes of fifth-century Greeks. Bureaucracies began to appear which drew on ancient traditions of Egypt and Mesopotamia, not those of the polis. The rulers themselves were given semi-divine honours, like the old Persian kings. In Egypt the Ptolemies revived the old cult of the pharaohs, and the first Ptolemy took the title of Soter, that is, 'Saviour'.

Still, the cities at least looked somewhat Greek. Their buildings were in the Greek tradition, and they had theatres, gymnasia, centres for games and festivals which were much like those of the past. Greek tradition showed in artistic style, too. Perhaps the best known of all Greek statues is that of Aphrodite found on the island of Melos and now in the Louvre at Paris (the '*Vénus de Milo*'); this is a Hellenistic work. As Greek style and fashion spread, so did Greek culture, even though the countryside remained almost untouched by it (Greek was not the native tongue of most people in the Successor states, though many came to speak versions of it). Soon Greek literature was being added to by writers in the new cities, who found audiences and patrons in an environment which was for a long time one of growing prosperity. Alexander's wars had released an enormous booty in bullion and precious objects which stimulated economic development as well as providing taxation to pay for standing armies and bureaucracies. The Hellenistic world was a much larger-scale affair than the old Greek world, and a broader stage for Greek culture.

The clearest indication of continuity with what had gone before came in one branch of intellectual activity, the study of science. Egyptian Alexandria was especially pre-eminent in science. Euclid, the man who systematized geometry and gave it a shape which lasted until the nineteenth century, lived there. Among other Alexandrians were the first man to measure the size of the earth and the first to use steam to transmit energy. Archimedes, who is famous for constructing war-machines in Sicily, as well as for his theoretical discoveries in physics, was probably Euclid's pupil, and another Hellenistic Greek, Aristarchus (from Samos, this time, not Alexandria), even arrived at the view that the earth moved round the sun, and not vice versa (this idea was not accepted by his contemporaries, though, because it did not square with Aristotelian physics). A body of knowledge and hypothesis such as this (and

A statue of Aphrodite found on the isle of Melos, and possibly the best-known work of sculpture surviving from the Hellenistic age, 'the Vénus de Milo'.

there was much more) was a major addition to the human toolkit. Hellenistic science was nonetheless held back because there was neither the inclination nor the apparatus to test some theories experimentally and because there was a bias towards the mathematical sciences rather than the applied. Moreover, though the existing state of technology may have made it difficult to make practical use of some ideas (though this view has been contested: some think that the Hellenistic artificers would have been able to build a steam-engine had they only conceived it), the triumphs of Hellenistic science may be thought somewhat to offset the loss of a tradition of self-government in politics and of a confident use of philosophy to seek answers to questions about the aims of life and the way men should behave. Yet the Hellenistic world succeeded in producing one important new ethical philosophy, Stoicism. It taught, roughly, that it was men's duty to be virtuous whatever the consequences for themselves. To be virtuous, it said, consisted above all in obeying the natural laws which ruled the universe and all men, not just Greeks. It was the first attempt to provide a philosophy for all humanity. It also produced the first condemnation of slavery, an extraordinary mental leap never achieved by the philosophers of classical Greece, and it was to be deeply influential for centuries among the elites of a new power, Rome.

THE RISE OF ROMAN POWER

Into a zone deeply influenced and even transformed by a culture emanating ultimately from Greece the power of a new empire was eventually to spread. It would go on, too, to incorporate much of mainland Europe and North Africa – zones hitherto never brought within the ambit of civilization, as well as the Hellenistic world. This was the empire of Rome. In a way she too was a Hellenistic successor state, but her history, both traditionally and in the light of modern scholarship, begins long before Alexander, after the expulsion of the Etruscan kings, around 510 BC.

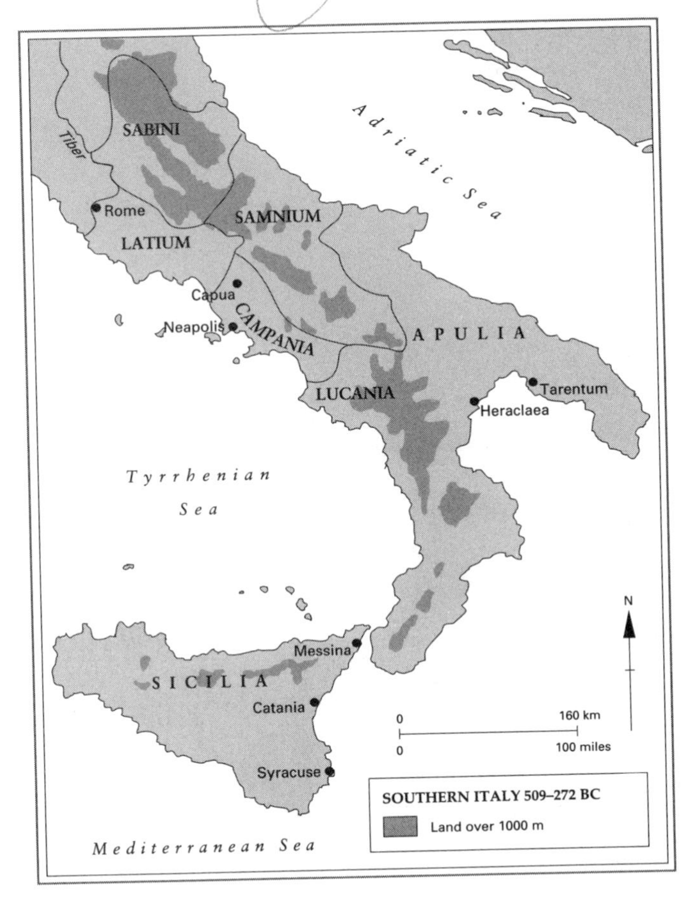

SOUTHERN ITALY 509–272 BC

Rome was then a republic. Her citizens were to go on insisting on their republican traditions even when they had long lived in what looked much more like a monarchy. Realistically speaking, the Republic can at most be reckoned to last about 450 years, until the middle of the first century BC, but that is a long time. It changed a lot over the years, but one change was more important than others, because it explains Rome's impact on later history. This was the spread of Roman power. In republican times it came to encompass the whole Mediterranean world in a system of domination and made a Roman empire which provided framework and cradle for much still shaping our lives today.

The early republic

During those centuries of empire-building, much changed at home. The first two centuries or so of the Republic were studded by violent political struggles, sometimes arising from the demands of the poorer citizens for a share in the power of the better-off and noble families, the 'patricians'. They dominated the Senate, which was the main governmental body, as the inscription carried on many monuments and on the standards of the army indicated: SPQR, the initials of the Latin words for 'the Roman Senate and People'. Yet somehow these struggles went on for a long time without mortal damage to the Republic. This says much for its institutions, which slowly changed as concessions were made to popular forces. Yet, though the poorer citizens won many victories and a bigger share of the spoils of power, Rome never became a democracy in the sense that they controlled the government for long.

For a long time the typical Roman citizen was a peasant farming his own small property, benefiting from the splendid climate and fertility which have always made Italy a rich country when well-governed, and sometimes when it is not. He already showed the industry and skill often demonstrated by later Italians. His farming was the basis of the early Republic; we must not think of the huge metropolis of later centuries, living on imported corn and swollen by huge numbers of immigrants, as typical of the Republic's early days. For a long time the typical Roman was a smallholder, and he was independent. Not until the second century BC did big estates owned by townsmen and relying on slave labour to grow cash crops of grain or olives (for oil) become at all common. To the very end of the story, Romans would look back sentimentally to the simple days of the early Republic as the times when the Roman virtues were upheld by the independent smallholding citizens.

A narrow agricultural base makes it harder to explain the first stage of Roman expansion. It would not be fair to say that Romans were always aggressive and anxious to make conquests. Often, Roman rule (like that of later empires) spread because of fear of neighbours and rivals rather than greed. Expansion was a slow business, too. Though Rome's territory doubled at the expense of her neighbours in the fifth century, and Roman supremacy replaced Etruscans in central Italy, this was not the beginning of an uninterrupted story of successful growth. In 390 BC Gauls from the north sacked Rome itself (this was the famous occasion on which, according to legend, the Capitol to which the Romans had withdrawn was only saved from a surprise attack by the honking of the geese who noticed what was going on). Still, by about 250 BC the Romans already dominated Italy south of the Arno; all of it by then belonged either to the Republic or to its allies, who were allowed to run their own internal affairs, but had to supply troops to the Roman army. In return their citizens enjoyed the rights of Roman citizens when they came to Rome. It was a little like the supremacy of Athens in the Delian League.

Even recently, the backbone of the population of Italy was its peasantry. These farmers were depicted on a bronze vessel used for the ashes of the cremated dead in a cemetery of the fifth century BC near Bologna.

Roman success was built on a number of advantages. The strategic position of the city helped and so did the long distraction of the Etruscans in struggles with the Greeks and other Latin cities, while Celtic tribes pressed them in the north. Another factor was a military system which made the best of Rome's manpower: every male citizen who owned property had to serve in the army if needed. This was no light demand; an infantryman had to serve sixteen years under the early Republic (though service was not for the whole year, since campaigns started in spring and finished during the autumn). This provided a military machine which became in the next few centuries the finest the world had seen. The pool of recruits on which the army drew grew steadily because of the obligations of allies to send contingents to it.

Even while taking over Italy, Rome had become involved farther afield. Some of the Greek cities had called in the king of Epirus to help them against the Romans in the early third century BC. He campaigned in the south and Sicily, thinking perhaps of building himself an empire in the west like Alexander's. He won battles all right, but at such a cost that we still talk of 'Pyrrhic victories' (his name was Pyrrhus) as ones which cost more than they are worth. At one time, too, it looked as if Ptolemaic Egypt might want Rome's alliance, and the first great preoccupations of the Republic outside Italy in fact did lie in Africa, though farther west than Egypt.

The Punic Wars

Carthage, in origin Phoenician, and richer than Tyre and Sidon had ever been, was a great naval power, with outposts in Sicily and Sardinia. It was at times allied to and at times at war with the Greeks of Sicily, and was a standing menace to the rich western coastal plains of Italy and the trade of their ports. In the end three 'Punic' (the name comes from the Latin word for Phoenician) wars were fought by Rome and Carthage. The first ended in 241 BC with the Carthaginians having to give up Sicily after more than twenty years of fighting (though it was not continuous). The Romans also took control of Corsica and Sardinia as a result, and they founded their first 'province' in western Sicily (mainland Italy was either governed directly as part of the Republic or was technically allied to it). These were the first Roman overseas territories. Wars are by no means always the most important events in a historic narrative but in this instance it is worthwhile to stick to the Punic wars for a little longer. Much was to follow from them. Before the second (which began in 218 BC) the Carthaginians had established themselves in Spain, settling at 'New Carthage' (the modern Cartagena). The Romans began to be alarmed when their power extended as far as the Ebro. A Carthaginian attack on

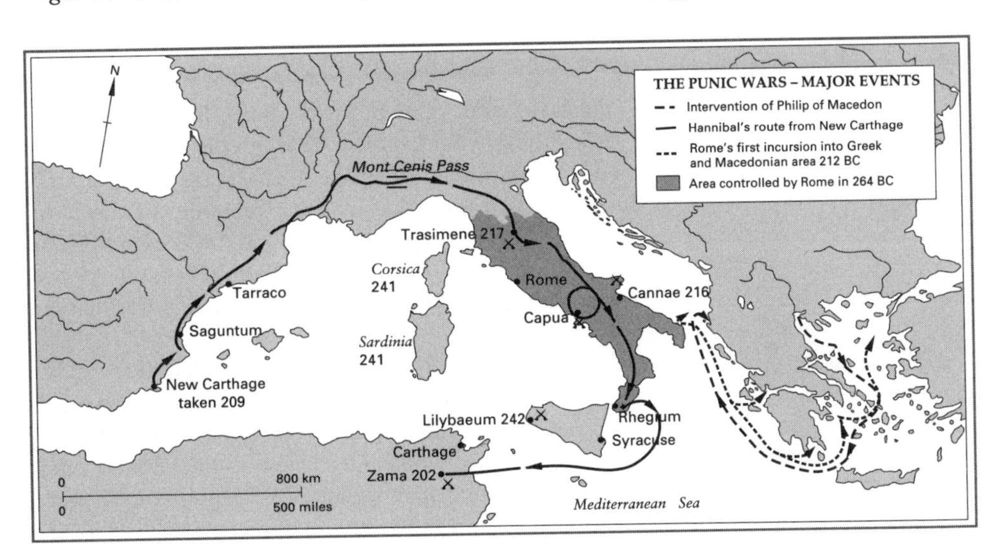

one of the few remaining independent cities on the Spanish coast was followed by the march of a Carthaginian army – complete with elephants – to Italy under Hannibal, the greatest Carthaginian general. A series of bad defeats for the Romans followed. Many of their allies changed sides. But the Romans hung on. In the end they recovered their grip. After twelve years in Italy the Carthaginians were starved and driven out. The Roman Senate gave their successful general, Scipio, permission to cross to Africa, and at Zama in 202 BC he broke the back of Rome's only serious rival in the West in one of history's decisive battles. The Carthaginians had to make a crippling peace. But many Romans still feared them mightily. A third Punic war did not break out for a long time – until 149 BC – but it then ended with so complete a defeat for the Carthaginians that their city was destroyed and ploughs were run over the site where it had stood.

By then the Roman empire was in being, in fact if not in name. The overthrow of Carthage had meant the end of Syracuse, the last independent Greek state in Sicily, because she had once more allied with the Carthaginians. All Sicily was now Roman and southern Spain, too, was conquered. Soon slaves and gold from Sicily, Sardinia and Spain were making some Romans aware that conquests might be profitable. Further east, Macedon had been allied with Carthage for a time, and so Rome had already begun to dabble in Greek politics. In 200 BC a direct appeal for help against Macedon and the Seleucids was made by Athens and the kingdom of Pergamon. The Romans were by then psychologically ready to become further involved in the East. The second century BC was crucial. Macedon was overthrown, the Greek cities were reduced to vassalage, and the last king of Pergamon bequeathed his land to Rome in 133 BC. A new province, called Asia (the western end of Anatolia) was set up in the same year. By then northern Spain had been conquered. Soon after, southern France (Gallia or Gaul) was taken. In the next century northern France followed and then further conquests in the East. This was an astonishing success story and it was not only Rome which drew benefits from it, nor only the conquered who paid a high price for it. So did the Republic.

The loser: a bronze head of a Gallic chief of the first century AD.

The decay of the Republic

The Romans liked to congratulate themselves for following what they called *mos maiorum* – 'the ways of our ancestors' would be a reasonable translation of this Latin phrase. They always showed a fondness for old traditions and liked to keep alive old ways of doing things. Roman religion was in large measure a matter of making sure that ancient ceremonies were kept up and carried out in the proper manner. Even when doing something new, the Romans liked to wrap it up in older packaging. One result was that the names of many of the Republic's institutions – and the idea that the state was a republic and not a monarchy – went on being used long after they ceased to be appropriate.

Roman citizenship is an example. The first citizens were all men and nearly always peasants. They had the right to vote and get justice before the courts, and the duty of serving in the army. Three important changes came about as the centuries went by. In the first place, rights of citizenship were gradually given to many people outside the original Roman territories. Secondly, the Punic wars impoverished the Italian peasant. Conscription took Roman soldiers away for longer from their homesteads and families, who often fell into poverty as a result, and the wars also did enormous damage in the Italian countryside. When peace at last returned, many former smallholders could not make a living there. On the other hand, men who had been able to make money out of the wars began to buy up land for farming in big estates. Slaves (part of the booty of conquest) were sometimes used to work them. The citizen-peasant tended to drift to the city to find a living there as best he could. He was on the way to being what the Romans called a 'proletarian' – someone whose only contribution to the state was to breed children. Both these changes affected politics. More poor citizens meant more voters to be bought, cajoled or bullied by politicians anxious to get into positions which would give them a chance to get at the rich prizes offered by conquest abroad.

The third change in the position of the citizen also stemmed from war: the army became more and more a full-time professional force, rather than one of citizens armed and brigaded for emergencies. One landmark was the ending of a property qualification for service. The pool of Roman military manpower had been showing signs of drying up. If the propertyless could serve, sufficient volunteers would come forward from among the poor who were willing to serve for pay, so that conscription now became hardly necessary. True, for some time the recruit still had to be a citizen, but in the end non-citizens were allowed to join up. They then received rights of citizenship as a reward for their service.

Gradually, in these ways, the Roman army grew apart from the Republic. The famed legions in which it was organized became permanent organizations, whose soldiers increasingly felt loyalty to their comrades and their generals. From the first century BC each legion carried 'eagles' – standards which symbolized the honour and unity of the legion, something of a combination of a religious idol and a regimental badge.

Impoverished citizens with votes to be bought; opportunities for politicians to get at wealth on a huge scale in the new territories to which they could be appointed as governors and generals; an army which was unbeatable (or almost) in the field and more and more loyal to itself and its leaders than to the Senate – these were slow but crucial political developments, and they went on for nearly two centuries, transforming the Republic under the surface even though much about it still looked the same. Meanwhile Rome was obviously getting richer. This was not just a matter of the loot and slaves available as a result of conquest to a few lucky enough to be in the right place at the right time. Poor citizens also benefited indirectly; when new provinces could be taxed, taxation ceased at home. Expensive 'games' were put on to amuse them. Some new wealth also went into the beautifying of Rome and other Italian cities, which sometimes reflected other changes as contact with the East became more common. This was especially true of that with the Greek cities whose cultural past was, after all, what educated Romans were brought up to respect as the roots of their own. But new fashions and standards came to the West, too, as the process of Hellenization spread farther. It showed in everyday things as well as in art and intellectual life; the Romans' passion for bathing has seemed to some people one of the most remarkable things about them, but in fact they took the fashion from the Hellenistic East.

Pax romana

In spite of growing corruption and violence in politics at home, and bad though some Roman provincial governors may have been, Roman power brought peace for longer periods to a larger area of the Mediterranean and Near East than ever before. Republican administration imposed order over many peoples and provided a common law. Many non-Romans who lived under Roman rule admired at least some of those who ran the system for their sense of justice, disinterestedness and the civilizing work they did. The practical outcome was very important for the history of the world, too. Before the Republic came to an end, it had created a political and military framework on a scale without parallel west of China and protected Hellenistic civilization. Many different cultures could live side by side within it and make their own contributions to the cosmopolitan whole.

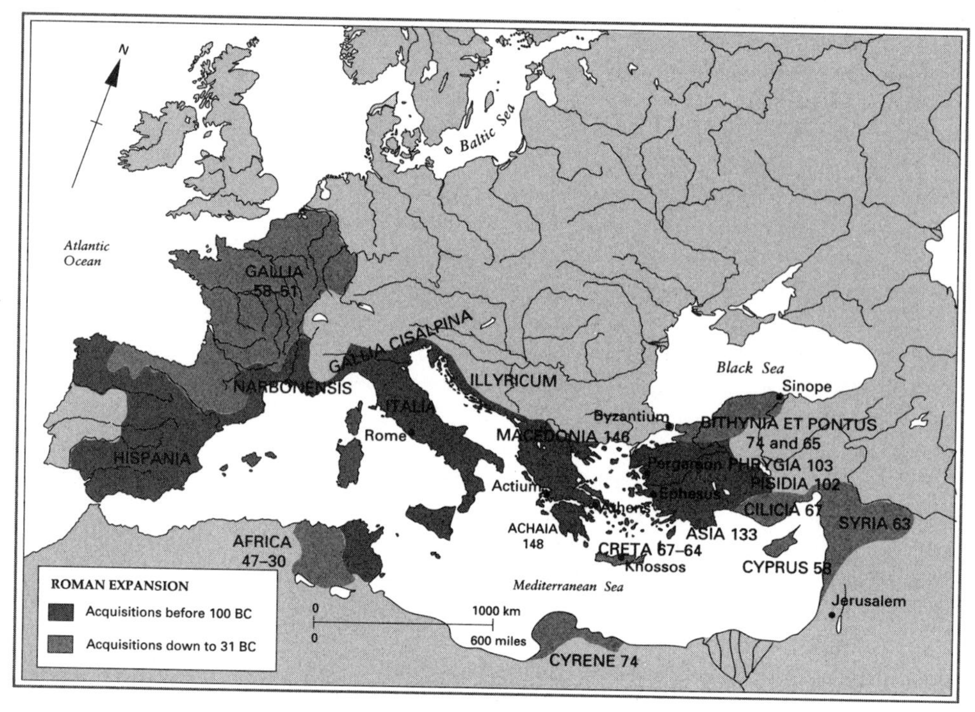

This structure went on growing for a long time. In 58 BC the Romans annexed Cyprus. In the next few years a young politician called Julius Caesar took command of the Roman army in transalpine Gaul (France) and finished off the independence of the Celtic peoples there. (He also led two reconnaissance expeditions across the Channel to the island Romans called Britannia, but did not stay.) These can be regarded as the last additions to the republican empire. By 50 BC all the northern coasts of the Mediterranean, all France and the Low Countries, all Spain and Portugal, a substantial chunk of the southern Black Sea coast and much of modern Tunisia and Libya were under Roman rule. By then, though, the Republic was on the verge of disappearance.

Why it should collapse has already been suggested, but the way things happened owed much to individuals and to chance, as it often does. In Italy restiveness among Rome's allies led first to war on them and then to the extension of Roman citizenship to virtually the whole of Italy – which made nonsense of the idea that the Roman popular assemblies (which only met at Rome) still had the last word in the Republic. More wars in the East threw up yet more war-lords with political ambitions at home. Round about 100 BC emergencies in Africa and southern Gaul led to the granting of exceptional powers to generals who were politicians at home and they used them against their political opponents as well as the Republic's enemies. Rome became a dangerous place – to political intrigue and corruption were now added murder and mob violence. People began to fear the emergence of a dictator, but were not sure where he was to come from.

Somewhat unexpectedly it turned out to be Julius Caesar, the conqueror of Gaul. His seven years there gave him three great advantages; he was away from Rome while other people were blamed for the increasing disorder, violence and corruption; he became enormously rich; and he won the loyalty of the best-trained and most experienced of the Roman armies. His soldiers felt that he was a man who would look after them, assuring them pay, promotion and victory.

Caesar has always been a fascinating figure. He has been seen both as a hero and as a villain, and his reputation has swung about. He did not have a very long career at the top, and it finished at the hands of his enemies, yet few have questioned his abilities. By writing his own accounts of his successful campaigns in some of the best Latin of his day, he helped to sustain a belief in them. He had great qualities of leadership, cool-headedness and determined patience. Though not cruel, he was ruthless. Whatever his aims and the morality of what he did, it can at least be agreed that he was no worse than most other politicians of his day and often showed himself better.

The end of the Republic

In January 49 BC Caesar struck. Claiming to be defending the Republic from its enemies, he crossed the river Rubicon, the border of his province, and marched on Rome with his army an illegal act. For four years he campaigned in Africa, Spain and Egypt, chasing his opponents who had armies in the provinces which they might use against him. He crushed opposition by force but also won over former enemies by mildness after success. He carefully organized his political support in the Senate and was made dictator for life. But some Romans feared that Caesar might re-establish a monarchy. In the end his enemies came together and in 44 BC he was murdered.

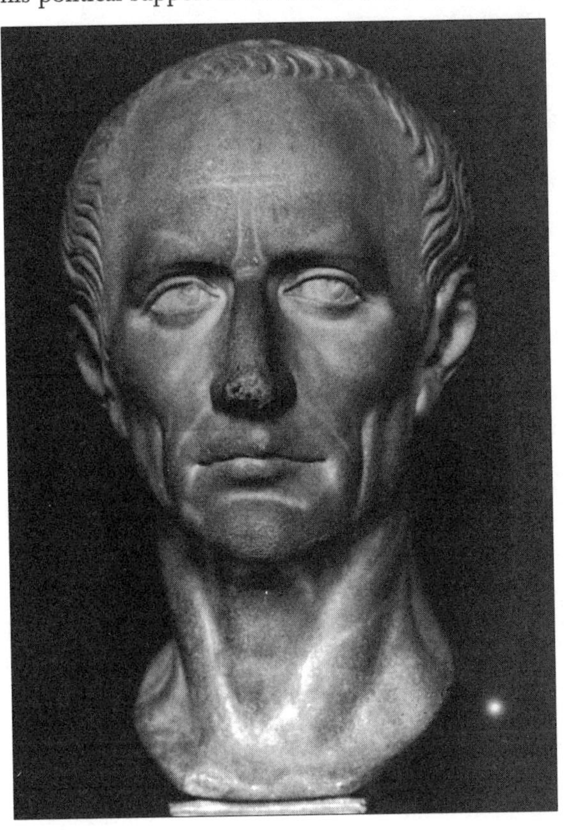

There survive many likenesses of Julius Caesar, several made soon after his death. Almost alone among the portraits of him, this one is believed to date from his lifetime.

The leader among Caesar's assassins (and reputedly a descendant of the founder of the Republic) was Brutus; like other politicians and rulers of the ancient world, he used coins as vehicles for propaganda. This one commemorates the day of Caesar's death as a day of liberation – a cap of liberty stands between two daggers – and may be the beginning of a tradition of idolizing republican tyrannicide which runs down to modern times.

In form, the Republic was still there. But the changes Caesar had already made in the direction of centralized power were left intact. Problems could not be solved by putting the political clock back. In the end his great-nephew and adopted heir, Octavian, made it clear that there had been an irreversible change and so began what we know as the Roman empire. Octavian first hunted down the politicians who had murdered Caesar. He then fought a civil war which took him as far afield as Egypt (it was duly annexed as a province, after the legendary suicides of Anthony and Cleopatra). When he returned to Rome supported by the loyalty of his old soldiers (and those of his great-uncle), he used his power carefully, getting the Senate to provide a cloak of republican respectability for everything he did. He was, formally speaking, only *imperator* – a title that meant he commanded soldiers in the field – but he was elected consul, the most important of the republic's executive officers, year after year. In due course he was given the honorary title of 'Augustus' and it is as Caesar Augustus that he has gone down in history. His power grew as more and more offices and honours were given to him, though he never ceased to insist that this was all within the old republican framework. He was *princeps* – first citizen not king. In reality, though, Augustus increasingly relied on the power given him by control of the army (he organized the first regiment for service in the capital itself, the Praetorian Guard) and on a bureaucracy of paid civil servants. He intended to be succeeded by a kinsman. This was only an adopted stepson (his own child was a daughter) but five Caesars in a row became *imperator* and *princeps* after him. And, after he died in AD 14, Augustus was declared a god as had been Julius Caesar.

In about AD 10, this gem was carved from onyx as a contribution to the imperial mythology then still in the making. Augustus, accompanied by the goddess Roma, is shown sitting in state, presumably contemplating the glory of his position while another figure places on his head a crown representing the civilized world. At the side, Tiberius steps from his chariot after his victory over the warlike Pannonians, while soldiers erect a trophy below. This remarkable work has been attributed to Augustus' official engraver and so no doubt expresses themes the emperor approved.

This was a big change. It settled the Roman state on a new course. It would be ruled in future by monarchs, though they would depend on, and therefore need to please, the army. By the time Augustus died the centuries-old domination of Rome by a relatively small class of politicians had ended with the triumph of one of the leading families among them, though the Caesars were not to enjoy an untroubled ascendancy. Augustus went to his grave to be remembered as the great bringer of peace and restorer of old Roman ways. But none of the three Caesars who followed his successor, Tiberius, died a natural death – and some have thought that Tiberius did not either. The empire (as we may now call it) was to bring great achievements and would spread Roman rule even farther, but it too was to fail in the end.

CHRISTIANITY

If historical importance is measured by impact on numbers of people, we can safely say that no single event in ancient times and perhaps none in the whole of human history is as important as the birth of the man whose name passed into history as Jesus. We can be fairly certain this was in Nazareth, in Palestine, and slightly less sure when it happened, though 6 BC seems the most likely date.

The whole of human history since shows how important it was. Quite simply, those later calling themselves Christians – the followers of Jesus – were to change the history of the whole globe. To find something which has had a comparable impact we have to look not to single events but to big processes like industrialization, or the great forces of prehistoric times like climate which set the stage for history. This has not prevented violent disagreement about Jesus and about what he was trying to do. But it can be seen easily enough that what gave the teaching of Jesus a much greater impact than that of other holy men of his age was that his followers saw him crucified and yet believed he later rose again from the dead.

They were Palestinian Jews, and to understand the story of Jesus we have to set him in the history of the Jews, his people. After the Babylonians carried off so many of them to Exile in 587 BC, and destroyed the Temple at Jerusalem, Jews had come to feel even more distinctive and even more unlike other Near Eastern peoples. Deprived of the Temple as a centre for their cult, they had turned to weekly reading of their scriptures, a practice which led in time to the appearance of the synagogue, a place of teaching and reading, not of sacrifice. Furthermore the prophets who had led some of the Jews back from Exile in 538 BC (after the Persian overthrow of Babylon) had preached a more exact and narrow observation of the Jewish law, in order to set Jews apart from other peoples, the 'gentiles', and had made sure the Temple was rebuilt. When Palestine was under Seleucid rule some Jews had taken to Hellenistic ways, but they belonged to an upper-class minority often distrusted and disliked by the people, who clung unquestioningly to their tradition – and indeed became even more tenacious of it. There was a great Jewish revolt in the second century BC against what we may call Hellenization. After this the Seleucid kings treated the Jews very cautiously.

The Jews in the Roman empire

The end of Seleucid rule in 143 BC was followed by a period of independence for about eighty years and then Judaea was taken by Rome. Two thousand years were to pass before there was again an independent Jewish state in the Middle East. By the time of Augustus, fewer Jews lived in Judaea than in the rest of the Roman empire. After the Exile the freedom of movement and trade offered first by the Hellenistic states, and then by the rule of Rome,

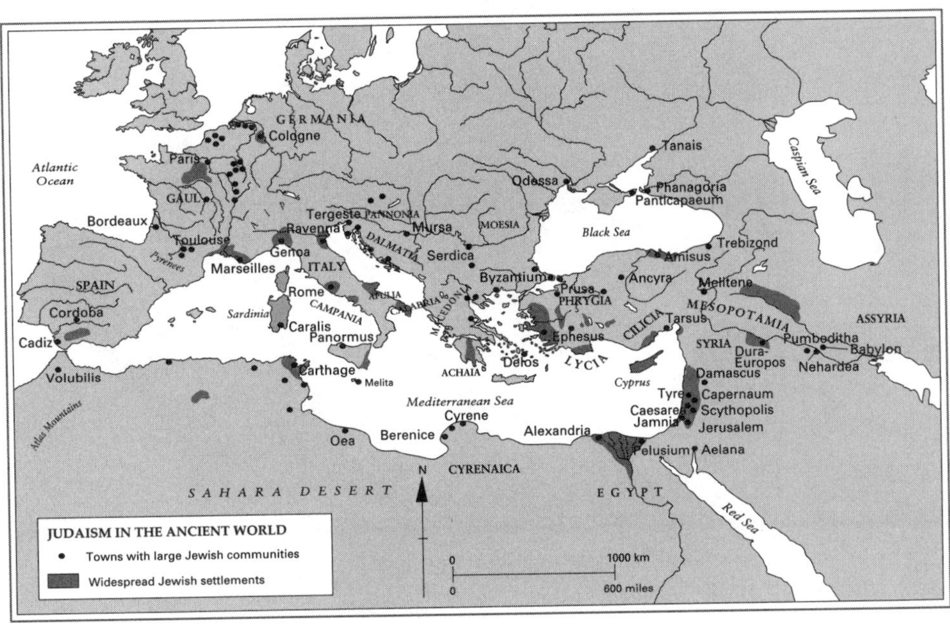

had spread Jews all round the Mediterranean coasts, into the Black Sea ports and into Mesopotamia. At Rome itself there may have been 50,000 and there was another great Jewish centre in Alexandria. This was the 'Dispersion' (diaspora). Some Jews had settled in the ports of Western India even earlier, by about 175 BC.

Jewish numbers grew slightly from the conversion of gentiles drawn to Judaism by its moral code, by religious ceremonies which centred round reading the scriptures and did not need shrines or priests, and above all because it promised human salvation. The Jewish view of history was clear and inspiring; it saw the Jewish people as one set apart by God, a chosen people who would be refined in the fire for the Day of Judgement, but who would then be gathered to salvation. It is difficult to see why this confident creed should have awoken resentment. Yet relations between Jews and their neighbours were often strained. Rioting was not uncommon and troubled the Roman authorities. Popular prejudices were easily aroused by Jewish distinctiveness and success.

Jesus of Nazareth

In AD 26 a new Roman governor, Pontius Pilate, was appointed at a bad moment in its history to the province of which Judaea was a part. It was especially disturbed. The Jews of Syria and Palestine hated one another, hated their Greek and Syrian neighbours, and hated most of all the Roman occupiers and their tax-gatherers. Some Jews belonged to a sect called Zealots who were in a sense a nationalist movement. Many Jews were waiting for a leader, a 'Messiah', one anointed by God, and a descendant of the line of David, who would take them forward to victory – whether military or symbolic there was much disagreement. Jesus of Nazareth was then about thirty years old. He had grown up amid these expectations. He knew himself to be a holy man and his teaching and the miracles reported of him awoke great

From the whole New Testament, no non-Jew is better remembered than Pontius Pilate, Roman governor of Judaea from AD 26 to 36. His historical existence has long been attested not only by the Gospels but by other literary sources; only in 1961, though, did this fragment of inscribed stone come to light and provide other evidence of his existence in the form of his dedication of a building to the emperor.

excitement. Of his life we have the Gospels as a record, the accounts written down after his death by followers, on the basis of the memories of those who had known him. The Gospels were written to show that they were right in thinking him a unique person – Messiah.

The uniqueness of Jesus was demonstrated for his followers by what happened at the end of his life. He was charged with blasphemy by the Jewish religious leaders and taken before the Roman governor, Pilate. Anxious to avoid further communal strife in a troubled city, Pilate bent the letter of the law somewhat and allowed him to be condemned. So Jesus was crucified, probably in AD 33. Soon afterwards his disciples believed that he had risen from the dead, that they had met him and talked to him after that, that they had seen him ascend into heaven and that he had left them only to return soon sitting at the right hand of God to judge all men at the end of time.

Whatever may be thought of the details of the Gospel records, it cannot plausibly be maintained that they were written by men who did not believe these things, nor that they did not write down what they were told by men who believed they had seen them with their own eyes. Clearly, too, Jesus' life was not so successful in a worldly sense that his teaching was likely to survive because of the impact of his ethical message alone. He had, it is true, especially attracted many of the poor and outcast, as well as Jews who felt that their traditions in the forms into which they had hardened were no longer wholly satisfactory. But these successes would have died with him had his disciples not believed that he had conquered death itself and that those who were saved by being baptized as his followers would also overcome death and live for ever after God's judgement. Before a century had passed, this message was being preached throughout the whole civilized world sheltered by the Roman empire.

St Paul

The new Jewish sect (for that was what it was) took root first in Jewish communities, whose distribution was of great importance in setting the pattern of early Christianity (a word taken from the Greek name soon given to Jesus – the 'Christ', or 'anointed one'). Soon, though, Jesus was preached also to the gentiles. This was the decision of a council of 'Christians' (the word was by then beginning to be used of his followers) held at Jerusalem in AD 49. Besides those who had known him personally (Jesus' brother James and his disciple, Peter, among them) there also may have been present a Hellenized Jew from Tarsus, Saul, known later as St Paul. After Jesus himself, he is the most important figure in the history of Christianity. Many gentiles were already interested in the new teaching, but it was Paul's missionary work and the decision of the Jerusalem Council that gentiles should not be asked to conform to the Jewish law – that is, accept the full rigour of the Jewish religion, and show it by undergoing circumcision and practising its dietary restraints – which released the most successful of world religions from the Jewish shell which had protected its earliest life.

Christianity not only now began to emerge from Jewish society, but came through Paul to distinguish itself from the world of Jewish ideas, though so far as we know, Jesus had at no time in his own teaching gone beyond the intellectual world of the law and the prophets, and had been scrupulous in his own religious observances. Paul, a Greek-speaker and an educated man, nevertheless put his view of Jesus' message into Greek and, in the process, into the language and ideas of Greek philosophy. Greek ideas of the distinction between soul and body, of the links between the visible, material world and the invisible, spiritual world were used by him to preach his message. He outraged orthodox Jews by preaching a Jesus who was God himself; such an idea could never have found a place within Judaism. It can be argued that Paul was the real maker of Christianity. Certainly most of the theology of the Christian Church has its roots in his interpretation of Jesus' teaching. Here there is room only to note that he seized the opportunity presented by a world at peace, protected by a framework of government and law in which men could travel easily and securely, a world in which the widespread Greek language made communication of ideas easy, to launch Christianity on its huge career of expansion. It is not surprising that Christians soon began to think that the Roman empire itself was somehow created by God to make the spreading of the Truth possible: it was divinely intended, some of them thought, to further Christianity. A more sinister reflexion also occurred to some of them as time passed: it was not the Romans, after all, but the Jews who had actually killed Jesus.

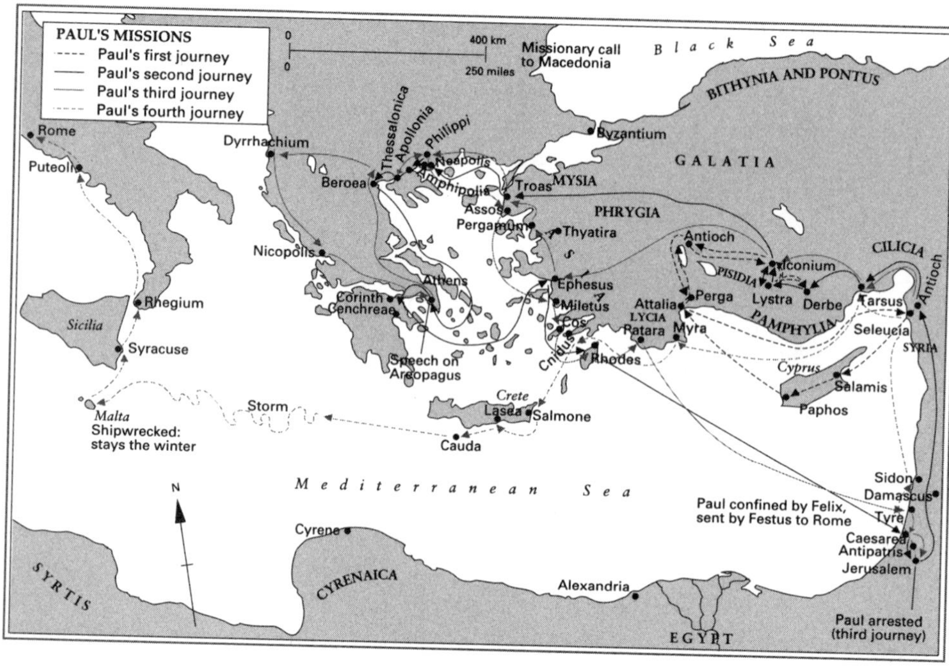

Almost the last thing we hear of Paul is that when he was accused by the Jewish leaders at Jerusalem of sedition and profanation of the temple, he used his rights as a Roman citizen to appeal from the judgement of the governor at Caesarea to the emperor at Rome. To the capital he made his way to await trial. What happened to him after that we do not know, though, according to early Christian tradition, he was martyred at Rome in AD 67. If he was, he had by then changed history.

THE ROMAN EMPIRE

Imperator, the title borne by Augustus and his successors, took a long time to come to mean the man at the top of the empire, what we call an 'emperor'. Much of the history of the empire was like this; as had happened under the Republic, institutions and ideas changed gradually and in ways almost unnoticed in the short term. In the century after Augustus' death there were twelve emperors, of whom the first four were related to him or his family, the last of them, Nero, dying in AD 68. At once the Empire dissolved in civil war; four emperors were proclaimed in a year. This showed that when an emperor was not able to assure a peaceful takeover for his successor, real power lay with the army; it became obvious in the 'Year of the Four Emperors'. There might even be more than one army to take into account; provincial garrisons sometimes supported different candidates, and the Praetorian Guard at Rome itself might sometimes have the last word, because it was on the spot. The Senate still appointed the first magistrate of 'the Republic', but could only manoeuvre and intrigue; in the last resort it could not defeat the soldiers. As for the emperors themselves, provided they kept the soldiers with them, their personal characters and abilities would decide what they could do.

A good emperor emerged in the end from the Year of the Four Emperors; Vespasian's worst fault seems to have been stinginess. He was not a Roman aristocrat (his grandfather had been a centurion turned tax-collector) but he was a distinguished soldier. The old Roman families had now clearly lost their grip on power, but Vespasian's own family – the Flavians – were not able to keep a hereditary succession going for long and second-century emperors went back to Augustus' solution of adopting heirs. Four of these, the 'Antonine emperors', gave the empire almost a century of good and quiet government, which later seemed a golden age. Three of them were Spaniards, one was a Greek. So now the empire did not belong to the Italians either.

Cosmopolitan at the top – as the origins of emperors showed – the empire was all the time breaking down barriers between peoples at the bottom. The Romanization of leading families in the provinces went ahead steadily. Young Gauls, Syrians, Africans and Illyrians all learnt Latin and Greek, wore clothes like those of the Romans and learnt to think of *Romanitas* – the Roman heritage – as something to be proud of. Meanwhile the civil servants and army held things together, respecting local feeling so long as the taxes came in regularly. When a decree gave the rights of citizenship to all free subjects of the empire in AD 212, this was the logical outcome of a long process of assimilation. By then even senators were sometimes non-Italian by birth. To be 'Roman' was by then not to have been born in a particular place, but to belong to a particular civilization.

The prestige of the emperors – or at least of the office they held – grew. Less and less did they resemble the 'chief magistrate', and more and more did they look like oriental kings, different in kind from their subjects. This was helped by the custom of regarding a dead emperor as a god. Julius Caesar and Augustus had both been deified after death; with Vespasian's son, Domitian, emperors began to be made gods while they were still alive. Particularly in the East, altars on which sacrifices had been made to the Republic or Senate were re-allocated to the emperor.

The Roman legacy

Those who regretted such changes could hardly deny that the empire was an astonishing achievement and one in which Romans could take pride. To provide regular, lawful government over a wider area than ever before, to black, white, brown 'Romans' equally, and to provide them also with the blessings of peace and the prosperity – all this was without precedent and remains the best ground for saying that the Romans did great things. Materially, they left great monuments, building and engineering behind. Centuries later, men explained Roman ruins as the work of long-departed giants and magicians, so amazing did they find them; an English seventeenth-century antiquarian said that Stonehenge was a Roman temple, because only the Romans could possibly have done anything so grand – but he was wrong too. Yet such mistakes are understandable and revealing. What the Romans left behind in brick, stone and concrete was of enormous impressiveness and long unrivalled in western Europe. Most of it had very practical aims. No legion was supposed to camp even for a night without digging itself into a properly planned and defensible camp with ramparts, so that the army got plenty of practice in surveying, engineering and building. Most Roman building, though, was in towns; they lived in an urban civilization. All over the empire public buildings and monuments showed what the Romans thought appropriate to civilized living. To service the towns, the Romans built the roads which linked them and provided the arenas, baths, drains and fresh-water supplies to make them comfortable. They liked magnificence and produced some vulgar things, but they were practical; they did not build anything so useless as the pyramids. Some of their tombs were very grand, though; centuries later, that of the emperor Hadrian in Rome became the castle of St Angelo.

The Romans used a very efficient but not very novel technology. They had better tackle than the Egyptians, as well as windlasses, cranes and iron tools which the builders of the pyramids had never had, but not much that was not already known to the Greeks. They used a wide range of materials, but most of them were already available, the exception being con-

MAJOR ROADS, CITIES AND GARRISONS OF THE EMPIRE IN THE AGE OF THE ANTONINES
Roads
N
North Sea
Hadrian's Wall
Baltic Sea
Black Sea
Caspian Sea
Atlantic Ocean
Mediterranean Sea
Persian Gulf
0
1000 km
0
600 miles
Eburacum
Deva
Lindum
Isca
Camulodunum
Londinium
Ulpia
Noviomagus
Novaesium
Vetera
Castra
Colonia Agrippina
Portus Itius
Bonna
Alauna
Augusta Treverorum
Mogontiacum
Divodurum
Augusta Vindelicorum
Vindobona
Avaricum
Vesontio
Basilia
Carnuntum
Augustodunum
Vindonissa
Lauriacum
Brigetio
Porolissum
Aventicum
Turicum
Virunum
Aquincum
Burdigala
Lugdunum
Poetovio
Brigantium
Vienna
Mediolanum
Siscia
Verona
Singidunum
Istropolis
Placentia
Asturica
Tolosa
Bononia
Sirmium
Viminacium
Nemausus
Genua
Ariminum
Ratiaria
Novae
Salonae
Oescus
Odessus
Narbo
Forum Iulii
Caesaraugusta
Narona
Naissus
Sinope
Philippopolis
Olisipo
Toletum
Tarraco
Byzantium
Sebastea
Roma
Amphipolis
Nicomedia
Emerita
Brundisium
Apollonia
Melitene
Capua
Valentia
Thessalonica
Caesarea
Corduba
Tarentum
Pergamum
Samosata
Gades
Carthago Nova
Nicopolis
Smyrna
Sardis
Tyana
Malaca
Messana
Athenae
Laodicea
Tarsus
Tingis
Rhegium
Patrae
Ephesus
Hippo Regius
Attalia
Antiochia
Cartenna
Saldae
Utica
Agrigentum
Corinthus
Salinus
Volubilis
Rusaddir
Caesarea
Syracusae
Rhodus
Ad Mercurium
Rapidum
Carthago
Pomaria
Castellum Tingitanum
Oppidum Novum
Thamugadi
Zarai
Thagaste
Sidon
Damascus
Capsa
Tyrus
Tacapae
Caesarea
Tamallenia
Turris
Philadelphia
Sabrata
Oea
Leptis Magna
Cyrene
Tubactis
Alexandria
Petra
Memphis
Aelana

crete, which they invented. It made possible building in new shapes. The Romans were the first architects to get away from the need to hold up broad spans of roof with lines of pillars: they invented the dome supported on vaults.

Among the most visible Roman works now left to us are roads. Occasionally they are still good enough to carry traffic, and even where they have disappeared, the routes they took are often still followed by modern roads. A special corps of surveyors kept up the skills which made possible the astonishing accuracy with which they went straight across hill and vale, and they were usually built by the legions. They gave the empire the communications which made possible the government of so wide an area. Between the age of the Caesars and that of the railway train there was no improvement in the speed with which messages and goods could be sent overland – and in some places communication got much worse in the next thousand years when Roman roads were not kept up.

Roman technology: a crane powered by treadmill in use for the building of a tomb.

Because Roman ruins provided huge quantities of ready-cut stone for later builders all over Europe, it is now difficult to imagine just how splendid much of the empire must have looked. Some great single monuments remain – the Pont du Gard in the south of France; the arena at Nîmes, not far away; the Black Gate at Trier; the aqueduct still bringing water to Segovia, in Spain; or the complex of baths at Bath, in England. At Pompeii there is a whole town to see. In many more places all over Europe, the Near East and North Africa there are fragments. Above all, there survives the astonishing wreckage of imperial Rome itself.

The Romans prided themselves on being tough and hardy, but they also liked comfort. Sometimes they overdid the business of self-indulgence (as the lists of what was served at the great feasts of the rich when big parties were fashionable show). Their enjoyment of bathing and central heating is more easily admired. They had great skill in all matters of plumbing and sanitation. Elaborate aqueducts brought drinking-water to the cities within which public baths and lavatories looked after outer and inner cleanliness. In private houses, steam-rooms and living-rooms were centrally heated from under the floor. Only in the twentieth century did the inhabitants of Britannia again get used to the idea that houses should be properly heated.

Away from engineering and hydraulics, the Romans' innovations were fewer. They made little contribution to pure science. In agriculture, watermills were just beginning to be introduced towards the end of imperial times; windmills had not made their appearance. Muscles, animal and human, remained the main source of energy and it has often been suggested that it was because large numbers of slaves were available that the Romans did not need to invent labour-saving machines. There may be something in this, but other explanations are possible. There was the enduring problem of turning a good idea into a practical invention, given the state of technology. Increasingly, too, the empire's history forced rural estates to try to be self-sufficient; they got by on what they could do for themselves, and did not try experiments. Finally, there was no stimulus from outside; China's treasury of technical skill was too

After almost nineteen centuries, the water-supply of the Spanish city of Segovia is still carried to it by this Roman aqueduct, borne by 128 arches which raise the water-channel to a hundred feet above ground at its highest point. Such monuments understandably made barbarians gape. The Romans were the greatest engineers yet to set about extending the human exploitation of the material world. Not for centuries was anything of such comparable scale and utility again to be built in Europe.

far away, and Rome's immediate neighbours had nothing very impressive with which to offer challenge to her.

The intellectual activity the Romans seem most to have admired is, characteristically, a practical one – the law, and the oratory that went with it. Rome did not stimulate philosophers like those of classical Greece (but neither did anybody else, the Chinese and Indians included). The Hellenistic philosophers were not such original thinkers as their predecessors, either. Roman culture, nonetheless, could point to some good expounders of Stoic philosophy, a few historians of note, and a galaxy of writers of Latin prose and verse, among whom Virgil, the epic poet, is unquestionably a giant figure even in world literature.

If it is easy to run down the Roman intellectual achievement by comparison with that of Greece, it should be remembered that to have produced for centuries such a succession of conspicuously able all-rounders suggests that the reliance of Roman culture on conservative

Reconstructed theatre, built 2nd century BC. Kourion, Cyprus.

ideas and the Greek tradition had much to be said for it. Roman politicians who reached the top were for a long time likely to have to act as administrators, generals, supervisors of building and engineering works, advocates and judges. Rome produced men in abundance who could do all this. And the kind of empire they ran was tolerant and cosmopolitan, in that even so revolutionary a creed as Christianity, with all its implications for future upheaval, could take root and flourish. There was some intellectual sophistication about it, too: it was a later, Christian empire that went back to trying people for blasphemy.

Christianity and the empire

Christian congregations and communities soon emerged all over the Roman world. Everyone recognized that the Christians at Jerusalem, where the first generation of the Church's leaders had actually known and heard Christ, deserved special respect. But the only links between all Christians were the rite of baptism (which was the sign of acceptance into the new faith), their belief in the risen Christ, and the ritual of the 'eucharist' – the special service which re-enacted and commemorated Christ's last meal with his disciples on the eve of his arrest, trial and crucifixion. Christians usually also believed that the end of the world was at hand, that Jesus would soon return to gather up those faithful to him, and would assure them salvation at the Last Judgement. If that was so, then clearly there was not much to do here and now except watch and pray. Running the churches was not therefore a very complicated business. Still, as they grew in numbers and wealth, there were administrative decisions to be taken,

The crushing of the Jewish Revolt of AD 66 was commemorated as a great triumph of Roman arms. On the Arch of Titus in Rome were placed reliefs, this one recording the bringing to Rome from the Temple of the Menorah, the table of the shewbread and the sacred trumpets, all of them spoils of war.

and so there appeared officers called bishops and deacons. As time went on, they were to take on more sacerdotal roles, concerning themselves more with the conduct of worship and questions of theology as well as with administration.

The first great change was Christianity's break with Jewry. Though it never shed its essential Jewish inheritance of monotheism, the Old Testament books of the Bible, and a view of human destiny as an extension of the special pilgrimage of a chosen people through history, and though Christian culture remains soaked in ideas and images drawn from the Jewish past, it nonetheless broke with Jewish society and the Jewish nation. The Romans long thought of Christians as just another Jewish sect, but the growth of gentile Churches made them distinctive. Jewish Christians meanwhile failed to convert their fellow-Jews to their view that the long-awaited Messiah of his people had come in Jesus. They could hardly go on attending the synagogue when it was known that they associated at common meals with gentiles who were uncircumcised, ate pork, and did not observe other features of the Jewish law.

Another turning point was reached with a great Jewish rising against the Romans in Palestine in AD 66 (the future emperor Vespasian was the local commander at the time). This was the worst Jewish rebellion the Romans ever had to master. After seven years' fighting, the reduction of Jerusalem by starvation to the point at which its inhabitants had turned to cannibalism to survive, and the destruction of the Temple rebuilt after the return from the Exile, the last Jewish forces committed mass suicide rather than surrender their stronghold at Masada in AD 73. Christians had not joined in the revolt, and this may have made the Roman authorities less suspicious of them. But neither had other Jews outside Palestine. Therefore, though Jerusalem was taken from the Jews after the revolt (Hadrian made it an Italian colony in 135, excluding all Jews from Judaea), elsewhere they were still left much to themselves under the government of their own religious authorities. Yet the revolt and its aftermath made the Jewish people even more self-conscious and reliant upon the observations of the strict Law, since the Temple was again no more. This made the position of Jewish Christians still more awkward.

Jews, moreover, had been the first persecutors of Christians. They had not only demanded the crucifixion of Christ, but had killed the first Christian martyr (St Stephen) and they had given St Paul his roughest moments. Some scholars have even blamed Jews at Rome for bringing down on the Christians the first Roman persecution, by pointing them out as scapegoats for a great fire at Rome in AD 64. Legend says that both St Peter and St Paul died in

Thirteen hundred feet above the Dead Sea, the rock of Masada was chosen by Herod the Great as a site for a fortress containing palaces, arsenals, barracks and great water tanks. It was virtually a personal refuge for him and his family. This was in about 30 BC. A century later, in AD 73, this fortress was the scene of the siege endured by Jewish rebels whose revolt against the Romans had by then been crushed everywhere else. In the end, faced with defeat, the entire garrison and their families destroyed themselves, only two women and five children whom they had hidden surviving, reports Josephus. Excavated in the 1960s, the scene of this heroic struggle is now a shrine of Israeli patriotism.

this persecution. Many Christians in the capital certainly perished horribly in the arena or were burned alive. Terrible as this was, though, it was unusual and local. Christians seem customarily to have enjoyed official toleration until well into the second century AD. Tales were told about them by the suspicious – they were said to practise black magic, cannibalism and incest, and some Romans disliked the way that their religion encouraged Jack (or Joanna) to think himself as good as his (or her) master in the eye of God and therefore to resist the traditional authority of husbands, parents and slave-owners. It was easy for the superstitious to think that the Christians were the reason for natural disasters – the old gods were angry that Christians were tolerated and so sent famines, floods, plagues, it was argued. But this did not much affect officials and authority only came officially into conflict with Christianity in the second century.

Persecution and progress

It was then discovered that some Christians would refuse to sacrifice to the emperor and the Roman deities. The Romans accepted a similar refusal from the Jews: they were a distinct people with customs to be respected. But by now most Christians were not Jews: why should they not carry out these acts of formal respect like other people? Condemnations followed – not for being Christian, but for refusing to do something the law commanded. This no doubt also encouraged unofficial persecution; in the second century there were pogroms and harryings of Christians in many parts of the empire, notably in Gaul.

Yet this century was also one of advance for the Church. The first of the great figures called the 'Fathers' – theologians and administrators who laid down the main lines of Christian doctrine so as to distinguish it more and more sharply from other creeds and to make more precise the duties and obligations of Christians – belong to this age. Among them, two were particularly important for the way in which they tried to connect Christian and Greek ideas (and therefore to help separate Christianity even more clearly from a mass of other oriental cults), St Clement of Alexandria and his pupil, Origen. The Fathers' intellectual and moral achievements were great, and some features of the age favoured them. A great search for new ways in religion was going on all over the Roman world in the second century, and Christianity profited from it. Moreover a new idea could spread quickly in a world held together by Roman law and order, where people could travel freely and everywhere find others speaking Greek.

By the end of the third century about a tenth of the population of the empire may already have been Christian, one emperor had been (at least nominally), and another seems to have included Jesus Christ among the gods honoured privately in his household. In many places the local authorities were by then used to dealing officially with the local Christian leaders, who were often prominent in their communities and, as bishops, played a large part in their affairs and represented them. What was more, the empire had concerns much more pressing than a religion of well-conducted believers.

THE FRONTIERS

Parthia and Persia

As long ago as 92 BC a Roman army had reached the Euphrates. For the first time the Republic was in direct contact with the Parthians, a people who were to play an important part in their affairs for the next three centuries. Nearly forty years later this became only too obvious when a Roman army invaded Mesopotamia and was within a few weeks wiped out in one of the worst military disasters of Roman history. Evidently the Parthians were not to be lightly interfered with. They were another of those originally nomadic Indo-European peoples from Central Asia, famous for their way of fighting on horseback, appearing to flee but then turning in the saddle to shoot arrows as they galloped off – hence the term 'Parthian shot'. They had chosen to settle just south-east of the Caspian, in an area later crossed by a major caravan route from Asia to the Levant – the Silk Road from China. In due course this would bring wealth to Parthian kings. There they lived, first under Persian rule, then under the Seleucids until in the middle of the third century BC the local Parthian governor had enough of Seleucid rule and decided to strike out on his own. This was the beginning of the independent Parthian kingdom which was to last for nearly five hundred years.

At one time in the following century a Parthian empire stretched from Bactria in the east to Babylonia and the Euphrates frontier with Syria (all that was left by this time of the Seleucid kingdom) in the west. Even Chinese emperors thought it worthwhile to open up diplomatic relations with Parthia (possibly, among other things, because of the fame of the splendid Parthian horses, much prized in China). The Parthian kings called themselves on their coins 'Great King' and 'King of Kings', traditional titles of the rulers of Persia, and claimed that they inherited the authority of the Achaemenids. Yet early Parthia was probably more like a coalition of great noblemen bringing their contingents of followers to the army of their overlord than a centralized and well-organized state. By the time the Romans met it,

Relaxations of a Persian monarch: possibly Shapur I (AD 240–271), a great antagonist of Rome, son of Ardashir, portrayed hunting, improbably mounted on the stag he despatches with his knife while another expires beneath his feet. One of many Sassanid dishes showing hunting scenes.

though, the Parthian army was a formidable machine. Besides its traditional mounted archers, it had an arm the Romans lacked, the heavy armoured horsemen called 'cataphracts', whose mounts, like them, were clad in chain-mail.

Rome and Parthia long tended to quarrel over Armenia, a frontier kingdom east of Anatolia which both thought of as falling into their sphere. There was a series of ding-dong struggles, each side sometimes victorious; once a Roman army actually occupied the Parthian capital, but the frontier did not change much. The disputed area was too far away for Rome to be able to hang on to conquests there without great effort and expense, and the problems of the Parthian kings at home were too distracting for them to think of expelling the Roman threat from Asia altogether. In about AD 225 the last Parthian king was killed by one of his vassals who was ruler of Fars, or 'Persia'. The man who overthrew him was named Ardashir

(Greek-speakers called him Artaxerxes). Ardashir's descendants would make the splendour and grandeur of the Achaemenids live again and were to restore Persian supremacy in much of the Near East.

The Sassanid Persian empire (it was named after one of Ardashir's ancestors, Sasan) was to become Rome's greatest antagonist. Its rulers strove to emphasize continuity with the past. They took up the traditional royal Persian titles and Zoroastrian religion. Ardashir also claimed all the lands ruled by Darius, the greatest Achaemenid. The traditions of Sassanid bureaucracy went back even further, to Assyria and Babylonia, and so did the royal claim to a divine authority. Not that this went uncontested. There were struggles between the monarchy and the great noble families who claimed descent from old Parthian chiefs who wanted to keep power in their own hands. Yet in the end, after centuries of conflict, the Sassanid empire was in fact, if not in form, to outlive the Roman. The threat Persia presented was all the more dangerous because it first appeared at a moment when Rome was handicapped by threats elsewhere and by confusion in her own internal affairs. This continued. Between AD 226 and 379, for instance, there were thirty-five Roman emperors, while only nine Sassanid kings ruled in Persia; long reigns and the stability that went with them were advantages. Shapur I (241–72) was perhaps the most outstanding Sassanid, at least until almost the end of his line. He once took captive a Roman emperor (Valerian, who, poor man, was said to have been skinned alive and stuffed by the Persians, though this may not be true), as well as conquering Armenia and invading the Roman provinces of Syria and Cappodocia on several occasions. After this there were longer periods of peace between Rome and Persia, but the two great powers never settled down to live easily together.

The Roman empire had reached its greatest extent long before, when the emperor Trajan died, in AD 117. It then covered an area about half that of the modern United States. Roman territories ran from north-west Spain to the Persian Gulf. Armenia had been annexed in AD 114, and this took Rome's frontier to the Caspian in the north-east. The big province of Dacia, north of the Danube, had been conquered a few years earlier. Some of these lands (notably those across the Euphrates) had almost at once to be given up, and even without them, this huge area posed big security problems. Though, it was only in the East that Rome was threatened by a great power – a state like Rome itself, capable of putting big armies into the field and carrying out long-term diplomatic and strategic plans – problems elsewhere grew harder to deal with as time passed. Africa was almost the only place where affairs stayed reasonably quiet after the acquisition of Mauretania in AD 42. This was largely because Rome had no African neighbours who much mattered; no major native populations lay just beyond the limits of Roman rule – there was only desert.

Europe: the limes

In Europe things were very different. All along the frontier from the Black Sea to the mouths of the Rhine were 'Germanic' peoples, with whom the Romans were often at war. Some of them had been thrown out of their original homelands by the Romans. They could be formidable opponents. Augustus had hoped to extend the limits of the empire to the Elbe, but it had become clear that this would not be possible – not least because of a great disaster in AD 9 when three legions were entirely wiped out in a defeat which affected Roman morale so badly that the numbers of the destroyed legions were never allowed to reappear in the army list. Because of the problem presented by these peoples, an elaborate frontier – or, as the Romans called it, *limes* – had been created.

The *limes* was not just meant to show where one government's responsibilities ended and another began, as do boundaries between most states today, but to protect what lay behind it and to separate it from something else. It separated two different states of culture, defining a 'Latin' Europe distinct from the northern and Trans-Danubian Germanic societies (the Slavs had not yet come on the scene). On one side of the frontier were Roman order, law, pros-

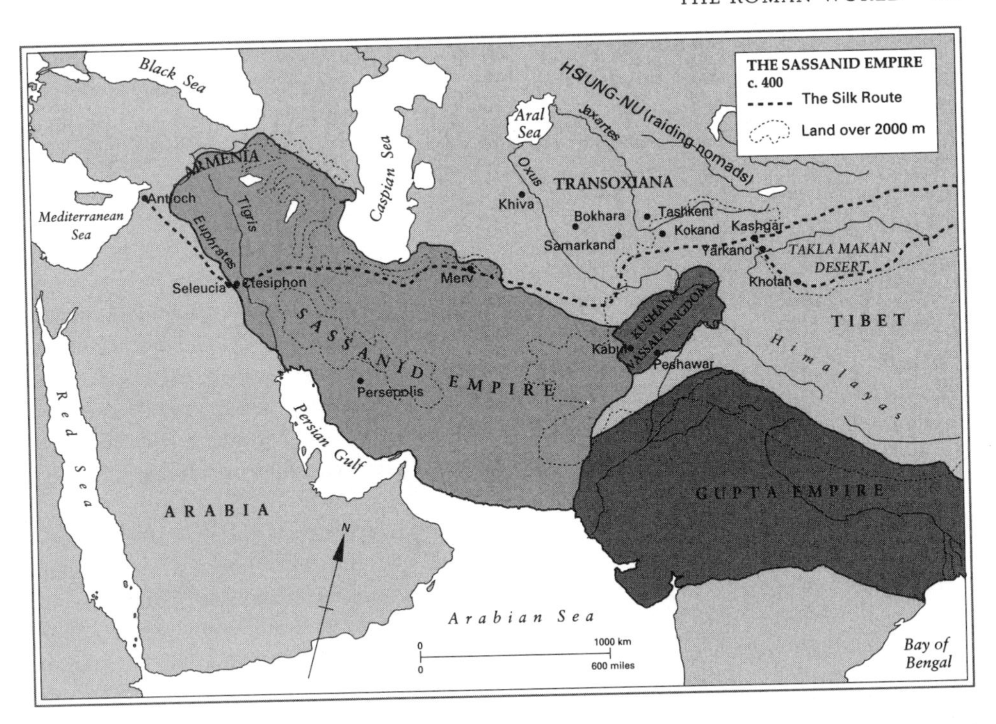

perous markets, fine towns – civilization, in short; on the other were tribal society, technical backwardness, illiteracy, barbarism. Of course, complete insulation was impossible and there was always coming and going. Still, the Romans saw the frontier as something they looked out from warily, not as a stage on a journey leading somewhere else, and where possible, they based it on natural obstacles. Much of it followed the lines of the Rhine and Danube. In the gaps between natural obstacles fortifications were built of turf, timber or sometimes masonry. Along it were dotted permanent legionary camps linked by signal towers and smaller strongpoints. Roads ran along the frontier so that troops could march quickly from point to point. One long stretch of works ran between the upper Rhine and Danube, and another in the Dobrudja ran down to the sea. The most remarkable of all was begun in about AD 122 in northern Britannia, between the Tyne and Solway Firth, and is still known as Hadrian's Wall, after the emperor who built it. It was eighty Roman miles long (about 120 km), a masonry wall, protected on both sides by ditches some ten metres wide and three deep, and sixteen forts, while smaller strongpoints were placed at intervals of a mile, with two turrets between each strongpoint. Its purpose, said Hadrian's biographer, was 'to separate the Romans from the barbarians'. As defences, such barriers were not effective unless properly manned. Twice, once at the end of the second century and once during the fourth, a temporary weakening of the garrison led to Hadrian's Wall being overrun and incursions of barbarian Scots and Picts far into the south, pillaging and destroying.

In continental Europe, the Rhine frontier, though comparatively short, was guarded by eight legions. By Augustus' day the army was already a long-service force, based on volunteer recruits, more and more of whom were drawn from the provinces. They were often barbarians, and served not only in specialist units with a local background – like the skilled slingers of the Balearic isles, or the heavy cavalry of the Danube provinces – but also in the legions of infantry which were the core of Rome's military power. Usually there were twenty-eight legions, about 160,000 men in all, all serving along the frontiers or in distant provinces like Spain and Egypt. (There were about as many troops again in auxiliary and specialist arms like the cavalry.) Long duty in the same areas tended to make the legions less mobile as time passed; the garrison towns contained large populations of dependants and families who could not easily move. But the internal network of roads still gave the empire's commanders great advantages in marching their forces swiftly from one place to another. Gradually the balance of the army's dispositions was changed to reflect changing strategic needs; by the early third century half the Rhine legions had been removed, while the army on the Danube had been doubled in size.

Barbarian pressure

Soon after AD 200 the Germanic peoples were pressing harder and harder on the frontiers, demanding to cross and settle within the empire's own lands. Some were no doubt attracted by civilization and wealth. But there were also more fundamental forces at work, pressure from other peoples to the east, who in turn were propelled westwards by changes in Central Asia, whether natural (such as climate) or political (such as the disturbance by the Han emperors of the Hsing-Nu, a people later to be remembered by Europeans as Huns). A sort of shunting was going on and at the end of the line the Germanic tribes were bound to bump into the Roman frontier.

The barbarians seem only to have been able to put a maximum of twenty or thirty thousand men into the field at once, but they were too strong for the third-century empire. Given strains and diversions elsewhere, it was impossible to hold them off for ever. First some of the Rhenish tribes were allowed to settle in Roman territory (where they were then recruited to help defend the frontier against later arrivals).

Then the Goths, another Germanic family of peoples, crossed the Danube in 251 (and killed an emperor in battle); five years later the Frankish peoples crossed the Rhine. Another group, the Alamanni, were soon raiding as far south as Milan. Meanwhile the Goths went on to Greece and then to harry Italy and Asia Minor from the sea.

This was a terrible time for Rome. While the barbarian onslaughts were going on a new period of civil war and disputed successions had begun. Several third-century emperors were killed by their own troops; one fell in battle to his own commander-in-chief, who was then slain by the Gauls after being betrayed to them by one of his officers. Crushing taxation, economic recession and soaring inflation meanwhile struck at people far removed from such exalted circles as these; local bigwigs began to be unwilling to serve as town-councillors and officials when such posts meant only that they would have to incur unpopularity by collecting heavier taxes – often in kind, as the monetary crisis grew worse. Another symptom was the rebuilding of many cities' defensive walls. They had not been needed under the Antonines, but now even those of Rome were put in order and towns which had never been fortified were given defences in the second half of the century.

DIOCLETIAN AND CONSTANTINE

At the end of the century Rome's luck changed, as once again a succession of able emperors came to the top. The first to turn the tide was an Illyrian, Aurelian, 'Restorer of the Roman Empire' as the Senate appropriately called him, though he was murdered as he was about to invade Persia. His successors were, nevertheless, like him, good soldiers. Nearly ten years after Aurelian's death Diocletian, another Illyrian, came to the throne and not only re-created (at least in appearance) the old power and glory of the empire, but actually transformed the way it operated. Of humble origins, Diocletian was very traditionally-minded and had a very exalted view of his role. He took the name 'Jovius', for example – Jove or Jupiter, the Roman name for the king of the gods, the old Greek Zeus – and seems to have seen himself as a god-like figure, supporting single-handed the civilized world.

By the time he turned to it, Diocletian had tried to provide other more practical remedies for the empire's troubles. An attempt to peg prices and wages and so halt inflation was a disaster. His most important step, though, was one whose full implications he may not have

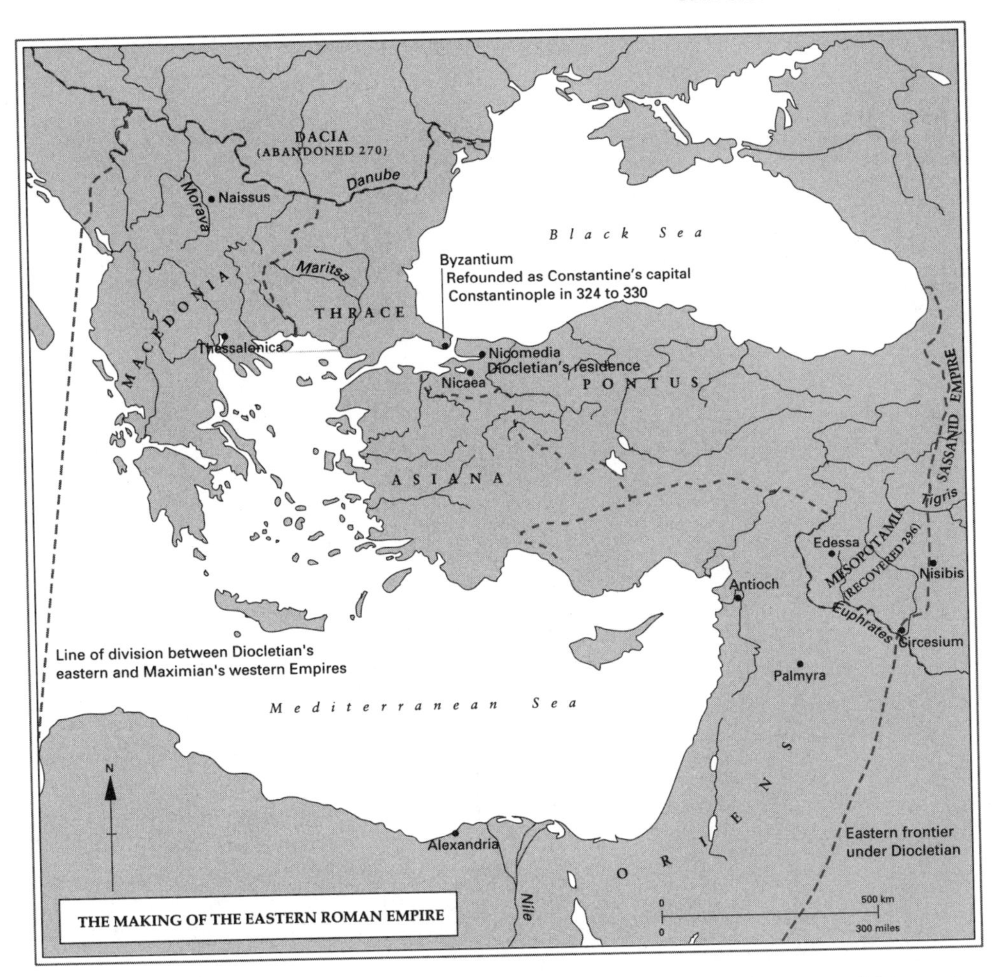

seen: more than any other single man, Diocletian opened the way to a division of the empire into two entities, East and West, which would go separate ways. Whether this outcome, or something like it, was inevitable has been much debated. Rome had welded together much of Alexander's empire in the Hellenized East with the western Greek world the great conqueror had never visited. Although there were always visible differences to remain between them, it was only in the difficult third century that strains appeared. It began to seem impossible to deal with the problems of the West when the resources of the richer East were needed against the barbarians and Persians. In AD 285 Diocletian tried the solution of dividing the empire along a line from the Danube to Dalmatia. He appointed a co-emperor to the western half who, like himself, had the title Augustus; each of them also had an assistant, nominated as his successor and called a Caesar. Other changes followed. The Senate's small remaining powers disappeared; to be a senator was now only an honour. The old provinces were divided into smaller units called 'dioceses', which were governed by imperial nominees. The army was regrouped and much enlarged; conscription was brought back, and soon there were about a half million men under arms.

Undoubtedly this re-arrangement helped for a time, but it had its own weaknesses. The machinery for assuring a quiet succession of Augusti only worked once, when Diocletian and his colleague abdicated in AD 305. (Diocletian retired to his enormous palace at Split, on the Croatian coast, whose ruins enclose much of the modern town there.) A bigger army meant more taxation, to be paid for out of a smaller population. Yet in the long term a very important step can be seen to have been taken. Though his successors did not stick to the division

A huge seated statue of Constantine, more than 30 feet tall, dominated his basilica in Rome. It represented a new kind of imperial iconography setting the emperor far above other men and foreshadowing the later exaltation of the emperor's image under the Christian empire of Byzantium.

of the empire on Diocletian's plan and there were again to be attempts to rule it as a whole, every future emperor had in practice to accept a large measure of subdivision.

Another part of the reform effort – and it must surely show that people were no longer taking the empire for granted or feeling loyalty to it in quite the same way – was an even greater emphasis on the unique, almost divine authority of the ruler himself, its oriental side in fact. This was to be very important for the future of Christianity, and boded ill for the old Graeco-Roman tradition of religious toleration. The question of Christian sacrifice to the imperial cult was revived. It was in 303 that Diocletian launched the last general persecution of Christianity; it did not long outlast his abdication two years later, though kept up a little longer in Egypt and Asia than in the West.

Paradoxically Christianity was just on the eve of its first great worldly triumph, thanks to the work of the emperor who may well be thought the most important of them all, Constantine. He was hailed as emperor by the army at York in 306 and in 324 he reunited the empire after two decades of civil war. He had soon decided to see if the Christians' god would help him. There is no reason to doubt his religious credulity or sincerity – he seems always to have hankered after a monotheistic creed and for a long time worshipped the sun-god whose cult was associated with that of the emperor. In 312, on the eve of an important battle and as a result of what he believed to be a vision, he had ordered his soldiers to put on their shield a Christian monogram by way of showing respect for the Christians' god. He won the battle. Soon afterwards toleration and imperial favour were re-extended to Chris-

Stilicho, the last general of the western empire, was the son of a Vandal, and twice saved Italy from barbarian invasion with the aid of armies composed of Visigoths, Alans and even Huns. When in the end he turned against the eastern empire, the emperor Honorius, his godson and former pupil, permitted Stilicho's murder. Three months later, the Goths sacked Rome.

tianity. Constantine went on to make gifts to churches (though his coins still for many years bore the symbol of the sun) and began to take part in their internal affairs by giving judgement at the request of the parties in an important ecclesiastical dispute. One can sense from his acts a Constantine moving only gradually towards personal conversion.

From 320 the sun no longer appeared on his coins and his soldiers had to go to church parades. In 321 he made Sunday a public holiday (though he said this was out of respect for the sun-god). He built churches and encouraged converts by giving them rewards and jobs. Finally, though he never formally disavowed the old religions and cults, he declared himself a Christian. Like many other early Christians, Constantine was not baptized until he was on his deathbed, but in 325 he presided over the first ecumenical council of the Church – one attended by bishops from the whole Christian world – at Nicaea. This founded a tradition that emperors enjoyed a special religious authority which was to last until the sixteenth century. Constantine also made another great contribution to the future when he decided to settle his capital at Byzantium, an old Greek colony at the entrance to the Black Sea. It was to be known as Constantinople. He wished to build there a city to rival Rome itself, but one unsullied by pagan religion. It was to remain an imperial capital for a thousand years and a focus of European diplomacy for another five hundred. But it was in making the empire Christian that Constantine shaped the future most deeply. He did not know it, but he was founding Christian Europe. He deserves his title – Constantine 'the Great' – though, as has often been said, because of what he did rather than why he did it, or what he was.

As the eastern empire seemed still in Constantine's day less threatened then the western, the effect of what he did increased the cultural division of east and west. The east was more populous, could feed itself and could raise more taxes and recruits; the west grew poorer, its towns slipping into decline, and depended on importing corn from Africa and the Mediterranean islands and, in the end, on barbarian recruits for defence. Gradually Constantinople came to rival Rome and even outshone it. More important still, Christianity helped to emphasize the separation of two zones: the Latin-speaking west had two great Christian communities within it, one Roman (presided over by the bishop, the pope of Rome) and one African. They increasingly diverged from the Greek-speaking Churches of Asia Minor, Syria and Egypt, all of which were more receptive to oriental influences and more influenced by Hellenistic tradition.

The end of empire in the West

Constantine's sons ruled the empire until 361. Soon after, it was divided again between co-emperors and only once more were east and west ruled by one man, the emperor Theodosius who in 380 finally forbade the worship of the old pagan gods, thus putting the empire's full force behind Christianity and in effect breaking with the old Roman past. But by his day things were already going downhill still faster in the west. By 500 the western empire had vanished.

It was not that society was suddenly engulfed as if by an earthquake. What disappeared was a machine, the Roman state in the west – or, rather, what remained of it. In the fourth century the western empire's administration had been seizing up. More demands were made on its dwindling resources. No new conquests could be made to help pay for defence. As taxes went up, more people left the towns and sought to live self-sufficiently in the country to avoid them. Less money meant a poorer army, and that meant more reliance on barbarian mercenaries – which cost still more money. Concessions to them had to be made just as pressure was building up from a new wave of migrations.

In the last quarter of the fourth century a particularly nasty nomadic people from Asia, the Huns, fell on the Gothic peoples who lived on the Black Sea coast and the lower Danube, beyond the Roman frontier. When the eastern empire bungled the peaceful settlement of refugees within the frontier one of these peoples, the Visigoths, turned on the Romans. In 378 they killed an emperor at the battle of Adrianople and soon cut off Constantinople by land from the west as more and more of them flooded into imperial territory. A few years later and the Visigoths were on the move again, but this time towards Italy. A Vandal general in the imperial service stopped them. From 406 the empire was employing barbarian tribes as 'confederates' (*foederati*) a word which meant barbarians who could not be resisted but who could

Pope Leo The Great turns back Attila the Hun from Italy and the holy city of Rome: a legendary encounter depicted one thousand years later by Raphael. The heroic and anachronistic treatment of the subject (here depicted on a tapestry) was embodied in a fresco in the Vatican, where visitors would at once catch the analogy with a later pope who sought vigorously to defend Italy from those he called `barbarians' – Raphael's patron, Julius II.

be persuaded to help. This was the best the western empire could now do for its defence, and soon it was clearly not enough.

Soon the barbarian peoples were wandering the length and breadth of the Latin West. In 410 Rome was itself sacked by Goths, an event so appalling that it led St Augustine, an African bishop and the greatest of the Fathers of the Church, to write a book which was to become one of the masterpieces of Christian literature, *The City of God*, in order to explain how God could allow such a thing to happen. The Visigoths eventually got as far as Aquitaine in south-west France before coming to terms with the emperor, who persuaded them to help deal with another barbarian people, the Vandals, who had by then overrun Spain. They pushed the Vandals in the end across the Straits of Gibraltar to settle in North Africa, making their capital at Carthage. There they remained, dropping across the Mediterranean in 455 to sack Rome a second time.

Terrible as such a raid was, the loss of Africa was more serious. The western empire's economic base was now shrunk to little more than part of Italy. It was hard to say exactly when the western empire ceased to be. The names and symbols were, like the Cheshire cat's smile, the last things to go. When the Huns were finally turned back from the West at a great battle near Troyes in 451, the 'Roman' army was made up of Visigoths, Franks, Celts and Burgundians – all barbarians commanded by a Visigothic king. When, in 476, another barbarian killed the last western emperor, he was recognized with the title 'patrician' by the eastern emperor. For all the forms, the reality was that the western empire had by then been replaced

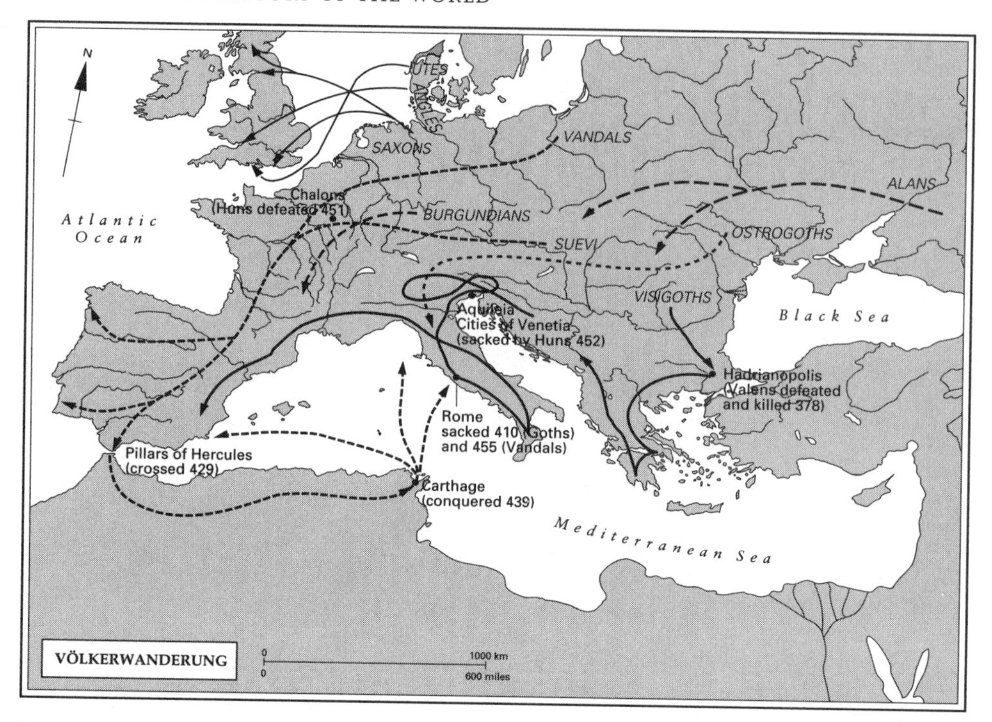

VÖLKERWANDERUNG

by a number of Germanic kingdoms. 476 is usually reckoned as the date at which a line can conveniently be drawn under the story of the empire which began with Augustus. But there are no simple endings in history, and many of the barbarians (some by this time educated by the Romans) saw themselves as the new custodians of a Roman authority which endured. They still looked to the emperor at Constantinople as their ultimate sovereign.

Changes and continuities

By the end of the fifth century many of them had settled down beside the old provincial gentry of Gaul, Spain and Italy, adopting Roman ways; some of them even became Christian. Only in the British Isles did the barbarians almost completely obliterate the old Roman past. In about AD 500, then, we are not at the end of the story of ancient civilization, whatever happened to the empire. Centuries earlier a Roman poet had remarked of one of Rome's conquests that 'Captive Greece took her wild captor captive'. He had meant, correctly enough, that, though the Greek states had gone under, the triumphant Romans had been captivated by Greek civilization. Something a little like that happened in the west as the Roman empire came to an end. Romans, therefore – and through them, Greeks and Jews – did not stop influencing history. There was to be a 'Roman' empire based on Byzantium for almost a thousand years and even in 1800 there was still in Europe something called the 'Holy Roman Empire'. Christian clergymen today still wear costume based on that of the Roman gentleman of the second century AD. Paris, London, Exeter, Cologne, Milan and scores of other towns and cities are all still important centres

Major Dates in the Last Centuries of the Western Empire

212 ad	Caracalla gives citizenship to virtually all free inhabitants of the empire
249	First general persecution of Christians begins
285	Diocletian's ordering of new imperial system: The 'Tetrarchy'
313	Edict of Milan restores Christian property and freedom of worship
330	Constantinople dedicated as capital
376	Goths cross the Danube
406	Vandals and Suevi cross the Rhine
409	Vandals, Alans and Suevi invade Spain
410	Legions withdraw from Britain: sack of Rome by Visigoths
412–414	Visigoths invade Gaul and Spain
420	Jutish and Anglo-Saxon landings in Britain
429–39	Vandals invade North Africa and conquer Carthage
455	Vandal sack of Rome
476	Deposition of last western emperor, Romulus Augustulus

of population, just as they were in Roman times, even if after centuries during which they had been much less prosperous than under the Antonines. Much of the map of Europe has still the shape the Romans had given it by their placing of garrisons and building of roads, and often their settlements reinforced the effect of natural divisions.

Today, though, the non-material continuities are perhaps the most obvious. First and foremost comes language: European languages are packed with words from Greek and Latin, the tongues through which the Bible first came to Europe. Our ways of counting and dividing time too come through the Graeco-Roman world, whatever roots they have believed that, in the Near East. It was Julius Caesar who took up the suggestion of an Alexandrian Greek that the Egyptian year of 365 days, with an extra day every fourth year, would be better than the complex traditional Roman calendar, and it was under Constantine that the Jewish idea of a Sabbath day of rest once in seven became accepted. And, of course, it is to early Christianity that we owe the distinction of BC and AD on which the whole Christian and most of the non-Christian world still works today (it was a little after 500 that a monk first calculated the date of Christ's birth; he was in error by a few years, but his decision has remained the basis of our calendar). Many more examples could be given – from Greek mathematics, Roman law, Christian theology, to take only a few – of ideas from antiquity still with great historical influence today.

As people came to look back in later times, they were much struck by what they owed to the civilization which produced them. But they also took something else from it. Most civilizations have classical ages from which they draw the standards by which they assess their own later achievements. Later Europeans sometimes exaggerated what the Greeks and Romans had done, but they found in it both an inspiration and a touchstone of their own performance. Classical antiquity became a myth of what civilization could and men ought to be. It is why, like their medieval predecessors, modern men too walk among the ruins of this great past and still find them amazing.

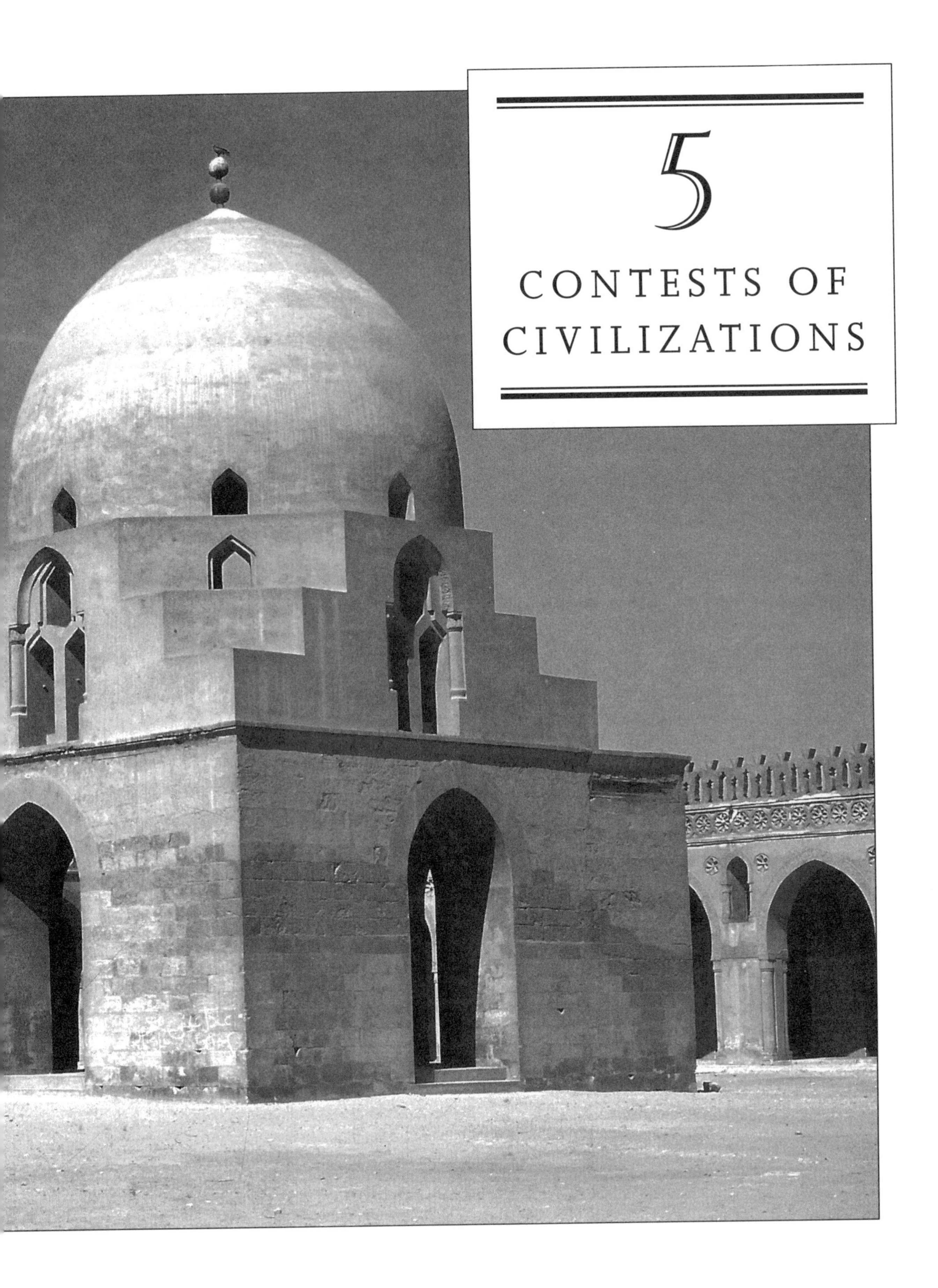

5

CONTESTS OF CIVILIZATIONS

THE BEGINNINGS OF BYZANTIUM

After its foundations were laid by Constantine, a Roman empire lived on after him in Constantinople for eleven hundred years. Its rulers always spoke of themselves as 'Romans', which was also what their enemies usually called them. They presided over half Christendom; the Christian world of the Mediterranean and New East. Looking back, there seem to be many reasons why its western and eastern halves should have gone different ways in the end. For a long time, though, it did not seem unavoidable – or, perhaps, even likely. Many gradual steps were needed before it could be taken for granted. First had come the slow shifting of the concerns of government towards the east in the third century and Constantine's great decisions that the empire should be Christian and that a new capital should be built on the Bosphorus at Byzantium (although he never lived there). The first helped to make the eastern and Greek-speaking provinces of the empire even more important; they had the largest Christian communities. The second speaks for itself. The fifth-century collapse of the western empire was the next obvious step towards separation; after this there could really no longer be any going back. The eastern empire could soon do no more than make the best terms it could with successful barbarians in the west. That at least kept alive the fiction that the empire was still united.

Justinian

Nonetheless, in 527 there came to the throne an emperor, Justinian, who still took that unity seriously and tried once more to govern the empire as a whole. In the end, though, he too helped to perpetuate the split, and most people think that it is with him that the story of a distinct Byzantine empire really begins. Justinian was a thoroughly unpleasant man: deceptive, ungrateful, suspicious and mean. But he was also ambitious, enterprising and brave, and he believed passionately in *Romanitas* and the empire as the guardian of civilization and true religion. He ruthlessly stamped out threats to his authority at home and briefly won back some Roman territory in the west. The Ostrogoths were driven out of Rome itself and Italy liberated for a while. Even the Visigoths of Spain were defeated and imperial government was again restored at Córdoba. The islands of Corsica, Sardinia and Sicily were recaptured. The cost was heavy. When people later looked back and blamed barbarians for devastating Italy, they were really lamenting what Justinian's armies had done, and it had only been done to

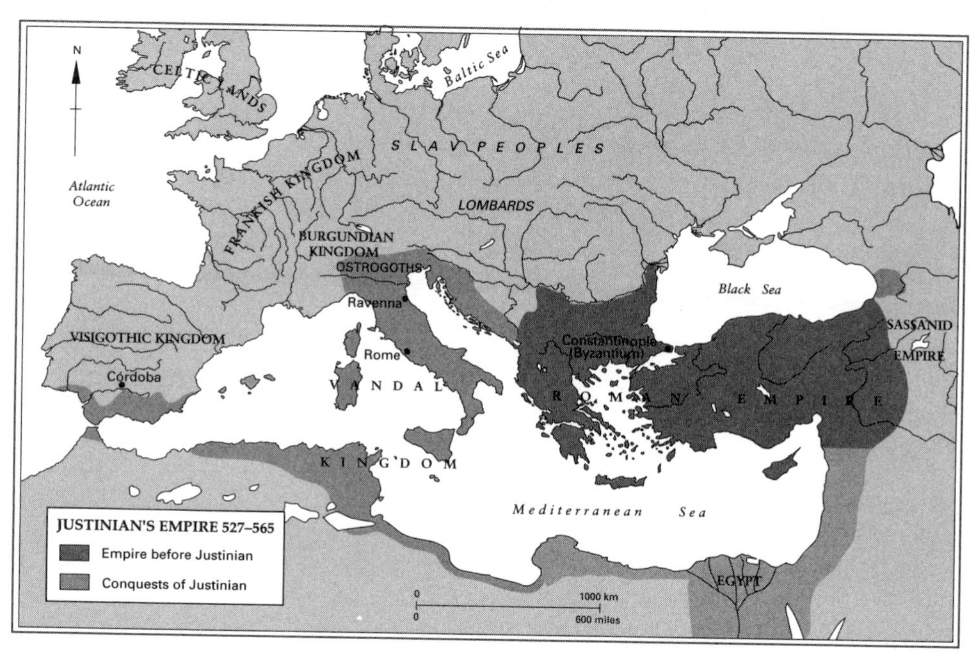

After Justinian's reconquest of Italy, much was done to beautify Ravenna, the last imperial capital there. This mosaic in the church of San Vitale dates from this period and shows the empress approaching the altar with offerings, attended by their suites. Theodora was a tiny woman, but in accordance with Byzantine convention she, like theEmperor, is shown taller than the attendants.

achieve a temporary success. The empire was always fighting on two fronts; expensive campaigns against the Persians soaked up men and funds. By the end of Justinian's reign, barbarians in Thrace again separated west and east Rome. The drift towards separate destinies for east and west had been only briefly interrupted.

Justinian's own contribution to ultimate division was considerable. Though he prided himself on speaking Latin and admired much of the Roman past, he did more than any other emperor to make Byzantium the seat of a distinctive political culture. One aspect of this was his attack on the huge confusion and complication of its law, some of which went back to the early Republic. The resulting consolidation of jurisprudence took only five years, but was to shape Byzantine and European history for centuries. Though apparently a conservative step, this was actually setting a new course. Immediately effective in the east, in the eleventh century the Roman Law of Justinian began to be accepted as the basis for good jurisprudence in western Europe too. It was powerfully biased towards seeing law as something made by rulers rather than (as in Germanic tradition) handed down in custom, and this appealed to many later princes, though not always to their peoples.

Other decisions, too, weakened old continuities. When Italy was reconquered, Justinian chose to make Ravenna the imperial capital instead of Rome. He abolished the Academy of Athens, which had survived from the age of its founder, Plato. He was determined to be a Christian emperor – or, at least, to rule over an empire seen to be Christian – and took away many of the special freedoms enjoyed by Jews, interfering with their calendar and worship and encouraging barbarian kings to persecute them. The old Hellenistic-Roman tradition of religious tolerance was abandoned, and abandoned towards some Christians, too. Justinian wholeheartedly backed up the Orthodox clergy who had at a number of important councils defined and denounced certain doctrines as heresy. There followed the harrying of groups with which heretics were identified – the Egyptian Copts, for example, and the Nestorians of eastern Syria, who were driven to take refuge in Persia. Those who did not flee remained to nurse their grievances and bitterness; in the long run this was to cost the empire dear.

Justinian could not, moreover, bring together the Latin Church of the west (more and more looking to the Pope of Rome for leadership) and the Greek Orthodox church, much as he wanted to do so, and this was an ideological obstacle to any reintegration of the old empire. The western Church would not accept the religious supremacy which he claimed for the emperor, even in matters of doctrine, a much more important point than might appear at first sight. The arguments about theology which Justinian entered upon with such gusto do not now seem very interesting but there was more to them than an imperial hobby. The western Church was always to assert that whatever duty might be owed by men to their earthly rulers, only the Church could tell them what their final duty was, for it was owed to God. Church and state in the west would, therefore, have to live together side by side, sometimes amicably, sometimes quarrelling, sometimes one being practically dominant, sometimes the other. From this tension would grow liberty. The eastern churches, on the other hand, held that both spiritual and earthly power belonged to the emperor himself. He had the last word in everything, for he was God's viceroy on earth. This was a view of government which was eventually to pass into the autocracy of the Russian tsars ('autocrat' was one Greek title for the emperor) and shape the future historical destiny of Russia.

From Justinian's time, the movement towards autocracy was never reversed, whatever concessions and weaknesses appeared in practice. What had once been the office of first magistrate of the Republic was finally orientalized. Byzantine monarchs were treated with the awesome deference given to a Persian king; their public acts were surrounded with elaborate ceremony and demonstrations of adulation. They were given ritual prostration (the kowtow) as a sign of respect. Byzantine art emphasized this. Almost entirely religious in its forms and subject-matter, it depicts emperors as God's agents, embodiments of divine power, and Christ as a kingly and conquering figure, rather than as the suffering, humiliated saviour of so much Catholic art. It was a style fed from Asia, as well as from old Rome. The emperors' images have around their heads the nimbus which the last pre-Christian emperors had borrowed from the sun-god, but some Sassanid rulers were shown with it, too.

The oriental cultural pressure on Byzantium is easy to understand. Many of the provinces of the empire lay in Asia, and it became more and more dependent upon them after 600 because it had little territory left in Europe. Officials had to speak Greek, but the empire was multi-racial and ethnic origin no more stood in the way of promotion – emperors came from Syria, Anatolia and the Balkans and Justinian himself was an Illyrian of Gothic descent – than it had done in the old days. The cities of Asia Minor might be Greek-speaking, but the countryside certainly was not. As time went by, the names of families from Anatolia loomed larger in Byzantine politics and administration – another source of Asian influence. When we add to this the inevitable coming and going across the frontiers which divided Byzantium from its Asian neighbours, it is not surprising that as the centuries go by it seems to resemble less and less either western Christendom or its own Hellenic heritage.

Religion and the state

The heart of Byzantium was its special Christian role and style. The empire had become a part of the machinery for the salvation of mankind and this was reflected in all it did. It had a huge propaganda and public relations dimension, too, and it is often impossible to see whether worldly or otherworldly considerations had priority. Justinian used Christianity and churchmen as a branch of diplomacy, standing god-parent to the baptized children of barbarian princes, and sending missionaries to convert the others. The riches and wealth of Constantinople – and of other parts of the empire, too, during his reign – helped to impress the neighbours. Its greatest physical monument is still the basilica he built, the Church of the Holy Wisdom, St Sophia. For centuries it was the greatest building of Christendom (though its huge dome collapsed once during Justinian's reign), in which the splendours of the imperial entourage at worship were paraded among silk and gold hangings and the blaze of mosaics and marbles of the Church.

Although the eastern empire lasted so long and underwent many changes, people who lived under it often pretended that nothing had changed at all. Its emperors went on calling

Hagia Sophia, the church of the Holy Wisdom, better known as Saint Sophia, was for centuries the greatest of Christendom's temples. It was Justinian's architectural masterpiece. The dome is an extraordinary feat of engineering (though it collapsed on one occasion) and its three aisles a huge theatre for the religious ritual of the court and the hierarchy of the Church. It was not a place of worship for the commonalty; as one scholar has put it, 'the audience was God'. Its original appearance, bedecked with hangings of silk and gold and blazing with the colour of its mosaics and marbles is hard now to recapture, for what the modern visitor sees is a building first 'purged' by iconoclasts, then redecorated and embellished again over the centuries until the Ottoman conquest which turned it into a mosque, adding minarets to the outside and Koranic inscriptions, pulpit, chandeliers and railings to the interior. Nowadays a museum, it became also in its last phase a model for Moslem architects and the inspiration of one oftheir finest works, the great Blue mosque about three-quarters of a mile away.

themselves *Augusti* right down to the end. Its religious essence did not change: it remained Christian and Christian in a special way, within the tradition called Orthodox. From this tradition there derive not only the churches of modern Greece and Cyprus, but also those of Russia, Bulgaria and some other Slav lands. 'Orthodoxy' – as it is sometimes called for short was in very many ways different from the Catholic Christianity which came to be dominant in western Europe. No Orthodox churchman had an authority like that of the Roman pope, for example; the patriarch of Constantinople, the acknowledged leader of the eastern Church after the seventh century, was in effect the emperor's appointment and, in return, gave the Church's blessing at an imperial coronation. The ordinary parish clergy were often married, whereas clergy in the western Church came to be celibate. This meant that the priesthood was not a society apart in Orthodox countries in quite the same way as it became in western Europe. Monks in the Greek Church were celibate, though, and here, too, lay a distinction:

Christian monasticism (which had begun in Egypt in the third century, when holy men had first retired to the desert to pray, meditate and discipline themselves against earthly temptation) remained closer to its original pattern under Orthodoxy than in its western forms, which increasingly threw up new ways of giving monks practical and communal as well as spiritual and individual roles.

The Greek Orthodox tradition also had to grapple with theological dispute and debate to a greater degree than the Roman Church in the early centuries. This in part reflects the presence of different religious traditions within the old Hellenistic world. Constantinople, Jerusalem, Antioch and Alexandria, the four great 'patriarchates' (or senior bishoprics), all stood for something a little different. Local interests and cultural traditions were liable to find expression in theological dispute, and this was one of the underlying forces at work for two or three centuries during which schism and separation abounded. During the same time, Antioch, Alexandria and Jerusalem all fell out of the orbit of imperial control (they were seized by Arab armies) and this gave more importance than ever among the eastern churches to the Greek Orthodox tradition of the Constantinople patriarchate.

Early Byzantine art, of the fifth or sixth century: St Simeon Stylites sitting on his column. A visitor brings his food (or perhaps a censer) and the bird may be the Holy Spirit appearing in the form of a dove, or a real bird again, providing the saint with sustenance.

Some of these disputes were subject to decision or clarification by general (ecumenical) councils of the whole Church but throughout these division between Rome and Constantinople persisted and even widened. The last general council recognized by both the Latin Catholic and Greek Orthodox churches to be valid was held in 787. By then the contrast between western and eastern Christianity was very visible, even in the strictest and most literal sense. An Orthodox church still looks very different from a western one, whether Roman or Anglican. One of the most obvious differences is the place given to images of the saints, the Blessed Virgin and Jesus Christ in Orthodox churches. These 'icons', as they are called, are displayed on a special screen and in shrines for veneration. They are not mere decoration, but are there to help focus devotion and teaching, as, it has been put, they are 'a point of meeting between heaven and earth'. Not surprisingly, this led to an emphasis on the painting of icons (or the making of them in mosaic) in Byzantine (and, later, Slav) art which produced the greatest masterpieces of those traditions. The popularity of icons was already established in the sixth century, though they could still give rise to bitter controversy. Until the imperial authority was put firmly behind the icons in the ninth century, 'iconoclasm', or attacks on icons, had been another influence making for divergence between Latin Catholic and Greek Orthodox Christianity. There was not to be a formal breach until 1054, but practically, Christendom had been already prised apart into a state of schism centuries before that. This was bound to accentuate cultural and political division between East and West as time went by.

Iconoclasts at work. *Whitewash is applied to an image of Christ in this ninth-century illustration from a manuscript.*

THE RE-MAKING OF THE NEAR EAST

In AD 500 and for centuries thereafter, civilizations and cultures were in conflict and contact at the edges of the western and eastern empires, and particularly in the Near East, as nowhere else in the world. Struggles began then which were to last centuries. A great English historian summed up one phase of them as 'the World's Debate', rightly recognizing the magnitude of the issues at stake. Yet for all the revolutionary impact of new forces in the region, its inhabitants shared much with one another. With relatively brief interruptions, great empires based in Iran had hammered away at the West for a thousand years before 500, but wars can sometimes bring civilizations closer, and in the Near East two cultural traditions had so influenced one another that their histories, though distinct, are inseparable. Through Alexander and his successors, the Achaemenids had passed to Rome the ideas and style of a divine kingship whose roots lay in ancient Mesopotamia; from Rome they went on to flower in Byzantium. Persia and Rome fascinated and, in the end, helped to destroy one another. By the time that happened 'Rome' itself had changed. It had come to mean the empire ruled from Constantinople which was, to all intents and purposes, one confined to Egypt, Palestine, Syria, Anatolia, Greece, and much of the Balkans up to the Danube, the great power representing Christianity in the Near East.

Persia and Byzantium

Like Byzantium, Persia the other Near Eastern great power in 500, was also the heir to traditions running back to the Achaemenids and even beyond, to ancient Mesopotamia. Her empire was about to be reconstituted for the last time, by conquests stretching from Armenia to the Yemen. Her rulers perhaps came to feel even more antagonistic to the Roman empire when it became Christian. Although Christians were supposed to enjoy toleration in Persia, the danger that they might prove disloyal in the continual wars with Rome made the fifth-century peace treaty which said this a dead letter. A group of Christians called Nestorians was the exception; they were tolerated by the Persians because they were persecuted by the Byzantines and therefore were thought by the Sassanids likely to be politically reliable. Religion tended thus to divide the region and weaken its earlier cosmopolitanism.

Around the two great centres of civilization clustered smaller states and satellites less advanced in their culture; beyond them lay straightforwardly barbarian lands. Byzantium's neighbours to the east included a few Christian outposts, such as the kingdom of Armenia. Far away beyond them was India, another centre of the civilization, but inaccessible beyond the mountains of Afghanistan and the plains of Indus. In Arabia there lived petty, semi-civilized Arab kingdoms and desert tribes, beyond the Oxus the nomadic peoples of Central Asia; to Byzantium's north were a newly-appeared group of tribes, the Slavs, settled along the lower Danube, and to its west lay the Germanic peoples.

The ever-renewed wars of Byzantium and Persia in the sixth century are for the most part a dull, ding-dong story. Yet they can be seen as the last round of the struggle of East and West begun by the Greeks and Persians a thousand years earlier, whose climax came at the beginning of the seventh century in the last world war of antiquity. The devastation caused may well have been the fatal blow to the Hellenistic urban civilization of the Near East. Chosroes II, the last great Sassanid, then ruled Persia. His opportunity seemed to have come when a weakened Byzantium – Italy was already gone and peoples from the Volga region, Slavs and Avars, were pouring into the Balkans – lost a good emperor, murdered by mutineers. Persian armies invaded Armenia, Cappadocia and Syria, ravaging their cities, and sacking Jerusalem in 615, bearing away the relic of the True Cross which was its most famous treasure. The next year Persian armies went on to invade Egypt; a year later still, their advance-guards were only a mile from Constantinople. They even put to sea, raided Cyprus and seized Rhodes from the empire. The empire of Darius seemed to be restored just as, at the other end of the Mediterranean, Byzantium was losing its last possessions in Spain. At this moment, the blackest in the long Roman struggle with Persia, a new soldier emperor came to the rescue. Heraclius, Armenian by descent, had come to the throne a few years earlier and now revealed his quality. He used sea-power to save Constantinople in 626, when the Persian army could not be transported to support an attack on the city by their Avar allies. In the following year Heraclius broke into Assyria and Mesopotamia, the old disputed heartland of Near Eastern strategy. The Persian army mutinied, Chosroes was murdered and his successor made peace. The great days of Sassanid power were over. The relic of the True Cross – or what was said to be such – was restored to Jerusalem. The long duel of Persia and Rome was at an end and the focus of world history was to shift at last to another conflict.

A NEW WORLD RELIGION: ISLAM

The Sassanids went under in the end because they had too many enemies. The year 610 had brought a bad omen: for the first time an Arab force defeated a Persian army. But for centuries Persian kings had been little preoccupied with enemies to their south. It was from that direction, though, that the fatal blows were to come, from the scorching, rocky, desert peninsula of Arabia. It had not always been like that. In the early Christian era inhabitants of irrigated lands there had sustained little kingdoms, trading through their seaports with India, the Persian Gulf and East Africa, carrying gums and spices up to Egypt, from which they would be passed into the Mediterranean. They were independent of the great empires, Roman and Persian, which never deeply penetrated the peninsula, and they prospered. But the irrigation system collapsed – we do not know why – the land became desert and barren, migrant tribes moved north from the south Arabian cities and the peoples of the peninsula reverted to nomadic, pastoral ways of a more primitive kind. From them emerged the armies of a new world religion.

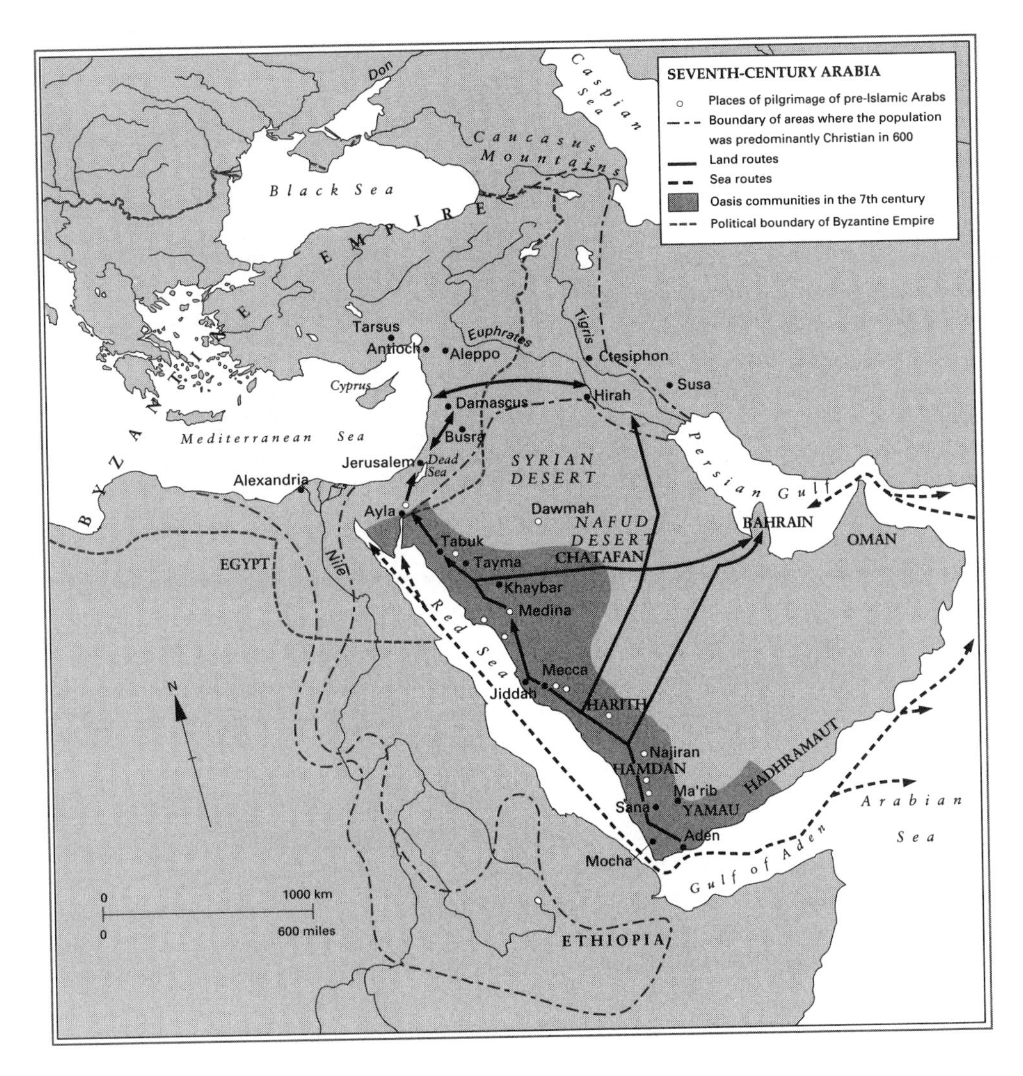

Islam – the religion founded by the Prophet Muhammad – is Christianity's only rival as a world religion in the vigour and range of its geographical spread. It springs ultimately from the same roots as Christianity, the tribal cultures of the Semitic peoples of the Near East. It is akin both to Christianity and its source, Judaism, in asserting that there is only one God. Nor do the three faiths share only monotheism; Moslems say that they worship the same God who is worshipped by Jews and Christians, though differently.

Muhammad the Prophet

Islam's story begins in a place so far hardly touched upon in this history. Muhammad was born in Arabia at Mecca, of poor parents who belonged to a minor clan of an important Bedouin tribe, in about 570. He was soon orphaned. How he grew up, we do not know, but it was in a place of some special note. Mecca was an oasis and a centre of pilgrimage. Arabs came to it from far away in order to venerate a black meteoric stone, the Ka'aba, a focus of pagan religion. Apart from a few who were Jewish or Christian, most Arabs were then polytheists, believing in nature gods, demons and spirits. But oases like Mecca, to which caravans came, and the little ports which were still in touch with the outside world also attracted a few outsiders and foreigners; some of them had brought knowledge of higher religions to Arabia by Muhammad's day, and Arabs already revered the god known to them as Allah, and worshipped by Christians and Jews.

A manuscript copy of the Koran, dedicated to a Sultan of Bokhara.

Muhammad seems to have been struck by signs that something was going wrong among his people. Commerce, population growth and foreign influence had begun to undermine their traditional and tribal arrangements. The old pastoral Arab societies had been organized around blood-ties; nobility and age were what made people respected in them, not money. Wealth did not always go along with noble blood and long years. Here was a social and moral problem. Muhammad began to reflect upon the ways of God to man. One day, as he contemplated in a cave outside Mecca, he heard a voice telling him to set down his vision of the word of God. For the next twenty-two years he spoke prophetically. What his followers wrote down as he did so was not put together until after his death but it became one of the great books of world history, the Koran. Like Jews and Christians, who treasured their own scriptures, Moslems were to be a people of a Book.

A new religion

First and foremost, the Koran laid out the fundamentals of what we now term 'Islam', a set of principles uniting a brotherhood of believers now world-wide. The word 'Islam' means submission, or surrender, and Muhammad saw himself as the mouthpiece through which God made known his will to men. Moslems were to believe that this had happened before; the great prophets of Israel and Jesus, too, had been true prophets, Muhammad taught, but he came to feel sure that he was himself the final Prophet, through whom God spoke his last message to mankind. That message set out a belief and a code of behaviour to meet the needs of Muhammad's own people, but it was to prove very acceptable to others. Its essence was the assertion that no God was to be worshipped but Allah; Islam was uncompromisingly monotheistic (one objection Moslems had to Christianity was that they believed it to be polytheistic, because it gave as much importance to Jesus and the Holy Spirit as to God the Father). Islam also prescribed a series of necessary religious observances of which the most important were regular prayer and the avoidance of pollution. This was all that was necessary for salvation.

As a creed, Islam was not only simple but revolutionary. it taught that those who clung to the old gods of Arab society would go to hell – a doctrine not likely to win it popularity among non-Moslem Arabs. The emphasis of the supreme importance of a brotherhood among believers was subversive, too, for it cut across tribal loyalties. Even some of his own kinsmen turned on Muhammad and in 622 he left Mecca with about two hundred followers and went north to another oasis about 250 miles away; it was to be renamed Medina – 'the city of the Prophet'. There he organized a new community and began to issue regulations on practical, daily concerns such as food, drink, marriage, war. On this, Islam was to be built as a distinctive civilization. This migration – the Hegira, as it is called – was, therefore, the turning-point in the early story of Islam, and has ever since been treated as the beginning of the Moslem calendar still in use all round the world. It was a break, even if not one whose potential was at once obvious, with traditional Bedouin society. Muhammad was founding a new sort of community.

He died in 632. Authority to interpret his teaching was inherited by a 'caliph'. The first of those who held this title were all related to the Prophet by blood or marriage and under them, the tribes of southern Arabia were conquered. Soon fighting spread to the north, to the Arabs of Syria and southern Mesopotamia. Soon, too, the 'patriarchal' caliphs of Muhammad's family were being opposed as exploiters; the caliphate, based on religious and doctrinal authority, seemed in a few years to be degenerating into a secular office. In 661 the last patriarchal caliph was deposed and killed. The office then passed to another family, the Ummayad, which held it for nearly a century. By the time they ceded it to another usurper, Islam had remade the map of the world.

A religion of conquest

Like Christianity before it, Islam in its first years looked pitifully weak in worldly terms. Any sensible man at the time of the deaths of Jesus or Muhammad would have thought the prospects of either religion bleak. Except to the faithful, neither could then have seemed likely to survive, let alone become a great force in world history. Yet both did, though in very different ways. Islam was from the start a religion of conquest. Even in Muhammad's day its military struggles had begun. He used Medina as a military base to subdue those who had opposed him at Mecca and the tribes nearby; those who submitted were welcomed into the umma, the brotherhood of believers which overrode tribal divisions. Yet Islam respected tribalism and the old patriarchal structure providing they did not interfere with its own rules, and confirmed the old dignity of Mecca as a place of pilgrimage. It was only those who resisted the new faith who were driven out of Medina – among them, the Arab Jews.

From this start, further military expansion followed soon after the Prophet's death when, in 633, Arab armies – 'Saracens', as the Byzantines called them – attacked the Sassanid and Byzantine empires – startlingly, at the same time. It took them five years to drive the 'Romans' from Syria and take Jerusalem, which remains to this day one of the holy places of Islam. Soon after this, the Persians lost Mesopotamia to them and Egypt was taken from Byzantium. An Arab fleet was formed; raids on Cyprus began (the island was later to be divided with the Byzantines) and by 700 Carthage had been occupied, the Berbers converted and turned into Arab allies, and the whole coast of North Africa was in Moslem hands. The Sassanid empire had long since collapsed, its last king driven out of his capital in 637 after a succession of defeats by Arab armies. He died near Merv in 651 after fruitlessly appealing for

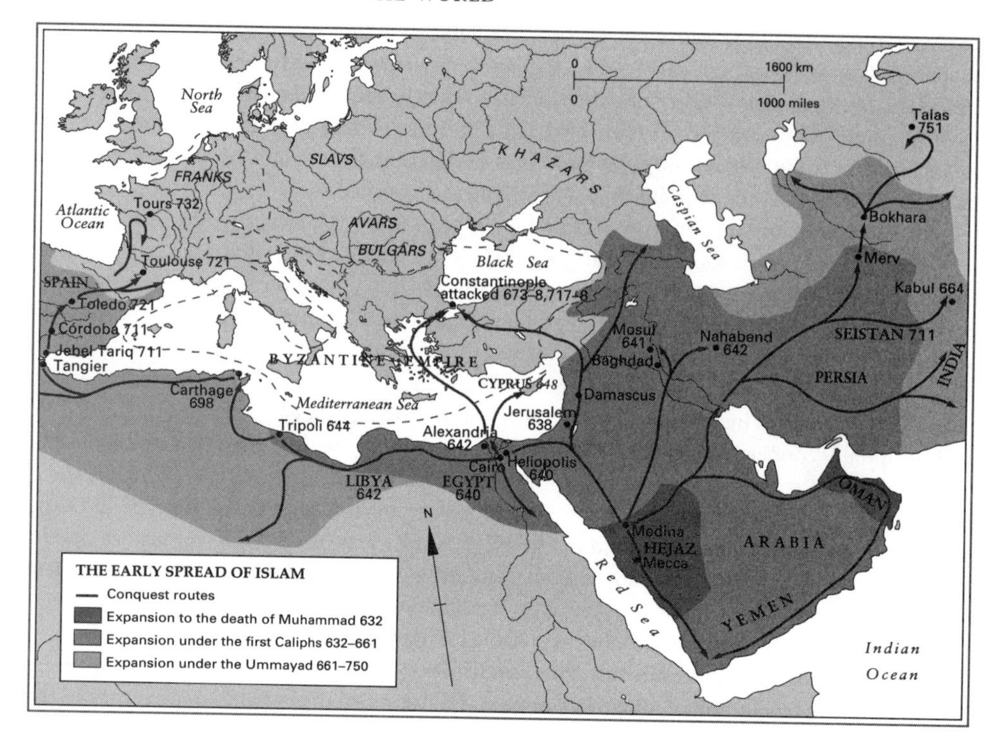

help to the emperor of China. Arab expeditions continued, pressing as far east as Kabul, capital of Afghanistan, and at the beginning of the eighth century crossed the Hindu Kush to invade India, and establish themselves for a while in Sind while ravaging Gujarat. At about the same time others crossed the Straits of Gibraltar and pushed into Spain, shattering the old Visigothic kingdom. In 717 they besieged Constantinople for the second time, though unsuccessfully, and had by then even penetrated the Caucasus.

This was nearly the high-watermark of Arab conquest. One army is reported to have reached China in the early eighth century, but whether or not it did, the Arabs won a last great victory in Asia over the Chinese high in the Pamirs, in 751, before being defeated by a people called the Khazars and settling down with a frontier across the Caucasus and along the Oxus. A few years before this, in 732, exactly a century after Muhammad's death, another Arab army had been turned back near Poitiers; this was to be the furthest penetration they ever made into western Europe, though they still had enough drive to raid deeply in the next few years. The tide had turned at last.

One reason for their astonishing record of success was that the Arabs' first opponents, Byzantium and Sassanid Persia, had been obliged to spend so much time fighting one another or dealing with their huge commitments on other fronts. The Byzantines had the Avars and Berbers to deal with, the Sassanids the central Asian invaders of whom the worst were the Huns. Within the Byzantine empire, too, there were disaffected peoples like those of Egypt, irritated by bad government and religious harassment from Constantinople and ready to welcome new masters. Arab success built on success, moreover: the overthrow of Persia meant that by the middle of the seventh century there was no great power left to hold off the Arabs anywhere west of China, except Byzantium. Nor had the Arabs much to lose; their soldiers were spurred on by the poverty from which they came and the belief that death on the battlefield against the infidel would be followed by entry to paradise. They believed that they were fighting for God. A great ideal inspired them – like some Christians and later revolutionaries – and made them brave, self-sacrificing and ruthless. The upshot was a record of conquest which made it look for a time as if Islam might rule the world.

THE ARAB ISLAMIC WORLD

Islam has never been a political unity. Its most important early political foci were the Ummayad and Abassid caliphates. The first came to an end in 750 when the last Ummayad caliph was overthrown by a usurper who prudently rounded off his triumph by a massacre of all the males of the defeated family. Thus began the Abassid caliphate, which soon came to resemble an ordinary dynastic monarchy. As a real political force, it lasted until the middle of the tenth century (though formally there was still an Abassid caliph in the thirteenth century) and under it Arab civilization reached its greatest glory. When the Arab ascendancy finally gave way, though, those who replaced it had already become Moslems, and under them Islamic civilization was to go on to new heights in Persia, India and elsewhere.

In early Abassid times Islam was still much marked by its Arabian origins. The most obvious symptom was the use of the Arabic language. Because the Koran was written in it, Arabic spread everywhere in the Islamic world. Indeed, 'Arabic' is a better description of the Abassid caliphate than 'Arab', for though it was Arabic-speaking, there were few other ways in which it was much like the raw seventh-century culture which had spread from the desert by conquest. In those early days Arab invaders had tried to keep apart from the peoples they conquered. They left local customs undisturbed – Greek and Persian went on being used until the eighth century as languages of government in Damascus and Ctesiphon, the former Sassanid capital, – and lived apart as a military caste in separate towns, supported by taxes levied on the neighbouring areas and neither trading nor owning land. This separation gradually gave way. Conversion to Islam became more frequent and garrison towns like Basra or Kufa gradually became true cities, engaged in trade. Once again, the Near East began to experience cosmopolitan empire, in which many different traditions could find a place in a culture upheld by an imperial regime. Some scholars have seen Abassid culture as the last flowering of Hellenism.

Caravan of pilgrims on the road to Mecca.

The Abassid caliphate

Particularly in the eastern (formerly Persian) provinces, non-Arab Moslems disliked Arab overlords; their discontent had helped to bring the Abassids to power, and the caliphate moved from the Ummayad capital of Damascus to Baghdad on the Tigris. Until then a little Christian village, Baghdad became a huge city, rivalling Constantinople, with perhaps half a million inhabitants, full of craftsmen and luxuries far removed from the simple lives of the first Arab soldiers of Islam. Islamic, Christian, Hellenistic, Jewish, Zoroastrian and even Hindu ideas mingled there, amid traders from many lands. Baghdad's prosperity was at its height under the legendary caliph Haroun-al-Raschid, supposedly the ruler to whom Scheherazade told her tales of *One Thousand and One Nights*.

For all this, though, and in spite of the fact that it came to look more like an oriental monarchy of a familiar kind, and less like a purely spiritual office, the caliphate was unswervingly Islamic. The new religion permeated the Arab empires, their offshoots and, later, their supplanters. Mosques in which the faithful met to pray towards Mecca each day were built all over the Islamic world; in all of them the same prayers were said and the same law was recited. The Islamic creed's resemblances to Christianity led some Byzantine clerics to treat Islam as a Christian heresy, rather than as paganism, but it was fundamentally and in every-day practice very different. It forbade wine and (like Judaism) pork to the followers of the Prophet, and commanded the performance of pilgrimage to Mecca (all Moslems are recommended still to make the pilgrimage once in their lives). It also imposed the duty of giving alms to the poor and allowed the keeping of slaves (but said they must be kindly treated). Above all, it enjoined the practice of regular prayer. But perhaps its indulgence of polygamy was to non-Moslems its most striking feature.

Facing page: Islamic mechanics built upon Hellenistic texts which transmitted the skills of such men as Archimedes and Hero of Alexandria. This elaborate machine, depicted in a fourteenth-century miniature (probably from Syria), is a clock. It indicated the passage of time by an elaborate series of movements by the little figures making it up and finally by the depositing of a little ball inside the elephant to mark each half-hour.

Women in Islam

Women did not attend prayers in Moslem mosques. According to Islamic tradition the Prophet himself had said that it was better that they should pray at home. It is still in its regulation of woman's role in society that Islam is most strikingly different from Jewry or Christianity. Surprising though it may seem to some, Jewish and Christian women have throughout their history always in principle enjoyed more freedom than Moslem women. The most obvious symbol is the veil still worn by many women in some Islamic lands and still likely to be imposed, at least by informal pressure, in lands where fundamentalist Moslems are in power. The Koran has, in fact, more to say on the subject of women than on any other social group. Partly because the main purpose of marriage was thought to be the raising of children, Islamic law allowed a man four wives and an unlimited number of concubines who did not share the legal status of the wives. The wives were all to be treated equally and each was given a dowry by the husband at the time of marriage which remained her own. This until very recent times was more favourable to women than was the law in many Christian countries, but, on the other hand, a Moslem wife could at any time be divorced by her husband at will, though she might not herself ask for a divorce on any grounds at all. Few Moslems may wish to use the full freedom which Islamic law gives them, but the Koran is quite clear about the Prophet's view of women's rights: 'Men have authority over women because God has made the one superior to the other', it says. 'Good women are obedient... As for those from whom you fear disobedience, admonish them, banish them to beds apart, and beat them'.

The Koran has always had a central authority in Islamic law, though it does not set out the whole of it, nor regulate every detail of Islam's complicated social arrangements. Like the reported opinions of the Prophet, it often only outlines the basis of good Moslem custom. Law-books soon began to appear in which teachers who claimed to have the authority of the Prophet's intention behind them collected judgements on specific problems. But the Ummayad conquests, which brought many different peoples and traditions under Islamic rule, demanded flexibility. The practice of the caliphs and their provincial governors led to the gradual softening and expansion of rules originally laid down for a simple, desert-dwelling, tribal society. The Abassids condemned the Ummayad caliphs for this, and under them Islamic law became much more rigid. There are some countries even today – Libya, Saudi Arabia, Pakistan for instance – where the law is still closely based on the Koran and in some of them its full rigour is still applied to criminals: the hands of thieves may be cut off, for example, and those guilty of adultery stoned to death.

Islamic learning and the arts

The Koran moulded Islamic culture in much else, too. The roots of Islam's emphasis on words and the art of words, both spoken and written, may well go back to the importance of the story-teller to tribal society, but the Koran made recitation and reading even more impor-

tant – indeed, it virtually created Arabic as a literary language. Educated Moslems often wrote verses and a high level of study was achieved in the Abassid caliphate; it was biased heavily towards theology (everything was looked at from that point of view) but it produced learned men. Literacy under the Abassid caliphate was probably more widespread than in any other contemporary civilization. Much translation into Arabic took place and, because of it, the texts of Hellenistic and Classical Greek were eventually to become much more widely known in Europe.

Translation helps to explain a great age of Islamic achievements in science and mathematics. Building on the work of the Greeks and drawing inspiration from Persia and India (the 'arabic' numerals we now use were in origin Indian, though set out by an Arab arithmetician), Arabic culture excelled in astronomy, medicine and mathematics. Textbooks (usually by Persians) of Arabic medicine were later used for centuries in Europe as standard teaching manuals. European languages themselves bear the mark of Arabic civilization. Such words as 'zero', 'cipher', or 'algebra' are all of Arabic origin, so, in music, are 'zither', 'guitar', and 'lute', while many terms of commerce – 'tariff', the French '*douane*' for a customs-post, and 'magazine' in the sense of a store – are all Arabic importations.

Not just because of Arab dress, a European (or for that matter a Chinese) traveller would also have found the appearance and style of the caliphates' cities and culture very different from what he knew at home. For a long time, Islamic painting emphasized calligraphy and intricate design rather than representations of things, because Islamic teaching forbade the making of likenesses of human forms or faces (only in the later Persian and Indian Islamic empires did the miniatures appear which are now so appealing to western eyes). In building, too, once the Roman invention of the dome had been adopted, Islam very quickly arrived at a distinctive and attractive style. It may have been inspired by a wish to build in a way which would set off the Arab conquerors from the people of the conquered territories. The first Islamic dome covers the Dome of the Rock, built at Jerusalem in 691, a shrine sacred to Jew and Moslem alike, for the first believe that on this spot Abraham was ready to sacrifice his son Isaac to God, while Moslems believe that from it the Prophet was miraculously taken up into heaven. Other Islamic domes were soon to follow and now the mosque and its accompanying minaret from which the Faithful are called to prayer are familiar sights in cities all over the world, and not merely in the Islamic lands.

ISLAM'S FURTHER BOUNDARIES

Islam's heartland has always been Arabia and the old Near East of the caliphates, the lands between Egypt in the west, and the Hindu Kush and Transcaspia in the east and north, but it quickly spread far beyond them. Conquest followed by the winning of converts soon carried the new religion along the African coast and into Spain, while other Arabs occupied parts of Sicily and some of the other Mediterranean islands. The conquests beyond Africa were not, though, to remain forever as Islamic lands.

In Spain, one branch of the Ummayad family never accepted the overthrow of their caliphate and set up an 'emirate', an independent kingdom, there in 756 at Córdoba. They

The great mosque of Córdoba remains one of the most remarkable buildings of theArab world but is almost all that is left of the splendour of a once-great Arab city which had no rival west of Constantinople. It was built as a forest of short columns, probably because the architect had many short Roman pillars at his disposal, but none tall enough for a high ceiling. Later the Spanish built a church in the middle of it. In spite of this and though it is now difficult to imagine it in its glory as the spiritual centre of western Islam,the mosque still remains a building of surpassing beauty and mystery.

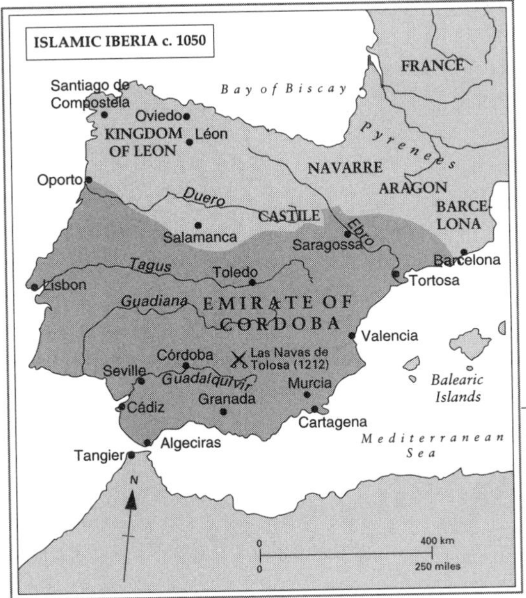

were to rule for centuries over much of southern Spain, where Islamic society was in some ways to reach even greater heights of cultivation and civilization than it did further east. In the tenth century, Arab Spain became the seat of a rival caliphate, too. But not long after this it was on the defensive and Christian kingdoms pressed in on it; Córdoba fell to them in 1236. Though two centuries more were needed to re-conquer the whole peninsula for Christianity, the brilliant Islamic civilization of El-Andalus finally went under in the fifteenth century.

Yet another new caliphate was set up in the tenth century in Egypt – a sign that the central Islamic lands were by then drifting apart politically; though they were never to be lost to Islam, they

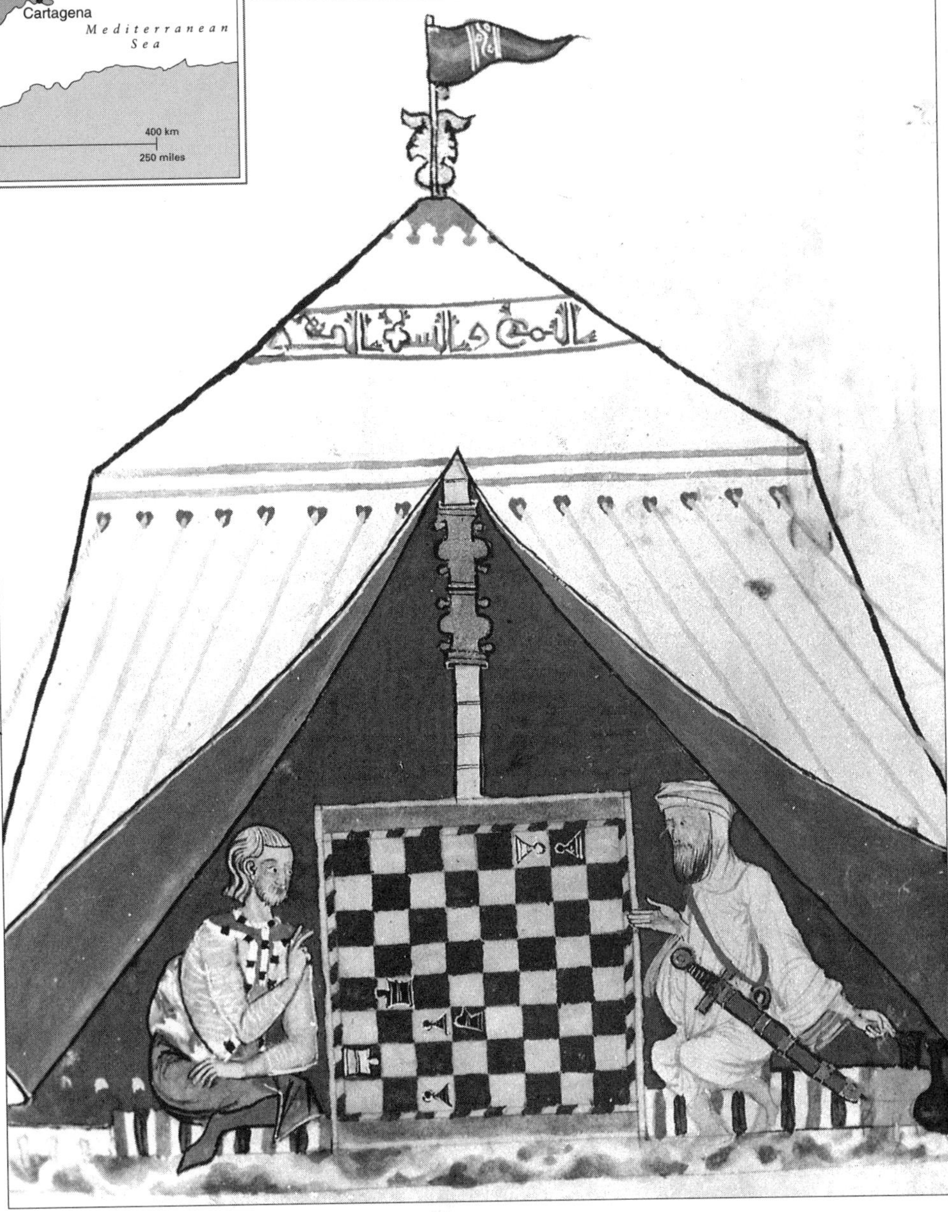

Alfonso the Wise, King of Castile, added Murcia to his kingdom and took Cadiz from theArabs, but is remembered as a great king less for that than for his patronage of scholars and artists, and for the splendour and eclecticism of Spanish culture during his reign. Something of its spirit is expressed by this illumination from a book on chess by the king: it shows an Arab nobleman and a Christian knight, his guest, peacefully at play.Christian culture in Spain was for centuries refreshed by currents from the Islamic world; the myth of the Reconquest was to obscure much of this from later Spaniards.

ISLAM BEYOND THE ARAB WORLD UNTIL 1800

- previously under a century or more of Moslem rule but no longer so by 1250
- Moslem rule but no longer so by 1300
- under Moslem rule c. 1250–1300
- coming under Moslem rule after 1300
- acquired and lost by Moslem rulers after 1300
- ★ substantial Moslem communities (c.1250) in areas never under Moslem rule
- ◗ areas continuing under Moslem rule until the present century with a majority or large minority of non-Moslems c.1250–1300

0 3200 km
0 2000 miles

have never since been united. Meanwhile, Islam penetrated Central Asia, India, the Sudan and the Niger basin between the eighth and twelfth centuries. Arab traders carried it to east Africa from the coast, and to west Africa by the trans-Sahara caravans. The dhows of the Arab merchants bore it also across the Bay of Bengal to Malaysia and Indonesia. Conquerors imposed Islam on much of India and by converting the Mongols, Islamic missionaries ensured that their faith would eventually reach China. All in all, it is an amazing success-story. While it unfolded, though, Islam outgrew by far any possibility of maintaining even such political unity as it had under the Arab caliphates. It evolved, too, in many ways beyond what the Prophet had taught. In short, it became the heart of a complex, diverse civilization, a standing challenge to other traditions and capable of rich interaction with them.

BYZANTIUM'S GREAT AGE: THE MAKING OF SLAV EUROPE

Historically, Byzantium was to be a shield for western Christendom. She held off first Persia and then for a long time the far more dangerous threat of Islam, as well as the continuing incoming tide of barbarian peoples further north. The eastern empire was a bulwark to the old Graeco-Roman world of the Mediterranean, even when pagan occupation of North Africa and Spain had turned its flank. Only after a thousand years did it at last give way. Yet for decades already in the seventh century the tide appeared to flow in favour of Islam. All the eastern empire's reconquests from the Persians were soon lost. For five years (673-678) Constantinople was itself besieged by Arab armies. Meanwhile, in Europe the tribes called Slavs pressed into Thrace and Macedonia and others, the Bulgars, crossed the Danube.

In 717 an Anatolian emperor, Leo II, came to the throne – a sign of the orientalized nature of the empire at the top. He was an outstanding ruler who regained much of the former imperial territory. He and his son stabilized the frontiers with the caliphate; Anatolia was recovered and securely held although there were still skirmishes and raids. An important alliance had been made with a nomadic people who had established themselves in south Russia, the Khazars; these served as a barrier to Islam's advance on that flank. True, in the west, nothing could be done; Ravenna went at last, and only a few toeholds were left in Italy and Sicily. But in the Balkans Leo and his family re-established the outline of Byzantine control at the periphery. Later, the Bulgar threat was to be contained. Altogether, the period down to the eleventh century was one of success and recovery. Cyprus, Greece and Antioch were all retaken; the struggle to win back North Syria continued. Meanwhile at home, there were dynastic troubles and religious controversy. But they did not prevent Byzantium from playing its great power role, and before the next great international crisis struck she had carried out another world historical task, the creation of Christian Slavdom. The world's history would have been very different had the Slav peoples never become Christian. That they did, was largely the work of Byzantium, which, in the process, gave them civilization, too. This double achievement may well have been Byzantium's most enduring legacy to the future.

The Slav peoples

Argument continues over where the Slavs came from in the first place; many of the Slav lands – notably in Russia – form part of the flat expanses where Europe and Asia meet; across them nomadic peoples had wandered for thousands of years and through them great folk-movements had pushed time and time again. Somehow, the Slavs successfully hung onto their lands in what are now Poland and Russia, though harried by Scythians, Avars, Goths and Huns. Between the fifth and seventh centuries, they were well-established both east and west of the Carpathians and were beginning to move south, no doubt under pressures similar to those stimulating other folk-wanderings. By the seventh century they were shielded somewhat on the eastern side, too, by other peoples who had settled there.

Gradually, Slav tribes spread over the Balkans. The first 'Slav' kingdom to emerge, though, was that of the Bulgars – who were not, in fact, really Slavs, but by origin (probably) Huns, some of whom were gradually 'Slavicized' through contact with the Slavs and intermarriage with them. When Byzantium recognized its existence in 716, though, it can be regarded as Slav. Its people had by then adopted many Slav ways. The Bulgars were to prove very troublesome. Not only did they occupy lands which had originally belonged to the empire, but they raided as far south as Constantinople itself, and the effort needed to guard against them always hampered Byzantium's efforts to recover its position to the west. One emperor is remembered as the 'Bulgar-slayer', but early in the ninth century the Bulgars managed to kill another in battle – the first to fall to barbarian arms for nearly five hundred years – and from his skull they made a cup for their king.

Byzantium sent to the Bulgars emissaries who mingled diplomacy with evangelism. The most important were two great ninth-century churchmen, the monks St. Cyril and St. Methodius. Cyril's name is still commemorated in the alphabet he devised (on the basis of Greek), for the Slav languages – Cyrillic; simplified versions of it are still used in Serbia, Bulgaria and Russia. Then a Bulgarian king, a tsar – or caesar, as the Byzantines called him – accepted Christian baptism. This was not the end of quarrels, though a very important step. One tenth-century Bulgarian king styled himself 'emperor of the Romans' and attacked Constantinople (unsuccessfully), and it was only in 1014 that the Bulgars were at last beaten so disastrously that they came to heel (the emperor who thrashed them blinded 15,000 of his Bulgar prisoners and sent them home by way of encouragement to their fellow countrymen). The Bulgar ruler of the day died (some say, of shock) and soon Bulgaria was officially a Byzantine province.

Kiev Rus

Meanwhile, Christianity had spread further north and east. On the upper stretches of many of the Russian rivers lived Slav communities ruled over by a fighting and trading aristocracy.

The other side of the long Byzantine struggle with the Bulgarians: the emperor Basil II, 'slayer of Bulgars'. At his feet grovel Bulgarian princes: Michael, standard-bearer of the heavenly host, steadies his spear and Gabriel places a crown upon his head. Above them all, the figure of Christ is seen, extending a heavenly crown to his champion. The psalter from which the painting is taken was made in Constantinople in the early eleventh century.

These were Norsemen – Scandinavians of the stocks in western Europe called 'Vikings', or 'Normans', but in Russia 'Varangians', who had conquered the Slavs after coming down the rivers from the north. They were tough, resourceful and brutal. They raided Constantinople, fought the steppe peoples and traded as far afield as Baghdad. One Varangian prince called Rurik settled at what is now Novgorod, on the Ilmen, in about 860, according to Russian tradition, and so long as there was a nobility in Russia, all its princes claimed descent from him.

The main centre of Varangian power and commerce shifted in the next few decades to Kiev, on the Dnieper. Kiev Rus, as it was called, was of great importance in Byzantine diplomacy. At worst, its warrior princes had to be held off; at best they might prove useful allies. But diplomacy was a slow business, and the Varangians too raided Constantinople. Eventually, though, in about 986, a Russian prince accepted Christianity for himself and his people – a momentous decision, almost as fateful for mankind as Constantine's, for it has shaped

Russia's history ever since. Two hundred years later, his Russian countrymen recognized this and he was canonized as St. Vladimir. We do not know how much his decision was one of conviction and how much one of calculation; the uncertainty is rather like that which surrounds Constantine's motives for conversion. But when Vladimir married a Byzantine princess the story of the greatest of the Slav nations had begun.

Kiev Rus may well have been much richer than most west European cities of the tenth century. It now became a Christian centre, as Vladimir and his successors set about imposing the new creed by force, helped by Bulgarian priests, overseen by the patriarch of Constantinople, who appointed the Metropolitan of Kiev. Russia adopted the Cyrillic alphabet and the liturgy of the Orthodox Church. Early in the eleventh century, Kiev was at the height of its splendour and influence. Iaroslav the Wise, its greatest ruler, then negotiated with rulers far away in western Europe as well as with Byzantium, and Kiev was thought by one western visitor to rival Constantinople. Educational foundations were promoted and the first code of Russian law promulgated. From this reign, too, comes the Primary Chronicle, one of the first works of Russian literature and a defence of the work of the princes of Kiev in unifying Russia under Christianity.

An emerging eastern Europe

A little after this great age, internal and external troubles were to make things more difficult for Kiev, and by then a Christian Poland had appeared, too. She was linked to western Christianity and not to Orthodoxy, though, because her ruler had chosen to accept the authority of the pope in Rome. Poland had to struggle with pressures from the German princes to the west as well as with the rivalry of Orthodoxy in the east and the Polish people have always had a particularly difficult historical destiny (often heroically and tragically fulfilled) as a nation Slav by descent and language, but western European and Roman Catholic in culture and religion, ground between aggressive and fearful powers on both sides.

Other Slav kingdoms emerged in Bohemia and Moravia. By the beginning of the twelfth century, there was a Christian Slav Europe. Yet it was not a unity, except in racial origin, a link which did not then count for much, It was divided between the two branches of the Christian Church – Roman and Orthodox as well as by the difficult terrain of the Balkans which split up the people who lived there into small nations. There was even a non-Slav people, the Magyars, who had come south of the Carpathians to settle in the Danube valley in the middle of the Slavs. Worse still, the Slavs had enemies east and west. Their relations with Byzantine might have settled down, but they were under pressure in the west from a drive eastwards by German settlers, headed by a religious order of soldiers – the Teutonic Knights who regarded war with the Slavs almost as a crusade. To the east, things were worse still. In 1240 a terrible thing happened, when Kiev itself was taken and sacked by a wild Asian people, the Mongols.

INNER ASIAN PEOPLES

The Mongols came from inner Asia, which might be defined not by geographical facts, but somewhat negatively, as the habitable region of Asia usually not effectively subject to any major state – such as China or Persia – nor much influenced by any of the major centres of civilization. The region's extent has in fact fluctuated, and a different definition might be

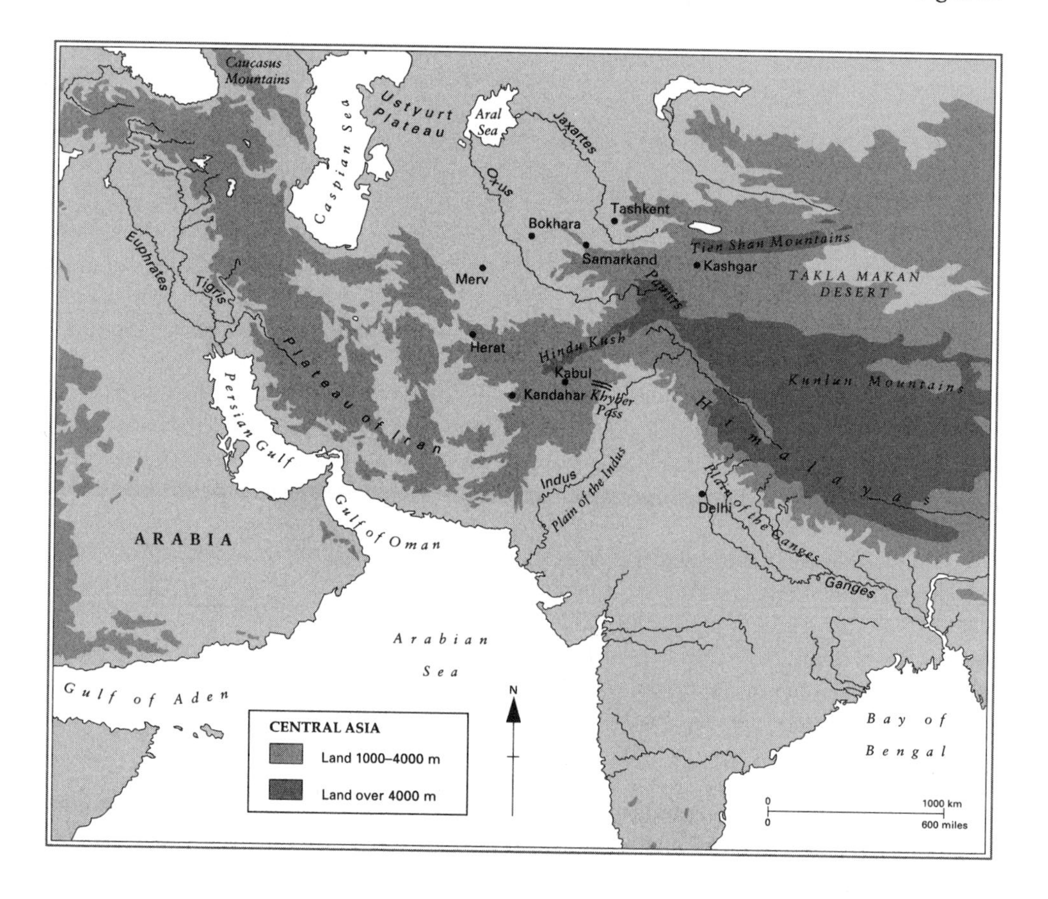

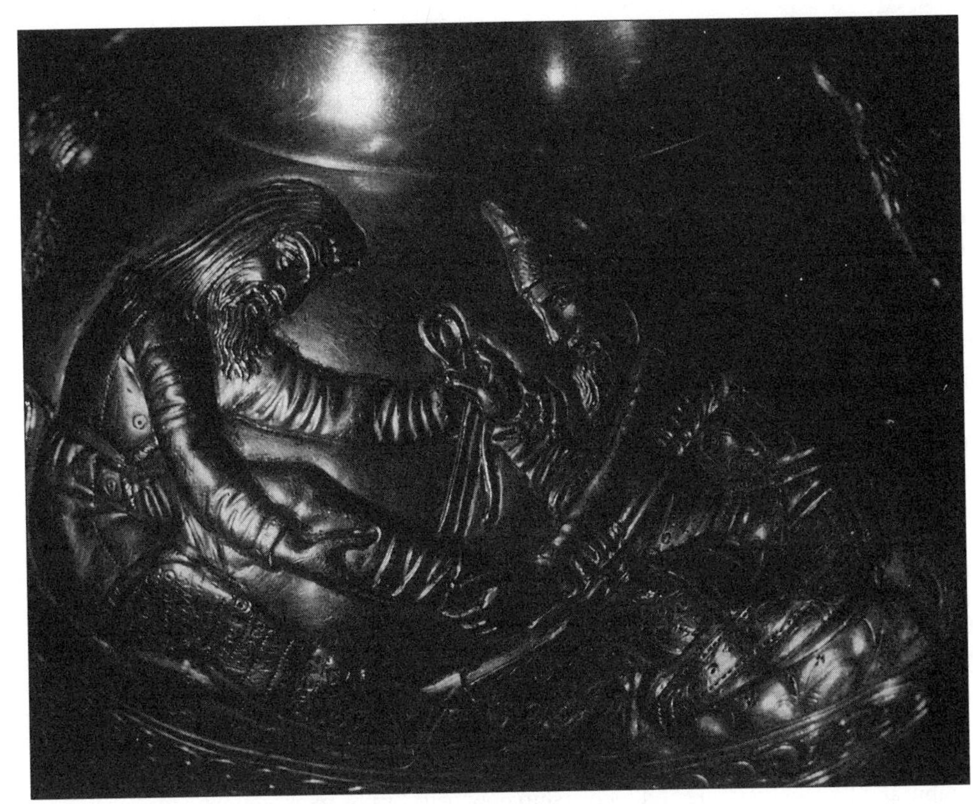

The Scythians were first mentioned in historical narrative by Herodotus and in the twentieth century were acclaimed by some Russian enthusiasts as the original stock on which their national tradition was founded. Much, though, remains to be discovered about them. Archaeology has already made it clear that contact with the Greek communities of the Black Sea zone stimulated them to a high level of artistry in precious metals. Much of our knowledge of this gifted people comes from objects laid in their tombs, such as the vessels on which these reliefs appear.

made in terms of the people who live there: Central Asia is that part of Asia which is suitable for nomads. For the most part it is a huge, high corridor between the Siberian forests to the north and the deserts, mountain ranges and plateaux running from the Caspian across the north of Iran and Afghanistan to the Hindu Kush and Tibet (which form its southern wall). Most of it is steppe, often arid, and very cold in winter. This helps to explain the long imperviousness of its peoples to outside influences until missionary religions evangelized them; then, Buddhists, Moslems and Christians all had some success among them. Such wealth as the region accumulated tended to crystallize at oases which sheltered populations always likely to arouse the antagonism and envy of the nomads. Some of the most famous – Samarkand, Merv, Bokhara – were stopping-places on the caravan routes from China to the west, the 'Silk Road'.

The nomadic steppe peoples learnt a specialized way of life which suited this environment. They were long to remain illiterate, but always had many skills which made them formidable opponents. They were often levers of world history. Because of climatic and political changes in the east, where even a small disturbance could mean life or death to a people dependent on pasture and set them on the move, they clashed from time to time with those who lived to the south and west of them. When this happened on a large scale, it made history. Among the steppe peoples, the Scythians, who had fascinated the Greeks and harried the Persians, were at the end of the line early in the Christian era, pushed from behind by other peoples who had behind them the Hsing-Nu, the forbears of the Huns. They in turn, had been set in motion by interference from the Chinese emperors. So, a huge historical shunting movement took place. By the fourth century AD the Huns themselves were as far west as Transcaspia, in the next they got furthest ever from home: western Europe seemed likely to fall to them, until a battle at Troyes turned them back in 451. Many other peoples took part in similar movements. Avars, Khazars, Pechenegs, Cumans all made their own impacts on world history at different times, though their differences are too complicated and their story is too long and confused to summarize briefly here. But one of them requires special attention: a clan of iron-workers who had been slaves of a Mongolian people called the Juan-Juan ('Avars' in Europe), the Turks.

The Turks

We first hear of the Turks in about 500. In the seventh century they accepted the nominal overlordship of the Chinese emperors. They were a loose dynastic connexion of tribes scattered from Mongolia to the Oxus river, with a head called a 'khan'. This web of influence from China to Persia has been called a Turkish 'empire', but it is difficult to be sure just how real it was. Still, some sort of a political relationship between its nomadic peoples spanned Asia for about a century, a considerable achievement for barbarians. At one time or another, rulers in China, Persia, Byzantium and India all found they had to take serious notice of the Turkish khans and do business with them; the Byzantines encouraged them to harass the north-eastern frontier of Persia, and the Sassanid empire in its last years seems to have allowed some of them to settle within its frontiers in return for help against Byzantium and the Arabs. The Turks benefited from such contacts and learnt the art of writing.

The Arab empires firmly sealed the Turks off again until the Abassid caliphate broke up into a jumble of successor states in the tenth century and the Turk peoples were able to get on the move again. By then, one clan among them, the Seljuks, had been converted to Islam. They pushed first into the Persian highlands and then on into Anatolia. Some of the Seljuks had already served as mercenaries in the Arab armies; together with their Moslem religion this brought them somewhat more under the sway of Arabic civilization. Like many barbarian peoples living on the edges of great centres of civilization, the Turks did not seek to destroy the elaborate life they had learnt to admire, but to share its benefits. Major works of Arabic and Persian literature and scholarship now began to be translated into Turkish.

Under the Seljuks, a true Turkish state at last came into being in Iran and Anatolia (where the Turks called their new province the Sultanate of Rum, since they saw it as part of the inheritance of Rome). This began the slow conversion of the once Christian Anatolians to Islam and helped to touch off a series of expeditions from western Europe which we call the Crusades, with the aim of driving back Islam. The Seljuk Turk empire did not survive long after 1200, even in Anatolia.

Meanwhile, another Turkish dynasty had taken over Egypt. The once-great Arab empires had by now long been thoroughly in decay, and for a time Christian recovery seemed assured. The western Crusaders set up new Christian kingdoms in the Levant, the Arabs had already lost Sicily, and, further west, the Christian reconquest of Spain had begun. It was thus on an already disorganized and weakened Islamic world that a new onslaught from the east fell in the thirteenth century.

Chinghis Khan

The people called Mongols brought Islam as near to disaster as it ever came. They emerged from the lands north of China now called after them, Mongolia. In the 1190s a young and somewhat embittered Mongol who had become khan of his people began to seek revenge for slights and injuries to them by the Tatars (another nomadic nation). Within a few years he was accepted by all the Mongol tribes as their 'universal khan', or, in their own tongue, 'Chinghis Khan'. Mispronounced in Arabic, this name became 'Genghis Khan' to Europeans. He proved to be the greatest conqueror the world has ever seen, terrorizing Europe and Asia alike. But Chinghis also built up something much more like a true empire than any other nomad chief, even though its only capital was the felt tents of his encampment.

Chinghis started by taking on a great power. He seized northern China in 1215, and then turned west. The Mongols were pagan, neither Moslem nor Christian, so Chinghis was impartial in his choice of victims. The first were the northern Islamic lands of Iran and the

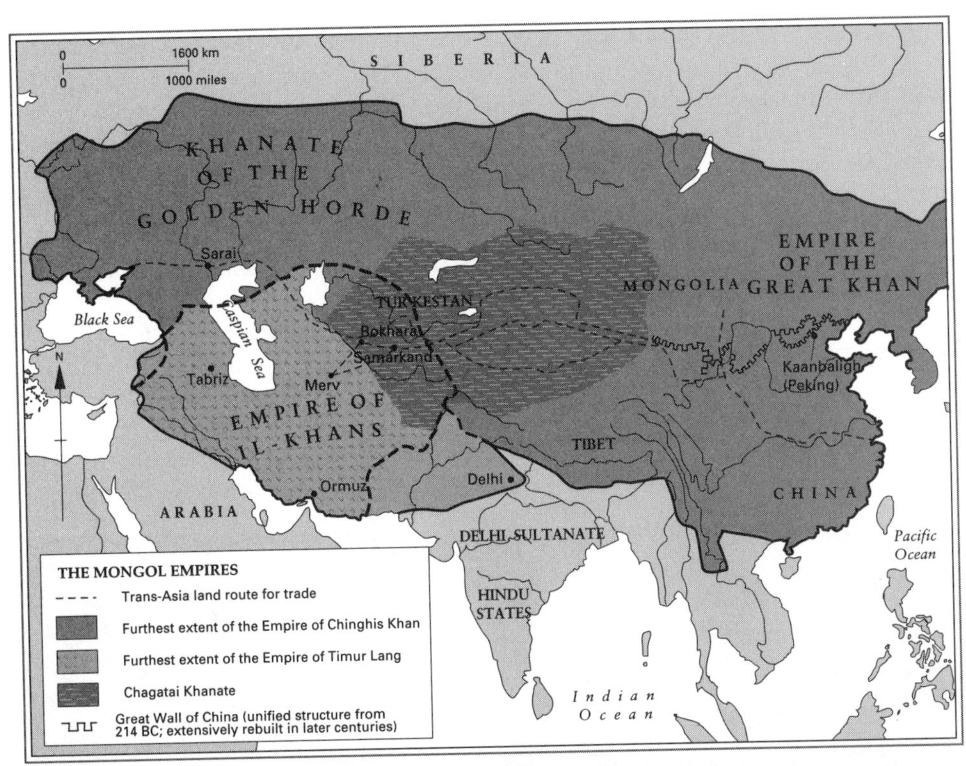

A Mongol warrior, plaiting the tail of his mount (from a fifteenth-century Persian painting).

rich oasis towns of Transoxiana. Resistance by the inhabitants was always followed by massacre. This encouraged later victims to surrender, and if they did they were usually spared. Soon after, Chinghis invaded southern Russia and levied tribute on the Christian princes there. Then he died, but his son and heir soon returned to the west and in 1236 the Mongols sacked Kiev. From there they raided deep into Europe, devastating Poland and Moravia, and pursuing the unhappy king of Hungary down through Croatia as far as Albania before they gave up. They only went home when the khan died and a new one had to be chosen. At his election among those present were an emissary from the pope, a Seljuk Sultan, a Russian Grand Prince and an envoy from the Abassid caliph.

Once Mongol internal affairs had settled down, the Mongol armies again fell on Islam. After defeating the Seljuks of Rum they turned on what was left of the Abassid caliphate, storming and sacking Baghdad. Superstitions about shedding a caliph's blood were met by rolling up the unfortunate fellow – the last Abassid – in a carpet and having him trampled to death by horses. Then a Mongol offensive was launched against Syria. The fate of Islam seemed to hang in the balance. But the Mongols were at last defeated by the Egyptian

At Samarkand stands the Gur-i-Mir, the burial place of Timur Lang. Commissioned by him as the tomb of a favourite nephew, it became his own and a part of a complex of buildings which grew into a family mausoleum. Craftsmen were brought from Isfahan to work upon this masterpiece of Timurid art, covered with coloured tiles and surmounted by a dome of unusal magnificence.

Mamelukes near Nazareth in 1260. Islam was saved, though a long Mongol ascendancy over much of Asia and Russia still lay ahead.

The unity of Chinghis Khan's empire had given way to a loose connexion of khanates ruled independently by Mongol princes, but with much in common. A sort of Mongol federation stretched at its greatest extent over something like a sixth of the old world's land surface. Its communications were good and well policed and the Mongols made an intelligent use of their conquered subjects. They enlisted them in their armies, while Chinghis used Chinese civil servants to run his taxation system and borrowed the Turkish script in order to write down the Mongol language. This was particularly important in shaping Mongol ideas. The story of Mongol China itself can be discussed elsewhere, but the Great Khans came to see themselves as somewhat like Chinese emperors. They expected other peoples to pay them tribute, not to negotiate with them as equals, and believed they exercised a universal monarchy on behalf of their own sky god. Yet they were tolerant in religion and the diversity of belief at the Mongol court impressed Christians. There was once talk of a khan being baptized, though nothing came of it.

At the end of the thirteenth century came an important change; the khan who ruled Persia became a Moslem. The danger to Islam had passed. Since then Persian rulers have always been Moslem. Peace and even some prosperity slowly returned to the old Islamic lands. There was only one more resurgence of the Mongol terror, under a conqueror as frightening as Chinghis himself, Timur Lang, by blood a mixed Turk and Mongol, who overran Persia in 1379. It had been beset by a succession of disputes and civil strife (Kublai Khan, 'Great Khan', had died at the end of the preceding century, and no overlord of all the Mongols was then appointed). Timur ravaged India (it was said that his advance on Delhi – which he sacked – was as swift as it was because of his army's wish to get clear of the smell of the piles of dead bodies left in its wake), fought other khans, ravaged Mameluke and Turkish lands and conquered Mesopotamia. But he was no empire-builder and his conquests hardly outlasted him. If he had an important impact on history it was negative: for a few decades his onslaughts stopped a Turkish people in Anatolia, the Ottomans, from going into the kill and extinguishing the Byzantine empire.

THE END OF BYZANTIUM

Byzantium's most fundamental weakness was always the need to fight against formidable opponents on more than one front at a time. After struggling against the Arabs and surviving, only for the Arabs to be replaced by fresh Moslem conquerors the Turks foremost among them – she had to grapple with Avars, Bulgars and Varangians to the north and north-east. At her back was the west – Christian, it is true, but increasingly separated from eastern Orthodoxy, and at times looking much like yet another enemy. Byzantium had not taken kindly to the crowning of a new claimant to the title of emperor in the west (he is remembered as Charlemagne) in 800. For this title to be taken, with the pope's connivance, by the king of a Germanic people called Franks seemed a challenge to the true inheritor of Rome. Byzantine officials tended afterwards to refer to all westerners as 'Franks' and this was what Europeans were to be called in the Near East and over much of Asia, as late as the twentieth century; people no more distinguished between Italians, Spaniards or Germans, than did Europeans make careful distinctions between different kinds of Turks. Things got worse when Franks and Greeks tried to co-operate against the Arabs. There were the long-enduring Byzantine claims in Italy, too, and the Normans' arrival in Sicily and their attempt to seize parts of Greece which belonged to the empire actually brought outright conflict. Byzantium grew weaker and her resources shrank from the eleventh century onwards, and as the old Byzantine ruling elites tended to be replaced by others drawn from Anatolian or Armenian aristocratic families, the cultural gap between two civilizations grew wider still.

A devastating blow was dealt the empire in 1072 when the Turks shattered the Byzantine army at the battle of Manzikert. After this, Asia Minor was to all intents and purposes lost to the empire, and with it went its most important recruiting ground and richest province. A little was recovered after the Mongol attacks on the Turks, but the territorial base of the empire was from this time very narrow. Two fresh difficulties had also loomed up. One was posed by a former Byzantine satellite, the Italian republic of Venice. In return for help in fending off the Normans from Greece, Venetians were allowed to trade freely throughout the empire as subjects, not foreigners. The cuckoo was admitted to the nest. In time, they were to secure a virtual monopoly of Europe's trade with Byzantium. Venetian naval strength soon overtook that of Byzantium, especially after a defeat of the Egyptian fleet by the Venetians removed the only important counterweight. Territorial gains followed commercial concessions (the Venetians even fought a war against Byzantium). By the thirteenth century a Venetian empire existed to which scores more islands and ports were to be added in the next three centuries.

The Crusades

The other threat also came from the west, in the movement called the Crusades. Through them the defence of Christendom in the Near East thus came to involve western Europeans whose interests were by no means the same as those of the Orthodox. In the twelfth century crusaders set up in the Levant four Latin states. Though they did not last, they were in formerly Byzantine territory. A dreadful revelation of what Frankish intervention in the area might mean came in 1204 when Constantinople was sacked by a Christian crusading army. It may have been a mortal blow. The empire survived, it is true, but Orthodox churchmen never forgave the Latin Christians. It was 1261 before the Byzantines recovered their own

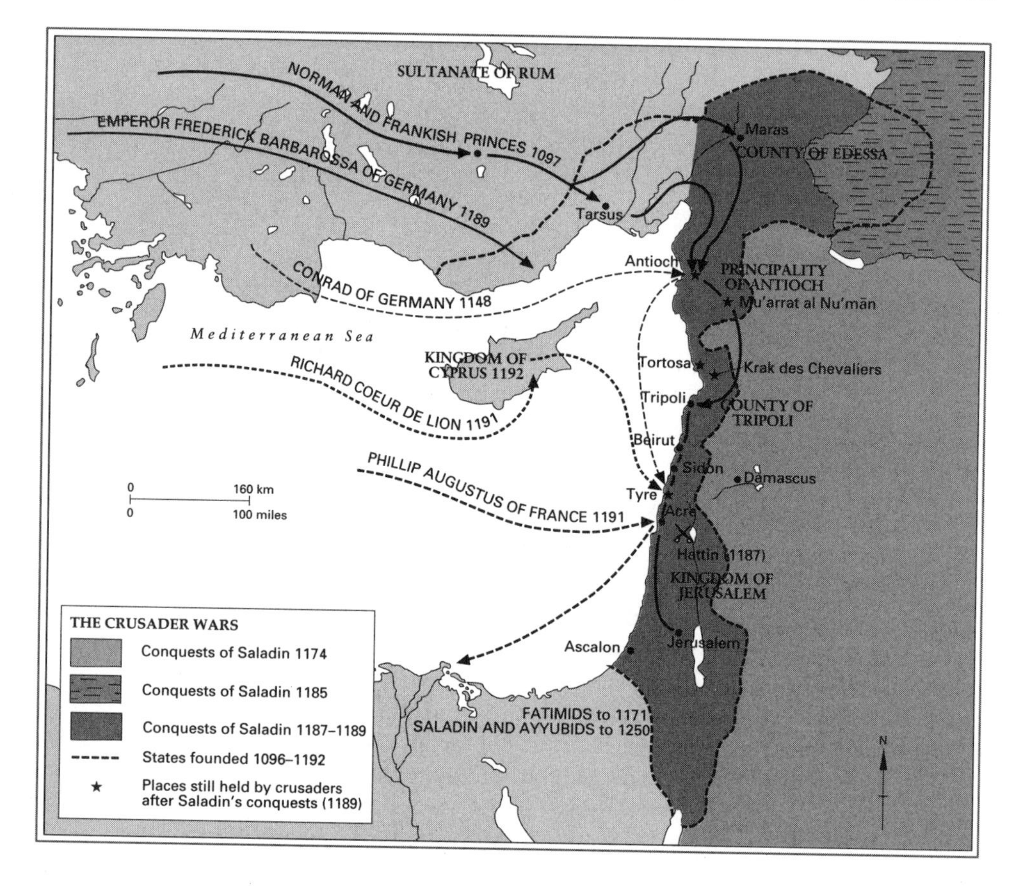

Mehmet II, the Ottoman conqueror of Constantinople, painted by the Venetian artist Gentile Bellini in1480.

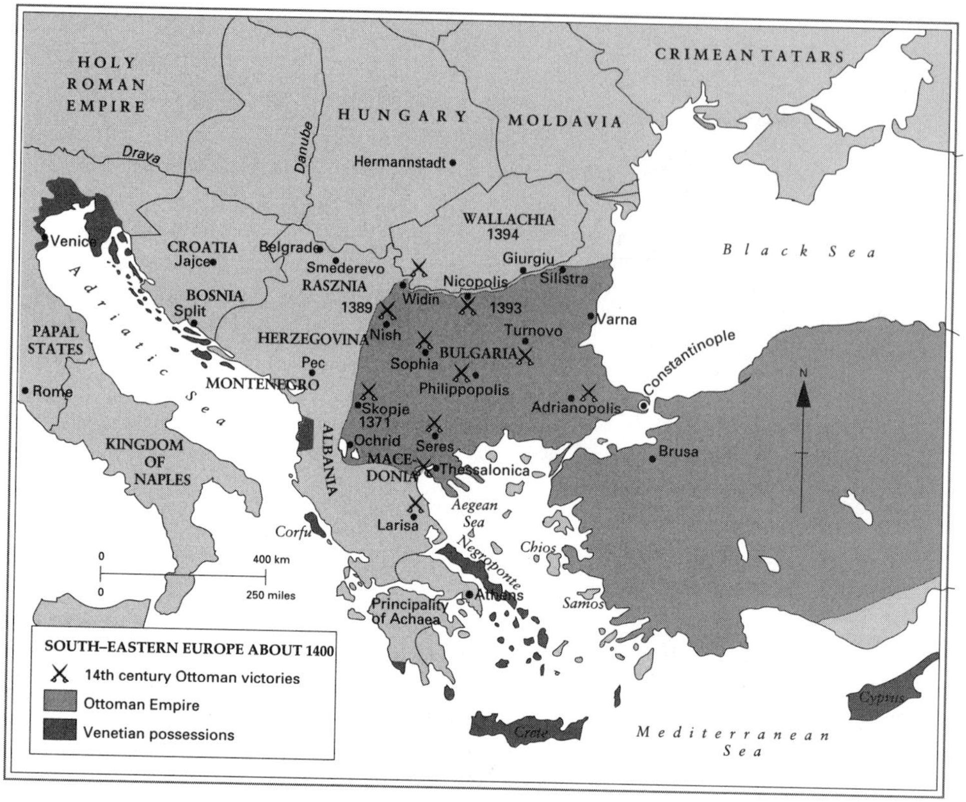

capital. By then the empire was little more than a fragment, a Balkan state with a fringe of Asian territory and a once-great capital. The Bulgarians were never again to be mastered, while the Venetians (and the Genoese, too) took over the whole Aegean island complex.

To deal with a threat from a Serb prince the fourteenth-century emperors had to call in the help of Moslems, the Osmanli Turks. This people (later known as Ottomans) had earlier helped the emperors to win back their capital from a 'Latin empire' installed there by the Crusaders. But they were to be Byzantium's final executioners. In spite of attempts to make up quarrels with the west, the Byzantines eventually had to face Islam once more alone. They resisted, sometimes by diplomacy (one Ottoman ruler married a daughter of an emperor) but in the end had to do so in arms.

By 1400 the Turks had eaten up much of the Balkans, conquering Serbia and Bulgaria, and thus once more establishing Islam deep in Europe. They had defeated another crusade and overrun Greece afterwards. They had already once besieged Constantinople. Only defeat by the Mongol Timur held them back for a little longer from going in for the kill. It seemed a second wave of Islamic conquest was sweeping the east, just as Islam was being expelled from Christendom in the west. When the Turks' advance was resumed, though they also pillaged the Venetian empire, their main aim was to take Constantinople itself. Early in April 1453, the final Turkish attack began. After nearly two months of siege, on 29 May, the emperor Constantine XI, eightieth in succession since his great namesake, went to Hagia Sophia, took communion, and then went out to die defending his capital on its last day as a Christian city. Soon it was all over and the Ottoman Sultan, Mehmet II, entered the city, went straight to the cathedral and there set up his triumphal throne.

'There has never been and never will be a more dreadful happening', wrote a terrified Greek scribe. No one was ready for it; western Christendom was aghast at the news. Not only had the Crescent triumphed over the Cross, but a thread which went back two thousand years in ancient Greece was snapped. Much more than a state, the Roman tradition itself was at an end, as well as a thousand years of Christian empire. It was also the most dramatic moment in a great historical change – the almost complete elimination of Christian states in the Near East, Balkans and the eastern Mediterranean and their replacement by a new world power, the Ottoman empire.

CHRISTENDOM'S NEW NEIGHBOUR

The Ottoman Turks were now the leaders of Islam and it is worth following their story just a little further. Perhaps because they had long been a border people, sandwiched between Christendom and the pagan Mongols, they seem to have been particularly fervent. They were among the fiercest soldiers of Islam (though willing to employ Christian renegades and mercenaries in their own army). Yet since the Crusades the Islamic Near East as a whole had become much more hostile to Christianity. The Prophet had said that Christians could be tolerated because they worshipped Allah, but after the Franks' behaviour to Moslems, and the welcome former Christian subjects of the Arabs gave to Mongol invaders, Islam became much fiercer towards Christendom.

In a couple of centuries the Ottomans built a Moslem barrier much more formidable than any earlier one between Western Christendom and Asia, holding off Orthodox Russia to the north and driving deep into the Danube valley. Their ships harried Mediterranean commerce and steadily mopped up the Venetian empire. Their armies were as good as the Europeans' perhaps for a long time rather better – and they had artillery. One monster cannon (built for the Turks by a Hungarian engineer) used at the final siege of Constantinople was so cumbersome that it needed a hundred oxen to move it and could only be fired seven times a day.

The Ottomans made a great difference to the Europe they ruled. In some places (Bosnia, for example) they permanently entrenched Islam by conversion. Elsewhere, they ruled Christian populations, who now looked to their Orthodox clergy for leadership. Christianity became the mark of nationality in the Balkans, preserving for many of the peoples of the

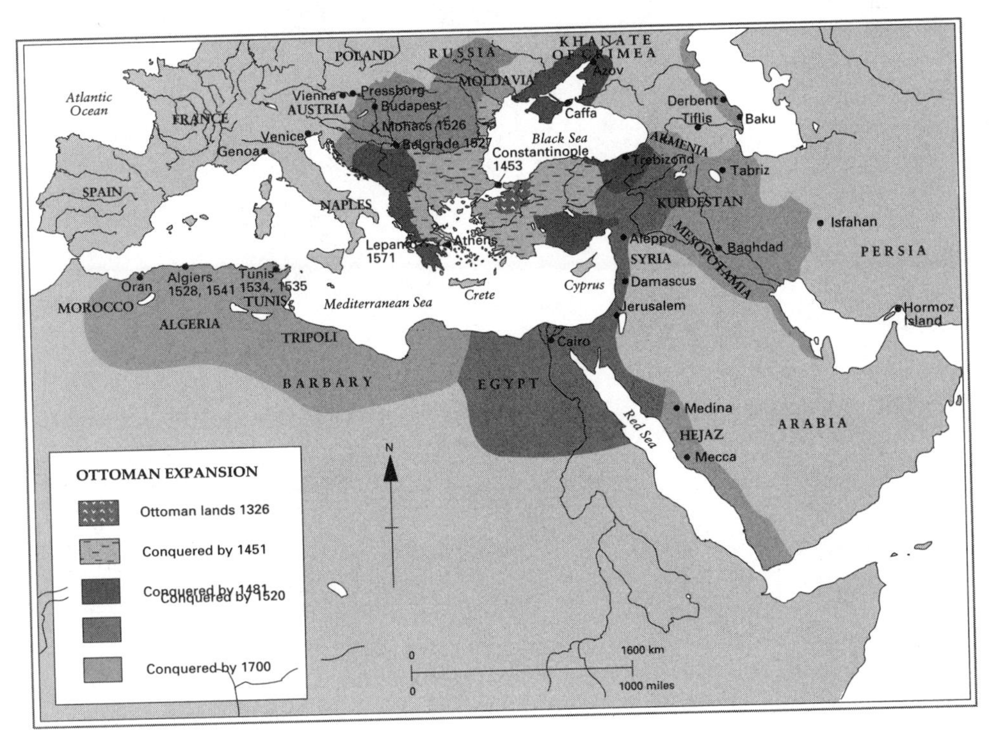

In 1526 Suleiman the Magnificent inflicted a shattering defeat on the Christian Hungarian army at Mohacs, south of Budapest. In this Turkish painting, Hungarian knights flee before the victorious Ottoman army.

region an important part of the Byzantine inheritance. It also divided Greek Orthodoxy – which still looked to the patriarch at Constantinople for leadership – from Russia, where the Church's presiding bishop finally settled in Moscow. Ottoman Turkey was a multi-racial, multi-religious society, though, even if Moslems were firmly on top. Non-Moslem subjects had internal self-government within their own community under their own religious leaders. Ottoman behaviour to Jews and Christians, therefore, was usually better than, say, Spanish Christians' behaviour towards Jews or Moslems.

No empire can be eternal, but Ottoman decline was to take a very long time – until 1922, when Ottoman power finally disappeared, in fact. In the sixteenth and seventeenth centuries decline did not look at all likely, as the Turks pushed deeper into Europe. In 1526 they wiped out the Magyar army in a defeat still remembered as a black day in Hungarian history and three years later they besieged Vienna. They also seized Cyprus, Crete and the rest of the east Mediterranean islands, took Syria and Egypt from the Mamelukes, Kurdistan and Mesopotamia from Persia, and sent an army as far south as Aden. In 1673 they were even back besieging Vienna for the second time and though this was on the eve of the turn of the tide and the beginning of the long retreat of Turkish power, the Ottomans still seized new possessions in the Mediterranean even as late as 1715.

It was a wonderful and frightening adventure, and it left a deep mark on European history, not only in areas the Ottomans occupied. Europe long quailed before Ottoman power and was fascinated by it for even longer. Coordinated resistance to it was never successfully arranged; in the end the burdens always fell on one or two princes: some others even at times looked to the Ottoman for help against fellow Christians. As for those eastern Europeans under Ottoman rule, the empire was another of those experiences which, from the drawing of the Roman imperial frontier onwards, have tended to separate Europeans east and west into two profoundly different groups of societies. But even if some nations felt it especially, the whole of European history was affected in one way or another by the Ottoman onslaught. Its deepest effect of all, perhaps, was to turn Europeans' thoughts away from contact with Asia overland via the Near East. Instead, they began to think harder still about something in the back of some people's minds since the end of the twelfth century, the idea of finding a way round Islam. The means were just becoming available as Constantinople fell. Portuguese ships had already been gingerly moving south down the Moroccan coast even before that, looking for a new route to the East and, possibly, an African ally to take the Turk in the flank. After 1453, they would look harder than ever.

6

WORLDS APART: THE GREAT ASIAN TRADITIONS

MAURYA INDIA

Though the religious pulse of Indian history was set by the end of the Vedic era, the future was then still malleable. New developments were still occurring. Monasticism, for example, an invention of Vedic times, provoked both ascetic experiment, and philosophical speculation. One very successful new cult which arose in reaction against the formalism of the brahmanical religion was Jainism, the creation of a sixth-century teacher who, among other things, preached a respect for animal life which made agriculture or animal husbandry impossible. Jains therefore tended to become merchants; today, the Jain community is one of the wealthiest in India.

Buddhism

Much the most important of the innovating systems, though, was the teaching of the Buddha, the 'enlightened one' or 'aware one' as his name may be translated. Siddhartha Gautama, its founder, was not a brahman, but a prince of the warrior class of the early sixth century BC. After a comfortable and gentlemanly upbringing in a state on the northern edge of the Ganges plain he found his life unsatisfying and left home. He then spent seven years in asceticism and self denial, before beginning to preach and teach. He set out an austere and ethical doctrine, whose aim was to achieve liberation from suffering by winning access to higher states of consciousness. The Buddha taught his disciples so to discipline or shed the demands of the flesh that nothing would prevent the soul from attaining the blessed state of Nirvana, union with the final reality or godliness which he believed to lie beyond life. Though self-annihilation, men could win freedom from the endless cycle of rebirth and transmigration which was the pattern of existence taught by the religion of his day and by later Hinduism.

The Buddha seems to have had great practical and organizing ability, unquestionable ethical integrity and a personality which quickly made him a popular and successful teacher. He sidestepped, rather than opposed, the brahmanical religion. The appearance of communities of Buddhist monks gave his work an institutional setting which would outlive him. He also offered a role to those not satisfied by traditional practice, in particular to women and to low-caste followers, for caste was irrelevant in his eyes. What became known as Buddhism, though, the religion developed by his followers from his teaching, may not really have expressed his own views, which were non-ritualistic, simple and atheistic. His doctrine soon underwent elaboration, and some would say, distortion by others, and like all great religions it assimilated much pre-existing belief and practice. Yet this helped it to retain great popularity and it was to spread, centuries after Gautama's death (in about 483 BC), to become the most widespread religion in Asia and a potent force in world history. It was the first world religion to spread beyond the society in which it was born.

Already when the Buddha was alive, an Indian style of civilization still living today and still capable of enormous assimilative feats stood complete in its essentials, a huge fact separating India from the rest of the world. Much of early civilization in India remains intangible, but the first reports based on a direct observer from the Hellenistic world, a Greek, sent as an ambassador to India by the Seleucid king in about 300 BC, come to us from only two centuries or so after the Buddha died. They tell us of an India beyond the Indus which Alexander's army never reached, for Megasthenes (that was the ambassador's name) travelled as far as Bengal and Orissa and met and interrogated many Indians. Though he tells tales of men who subsisted on odours instead of food and drink, of others who were cyclopean or whose feet were so large that they used them to shelter from the sun, of pygmies and of men without mouths, he also describes the India of a great ruler, Chandragupta, founder of a dynasty named the Maurya. Something is known about him from other sources. Some believed that he had been inspired to conquest by having as a youth seen Alexander the Great during his invasion of India. However this may be, Chandragupta had built a state which encompassed not only the two great valleys of the Indus and Ganges, but most of Afghanistan (taken from the Seleucids) and Baluchistan, with its capital at Patna. It seems from Megasthenes' account that its peoples were already divided, broadly speaking, into two religious traditions – one the old Brahmanical religion which was the root of Hinduism, and the other, apparently, Buddhist. Chandragupta himself is said to have spent his last days in retirement with the Jains, near Mysore, ritually starving himself to death.

A sixth-century bronze statue of the Buddha, who was more and more the object of quasi-idolatrous devotion as the Gupta period went on.

Asoka

Chandragupta's son and successor began to extend this empire farther south. But it was the third Mauyra emperor, Asoka, who completed the process and finished by ruling over a larger area of India than was ever to be under one government again until the height of British power in the nineteenth century. Under Asoka, too, we at last begin to have fairly full documentation. He left many inscriptions, recordings, decrees and messages to his subjects, and these suggest Persian and Hellenistic influences; India was more closely in touch with the outside world at that time than China. At Kandahar in Afghanistan (one of the many cities named after Alexander the Great) Asoka left inscriptions in both Greek and Aramaic.

Although this was probably a more formally organized government than any hitherto, what mattered more by Asoka's day was that India was by then firmly organized on caste lines. Over it ruled Asoka's bureaucracy, helped out, it seems, by a large secret police or internal intelligence service. Besides the duties we might expect it to carry out – gathering taxes, keeping public order, supervising irrigation, for example – this government machine had the task of promoting a set of beliefs, an ideology, we might say. Asoka himself was a Buddhist (he is said to have been converted after witnessing and being horrified by an especially bloody battle), but the message proclaimed on many of the inscribed pillars he left behind to link, symbolically, heaven and earth, is not merely Buddhist. The ideas on them are summed up in the word Dhamma, a derivation from a Sanskrit word meaning 'Universal Law'. They recommended religious toleration, non-violence and respect for the divinity of all men. This is a very surprising set of ideas for an ancient empire and has been anachronistically lauded by modern Indians. Probably these precepts – they were not set out as laws or decrees to be obeyed and enforced – were part of an attempt to make it easier to govern such a huge ramshackle collection of many peoples, creeds and languages. 'All men', says one of Asoka's inscriptions, 'are my children'. It would certainly have made government much easier if everybody could have been brought to agree, and it is notable that Asoka only started to put up his pillars towards the end of his reign (after about 260 BC), when, of course, his conquests were already complete.

Asoka also carried out public works likely to benefit all his subjects. Reservoirs were built, wells were dug, rest-houses put up at regular intervals along the roads of the empire and banyan trees were planted to give shade to travellers. But this does not seem to have helped much in overcoming India's divisions. Instead, some of her most important religious characteristics seem to have deepened and hardened during Mauryan times. Religious ideas and literature took a step towards the crystallization of Hinduism, because this was when two great Indian epic poems, the *Mahabharata* and *Ramayana*, tales of gods, demons and historical derring-do, began to take their final form. To the first was now added the *Song of the Lord* (*Bhagavad Gita*). This was to become the central document of Hinduism, as important to it as the New Testament is to Christianity. But more popular and superstitious cults also flourished in

Mauryan times, and one of them was connected with the *Bhagavad Gita* very closely, for it was the cult of the most popular of all the Indian gods, Krishna, who is also the subject of the poem. Buddhism prospered too under Asoka, perhaps in the main because the emperor supported it. He even sent missionaries aboard. Those in Egypt and Macedonia did not do so well as others in Burma and Ceylon; in that island Buddhism was to remain the dominant religion from this time.

Many of these developments outlasted the Mauryan empire, which began to break up soon after Asoka's death. Why this happened is not clear, but perhaps the simplest explanation is that it outgrew its resources. Like all other ancient empires, the Maurya was essentially parasitic on an agriculture capable of only limited expansion. Its bureaucracy can only have been elementary: it was probably always likely to decay into favouritism and corruption, because there was no system of recruitment and control on rational lines. In any case Indian society was likely to remain in large measure pretty independent of political regimes. It was organized around family and caste, and what happened above their level did not much concern the average Indian. Like the Chinese, Indian empires could not fall back on much loyalty if they ceased to deliver the goods in terms of order and well-being.

Religion, caste, and family, then, provide the long continuities in Indian history. The economy did not alter much, either. In many ways the life of the Indian peasant can have changed very little between Mauryan times and the arrival of Europeans in the sixteenth century AD. Perhaps this was inevitable given the climate and the routines it imposed. Yet there were some important developments in Indian society about whose origins we do not know much, but which must lie at least as far back as Mauryan times. One is the growth of trade, which gave government a revenue from tolls – hence, perhaps, the attention to road-building. A growth of commercial and industrial guilds accompanied this and went so far that some guilds were regarded as threats to the authority of the king. Foreign trade too, with both Africa and the Roman empire, was growing all the time.

HINDU INDIA

The last Mauryan emperor was assassinated in about 184 BC. Indian history then dissolves into a very confused story for about five hundred years. One or two things stand out in the muddle. First in importance is a series of invasions from the north-west. The Bactrians were descendants of the Greeks left behind by Alexander the Great on the upper Oxus, where they established an independent kingdom between India and the Seleucid state in the third century BC. They were soon drawn to India, and pushed into the Indus Valley in the first century BC, to be followed by other peoples who at different times set themselves up in the Punjab, Parthians and Scythians among them. One of the most interesting but mysterious of these invading peoples were the Kushanas. They had migrated all the way from the borders of China and always seem to have had their main interest focused on Central Asia, but they ruled an empire which at one time stretched from the steppes to Benares on the Ganges. The Kushanas were keen Buddhists. It was in their time that sculptured images began to be made of the Buddha (often in a style which shows Greek influence). This was one illustration of the way in which Buddhism was coming down to earth, to be a religion like other religions. But many changes were taking place at the same time, and all the Indian religions and sects show interplay with one another.

The Kushanas went under in due course, and India again dissolved into a jumble of kingdoms. Political unity did not reappear until a new empire, the Gupta, was founded in AD 320. What can be distinguished amid the confusion of the centuries before that, when the Roman empire was at its height, is continuing disturbance from the north-west. The newcomers brought new influences (it is likely that Christianity appeared in the first century AD) yet never overcame the continuing and growing power of Indian tradition. Nor did invaders ever really penetrate the south. After Mauryan times, indeed, that area was not to be united again to the north in a political unity until the British Raj. The Deccan was to remain the region where Hinduism was most strongly entrenched and conservative. It kept its own Dravidian, non-Aryan rulers, and was in many ways a world apart from Hindu northern India, for all its formal sharing of its religion and beliefs. In any case, north and south the broad patterns of Indian life went on largely undisturbed by the coming and goings of rulers. Most Indians lived then (as now) in villages which were more or less self-sufficient and unaffected by events outside. After each new inrush of invaders from across the mountains, or the passing of a more successful conqueror and consolidator than usual, there sooner or later reappeared a scatter of kingdoms built around old local centres and communities. These tended to survive across the centuries, however much disturbed from time to time.

The Gupta

The first Gupta emperor, like his Maurya predecessors, had his capital at Patna and his dynasty ruled a united northern India from the Ganges valley. The peace and freedom from invasion which they provided later led many Indians to look back to the Gupta age as a golden one of peace and good government, a classical period when many of the arts came to fruition for the first time. It is from the Gupta era that there survive the first of great numbers of stone temples, richly embellished with sculpture, which are as important to the history of Indian art and architecture as are the Gothic cathedrals of the Middle Ages to the development of European. Literature flourished; under the Guptas there began the long tradition of Indian popular drama based on stories contained in the great Sanskrit epics, which are still very popular with Indian cinema-goers today. It was also a time of scholarship and philosophical advance. In the fifth century Indian arithmeticians invented the decimal system, an invention of huge importance to mankind, and passed later to the West by the Arabs of the caliphates.

The Spread of Buddhism

c. 563–483 bc	Life of Gautama
c. 370	Council registers first major divisions within Buddhism
255	Emperor Asoka converted to Buddhism
c. 240	Buddhism adopted in Ceylon
61ad	Traditonal date of arrival of Buddhist missionaries in China
c. 120	Buddhist council in Kashmir establishes major texts of Mahayana Buddhism
c. 120–162	Kaniska, Kushana King, promotes Buddhism in Gandhara, Punjab, Sind
c. 200	Buddhism penetrates Indonesia
c. 400–500	Buddhism penetrates Burma
c. 550	Buddhism arrives in Japan
844	Buddhism arrives in China
c. 1300	Buddhism accepted in Thailand

Some of the most important developments under the Guptas had nothing to do with the dynasty; they were continuations of what had already been going on long before and would outlive it. They can be summed up as the gradual but final emergence of classical Hinduism. We do not know exactly when they took their fixed form, but after Gupta times the complicated Hindu social arrangements (which are still so important in shaping Indian life) and the beliefs associated with them are clearly in place. Hinduism's roots go very far back into the past, perhaps beyond the Aryan invasions, for even in the Indus Valley civilizations gods were worshipped who may have been forerunners or 'ancestors' of the Hindu Shiva. But now Hinduism overflowed the mould of Old Brahman and Vedic religion which had long enclosed it.

Belief and society

In Gupta times something very like the Hindu society of later India was already in existence. Its basis was the caste system, by now already evolved far beyond the old four-class division Vedic society. Religious belief had changed too. Hindu religion is difficult to describe, though, for it is not a matter of creeds or statements about what has to be believed. Nor indeed is Hindu religion to be thought of as something apart, or a distinct side of life. It is, rather, a way of looking at the world as a whole (unseen as well as seen) and living in it. If there is a central practical principle to Hinduism, it is to live your life in accordance with your place in the scheme of things. For the peasants – and most Indians are, and always have been, peasants – this might mean only superstitious attempts to secure the goodwill of the deities at the local temple, respect for caste and its practical restraints, and taking part in popular festivals such as those for which great wheeled cars called juggernauts, painted and carved with demons, gods, goddesses and monsters, are still trundled through village streets today. More specialized cults of major gods and goddesses such as Shiva or Krishna also existed. But so did a pure philosophical Hinduism far removed from the crudities of animal-sacrifice and image-worship which went on at the popular level (rather as there was for many centuries a lot of magic and superstitious praying to saints in popular Christianity). Its most developed form was called Vedanta, an abstract belief which (somewhat like versions of Buddhism) stressed the unreality of the actual, material world. It taught that men ought to seek disengagement from that world by winning a true knowledge of reality, or brahma.

Hinduism had something to meet all needs so far as doctrine went. But the way it worked in daily life tended to become more rigid and narrow. Child marriage and the introduction of the practice called suttee, which forced widows to submit to being burnt to death on their husbands' funeral pyres, go along with many other signs in the fifth and sixth centuries that women had to accept a much lower place in society as time went by. In early times the Brahmans had allowed them access to knowledge of the Vedic scriptures, but this stopped.

Buddhism too was undergoing important developments for centuries before and after Gupta times. Instead of being remembered mainly as a great teacher, the Buddha was increasingly seen as the outstanding example of what were called the bodhisattvas, human saviours who renounced the goal of self-annihilation for themselves in order to remain in the world and teach the way to salvation. The most important change was the appearance of what is called 'Mahayana' (or 'great vehicle') Buddhism. This in its most straightforward version amounted to little more than worshipping the Buddha himself as a divine saviour, one manifestation of a great, single, heavenly Buddha. It became much more popular than the tough practices of personal mortification and austerity which had been taught by Gautama. He had forbidden idol-worship, but from the first century onwards increasing numbers of images of him were made and believers worshipped them in temples. In the end Mahayana Buddhism became the dominant form of this religion, which flourished in Nepal, Tibet, China and Japan, while the older tradition maintained itself better in Indonesia and Malaysia.

There is a slight convergence in the way the two major Indian religions were evolving. The Buddha was more and more thought of as almost a god, though a vague one; this was somewhat akin to the idea Hindus had of a soul at the heart of all things. Both religions came to favour contemplation, passivity, the performance of well-known duties, fitting in with the scheme of things rather than attempts to change the world, whether by conversion or by decisive acts. Both went well with a view of time as the endless recurrences of cycles which individuals could not break. Both fed a very different outlook from that of Christianity or Islam. That must have made a lot of difference to the way Indian civilization developed even at a very worldly level. Indian social institutions and above all caste imposed economic burdens; when birth decided what your economic role was to be in life, enterprise and aptitude were narrowly channelled, and so was the ambition of any but the warrior – and his ambition was likely to be satisfied in a somewhat destructive way.

ISLAMIC INDIA

Towards the year 500, the Guptas were showing signs of losing control and before long India dissolved again into little princedoms. Soon it was again being harried by invaders from the north-west – this time a branch of the Huns who managed to wipe out Buddhism in

At Mahabalipuram is a collection of shrines and temples from the seventh and eighth centuries. Beside them stands a gigantic piece of sculpture, carved from a single huge boulder and depicting the Descent of the Ganges from the Himalayas. It incorporates the finest evocations of animal forms to be found anywhere in the Eastern world in one colossal celebration of thankfulness for Shiva's gift of water.

Afghanistan though they left little lasting impact in the Indus Valley. Nor did the Arabs, who came next and conquered the Punjab for a while in the eighth century. Their arrival, though, opens the story of Islam in India.

A second, more powerful (though still not lasting) influx of Moslems came in the eleventh century, in the form of Turkish peoples who intended to stay. Within a few decades they established Moslem rule from Delhi over the whole Ganges valley. Some Turks were anxious to make converts, and persecuted Hinduism, destroying its temples. But they did not last: in 1398 Delhi was sacked by Timur Lang, and that was the end of the Turkish Sultans, the first Moslems to rule much of India. Yet this was not the end of Islam in the subcontinent. It was by then firmly rooted there. It was to prove a much greater challenge to India's powers of digestion than any other creed. With Islam went other influences, too. The cultures of the Indian courts were from this time strongly marked by Persian style and fashion.

The first Moghul emperors

The man who restored Islamic empire once more in India saw himself very much in the Persian tradition. Babur was descended on his father's side from Timur Lang, and on his mother's from Chinghis. He was therefore part Mongol, and the empire he founded was to be called the 'Moghul' empire (that being the Persian word for Mongol), though he did not use the term. He stemmed from those descendants of Timur who had settled down in Persia and were really more Turkish than Mongol. Babur was a lover of Persian poetry and of the gardening in which the Persians delighted (he introduced grapes, melons, bananas and sugar-cane to Kabul). He was also a great book-collector and a highly literate man. He was a poet, wrote a remarkable autobiography as well as a forty-page account of Hindustan at the time of his conquest, noting

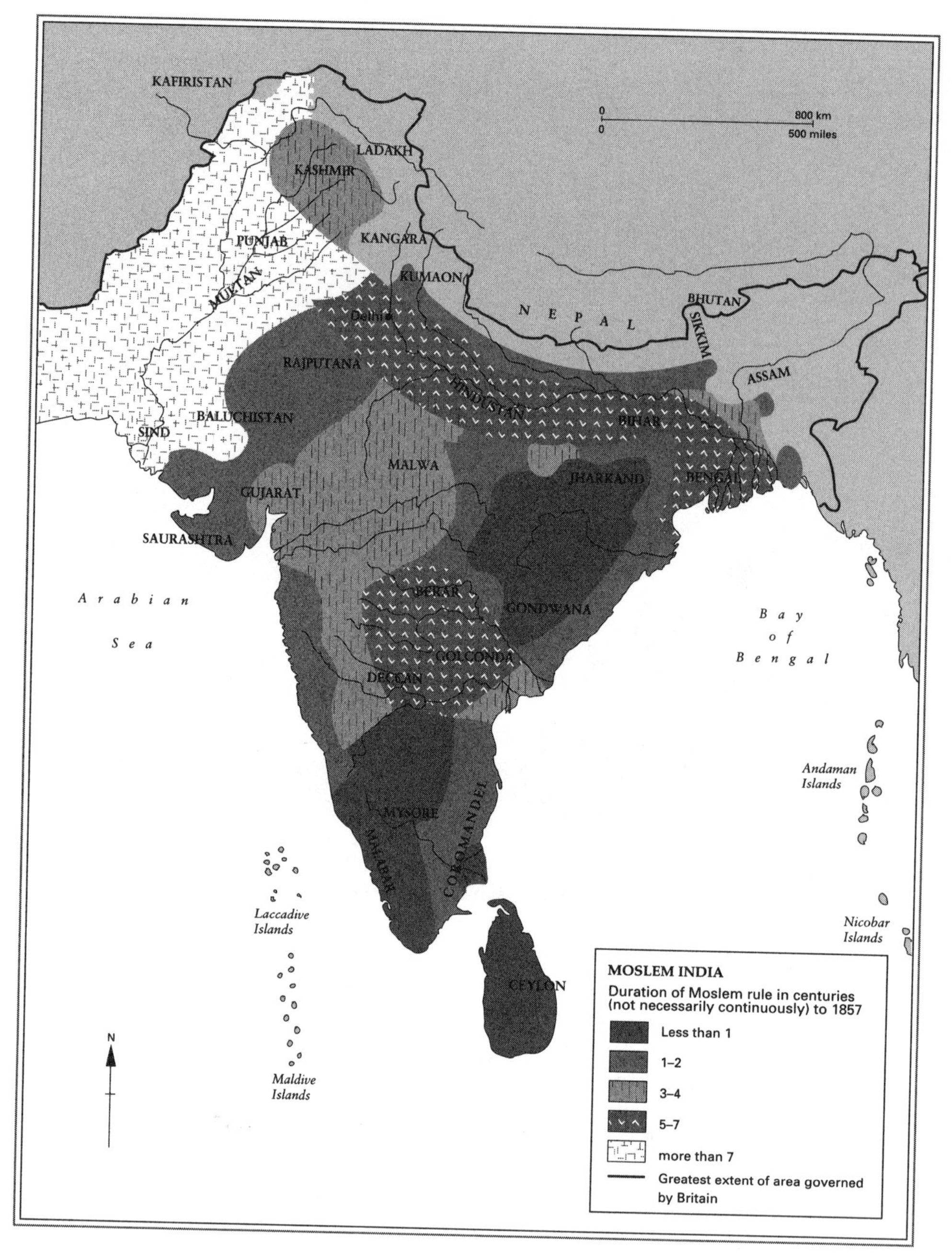

not only its customs and caste-structure but its wild life and flowers. Above all, Babur was an outstanding soldier, who had begun a military career by riding beside his father in battle at the age of ten. He had shown what he was made of by capturing Samarkand at the age of fourteen in 1497. His own base was Kabul, and from there he was invited into India in 1525 by some of its discontented and disloyal Moslem princes. After over-running the Punjab – which the princes had reckoned on – Babur went on to take Delhi, killing its Sultan in battle and then turning on those who had invited him to India – which they had not. He next subdued Hindu princes who had taken advantage of the quarrels of the Moslem rulers, and by 1530 (the year of his death) he ruled an empire stretching from Kabul (where, significantly, he was buried) to the borders of Bihar.

A battle between the forces of Persia and the Moghuls in an illustration of the mid-seventeenth century.

This was enlarged still further by his grandson, Akbar, whom awed Europeans called simply 'the Great Moghul'. He, too, was a brave soldier, and he once killed a tiger in single combat with his sword. Even as a boy he liked riding his own fighting elephants and preferred hunting to lessons – with the result that he, unlike all other Moghul emperors, was almost illiterate. But in the best traditions of his line he admired learning and art, collecting books and paintings and maintained a department of court painters. Both Moghul architecture and painting under his rule reached their peak.

Akbar reigned from 1555 to 1605 (thus overlapping at each end the almost equally long reign of Queen Elizabeth I of England). He showed great skill in handling the religious differences of his subjects, and one of his first acts as a ruler was to marry a Rajput (and therefore Hindu) princess. Even before this he allowed the Hindu ladies of his harem to practise their own religion. He was no persecuting Islamic missionary. Before long he abolished a poll-tax on non-Moslems levied by his predecessors, and his finances were managed by a Hindu minister. Akbar also showed great clemency towards defeated Hindu princes. His Hindu wife was the daughter of the greatest of the Hindu kings of Rajputana, the ruler of what is now Jaipur so marriage also had a place in his conciliatory diplomacy. Altogether the impression given by Akbar's reign is one of well-directed tolerance and, in consequence, of the easier government of so diversified an empire.

When Akbar died, his dynasty was more solidly established than any earlier one which had ruled so much of India, not least because a new land-tax provided money to keep the empire going without further expansion by conquest. (It also seems actually to have led to increases in agricultural production.) Some of the Moghul empire's administrative innovations were to last well into the times of British India. There were even signs that the Moslem ruling class were softening somewhat in their hostility to Hinduism, which they should have regarded as idolatry. A new language common to members of both religions had already made its appearance – Urdu, the tongue of the camp, used by the Moslem conquerors to their Hindu subjects as a lingua franca. It combined a Hindi structure with a Turkish and Persian vocabulary and was another institution which lasted right down to the twentieth century. Not only was Urdu the general language of the Indian army under the British, but it is still widely spoken today.

Akbar's reign brought another portent. Even in Roman times India's western ports had traded with the Mediterranean. In the sixteenth century began the first regular contacts with Atlantic Europe. The Portuguese were the first Europeans to arrive. They appeared on the Malabar coast just before 1500, and slowly moved round to the Bay of Bengal in the second half of the sixteenth century, setting up trading posts there. Akbar invited some of them to send missionaries learned in their faith to his court to dispute with Moslem divines, and three arrived in 1580. The day of the British was then still to come, but on 31 December 1600, the last day of the sixteenth century, some of Queen Elizabeth I's subjects founded the first English East India Company in London. Three more years passed before the first English emissary arrived at Akbar's court, but this was a milestone in Indian history as well as English. Europeans were from now on to arrive in growing numbers – and India would never again be able to ignore them.

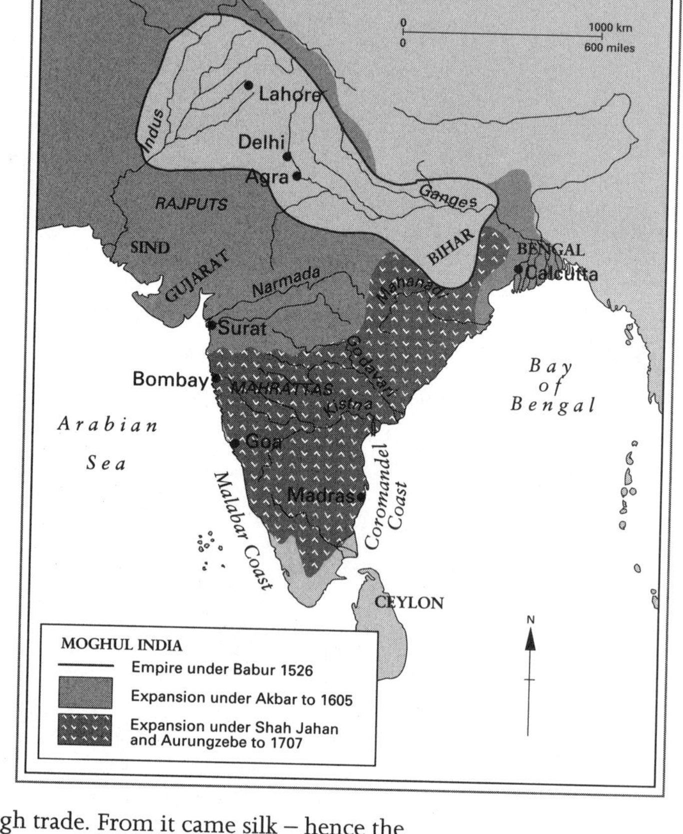

IMPERIAL CHINA

Of all the strange but not entirely unknown lands which lay beyond Rome's frontiers China was the most mysterious and the least known. With it, the Romans were only just and indirectly in touch through trade. From it came silk – hence the Latin name the Romans gave China: 'Serica', or 'silken'. Her culture continued for centuries to be protected by her isolation. China had complicated and close relations with the people of central Asia; yet, once unified, she had for many centuries on her borders no great states with whom relations had to be carried on. This isolation was, if anything, to increase as the centre of gravity of European civilization moved west and north and as it was more and more cut off from East Asia by the inheritors of the Hellenistic legacy: Byzantium, Sassanid Persia, and then Islam.

China was thus to remain remote, inaccessible to most of the currents changing other parts of the Eurasian landmass, and far from sources of disturbance in other great civilizations. Even Islam, when it came, made much less difference than it did elsewhere. China was also endowed with a huge capacity to absorb alien influence when it did arrive. Her assimilative power rested on the culture of an administrative élite which survived dynasties and empires and kept China on the same course. Because it kept up written records from very early times which provide an incomparable documentation, crammed with often reliable facts (though chosen by a minority whose preoccupations they reflect), we owe much of our knowledge of China to its scribes. Their histories emphasize continuity and the smooth flow of events. Given the needs of administration in so huge a country this is perfectly understandable; uniformity and regularity were clearly to be desired. Yet such a record leaves much out; even in historical times the real concerns and life of the vast majority of Chinese are hard to discern.

The narrative of the state is a little easier to establish. China's history after the end of the period of Warring States has a backbone of sorts in the waxing and waning of dynasties. Dates can be attached to these, but there is danger of an artificial, or at least of a misleading,

A section of the Great Wall near Peking.

authoritativeness to be overcome in using them. It could take decades for a dynasty to make its power a reality over the whole empire and even longer to lose it. With this reservation, though, the dynastic reckoning is helpful. It gives us major divisions of Chinese history down to this century. The first which need concern us are the Ch'in, and Han periods.

The Ch'in

The Ch'in came from a western state still looked upon by some Chinese as barbarous as late as the fourth century BC; in this and in the timing, too, there is a suggestive similarity to the rise of Macedon. They prospered, perhaps in part because of a radical reorganization carried out by a legalist-minded minister in about 356 BC; perhaps also because of their soldiers' use of a new long iron sword. After swallowing Szechwan, the Ch'in achieved the status of a kingdom in 325 BC. The climax of Ch'in success was the defeat of their last opponent in 221 BC and the unification of China for the first time in one empire under the dynasty which has given the country its name. It was a great achievement. China may be considered from this time the seat of a single, self-conscious civilization. There had been earlier signs that such an outcome was likely and by the end of the Warring States Period some parts of China already showed marked similarities which offset the differences between them. The political unity achieved by Ch'in conquest over a century was in a sense the logical corollary of a cultural unification already well under way. Some have even claimed that a sense of Chinese nationality can be discerned before 221 BC; if so, it must have made conquest easier. Yet the Ch'in were displaced after less than twenty years.

The Han

It is easy to exaggerate the extent and effectiveness of government in early China, and conjectural boundaries do not mean much, but for over four hundred years, from 206 BC to AD 220, China was ruled by emperors of two dynasties with the same name – the Han. For a short period (AD 9-23) the 'former' Han ceased to reign and the 'later' Han did not at once succeed them, but the two eras can for many purposes be thought of as a whole. Having said that, some individual emperors were much more vigorous and effective than others. Drawing the boundary of the Han empire on the map is a pretty theoretical affair; control was certainly not achieved over the whole area within it even in the sense that Rome ruled her

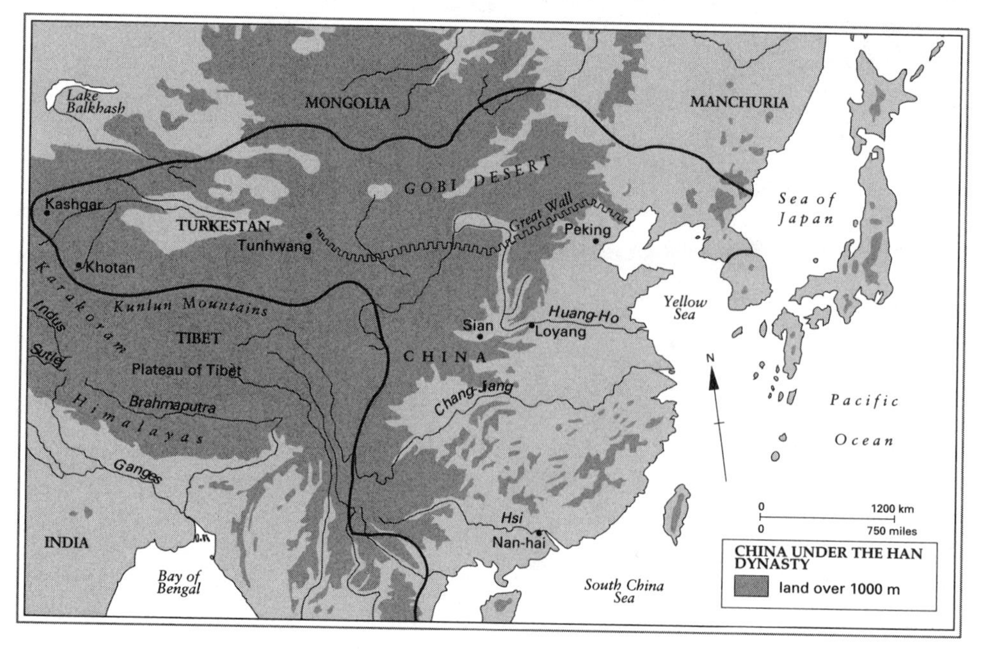

empire at that time. Nor did Chinese civilization permeate what is now thought of as China in the way Hellenistic civilization permeated western Europe, the Mediterranean and the Near East. Chinese script had only been standardized under the Ch'in (just before Han times). A tribal society in which only a few leading men were 'sinicized' (that is, had taken to Chinese ways) was normal over much of the area politically subjected to the Han, especially in the south. There was still not much everyday cement for the regime.

Nevertheless, Han emperors extended China's claims to political domination further than any of their predecessors. In theory at least, the Han empire was, at its greatest extent, as big as the Roman. The emperor Wu Ti, the 'Martial Emperor' who reigned 141-87 BC, was especially acquisitive. Under him a big area of Central Asia, the Tarim Basin, was taken into the empire, as well as southern Manchuria – land north of the Great Wall – and much of the south-eastern coast of China. The Thai peoples of the Mekong Valley were also subjugated and Annam accepted the overlordship of the Han. Later, Mongolian peoples called the Hsing-Nu were driven north of the Gobi Desert and so sent off on the beginnings of a long march which made them a force in world history and better known as 'Huns'.

Expansion increased China's contacts with other parts of the world. Those with the Mediterranean remained only indirect, because most of China's trade was by land. The most desirable commodity she produced was silk and from about 100 BC it was being sent to the west by caravans along the 'Silk Road' of Central Asia. Perhaps the new contacts of this era with the nomadic horsemen of the deserts explain the appearance of the beautiful bronze horses which began to be cast in Han times.

Religion in China

Yet, in spite of such wider contacts, China remained remarkably isolated and untouched by outside influence. Nothing like the effect Judaism had on Graeco-Roman civilization through Christianity can be detected. Islam penetrated Turkestan and the corners of the empire and flourished but it did not bite deeply. The one possible challenger to tradition was Buddhism, which appears to have made its way to China during the first century AD along the trade-routes from Central Asia. It may well have been China's most important cultural import before the nineteenth century, and was in principle much more alien to Chinese religion than any other before Christianity, for it emphasized other-worldliness, rather than the fulfilment of duties to society. Perhaps it was launched on the road to China by the Kushanas, for they spread it to Central Asia; thence Buddhism travelled to North China, mainly in its Mahayana form. The comforting idea of a Buddha who was a saviour, to whom the believer could look in faith for help, must have seemed very attractive at a time of upheaval and social disintegration. Buddhism could be all things to all men – superstition to the humble, stimulating new philosophical ideas to the educated, and its artistic style was appealing.

Buddhism gradually spread downwards through society. Students and monks began to go to and fro between China and India in search of Buddhist instruction. Buddhism's greatest success came between the sixth and ninth centuries. Yet during this time it was changing and often turned into something very unlike what Gautama had taught. Though Buddhism everywhere was then something of a jumble – often of conflicting and contradictory doctrines – in China it underwent special developments. Sectarian movements appeared (one of them the meditation movement later known under its Japanese name of 'Zen'). Celibacy, a central tenet of classical Buddhist teaching (and especially unwelcome to the Chinese, who attached great importance to assuring the continuity of the family and the cult of ancestors), was abandoned by some Buddhist clergy.

The state, too, took a hand in regulating Buddhism, notably by limiting the number of monks and monasteries (this was to prevent their riches from escaping the taxation system) though it proved hard to enforce, and there were occasional outbreaks of persecution. The worst came in the ninth century when all alien religions were banned. Official sources say that more than 4600 monasteries were destroyed and over a quarter of a million Buddhist monks and nuns lost their tax exemption. This happened while Confucianism was recovering its grip on the educated and it marked the beginning of the decline of Buddhism in China. It was never again so strong. It had not remade Chinese civilization, but had only contributed some more elements to it.

Buddhism in fact hardly troubled the religious attitudes of most Chinese. Chinese official religion was not harsh towards it or other creeds; the state was only interested in persecuting when a new faith seemed in some way a political or social threat. Chinese tradition stressed little more than that the proper persons should carry out ritual sacrifices and that ancestors should be venerated; it was very undogmatic. Confucianism reinforced this, with the result that the educated Chinese were remarkably tolerant (something for which Europeans were later much to admire them) and later, under the T'ang, one emperor issued an edict permitting the preaching of Christianity (which had reached China through Nestorian missionaries). Because imperial China was unlikely to show much resistance to new religious ideas from the outside, it may have seemed probable that they would prosper, but this was not so. Although, the breakdown of much of the traditional society during the troubles of the Han collapse and its aftermath made people cast about for new cults and beliefs – rather as happened in the declining Roman empire – the beneficiaries were popular cults and developments of Taoism, which became a mish-mash of faith-healing, superstition and Buddhist ideas.

This celebrated bronze horse buried at Kansu in north-west China in the second century AD may have been meant to be a magical creature; it seems to be galloping through the air. Similar models were often made of real horses and were buried with their owners.

Chinese civilization

In spite of foreign influences, then, educated Chinese in Han times found it easy to think of their country as the centre of the world and the seat of true civilization. This intellectual assurance no doubt explains much of their indifference to what was going on elsewhere. But other factors must have helped. Geographical remoteness, for example, was always a factor; anywhere that might have exercised a dynamic stimulus (culturally speaking) was a long way away, and this both kept out potentially disruptive forces and narrowed China's chance of experience which might have made her rulers more curious about the outside world. China was, too, virtually economically and technologically self-sufficient. She had rich natural resources, and by Han times her agriculture and technology were able to exploit the environment successfully. Nothing obviously superior was available elsewhere. Rice, introduced in very early times from either south-east Asia or India, was the last really radical innovation in material life from outside before modern times.

The Han era brought further refinements and inventions. Han scientists produced the first magnetic compass, with dial and pointer (though it was not used for navigation, but for laying out temples with the correct geomantic orientation) the first system of cartography based on a grid system, machines to record earthquakes and callipers with decimal graduation for craftsmen. In retrospect, though, of all the innovations of the period, the most striking is the discovery of how to make paper (announced from the imperial workshops in AD 105). This was to be of enormous importance to the whole human race, even if knowledge of paper-making was not to reach the west for several centuries. It was cheaper than papyrus or parchment (though it deteriorated much more rapidly than the latter) and easier to make.

Transport and therefore communications also improved during Han times. The rudder attached to the stern of a ship (as opposed to a big paddle, hung over one side) appeared in the first century BC; European ships had to wait another twelve hundred years or so for it. It was also under the early Han that a breast-strap harness was developed for horses; far heavier loads than before could now be drawn. A little after the end of the dynasty, the Chinese were to introduce the stirrup – an invention of enormous importance in warfare because of the greater security and control it gave to the rider. The new crossbow, on the other hand, was invented under the Han. It was a major technological achievement, more powerful and more accurate than the bows of the barbarians, and beyond their power to imitate, since non-Chinese long lacked the technique of making the bronze locks the new weapons required.

An example of grave-goods from a tomb of the lower Han period, an earthenware figure of a fishmonger.

Such innovations bear witness to the richness of Han civilization, and in many ways it was only the beginning of a glorious period; Chinese science and mathematics in the next thousand years were to throw up far more new ideas than European. For the rulers and the rich, too, Han China at its peak must have been a splendid society. Some of its loveliest creations in silk, wood and paint have perished; when palaces began to be burnt in the troubled last decades of the dynasty, priceless collections were destroyed. Still, many beautiful things survive because of the Han practice of burying the rich and noble with many of their possessions, or with models of them. One particularly notable recent discovery was of the elaborate jade suits in which a former Han prince and princess were buried. Under the later Han, bronze objects, especially models of horses, show a new development of one of the oldest Chinese arts, bronze-casting, while new coloured glazes were invented by the potters.

In much of Han art Chinese civilization already seems deliberately to be looking backwards rather than to the future. This was true of the life of the mind too. The scholarly writing of dynastic histories began under the Han, when the greatest of Chinese historians, Ssu-ma Ch'ien, wrote his *Historical Records*, highly esteemed by specialists in these matters. But the most important cultural development was the establishment of Confucianism as the official ideology of the state. This triumph for the sage's teaching (or what was taken to be his teaching) came about in part because scholars naturally wished to restore the ravages caused to Chinese learning and libraries under the Ch'in, by whom they had been deeply offended. Though a few of them had been favoured and gave the dynasty advice, there had been a nasty moment in 213 BC when the emperor turned on critics of the despotic and militaristic character of his regime. Books were burned and only 'useful' works on divination, medicine or agriculture or those glorifying the dynasty were spared; more than four hundred scholars perished. As the Han scholars worked, they rediscovered Confucian texts lost under the Ch'in. What was really at stake is not exactly clear, but the Han emperors were willing to conciliate the intellectuals. They established professorships of Confucian studies, ordered regular sacrifices to Confucius to be made in all government schools and began to admit recruits to the civil service on the basis of examination in the Confucian classics. This led to the formalization of Confucian doctrine into a long-lived orthodoxy. The canonical texts were established soon after 200 BC. Official Confucianism was to go on to absorb something from other schools of thought, but its ethical precepts henceforth remained dominant in the philosophy which formed China's future rulers. In AD 58 sacrifices to Confucius were ordered in all government schools. Eventually, under the T'ang, administrative posts were confined to those trained in this orthodoxy. For over a thousand years it provided China's governors with a set of moral principles and a literary culture doggedly acquired by rote-learning. The result was one of the most effective and ideologically homogeneous bureaucracies the world has ever seen.

Yet scholarly support was not enough to ensure the survival of the Han dynasty. It faced severe internal and external challenges. The most serious were more frequent peasant rebellions. As population grew, more peasants became landless and unable to find money for taxes and food. There were renewed barbarian attacks from the outside, too. Warlords who controlled the professional soldiers on which the dynasty relied tended to seize power at home; barbarians, allowed within the frontier in the hope of converting them to Chinese ways, turned on those who had brought them in. It was to the son of the greatest of the warlords that the last Han emperor resigned his throne in AD 221. When he did so, China fell apart again.

Much of the evidence of the brilliant Han culture was dissipated or destroyed during the fourth and fifth centuries, when barbarians returned to harass the frontiers. China then once more dissolved into a congeries of kingdoms, some under barbarian dynasties. In this crisis, nonetheless, there is still observable China's striking power of cultural digestion.

Gradually, barbarians were seduced by Chinese ways; they lost their own identity, took to Chinese clothes and language, and became simply other kinds of Chinese. The prestige which Chinese civilization enjoyed among the peoples of Central Asia was already very great. There was a disposition among her uncivilized neighbours to see China as the centre of the world, a cultural pinnacle, somewhat in the way in which the Germanic peoples of the West had seen Rome. One Tatar ruler actually imposed Chinese customs and dress on his people by decree in 500. The central Asian threat was not over; far from it, it was in the fifth century that the first Mongol empire appeared in Mongolia. Nonetheless, when the T'ang, a northern dynasty, came to receive the mandate of heaven in 618 China's essential unity was in no great danger, and disunity and barbarian invasion had not damaged the foundations of a Chinese civilization which now entered its classical phase.

Under the T'ang the institutional and cultural aspects of Chinese civilization cannot be separated. To the family and the state the Chinese looked for authority; those institutions were unchallenged, for in China there were no entities such as Church or communes to confuse questions of right and government so fruitfully as in Europe. The essential characteristics of the Chinese state, too, were all in place by T'ang times. They were to last until this century and the attitudes they built up linger on still. In their making, the consolidating work of the Han had been especially important, but the office of the emperor, holder of the mandate of heaven, had been taken for granted even in Ch'in times. The comings and goings of dynasties did not cast discredit on the office since they could always be ascribed to the withdrawal of the heavenly mandate. Meanwhile, the power inherent in emperor's position grew.

Gradually, a ruler who was essentially a great feudal magnate, his power an extension of that of the family or the manor, became one who presided over a centralized and bureaucratic state. This had begun a long way back. Only a potent state could have organized and deployed the human resources the first Ch'in emperor had already used to link together the existing sections of the Great Wall in 1400 miles of continuous barrier against the barbarians (legendarily, an achievement which cost a million lives). The Han emperors had been able to impose a monopoly of coining and standardized the currency. Under them the civil service had been founded, too.

An imaginary and conventional seventeenth-century representation of Confucius, whose mystical status and supposed teaching had by then hardened into an official ideology.

Bureaucracy

Territorial expansion had required more administrators. An enlarged bureaucracy was to survive many periods of disunion (a proof of its vigour) and remained to the end one of the most striking and characteristic institutions of imperial China. It was probably the key to China's successful emergence from the era when collapsing dynasties were replaced by competing petty and local states which broke up the unity already achieved. It linked China together by ideology as well as by administration. Trained and examined in the Confucian classics, the civil servants assured that literacy and political culture were wedded in China as nowhere else.

The officials were in principle distinguished from the rest of society only by education (the possession of a degree, as it were). Most of them came from the land-owning gentry, but they were set apart from it. Once selected for office by the test of examination, they enjoyed a status lower only than that of the imperial family, as well as great material and social privileges. They had two crucial annual tasks, the compilation of the census returns and the land registers on which Chinese taxation rested. Their other main duties were judicial and supervisory, for local affairs were very much left to local gentlemen acting under the oversight of the two thousand or so district magistrates from the official class. Each of these lived in an official compound, the yamen, with his clerks, runners and household staff about him. Over them, there watched a state apparatus of control, checking and reporting on a bureaucracy which at its greatest extent ruled an area much larger than the Roman empire.

This structure had huge conservative power. Thanks to the examination system, governmental practice was shot through by agreed ideals. Moreover, though it was very hard for anyone not assured of some wealth to support himself during the long studies necessary for the examination – writing in the traditional literary forms itself took years to master – the principle of competition ensured that a continuing search for talent was not quite confined to the wealthier and established gentry families; China was a meritocracy, even if a restricted one. From time to time there were instances of corruption and the buying of places, but such signs of decline usually emerge into the records only towards the end of a dynastic period. For the most part, the imperial officials showed remarkable independence of their background. They were in theory the emperor's men; they were not allowed to own land in the province where

An eighteenth-century album contains this painting of an examination, a traditional procedure long unchanged in China. Magistrates are here shown at their re-examination in the Confucian classics. The emperor himself (a T'ang monarch of the eighth century) assesses the results, on which promotion or dismissal was based.

they served, serve in their own provinces, or have relatives in the same branch of government. They were not the representatives of a class, but a selection from it, an independently recruited élite, renewed and promoted by competition. They made the state a reality. To those among them who had risen through the official hierarchy to its highest levels and had become imperial counsellors, the only rivals of importance were the court eunuchs. These creatures, often trusted with great authority by emperors because, by definition, they could not found families, were the only political force escaping the restraints of the official world.

China had little sense of the European distinction between government and society. Official, scholar and gentleman were usually the same man, combining many roles which in Europe were increasingly to be divided between governmental specialists and the informal authorities of society. He combined them, too, within the framework of an ideology which was much more obviously central to society than any to be found elsewhere than perhaps in Islam. The preservation of Confucian values was not a light matter; it was not satisfiable by lip-service. The bureaucracy maintained those values by exercising a moral supremacy somewhat like that long exercised by the clergy in the West – and in China there was no Church to rival the state. The ideas which inspired it were profoundly conservative; the imperative administrative task was the maintenance of the established order; the aim of Chinese government was to oversee, conserve and consolidate, and occasionally to carry out large public works. Its overriding criteria were regularity and the maintenance of common standards in a huge and diverse empire, where many district magistrates were divided from the people in their charge even by language. In achieving these aims, the bureaucracy was very successful.

Each dynasty went through a cycle of rise and decline. The start of a period of decline was usually marked by its inability to protect the frontiers against fresh barbarian invasions, and by hard times, famine and peasant revolt at home. These led to the breakdown of normal taxation and often of law and order. Local warlords running little kingdoms of their own tended to emerge. Yet these periods of disorder never got so badly out of hand after the tenth century that they led men to question the fundamental principle that China ought to be governed by the emperor as one country. Though, under the later Sung China was divided into a northern state ruled by barbarians and a southern ruled by Chinese, it never broke up into very small units again until the twentieth century.

Continuity

The historical record is therefore less scrappy and muddled than it looks at first sight. For a thousand years, many things went on almost without interruption, uproar and invasion from time to time notwithstanding. Rulers and dynasties might come and go, but the imperial throne itself continued to be respected, even if the mandate of heaven could be withdrawn from the particular individual or dynasty who held it at any one time. The position of the civil service was practically unquestioned, and, increasingly recruited by merit and talent, was unparalleled anywhere else in the world in its efficiency and intelligence. The Sui and T'ang emperors restored the entry examinations started by the Han, and, through them, the bureaucracy was ensured a very conservative cast of mind. In the longest run this was a terrible handicap, but the world outlook it gave educated Chinese lasted remarkably unchanged until only a century or so ago, which suggests it was for a long time not very remote from China's needs.

The examination system also drove even deeper the division of the Chinese people into an educated ruling elite and a non-educated mass who were ruled. Yet as the centuries passed, China's social structure was bound to change somewhat. There was, broadly speaking, a long decline in the importance (though not the wealth) of the aristocracy as the empire grew older; they had to give up more and more political and administrative power to the professional civil servants. There were growing numbers of merchants to be taken into account, as trade developed and towns increased in both number and size. Other creeds than the Confucian orthodoxy of the officials and gentry were from time to time important, too. Even some who were high in the social scale turned to Taoism or Buddhism when, after the Han collapse, disunity gave the latter an opportunity to penetrate China. In its Mahayana version it posed more of a threat than any other ideological force before Christianity, for, unlike Confucianism, it posited the rejection of worldly values, and it was never to be eradicated altogether, in spite of persecution under the T'ang (whose attacks on it were probably mounted for financial rather than ideological reasons). In spite of much material damage inflicted on Buddhism, Confucianism had to come to terms with it. No other foreign religion influenced China's rulers so strongly until Marxism; even some emperors were Buddhists.

Such religious and philosophical traditions as Taoism and even Buddhism, important as they were, touched peasant life very little. Amid the insecurities of war and famine, the peasant turned to magic or superstition. What little we can see of his life suggests that it was often intolerable, sometimes terrible. Peasant rebellion, first noticeable under the Han, became a major theme of Chinese history, punctuating it almost as rhythmically as the passing of dynas-

On the site of an ancient shrine, the wooden Temple of Heaven of Peking as it now appears is really the outcome of an eighteenth-century repair programme. Its basic design, though, goes back to pre-Han times, an example of the conservatism of Chinese architecture exemplified in the ground plan of the 'forbidden city' of Peking itself which followed the gridiron street-plan prescribed by the ancient books of architecture. Properly speaking, the 'forbidden city' was the Palace compound only; the Temple of Heaven lay outside this central area.

ties. Oppressed by officials acting either on behalf of an imperial government seeking taxes for its campaigns abroad or in their own interest as grain speculators, the peasants turned to secret societies, another recurrent theme. Their revolts often took religious forms. A millenarian, Manichaean strain has always run through Chinese revolution, bursting out in many guises, positing a world dualistically divided into good and evil, the righteous and the demons. Sometimes this threatened the social fabric, but the peasants were rarely successful for long.

Chinese society therefore changed very slowly. For all the important cultural and administrative innovations, the lives of most Chinese altered little in style or appearance over the centuries. The comings and goings of the dynasties were accounted for by the notion of the mandate of heaven. Although it could point to great intellectual achievements, China's civilization at an early date seems self-contained, self-sufficient, stable to the point of immobility. Yet slow changes there were. One was the continuing growth of commerce and towns which made it easier to supplement the obligatory labour service of a peasant by taxation, for such new resources could be tapped by government. The work of the Ch'in on the Great Wall was followed by that of later dynasties which extended it, and sometimes rebuilt portions of it. It still astonishes the observer and far outranks the Roman *limes*. Just before the inauguration of the T'ang, too, a great system of canals was completed which linked the Yangtze valley with the Yellow River valley to the north, and Hangchow to the south. Millions of labourers were employed on this and on other great irrigation schemes, works comparable in scale with the Pyramids and the great cathedrals of medieval Europe.

THE LATER DYNASTIC STORY

A civilization with impressive achievements already to its credit entered a new and mature phase in 618. For the next thousand years, as for the previous eight hundred, the formal framework of its evolution can be set out most easily as a dynastic sequence. After the Han, disorder divided China for more than 350 years. A general of mixed Chinese and barbarian blood then reunited the country in 581. The Sui dynasty he founded lasted only some thirty years before another general (also of mixed descent) seized the throne. This inaugurated the T'ang dynasty, under which China was again a unity for nearly three and a half centuries. Another period of disorder then followed, but this time it lasted only fifty years before the Sung ascended the imperial throne in 960. Though the Sung lost control of northern China to peoples from Manchuria in the twelfth century, they hung on in the south until 1279. In that year Kubilai, the grandson of Chinghis Khan, completed the Mongol conquest of China. He adopted the Chinese dynastic name of Yuan and his successors ruled China until 1368, from the new capital at Peking, when they were replaced by a dynasty founded by a rebel Chinese commoner, the Ming, which lasted until 1644.

Important themes overrun such divisions into alternating order and disorder. One is population history. There was a shift of demographic weight towards the south during the T'ang period; henceforth most Chinese were to live in the Yangtze valley rather than the old Yellow River plain. The devastation of the southern forests and exploitation of new lands to grow rice fed them, but new crops became available, too. Together they made possible an overall growth of population which further accelerated under the Mongols and the Ming. Perhaps eighty million Chinese in the fourteenth century more than doubled in the next two hundred years, so that in 1600 there were about 160 million subjects of the empire. Given populations elsewhere this was a huge number.

As numbers gradually rose, all cultivatable land came to be occupied. It was farmed more and more intensively in smaller and smaller plots. More and more peasants were landless. The one way out of the trap of famine was rebellion. At a certain level of intensity and success this might win support from the gentry and officials, whether from prudence or sympathy. When that happened, the end of a dynasty was probably approaching, for Confucian principles taught that, although rebellion was wrong if a true king reigned, a government which provoked rebellion and could not control it ought to be replaced, for it was *ipso facto* illegitimate. Thus for many centuries, population pressure was a major lever of China's history. Yet it tended to make itself felt to the authorities only in indirect and obscured ways – when famine or hunger drove men to rebellion. A much more obvious threat came from the outside, though China, which had been thought of as a possible ally by Byzantium, which had sent armies to fight the Arabs and received ambassadors from Haroun-al-Raschid, was a great world power. Essentially her problem was that of Rome, an overlong frontier beyond which lay barbarians. T'ang influence over them was weakened when central Asia succumbed to Islam. Like their Roman predecessors, too, the later T'ang emperors found that reliance on soldiers could be dangerous. Hundreds of military rebellions took place under the T'ang and any rebellion, even if short-lived, had a multiplier effect, tending to disrupt administration and damage the irrigation arrangements on which food (and therefore internal peace) depended.

In the end, unable to police their frontier effectively and troubled at home, the T'ang went under in the tenth century, and China collapsed again into political chaos. During it, the continuity and recuperative power of the bureaucracy and her social institutions kept China going. After dynastic change, the inheritors of power, even if from outside, usually turned to the relatively small number of officials in post and thus drew into the service of each new government the unchanging values of the Confucian system. Confucian teaching made it easy for the dynasty to change without compromising the deepest values and structure of society. A new dynasty was almost bound to turn to the officials for its administration and to the gentry for most of its officials who, in their turn, could get some things done only on the local notables' terms when central government was weak.

Classical China

Recurrent disunity did not, therefore, prevent China's rulers, sages and craftsmen from bringing Chinese civilization to its peak in the thousand years after the T'ang inauguration.

Some have placed its classical age as early as the seventh and eighth centuries, under the T'ang themselves, while others discern it under the Sung. T'ang culture reflected the stimulus of contacts with the outside world, but especially with central Asia. The capital was then at Ch'ang-an at the end of the Silk Road, in Shensi, a western province. Ch'ang-an means 'long-lasting peace' and to it came Persians, Arabs and central Asians who made it one of the most cosmopolitan cities in the world. It contained Nestorian churches, Zoroastrian temples, Moslem mosques, and the objects which remain to us suggest it may have been the most splendid and luxurious capital of its day. Many of them reflect Chinese recognition of styles other than their own – the imitation of Iranian silverware, for example – while the flavour of a trading entrepôt is preserved in the pottery figures of horsemen and loaded camels which reveal the life of central Asia swirling in the streets of Ch'ang-an. These figures were often finished with the new polychromatic glazes achieved by T'ang potters; their style was imitated as far away as Japan and Mesopotamia. The presence of the court stimulated both such craftsmanship and visits of merchants from abroad. From tomb-paintings something of the life of the court aristocracy can be seen. The men relax in hunting, attended by central Asian retainers; the women, vacuous in expression, are luxuriously dressed and, if servants, are elaborately equipped with fans, cosmetic boxes, backscratchers and other paraphernalia of the boudoir. Great ladies of the T'ang era, too, favoured central Asian fashions borrowed from their domestic staff.

A characteristic pottery figure of T'ang times, an armoured cavalryman.

The history of women, though, is the history of another of those other Chinas always obscured by the bias of the documentation towards the official culture. Their lot was probably a hard one. We hear little of them, even in literature, except in sad little poems and love stories. Yet presumably they must have made up about half of the population, or perhaps slightly less, for in hard times girl babies were exposed by poor families to die. That fact, perhaps, characterizes women's place in China until very recent times even better than the more familiar and superficially striking practice of foot-binding, which produced grotesque deformations and could leave a high-born lady almost incapable of walking. Even by the savage standards of other cultures in more barbaric places, Chinese women seem until the twentieth century to have been a much oppressed class. Peasant women carried out much of the heavy work of the fields, and upper-class women enjoyed little freedom.

Official culture also excluded most of the tenth or so of the Chinese population who lived in the cities, some of which were the biggest in the world. Ch'ang-an, when the T'ang capital, is said already to have had two million inhabitants. No European city was so big, and contemporary Canton or Peking were even larger. Such huge cities housed societies of growing complexity. Their development fostered a new commercial world; the first Chinese paper money was issued in 650. Prosperity created new demands, among other things for a literature which did not confine itself to the classical models and written in a colloquial style far less demanding than the elaborate classical Chinese. City life gradually secreted a literate alternative to the official culture, and (because it was literate) it is the first unofficial China to which we have some access. Popular demand could be satisfied because of the invention of paper. By AD 700, this had been followed by printing, whose origins lay under the Han in the taking of rubbed impressions from stone. Printing from wood blocks followed and movable type appeared in the eleventh century AD. Soon after this large numbers of books began to be published in China, long before they appeared anywhere else.

The culture of Ch'ang-an never recovered from its disruption by rebellion in 756, only two years after the foundation of an Imperial Academy of Letters (about nine hundred years before any similar institution in Europe). Yet the Sung ascendancy was to produce more great pottery; the earlier, northern phase of Sung history was marked by work still in the coloured, patterned style, while southern craftsmen came to favour monochromatic, simple products. Significantly, they attached themselves to another tradition: that of the forms evolved by the great bronze-casters of earlier China. For all the beauty of its ceramics, though, Sung is more notable for some of the highest achievements of Chinese painting, their subject-matter being, especially, landscape. Above all, though, the Sung era is remarkable for a dramatic transformation of the economy.

The Sung mystery

In part this can be put down to technological innovation – gunpowder, movable type and the sternpost can all be traced to the Sung era – but that may have been as much a symptom as a cause of a surge in economic activity between the tenth and thirteenth centuries. It led to a shift southward of China's economic centre of gravity, and the rise of new ports such as Canton and Fuchow. It appears to have brought to most Chinese a real rise in incomes in spite of continuing population growth, an astonishing fact. For once economic growth in the pre-modern world seems for a long period to have outstripped demographic trends. One change making this possible was certainly the discovery and adoption of a rice variety which permitted two crops a year to be taken from well-irrigated land and one from hilly ground only watered in the spring. Rising production in a different sector of the economy is dramatically suggested by one scholar's calculation that within a few years of the battle of Hastings, China was producing nearly as much iron as the whole of Europe six centuries later. Textile production, too, underwent rapid development (notably through the adoption of water-driven spinning machinery) and it is possible to speak of Sung 'industrialization' as a recognizable phenomenon. Why this happened, and why it did not continue, are questions still vigorously debated.

Undoubtedly there was a real input to the economy under the Sung economy by governmental investment in public works, above all, communications. Prolonged periods of freedom from foreign invasion and domestic disorder also must have helped, though the second benefit may be explained as much by economic growth as the other way round. The main explanation, though, seems likely to be an expansion in markets and the rise of a money economy which owed something to factors already mentioned, but which rested fundamentally on rising agricultural productivity. So long as this kept ahead of population increase, all was well. Capital became available to utilize more labour, and to tap technology by investment in machines. Real incomes rose. It is much harder to guess why economic expansion, once started, did not go on. Instead, average real incomes in China stabilized for something like five centuries, as production merely kept pace with population growth. After that time, incomes began to fall, and continued to do so to a point at which the early twentieth-century Chinese peasant could be described as a man standing neck-deep in water, whom even ripples could drown.

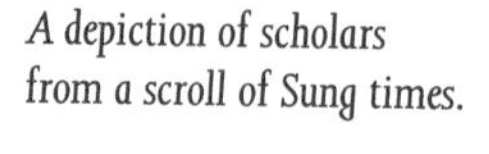

A depiction of scholars from a scroll of Sung times.

China did not go on from the success of Sung days to produce a dynamic, progressive society, then. In spite of printing, her masses remained illiterate down to the present century. The empire's great cities, for all their size and commercial vitality, produced neither the freedom and immunities which sheltered men and ideas in Europe, nor the cultural and intellectual life which in the end revolutionized European civilization, above all, they do not seem to have stimulated effective questioning of the established order. Even in technology, where China achieved so much so soon, there is a similar strange gap between intellectual fertility and revolutionary change. The Chinese often showed their inventiveness, but once Chou times were over, it was the cultivation of new land and the introduction of new crops, rather than technical change, which raised production. Masterpieces had been cast in bronze in the second millennium BC and the Chinese were casting iron fifteen hundred years before Europeans, yet much of the engineering potential of this metallurgical tradition was unexplored even when iron production rose so strikingly. What he called 'a sort of black stone' was burnt in China when Marco Polo was there towards the end of the thirteenth century; it was coal, but there was to be no Chinese steam engine. Chinese sailors in Sung times already had the magnetic compass, but though naval expeditions were sent to Indonesia, the Persian Gulf, Aden and East Africa in the fifteenth century, their aim was to impress those places with Chinese power, not to accumulate information and experience for further voyages of exploration and discovery. They were soon given up, in any case.

Such a list could be much lengthened. Perhaps Chinese civilization was just too successful in pursuit of different goals, the assurance of continuity and the prevention of fundamental change. Neither officialdom nor the social system favoured the innovator and even merchant families were happy, with prosperity, to integrate themselves with the official class. Pride in the Confucian tradition and the confidence buttressed by great resources and remoteness made it difficult to learn from the outside. The Chinese were not intolerant. Jews, Nestorian Christians, Zoroastrian Persians, and Arab Moslems long practised their own religion freely, and the last even made some converts, creating an enduring Islamic minority. Contacts with the West multiplied, too, under Mongol rule. But formal tolerance never led to much receptivity in Chinese culture.

Mongol China

By the end of the thirteenth century, all China had been overrun by the Mongols. Like earlier invaders, the Mongols once more showed China's continuing seductive power over its conquerors. The first blow was very heavy. Something like thirty million lives may have been lost during the conquest, or well over a quarter of China's whole population in 1200. Yet, the Mongol empire under Kubilai, the last of the Great Khans, moved its centre from the steppes to Peking. From this time Mongol China can be considered Chinese, not Mongol; Kubilai adopted a dynastic title in 1271. He broke with the old conservatism of the steppes, the distrust of civilization and its works. He was to spend nearly all his life in China, though his knowledge of Chinese was poor, and his followers slowly succumbed to Chinese culture in spite of their initial distrust of the scholar officials. China changed the Mongols more than the Mongols changed China, and the result was the magnificence reported by the amazed Marco Polo.

Nonetheless, the Mongols sought by positive prohibition to keep themselves apart from the natives. Chinese were forbidden to learn the Mongol language or marry Mongols. They were not allowed to carry arms. Foreigners, rather than Chinese, were employed in administration where possible, a device paralleled in the western khanates of the Mongol empire: Marco Polo was for three years an official of the Great Khan; a Nestorian presided over the imperial bureau of astronomy; Moslems from Transoxiana administered Yunan. For some years, too, the traditional examination system was suspended. Some of the persistent Chinese hostility to the Mongols may be explained by such facts, especially in the south.

Yet, the Mongol achievement was very impressive. China's potential, once united, as a great military and diplomatic power was again shown. The conquest of the Sung south was not easy, but once it was achieved (in 1279) Kubilai's resources were more than doubled. He assembled a major fleet and began to rebuild the Chinese sphere of influence in Asia. In the south, Vietnam was invaded (Hanoi was three times captured) and Burma was occupied for a time after Kubilai's death. These conquests were not, it is true, to prove long-lasting and they resulted in tribute rather than prolonged occupation.

Mongol cavalry of the Ming era: a mounted archer and his stocky pony, whose plaited tail is characteristic of inner Asia.

Success was also qualified in Java; a landing was made there and the capital of the island taken in 1292, but it proved impossible to hold. But only towards Japan was the Mongol thrust wholly unsuccessful, while the maritime trade with India, Arabia and the Persian Gulf which had been begun under the Sung was further developed.

Since it failed to survive, the Mongol regime cannot in the end be considered a success, but this does not take us far. Much that was positive was done in just over a century. Foreign trade flourished as never before. Marco Polo reports that the poor of Peking were fed by the largesse of the Great Khan, and it was a big city. A modern eye finds something attractive, too, about the Mongols' treatment of religion. Only Moslems were hindered in the preaching of their doctrine; Taoism and Buddhism were positively encouraged, for example by relieving Buddhist monasteries of taxes (this, of course, meant heavier impositions on others, as any state support for religion must; peasants often pay for religious enlightenment). Yet in the fourteenth century, natural disasters already combined with Mongol exactions to produce a fresh wave of rural rebellions, the telling symptom of a dynasty in decline. Secret societies began to appear again; one of them, the 'Red Turbans', attracted support from gentry and officials. One of its leaders, a monk called Chu Yan-chang, seized Nanking in 1356. Twelve years later he drove the Mongols from Peking and the Ming era began.

The Ming

Like many other Chinese revolutionary leaders Chu Yan-chang gradually became an upholder of the traditional order. The dynasty he founded, though it presided over a great cultural flowering and managed to maintain the political unity of China, only confirmed China's conservatism and isolation. In the early fifteenth century an imperial decree forbade Chinese ships to sail beyond coastal waters or individuals to travel abroad. Soon, Chinese shipyards lost the capacity to build the big ocean-going junks; they did not even retain their specifications. The great voyages of the eunuch Cheng Ho, who might have been a Chinese Vasco Da Gama, were almost forgotten. At the same time, the merchants who had prospered under the Mongols were harassed.

Meanwhile the Ming dynasty ran to seed, its decline registered by a succession of emperors virtually confined to their palaces while favourites and imperial princes disputed the enjoyment of the imperial estates. Except in Korea, where the Japanese were beaten off at the end of the sixteenth century, the Ming could not maintain the outskirts of Chinese empire. Indo-China fell away from the Chinese sphere, Tibet went more or less out of control and in 1544 Mongols came back to burn the suburbs of Peking. In the following century the Ming were threatened by a people from the north of the Great Wall, the Manchu, living in a province which they later gave its name, Manchuria. When they intervened in China's internal troubles it was the end of the Ming. In 1644 a Manchu dynasty, the Ch'ing, was placed on the throne, and its emperors were to remain there until the twentieth century.

THE CHINESE SPHERE AND JAPAN

The ebbings and flowings of Chinese imperial power did not much affect the enduring influence of Chinese civilization and culture throughout that great East Asian region over which China was so often and so long the paramount power. For most of recorded history, something like a third of the human race has always lived in this huge zone, and the persistent ascendancy of Chinese cultural influence in it has been an enormously important fact of world history. The area is immensely varied, both in climate and terrain, and hard to categorize. Burma, for example, which is from many points of view part of the Indian sphere, nonetheless is inhabited by people whose language is (like those of the Thais and Vietnamese) 'Sinitic' – of the same family as that of the Chinese. Further east, for all the still-obvious influence of China in Japan and Korea, those countries have traces in their languages of the non-Sinitic 'Altaic' group (those spoken by the Turkish peoples). The whole region tends to be very densely populated, a fact of which a partial explanation is the capacity of many places in this hot, moist region to produce two crops a year given intensive labour. It is therefore an area of rich variety in custom and culture, and yet all the East Asian peoples are similar in seeming to show great industry and enterprise, and a willingness to accept a marked subordination of the individual to the group.

China's formal, as distinct from cultural, relations with her immediate neighbours always fluctuated with her military power. The Sui established control over northern Vietnam (Annam), made conquests in Tibet and brought to heel the eastern branch of the Turk peoples. The T'ang emperors had to deal with the Turks again later, but also carried the boundaries of the empire far to the west again, to the Pamirs, taking in the Tarim basin. Korea became a vassal state in T'ang times and a Korean general then led an army across the Pamirs to prevent a junction of the Arabs and Tibetans, though this campaign ended in disaster at the hands of the Arabs at Talas, in 751.

This was an important turning-point. After this Chinese military power waned in Central Asia and this was the main reason why Islam was able to establish itself there. The Sung, in due course, were less successful in dealing with their barbarian neighbours than their predecessors and actually had to pay tribute to some of them. Nor could they win back Annam. It was a Manchurian people which eventually seized North China from them and, in due course, Mongols who rebuilt the unity of the empire. In the end, though, it was not the management of vassal and tributary neighbours which provided imperial China with an insoluble problem, but a barbarian race which came from far away. It was in 1557 that a handful of Portuguese set up the first permanent European settlement in China. They were to be there for a long time.

Japan

At about the same time, other Portuguese succeeded in getting into Japan, the most important of the lands with distinct cultures within the Chinese sphere of civilization, and one for a long time even less known to non-Asians than others. Because Japan is an island country, the sea has protected her – she has never been successfully invaded – and has helped to feed her people. Only recently has fishing ceased to provide the bulk of the protein consumed by Japanese. For a long time Japan was able to take from outside what she wanted while keep-

Minamoto Yoritomo, the unscrupulous, brave and resourceful warrior who established in Kamakura a power which was the base of the first bakufu; he was the first of the shoguns. This is a contemporary portrait; we have no such lifelike likenesses of the feudal lords of early medieval Europe, some of whom must have been not dissimilar in their ambitions for their families – however inferior in their successes.

ing out what she did not, because of the sea, which made the Japanese sailors too, though for a long time this showed in successful piracy and fishing (like the sea-faring of English west-countrymen) rather than in distant enterprises.

Korea is the Asian mainland nearest to Japan, and the Japanese have always been very sensitive about that country. At one time, in the eighth century AD, Japan's rulers held territory there and for much of the twentieth century they dominated it. But China was for much longer the nominal overlord of Korea and China was always the foreign power whose behaviour mattered to Japan more than any other. From very ancient times she influenced Japan deeply. Though their languages are different, both the Japanese and Chinese peoples are Mongoloid (though there are a few Caucasians in the north of the Japanese islands, the descendants of the aboriginal Ainu). In pre-historic times bronze technology seems to have passed from China to Japan. Then, after the Han collapse, when the Japanese began to show much more interest in Korea, contacts with the great mainland civilization multiplied rapidly. The title of 'emperor' given to Japan's ruler, together with Confucianism, Buddhism and a knowledge of iron-working, all passed from China to Japan. Chinese potters came to Japan at an early date, setting up kilns there and inter-marrying with the natives, and from this sprang much of Japan's later artistic achievement. Chinese script was adapted for writing the Japanese language, and government began to show traces of Chinese influence too. In the sixth and seventh centuries, when Chinese influence was at its height, reforming Japanese statesmen made great efforts to set up a centralized government with a civil service on Chinese lines, based on merit, not birth, and an emperor who was a real ruler and not just the head of the most respected clan.

All this may make it appear that Japan borrowed from abroad all that made her a civilized nation. Given the impressiveness of T'ang China and the frequent evidence of Buddhist influence on Japanese art, it would be easy to understand if this were true, but it is in fact not so. The roots of Japanese civilization and government lay at home.

The first Japanese chronicles (compiled in the eighth century) explain how the land and people of Japan were made by the gods, but the earliest firm chronology comes from Chinese and Korean sources three centuries earlier. It shows that at the beginning of the fifth century government was already centred on an emperor. He was supposed to be a descendant of the sun goddess and exercised a general headship over the Japanese national family from his ancestral domains in what was later the province of Yamato. That national family was organized in clans, the main units of Japanese as of earlier Chinese society. From time to time one clan rose to greater power than the others, usually by influencing or even controlling the emperors. But the imperial family itself went on in an unbroken line (though sometimes helped out by adoption); the present emperor still traces his succession directly back to the first, a remarkable claim to continuity.

Between 500 and 1500 there were two important periods when individual clans dominated Japan. In the eighth century the Fujiwara came to the top. For the next two or three centuries they effectively controlled the emperors through marriage alliances and the relationships that followed. During the Fujiwara era the imperial capital was at Heian, the modern Kyoto. The emperor lived there, carrying out his heavy ceremonial and religious duties. But the power of the Fujiwara ebbed. There was fighting between the clans, and a ruthless and very able general, Minamoto Yoritomo, took power in 1185; this was the beginning of the ascendancy of the Minamoto (usually described as the Kamakura era, from the district where the main Minamoto lands were to be found). The Minamoto themselves gave way in the fourteenth century, and after that Japan dissolved into violent and bloody civil wars until the sixteenth century.

The shogunate

At first sight this does not look very interesting. It seems to be on about the same level as the struggles of barons and great families in medieval Europe. In fact it was an important age, in which Japan developed on very individual lines. To begin with, in spite of some attempts to create a strong civil service recruited by merit at the beginning of Fujiwara times, imperial power dwindled away. Power remained in the hands of the nobles. In the end this was to make it very difficult for other interests to be taken into account, and Japanese society tended to freeze into immobility. Offices became hereditary, and rights to levy imperial taxes were granted to those enjoying Fujiwara favour. This eclipse of the emperors was completed in the 'Kamakura' period (1185-1333) when effective government passed to the Minamoto 'shogun', or commander-in-chief, who ruled in the emperor's name but in fact independently and in the interest of his own clan, whose lands were not in the area of Hsian but of Kamakura. Yet there was a steady progress from the erosion of the power of the emperor, to the erosion of any idea of central authority at all. It was very unlike China, and the consequence, often, and repeatedly, was anarchy and even civil war between magnates.

All this would have been very dangerous (and might not have happened) without the protection of the sea. But for most of this era the Japanese had only had one 'national' military problem, the containment of the Ainu barbarians, and Japanese society was not developed enough in other ways to make people demand a more centralized government. Most Japanese seem to have been content to accept the authority of clan and family and the national cult, Shinto.

Another important trend in these centuries was towards a much more military society in which the martial virtues of loyalty, endurance and bravery came to be held in great esteem. In part this was because the lesser nobles and country squires became more independent as the Fujiwara era drew to a close. The civil wars, in which warriors bound themselves to serve lords as retainers, greatly strengthened this. In a much exaggerated form it was not unlike what was going on at about the same time in Europe within the way of ordering society which has been called 'feudal'. In Japan, there gradually emerged as the most respected class below the great nobility the 'samurai', whose knightly ideals have ever since been an inspiration to Japanese patriots – and have also often helped to make Japanese society very violent.

Swordsmanship was highly esteemed in Japan, as were the great fencing-masters, while the beautiful weapons forged by the Japanese smiths remained treasured and revered possessions.

Japanese admiration for the warrior went with a developing sense of Japanese superiority and military invincibility which owed much to the successful resistance to two attempted Mongol invasions, the first in 1274, the second in 1281. There were huge undertakings by well-equipped expeditions (among other things the Mongols made use of Chinese technology and used catapults to throw bombs which burst in the air). The second was effectively destroyed by a storm which wrecked the Mongol fleet – the kamikaze, or 'divine wind', which was seen as a heavenly intervention on behalf of Japan.

Against this background the lot of the ordinary Japanese the peasant – changed only for the worse. For a long time the economy did not much thrive; farming remained what it had always been technically and there was no town growth like that in China. Japan managed slowly to grow more food but largely through the gradual increase in the size of estates and therefore of farmed area rather than through technical advance. The peasant paid heavy taxes, usually to his lord, to whom the right of levying them had been granted by the shogunate, and raised the rice crop which provided most of Japan's food. In the fifteenth century things rapidly worsened. There were plagues and famine, peasants formed leagues to protect themselves, with unemployed warriors as their leaders, and risings followed.

On this poverty there nonetheless rested a brilliant civilization. Under the Fujiwara its brilliance was restricted to the imperial court circle, but later it came to be shared by the whole ruling class. Japan gradually shook off Chinese cultural influences or remoulded them to its own needs. The first Japanese literature and the No drama – a unique combination of poetry, mime and music carried out in elaborate costumes and masks – made their appearance. Interestingly some of the most famous Japanese books were by ladies of the Heian court; it seems to have been felt that, as they were probably not up to anything very serious, women could properly write in Japanese although men should still use Chinese for the serious works of art or learning they produced (much as educated Europeans long went on using Latin as the language of scholarship).

Some of the most beautiful works of art which have ever been made came from this culture. Japanese artists have always emphasized scale, simplicity, and perfect craftsmanship and have shown it in pottery, painting, lacquerwork, and silk-weaving as well as in arts which are much more especially Japanese, such as flower-arrangement, landscape-gardening, or the production of beautiful swords by armourers. Great artists enjoyed widespread fame and admiration. All the arts came to high perfection, too, during the most anarchic period of Japan's history, despite the social and economic damage caused by civil strife.

Some of the beautiful things made by the Japanese eventually began to find their way to markets abroad. In the fifteenth century China was an important customer and Buddhist monks played an important part in her trade with Japan. Inevitably, interest was awoken, and sooner or later people would want to know more about the strange, remote island empire from which such treasures came. Among the curious would be Europeans, the first among whom to arrive were the Portuguese, probably in 1543. Others soon followed. Japan's internal conditions at that time put no obstacle in their way. Nagasaki, a little village, was opened in 1570 to the Portuguese by one magnate who was already converted to Christianity. Besides their faith, though, the intruders had also brought firearms, whose first impact on Japanese society was to inflame still more its appetite for internal strife. The eagerness with which Japanese adopted the new weapons seems in retrospect a portent of what was to come, the most considered and well-motivated of all processes of deliberate modernization by a non-European people, albeit one that lay two and a half centuries ahead.

L:ANGLI ET FRA

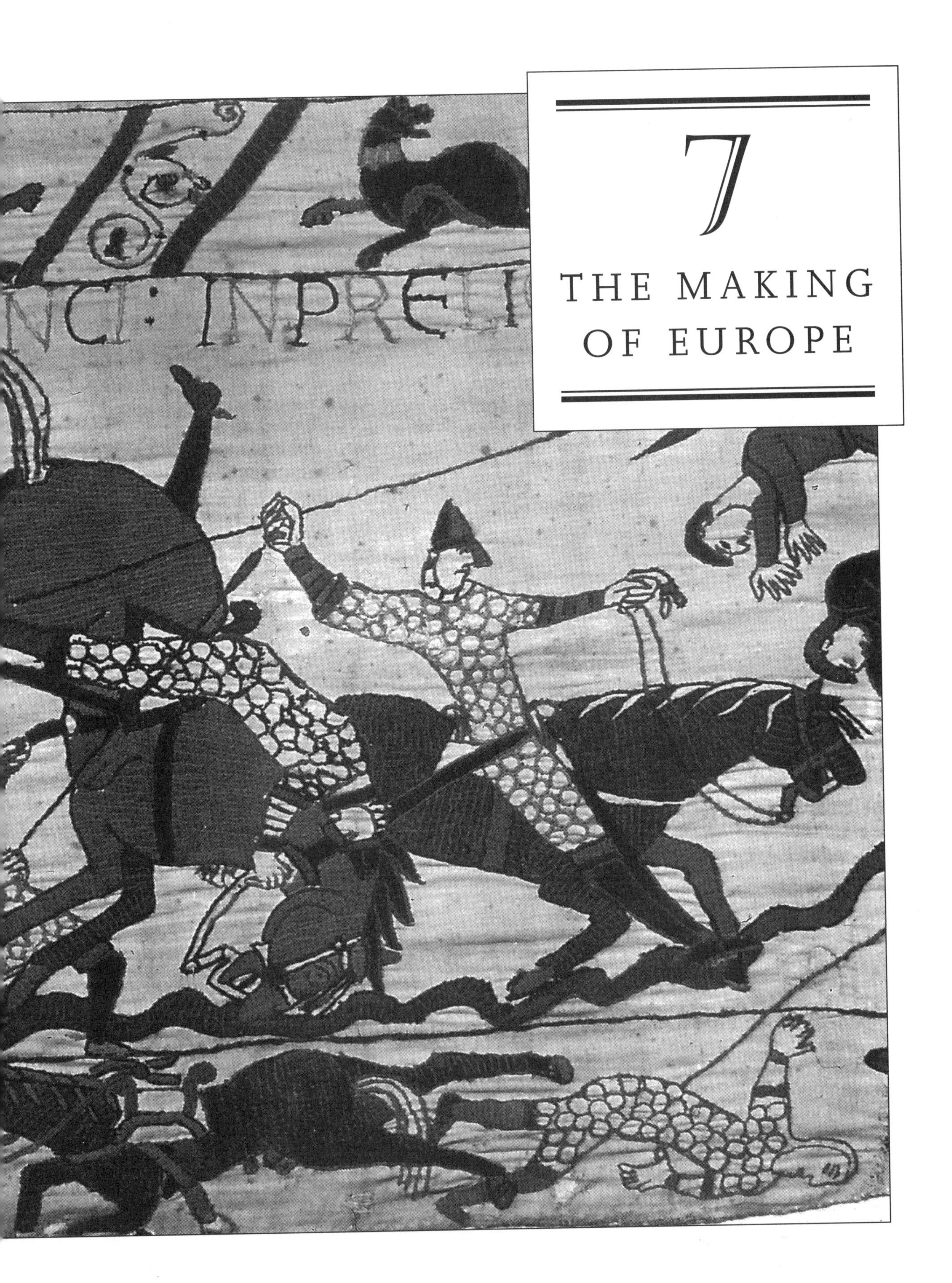

7 THE MAKING OF EUROPE

MEDIEVAL CHRISTENDOM: THE WEST

The centuries between the end of classical antiquity and the year 1000 or so now look very much like Europe's age of foundation. Certain great markers then laid out the patterns of the future, though change was slow and its irreversibility uncertain. Then, in the eleventh century, a change of pace can be sensed. New developments become discernible and, as time passes, it is clear that they are opening the way to something quite different. An age of adventure and revolution was beginning in Europe, and it was to go on until European history merged with the first age of global history. As a result, much which we can recognize as part of our idea of Europe was in place in, say, 1500 and it had not been there a thousand years earlier. By then towns and cities of which many were to become the great urban agglomerations of our own day were scattered scross the continent. Over much of western Europe men and women were by then already getting a living from other sources than agriculture. Complicated patterns of trade, some of it over long distances, and much of it by sea, can be made out. Technological skills were already producing change faster than anywhere else in the world. And most Europeans were aware that, whatever they might call it and however they imagined it, the world outside was a very different place from the one they lived in. It was, above all, a non-Christian world: Europe was distinctive because it was Christian and, by the end of the fifteenth century, cradled a Christian civilization. Indeed, if people thought about it at all in abstract terms, it still was as 'Christendom'. With this fact, the story begins.

The origins of the Christian West

We should be wary of the associations of the word 'government' when thinking of what succeeded it. Formally, it was a number of barbarian kingdoms, all of them set up by the Germanic peoples who had flooded into the western Empire from the third century onwards; they were really matters of tribal organization and the personal ascendancy of powerful men.

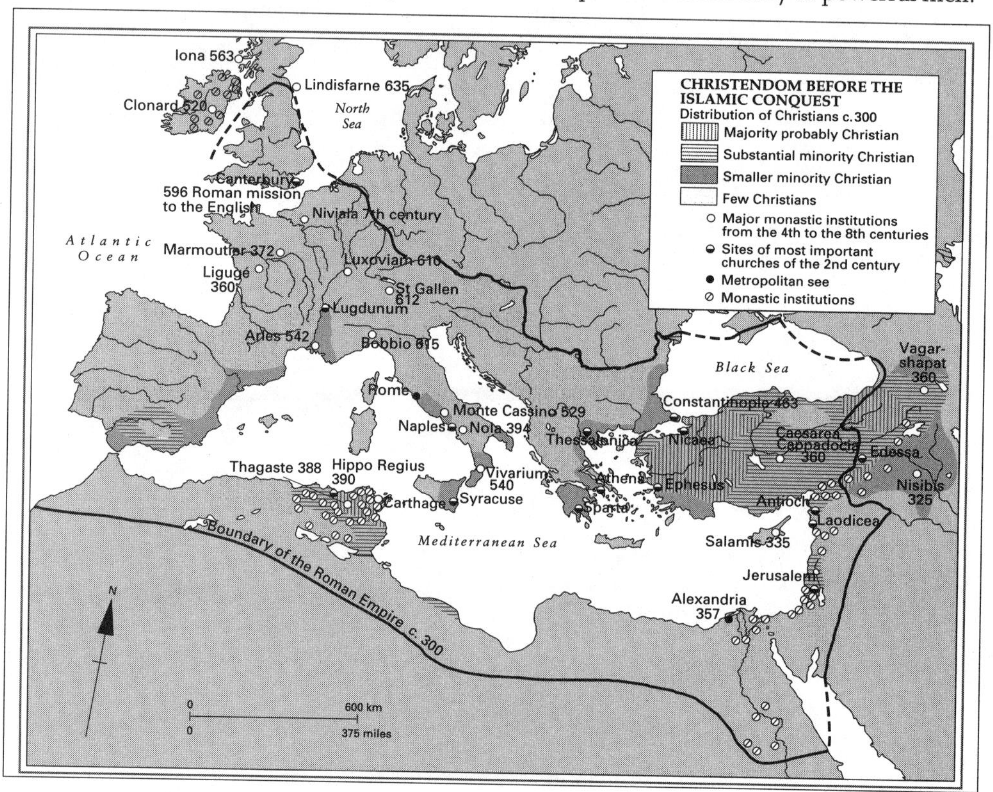

They were somewhat Romanized, and gradually Christianized, but for centuries, even the greatest European kings were hardly more than barbarian warlords to whom men clung for protection and in fear of something worse. The foundations of a new civilization had to be laid in barbarism and backwardness. For centuries, no city in western Europe, not even once-great Rome, could approach in magnificence Constantinople, Córdoba, Baghdad or Ch'ang-an, nor could architecture in Europe compare with that of the Hellenistic past, of Byzantium or of the Asian empires. When such architecture emerged, it did so by borrowing their styles. For just as long, western Europe could produce no science, no school to match those of Arab Spain or Asia. It was a cultural backwater. Its inhabitants grew used to privation rather than opportunity and huddled together under the rule of the warriors whom they needed for their protection.

Obviously, the Roman heritage had not disappeared overnight. The barbarian kingdoms employed scribes and clerks when they wanted to keep records, took up Roman ways and gave Roman titles to their own dignitaries. They had for a time to do business locally with the notables and officials of the old imperial days, too. Although it is very difficult to measure, something of the Roman past lived on west of the Rhine and the upper Danube. Language is one indicator of survival. The 'Latin' or 'Romance' languages are still spoken in Italy, France, Spain and other places because the influx of barbarians was not great enough to swamp the forms of Latin which were widespread in those countries. Many of the most important men among the barbarians began before long to speak it themselves – and to write it, if they knew how to write. In the British Isles, on the other hand, Latin virtually disappeared and the Germanic tongues of the invaders replaced the language of the Romano-British, while Latin had never established itself beyond the *limes*.

By far the most important institution left behind by the empire was to turn out to be the Church. All over western Europe its bishops – sometimes men of important families, with wealth and powerful connexions to back them up – were key figures in local affairs, taking on tasks previously carried out by imperial officials. The Church came more and more to stand for what Rome itself had long stood for – civilization. The line between Christianity and paganism was also the line between Roman civilization and barbarism.

The papacy

One bishop was especially important, the pope of Rome. There were several reasons why his standing and influence should grow as the centuries passed. Unquestionably the most important was the growing separation from the east. Rome, where St Peter and St Paul had been martyred, and where there had long been a big Christian community, was the most important bishopric in the west and the natural leader of the Church there. Later, when the Arabs swept through North Africa and its old Christian communities disappeared, this leadership was even more unquestioned. Rome was also a place to which a profitable pilgrim trade came; the bones of St Peter himself were to be found there, men believed. The fact that Rome was for so long the old imperial capital and that the pope had long been used to dealing with high imperial authorities also helped when a breakdown of government followed the barbarian invasions of Italy. As the imperial administration collapsed, so, in many places, the pope's men had virtually to take over local government. One pope went to see the Huns' leader, Attila, and men said that it was his personal intervention which had turned the barbarian back. In many ways, then, the papacy rose steadily in importance.

Inevitably, much of the Church's history is the history of the papacy, Christianity's best-documented institution. The division of the old empire meant that if there was anywhere in the West a champion of the faith, it had to be Rome, whose claim to primacy rested on the guardianship of St Peter's bones and it was indisputably the only apostolic see in the West. After the tenure of pope Gregory the Great (590-604), the founder of papal supremacy in the western Church, it was implausible to maintain the theory of one Christian Church in one

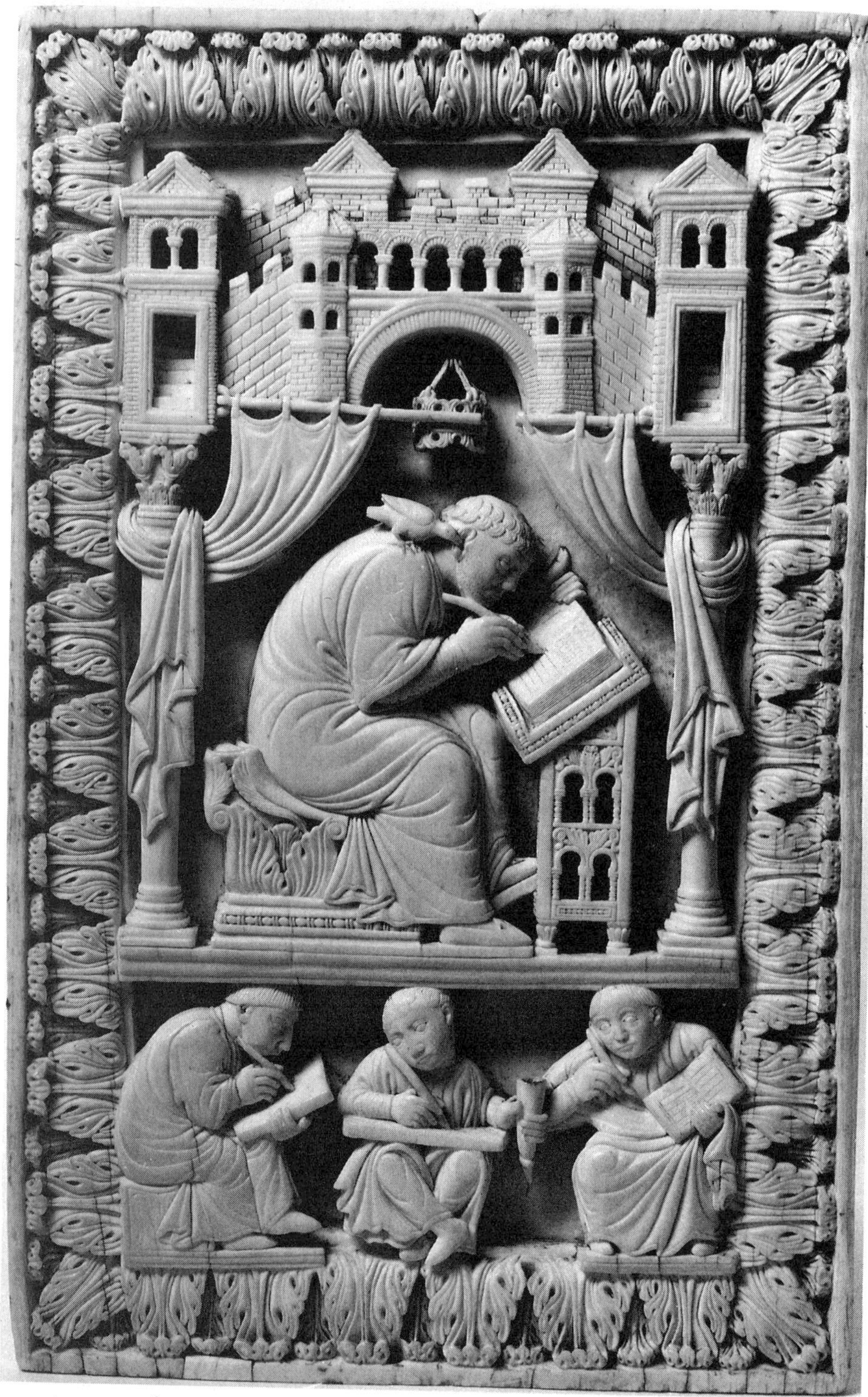

The tenth century Rhenish artist who made this ivory chose for his subject St Gregory I (the Great), writing under the influence of the Holy Spirit, here shown as a dove on his shoulder. Below him are the scribes of a monastic copying-school, one holding the ink horn for the others. Gregory left a large number of writings which were to influence the future developments of the church after his death as much as he had done through the great decisions of his pontificate.

empire, even if an imperial official hung on at Ravenna as the representative of the emperor in Byzantium. The last emperor who visited Rome did so in 663 and the last pope to go to Constantinople went there in 710.

Papal diplomacy helped a pattern of Christian kingdoms to emerge slowly from barbarian Europe. There were difficulties; not only were many of the Germanic peoples pagan,

Eight miles off the west coast of Ireland lies the seven-hundred-foot-high rock of Skellig Michael. Some beehive cells for monks and a tiny church are the remains of an old Celtic monastery established there at an unknown but early date. From such precarious footholds at the edge of the civilized world came the Irish monks who were the first Christians to evangelize the former Roman province of Britain and the Celtic peoples of Scotland.

some were Christian heretics – Arians, who denied the divinity of Christ. Their background could not be eliminated overnight. So the Church often acted very cautiously and prudently, 'christening' old pagan shrines, for example, by associating them with a Christian martyr or hermit, and adopting heathen feast-days and turning them into Christian ones. A lot of near-magic and simple superstition was thus accepted as a part of Christianity and so it long remained. But on the big questions – condemning blood-feuds, for example, or upholding the principle of Christian marriage, which was monogamous – the Church stood firm. And so, in the end, it slowly civilized the peoples from whom were to emerge the first Europeans.

The only part of western Europe in which Christianity had been virtually obliterated was the old Roman province of Britannia. Christianity survived in Ireland and on the fringes of Scotland, where it had been driven by barbarian invasions. From this perimeter missionaries came back in the sixth century to what was to become England, in particular under the leadership of St Columba, who founded a monastery on Iona in 563. But England was also to be the target for Rome's first major missionary effort. Assisted by the Celtic Church (with which, however, they were in rivalry) the Roman missionaries who worked from Canterbury re-established Christianity in England in the seventh century by converting the kings of the little kingdoms there which had emerged from the boatloads of Saxons, Jutes and Angles who had been arriving since the legions left in 407.

Monasticism

The papacy was not the only institution which played a crucial part in making western Europe Christian. Another was monasticism, which had first appeared in Egypt and the Near East. In the West, though, it took a notably new form in the sixth century, largely because of the work of one remarkable man, St Benedict. From a community he founded in 529 at Monte Casino in southern Italy, he had a huge effect on the nature of monasticism throughout the western Church. His 'Rule', a set of standing orders for his monks, was the seed of a general reform. It prescribed that besides attending to their devotions, monks should both work and study. Even the Irish monastic rule of St Columba, at first a serious rival, gave way to it. The 'Benedictines', following Benedict's Rule exactly, and after them other new orders of monks who used it as a model, built all over Europe communities in which men worked and prayed together, preaching to and evangelizing the neighbourhood. They were centres of education, art, and even farm management. From these monasteries more and more monks went out as the centuries passed to advise kings and become bishops and to put backbone into the ordinary clergy of the parishes – the 'secular' clergy as they are sometimes called, to distinguish them from the 'regular' clergy of the monastic orders. One monk became pope Gregory the Great.

There were communities of female 'religious' too. Christianity was certainly not alone among world religions in finding a role for women, and other countries have had holy women, priestesses and what have been called, for want of a better word, 'nuns'. But the female religious orders of Christendom, modelled on those of the monks, were quite new both in the scale of their foundations and the opportunities they provided to women. Hundreds of thousands of women found a place in Christian religious communities as the centuries passed. Christianity has been accused of rating women too lowly, but about its religious women it has no need to be defensive. There was until the Christian era virtually no respectable profession open to a woman in any society, but Christianity opened the way to a future in which some of them nursed, some of them studied, some of them administered estates and some devoted themselves to the contemplation and prayer which they thought their highest task. No other society gave such opportunities to its women, though they were opportunities purchased at the price of vows of obedience and chastity, and of acceptance of a life withdrawn from the world.

A NEW EUROPEAN STRUCTURE

By AD 1000 a new western Christendom is visible consisting of about half the Iberian peninsula, modern France, Germany west of the Elbe, Bohemia, Austria, the Italian mainland and England. On its fringes lay Christian Ireland and Scotland, and, just coming into its ambit, the Scandinavian kingdoms. Outside this area, the Arabs were established in much of Spain as well as in Sicily, Corsica, Sardinia and the Balearics. To the east Slav Christendom and Byzantium were both culturally worlds apart from the West, though they provided a Christian cushion which sheltered it from the full impact of eastern nomads and of Islam. Western Europe was all but landlocked; though the Atlantic was wide open, there was almost nowhere to go in that direction once Iceland had been settled by the Norwegians, while the western Mediterranean, the highway to other civilizations and their trade, was an Arab lake.

Looked at more closely, the importance of certain great changes since the Dark Ages which followed the western Roman empire's collapse is clear. There had been a cultural and psychological shift away from the Mediterranean, the focus of classical civilization. The centre of European life, in so far as there was one, moved to the valley of the Rhine and its tributaries. That was where its main concentrations of wealth, such as they were, were to be found. A second change was more positive, a gradual advance of Christianity and settlement in the east. By 1000, the advance guards of Christian civilization were well beyond the old Roman frontiers. A third change had been the slackening of barbarian pressure, after three or four dangerous centuries. It is true that two hundred years later, when the Mongols menaced her, it must still have been difficult to feel that Christendom was no longer to be a prey to outsiders. Yet it was true. She had ceased to be wholly plastic. Not only had the pressures upon her begun to relax by that date, but some of the lineaments of a later, expanding Europe were already to be seen. In the central area of western Christendom a future France and a future Germany were in the making. To the south lay the beginnings of a new west Mediterranean littoral culture, embracing at first Catalonia, the Languedoc and Provence. With time and the recovery of Italy from the age of barbarism, this had extended itself further to the east and south. A third distinct region was constituted by the somewhat varied periphery in the west,

A tenth-century history of the councils of Spain, the 'Codex Vigilanus' (or codex of Vigila) described the history of Spain, as the chroniclers of other former barbarian peoples described theirs, as a story of Christian monarchy. The Codex ends with a page showing five kings, three Visigothic, two kings of Leon, with the wife of the one who commissioned it. The bottom row depicts some of the very few individual artists who emerge from the anonymity of early medieval art, the scribe Vigila himself (in the centre) with an assistant and a pupil.

north-west and north where there were to be found the Christian states of northern and post-Visigothic Spain, England, with its independent Celtic and semi-barbarous neighbours, Ireland, Wales and Scotland, and, lastly, the Scandinavian states. One could argue about allocating some areas (such as Aquitaine, Gascony and sometimes Burgundy) to one or the other of these three regions but these distinctions are real enough to be useful. Historical experience, as well as climate and race, differentiated these regions, little as most men living in them would have known it.

The Franks

Men, as well as geography, had done much to bring about this degree of definition. In the fourth century there had come to be settled as *foederati* within the Roman frontiers a Germanic people called Franks. They lived between the Scheldt and Meuse, in lands now Belgian, another branch of the same stock remaining east of the Rhine. Among them was one kinship group remembered as Merovingians, under whom the Franks of the fifth century turned on their erstwhile hosts, the Roman governors of Gaul, and conquered much of the country to the west and down to the Loire.

One Merovingian prince, Clovis, won acceptance as ruler of both western and eastern Franks. He married (traditionally, in 496) a Burgundian princess who was a Catholic and then underwent a battlefield conversion somewhat like that of Constantine, to Christianity. This gave him the support of the Gaulish Church and the pope, and with it he went on to win an ascendancy throughout Gaul from the Visigoths and Burgundians. He moved his capital to a little town on an island in the Seine, called by the Romans Lutetia. Its inhabitants were Christian (there had been bishops there at least since 250) and known as *Parisii*, and it is now called Paris. Clovis was eventually buried there as a Christian, the first Frankish king so to be interred. When he died, his lands were divided in the customary Frankish manner.

The heritage was later put together again (his descendants adding to it the lands of the Ostrogoths north of the Alps), then divided once more. What survived, nevertheless, was a clear Frankish supremacy over much of north-west Europe and an important new ally for the Church. From a barbarian people, a new power had emerged. This was to be the key to a Frankish shaping of a new order in the west, and therefore in the end, of a New Europe. The heartland of the medieval West was to be the Frankish heritage. It was neither wealthy nor

The Frankish warrior. Taken from a Saxon funeral bottom stele of about 700, this is one of the earliest naturalistic renderings of a barbarian which survives.

No authenticated portrait of Charlemagne exists, but this lively scene of him with his son Pepin and attendant scribe, ink horn in hand, may well give a fair impression of his ordinary dress and bearing.

culturally very advanced. With fewer towns than the south its life centred on the soil. Its rulers were successful warriors turned landowners. Yet from this base, the Franks began the colonization of Germany, protected the Church, and hardened and passed on a particular tradition of Christian kingship to the future.

Though there was dynastic continuity after Clovis, a succession of impoverished and therefore feeble kings had conceded more independence to Frankish aristocrats, and one family among them, from Austrasia, the Frankish lands east of the Rhine, came to overshadow the Merovingian royal line. It produced Charles Martel, the soldier who turned the Arabs back at Tours in 732 and the supporter of St Boniface, the missionary who evangelized Germany. This is a considerable double mark to have left on European history and it confirmed the alliance with the Church. After Martel's second son, Pepin the Short, was chosen king by the Frankish nobles in 751, the pope came to France and anointed him king as Samuel had anointed Saul and David, and gave him the title of Patrician of Rome. The papacy needed its powerful friend. Lombards were terrorizing Rome, and the eastern emperor would not help (and in Catholic eyes was, at that moment, almost a heretic). Pepin defeated the Lombards and in 756 established the Papal States of the future by granting Ravenna 'to St Peter'. It was another landmark in European history, the beginning of eleven hundred years of the papacy's Temporal Power, during which the pope ruled his own dominions like any other monarch. There also followed the reform of the Frankish Church, further colonization and conversion in Germany (through wars waged against the pagan Saxons), and the throwing back of the Arabs across the Pyrenees. The Roman-Frankish axis was good for Catholicism.

Charlemagne

Pepin's lands were once more divided at his death but the whole Frankish heritage was united yet again in 771 in his elder son Charlemagne, the greatest of the Carolingians, as his line came to be called, and soon a legend. He was obviously still a traditional Frankish warrior-king; he conquered, and his business was war. It was more unusual that Charlemagne took very seriously the Christian sanctification of this role. He sought to magnify the grandeur and prestige of his court by filling it with evidence of Christian learning and patronizing learning and art. Territorially, he was a great builder, but this had its religious aspects, too. He overthrew the Lombards, the subjugators of Italy and tormentors of the Popes, and took their crown and their Italian lands. For thirty years he hammered away in campaigns in the east and achieved the conversion of the Saxon pagans by force. Fighting against the Avars, Wends and Slavs brought him Carinthia and Bohemia and, perhaps as important, an open route down the Danube to Byzantium. To master the Danes, the Dane Mark (March) was set up across the Elbe. Early in the ninth century Charlemagne pushed into Spain and set up a corresponding Spanish March across the Pyrenees and running down to the Ebro and the Catalonian coast.

Historians have been arguing almost ever since about what this agglomeration really was, as they have also argued about what Charlemagne's coronation in Rome by the pope on Christmas Day, 800, and his acclamation as emperor, actually meant. An emperor whom

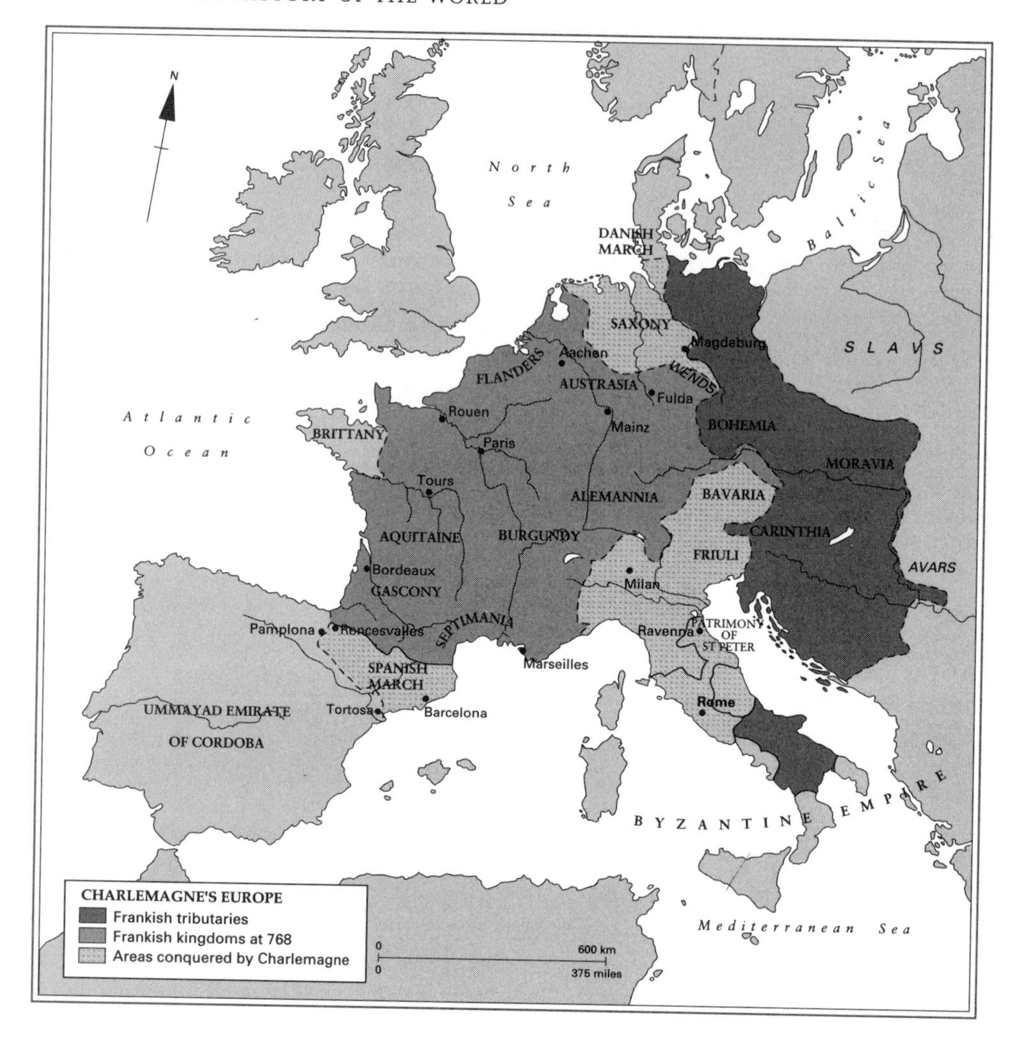

everybody acknowledged to be such lived in Constantinople: were there now to be two emperors of a divided Christendom, as in later Roman times? Clearly, empire was a claim to authority over many peoples; by taking the title, Charlemagne said he was more than just a king of the Franks. Perhaps at first Italy mattered most in explaining it, for among the Italians a link with the imperial past might be a cementing factor as nowhere else. An element of papal gratitude – or expediency – was involved, too; Leo III had just been restored to his capital by Charlemagne's soldiers. Yet Charlemagne is reported to have said that he would not have entered St Peter's had he known what the pope intended to do, and he may have disliked the implied arrogation of authority by the pope.

Whatever his motives, Charlemagne's seal bore before long the legend 'Renewal of the Roman empire'. This was conscious and ostentatious reconnexion with a great past, and it is hardly surprising that though his title was for a few years later recognized at Constantinople as valid in the west, the new emperor's relations with Byzantium were never easy. Perhaps significantly, Charlemagne had with the Abbasid caliphate somewhat formal but not unfriendly relations; the Arabs menaced Byzantium. With the Umayyads of Spain he got on less well; they were near enough to be a threat.

To protect the faith from pagans was a part of Christian kingship but this did not mean independence for the Church. Charlemagne firmly subordinated it to his authority, using it as an instrument of government and ruling through its bishops. He presided over Frankish synods, pronouncing upon dogma as authoritatively as had Justinian, and seems to have hoped to reform both the Frankish Church and the Roman, imposing the Rule of St Benedict. Charlemagne foreshadowed the later idea that a Christian king should be responsible not only for the protection of the Church but for the quality of the religious life within his dominions.

His court at Aachen was perhaps primitive by comparison with that of Byzantium – and possibly even in comparison with those of some of the early barbarian kingdoms – but when Charlemagne's men brought materials and ideas from Ravenna to beautify Aachen, Byzantine art began to fertilize the north European tradition. Classical models, too, influenced his artists. But it was as an intellectual centre and through its scholars that the court was most successful and important. Scribes copied texts in a new hand called Carolingian minuscule which was to be one of the great instruments of culture in the West. Charlemagne had hoped to use it to supply an authentic copy of the Rule of St Benedict to every monastery in his realm, but its greatest importance lay in the making of copies of the Bible. It was to be the major text in the monastic libraries which now began to be assembled throughout the Frankish lands. This had a more than religious significance: the Jewish history of the Old Testament was full of examples of pious and anointed warrior-kings and these were models it would be helpful to publicize. Copying and the diffusion of texts went on for a century after the original impulse had been given at Aachen and were the core of what has been called 'the Carolingian Renaissance'. The term had none of the later pagan connotations of the noun; it was emphatically Christian. Its purpose was the training of clergy to vitalize and leaven the Frankish Church and carry the faith as missionaries further to the east. There were several Irishmen and Anglo-Saxons in the palace school at Aachen, among them an outstanding cleric from York, a great centre of English learning, Alcuin. His most famous pupil was Charlemagne himself, but he had several others and managed the palace library. At his school at Tours, where he became abbot, he expounded classical authors to the men who would govern the Frankish church in the next generation.

Alcuin's is as striking an example as any of the shift in the centre of cultural gravity in Europe, northward and away from the classical world. But besides Franks, there were Visigoths, Lombards and Italians too who contributed to teaching, copying and founding the new monasteries which spread outwards into both east and west Francia. One of them, Einhard, wrote a life of Charlemagne from which we learn that he was sometimes garrulous, that he was a keen hunter and passionately loved swimming and bathing in the thermal springs, which explain his choice of Aachen as a residence. Charlemagne comes to life in Einhard's pages as an intellectual, too, speaking Latin as well as Frankish, and understanding Greek. This is made more credible because we hear also of his attempts to write, keeping notebooks under his pillow so that he could do so in bed, 'but,' Einhard says, 'although he tried very hard, he had begun too late in life'.

A vivid picture can be formed of Charlemagne as a dignified, majestic figure, striving to make the transition from warlord to ruler of a great Christian empire, and having remarkable success in so doing. Clearly his physical presence was impressive (he probably towered over most of his entourage), and men saw in him the image of a kingly soul, gay, just and magnanimous, as well as a heroic paladin of whom poets and minstrels would for centuries sing. He gave a new majesty to Germanic kingship. When his reign began, his court was still peripatetic, normally eating its way from estate to estate throughout the year. When Charlemagne died, he left a palace and a treasury established at the place where he was to be buried. He had been able to reform weights and measures, and had given to Europe the division of the pound of silver into 240 pennies (*denarii*) which was to survive in the British Isles for eleven hundred years. But his power was also very personal. In the last resort, even a Charlemagne could only rely on a monarchy based on his own domain and its produce and on the big men close enough to him for supervision. These vassals were bound to him by especially solemn oaths, but even they began to give trouble as he grew older.

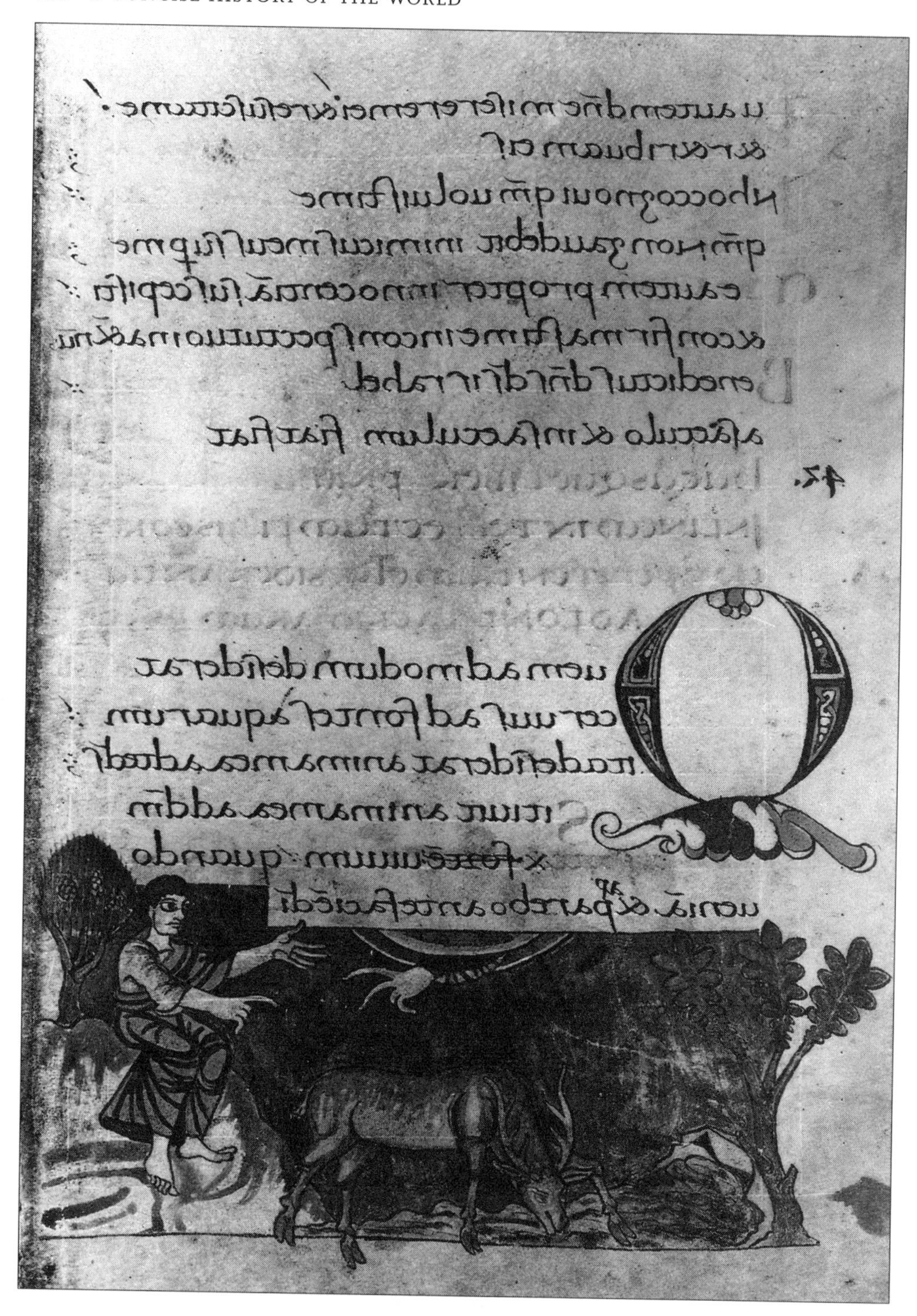

A psalter copied at a French Abbey in about 825, written in the new, easily-read hand of the Carolingian copyists.

The beginnings of France and Germany

Charlemagne's successors had neither his authority nor his experience. Partition was not long delayed after his death. Regional loyalties formed around individuals and in the end three of Charlemagne's grandsons divided up the heritage of the Franks in the Treaty of Verdun of 843, a settlement of great importance in defining Europe's future map. It gave a core kingdom of Frankish lands centred on the western side of the Rhine valley (containing Aachen) to Lothair, the reigning emperor (thus it was called Lotharingia) together with the 'kingdom' of Italy. To the east, the lands of Teutonic speech between the Rhine and the German Marches went to Louis the German. Finally, in the west, a tract of territory including Gascony, Septimania and Aquitaine, went to a half-brother of these two, Charles the Bald. This settlement effectively made the first political distinction of France and Germany; Lotharingia had much less linguistic, ethnic, geographical and economic unity than either of the other two kingdoms for it was there in large measure because three sons had to be provided for. Much future Franco-German history was going to be about ways of dividing it.

Carolingian kings grew weaker as time went by. In west Francia they lasted just over a century after Charles the Bald. By the end of his reign Brittany, Flanders, and Aquitaine were to all intents and purposes independent. The west Frankish monarchy thus started the tenth century in a weak position. In 911 Charles III, unable to expel the Norsemen, conceded lands in what was later Normandy to their leader, Rollo. Baptized the following year, Rollo set to work to build the duchy for which he did homage to the Carolingians; his Scandinavian countrymen continued to arrive and settle there until the end of the tenth century, yet somehow they soon became French in speech and law. Meanwhile, from confusion over the succession in west Francia there emerged a count of Paris who steadily built up his family's power around a domain in the Ile de France. This was to be the core of the later France. When the last Carolingian ruler of the west Franks died in 987, Hugh Capet, son of the count of Paris, was elected king. His family was to rule for nearly four hundred years. For the rest, the west Franks were by then divided into a dozen or so territorial units ruled by magnates of varying standing and independence.

The Holy Roman Empire

800 ad	Coronation of Charlemagne
840–3	Division of Carolingian empire on death of Louis the Pious. Lothair I takes title of emperor (together with Italy and Lotharingia).
955	Battle of Lechfeld: Otto I (the Great) finally removes the Magyar menace with this victory.
966–72	Otto I's third expedition to Italy: deposition of one pope, restoration of another, nomination of a third.
998	Otto III deposes a pope.
1046	Henry III deposes three rival popes and reaffirms right of nomination to the papacy.
1075–1122	Investiture struggle, formally ended by Concordat of Worms.
1125	Elective principle for selection of emperors established with accession of Lothair II.
1138	Hohenstaufen dynasty of emperors begins with Conrad III. Prolonged struggle with papacy follows.
1152–90	Frederick I (Barbarossa) began use of style 'Holy Roman Empire'.
1183	Peace of Constance (between emperor, pope and Lombard cities) opens way to divergence of Germany and Italy under formal suzerainty of the emperor.
1245	Frederick II deposed by Pope Innocent IV at Synod of Lyons.
1268	Last Hohenstaufen prince murdered.
1356	The 'Golden Bull' of Charles IV settles constitution of the Holy Roman Empire until 1806.

The German emperors

Across the Rhine, when the last Carolingian king died in 911 a political fragmentation followed which set political patterns in Germany for nearly a thousand years. The assertiveness of local magnates combined with stronger tribal loyalties than in the west to produce a half-dozen powerful dukedoms. The ruler of one, Conrad of Franconia, was chosen as king by the other dukes, who wanted a strong leader against the Magyars, and at his coronation the bishops anointed him, the first ruler of the east Franks so to be treated. But Conrad was not successful against the Magyars. When he strove, with the support of the Church, to exalt his own house and office the dukes gathered their peoples about them to safeguard their independence. The four which mattered most were the Saxons, the Bavarians, the Swabians and the Franconians (as the east Franks became known). Conrad nominated one of the rebels his successor and the dukes agreed. In 919, Henry 'the Fowler' (as he was called), Duke of Saxony, became king. He and his descendants, the 'Saxon emperors', or 'Ottonians', ruled the eastern Franks until 1024.

Henry the Fowler avoided ecclesiastical coronation. He had great family properties and the

tribal loyalties of the Saxons on his side and brought the magnates into line by proving himself a good soldier. He won Lotharingia from the west Franks, created new Marches on the Elbe after victorious campaigns against the Wends, made Denmark a tributary kingdom and began its conversion, and defeated the Magyars. His son, Otto I, thus had a goodly inheritance and made good use of it.

He continued his father's work in disciplining the dukes, and in 955 he inflicted on the Magyars a defeat which ended for ever the danger they had presented. Austria, Charlemagne's east March, was recolonized. Though he faced some opposition, Otto made a loyal instrument out of the Church; in Germany churchmen tended to look with favour to the monarchy for protection against predatory laymen. With Otto ends, it has been said, the period of mere anarchy in central Europe; under him, there is the first inkling of something we might call a self-conscious Germany.

The emperor Otto III receives the homage of the four parts of the empire, Sclavonia, Germania, Gallia and Roma. Like the leading personage in a Byzantine painting, he is drawn larger than the accompanying figures.

In 936 Otto was crowned at Aachen, accepting the anointing which his father had avoided, German dukes served him as his vassals in the old Carolingian style at his coronation banquet. Fifteen years later he invaded Italy, and assumed its crown. The pope refused him an imperial coronation for ten years, and then crowned him in 962, so uniting again the German and Italian crowns. It was the beginning of what would one day be known as the Holy Roman Empire and would last nearly a thousand years. Yet it was not so far-flung an empire as Charlemagne's, nor did Otto dominate the Church as Charlemagne had done. Nevertheless, it was a remarkable achievement. His successors successfully maintained the tradition he established of exercising power south of the Alps. Otto III, his grandson, made a cousin pope (the first German to sit in the chair of St Peter) and followed that by appointing the first French pope. Rome seemed to captivate him and he settled down there. Like both his immediate predecessors, he called himself 'Augustus' but in addition his seals revived the legend 'Renewal of the Roman Empire' – which he equated with the Christian empire. Half Byzantine by birth (his mother was a Byzantine princess), he saw himself as a new Constantine and dreamed of a Europe organized as a hierarchy of kings under the emperor, an eastern notion. After his death in 1002, his body was taken to Aachen, as he had ordered, to be buried beside Charlemagne.

Otto left no heir, and his elected successor, Henry II, was at heart a German ruler, not an emperor of the West. His seal's inscription read 'Renewal of the kingdom of the Franks' and his attention was focused on pacification and conversion in east Germany. Though he made three expeditions to Italy, Henry relied there on playing off of factions against one another, and with him the Byzantine style of the Ottonian empire began to wane. Yet the eleventh century opened with the idea of western empire still capable of beguiling monarchs, though the Carolingian inheritance had long since crumbled into fragments. Germany was a reality, even if still inchoate and conceptually barely existent. Meanwhile, in France, too, the main line of the future was settled, though it could not have been discerned at the time. Though the suzerainty of the Capetians was for a long time feeble over the fragments of West Francia, they had on their side a centrally placed royal domain, including Paris and the friendship of the Church.

Southern Europe

The other major component of the Carolingian heritage had been Italy. Since the seventh century it had been evolving away from the possibility of integration with northern Europe and back towards re-emergence as a part of Mediterranean Europe. Even after Charlemagne had overthrown the Lombards, though, the popes were often in danger. They had to face both the rising power of the Italian magnates and their own Roman aristocracy. The western Church was at its lowest ebb, lacking cohesion and unity. The Ottonians' treatment of the papacy showed how little power it had. An anarchic Italian map was another result of this situation. The north was a scatter of feudal statelets. Venice was already a successful independent republic and had been pushing forward in the Adriatic; her ruler had just assumed the title of duke. She is perhaps better regarded as a Levantine and Adriatic rather than a Mediterranean power. City-states which were republics existed in the south, at Gaeta, Amalfi, Naples. Across the middle of the peninsula ran the Papal States. Over the whole fell the shadow of Islamic raids as far north as Pisa, while emirates appeared at Taranto and Bari in the ninth century. These were not to last, but the Arabs completed the conquest of Sicily in 902 and were to rule it for a century and a half with profound effects.

The Arabs shaped the destiny of the other west Mediterranean coasts of Europe, too. Not only were they established in Spain, but even in Provence they had more or less permanent bases (one of them at St Tropez). The inhabitants of the European coasts of the Mediterranean had, perforce, a complex relationship with the Arabs, who appeared to them both as freebooters and as traders. Southern France and Catalonia were areas in which Frankish had followed Gothic conquest, but the physical reminiscences of the Roman past were plentiful in these areas and so was a Mediterranean agriculture. Another distinctive characteristic was the appearance of a family of Romance languages in the south, of which Catalan and Provençal were the most enduring.

The Norsemen

Pagan Norsemen had long been shaping the history of the British Isles and the northern fringe of Christendom. Probably because of over-population, the Scandinavians had begun to move outwards from the eighth century onwards. Equipped with longboats which oars and sails could take across seas and up shallow rivers and tubby cargo-carriers which could shelter large families, their goods and animals for six or seven days at sea, they thrust out to implant a civilization which was to stretch from Greenland to Kiev. The Norwegians colonized. The Swedes who penetrated Russia and survive in the records as Varangians were much busier in trade. Danes did most of the plundering and piracy. But no single people had a monopoly of any one of the themes of the Scandinavian migrations, and those who took part in them are traditionally called, indiscriminately, 'Vikings', a word that meant 'sea-rover', or 'pirate'.

Colonization was their most spectacular achievement. Norsemen wholly replaced the Picts in the Orkneys and the Shetlands and from them extended their rule to the Faroes (previously uninhabited except for a few Irish monks and their sheep) and the Isle of Man. Such settlement was more lasting and profound than that on the mainland of Scotland and Ireland, where it began in the ninth century. Yet the Irish language records their importance by its adoption of Norse words in commerce, and Dublin was founded by the Vikings as a trading-post. The most successful colony of all was Iceland. Irish hermits had anticipated Norsemen there, too; it was not until the end of the ninth century that they came in large numbers. Soon there may have been 10,000 Norse Icelanders, living by farming and fishing, in part for their own subsistence, in part to produce the salt fish which they might trade. In that year the Icelandic state was founded and the Thing (a council of the big men of the community rather than the first European 'parliament' as it was later seen) met for the first time.

One of the most beautiful machines of its age, the Viking ship. This one, buried as the bier of a prince, may have originally belonged to a king, for it was exceptionally finely carved and decorated.

Colonies in Greenland followed in the tenth century. Norsemen were there for five hundred years before they disappeared. Of discovery and settlement further west we can say much less. The Sagas, the heroic poems of medieval Iceland, tell us of the exploration of 'Vinland', the land where Norsemen found the wild vine growing, and of the birth of a child there (whose mother subsequently returned to Iceland and went abroad again as far as Rome as a pilgrim before settling into a highly sanctified retirement in her native land). There are reasonably good grounds to believe that a settlement discovered in Newfoundland is Norse. But we cannot at present go much further than this.

In western European tradition, the colonial and mercantile activities of the Vikings were from the start obscured by their horrific impact as marauders. They had some very nasty habits, spread-eagling among them, but so did most barbarians. Churchmen were doubly appalled, both as Christians and as victims, by attacks on churches and monasteries; the Vikings found the concentrations of precious metals and food so conveniently provided by such places especially attractive targets. Nor were the Vikings the first people to burn monasteries in Ireland. Nonetheless, their impact on northern and western Christendom was indisputably terrifying. They first attacked England in 793, Ireland two years later. In the first half of the ninth century the Danes began to harry Frisia year after year, the same towns being plundered again and again. The French coast was then attacked; in 842 Nantes was sacked with a great massacre. Within a few years a Frankish chronicler bewailed that 'the endless flood of Vikings never ceases to grow'. Towns as far inland as Paris, Limoges, Orleans, Tours and Angoulême were attacked. Soon Spain and the Arabs suffered; in 844 the Vikings stormed Seville. In 859 they even raided Nîmes and plundered Pisa.

These terrible incursions changed those on whom they fell, especially the West Franks. Viking raids on coastal sites could not be effectively anticipated or often resisted. Their ravages threw new responsibilities on local magnates; central and royal control crumbled away and men looked more and more towards their local lord for protection. Not all the efforts of rulers to meet the Viking threat were failures, though. The Vikings could be (and were) defeated if drawn into full-scale field engagements and the main centres of the Christian West were on the whole successfully defended, but rulers often paid what the English called Danegeld, a tribute to buy them off.

England

England had soon become a major target. Its pagan kingdoms had largely been Christianized by the ninth century. The Roman mission which had established itself in Canterbury in 597 had competed with the older Celtic Church until 664, a crucial year when a Northumbrian king at a synod of churchmen held at Whitby pronounced in favour of adopting the date of Easter set by the Roman church – a symbolic choice, determining that the future England would adhere to the Roman traditions, not the Celtic. From time to time, one English kingdom or another had been strong enough to have some sway over the rest, but none could stand up to a wave of Danish attacks from 851 onwards. They led to occupation of two-thirds of the country and the establishment of substantial areas of settlement. At this juncture, the Anglo-Saxons were saved from relapse into paganism and incorporation in the Scandinavian cultural sphere by the king of Wessex.

Alfred is the first Englishman whose name must appear in this book. He is also the first English national hero. As a child of four, he had been taken to Rome by his father and was given consular honours by the pope. The monarchy of Wessex was indissolubly linked to Christianity. Alfred saw

According to Germanic tradition, a king should reign with counsel from his great men. In Anglo-Saxon England the witanagemot joins the king in judging a criminal.

himself as much as a defender of the faith against paganism as of England against an alien people. In 871 he inflicted the first decisive defeat on a Danish army in England. Significantly, a few years later the Danish king agreed not only to withdraw from Wessex but to accept conversion as a Christian. This registered that the Danes were in England to stay but also that they might be divided from one another. Soon Alfred was leader of all the surviving English kings; eventually, none was left but he. He recovered London and when he died in 899 the worst period of Danish raids was over and his descendants were to rule a united country. Their rule was accepted even by the settlers of the 'Danelaw', the area defined by Alfred and marked to this day by Scandinavian place-names and fashions of speech.

Alfred had also founded a series of strongholds ('burghs') as a part of a new system of national defence by local levies. They not only made possible the further reduction of the Danelaw but set much of the future pattern of English life; on them were built towns whose sites are still inhabited today. Finally, with tiny resources, Alfred deliberately undertook the cultural and intellectual regeneration of his people. The scholars of his court, like those of Charlemagne, proceeded by way of copying and translation: the Anglo-Saxon nobleman and cleric were intended to learn of, above all, the Bible in their own tongue. These innovations marked the beginning of a great age for England. A surge of monasticism revivified the Church. A local government structure of 'shires' appeared (some of its boundaries endured to 1974). The Danes were held in a united kingdom through a half-century's turbulence and only when ability failed in Alfred's line did the Anglo-Saxon monarchy come to grief. A new Viking offensive took place and huge sums of Danegeld were then paid. One Danish king (this time a Christian) overthrew the English king and then died, leaving a young son to rule his conquest. This was the celebrated Canute, under whom England was briefly part of a great Danish empire (1006-35). There was a last great Norwegian invasion of England in 1066, but it was shattered at the battle of Stamford Bridge.

Scandinavian civilization left Europe important legacies: the duchy of Normandy and the literature of the Sagas, for instance. Yet it was in the end only an episode. In settled lands, the Norsemen gradually merged with the rest of the population and by Canute's day even the Danes of the Danelaw were speaking English and accepting English law as their own. When the descendants of Rollo and his followers turned to the conquest of England in the eleventh century they were really already Frenchmen; the war-song they sang at Hastings was about Charlemagne the Frankish paladin and the men of the Danelaw they conquered were by then English (the Vikings had similarly lost their ethnic distinctiveness in Kiev Rus and Muscovy). The 'Normans', as we remember them, installed a new royal family and a new aristocracy in England in 1066, but within a few centuries their royal house and aristocracy became English, too.

THE TEMPERING OF THE MEDIEVAL CHURCH

The leaders of the Church in the centuries between the end of the ancient world and the eleventh or twelfth century often felt isolated and embattled. Increasingly at odds with, and finally almost cut off from, eastern Orthodoxy, Catholicism developed an aggressive intransigence almost as a defensive reflex. It was a sign of its insecurity. Even if Charles Martel had turned back Islam in the west, the Arabs remained a formidable threat, and the Vikings were sheer heathens. At home, too, Churchmen felt at bay and beleaguered. In the middle of still semi-pagan populations they had to make what they could of a culture with which they had to live, judging nicely what concession could be made to local practice or tradition. All this had to be done by a body of clergy of whom many, perhaps most, were men of no learning, not much discipline and dubious spirituality. Meanwhile, kings and lay magnates prowled about them, sometimes helpfully, sometimes hopefully, always a potential and often a real threat to the Church's independence of the society it had to strive to save.

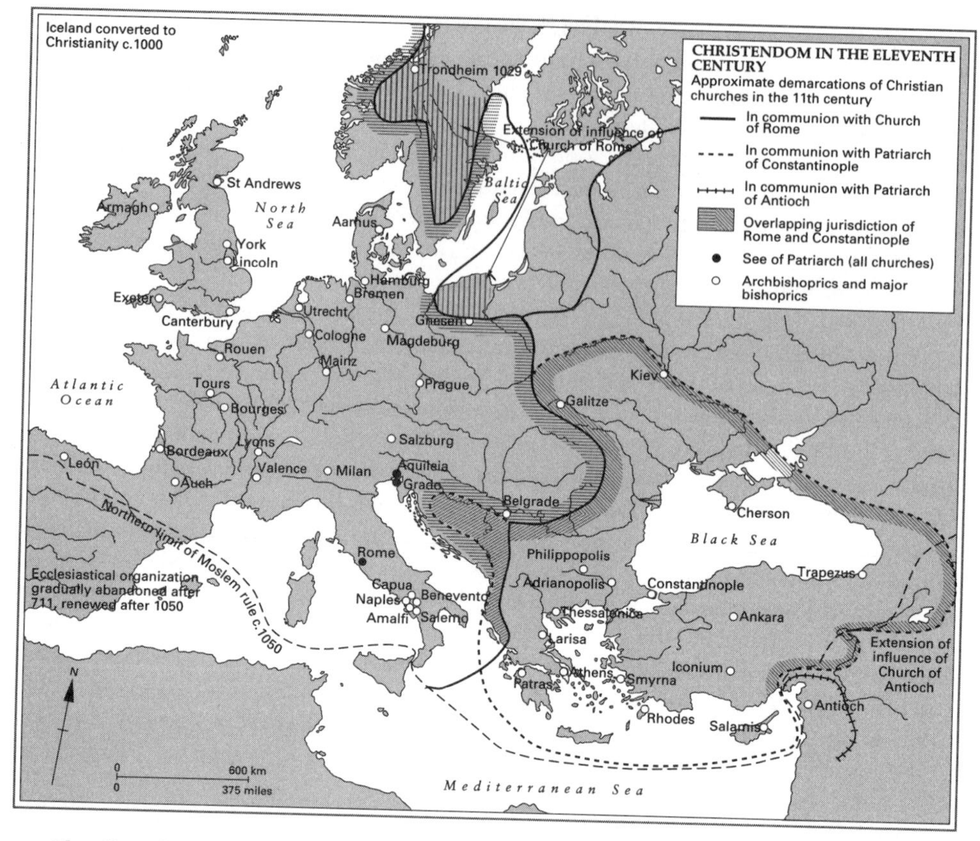

The Church was used to seeking protection. When Ravenna had fallen to the Lombards, Pope Stephen set out for Pepin's court, not that of Byzantium. There was no desire to break with the eastern empire, but Frankish armies could offer protection no longer available from the east. Protection was needed for the papacy, too, from the Arabs who menaced Italy from the beginning of the eighth century, and, increasingly, from the native Italian magnates who became obstreperous in the ebbing of Lombard hegemony. There had been some very bad moments in the two and a half centuries after Pepin's coronation. Rome seemed to have very few cards in its hands and at times only to have exchanged one master for another. For a long time the popes could hardly govern effectively even within the temporal domains, for they had neither adequate armed forces nor a civil administration. As great Italian property-owners, they were exposed to predators and blackmail, and emperors had sometimes been great makers and unmakers of popes.

The sources of ecclesiastical power

Yet there was another side to the balance sheet. Pepin's grant of land would in time form the nucleus of a powerful papal state. In the pope's coronation of emperors there were veiled claims, too, perhaps to be the identifier of rightful emperors; perhaps, some thought, the pope conferred the crown and the stamp of God's recognition on the emperor only conditionally, not irrevocably. The papal coronation of Charlemagne, like that of Pepin, may have been expedient, but it contained a potent seed.

Immediately and practically, the support of powerful kings was needed for the disciplining of local churches and the support of missionary enterprise. From the Papal-Frankish entente of the eighth century there emerged gradually the idea that it was for the pope to say what ecclesiastical policy should be and that the bishops of the local churches should not pervert it. An instrument of standardization was being forged out of Pepin's use of his power as king to reform his countrymen's Church and to do so on lines which brought it into step with Rome and further away from Celtic influences.

Papal power ebbed and flowed as the centuries went by. In the eighth century, a famous forgery, the 'Donation of Constantine', purported to show that Constantine had given to the Bishop of Rome the former dominion exercized by the empire in Italy; a hundred years later a pope addressed kings and emperors, it was said, 'as though he were lord of the world', reminding them that he could appoint and depose. What was more, he used the doctrine of papal primacy against the emperor of the East, too, in support of the Patriarch of Constantinople. This was a peak of pretension hard to sustain in practice. In the collapse of papal authority in the tenth century, the throne became the prey of Italian factions. The day-to-day work of safeguarding Christian interests in hard times had to be left in the hands of the bish-

St Etienne at Nevers: one of the most magnificent and unspoiled Romanesque churches of France, built between 1063 and 1097 to serve as a chapel for a Cluniac priory.

In the second half of the tenth century a school of copying and illumination flowered at Winchester under the influence of currents from continental Europe. The newly founded Benedictine abbey – the New Minster – received a charter which is the oldest manuscript made for the house and contains this miniature showing King Edgar offering to Christ the charter he has granted to the monks.

ops of the local Churches. Needing to respect the powers that were, and seeking the protection and help of kings and princes, they often became royal servants, as much under the thumbs of their secular rulers as, often, the parish priest was under the thumb of the local lord. This could be a humiliating dependency.

Cluniac reform

In the tenth century, there began a great creative movement of reform. Its heart was a renewal of monastic ideals; a few noblemen founded new houses which were intended to recall a degenerate monasticism to its origins and observance of the Benedictine Rule. Most of them were in the old central Carolingian lands, the area from which the reform impulse radiated outwards, and the most celebrated was the Burgundian abbey of Cluny. Founded in 910, for nearly two and a half centuries it stimulated reform in the Church. Its monks followed a revision of the Benedictine rule and evolved something quite new; the Benedictine monasteries had been independent communities, but the new Cluniac houses were subordinate to the abbot of Cluny himself; the general of an army of (eventually) thousands of monks who only entered their own monasteries after a period of training at the mother house. At the height of its power, in the middle of the twelfth century, more than three hundred monasteries some as far away as Palestine, looked for direction to Cluny, whose abbey contained the greatest church in western Christendom after St Peter's at Rome.

Learning and culture

It is easier to speak of institutions, structure and law than of other sides of religious life in the early Middle Ages. Religious history tends to lose something of its spiritual dimension in the records of bureaucracy. Yet that spiritual dimension was unchallenged, unique, and pervaded the whole fabric of society. The Church had a monopoly of culture, too. Terribly damaged and narrowed by the barbarian invasions and the intransigent other-worldliness of early Christianity though the classical heritage had been, what was preserved of it had been preserved by churchmen; by the tenth century the Benedictines and the copiers of the palace schools had ensured that not only the Bible but Latin compilations of Greek learning were available to their contemporaries. Through their version of Pliny and Boethius a slender line connected early medieval Europe to Aristotle and Euclid. Literacy, though, hardly went beyond the clergy. Well into the Middle Ages, even kings were normally illiterate. The clergy controlled virtually all access to such writing as there was. In a world without universities, only a court or church school offered the chance of letters beyond what might be offered, exceptionally, by an individual cleric-tutor. The effect of this on all the arts and intellectual activity was profound; high culture was not just related to religion but took its shape only in the setting of overriding religious assumptions. The slogan 'art for art's sake' could never have made less sense than in the early Middle Ages. History, philosophy, theology, illumination, all played their part in sustaining a sacramental culture. However narrowed it might be, though, the legacy they transmitted, in so far as it was not Jewish, was classical.

This is all very general. We must again recall that we can know very little directly about what must be regarded as both theologically and statistically much more important than this (and, indeed, as the most important part of all that the Church did) – the day-to-day business of exhorting, teaching, marrying, baptizing, shriving and praying, the whole religious life of the secular clergy and laity which centred about the provision of the major sacraments. The Church for centuries deployed powers which often cannot have been distinguished clearly by the faithful from those of magic. It used them to drill a barbaric world into civilization. It was enormously successful in doing so and yet we have little direct information about how it was done.

GETTING A LIVING

Of the social and economic reality of the Church we know much more, and that reality was vast. The Church, a great landowner, controlled much of society's wealth. Its income came above all from its land and an individual monastery or chapter of canons might have very large estates. All this helped to generate documents from which information can be culled about the economy of the day. For a long time it was very primitive. At the end of antiquity economic life in western Europe had everywhere declined and sometimes collapsed. Not everyone felt the setback equally. The most developed economic sectors went under most completely. Barter replaced a money economy; when silver began to circulate again, there was still not much coin – particularly of small denomination – in circulation. Spices disappeared from ordinary diet; wine became a costly luxury; most people ate and drank bread and porridge, beer and water. Scribes turned to parchment, which could be obtained locally, rather than papyrus, now hard to get (this was to turn out to be an advantage, for minuscule was possible on parchment, and had not been on papyrus, which required large, uneconomical strokes). Recession ruined the towns and the universe of trade disintegrated. Contact was maintained with Byzantium and further Asia, but commerce in the western Mediterranean dwindled during the seventh and eighth centuries as the Arabs took over the North African coast. Later, thanks again to the Arabs, it was partly revived (one sign was a brisk trade in slaves, many of whom came from the Slav peoples who thus gave their name to a whole category of forced labour). In the north, too, there was a certain amount of exchange with the Scandinavians, who were great traders. But for most Europeans getting a living meant agriculture.

Subsistence was for a long time almost all that could be hoped for. Animal manure or breaking new ground were the only ways of improving a return on seed and labour which was by modern standards derisory. Only centuries of laborious husbandry would change this. The animals who lived with the stunted and scurvy-ridden human tenants of a poverty-stricken landscape were themselves undernourished and undersized; for fat, the luckier peasant depended upon the pig, or, in the south, on oil. Only with the introduction in the tenth century of plants yielding food of higher protein content did the energy return from the soil begin to improve. Then came some technological advance. Mills became more numerous and a heavier, wheeled plough was able to bring under cultivation more of the more fertile but heavy, damp, sticky soils of the northern plains. Because the new ploughs needed big teams of oxen to pull them, there followed the adoption of a system of rotations. The little strips worked by individuals were scattered over two or three fields belonging to the local community. All the strips in each field were put under the plough (or left fallow) at the same time to make the best use of plough-teams. It was more efficient than any method of cultivation known hitherto in Europe. More varied crops were another result. Among them were the extra oats needed to feed horses, which could therefore be used instead of oxen. Soon this led to the devising of better harness, and, notably, the invention of the whipple-tree. Agriculture was beginning to show signs of major change by 1000 BC.

It then fed and maintained a population probably smaller than that of western Europe in Roman times though even approximate figures are almost impossible to establish. At any rate, there is no evidence of more than a very slow population growth until the eleventh century. The population of western Europe may then have been approaching forty million fewer than live in the United Kingdom today. In that sparsely-populated world, possession of land or access to it was the supreme determinant of the social order. Somehow, slowly, but logically, the warriors of barbarian societies became landowners too. With the dignitaries of the Church and their kings, they were the rulers. Land-ownership generated not only rent and taxation, but jurisdiction and labour service, too. Landowners were the lords, and gradually their hereditary status was to loom larger and their practical prowess and skill as warriors was to be less emphasized. Some of them were granted lands by a king or great prince. They were expected to repay the favour by service. To grant exploitable economic goods in return for specific obligations was very common, and this idea lay at the heart of what later men, looking back at the European Middle Ages, called 'feudalism' – no medieval man used the word, for it had not been invented then.

Feudal dependences

Roman and Germanic custom both favoured dependency and in the later days of the empire, or the troubled times of Merovingian Gaul, it became common for men to 'commend' themselves to a great lord for protection; in return they offered him a special loyalty and service. This easily fitted into the practices of Germanic society. Under the Carolingians, the practice began of 'vassals' of the king doing him homage; that is to say, they acknowledged with distinctive ceremonies, often public, their special responsibilities of service to him. He was their lord; they were his men. Vassals, though, had vassals of their own and one lord's man was another man's lord. A chain of obligation and personal service could in theory stretch from the king down through his great men and their retainers to the lowest of the free. And, of course, it might produce complicating and conflicting demands. Even a king could be another king's vassal in respect of some of his lands. At the bottom of the pile were the slaves, more numerous perhaps in southern Europe than in the north but everywhere tending to decline in numbers and to evolve marginally upwards in status to that of the serf – the unfree man, born tied to the soil of his manor, but nevertheless, not quite without rights.

Few medieval men or women can often have needed to form a view of the world as a physical whole, and fewer still would have had much empirical knowledge on which to base one. This must have made it easier to hold to the view that Jerusalem was the world's geographical centre – as in this thirteenth-century map.

Though much of the land of Europe was divided into fiefs – the *feuda* from which feudalism takes its name – which were holdings bearing obligation to a lord, there were always important areas, especially in southern Europe, which were not feudal in this sense. There were also always some freeholders, more numerous in some countries than others, who owed no service for their lands but owned them outright. Nevertheless, contractual obligations based on land set the tone of medieval civilization. Corporations, like men, could be lords or vassals; a tenant might do homage to the abbot of a monastery (or the abbess of a nunnery) for the manor he held on its estates, and a king might have a cathedral chapter or a community of monks as one of his vassals. There was much room for complexity and ambiguity. But the central fact of an exchange of obligations between superior and inferior ran through the whole structure and is one key to understanding medieval society.

It justified the extraction from the peasant of the wherewithal to maintain the warrior and build his castle. From this grew the aristocracies of Europe. The military function of the system long remained paramount. Even when personal service in the field was not required, that of the vassal's fighting-men (and later of his money to hire fighting-men) would be. Of the military skills, the one which long mattered most was that of fighting in armour on horse-back. At some point in the seventh or eighth century the stirrup was adopted in Europe; from that time the armoured horseman had it for the most part his own way on the battlefield until the coming of weapons which could master him. From this technical superiority emerged the knightly class of professional cavalrymen, maintained by the lord either directly or by a grant of a manor to feed them and their horses, another source of the warrior aristocracy of the Middle Ages and of European values for centuries to come. But for a long time movement into the knightly class was not difficult.

Medieval kings

A king might have less control over his own vassals than they over theirs. The local lord, whether lay magnate or bishop, must always have loomed larger and more important in the life of the ordinary man than a remote and probably never-seen king or prince. In the tenth and eleventh centuries there are everywhere examples of kings obviously under pressure from great men. The country where this seemed least to present a problem was Anglo-Saxon England, whose monarchical tradition was the strongest of any. Kings everywhere had some strengths of their own, though. Their office was unique, its sacred, charismatic authority confirmed by the anointing by the Church. Kings were set apart by the special pomp and ceremony which surrounded them and which played as important a part in medieval government as does bureaucratic paper in ours. If in addition a king had large domains of his own, then he stood an excellent chance of having his way.

Not always in the technical and legal sense, but in common sense, no-one except kings and great magnates had much freedom in early medieval society. Everyday life was cramped and confined by the absence of much that we take for granted. There was nothing much for laymen to do, after all, except pray, fight, hunt, farm or run an estate so that others farmed for you; there were no professions for men to enter, except that of the Church, and small possibility of innovation in the humdrum style of daily life. Choices were even more restricted as one went further down the social scale, and women's were always narrower than those of their menfolk. Only with the gradual revival of trade and urban life as the economy expanded was all this to change.

A royal castle: Caerphilly, one of a chain of fortresses making up the Plantagenet security system for Wales after its final subduing in the late thirteenth century.

THE HINGE OF THE TWELFTH CENTURY: INVESTITURE

It was a paradox that the very eminence of the Church in the High Middle Ages was itself a source of danger. As churchmen were more and more involved in affairs of state, there was a danger that it would be diverted from its true purpose – the worship of God and the salvation of souls – and would simply be used as a force to back up lay rulers. This danger lay behind a great quarrel remembered as the 'Investiture Contest', which came to a head in a clash between popes and lay rulers over who had the right to 'invest' – that is, confer authority upon – a bishop in his diocese. In one way it was never settled: it was an early round in a long-lived contest running through western European history down to this century about whether State or Church should have the last word. It was a conflict from which Europe greatly benefited, for it gradually led some people to think that it might perhaps be best if neither had the last word and that some things were too important to be left to anyone except the individuals it affected. Once that idea can be seen to have taken root, liberalism's central idea of the autonomy of the individual may be said to be born.

The murder of Becket, archbishop of Canterbury, by royal servants an incident in the long medieval struggles of kings and churchmen, and a martyrdom which made Canterbury a place of pilgrimage for western Christendom.

Nonetheless, the central episodes of the Investiture Contest lasted only a half-century or so and the issue was by no means clear-cut. The very distinction of Church and State implicit in some aspects of the quarrel was in anything like the modern sense still unthinkable to medieval man. Many specific practices at issue were by and large quite soon the subject of agreement and many clergy felt more loyalty to their lay rulers than to the Roman Pope. Much of what was at stake, too, was very material, the sharing of power and wealth within the ruling classes who supplied the personnel of both royal and ecclesiastical government in Germany and Italy, the lands of the Holy Roman Empire.

The most famous battle of the Investiture struggle was fought just after the election of Pope Gregory VII in 1073. Hildebrand (Gregory's name before his election: hence the adjective 'Hildebrandine' sometimes used of his policies and times) was a pope of great personal and moral courage, if deeply unattractive as a person. He had fought all his life for the independence and dominance of the Papacy within western Christendom and when reform became a matter of politics and law rather than of morals and manners Hildebrand was likely to provoke rather than avoid conflict. Perhaps strife was inevitable.

Many churchmen for many years had been sure the Church was in need of reform. They thought the clergy should live lives different from laymen's lives: they should be a distinct society within Christendom. They attacked simony (the buying of preferment), the marriage of priests, and interference with them. Perhaps, too, the emperors were bound to find themselves in conflict with the Papacy sooner or later. In Italy the empire had allies, clients and interests to defend. Since the tenth century, both the emperors' practical control of the Papacy and their formal authority had declined. A new way of electing popes left the emperor with a theoretical veto and no more. The working relationship of the two powers had deteriorated.

Gregory VII took his throne without the customary imperial assent, simply informing the emperor of the fact. Two years later he issued a decree on lay investiture. What it actually said has not survived, but it is known to have forbidden any layman to invest a cleric with a bishopric or other ecclesiastical office and excommunicated some of the emperor's clerical councillors on the grounds that they had been guilty of simony in purchasing their preferment. To cap matters, Gregory summoned the emperor to Rome to appear before him.

Henry IV responded by getting a German synod to declare Gregory deposed. This earned him excommunication, which would have mattered less had he not faced powerful enemies in Germany who now had the pope's support. In the end, Henry had to give way. He came in humiliation to Canossa, where he waited in the snow barefoot until Gregory would receive his penance in one of the most dramatic of all confrontations of lay and spiritual authority. But Gregory had not really won. His assertion that kings could be removed when the pope judged them unfit or unworthy was almost unthinkably subversive to men whose moral horizons were dominated by the idea of the sacredness of oaths of fealty; it was bound to be unacceptable to any king. As a result, investiture ran on as an issue for the next fifty years. Yet Gregory had separated clerics and laymen as never before and had made unprecedented claims for the distinction and superiority of papal power. More would be heard of them in the next two centuries, as other popes steadily pressed papal claims and built up the Church's administrative machine, the Curia, a Roman bureaucracy like the household administrations of the English and French kings. Through it the papal grip on the Church itself was strengthened while papal jurisprudence and jurisdiction ground away, drawing more and more legal disputes away from the local church courts to papal judges.

Papal monarchy

As the investiture contest receded, secular princes became better-disposed to Rome. Though a spectacular quarrel in England over clerical privilege and immunity from the law of the land led to the murder (and then the canonization) of an archbishop of Canterbury, the position of the Church was not much challenged. Yet even some clergy strongly disliked the affirmation of Papal authority. They believed that councils of the whole Church had more right than the pope to lay down its laws, and this was to lead to a major upheaval – the Conciliar movement – in the fifteenth century. Others began to say that on many points the authority of the Pope could be set aside if the Bible (or some other authority) gave grounds for doing so. From these and many other ideas sprang in the end both reform and heresy.

One expression of such stirrings was the appearance of a new kind of regular clergy called friars. They belonged to religious orders bound by a Rule like that of monks, but went about in the world, preaching and teaching, instead of living in a monastery. They were called 'mendicants' (that is, beggars) because they relied on alms for their support, and the first two mendicant orders were the most important: the Franciscans (named after the Italian, St Francis) and the Dominicans (named after the Spaniard, St Dominic). Franciscans specifically undertook a mission to the poor; some of them came near to advocating social revolution. The Dominicans concentrated on war against heresy by study (they were great scholars), by preaching (they are still called the Order of Preachers) and by pursuing heresy through a new machine of investigation, the Inquisition. It was not always easy to decide which was the true face of the Dominicans – that of the Sicilian St Thomas Aquinas, who was the greatest Christian philosopher of the Middle Ages, reducing all knowledge to a system intelligible on Aristotelian lines, or that of the Spanish Torquemada, the judge who harried heretics and Jews in fifteenth-century Spain.

Honorius III, a vigorous and intelligent pope, carried on the good work of his predecessor Innocent III in asserting the authority and role of the Roman see. He also approved the foundation of the Dominican order and confirmed by a bill the revised rule of the Franciscans, founded under Innocent III, a ceremony commemorated in this sculpture.

One other new and important institution to emerge under the wing of the Church at about the same time was the university. Bologna, Paris and Oxford were the first; by 1400 there were over fifty in western Europe. They rapidly improved the general educational standard of the clergy, but also produced administrators, and they were sometimes favoured with special privileges and grants by rulers because of this. When laymen came at a later date to seek training at the universities, these institutions would still be – as they were to be for a long time – usually under the control of the Church, which prescribed what would be taught. The result was that the universities were another great force making for the permeation of medieval society by religion. For hundreds of years educated men in Europe could not think except along the lines laid down by the Church, just as those of China could not escape from Confucianism.

Europe's first lunges overseas: the Crusades

Long before the Ottoman onslaught revealed how dangerous Islam still could be, the 'Franks' had taken the offensive in the Near East. Even in the eleventh century the Normans won back Sicily and South Italy more or less completely from the Arabs. Soon after this – in 1095 – the pope launched the idea of the first of the great expeditions called Crusades. His aim was to deliver the Holy Land – Palestine – from Islam and there were to be four really important Crusades. The first succeeded in recapturing Jerusalem, which was not retaken by the Arabs for nearly seventy years. Before that there had been a Second Crusade (1147-49) in which a German emperor and a French king took part. It began badly with a massacre of Jews in the Rhineland and ended in disaster. The Third Crusade (1189-92) sought to recover Jerusalem, which had been by then recaptured by Saladin, a great Moslem soldier, in 1187. It failed, although this time the English king (Richard Coeur-de-Lion or Lionheart) joined the French king and the German emperor (who was drowned while still on his way to Palestine). Finally a Fourth Crusade left (this time without kings) in 1202. It was financed by the Venetians and at once got diverted into meddling with the internal politics of Byzantium; the upshot was the terrible sack of Constantinople by the Crusaders in 1204 and the establishment of a 'Latin

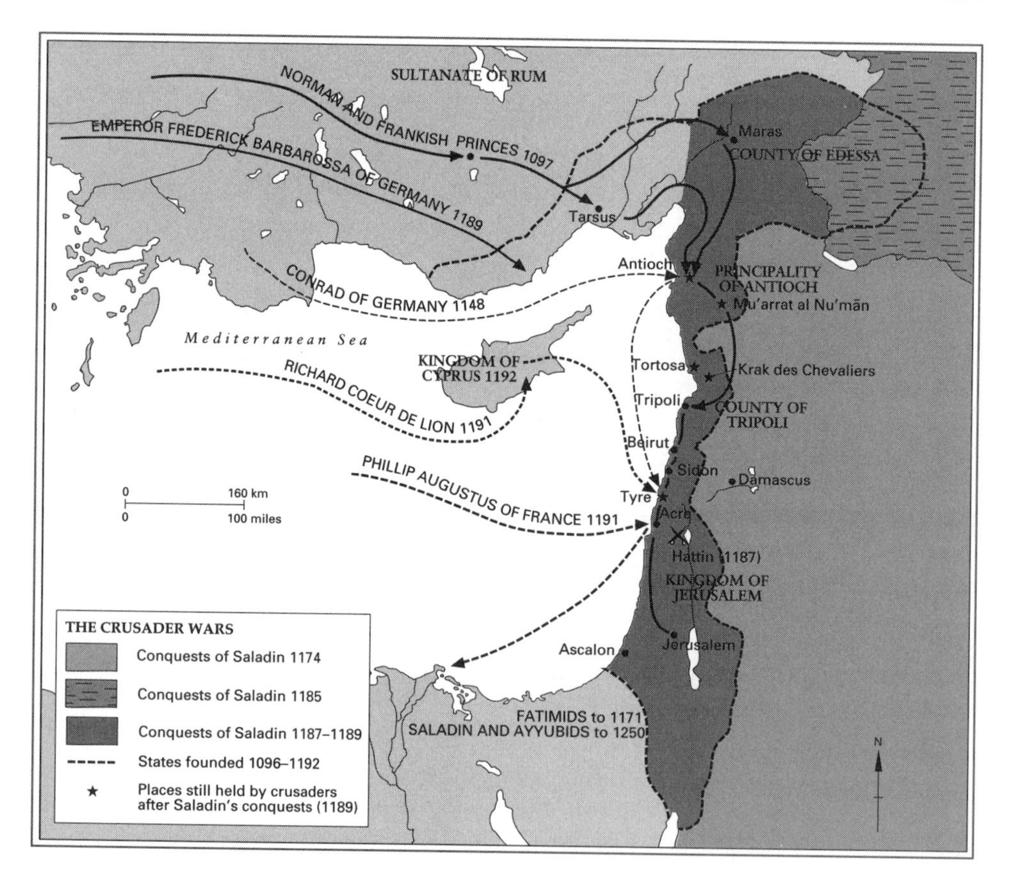

Under Moslem rule, Spanish Christians produced what was later called Mozarabic or 'arabized' art. One of its most astonishing productions is a collection of illuminated manuscripts of a popular work of devotion and scholarship, a commentary on the *Apocalypse* by an eighth-century monk, Beatus of Liebana. About twenty copies survive and this illustration of the text 'Behold, he cometh with clouds; and every eye shall see him' is from an eleventh-century version.

Empire' there for the next few decades. After this disgraceful episode – after all, the Byzantines were Christians, and if the Crusading movement meant anything it was surely the upholding of the Faith – there were several more crusades, but their great age was over. They never again looked like winning back Palestine.

What the Crusades revealed matters more than what they did or did not achieve. They were the most striking expression of a new temper and outlook in western Christianity. They were sustained by real religious fervour; thousands of humble people set out on them – often to perish miserably because of their ignorance and lack of preparation for what awaited them.

They were part of the religious revival of the eleventh century, and illustrate the divergence of east and west Christendom. For the eastern empire, struggling with Islam was not just about religion – it was also about power, politics and territories which had previously belonged to Byzantium. Empire and caliphate treated one another as great powers, just as Rome and Persia had done, and religious differences between them were often subordinate to others.

Not that religion alone recruited crusaders. Many of them had no very respectable motives at all – they were out for booty or, preferably, lands. Just as Norman knights set out for Anglo-Saxon England and Saracen Sicily to get estates for themselves, so many of the Crusaders thought that the four 'Latin Kingdoms' founded after the first Crusade or the 'Latin Empire' which emerged in 1204 would assure their futures. In this light, the Crusades were the first example of the rapacity of European overseas imperialism. As often in later times, noble and ignoble aims were mixed up in the minds of men who tried to set up western institutions in remote, exotic settings. They did so with clear consciences because their adversaries were infidels who had seized Christianity's most sacred shrines. 'Christians are right, pagans are wrong' was the way a famous medieval poem, *The Song of Roland*, summed up this kind of thinking.

When Islam recovered, the Crusading states speedily collapsed. The enmity of Moslem and Christian intensified, though Europe had not rejected everything Islam culture had on offer. Through Spain and Sicily she absorbed much of what Islam had to give, whether it was

The Crusades

Conventionally, the 'Crusades' is the name given to a series of expeditions directed from western Christendom to the Holy Land whose aim was to recover the Holy Places from their Islamic rulers. Those who took part were assured by papal authority of certain spiritual rewards indulgences (remission of time spent in purgatory after death) and the status of martyr in the event of death on the expedition. The first four Crusades were the most important and made up what is usually thought of as the Crusading era.

1095 ad	Urban II proclaims the *First Crusade* at the Council of Clermont. It culminated in
1099	The capture of Jerusalem and foundation of the Latin Kingdoms.
1144	The Seljuk Turks capture the (Christian) city of Edessa, whose fall inspires St. Bernard's preaching of a new Crusade (1146).
1147–49	The *Second Crusade*, a failure (its only significant outcome was the capture of Lisbon by an English fleet and its transfer to the King of Portugal).
1187	Saladin reconquers Jerusalem for Islam.
1189	Launching of *Third Crusade* which fails to recover Jerusalem, though
1192	Saladin allows pilgrims access to the Holy Sepulchre.
1202	*Fourth Crusade*, the last of the major crusades, which culminates in the capture and sack of Constantinople by the crusaders (1204) and establishment of a 'Latin Empire' there.
1212	The so-called 'Children's Crusade'.
1216	The *Fifth Crusade* captures Damietta in Egypt, soon again lost.
1228–9	The emperor Frederick II (excommunicate) undertakes a 'crusade' and recaptures Jerusalem, crowning himself king.
1239–40	'Crusades' by Theobald of Champagne and Richard of Cornwall.
1244	Jerusalem retaken for Islam.
1248–54	Louis IX of France leads crusade to Egypt where he is taken prisoner, ransomed, and goes on pilgrimage to Jerusalem.
1270	Louis IX's second crusade, against Tunis, where he died.
1281	Acre, the last Frankish foothold in the Levant, falls to Islam.

There were many other expeditions to which the title of 'crusade' was given, sometimes formally. Some were directed against non-Christians (Moorish Spain and the Slav peoples), some against heretics (e.g. the Albigenses), some against monarchs who had offended the Papacy. There were also further futile expeditions to the near East. In 1464 Pius II failed to obtain support for what proved to be a last attempt to mount a further Crusade to that region.

Greek learning passed on through the great Arabic scholars of Córdoba, or the luxuries of silk clothes, perfumes, new foods and taking more baths (which Europeans must have badly needed to do) brought back by Crusaders. But Islam hardened. The Christian subjects who had shown their Moslem rulers disloyalty suffered once Moslem rule returned in the Levant, and the long-lived Christian communities of the region began to decline and for the most part disappeared.

The Crusades had changed the outlook of Christians, too. The new militancy and determination they showed to go out and conquer in the name of the Cross was to be one psychological root of the confidence with which later Europeans went out and took the world. Like the crusaders, those later men knew in their hearts that they had right on their side and should use force to make it prevail. This spirit showed itself, too, in Spain. Geography, climate and Moslem division had all helped Christianity to survive in the peninsula after Islamic conquest. In the Asturias and Navarre Christian princes or chieftains had hung on. Later, aided by the establishment of the Spanish March by Charlemagne and its subsequent growth under the new Counts of Barcelona, they began to nibble away at Islamic Spain. A kingdom of Leon emerged in the Asturias to take its place beside a kingdom of Navarre. When, in the tenth century the Christians fell out with one another the Arabs again made headway against them, above all at the very end of the century when a great Arab conqueror, Al-Mansur, took Barcelona, Leon, and in 998 the shrine of Santiago de Compostela itself. Yet within a few decades Christian Spain had rallied and Islamic Spain fell into disunion. For Spain, above all, Christianity was the crucible of nationhood, and a confrontation of civilizations supplied the energy to fire it. The centuries of the Reconquest which followed were for Spaniards really centuries of Crusade.

EUROPE'S EAST

On Europe's eastern frontiers, the same aggressive, thrusting spirit was at work as in the Levant and Spain. It was, though, often also deployed against other Christians and it was backed up by a major folk-movement. The story goes back as far as Charlemagne's conquest. For much of his life he was on campaign. The great emperor's position had depended very much on expansion. Lands were needed to supply the royal treasury and estates had to be given to the Frankish nobles to keep them happy. Some of Charlemagne's wars brought neighbouring Christian peoples under his control, but others were against pagan peoples further away. The Saxons who lived between the Elbe and Rhine endangered Frankish control of Bavaria and Frisia and so were conquered – and christianized. This in effect took Germany's eastern frontier to the Elbe. The heathen Avars lived on the Hungarian plain and were more of a menace as raiders. They too were brought to heel.

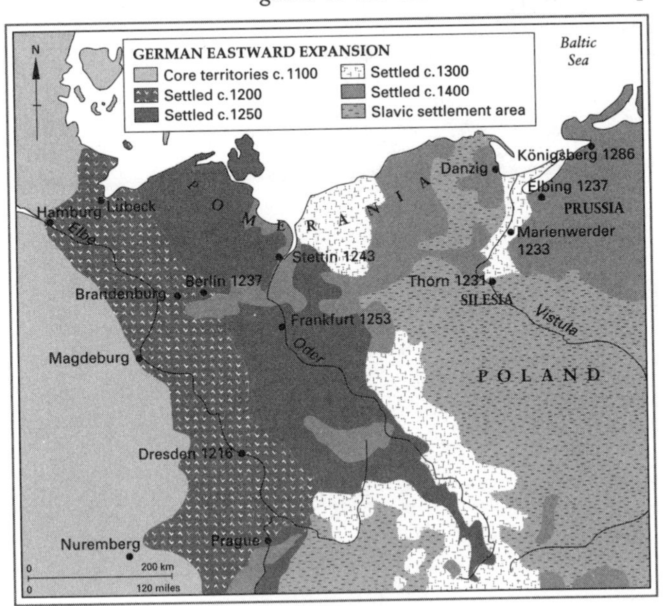

Between 1100 and 1400 the Frankish push to the east became a great folk-movement. Germans colonized and built towns on the Prussian and Polish marches, transforming the racial, economic and cultural map there. They pushed steadily into Slav lands, and so began a struggle between Slav and Teuton in eastern Europe which went on until this century and may not yet be settled. Its most recent landmark was the expulsion of huge numbers of Germans from Polish territory at the end of the second World War (some seven million left between 1944 and 1952), a reversal of much of the great medieval advance when for many centuries the Germans had their own way. An important part in that was played by the Teutonic Knights, a religious order of soldiers, taking monastic vows, and doing their Christian service on horseback by slaughtering Prussian and Baltic pagans – and Christian Slavs.

The rise of Muscovy

For a long time, the Germans benefited from Slav weakness. Not only had Kiev Rus broken up in the eleventh century, but its successor states had for a long time had their backs to the wall. As Byzantium declined, their greatest Orthodox ally was no longer available and the Russian princes had to face the Mongols alone. After the sack of Kiev in 1240, Muscovy, the major principality, paid tribute to them and their Tatar successors of the Golden Horde for two and a half centuries. Although a thirteenth-century prince of Muscovy, Alexander Nevsky, successfully beat off the Teutonic knights, the western threat continued to embitter Russian feelings towards the Germans whose intransigent Catholicism, with its aggressive assertion of the supremacy of the Pope, was completely unacceptable to the Orthodox. So Russian civilization drew even further away from the west. Its centre of gravity shifted to Moscow, whose princes were more despotic than those of other Russian states. Perhaps because of this they were better able to survive, and to extend their power, as they did, over their neighbours. Another change in the east was the emergence of a new Catholic Slav state, huge in area, which incorporated Poland and Kiev, the duchy of Lithuania. Though Catholic, the Lithuanians, too, fought the Germans and it was their troops who inflicted on the Teutonic Knights their most shattering defeat, at Tannenberg in 1460. Yet the Lithuanians also harassed Muscovy, which somehow hung on, playing politics with the divisions among its neighbours, Tatar and Catholic alike.

The heart of an empire in the making: the Moscow Kremlin. The buildings are the cathedral of the archangel Michael and the Belfry of Ivan the Great (which, in spite of its name, was not completed until 1600).

The fall of Byzantium was an event in Russian history. From this time, the undoubted centre of Orthodox Christendom was Moscow. Russian churchmen began to say this was divinely ordained: Moscow was, as they put it, the 'Third Rome'. The moment was a good one, for both the Lithuanians and the Tatar Horde were rent by internal quarrels. At this juncture, in 1462, there came to the throne of Muscovy a new prince, Ivan III. He was to expel the German merchants who tempted the German Baltic cities to dabble in Muscovy's affairs, beat off a last Tatar onslaught and absorb Novgorod, a Russian state with a more republican tradition than Muscovy. He is rightly remembered as the first national monarch of Russia, giving it something of the cohesion already won for France and England by their kings. But Ivan also built what was quickly recognized to be a different sort of monarchy from those in the west. The Russian ruler was (to quote his title) 'autocrat by the grace of God', a conscious use of a Greek term. Ivan was the first Russian ruler to be called Tsar, another conscious claim to the heritage of the Caesars (and he married a niece of the last Greek emperor). He was the real creator of a Russian monarchy which was to last until 1917, and his successors continued to use the Byzantine double-headed eagle which he adopted as part of the imperial insignia. Rightly, he has gone down in history as 'Ivan the Great'.

Ivan IV, the 'Terrible', tsar at the age of three, who reigned for over a half-century.

Under Ivan's successors came the first victories over the Lithuanians. Muscovy had always had strategical potential. It lay at a focus of river systems and therefore of Russian communications. From the fifteenth century, too, it was the main centre of Russian population. To its immediate south lay the 'Black Earth' zone, celebrated for the richness of its agricultural land. Moreover, between 1389 and 1598, over more than two centuries, there were only six Muscovite rulers in succession, and long reigns gave government great stability. There were many advantages waiting to be used by energetic and able rulers – and Muscovy was to continue to have them after Ivan, even if some of them were also mad. 'Ivan the Terrible' (Ivan IV, who came to the throne in 1533) certainly did so, yet by the middle of the sixteenth century Muscovy's 400,000 square kilometres had grown to nearly three million; by 1600 the new Russia was to be as big as the whole of the rest of Christian Europe put together. In addition, Ivan IV had further entrenched Russian monarchial power by building a royal domain consisting of something like Russia's total area; it was to crumble again in times of trouble under his successors but it emphasized in another way the distinction of the Muscovite style in government.

NATION-MAKING

Many now think that a state ought to exist only if there is behind it a people with a sense of nationality. This is the political doctrine we call 'nationalism'. Many modern states pretend to be like this and a few are. Until recently, though, this would have been thought unusual, or even nonsense. States existed long before people thought on nationalist lines, and often it was the actual existence of a state which for the first time gave people the feeling that they belonged to the same nation; states have usually created nations, in other words, not (as is now claimed ought to be the case) the other way round. The later Middle Ages shows this very clearly, for there then came into view some of the major states and nations of western Europe – by 1500 England, Scotland, France, Spain, Portugal all had shapes not very different from those they still have today; many of their inhabitants felt a sense of nationhood, too. Elsewhere, though some people certainly felt they were 'Germans' or 'Italians' because they spoke the same language, that had nothing to do with their governments. Italy was only a 'geographical expression' (as an Austrian statesman remarked as late as the nineteenth century). The pope had a very special European position as head of the Church, but among the Italian princes, he was just another ruler, one among many. He had his temporal power (distinct from his spiritual power as ruler of the Church) over his own realm, which was meant to be as independent as the republic of Venice or the little duchy of Parma. Germany was even more complicated. The Holy Roman Empire, whose princes, cities and 'imperial knights' were the feudal dependents of the Emperor, included most of it, but much of the Empire was not German. Periodic attempts to turn the Empire into a more centralized structure, in which the Emperor would be more of a king than a feudal lord, always failed. So Germans remained people who spoke versions of German but were subjects of the Archbishop of Mainz, or the Hanseatic trading cities of the north, or of the 'Elector' of Bavaria, or of (literally) one of hundreds of other sovereign units, some tiny.

England and France

The first major national monarchies were the English and French. Alfred is the beginning of the story of the one, and Hugh Capet of the other. His descendants were to rule France for four hundred years. As the centuries went by the kings of France slowly built up their territory and authority around the core of the Capet family estates. France long remained more 'feudal' than England, in that there was no tradition of a national monarchy standing over all men to which everyone owed allegiance, whatever his obligations to other lords. Still, the Capetians were very successful. They had by 1300 taken Normandy and other feudal dependencies from the kings of England (who since the Norman Conquest had held domains and fiefs in France). The kings of England long kept up their claims and the result was often war between the two monarchies. Between 1337 and 1453 England and France engaged in what is called the 'Hundred Years' War. It was not a period of continuous fighting, but was very important in strengthening kings in each country (particularly, those of France) and in making the French and English feel much more patriotic – or, at least, anti-foreigner.

England was the long-run loser, in spite of her great victories at Agincourt and Crécy. By 1500 her king had little land left on the continent. But France

The Hundred Years War

The name conventionally applied to a period of intermittent Anglo-Fench struggle in pursuit of English claims to the French crown. After performing homage for his lands in Aquitaine to the King of France, the English King, Edward III, quarrelled with his overlord which led to open hostilities and in

1339 Edward III proclaimed himself King of France, in right of his mother. There follow

1340 English victories at Sluys (naval, 1340) and Crécy (1346), and the capture of Calais (1347).

1355–6 Raids by the Black Prince across France from south-west and French defeat at Poitiers.

1360 Treaty of Bretigny ends first phase of war. Edward given an enlarged, soverign duchy of Aquitaine.

1369 The French re-open the conflict, the English fleet defeated at La Rochelle (1372) and lose Aquitaine. Steady decline of English position follows.

1399 Deposition of Richard II (married 1396 to daughter of Charles VI of France) renews French hostility.

1405–6 French landing in Wales and attack on English lands in Guienne.

1407 Outbreak of civil war in France, exploited by English.

1415 Henry V re-asserts claim to French throne. Alliance with Burgundy and defeat of French at Agincourt, followed by re-conquest of Normany (1417–19).

1420 Treaty of Troyes confirms conquest of Normandy, marriage of Henry V to daughter of the King of France and his recognition as regent of France.

1422 Death of both Henry V and Charles VI of France. Infant Henry VI succeeds to the English throne; continuation of war successfully by English until

1429 Intervention of Jeanne d'Arc saves Orleans; Charles VII crowned at Reims.

1430 Henry VI crowned King of France.

1436 Loss of Paris after collapse of Anglo-Burgundian alliance.

1444 Treaty of Tours: England concedes duchy of Maine.

1449 The treaty of Tours is broken by the English, resulting in the collapse of English resistance under concerted French pressure.

1453 English defeat at Castillon ends English effort to reconquer Gascony; English left with only Calais and the Channel Islands and the struggle peters out in their abortive expeditions of 1474 and 1492.

1558 Loss of Calais to France (but the title of King of France is retained by English kings down to George III – and the French coat of arms is displayed in the *Times* newspaper's device until 1932).

suffered more, for it was on her territory that the campaigning took place. In the second half of the fifteenth century though, when untroubled by English armies, her kings could at last settle down to putting their house in order. Once there was no danger of foreign invasion and support for the discontented, they could deal much more firmly with an awkward nobility, and doing just this was to be much of the story of the French monarchy for the next two centuries.

One famous French national heroine of the fifteenth century is Joan of Arc. At the time, it is not likely that many Frenchmen knew about her, but in the long run she was to become a saint and a legend. The later Middle Ages provided in this way, too, materials for a strengthened sense of national identity. In the fourteenth century St. George (a somewhat obscure figure, of whom little is known and who probably did not kill anything like dragons at all) became the patron saint of England, whose soldiers put his red cross on their coats. People began to write national histories at about that time and to discover heroes in their past. One was King Arthur, who, if he existed, was probably a fifth-century Roman-British aristocrat, but was more or less invented by a Welshman in the twelfth century.

Some of the legends and stories now beginning to become current were written down in languages which were the common speech of the people and not in Latin, the traditional scholarly tongue. 'Vernacular' literature as it is called is an indicator of the subdivision of the peoples of Europe into the old barbarian tribes and of their evolution into nationhood. It was to be given enormous stimulus by printing. Works began to appear written in languages similar enough to to modern ones to be read without much difficulty. The huge poem written by the Florentine poet Dante, *The Divine Comedy*, is a landmark in the history of the Italian language, just as are *The Canterbury Tales* of Chaucer in that of English.

Spain

One of the first works of Spanish literature is a long poem about a national hero, called the Cid. His nation's story was a very special one. Spain had only developed very slowly so far as its state structure was concerned (not even in 1500 was it legally one kingdom) but rapidly and strongly at the level of popular feeling. This was because Spain was made by war, the *Reconquista*, the reconquest of Spain for Christianity. The balance between Christians and Arabs swung back and forward for a time but the trend was against the Arabs, who had internal divisions of their own to cope with. By the mid-twelfth century Toledo, the major Arab city of Spain, was again Christian, and Portugal had emerged as a Christian monarchy. In the middle of the thirteenth century, Seville, a great centre of Arab culture, fell to the King of Castile and the kingdom of Aragon had taken the Arab city of Valencia. In 1469, the king of Aragon, Ferdinand, and the queen of Castile, Isabella, were married. Under them – '*los reyes Catolicos*', the Catholic monarchs, as they have gone down in Spanish history – the reconquest was completed. In 1492 its last stronghold, the beautiful city of Granada, fell to the Spanish. The Cross had triumphed in Spain and within a few years the expulsion of Jews and Moors (as the Arabs and Berbers had come to be called) was to show that Spanish nationality was tied to the idea of Christianity and ideological purity in a much deeper way than that of any other nation; often this was to have tragic results.

War and power

Clearly, fighting played a great part in nation-making in western Europe. From Alfred onwards, European nations were made by people strong enough to make them hold together, and often by making them join together against a foreign invader. Before the coming of gunpowder, fighting was mainly done by the armoured horsemen whose victories were secured by building castles for the longer term. Noblemen as well as kings built castles, and this gave some of them the chance to be very independent. Then, in the fourteenth century, came two changes. The English and Welsh longbowmen at Crécy (and at Agincourt

Longbow and crossbow gave way only slowly to what Shakespeare called 'villainous saltpetre', but the coming of the first guns heralded the end of a long era of warfare during which men could rely only on the extra power given to missile weapons by the tension of cord, wood or steel, or the acceleration of sling or throwing-stick. Early guns,like the mortar shown in this fifteenth-century manuscript, were crude, unreliable and inaccurate, but they rapidly improved and by 1500 were essential elements in any royal armoury.

in the next century) showed that the armoured horseman was not unstoppable. But to hire bowmen in large numbers so as to produce the clouds of arrows which could halt a cavalry charge was expensive. Kings had more money than most magnates with which to do it, and professional soldiers, fighting for pay, came into being to serve them. Often they moved about Europe from paymaster to paymaster. The other change came with the first guns. They were crude, feeble things. But soon some were being made whose shot could penetrate armour and knock down walls, and it was not long before European guns were the best in the world (and, almost incidentally, gun-smiths and metal-workers were multiplying in numbers which represented a significant addition to the still small pool of technologically skilled labour). The day of the old independent aristocracy as a military caste was now coming to an end. Magnates might still exercise power, but it would have to be by working through kings or by getting power delegated to them. The process was slow and drawn-out, but gradually the baronial revolts which had so often troubled and sometimes tamed kings (the English owed the great constitutional document called 'Magna Carta' to one) became rarer. That, though, is a story running into later times.

By 1500 some countries had gone much further along the road towards creating true states than had others. If centralized government is the test, the most advanced kingdom in Europe by that date may have been England. Though even in the fifteenth century, men could still look on warfare as a good investment for noblemen hoping to make some money by taking prisoners for ransom, and the rules of chivalrous conduct still limited somewhat the frightfulness of the struggle for the knightly class, those days were fast ebbing. War was becoming a national business and that meant it increasingly demanded the resources of a large state to finance it.

LATE MEDIEVAL SOCIETY

Most people at the end of the Middle Ages would have thought very little about nations and states. For them it was the more local focus of power and authority that mattered. This was one reason why a new vigour in the growth of towns after 1100 is important. It was one aspect of a general quickening of economic life.

Kings at first liked towns for encouraging trade, which could be taxed. Often, they gave them charters which ensured economic rights and privileges (to hold markets, for example) and made it difficult for the aristocracy to interfere with them. Towns did not fit into the feudal way of doing things. They were free. Within them new kinds of organization appeared, and a new breed of men, whose living came not directly from land, but from the professions, trade or manufacture. They were burghers – those who lived in what the English called a 'borough' and Germans a *burg*. The French called them *bourgeois*. Their leaders were drawn from the 'guilds' of those practising a particular trade or employment.

Thus, European towns had come to stand apart in important ways from the rest of society (and this does not seem to have happened to the towns of other civilizations). They tempted to them people who disliked a countryside where the landlord had the last word; 'town air makes you free', said a German proverb.

But no general statement holds good everywhere in medieval Europe. It was an enormously varied place. Whatever broad similarities existed, local differences were so great that every statement has to be qualified. Moreover, although society began to grow more complex in the twelfth century, and although there began to be many more people about who got their living as lawyers, or shopkeepers, or weavers, the economy as a whole remained in the last resort dependent on the land. Agriculture overshadowed everything else. Wool, hides, grain for brewing, wood for building were industrial materials for such manufacturing as there was. Every town fed itself on the produce of the neighbourhood. Even politics was ultimately about who should own land or have the right to a share of the surplus it produced. Survival or starvation depended on the state of the harvest; it was a long time before food could be brought far to meet a local scarcity.

The Black Death and after

The slow but steady rise in agricultural production in northern Europe and the clearing and cultivating of extra land had already achieved much by 1100. Even then there were still huge areas of wasteland and woodland, too, and it was still mainly by finding fresh land that agriculture was to go on growing. In the fourteenth century, though, demand and supply came much more into balance; there was little land left which could easily be cultivated with the methods then available, and this was when a wave of epidemics fell on Europe. They seem to have started in an outbreak of bubonic plague in China in 1333. Rats (on which lived fleas whose bodies contained the germs of the plague) then carried the disease along trade-routes to the Middle East and Russia, first in caravans, then in ships. In 1348 it appeared in western Europe – and in the same year entered England through a little Dorset seaport. Its effects were everywhere terrible, both because of the very nasty symptoms accompanying some versions of the disease and because of the huge toll it took of lives, and it is remembered as the Black Death. In Bristol, a third of the inhabitants died and this may have been about the same proportion as in England as a whole. On the continent of Europe things were even worse in some places, for the Black Death came back again and again, other diseases killing those who had escaped that of plague, and then starvation after that, because food production went down and trade dwindled. People became so frightened that some of them believed that even a glance from a sick person could kill. The last bad attacks came in 1390. By then the Black Death had brought havoc and change all over Europe.

The fall in population led to a continent-wide collapse of food production as there ceased to be labour to till the fields. That crisis, in its turn, was followed by social struggle – peasant revolts and risings. They were provoked by landlords trying to make a reduced population provide labour as in the old days, although the serfs knew full well that a shortage of labour meant they could ask for high wages to work as paid labourers. Landlords tried to enforce their old legal rights, but economic facts increasingly forced them to hire labour instead of taking it in kind as part of their tenants' rent. This was one of the first signs of the beginning of the replacement of an old kind of society by one founded on wage-labour. The Black Death thus accelerated an important long-term change.

It also led to much property changing hands. As landlords ceased to be able to farm their own lands, some of them divided their large estates and sold farms to tenants; in England, for example, there were fewer very large estates in private hands and more medium-sized ones after the plague. There was some inflation too, because there was still as much cash about, but it was spread between fewer people. Some classes suffered more than others in the toll the plague took, especially town-dwellers (and this also meant that universities suffered, for they were in towns; some European universities closed down for ever).

Technological innovation

Although changes in ownership may have led to land being used in more efficient units, there does not seem to have been any technical change in agriculture comparable to what had been achieved by 1100 or so. In other activities technical advance was sometimes faster. Throughout the Middle Ages, though, the typical technologist was a craftsman who had mastered the hand tools of a particular trade. There was no sudden breakthrough in mechanical methods or understanding, but rather a slow accumulation of a pool of skills and techniques, especially in metallurgy. In the thirteenth century, more than a thousand years after the Chinese, Europeans at last learnt how to cast iron by pouring it molten into moulds. Yet the progress they achieved was won not by inventing things quite anew, but by adapting known methods.

Churches and cathedrals were, other than war, the greatest capital investments of medieval Europe. They also expressed its greatest artistic and engineering triumphs. Ely cathedral, the outstanding example of the Romanesque style in England, was completed in 1189.

In 1500, and for another couple of centuries after, there was for all practical purposes, still only wind, running water or muscle-power available as power to drive machines. These forces were, nevertheless, put to use much more efficiently than a few centuries earlier. Wind used to drive a windmill made an enormous difference to work, particularly in the countryside, which had previously had to be done by hand (above all grinding corn). Water-mills could more easily be used for industrial work, too, and were to provide most of Europe's machine-power (and, indeed, the world's) for centuries yet. In the thirteenth century they began to be used to drive bellows in forges and, soon, for 'fulling' (cleaning and thickening) cloth, for working hammers, and for cutting lumber in saw-mills. But for a long time most work was still done by the muscles of animals and men. Changes only came as people discovered more efficient ways of using those muscles, through better ways of harnessing horses or oxen, or finding new uses for iron tools, which needed less effort behind them than wooden or bone ones.

The biggest and most spectacular medieval industrial process was building, but this was very specialized. Medieval builders did not produce such great works of public engineering as the Romans, but they built an astonishing series of buildings of which the great Gothic cathedrals were the finest, and these required high skill both in engineering and in stone-cutting. This enlarged a pool of craftsmen whose growing importance can also be seen in the appearance of specialized manufacturing regions. These were areas in which there were large numbers of individual workmen and small workshops, not places with big factories: the same pattern can still be seen in parts of Asia today. The most obvious and richest specialized in textiles. They appeared first in Italy; later came the rich textile towns of Flanders, to which wool from England became an important export. But textile specialization is also commemorated in the magnificent parish churches of the rich English wool and weaving regions where merchants invested their money in building to the glory of God and the insurance of the life to come. Building, and the sculpture which went with it, was one of the great medieval art forms, but the development of others also shows a growing wealth. Illumination and calligraphy became even more elaborate as time passed. Jewellery was prized and there was always a market for it among the rich, and so was work in glass, though the finest stained glass was to be found in churches. Most medieval windows, too, were not glazed; glass was too expensive. Painting too was for a long time almost entirely concerned with religious subjects, but gradually began to liberate itself from the Church. Heraldry, a late-comer among the arts, was an important employer of painters, sculptors and scriveners.

Daily life

Growing wealth did not mean that the mass of Europeans lived much more comfortably, nor even that they were well-fed. Like most people during most of history, they ate rather dull and boring food as a rule. People could only eat what was easily available where they lived, and this usually came from grain, often wheat or barley, made up into different kinds of porridge or gruel – it was quite good food, so far as filling you up went, but cannot have tasted very nice. To flavour such a diet there was rarely anything except a little honey or some curds or sour milk. To wash it down there were various kinds of beer (made from fermenting the grain), milk or plain water. Europeans living in the right place might get occasional fresh fish or some other seafood, or meat by hunting. In some places there were plentiful vegetables, but in northern Europe for a long time people had little but nuts, berries and wild fruit as extras in their diet. On the shores of the Mediterranean, people could grow olives and grapes. Though more plentiful in Europe than elsewhere, meat, butter and cheese were for a long time only eaten by the rich, who must often have been stronger than the poor. Even the well-off, though, were often not really very healthy. Broadly speaking, most people did not get enough fats or vitamins to stand up well to disease or the very hard lives they lived. They tended to die younger, to give way to illness more easily and to suffer from skin diseases and bad teeth (often caused by bad food) much more than people do today in any but the poorest parts of the world. Women too faced additional hazards in that they were often weakened by frequent pregnancies and heavy work in the fields, while the lack of medical skill made childbirth itself a major danger.

Nothing of much value was known for a long time about the conditions which led to disease spreading or about how to treat it; people did not therefore take steps which would be thought obvious today. No European cities in the Middle Ages, for example, had drains as good as those of many Roman towns centuries before, or even as those of Mohenjodaro. Rubbish and dirt were allowed to pile up in the streets, breeding flies, smells and infection until the rain washed them away. People who had diseases were rarely kept away from others, although the Black Death resulted in the first attempts at quarantine of ships from the east. Sometimes, when a disease was carried from one place to another, it had especially fierce effects because immunities were very localized.

December was the month for killing pigs, whose salted or smoked meat was the best that could be had during the long winter months, and whose blood and offal could be turned into sausages and stored. In the margin of this fifteenth-century picture can be seen a little of a lighter side of medieval life – little boys playing some kind of tug-of-war on their sledges.

In the end people's chances of living to old age usually increased because they found ways of growing more food. Previously all that could be done was to turn to medicine after an illness had been caught; in Europe this was for the most part only a mixture of magic, superstitious hocus-pocus and a few practical observations about the way certain herbs or drugs worked. Most of what was soundest in European medicine seems to have been learnt from the Arabs. It was a long time before a doctor's attendance at a sickbed was thought as useful as that of a priest – and considering what was known about medicine, this may not have been a bad judgement. The only major success of medicine was that leprosy was almost stamped out; leper hospitals ('lazar-houses') were built outside town walls, where the unhappy lepers were isolated from their fellow men.

The heart of a civilization

This was a society which, if it had made enormous strides since Late Antiquity was nonetheless still almost unimaginably different from our own. For all its potential, only here or there could a hint have been gathered of further change to come. Above all, the central place in it of religion made it unlike our own society. Europe was still, for most Europeans, Christendom, and more consciously than ever after 1453. Almost the whole of its life was defined by religion. The Church was for most men and women the only recorder and authenticator of the great moments of their existence – their marriages, their children's births and baptisms, their deaths.

Many of them gave themselves up to it; there were proportionately many more religious, both male and female, than today, but though some of them sought withdrawal to the cloister from a hostile everyday, what they left outside was no secular world such as ours. Learning, charity, administration, justice and huge stretches of economic life all fell within the ambit and regulation of religion. Even when men attacked churchmen, they did so in the name of the standards the Church had itself taught them and with appeals to the knowledge of God's purpose it gave them. Religious myth was not only the deepest spring of a civilization, it was still the life of all men. It defined human purpose.

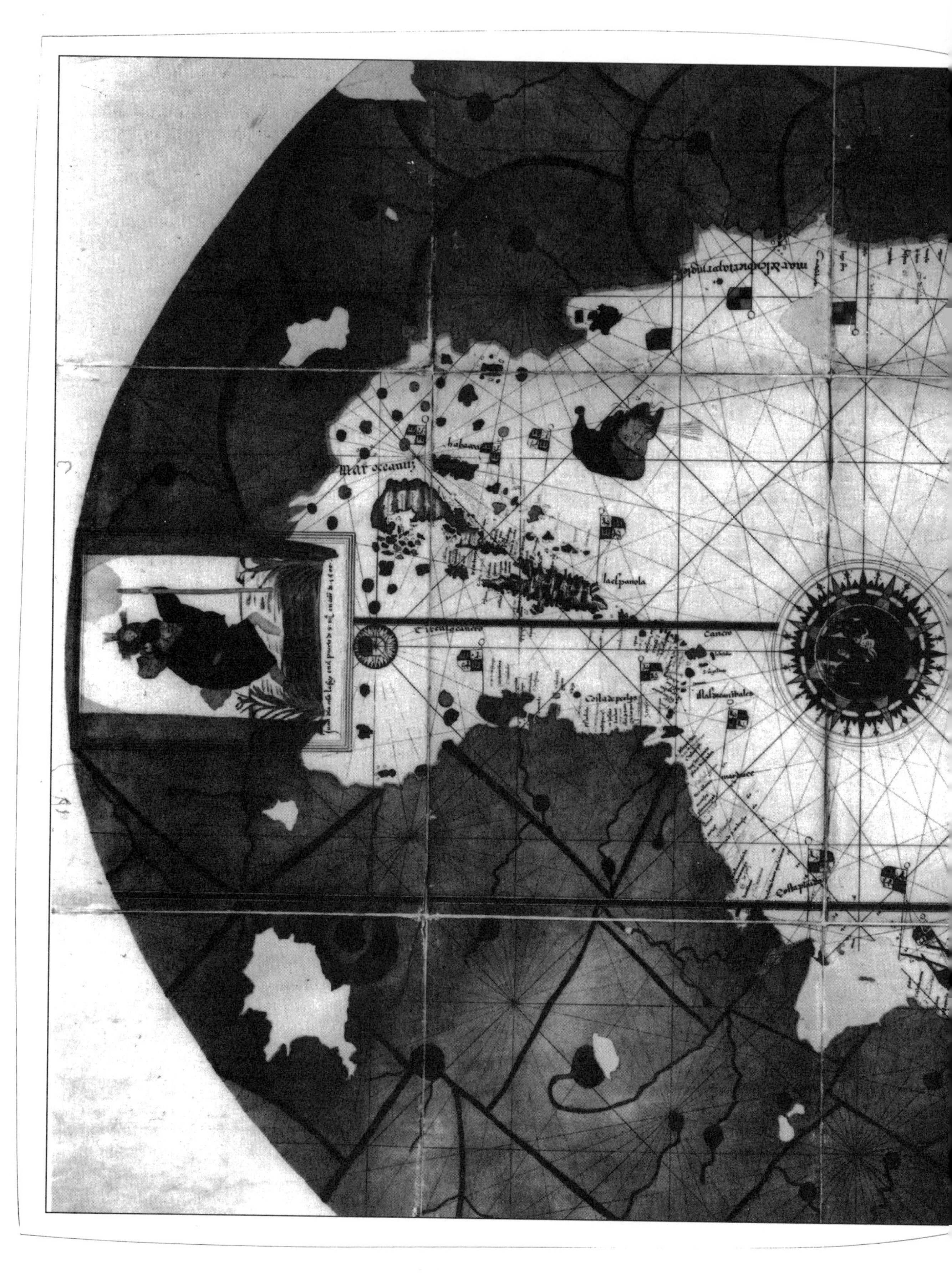

Mar oceanuz
la española
Cancro
Costa de perlas

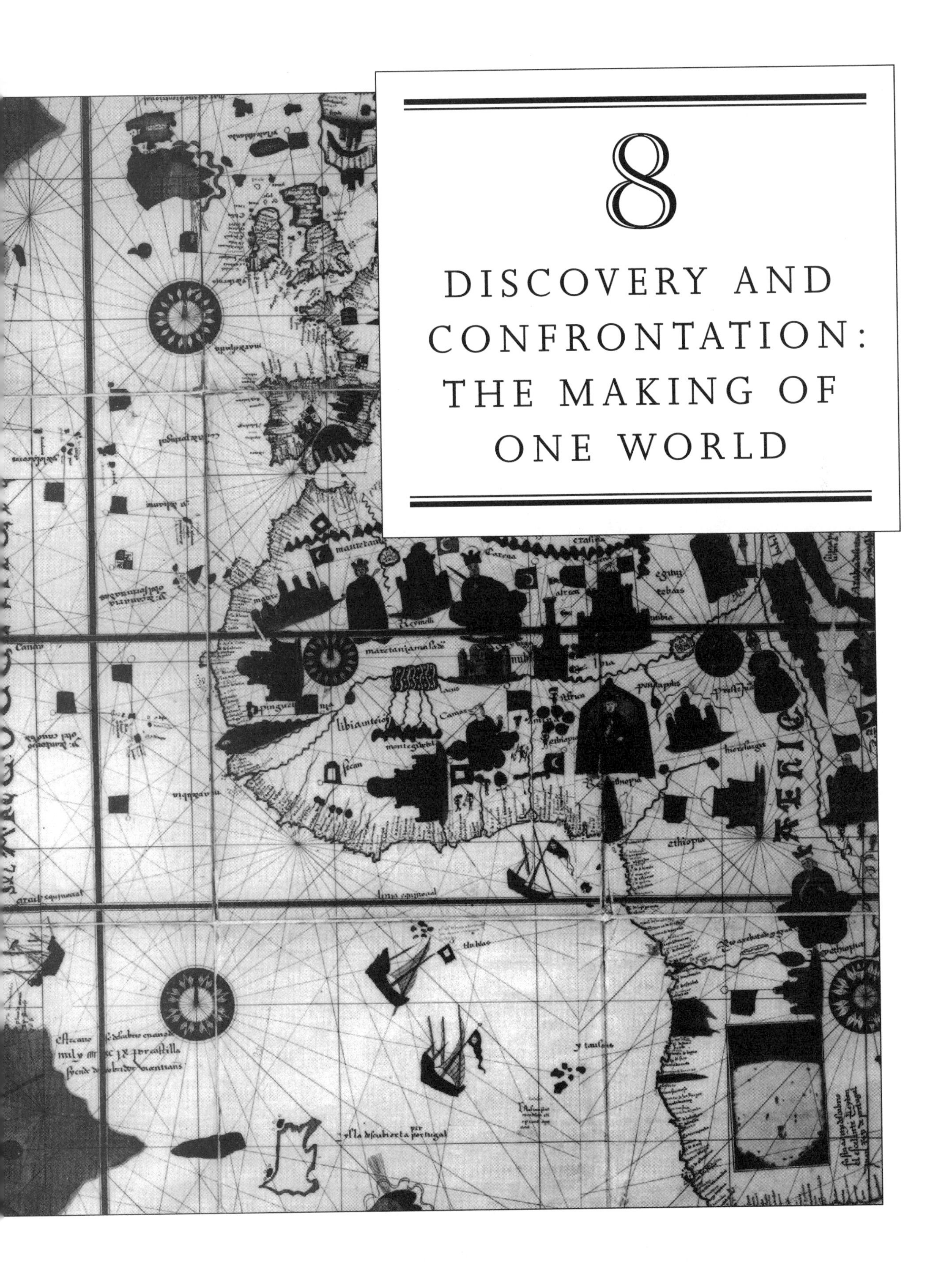

8

DISCOVERY AND CONFRONTATION: THE MAKING OF ONE WORLD

THE EUROPEAN INITIATIVE

Besides being a good round number, AD 1000 has a certain special interest as a date. As it got nearer many people in Christian Europe thought it must bring the end of the world and Judgement Day. Except for those few who thought that it would be a good idea if the world were to come to an end and felt no unease about meeting their Maker, Europe in that year was hardly a place to encourage optimism. It was poor, only just shaking off its sense of being beleaguered by Huns, Avars, Vikings and Arabs, and over much of it law and order were barely established. By 1500 or so, that had begun to change. Europe was still poor (by modern standards) but was immensely more wealthy than five centuries earlier, with bigger and more prosperous towns, more interchange and trade between them, more artistic and scholarly work going on, new and more effective governments, and a new outlook on the world outside. There is evidence of stirring, of enterprise, of excitement over new horizons.

The Renaissance

Something of this – the blossoming of arts and scholarship between the fourteenth and sixteenth centuries – is sometimes referred to in a kind of shorthand as the 'Renaissance',originally a French word which means 'rebirth'. Every European country west of Russia felt in some measure its influence and most of them contributed something to it. Italy, though, was its real centre and heart. From about 1350 to 1450 many more scholars, artists, scientists and poets lived in the cities of Italy than in any other country. The rest of Europe went to school to Italy, as it were, to learn how to copy the beautiful and clever things they could find there; the Italians looked to the classical past of Greece and Rome.

The Renaissance had its roots in discovering again part of Europe's past which had been over-shadowed by Christian civilization during the Middle Ages. Raphael glorified the great philosophers of Greece in paint and humanist writers imitated the style of the Roman Cicero

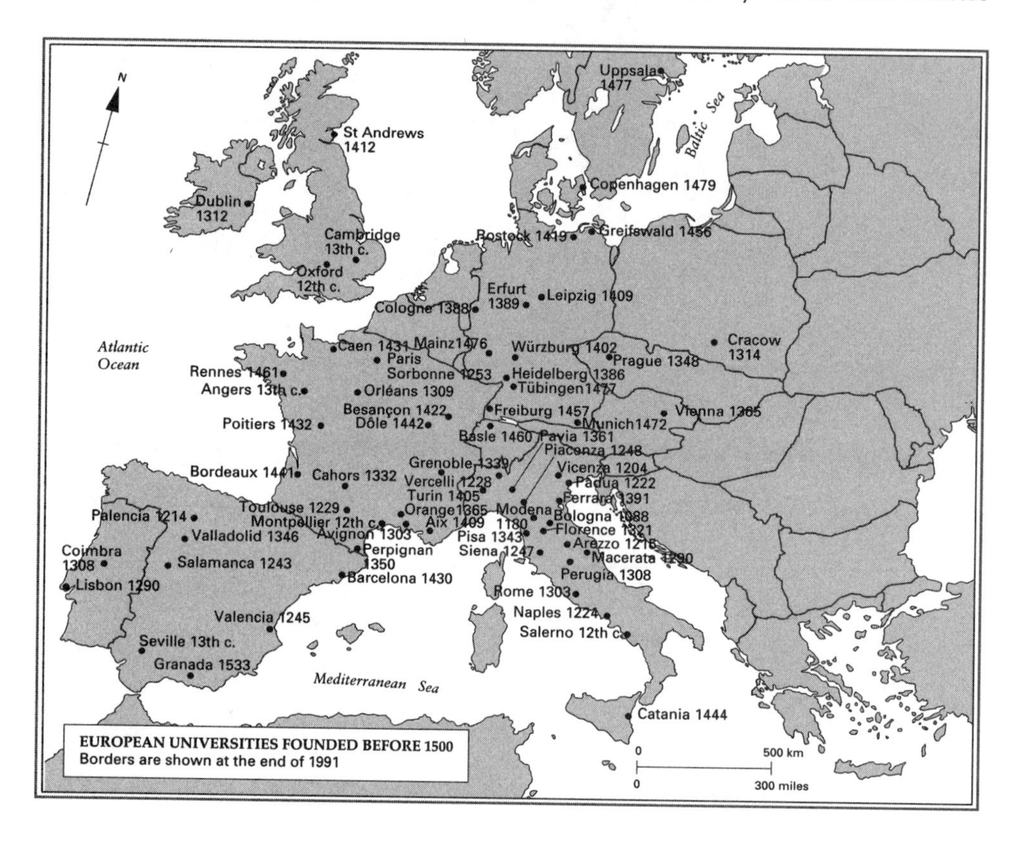

EUROPEAN UNIVERSITIES FOUNDED BEFORE 1500
Borders are shown at the end of 1991

in order to write elegant Latin. The 'rebirth' of classical learning, indeed, was what gave the Renaissance its name. Yet the most striking evidence of what the Renaissance did is still its art. In painting, sculpture, engraving, architecture, music and poetry it left behind a vast number of beautiful creations which have for centuries shaped men's ideas of what beauty should be. This art came to a climax in the late fifteenth and early sixteenth centuries, the age of – among others – Michelangelo, sculptor, painter, architect and poet; of Raphael, painter and architect; and of Leonardo da Vinci, painter, engineer, architect, sculptor and scientist. The men of the Renaissance admired such all-rounders. They gave people a new idea of human excellence. Man came to be seen as a creature of greater potential here on earth than the Church had taught. In Michelangelo's painting of the Creation of Adam, the father of the human race is a gigantic, heroic figure, dwarfing in power and dramatic effect even his creator, whose finger gives him life.

It was Renaissance scholars who first began to talk of the 'Middle Ages' (or, rather, 'Age', for they first spoke of it in the singular) as something standing between them and the classical past of whose importance they were so conscious. Yet the most important fact about Europeans had not changed much. European civilization in 1500 still had a religious heart – and, indeed, more than that, for it was visibly clothed and presented itself to the world in religious forms. The first printed European book had been the Bible, the sacred text of this civilization. By 1500 it was already being printed in German, Italian and French translations (the English had to wait for their printed version until 1526) and more people than ever before were reading it by 1500. After the fall of Constantinople, many Europeans felt that perhaps they were really the only people who were Christians (for too little was known about the Muscovites to take them seriously) and this was an exciting idea. Perhaps, after all, Europe might be seen as the centre of the world as Jerusalem had once been.

Discoveries

This changed atmosphere explains much of what has been called the 'Age of Discovery', though perhaps it is better to qualify that as the 'European Age of Discovery' (few non-Europeans discovered much hitherto totally unknown in the fifteenth century or after). It was not just because Ptolemy's world map had been brought to the west in 1400 and (after its first publication in print in 1477) had given Europeans new ideas, Ptolemy was, in fact, just about to be finally out-dated by newly acquired knowledge, particularly in two respects – the facts that lands unknown to him existed to the west, across the Atlantic, and that Asia could be reached by sea round Africa. Nor can the age be explained simply by technical advances in ship-building and seamanship, important as they were. The Chinese had long had magnetic compasses and had built big ocean-going junks, while Arab dhows moved incessantly to and fro across the breadth of the Indian ocean, and, far away, Pacific islanders made mysterious long voyages in open canoes with great navigational skill. A central problem remains: why was it Europeans who carried out the uniting of the globe through a great series of enterprises on land and sea which only came to an end with the great arctic and antarctic expeditions of the twentieth century? Why did Arabs or Chinese not reach the Americas first? No simple explanation is satisfactory. The best answers rest on an accumulation of considerations and it is safest not to give any one a decisive pre-eminence.

Among them, clearly, improvement in the design and building of its ships had certainly helped to prepare Europe for a new world role. The stern-post rudder was being used by Europeans in 1300. With improved rigging as well, ships became more manoeuvrable, safer and faster. By 1500, the tubby 'cog' of medieval sailors in northern Europe had given way to a little three-master with mixed sails, some square-rigged, some fore-and-aft; it was in its essentials the sailing-ship design which was to dominate the seas for the next three hundred and fifty years. Important advances had also been made in navigation. Centuries earlier, the Vikings had been outstanding seafarers, making long oceanic voyages out of sight of land because they had known how to sail along a line of latitude, using the sun's height above the horizon at midday to keep them on course. Then, in the thirteenth century, the compass arrived in the Mediterranean (it may have come from China, but there is no direct evidence that it did) and began to be used. In 1270 comes the first reference to a ship's use of a chart. They made it easier to improve European geographical knowledge and to spread it at an increasing rate in the next two centuries.

Another thread in the story is the appearance of new incentives. One, very important, was the hope of commercial gain; gold and pepper, it was known, came from south of the Sahara – perhaps their source could be found? Then there was the failure of the Crusades and the resurgence of Islam; though on the retreat in Iberia, it was advancing in the eastern Mediterranean and the Balkans, as well as India. The growing Ottoman threat spurred Europeans to dream of finding a way round it or allies to use against it. There was Christian missionary zeal, too. The first discoverers were medieval men who saw the world in religious terms; perhaps, they thought, they could find Prester John, the legendary Christian king of Ethiopia, as well as winning souls for Christ. Finally, there was simple curiosity. Whichever motive came out top in any particular instance, the outcome was that more and more western Europeans became interested in the possibility of oceanic exploration in the thirteenth and fourteenth centuries.

Vasco da Gama, explorer, whose achievements were rewarded by wealth, honours, and the vice-royalty of Portuguese India, where he died.

The Portuguese

Outstanding among them was a brother of the king of Portugal, Prince Henry – the 'Navigator' as he was later called. Like the men of Devon and Cornwall, the Portuguese seem to have had salt-water instead of blood in their veins. The coast of their country is wholly Atlantic. From its little harbours they sent out on it hundreds of fishing and trading vessels. These trained the seamen who carried the first European colonists into the Atlantic, the Portuguese (and a few Spanish) settlers who went to look for land in Madeira and the Canaries. As for trading enterprise, the Portuguese had to look out to the Atlantic for that, too, for they were landlocked by Spain and barred from the Mediterranean trade by the fierce opposition of the Genoese and Venetians who wanted to keep it for themselves, and, although they conducted crusades of their own in Morocco, they did not make much headway there.

A modern way of putting it would be to say that Prince Henry subsidized research. He organized expeditions southwards, down the coast of Africa. In 1434 the Portuguese first rounded Cape Bojador; ten years later they established themselves in the Azores and had got as far down the African coast as Cape Verde. In 1445 they reached Senegal and soon afterwards built a fort there. In 1473 they crossed the Equator and in 1487 reached the tip of Africa, the Cape of Good Hope. Ahead lay a huge prize, the spice trade across the Indian Ocean so long monopolized by the Arab dhows. The existence of that trade meant that Arab pilots were available. So, almost at the end of the century, a Portuguese captain, Vasco da Gama, was commissioned by his king to find a route to India. After picking up an Omani pilot in east Africa, he sailed east and dropped anchor at Calicut, on its western coast, in May 1498.

The New World

Six years earlier, a Genoese sailor, Christopher Columbus as he is called in English, had taken another and still greater step. After unsuccessfully asking the king of Portugal for support for a voyage of discovery which (he believed, basing himself on Ptolemy) would take him to Asia by sailing westwards across the Atlantic, he succeeded in 1492 in persuading Isabel of Castille, the 'Catholic Monarch', to give her support and at last set sail. After 69 days his three tiny little ships made a landfall in the Bahamas. A fortnight later he discovered Cuba, which he named Hispaniola. The following year he came back with a much better-equipped expedition and explored the islands known ever since as the West Indies. Without knowing it, he had discovered a New World – a name first used of it in 1494 – and his leap in the dark had changed world history. Unlike the Portuguese navigators who bravely and resourcefully – but systematically – navigated round a known continent to a known destination, Columbus had stumbled upon two whole continents not hitherto known to exist; they were not only unknown, but unexpected. Thus they were truly 'new'. In 1495 the first map appeared showing his discoveries, with Cuba marked as an island instead of (as Columbus had made his crew swear) part of the Asian mainland. He had refused to admit the possibility of a new continent and until his dying day insisted that he had discovered the offshore islands of Asia.

One sequel must be mentioned. In 1502 an Italian in a Portuguese vessel struck southward from the coast of what is now Brazil to sail as far south as the river Plate. His voyage demonstrated conclusively that a whole continent lay to the south of the Caribbean region

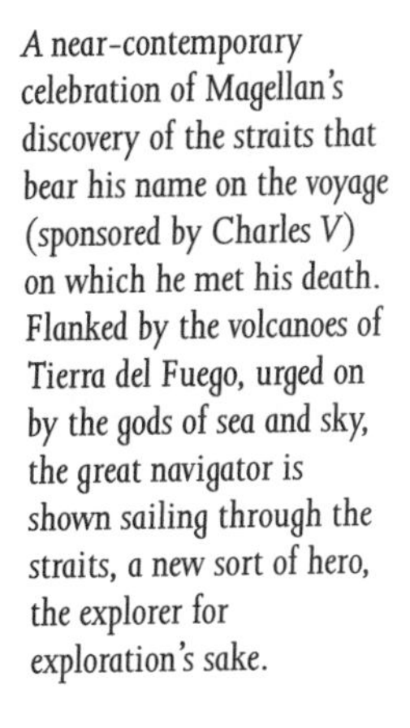

A near-contemporary celebration of Magellan's discovery of the straits that bear his name on the voyage (sponsored by Charles V) on which he met his death. Flanked by the volcanoes of Tierra del Fuego, urged on by the gods of sea and sky, the great navigator is shown sailing through the straits, a new sort of hero, the explorer for exploration's sake.

where the first great discoveries had been made. The Italian's name was Amerigo Vespucci. In his honour a German geographer named the new continent after him five years later – America. Later the name was applied to the northern continent, too.

Exploration was thus the last of a complex of forces leading Europeans to see their relationship to the rest of the world in a new way. World discovery was next to be followed by world transformation – also the work of Europeans. It was increasingly spurred by a confidence which grew as success was built on success. But great successes were already there to be built on by 1500, when Europeans stood at the beginning of an age in which not only their confidence but their energy were to grow, for a long time seemingly without limit. The world had not come to them, they had gone out and taken it, and had done so with much more success than their Crusading predecessors. To understand why and how that happened, we need to turn to look at the world the Europeans were discovering. That, too, was the product of long histories – but histories very different from those of the discoverers and conquerors.

The Age of the Major Discoveries

1445 Portuguese land on Cape Verde Islands.

1455 Papal bull *Pontifex Romanus* recognizes Portuguese monopoly of African exploration.

1460 Death of Prince Henry 'the Navigator'.

1469 Afonso V of Portugal leases a monopoly of West African trade in return for continued exploration.

1479 Spain agrees that Portugal should have monopoly rights in trade with Guinea.

1481 Fort founded at Elmina (in modern Ghana) as base for Portugal's African trade.

1482 Portuguese reach the Congo.

1488 Bartolomeu Diaz rounds Cape of (Good) Hope.

1492 Christopher Columbus reaches the West Indies.

1494 Treaty of Tordesillas giving Spain exclusive rights of exploration W of a line drawn N–S across the Atlantic. Portugal had similar rights E of this line.

1496 First voyage of discovery by Italian John Cabot, commissioned by Henry VII of England.

1497 Cabot reaches Newfoundland on his second voyage.

1498 Vasco Da Gama arrives at Calicut, having discovered the sea route to India.

1499 Under Spanish flag, Florentine Amerigo Vespucci discovers South America.

1500 Portuguese Pedro Alvares Cabral discovers Brazil.

1507 Term 'America' used to denote the New World.

1508 Cabot sets out to find the Northwest Passage.

1513 Balboa crosses the isthmus of Darien to reach the Pacific.

1519 Portuguese Ferdinand Magellan and Juan Sebastian del Cano sail westwards in search of Spice Islands.

1522 Del Cano returns to Spain, having circumnavigated the globe.

AFRICA BEFORE MODERN TIMES

Africans and Africanists alike rightly dwell upon Africa's importance in pre-history. Most of the evidence for the life of the earliest hominids is African and Africa is where the human story began. If indeed the first man and woman appeared there, then more of us than we usually think are Africans by descent, and if humanity did not evolve anywhere else independently, but spread out from that continent, then we are all Africans in the last resort. But we are hardly in any degree Africans in any important sense. The major cultures of the world (with a few exceptions in North and South America) owe little to Africa, whose contribution to the accumulation of the cultural capital of civilization has been smaller than that of other continents. With the Upper Palaeolithic and the Neolithic, the focus of prehistory moved away from its African cradle. Much that was of importance continued to happen in the continent after that time but its greatest era of creative influence on the rest of the world was over. The Nile valley, which produced the one African civilization to have much influence ouside the continent, mattered less than Sumer or the Aegean. Egypt's major cultural influence went little beyond its geographical limits and, if we exclude that area, Africa is for most of historical times and until very recently a continent which has had little to offer the world except its natural resources. It has been the home of peoples to whom things have happened, rather than the source of ideas or techniques which changed life elsewhere. Even the most important changes in African history have been produced by forces working from the outside.

Why Africa's role in civilization even in early times was so muted we cannot say but there is a strong possibility that the primary force may have been a prehistoric change of climate

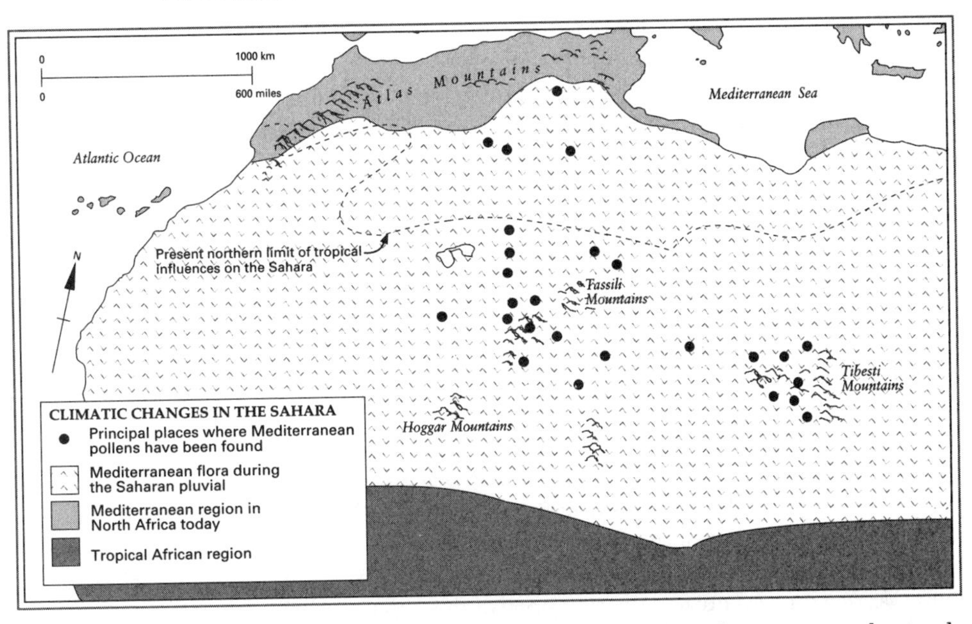

which left much of it hard to live in. As late as 3000 BC or so, the Sahara supported animals such as elephants and hippopotami which have long since disappeared there; it was the home of pastoral peoples herding cattle, sheep and goats. In those days, what is now often desert and arid canyon was fertile savannah intersected and drained by rivers running down to the Niger and by another system seven hundred and fifty miles long, running into Lake Tchad. The peoples who lived in the hills where these rivers rose have left a record of their life in rock painting and engraving very different from the earlier cave art of Europe which depicted little but animal life and only an occasional human. This record suggests also that the Sahara was then a meeting-place; negroid and what some have called 'Europoid' peoples (those who were, perhaps, the ancestors of later Berbers) as well as the Hamitic Tuaregs were in contact there. One of these peoples seems to have made its way down from Tripoli with horses and chariots and perhaps to have conquered the pastoralists. They do not seem to have belonged to the Indo-European family but whether they did so or not, their presence, like that of the negroid peoples of the Sahara, confirm that Africa's vegetation was once very different from that of later times: horses need grazing. Yet by historical times the Sahara is already desiccated; the sites of a once prosperous people are abandoned, the animals have gone.

The African peoples

Another difficulty in assessing Africa's true place in history is that, except in Egypt, it produced until recently few written records. There are a few references to the rest of the continent in the Egyptian records of government. Roman and Byzantine archives provide more material, but tell us about little except North Africa and the Sudan. Apart from this, until the appearance of Islam, we have to make do with legend and travellers' tales. When the Greek historian Herodotus came to write about Africa in the fifth century BC, he found little to say about what went on outside Egypt (whose records he could not read). His Africa was defined by the Nile, which he took to run south roughly parallel to the Red Sea and then to swing west along the borders of Libya. South of the Nile there lay for him in the east the Ethiopians, in the west a land of deserts, without inhabitants. He could obtain no information about it, though he heard of dwarfish people who were sorcerers. Given his sources, Herodotus was not talking topographical nonsense, but he grasped only a third or a quarter of the ethnic truth. The Ethiopians, like the old inhabitants of upper Egypt, were members of the Hamitic peoples who make up one of three racial groups in Africa distinguished by modern anthropologists at the end of the Stone Age. The other two were the ancestors of the modern Bushmen, inhabiting, roughly, the open areas running from the Sahara south to the Cape, and the negroid group, eventually dominant in the central forests and West Africa. (Opinion is divided about the origin and distinctiveness of a fourth group, the Pygmies.)

To judge by the record of their stone tools, cultures associated with Hamitic or proto-Hamitic peoples seem to have been the most advanced in Africa before the coming of farming, whose emergence was, except in Egypt, slow. In Africa, prehistoric hunting and gathering cultures were to coexist with agriculture right down to modern times. In due course, though, demographic growth followed the production of food in greater quantity, and this changed African population patterns. After making possible the dense settlements of the Nile valley which were the necessary preliminary to Egyptian civilization, farming built up in the second and first millennia BC the negroid population south of the Sahara, in the grasslands separating desert and equatorial forests. Agriculture seems to have spread southwards, rather than being discovered in more than one place. Nutritious crops better suited to tropical conditions and other soils than the wheat and barley which flourished in the Nile valley were in due course to be found in the millets and rice of the savannahs. The forest areas, though, had to wait for exploitation until yet other plants suitable to them arrived from South-East Asia and eventually America; none of this happened before the birth of Christ.

Iron

The divergence of cultural trends within the continent was accentuated by the coming of metallurgy. Copper seems to have been worked in the Sahara late in the second millennium BC, the raw material possibly being drawn from mines in what are now Mauretania and Senegal. By the sixth century BC, copper-mining was going on in Katanga. Iron came to Africa first from the western Asia peoples through Egypt at the end of the second millennium, but it was a long time before it began to be worked there. When this happened, it was in some parts of the continent the first metallurgical skill to appear; some Africans moved straight from Stone to Iron ages without a Bronze (or Copper) age in between. Iron began to be smelted in what is now upper Nigeria in the fifth century BC, and the techniques may have originally crossed the Sahara from the Phoenician cities of the North African coast.

Pottery head from northern Nigeria, dated between the first and fifth centuries BC.

Iron had a very great impact. One of the earliest appears to have been political. The first exploitation of African ores of which we hear comes in the first independent African polity other than Egypt of which we have information, the kingdom of Kush, high up the Nile, on the frontier zone of Egyptian activity. After Nubia had been absorbed the Sudanese principality to its south was garrisoned by the Egyptians, but by about 1000 BC it was an independent kingdom, and one deeply marked by Egyptian civilization. Probably its inhabitants were Hamitic people and its capital was at Napata, just below the Fourth Cataract. By 730 BC Kush was strong enough to conquer Egypt. Five of its kings ruled as the pharaohs known to history as the Twenty-Fifth or 'Ethiopian' Dynasty. They could not arrest the Egyptian decline and when the Assyrians fell on Egypt, the Kushite dynasty ended. Though Egyptian civilization continued to influence the kingdom of Kush, a pharaoh of the next dynasty invaded it in the early sixth century BC. After this, the Kushites, too, began to push their frontiers further to the south and in so doing their kingdom underwent two important changes. It became more negroid (its language and literature show a weakening of Egyptian trends) but it is at this point that iron begins to influ-

Probably the first civilisation to influence Black *Africa* was that of Egypt, which spread southward through the Sudanese kingdom of Meroe. This lion-headed god on a native tablet inscribed in Meriotic is one relic of that.

ence its destinies. Kush extended its territory over new territories which contained both iron ore and the fuel needed (in considerable quantities, given the techniques available) to smelt it. The art of smelting had been learnt from the Assyrians in the seventh century. The Kushite capital at Meroe now became the metallurgical centre of Africa. Iron weapons gave the Kushites the advantages over their neighbours which northern peoples had enjoyed in the past over Egypt, and iron tools extended the area which could be cultivated. On this was to rest some three hundred years of prosperity and civilization in the Sudan, though later than the age we are now considering.

Before the Christian era iron-working had spread south of the Sahara into central Nigeria; it took 1200 or so years to reach the south-eastern coasts. It must have helped the spread of agriculture into previously unworkable or inaccessible parts of Africa and therefore to have stimulated population growth, though indirectly and not sharply (even at the start of the Christian era there were probably less than 20 million Africans in all), because Africans long tended to farm as migrants, clearing and exhausting land and then moving on. They neither discovered nor adopted the plough until much later. One reason may be that in many places disease made it difficult to breed draught animals to pull one. The highlands of Ethiopia were almost the only place in Africa where horses were actually bred.

Iron-working and agricultural techniques (new food crops from Asia early in Christian times, for example) were the first of a long list of the importations to Africa which have made it easier for its inhabitants to live in sizable numbers away from the Nile Valley and Mediterranean coast, though southern Africa remained in the Stone Age more or less until the coming of the Europeans. Innovation nonetheless provided even there the first means of overcoming the huge handicaps and barriers which climate, terrain and disease always put in the way of civilization. They are the start of a story which runs into modern times and the arrival from the outside of medicine, hydro-electric dams and air-conditioning. But for a long time Africa south of the Sahara remained tied to a shifting agriculture, and lagged behind in pottery, milling and transport because it lacked the wheel. Nor was much of Africa literate until modern times.

Facing page: The menacing European: a Benin ivory salt-cellar of the seventeenth century, a ship surmounting it, and Portuguese soldiers gathered menacingly about its base.

Early cultural divisions

In spite of the long absence of written sources other than those provided by the Arabs or the Ethiopian Copts, though, it is possible to discern the main currents of African history fairly easily. Africa can now be divided culturally very roughly into an Islamic north and a non-Islamic south (this is by no means the same division as that between negroid and non-negroid Africa). Outside this scheme lie the highlands of Ethiopia, inhabited by non-negroid peoples speaking Amharic. We know that in about 300 BC the Ethiopians overthrew the kingdom of Kush. Later, in the fourth century AD, Ethiopia was to become one of the first Christian kingdoms in the world, when Coptic Christians from Egypt converted its rulers. They remained directly in touch with the rest of Christendom only briefly after that, though, because Arab invasion of Egypt put an Islamic barrier between them and it. After this Ethiopia was for centuries the only Christian nation in Africa and its only literate non-Islamic society. But she had little to do with the outside world until five or six hundred years ago.

Meanwhile, in North Africa, the Maghreb Christian communities established in Roman times were in due course all but extinguished by Islam. Only those in Egypt remained numerous. The Arabs spread by military conquest along the whole northern coast, converting the Berber and Moroccan peoples as they went. To the south-west, across the Sahara, the Berber tribes of the desert had long had contacts with negroid peoples which had brought West Africa, even in the second millennium BC, into economic relationships with the Mediterranean world. But scholars still argue about what the reality of these was. After the Arab conquests in the north Islam was carried across the Sahara by caravans of Arab explorers and merchants looking for the source of the gold and slaves already filtering north. By the end of the eleventh century Islam was established in the Niger Valley and West Africa. By then, too, Somalia in the east had been Islamicized.

Ghana and Mali

The arrival of Islam is thus enormously important for historians. Arab travellers provide our first direct written evidence based on the actual observation of black Africa. They were sometimes shocked by what they saw – the nakedness of African girls, for example – but they recorded much that is helpful. One West African polity they tell us about has a now familiar name: Ghana, a kingdom ruled by a Berber dynasty, it seems, as early as the fourth century. Ghana was clearly important once this dynasty was driven out in the eighth century, for it was the 'land of gold', as an Arab writer described it. Gold came up from Ashanti and Senegal to Ghana's traders, who passed it on to the Arab caravans from which it made its way to the Near East, together with salt and slaves. At its greatest extent Ghana stretched from the Atlantic to the upper Niger, and it appears to have flourished from the eighth to the mid-eleventh centuries AD. One of the matters still discussed by specialists is the extent to which it continued to owe much of its government (and perhaps its ruling class) to its former Berber rulers from the north. Whether it should be seen in such terms or not, there was a reversion to Berber rule in the eleventh century under kings from the Islamic Maghreb.

The eventual destroyer of Ghana, and one of the states which succeeded it after its break-up, was Mali (another name revived by a modern African state), a much bigger and Islamic kingdom, covering the whole Senegal Basin. Its king was so rich that he was said to have ten thousand horses in his stables. His wealth caused an enormous sensation in the Arab world

when he made a pilgrimage to Mecca in 1307. But this empire also broke up in the fifteenth century, when the trade across the Sahara came under the control of another empire, the Songhai, whose supremacy lasted until the end of the sixteenth century. Western sub-Saharan Africa also was by then almost entirely under Moslem chiefs and kings, as much of it still is. The Islamic conversion of black Africa took place from the top of society downwards, and many pagan practices long persisted even when these countries were, formally speaking, Moslem. Farther south Islam did not penetrate, except where the Arabs touched the coastal regions, and the story of southern Africa is even harder to get at than that of the north.

Southern Africa

The fundamental movement in the history of the south was a long migration of peoples speaking languages of a family called Bantu at about the beginning of Christian times. They came from eastern Nigeria, spreading through the Congo Basin and over most of southern Africa. This laid out a pattern of settlement which has lasted until the present, though complicated and added to by later migrations. Eventually some of these migrants reached the eastern coast where, once again, Black Africa met the Arab world. Traders who arrived on the coast from the Red Sea and the Persian Gulf from the eighth century onwards called East Africa 'Zanz' (from which later came the name 'Zanzibar'). They founded the coastal towns which began the urbanization of this part of Africa and bought gold, copper and iron from the inhabitants. Indonesian visitors may have turned up too, for some of them had settled in Madagascar, to which they had brought plant species from Asia to add to its crops. As for indirect links with the outside world, they stretched even farther afield. Chinese products have been found in East Africa and rich Cantonese in the twelfth century were reported to own large numbers of African slaves.

It is not easy to discover much about the way the southern African kingdoms were run. Non-literate, they can hardly have had bureaucracies and their kings probably ruled them within the restraints of custom and respect for tradition. Some of these kingdoms were big, but they had no developed major religion. We hear from the Portuguese of one at the end of the fifteenth century on the lower Congo called the kingdom of the Bakongo. Its rulers asked for missionaries, sent an embassy to Lisbon and welcomed the Europeans. The king was baptized as Alfonso I in 1491 (though authorities disagree about whether he soon reverted to paganism or lived and died a model Christian king). But at this moment a new age was opening; three years before the Portuguese had first sighted the Cape of Good Hope and in the next era of African history Europeans were to be the ultimate cause of most of its crucial developments.

Soon, the Portuguese could report the discovery of another major state in East Africa, ruling a big area in the Zambezi Valley. This had followed an earlier culture, in what was later to be Rhodesia, which archaeologists have called 'Azanian', and which has left traces of elaborate mining, canal-digging and well-building. Such activities began the exploitation of this region's mineral wealth and it has gone on ever since. The gold supply supported a kingdom which must have lasted at least four centuries and which has left ruins (probably of the fifteenth century) of the only big stone buildings in southern Africa. The most famous are at 'Great Zimbabwe', where there was a royal capital and burial place whose earliest buildings date from the eighth century, though the most impressive of them were probably built in the sixteenth or seventeenth. They consist of about 80 hectares in all of enclosures, some with massive walls and towers built in dry masonry from shaped stones laid with great exactness. The first Europeans to discover Zimbabwe found it hard to believe that Africans could have created anything so elaborate and impressive (rather as archaeologists once thought that Mycenaeans were needed to explain Stonehenge), but it is now clear that they are African work. The Great Zimbabwe site, like the beautiful bronze work of Benin, shows the artistic capacity of black Africa, but also its limitations.

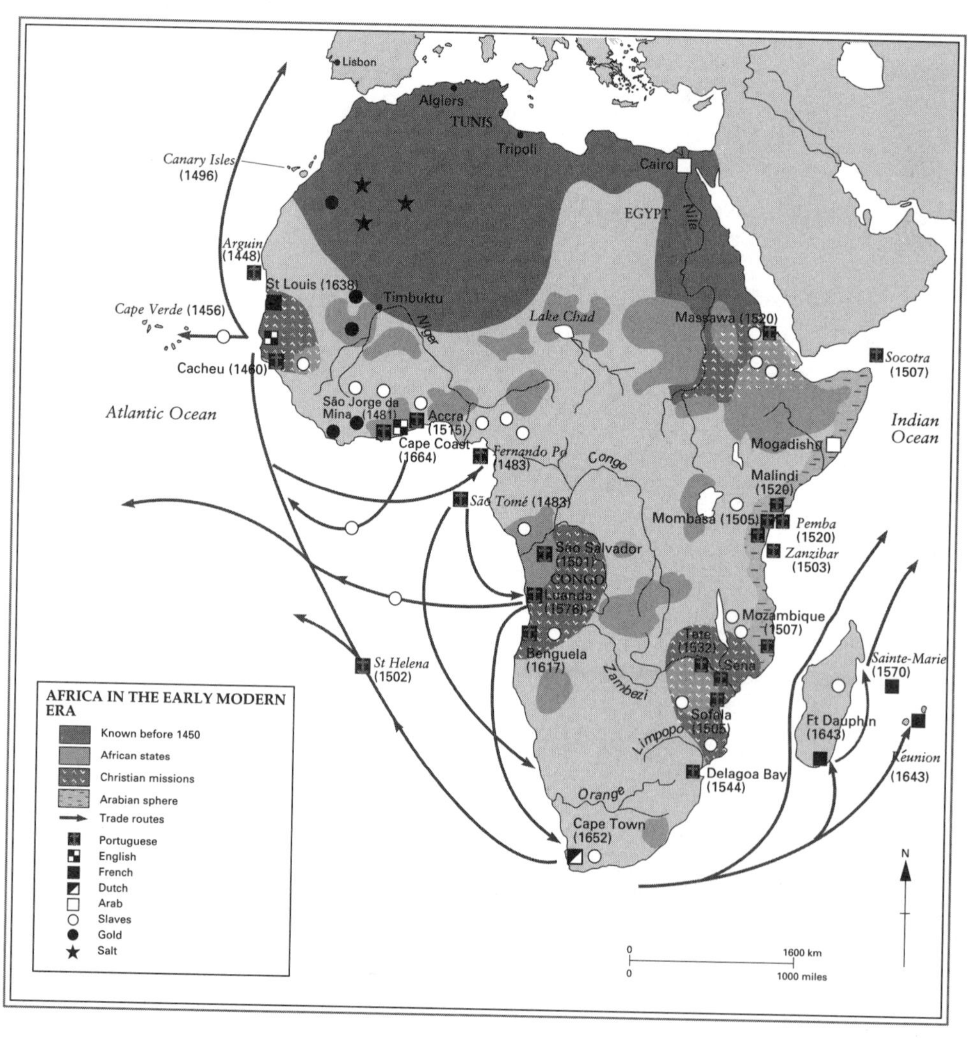

In 1500, most of Africa was still untouched by the Arab and Christian outsiders. Yet they had already brought literacy and other techniques of high civilization to some of its most advanced cultures. For the interior of sub-Saharan Africa, effective contact with such influences still lay a long way in the future even in 1500. Yet such contact as there was had already revealed itself as liable to have some very unpleasant aspects. Arab slave-traders had by then been at work for centuries collecting from their compliant rulers columns of black men, women and children to be marched as slaves either north to the Nile valley and the Near East, or to the coast and the dhows which waited there to carry them to Oman, Persia, India, and even Canton. On the west coast, the Portuguese had in 1441 captured and brought home black men they described (inaccurately) as Moslems; one year later the first sale of African slaves was held. By 1500, perhaps 150,000 black slaves had been taken from Africa by the Portuguese; there are better European records for guessing numbers than there are Arab.

THE AMERICAS BEFORE THE EUROPEANS

The history of man in the Americas is much shorter than that in Africa or, indeed, than in any other continent except Australia. Something like thirty thousand years ago, Mongoloid peoples crossed into North America by land from Asia: the continent has always been peopled by immigrants. Over the next few thousand years they filtered slowly southwards. The Americas contain very varied climates and environments and archaeological evidence shows they threw up almost equally varied patterns of life, based on different opportunities for hunting, food-gathering and fishing. Some early Americans arrived at agriculture independently of the Old World, though there is still disagreement about when precisely this happened. It was, nevertheless, a change which came later than in the Fertile Crescent. Maize began to be cultivated in Mexico in about 5000 BC, but had been improved by 2000 BC in Meso-america into something like the plant we know today. Large settled communities became possible. Farther south, potatoes and manioc (another starchy root vegetable) also begin to appear at about this time and a little later there are signs that maize has spread southwards from Mexico. Everywhere, though, change is gradual; to talk of an 'agricultural revolution' is even less appropriate in the Americas than in the Near East. In the northern continent, only a few Americans ever got so far as agriculture before the coming of Europeans, though they were well adapted to the life of hunting and gathering. The Plains Indians lived contentedly until the destruction of their habitat by the white American, and the Esquimaux found ways to survive in very inhospitable conditions.

Further south, agriculture led in due course to civilization. But American civilization was always different from that elsewhere, because of its long isolation. If there were occasional and perhaps accidental visits from Polynesian and other Pacific peoples to the west coast of America, no one has yet shown that they had any general effect on the culture of the Americas. In any case, it was America's remoteness from other major centres of civilization which really mattered. Metal-working could spread from Mesopotamia to ancient Egypt, and Christianity could be carried from the Mediterranean into China by way of Central Asia, but there was no persisting contact between the Americas and the other great centres of civilization until after 1492. Viking settlements made in Greenland and Labrador in the ninth century disappeared long before this, perhaps wiped out by Esquimaux.

The American civilizations have therefore very special and well-defined characteristics. No doubt the most important (their basic dependence on maize, for instance) derive from possibilities of the regions where they emerged, the particular geography and climates which made it easier for them to develop in some directions rather than others. There were three major areas: the Andes, the heavy rain-forests of Yucatán, Guatemala and Honduras, and the central valley of Mexico. The latest civilizations they produced were all well alive when the first Europeans arrived, so that something is known about them from their discoverers' accounts of what they found, as well as from archaeology. Through them, too, something can be seen of their predecessors.

Olmec culture

The first recognized American civilization is that of the Olmecs of the eastern Mexican coast, and it was of major importance. It was focused, it seems, on important ceremonial sites with large earth pyramids, where colossal monumental sculpture and fine carvings of figures in jade have been found. The Olmec style is highly individual and for several centuries after 800 BC seems to have prevailed right across Central America as far south as what is now El Salvador. But it retains its mystery, appearing without antecedents or warning in a swampy, forested region which makes it hard to explain in economic terms. We do not know why civilization which elsewhere required the relative plenty of the great river valleys should in America spring from such unpromising soil. Yet the gods of the later Aztecs whom the Spanish found dominating Mexico were to be descendants of those of the Olmecs, and Olmec civilization established much else which remained basic to Mesoamerican life: monumental sculpture, city-planning, jade-carving of small objects. It may also be that early hieroglyphic systems which appeared in Central America originate in Olmec times, though the first survivals of the characters of these systems date from only a century or so after the disappearance of Olmec culture in about 400 BC. Again, we do not know why or how this happened. Much further south, in Peru, a culture called Chavin (after a great ceremonial site) survived a little later than Olmec civilization to the north. It, too, had a high level of skill in working stone and spread vigorously only to dry up mysteriously.

A Mayan relief once forming part of the decoration of the lintel of a house. It shows a form of ritual self-torture or, perhaps, penance. Before a priest-like figure kneels a suppliant who draws through his or her tongue a cord studded with thorns.

The Maya

Important for the future though these early lunges in the direction of civilization were, they came millennia behind the appearance of civilization elsewhere. When the Spanish landed in the New World nearly two thousand years after the disappearance of Olmec culture they would still find most of its inhabitants working with stone tools. They would also find complicated and living societies (and relics of others past) which had achieved prodigies of building and organization far out-running, for example, anything Africa could offer after the decline of ancient Egypt. One of these, Maya civilization, was long past its zenith when the Europeans arrived. Much of the area where it thrived (the Yucatán peninsula, Honduras, and Guatemala) is unpromising for human settlement, and probably always was. Though there are some mountainous and temperate regions, most of it is low-lying and covered in tropical rain-forest which fosters ferocious insects and animals, and breeds disease. Yet there the Mayas built temples and pyramids almost as big as those of ancient Egypt. What is more, they did this solely on the basis of an agriculture of clearing and burning (which may have been in fact the one best adapted to the particular conditions of rain-forest soil, but which offered small returns). Many Central American cultures seem to have shared the same gods, derived from earlier times, and to have given similar importance to the calendar and the maintenance of large ceremonial sites. Among them, though, Mayan is undoubtedly the most impressive.

Though Mayan civilization has been traced back to the second millennium BC, its earliest substantial traces come from about AD 100, and its finest period of creativity lay between 600 and 900. It has not left urban ruins; the Mayas lived in small hamlets. But we have temples, pyramids, tombs and courts, the remains of great ceremonial centres of a religion which tried to impress the beholder with his remoteness from the gods and the importance of those who climbed the long flights of steps up the pyramids to communicate with them. Maya society in this classical era seems to have been ruled by a noble class of warriors and its hereditary priests. Religion was a matter of carrying out ritual and ceremonial in accordance with a calendar based on astronomical observation. This has struck some scholars as the best evidence of the high level of Maya culture, for it rested on major skill in mathematics. The Mayas counted in twenties and had a system of numbering like ours, in which the same symbol can stand for different values according to its position – for example, 10, 1, 0.1, 0.01 ... and so on. The religious leaders of the Mayas had an idea of time much vaster than any other civilization of their day. They thought of an antiquity of hundreds of thousands of years and may even have arrived at the idea that time has no beginning. Three of their books have survived, written on bark paper folded like a screen. They are books of ritual, and from them something has been learned about Maya chronology; the rest has to be worked out from radiocarbon dates, archaeology and stone inscriptions in a hieroglyph only now beginning to be deciphered. The indications are that writing was used for other purposes too, but no books of history or prophecy have been found.

The Maya achievement was specialized. Though they had skilled craftsmen and exported beautiful objects carved in jade to other parts of Central America, the Mayas never discovered the wheel or the arch and believed in a very crude set of gods. This of course makes it all the more surprising that they should have invested such huge resources in the building of their great ceremonial sites – an unusually unproductive activity in economic terms since it does not seem even to have had any useful technological spin-off. Whatever the strains, Maya civilization began to decline from the tenth century onwards, when a great earthquake or eruption led to the abandonment of many of the central sites. Shortly afterwards came invasions of Maya territory by peoples from the Mexican plateau, metal-users. Chichen Itza, the greatest of their centres, was abandoned in the thirteenth century, and Maya society seems to have dissolved into a scatter of little statelets, though only at the end of the seventeenth century did its last stronghold in Yucatán fall to the Spanish. But by then Maya civilization had really come to an end, leaving no important tradition or technique to the future, only a fascinating series of ruins and a language still spoken today by two million people.

Inca Peru

Peru on the eve of the Europeans' arrival in the Americas was incomparably the most advanced site of civilization in the western hemisphere. The Peruvians of that day had taken over and advanced further techniques drawn from earlier peoples. They mined gold and silver and had great skill in working them. They used bronze-bladed hoes in their agriculture (but they had no ploughs, nor draught animals). Besides building with remarkable skill in dry masonry with huge, perfectly fitted blocks, the Peruvians were outstanding weavers;

some judges have thought them the best in the world of their day. They also had skilled surgeons, capable of carrying out difficult and dangerous operations on anaesthetized or hypnotized patients. They kept records, though not in writing, but with a code of knots in coloured cords called quipu. But the most remarkable feature of their society was its organization. It had been built up through conquest by a people called the Incas, whose centre was at Cuzco from about 1200 (the traditional list of Inca emperors goes back this far). At the end of the fifteenth century they had some sort of claim to rule an area of about 720,000 square kilometres from northern Ecuador to central Chile, a huge achievement, considering the difficulties presented by the terrain, though they had only really been expanding fast for about seventy years.

Within this empire, which was ruled as a despotism by the chief Inca and the Incas who were its dominant caste, communications were provided by about sixteen thousand kilometres of roads, travelled by chains of runners in all weathers. Along them were built rest-houses for travellers on official business and, where necessary, suspension bridges to cross gorges. The population was grouped in units of ten households. Travel or departure from the local community was not allowed, though the Incas moved newly conquered peoples away from their ancestral lands, replacing them with more docile immigrants when they wanted to make sure of new parts of the empire. There was no private property or money and only local trade by barter in handicrafts. Wild animals were considered public property and were hunted in great game drives. This occasionally provided meat; for the most part the state apparatus reallocated the agricultural produce it collected and doled out manufactured goods in exchange.

Eight thousand feet high in the Andes, the Inca city of Machu Picchu was never reached by the Spaniards and was only discovered in 1911. No written records survive to explain its history, but it seems to have been a religious centre, built in the fifteenth century AD.

In this carefully controlled system the ordinary Peruvians can hardly have found life very exciting; perhaps the occasional stint of obligatory labour in the mines or on public works may even have been an agreeable change from the narrow routine of daily life. Though a man could choose his own wife, he could do so only within his own community, because of restraints on travel. He could not buy or sell because there was no money. As for the children of leading members of conquered peoples, they were taken to Cuzco and educated there in such a way as to give them the outlook needed to uphold Inca rule. In the background the Inca army was always there to deal with rebellion. Yet, though Inca government worked efficiently, it did not wipe out discontent among the Incas' subjects (as the first Europeans to arrive discovered). It remains nonetheless a remarkable example of a society run on totalitarian lines. The individual was firmly kept in second place to the collective – something true of most human societies which have ever existed until recent times, but a characteristic imposed by the Incas with an efficiency unusual when none of the technological advantages of modern governments existed. Some people have always found the Inca system appealing, and sixteenth-century Europeans were soon being given fulsome accounts of its seeming justice and efficiency.

Mexico

In about 1100 the valley of Mexico and some of the old Maya lands were dominated by a people called the Toltec. Their capital, at Tula, was a substantial city, with irrigation to sustain the farms which supplied it. They were a military people, who seem to have been able to rely on enslaving their neighbours whose forced labour they used to carry out and maintain major building works. Shortly afterwards their domination gave way to a confused period, which ended with the emergence of new top dogs by about 1350. The new ruling caste was what was long called Aztec, though the name Mexica is now often preferred. On the site of a vil-

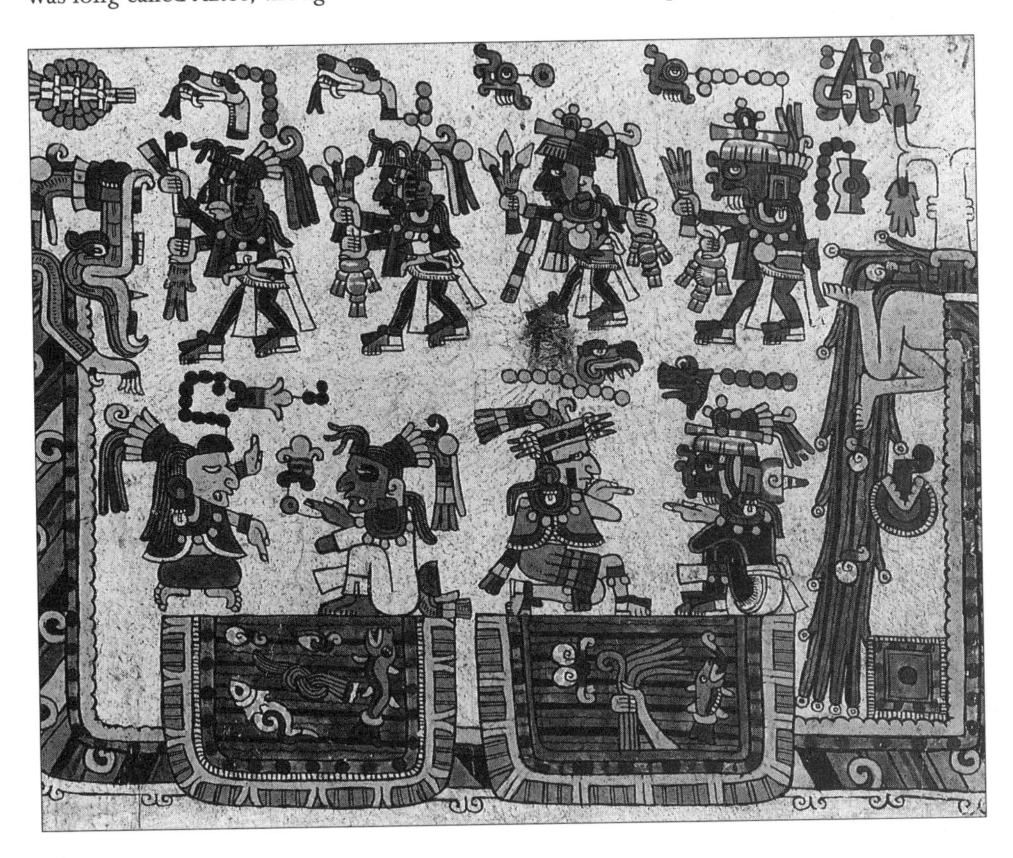

Among the Indian peoples who fell under the sway of the Aztecs were the Mixtecs. This example of their manuscript was sent to the emperor Charles V by Cortes.

lage at the edge of lake Tezcoco they founded the capital of an empire which in the next century and a half expanded (mainly after the accession of a very vigorous chief in 1429) to take in the whole of central Mexico. At its most impressive, this city, Tenochtitlan, was described by the excited Spaniards who first saw it as outdoing Rome or Constantinople in magnificence. An aqueduct brought good water to it from the springs at Chapultepec, five kilometres away. It was filled with temples and dominated by huge pyramids.

These seem to have been built with the skills and in the styles of the peoples the Aztecs (or Mexica) conquered. Like the Toltecs before them, they ran a military, tributary empire, maintained by their subjects. They themselves seem to have been very uninventive. No single major invention or innovation in Mexican civilization can be confidently dated later than the Toltecs. The greatest Mexican religious centre, Teotihuacán, thirty kilometres or so from the Aztec capital, was a built-up area five and a half kilometres long and three kilometres wide. All of it was there before AD 600 and had been occupied for perhaps a thousand years before that. It may have had 200,000 inhabitants when at its peak. It was the centre of a religious ritual based on pyramids, like those of other parts of America. Much of what made Aztec society so impressive to the first Europeans who discovered it was not Aztec at all, but had deep roots, even in Olmec times, and this is probably true of Teotihuacán.

Though derivative, the Aztec empire, still expanding when the Europeans arrived, was impressive, strange and even terrifying to European eyes. Its elected chief and religious leader directed a state organized in twenty clans, using pictographic records, and allocating annual subsistence from communal property held by the clans in return for compulsory labour and military service. There seems to have been a highly skilled agriculture, and it grew cotton. The use of the wheel for transport was unknown, though Mexico had craftsmen who were fine potters, jewellers, weavers and workers in feathers; they were also good at working in copper and gold, though they knew nothing of iron. For their best cutting edges the Aztecs relied mainly on obsidian, a volcanic glass. The warriors' skills were the most highly esteemed of Aztec society, and outstanding warriors could join organizations somewhat like European orders of knighthood, which performed special dances and rituals.

After they had taken in its wealth and magnificence, the first Europeans to see Aztec society were impressed above all by its cruelty. Its religion demanded human sacrifices and these were performed with revolting accompaniments: decapitations, flayings and the tearing out of victims' hearts while they were still alive. (One important Aztec art-form was the stone box used for burning and storing human hearts.) No less than 20,000 people are said to have been sacrificed at the dedication of the great pyramid of Tenochtitlán in 1489; Aztec mythology said that the gods had been obliged to sacrifice themselves to give the sun the blood it needed for its food, and these grisly rituals were re-enactments of this myth. Because new victims were needed all the time, the Aztec state could never be at peace. The Aztecs did not therefore really mind that their subjects were only very loosely controlled and that revolts were frequent, for this provided an excuse for rounding up more prisoners for sacrifice. But it meant also that, when the Europeans arrived, they found ready allies and helpers among the non-Aztecs.

The major American civilizations of the age of Conquest all shared features which now make them seem rather depressing. In a way they seem to have reached dead ends, limited by their low level of technology and religious and social ideas. Individual artistic achievements stand out – Aztec pottery, for instance – but made no impact on other civilizations. No doubt isolation is the main reason. The contributions of the Americas to the common life of mankind were not to be made through the inventions of these developed cultures, in fact, but from the things they passed on unwittingly – the obscure, unrecorded discoveries of the anonymous primitive cultivators who had first found out how to farm the ancestors of maize, potatoes and squash, for example. They, without knowing it, were adding hugely to the future resources of mankind. The glittering civilizations of Incas, Mayas and Aztecs, for

The greatest structure of Chichen Itzá is a temple pyramid dedicated to the legendary and semi-historical Kukulean, a possible leader of the Toltec invaders who overthrew the old Mayan hierarchies in the tenth century AD. The site is strongly marked by northern and Mexican, non-Mayan influences.

all their fascinating relics, matter less and are really little more than beautiful curiosities in the margin of world history. What survives from them as continuing influence can now be seen best in humbler, but persistent traits of daily life in the New World – chocolate, or the making of the tortilla of unleavened maize flour, or the Maya language, still spoken by some peasant communities.

THE BEGINNINGS OF EUROPEAN COLONIALISM

What is now called European imperialism made its first lasting impact on other peoples in the Americas and Africa. 'Imperialism' is a word which has acquired very broad meanings. It has been used to refer to almost any kind of continuing domination – political, economic and even cultural exercised, whether consciously or unconsciously, by one group of human beings over another. Historians have to look more closely at that vague set of ideas, if they are to understand how things came about, and what agencies were at work at different times. One part of the phenomenon of imperialism which is easiest to trace is the story of winning outright physical possession – a freehold, as it were – in foreign lands, whether by a government in Europe, or by settlers in lands with native populations (the Vikings of the ninth and tenth centuries were the first European colonists of this sort in the Americas), or even by the setting-up of new political units dominated by colonists (such as, for instance, the short-lived Crusader states of the Levant had been).

The age of the discoveries opened up new opportunities for such overseas conquest, above all to nations with easy access to the Atlantic. By then, landward expansion was at an end in the German east, the Reconquest of the Iberian peninsula was drawing to a close, and any dream of reviving the old crusader states was impossible, given the strength of the Ottoman advance. So it was the men of Portugal and Spain, and to a lesser extent, England and France, who launched the first great wave of the expansion of European rule and Europeans' settlement overseas. By 1600, only the Iberians had achieved very much. The Portuguese had a chain of harbours and forts which guarded a trading hegemony stretching, at its end, to China and Japan, a few areas in Africa later to be settled and developed as plantation colonies, and Brazil. The Spanish meanwhile had built up in the Americas the most impressive agglomeration of territory, at least notionally, ever to be acquired by one kingdom (Castile) down to this time. The French and British could by 1600 point only to a few stabs at exploration and settlement in North America (the creation of a French viceroy of Canada, Newfoundland and Labrador in the 1540s was mere window-dressing) and a quickly-established and successful tradition of piracy and free-booting at the expense of the Spanish in the western hemisphere. Their major American settlements, like those of the Dutch, were only to be made later.

Spanish empire

It follows from the chronology (and the process of settlement begins on the Atlantic islands in the fourteenth century) that the men who made the first European empires were medieval men. That is to say, their thinking was set in the mould provided by the classical legacy of Greece and Rome and, more important still, Christianity. This was the legacy which shaped those whom the Spanish called the *conquistadores* – the 'conquerors' – who appear in the 1490s first as settlers in the Caribbean islands and next as explorers making tentative stabs at the mainland in the area of what was later called Venezuela. In 1513, some of them crossed the isthmus of Panama, settled down, built huts and sowed crops, a sign that they were there to stay; the first mainland Spanish jurisdiction was then established in the region of Panama. By then there were already many settlers in the Caribbean islands and African slaves had been introduced there to provide labour. The Americas grew more and more attractive to poor, land-hungry Spanish squires and soldiers of fortune as more became known about them.

The most famous of the *conquistadores* was a half-heroic, half-piratical figure, a Spanish officer called Hernan Cortés. In 1518 he left Cuba for Mexico, broke free of the control of his superiors by burning his boats on landing there, founded the town of Vera Cruz and then led his men inland to the high plateau which was the heart of the Aztec empire. Within a few months he had conquered it. A few years later, in 1531, another Spaniard, Pizarro (an even more ruthless adventurer than Cortés), marched through the Andes to the Inca capital and destroyed the Inca regime. Thus Mexico and Peru both passed to the Spanish crown, to be added to the territories already acquired in what is now Venezuela and Central America.

Legends about fabulous wealth to be won in the Americas filtered home to Spain and drew more Spaniards like a magnet to the 'Indies'. The motives of those who came were very mixed. They wondered at the wealth of the native civilizations, and were undoubtedly most excited by the possibilities of carrying off their treasures. For a long time men were to explore South America tirelessly, looking for the legendary city the Spanish called El Dorado – the golden place – because of its fabulous wealth. The *conquistadores* had other motives, too, though. Many of them sought land for estates, or slaves to work the ranches and farms they had already built up on the islands. This did not make them gentle to the Indians. Though they might sometimes wish to bring to the Indians the Gospel of Christ, and although the Spanish clergy sought to restrain them, the settlers had behind them the militant Christianity which had won back Spain from the Moors in the 'Reconquest' and it was not a creed respectful of cultural difference. Furthermore, many of the Spanish were sincerely horrified at such practices as the Aztec human sacrifices (however hard it may be for us to understand why men used to the idea of burning Christian heretics should have been so offended).

Americans old and new

The effects on the native populations of the arrival of the Spanish were almost uniformly disastrous, but not all of them were the responsibility of the new arrivals (unless it is argued that they should not have gone to America at all). The diseases they brought with them (smallpox

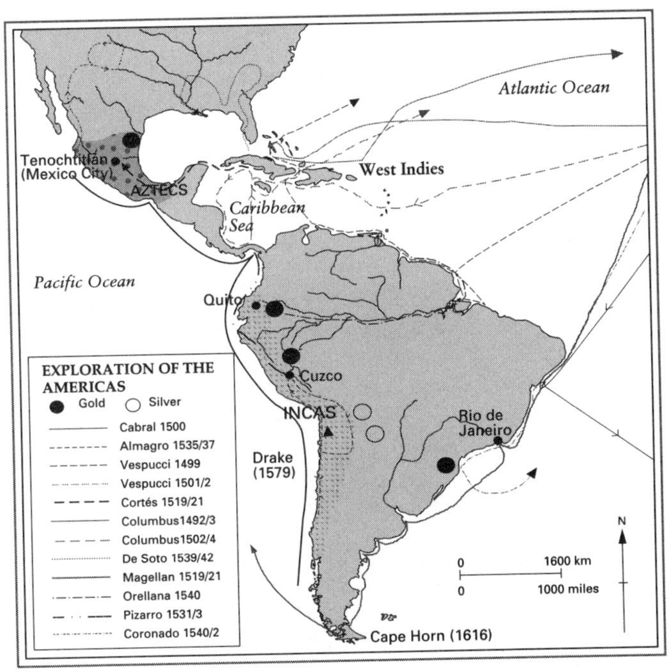

was the worst) had a catastrophic demographic effect, first in the islands and then on the mainland. It was possibly reinforced by the psychological shock of the Spaniards' power and seeming irresistibility. The Indians, for example, had never seen horses; Cortés had brought sixteen with him and the Aztecs could not at first believe their eyes when they saw men dismounting from them, for they seemed to be watching beasts dividing themselves in two.

In the islands there was soon insufficient labour for the settlers. They ruthlessly exploited what there was and though churchmen fought against their cruelties, they could not do much to protect the Indians. The usual arrangement was that a Spanish settler was allotted the labour services of a native community in return for government and protection. As disease and overwork took their toll, it became all the more important to royal officials and settlers alike to prevent labourers from leaving the plantations where they worked and so the system was tightened up even more. Over large areas of South America even today the word familiarly used for a peasant is *péon* – the Spanish word for a pawn in a game of chess, the lowest-valued piece on the board.

American populations in the Spanish dominions slowly rebuilt themselves by natural increase and immigration from Europe over the next two centuries or so. The outcome was a number of Ibero-American societies in which upper and middle classes of mainly European blood ruled over a predominantly Indian population. Although neither the Spanish nor Portuguese much minded inter-marriage (the Spanish had long lived in a multi-racial society), colonial society respected European blood; the more you had the more likely you were to be fairly well off and in a position of power. The people of predominantly European blood born in the Americas (*Creoles* was the Spanish name for them) were the rulers and landlords of the Indian survivors of the old civilizations, almost all of whose spectacular past achievements disappeared. Many Indians came to speak a kind of Spanish and became Christian at least in name.

Institutions and government

The story was much the same in Portuguese Brazil, except that there was almost nothing in that country in the way of civilization to be displaced as there had been in Mexico and Peru. Moreover so many African slaves were brought there to work on sugar plantations that the black cultural heritage of Africa was soon as important in Brazil as the Indian. As in the Spanish colonies, so in Brazil Christianity was one of the most conspicuous ways in which European civilization was implanted in a non-European setting; most of the oldest buildings in modern Brazil are churches. Slowly, too, other European imports took root in Central and South America. Forms of government based on laws, traditions and institutions which went back deep into the European past and had no logical connection with American society at all were taken for granted by settlers and governments in Spain and Portugal alike and they imposed them. So, after the empires passed away, a system of states on European lines, with civil servants and law courts like those of Europe, would survive in South America, as would the domination of European languages.

The Spanish and Portuguese American empires were huge but thinly populated. For a long time there were only a few European immigrants to exploit them and the Indian num-

The Japanese have always been close observers of strangers and carefully recorded them. In this case, the Europeans shown are some of the first Portuguese (one of them accompanied by a black slave) who arrived in Japan in the 1540s. The Portuguese were to remain for nearly a century, until 1638, virtually monopolizing European access to the mysterious country to which they had brought Christianity and firearms.

bers fell; in 1600 they may only have had about 10 million people between them. In theory at least, the Spanish governed by 1700 an area running from the River Plate in the south to the Colorado in the north. It included almost the whole length of the Pacific coast from southern Chile to northern California (where the name San Francisco commemorates Spanish sovereignty), much other territory north of the Rio Grande, and Florida. These lands were very sparsely inhabited even in 1800; over most of them there was no Spanish presence except a few mission stations and a fort or two, though some of these were to be the sites of cities very important in later times. The rest of the northern viceroyalty of 'New Spain' consisted of Mexico, which was thickly settled, and the lands of the isthmus. Peru and some of the larger Caribbean islands were by 1800 also centres of fairly large populations, with major cities. The 'Indies' were governed in theory as sister kingdoms of Castile and Aragon by viceroys, but in practice they had to be run with a fair degree of independence. Yet they were part of a world-wide empire, linked by way of Acapulco and Panama with the Philippine islands and Spain.

North America

For a long time European colonization in North America looked very much less impressive than that further south, but by 1700 the eastern seaboard of the continent was studded with English colonies. One of them, Virginia (so named after Queen Elizabeth I, who did not marry), had been the site of the first unsuccessful attempts to 'plant' a colony in the 1580s (early colonies were often spoken of as 'plantations'). Further attempts followed early in the next century but for a long time North America ran a poor second in economic attractiveness to the Caribbean and by the 1620s English settlers on islands in the West Indies had much more success to show than their cousins on the mainland.

The seventeenth century finally brought settlement to North America on a scale sufficient for success, and progress speeded up. By 1700 about 400,000 people of imported stocks (mainly British) lived in North America in twelve English colonies. The first settlement to survive had been set up at Jamestown, in the modern state of Virginia, in 1607 (just a year before a French explorer, Champlain, built a small fort at Quebec, and only a few before Dutch settlements appeared on what is now the site of New York City). There was plenty of land in North America which could be cultivated on European lines, and the English began the practice of transplanting whole communities – men, women and children – who set to work as farmers. Like the colonies of the ancient Greek cities, this made them substantially independent of the mother country. The next thing was the discovery that tobacco-raising – first in Virginia – provided a trading product for export. Tobacco was all-important in the early history of Virginia and that of the later colony of Maryland too; it was even used instead of money for accounting debts. It was the first of several 'staple' products – cotton, rice, indigo were others which gave the colonists the wherewithal to buy things they wanted from the mother-country. Others came from fishing and its associated activities. Canada never had a product of equal importance; the fur trade was not comparable as a support for population, and this is one reason why French settlement grew only very slowly. In 1661, there were only about three thousand French people in Canada.

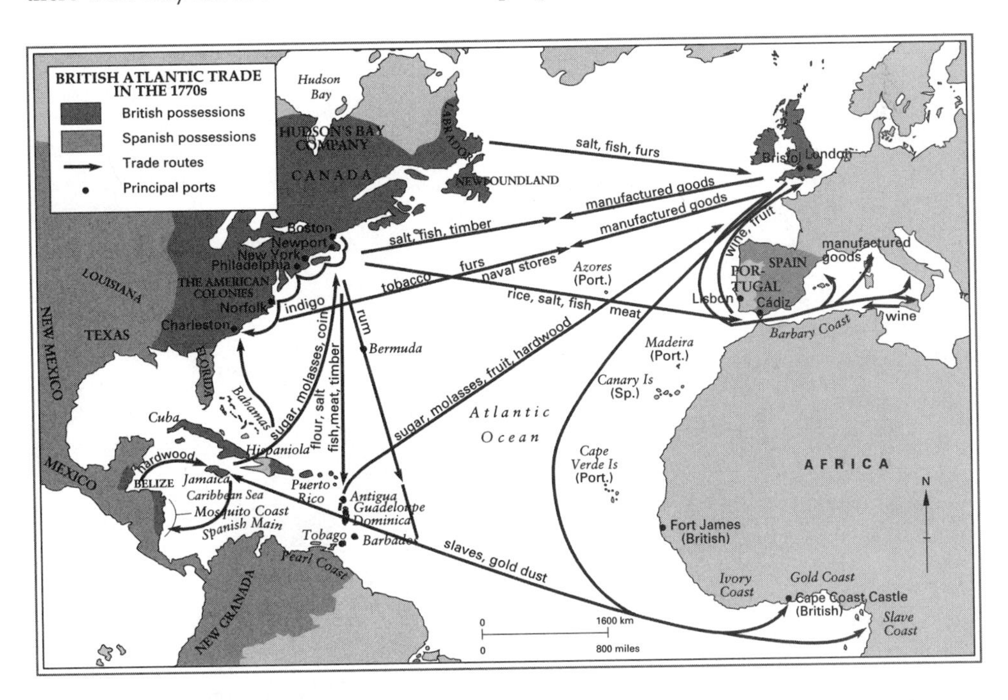

English colonies developed from the first on differing lines, shaped by climate and geography. Another source of distinctiveness was that 'New England' (as the northernmost group of colonies were soon called) began to attract men and women with strong and special religious views, often reflecting the more extreme 'Protestant' doctrines of those called Calvinists; they usually had very rigorous ideas about behaviour and disliked ritual and ceremony in religious worship (though they were sedulous in inflicting their own forms of address and behaviour in a ritualistic way). In England such people were called 'Puritans', and though many of them still thought of themselves as members of the established Church of England when they first came to America, they tended to break away from it when they had three thousand miles of Atlantic between them and home. They were not, usually, a very jolly lot; some of them became very irritated when an immigrant not so strait-laced as they set up a maypole. On the whole, New England became known as a place for those who wished to break with old ways, while those who had more affection for the customs of the old country went to the colonies farther south, like Virginia and the Carolinas.

Travellers' tales have always been popular. Captain John Smith (depicted here) was among the founders of the first successful English colony in North America and one of the first English authors to cater for a new demand for tales from the Americas.

The Puritan settlers who landed in 1620 from the Mayflower and founded a colony at Plymouth, Massachusetts, were to become legendary as the 'Pilgrim Fathers', and with their story there came also to be associated a tradition of self-government. This did not necessarily mean democracy. In Massachusetts government tended to fall into the hands of a very narrow circle of the well-to-do and the Calvinistic clergy, but in other colonies – Connecticut, for example, or Rhode Island – more democratic forms of government appeared. Almost everywhere, though, the realities of travel in the age of the sailing-ship and horse, and the conditions of the New World, tended to erode the actual power of government in England to control its colonists. Local self-government was soon a reality in the Anglo-Saxon colonies, whatever the arrangements under which they had first been set up.

The quaker William Penn founded the colony of Pennsylvania. His unusually scrupulous relations with Indians became near-legendary and are commemorated in this picture showing an agreement of a treaty with them.

None of the North American colonists had to deal with complicated, rich native societies like those of Mexico and Peru. North American 'Indians' in the seventeenth century were often only just entering the agricultural phase of their existence; their technology was at best neolithic. Nonetheless they could offer valuable advice to the white settlers. In their early days Massachusetts colonists were actually saved from starving by food provided for them by Indians. Unhappily this did not bring the Europeans to treat them any better in the long run. Gradually the white settlements encroached on traditional hunting lands and there began a long-drawn-out era of conflict which was to end with the virtual extinction of many aboriginal peoples. Those who survived did so by moving further west. This was one of the costs of the opportunities which English America more and more obviously offered to thousands of poor Europeans. Drawn by its attractions, Germans, Huguenots and Swiss began to arrive at the end of the seventeenth century; the Dutch had been earlier arrivals. North America was by 1700 already proving a 'melting-pot' of different stocks.

THE ASIAN SPHERE

Conquest and settlement were not the whole story of the European impact on the world. The same drive to make themselves rich, the same conviction of spiritual supremacy and the innate rectitude of their cause, the same pressures of rivalry, all operated east of the Cape of Good Hope and west of Acapulco just as they did in the Americas or on the slaving coasts of Africa. But there were important differences, too. In the first place, Asia had traditionally been a source of much-prized commodities, high in value and small in bulk (the outstanding examples were spices). They had to be bought or bartered for. Secondly, much of Asia with which Europeans were in contact was ruled by empires and powers which commanded great military resources (including firearms and cannon), had long traditions of settled government, and had, in at least some cases, claims to respect as great powers. Thirdly, some of them could demonstrate cultural and artistic achievement on a scale which left many Europeans uneasily conscious of their inferiority; China, at least, went on being an object of adulation to some European thinkers until well into the eighteenth century. Finally, there were a

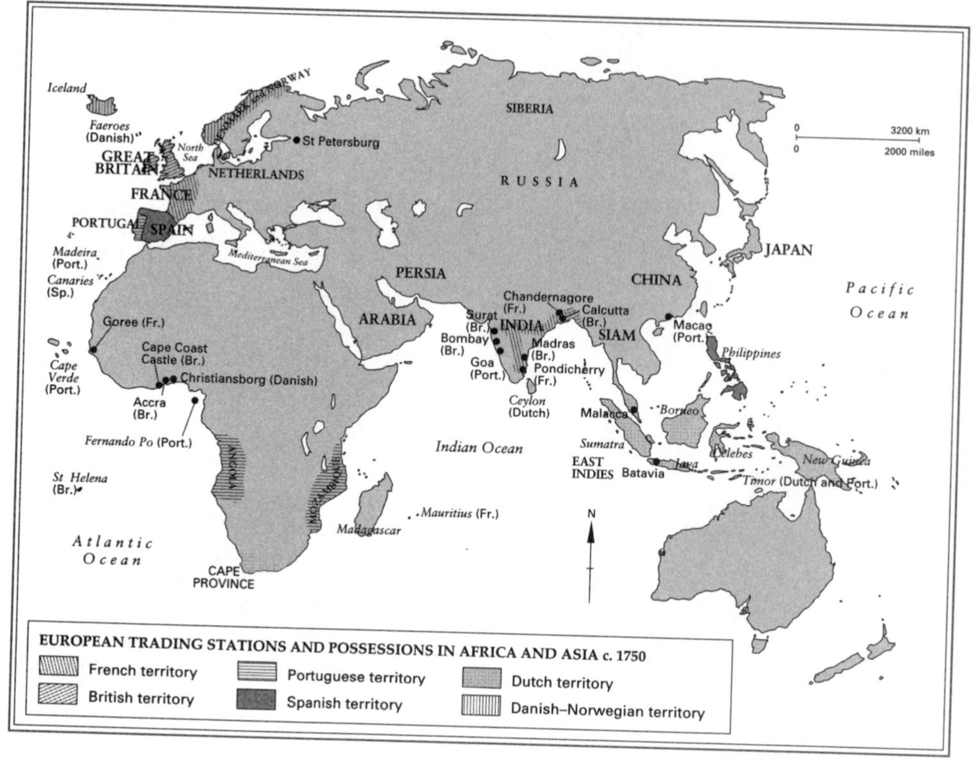

great many Asians and very few Europeans – and, it may be remarked, it was the Europeans who tended to succumb to disease for which they were not prepared, so that settlement was hardly an option.

The outcome, by and large, was the establishment by some Europeans – Portuguese, Dutch, English, French – of what can be called trading rather than colonial empires. They consisted of scattered depots, usually ports, valuable for the conduct or protection of commerce, and of certain treaty or customary rights and privileges enabling European merchants to do their business. This left the peoples of most Asian countries virtually unaware of the European presence.

Europeans and China

China was the most striking example. After a series of piratical exploits which aroused the rage of the empire, the Portuguese were turned out in 1522. They then successfully established themselves at Macao, but they were not allowed up-river to Canton, and this state of affairs continued under both Ming and Manchu. Meanwhile, China seemed as unchallenged and unassailable as ever. For all its heavy cost in lives (some 25 million died) Manchu conquest seemed to restore China's imperial power and inaugurate another flowering of the arts. K'ang-hsi, who reigned from 1662 to 1722 and was the greatest of Ch'ing emperors, inaugurated conquests which were to continue during the eighteenth century. He took over Formosa, occupied Tibet, mastered the Mongols – something of a turning-point, for from this time the nomadic peoples of Central Asia at last began gradually to fall back before the settler. Further north, in the Amur valley, another new historical chapter opened when, in 1685, a Russian post was destroyed and China made her first treaty with a European power to secure her frontier. Chinese relationships to the outside world were developing faster, perhaps, than any Chinese knew. Later, in the eighteenth century, Tibet was again invaded and vassal status reimposed on Korea, Indo-China and Burma. Meanwhile, at home, peace and prosperity produced a silver age of the high classical civilization which some scholars believe to have reached its peak under the later Ming. Much beauty and scholarship were nevertheless still produced under the Manchu, and as early as K'ang-hsi's reign the imperial kilns began a century of technical advance in enamelling which produced exquisite glazes.

Yet Manchu China's civilization was still the civilization of an élite. It was as much the property of the Chinese ruling class as it always had been, a fusion of artistic, scholarly and official activity and profoundly conservative. It strove to imitate and emulate the best, but the best was always past. The practical outcome seems starkly apparent by the eighteenth century. Paradoxically, for all her early technological achievement China had not arrived at a mastery of nature which could fit her to resist Western intervention. Gunpowder is the most famous example; the Chinese had it before anyone else, but could not make guns as good as those of Europe, nor even employ effectively those made for them by European craftsmen. Chinese sailors had long had the use of the mariner's compass and a cartographical heritage which produced the first grid map, but they were only briefly exploring navigators. They neither pushed across the Pacific like the more primitive Melanesians, nor did they map it, as did the later Europeans. For six hundred years or so before Europe had them, the Chinese made mechanical clocks fitted with the escapement which is the key to successful time-keeping by machines, yet Europeans brought with them to China an horological technology far superior to the Chinese. The list of unexploited intellectual triumphs could be much lengthened.

China was, therefore, dangerously cramped in her possibilities of response to the outside world and there were more dangerous threats on the way. The Russians were by 1700 installed in Kamchatka, were expanding their trade on the caravan routes and were soon to press on into the Trans-Caspian region. Even peace and prosperity had a price, for they brought faster population growth. By 1800 there were over three hundred, perhaps even four hundred, million Chinese.

By then, too, Europeans' awe at the size and grandeur of the empire had begun to ebb. Probably the high point of their respect for it had been reached in the seventeenth century. A Roman Catholic mission of Jesuits had then been allowed to establish itself at Peking. Its members were well-treated (the Chinese were interested in the clock-making, architectural and astronomical skills of the Jesuits) and began to hope that the conversion of the empire was in sight: perhaps a Chinese Constantine would be found. But this mission came to an end when the papacy condemned the concessions its members had made to Confucianism. It was a sign that European values were likely to be less open to sinicization than those of other barbarians who had come to China. But admiration for Chinese art and style continued, leading to a craze in Europe for chinoiserie in the eighteenth century, when many European merchants (notably British) first were able to settle in Canton to satisfy it with Chinese goods.

Japan

Japan, meanwhile, remained much more tightly sealed against Europeans. The period after 1603, when the old title of shogun was revived, is known as the 'great peace', and for two and a half centuries the emperor was firmly kept in the wings of Japanese politics. The shoguns themselves changed from being outstandingly important feudal lords to being in the first place hereditary princes and in the second the heads of a stratified social system over which they exercised viceregal powers in the name of the emperor and on his behalf. This regime was called the *bakufu* – the government of the camp. The key to its structure was the power of the Tokugawa house which controlled the shogunate. The feudal lords became in effect Tokugawa vassals. They were carefully watched. The lords lived alternately at the court or on their estates; when they were on their estates, their families lived as potential hostages of the shogun at Edo, the modern Tokyo.

Japanese society was strictly separated into hereditary classes. The noble samurai were the lords and their retainers, the warrior rulers who dominated society and gave it its tone as the gentry bureaucrats gave one to the Chinese. The original links of the retainers with the

Japanese castle-building: Himeji, Kyoto, built in 1609.

land were virtually gone by the seventeenth century and they lived in the castle towns of their lords. The other classes were the peasants, the artisans, and the merchants, the lowest in the social hierarchy because of their non-productive character, in spite of the vigour of Japanese trade. The aim of the whole system was stability, attention to the duties of one's station and confinement to them was determinedly enforced. Hideyoshi himself, the first shogun, had supervised a great sword hunt whose aim was to take away these weapons from those who were not supposed to have them – the lower classes. Japan's society accordingly came to emphasize the things that could ensure stability: knowing one's place, discipline, regularity, scrupulous workmanship, stoical endurance.

It was a weakness of this system that it presupposed isolation. It was for a long time threatened by the danger of a relapse into internal anarchy; there were plenty of discontented nobles and restless swordsmen about in seventeenth-century Japan. But there was also an obvious external danger: the Europeans. They had already brought to Japan imports which would have profound effect, above all, firearms. There was also Christianity, at first tolerated and even welcomed as something tempting traders from outside. In the early seventeenth century the percentage of Japanese Christians in the population was higher than it has ever been since. Soon, it has been estimated, there were over half a million. Nevertheless, once Christianity's great subversive potential had been grasped by Japan's rulers, a savage persecution began. It brought trade with Europe almost to an end. The English, Spanish and Portuguese left within a couple of decades. Other steps followed. Japanese were forbidden to go abroad, or to return if they were already there, and the building of large ships was banned. Only the Dutch, who promised not to proselytize and were willing, symbolically, to trample on the cross, kept up Japan's henceforth tiny contact with Europe from a trading station on an island in Nagasaki harbour.

A changing country

After this, there was no real danger of foreigners exploiting internal discontent. But in the settled conditions of the 'great peace', military skill declined and military technology grew out-of-date. When the Europeans came back, Japan's military forces would be technically unable to match them. There were other difficulties too as a result of the general peace in which internal trade prospered. The Japanese economy became more dependent on money.

Old relationships were weakened by this and new social stresses appeared. At the same time, merchants did well. Gradually the warriors became dependent on the bankers. Towns were growing, too, and by 1700 Osaka and Kyoto both had more than 300,000 inhabitants, while Edo may have had 800,000. Other changes were bound to follow such growth.

While Japan's rulers slowly came to show less and less ability to contain new challenges to traditional ways, those challenges stemmed from a fundamental fact: an economic growth which in historical perspective now appears the dominant theme of the era. Between 1600 and 1850 Japan's agricultural production approximately doubled, while her population rose by less than half. The explanation of what seems to have been a successful stride to self-sustaining economic growth is still debated. It must have helped that the seas around Japan kept out invaders such as the steppe-borne nomads who time and again harried the wealth-producers of mainland Asia. The great peace was another bonus. There were positive improvements to agriculture stemming from more intensive cultivation, better irrigation, the exploitation of new crops brought (originally from the Americas) by the Portuguese. But for all its aspirations to regulate society, too, the government of the bakufu in the end probably favoured economic growth because it lacked power. Instead of an absolute monarchy, it came to resemble a balance-of-power system of the great lords, sustainable only so long as there was no foreign invader to disturb it. As a result it could not obstruct the path to economic growth and did not divert resources from producers who could usefully employ them. Indeed, the economically quasiparasitical samurai actually underwent a reduction in their share of the national income at a time when producers' shares were rising. It looks as if by 1800 the per capita income and life expectancy of the Japanese was much the same as that of their British contemporaries.

Much of this has been obscured by more striking features of the Tokugawa era at a different level. The new prosperity of the towns created a clientele for printed books and the coloured wood-block prints which were later to excite European artists' admiration. It also provided the audiences for the new kabuki theatre. Yet brilliant though it often was, and successful, at the deepest economic level (if undesignedly), as it was, it is not clear that the Tokugawa system could have survived much longer even without the coming of a new threat from the West in the nineteenth century. Towards the end of the period there were signs of uneasiness, but Japan had already made for herself a unique historical destiny and it would mean that she faced the West in a way very different from the subjects of Manchu or Moghul.

Moghul India in decline

By then, another great centre of Asian civilization was showing much more marked signs of concession to the Europeans than either China or Japan. Akbar's court empire was one of the most powerful and his court one of the most sumptuous in the world. His successor was contemptuous of the presents sent to him by James I, the ruler of an impoverished England, a few years later. Yet the future of India lay with those who brought them. The Moghul emperors were to continue in direct descent, though not without interruption, until the middle of the nineteenth century, but with a long period of declining power. Under the three rulers who followed Akbar the empire grew to its greatest extent in the first half of the seventeenth century and then began to decay in the second.

Shah Jahan, Akbar's grandson, began the piecemeal acquisition of the Deccan sultanates and tried without success to drive the Persians from Kandahar. During his reign there was a weakening of the principle of religious toleration, though not sufficiently to place Hindus at a disadvantage in government service; administration remained multi-religious. At Agra, a lavish and exquisite court life was kept up. The emperor built there the most celebrated and the best-known of all Islamic buildings, the Taj Mahal, a tomb for his favourite wife. It is the culmination of the work with arch and dome which is one of the most conspicuous Islamic legacies to Indian art and the greatest monument of Islam in India.

India's most famous building and the supreme achievement of Islamic architects in the subcontinent, the Taj Mahal, built by Shah Jehan as a tomb for his favourite wife.

Below the level of the court, life in Moghul India was becoming less attractive. Local officials had to raise more and more money to support not only the household expenses and campaigns of Shah Jahan but the social and military élites who were parasitic on the producing economy. Without regard for local need or natural disaster, a rapacious tax-gathering machine may at times have been taking from the peasant producer as much as half his income and virtually none of it was productively invested. Yet even Shah Jahan's demands probably did the empire less damage than the religious enthusiasm of his third son, Aurungzebe, who set aside three brothers and imprisoned his father to become emperor in 1658. He combined, disastrously, absolute power, distrust of his subordinates and a narrow religiosity. Revolts against Moghul rule owed much to Aurungzebe's attempt to prohibit the Hindu religion and destroy its temples and to his restoration of the poll-tax on non-Moslems. A Hindu's advancement in the service of the emperor became less and less likely; conversion was necessary for success. A century of religious toleration was cancelled and this weakened old loyalties and cooperation.

It also made it impossible to conquer the Deccan, which has been termed the ulcer which ruined the Moghul empire. As under Asoka, North and South India could not be united. The Mahrattas, the hillmen who were the core of Hindu opposition, constituted themselves under an independent ruler in 1674 and allied with the Deccan sultans to resist the Moghul armies in a long struggle which threw up a heroic figure, Shivagi, who has become something of a paladin in the eyes of modern Hindu nationalists. He it was who built from fragments a Mahratta political identity which soon enabled him to exploit the taxpayer as ruthlessly as the Moghuls had done. Aurungzebe was continuously campaigning against the Mahrattas down to his death in 1707. His three sons disputed the succession and the empire almost at once began to break up and a much more formidable legatee than the Hindu or local prince was waiting in the wings – the European.

European arrivals

Since Akbar's day, the Europeans had been allowed to establish toeholds and bridgeheads. The Portuguese were first. The English won their first west-coast trading concession early in the seventeenth century. Then, in 1639, on the Bay of Bengal and with the permission of the local ruler, they founded at Madras the settlement which was the first territory of British India, Fort St George. Though they fell foul of Aurungzebe, they got further stations at Bombay and Calcutta before the end of the century. Their ships maintained a paramountcy in trade they had won from the Portuguese, but by 1700 a new European rival had appeared. A French East India Company founded in 1664 soon established its own settlements.

Europeans long associated the name of Madras with the cottons it exported to them. This painted hanging provides an example of the seventeenth-century Indian's vision of the European.

A century of conflict between French and English lay ahead. Politics began to complicate trade amid the uncertainties aroused when Moghul power was no longer as strong as it once had been. Relations had to be opened with the emperor's opponents as well as with him. By 1700 the English were well aware that much was at stake and already by then India was being caught up in events not of her own making, the era of world history, in fact. Little things show it as well as great; in the sixteenth century the Portuguese had brought with them chili, potatoes and tobacco from America. Soon maize, pawpaws and pineapple were to follow. Even Indian diet and agriculture were already changing.

It is likely that it was not the coming of the European which ended the great period of Moghul empire; that was merely coincidental, though the newcomers could reap the advantages. No Indian empire had ever been able to maintain itself for long and the diversity of the subcontinent and the failure of its rulers to find ways to tap indigenous popular loyalty are probably the main explanation. India remained a continent of exploiters – the ruling élites – and producers – the peasants – upon whom they battened. By the end of the seventeenth century, India was ready for another set of conquerors.

AR

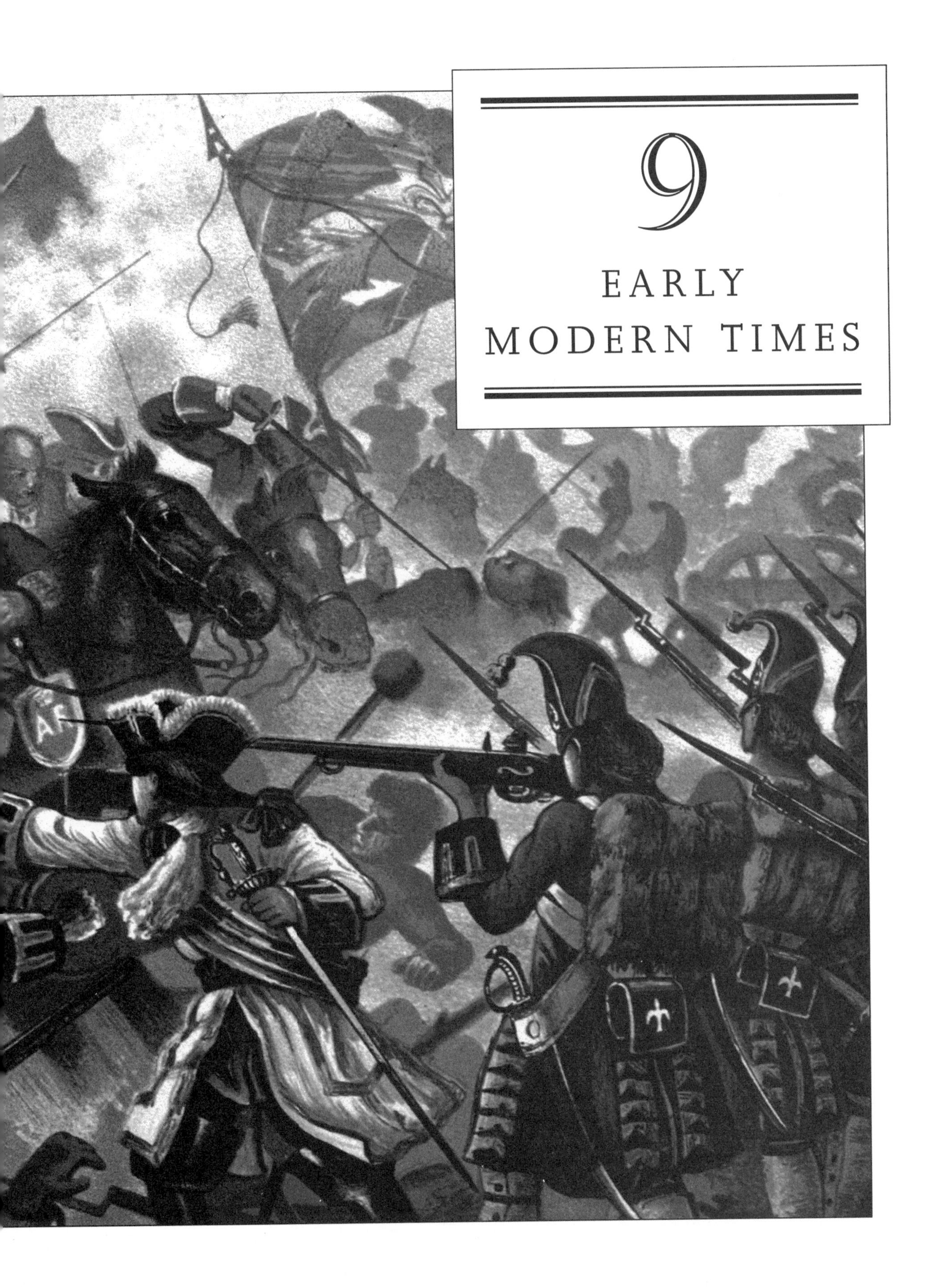

9 EARLY MODERN TIMES

THE FIRST SIGNS OF GLOBAL HISTORY

Modern history begins in Europe, where the forces and processes which were to make a unity of world history – a huge world-wide complex of inter-relating events and reciprocal movements – came first into being, and can first be seen at work. Perhaps the first sign at all of this was simply the new knowledge of the natural physical shape of the world and its continents, all of which (except for Antarctica which was as yet undiscovered) were at least known to exist in 1500, even if their form was not clear.

There was coming into being in Europe, too, the first faint idea of the huge variety of the world which might remain to be discovered and of peoples and cultures even more unlike those of Christian Europe than were those of Islam or Orthodoxy. Something else which could not then yet be known, though, was anything about the numbers, structure and distribution of mankind. Even now we can do little more than make careful guesses about population in the sixteenth century. Governments and scientists had not begun to collect statistics in a systematic way and, indeed, there was strong public feeling in European countries against taking a census even as late as the eighteenth century. Counting people almost always foreshadowed higher taxes; what was more, there were Biblical precedents against such things. As a result, any estimates must be treated with much caution. There happen to exist quite a lot of figures for Italy's population in 1500, but scholars still come up with totals which vary from five to ten millions.

Making the best of such information as we have, it looks as if the total world population in 1500 was about 425 millions. The largest part of this in any one continent was in Asia. The greatest single political unity in terms of numbers was (as it has always been ever since the end of the Roman empire) China, which can hardly have had much less than 100 million inhabitants. Probably the next biggest Asian country was India, though estimates of its population are very vague. Europe, including Russia, may have had eighty million inhabitants. After the enormous set-back caused by the Black Death, the population in some parts of Europe had not got back to levels of the early fourteenth century by 1500. France was then the biggest single European country. She had something like sixteen million inhabitants, and was undergoing a burst of rapid growth. As for America, at the time of its discovery by Europeans, there must have been only about a million Indians and Eskimos in the whole of the vast spaces of North America – the largest single area still sheltering the old pre-agricultural

The Plague at Tournai, 1349.

way of life – though in contrast there may have been fourteen million Americans south of the Rio Grande in 1500, about five million of them in central Mexico.

We can now see, though, that populations in some European countries had already in 1500 begun to grow in a new, uninterrupted way which has continued to the present. What is more, though there were big differences between them (different countries grew at very different rates at different times), they grew much faster than in earlier times. This began to change the balance between continents. By 1800 there seem to have been about 900 million people in the world, about twice as many as three centuries before. This was a slow but very large increase. It is not easy to explain, though the basic cause may be a spell of climatic improvements and better harvests. But of these more than a fifth – about 185 millions were Europeans, a larger proportion than ever before. Both the new rapidity in population growth and the removal of earlier restraints upon it in a few areas were trends which were to become world-wide in our own times.

THE REVOLUTION IN AGRICULTURE

Population rose because there was more food about. But the reason for that was for a long time operative only in Europe and for a long time, it would not have been easy to see. For most of these three centuries there was a very broad similarity – as there had been throughout the history of civilization – between what people ate the world round. It was nearly always bread or a cooked grain – wheat, maize, rice and rye were some of many varieties used in different parts of the world. Growing grain is a more efficient way of getting calories out of a given area than raising stock on it. Although even in the Middle Ages meat-eating had been more common in some European countries than elsewhere in the world, in 1800 most Europeans still rarely tasted meat. Like the inhabitants of other continents, they supplemented grains with chestnuts, beans and other vegetable foods and with eggs and fish. Hard times could recur in Europe even in the eighteenth century; there were famines in France in the 1770s and 1780s. Elsewhere in China, Russia, India, and in parts of Africa, they have gone on down to the present day (though not usually with such prolonged effects as in former times, because food can now usually be brought quickly from some other part of the world).

Europe's improvement

In spite of all such similarities, though, Europe already differed fundamentally from elsewhere in 1800 in that much more food per head was being grown there than three hundred years earlier. There had been plenty of room for improvement. Even a well-run and efficient medieval farm would seem a poor sort of place to a modern European farmer. Given the amount of labour put into it, its output of crops was very low; seed corn usually yielded a crop only five times its own weight. Nothing like so much per hectare could be grown in 1500 as the same farm would grow today. Very little departure from traditional ways was possible, even if anyone could think of such a thing. In short, much medieval farming in Europe was like what still goes on in parts of Asia and Africa today. Yet a change was coming and by 1800 it was irreversible. The advance in European agriculture in these three hundred years revolutionized human development more than anything since the invention of agriculture itself.

Europeans had always had natural advantages. Thanks to good rainfall much of Europe's land surface can be cultivated. Fish in its coastal waters provide plenty of easily harvested food. Under the surface lay important mineral deposits, including some of the richest iron and coal fields in the world. Before these could be properly tapped, there was abundant wood for fuel and building. Nonetheless, in 1500, most Europeans still lived as subsistence farmers – that is to say, they grew enough for their own needs, and only a few of them regularly produced a surplus to sell to those who did not live in the countryside. Even when they did, their market was usually very local. In spite of trade between countries in wine, wool, hides and a little grain, most daily food was grown and raised close by the place where it was consumed.

What was grown (and ways of growing it) began to be more varied, too. For many centuries now a few big regional divisions in Europe have set the basic patterns of farming. Setting aside for a moment the Scandinavian peninsula, Europe can be seen at its simplest as two zones; a broad plain and a corresponding stretch of high, often mountainous country to the south of it. With almost no disturbance except a slight rise just west of Moscow, the great European plain rolls westwards for more than 4000 kilometres almost uninterrupted by mountains or even high hills. It begins with the broad expanses of Russia, starts to narrow somewhat south of the Baltic, in Poland and western Germany, and then fans out again round the Ardennes and French Massif Central to taper away to the Pyrenees. Across the North Sea, England is part of it too, and it peters out in the foothills of Wales and Scotland. In modern times, this has been Europe's historic grain-growing area. Grain has long provided Europeans with both food and drink. Beer made from barley, and spirits distilled from grains – whisky or vodka, for example – are the traditional alcoholic drinks of this region. It has well-defined boundaries. In Russia there is the line of the northern coniferous forest. The sea provides a northern limit farther west and the southern flank is protected by the mountain-walls of the Carpathians, Alps, Massif Central and Pyrenees.

South of these mountains the ground is generally high-lying except for a few river-valleys of which the Danube, Rhône, Po and Ebro are the most notable and important. Grain is also grown on a large scale in this southern area in some places (the Danube valley is one, the plateau of Castile another), but the high ground is often farmed by stock-raising and pasturing. It is the land of the vine: wine and grape spirits provide its alcoholic drinks. Finally, around the Mediterranean coasts and over much of Spain, it is the land of olives, which provide oil.

Dividing Europe in a different way, the Elbe is not a bad marker of the northern plain's separation into east and west. History has frequently taken different roads either side of a line from the mouth of the Elbe to the head of the Adriatic. Roughly speaking the same line runs along the January 'isotherm' of 0 degrees Centigrade – the line which links places having zero temperature in that month. The west, warmed by the air and water currents we call the 'Gulf Stream', is warmer than the east, swept by belts of cold air from the Arctic and the landmass of Asia. The Sea of Azov, for example, lies as far south as the French city of Lyons, but is often frozen in winter, while the Rhône at Lyons continues to flow. This division has had big consequences for east and west Europeans and the way in which they get their living.

One was that they raised different grains. In eastern Europe until very recent times the hardy rye was the usual grain for human consumption, while wheat or maize (the 'Indian corn' introduced from America in the sixteenth century) were commoner farther west. Another striking difference was that (broadly speaking and with many local variations) most peasants west of the Elbe in 1800 were either freemen owning little plots of land or tenants paying rent in cash or kind. In the east they were much more likely even at that late date to be serfs, 'tied' to the soil of the manor on which they lived, unable to leave without permission. This difference had become much more marked after the seventeenth century, when serfdom entrenched itself more deeply still in the east just as it was dying in the west.

In both east and west Europe many more specific local differences sprang from the needs of husbandry in particular areas; soil, climate, knowledge and local markets all give variety to the picture. They gradually led to specialization, which had other far-ranging effects. Even in the sixteenth century, for example, grain grown in the lands south of the Baltic was shipped to western Europe, and this meant growth in the shipping industry and new profits for the old German towns of the league of sea-ports called the 'Hansa'. In fifteenth-century

An agriculturist and loyal subject presents his king, William IV, with a prize specimen of a bull. In the background, Tredegar Castle, Monmouth.

England East Anglia was already specialized in barley-growing and sheep-raising, while the Thames valley produced wheat, and the northern and western counties grazed cattle. Even the appearance and special qualities of animals were different in different places. The Merino sheep (later to spread world-wide) was suited to the dry pastures of Spain; it looked somewhat goat-like to English eyes, but it gave the best wool. Sheep raised on England's greener pastures, on the other hand, had coarser fleeces but carried more meat. Such variations meant that levels of well-being and comfort differed from country to country. Foreigners noticed that peasants and craftsmen in seventeenth-century England wore woollen cloth, whereas their continental equivalents long wore coarse linens made from flax.

It would be easy to go on listing such differences, for there were many, but though interesting it would not do more than confirm the basic point: backward by modern standards as European agriculture may seem to have been in 1500, it was already pretty diversified. What is more (since nothing happens neatly in history), some of its variety was then already reflecting the first beginnings of the big transformation that was coming, the 'Agricultural Revolution', as it used to be called. Eighteenth-century Englishmen preferred the word 'Improvement', and, although there was a change which was indeed revolutionary, because it transformed the world, it came about only gradually and slowly. Why it should have happened in Europe is still much of a mystery. The basic explanation may be the slow accumulation of wealth and resources – particularly shown in the growth of towns – going on there since the twelfth century. But it is odd that something similar did not happen in, say China, where a great growth of cities also occurred and where intensive labour (required for rice-growing) and manuring (contracts for removing human waste – 'night-soil' as it was called – from towns were very valuable in both China and medieval Europe) were employed.

New methods

Improvement almost always meant specialization. Individual farmers stopped trying to grow everything and concentrated on the things they could do best, buying their other needs elsewhere. It was also always accompanied by technical betterment. This might mean new 'rotations' (that is, using fields for different crops each year in such a way as to rest and improve the soil rather than exhaust it), new products (potatoes and maize from America were outstanding examples), new treatments (liming, for example, for the soil), new varieties of familiar crops (special grasses for pasture), new care for the soil (building drains and hedges), new machinery (though this came in more slowly), or simply the enclosing of land which was formerly 'common' so as to make it the property (and therefore the interest) of one man. In the end all these things led to more being extracted from the land, and greater production meant more food and cheaper clothes.

Some of these changes first appeared in parts of Italy and in Flanders in the fourteenth and fifteenth century. They were pushed farthest in the Low Countries, from which they spread to England in the sixteenth and seventeenth centuries. Among the first results there were the enclosing of land for sheep, the bringing-together of the scattered medieval strips

For most of human history, women have worked on the land. These Flemish haymakers, depicted by Pieter Brueghel the Elder in the early sixteenth century, were probably better dressed and shod that many peasant women of today, and certainly better than the Asian and African women who still make up the majority of the agricultural labour force in many countries.

of an individual holding into compact fields, the draining of local land (especially in the Fens), and the making of new land from marsh and sea (as the Dutch had done). This laid the foundations for a tremendous technical advance in English farming in the eighteenth century, when it became the best in the world. New breeds of animal and varieties of crops multiplied, and the first important innovations in machinery since the coming of the wheeled plough were made with mechanical drills, horse-drawn harrows and threshing-machines.

Visitors came from all over Europe to see English farming and the new methods spread back to the continent, especially to Germany and the east. Here, where the soil was often poor, it was doubly important to take advantage of every possible way of improving it. This led, paradoxically, to landlords hanging on firmly to one aspect of the past. Since what was needed above all in eastern Europe to improve productivity was labour, landlords resisted all attempts to break up the old manorial system. Serfdom had virtually been replaced by wage-labour in England by 1500, but in the next two hundred years it became much more common in Germany and Poland (to say nothing of Russia). The Junkers of East Prussia, as noblemen there were called, got as much work as possible out of their serfs, tying them more firmly by legal rules to the manor so that they could do so. In 1800 it was still quite normal for a peasant on an east German estate not to be able to leave it for a job elsewhere or to marry without permission, or to attend to his own patch of garden before he had done work he owed his landlord. (Nor was this work always due in the field; the serf's children and womenfolk might have to work in the house for the lord, too.) In Russia things were even worse – and were to get worse still – though technical improvement there was slight. Certainly improvement was far from being the only reason why serfdom persisted in eastern Europe while disappearing in the west, but it is part of the explanation. It was very convenient to tighten up the demands made on serfs if you wanted to improve your estate. In some places (Poland was the worst) the outcome was that peasants were reduced to near-slavery.

General progress often makes many individuals miserable, but it is hard to argue that the overall and long-term effects of improvement were not good. By 1800 there were still many hungry people in Europe, but in some countries they were far fewer than three centuries before; a corner in history was being turned because agriculture was the mainspring of the economy. Apart from minerals and fish-products, most manufactures and commerce depended on and dealt in things grown or raised on the land – hides for shoes, wool for cloth, grapes and hops for wine and brewing.

RULERS AND RULED

In 1500 Europe outside the Ottoman Turkish lands was almost entirely Christian. Well to the east, Christendom was divided by the line at which Roman Catholicism gave way to Orthodox Christianity; there were borderlands in Hungary, the Ukraine and what is now former Yugoslavia, where these denominations were rather mixed up. Roman Catholic bishoprics could be found as far east as Vilna and Dniester. As for the Europeans under Moslem Turkish rule, they usually belonged to one of the Orthodox Churches. In the next few centuries Islam was to advance with Ottoman rule still farther in Europe, by conversions among the Balkan peoples, but by way of balancing this the many Moslems who still lived under Spanish rule in 1500 disappeared. At that date there were also Jews in almost every European country; in some Jewish numbers were very small, but many were to be found in the borderlands of Poland and Russia, to which Jews had fled from persecution in western Europe during the Middle Ages. For the most part, geographical Europe was nearer to being the same thing as what was called 'Christendom' – the part of the world lived in only by Christians – than ever.

Political and legal description is much more difficult. Spain, Portugal, England and France, it is true, looked somewhat like their modern equivalents in 1500. Each of them had good natural boundaries and that helped. The Pyrenees, the Atlantic and the Mediterranean cut off the Iberian peninsula; once the Moors were defeated, it was not easy for outsiders to interfere there. But Portugal had its own king, and Spain, though united under the same rulers, was legally divided into the kingdoms of Castile and Aragon, each of which had separate laws and customs. Up in the north too there was the little independent kingdom of Navarre. As for England, she was almost an island and her kings had conquered Wales long before. Yet she still had an independent neighbour in Scotland; though the two kingdoms shared a king after 1603, they were not brought together into one state, 'Great Britain', until 1707 (and even then many of their laws remained different). Ireland, though an island, was a conquered province ruled by an English viceroy until the eighteenth century. At that time English kings still called themselves kings of France; this was by then pure boasting or antiquarianism, but in 1500 England still held a tiny patch of land around Calais, though the French kings were effectively the overlords of most of modern France. Some eastern areas, notably much of Burgundy, Savoy, Alsace and Lorraine, had not yet been brought under their rule, though, and right inside France a few little 'enclaves' belonged to foreign rulers; the most notable of them being Avignon, under the pope.

Dynasticism

Most rulers had claims on lands elsewhere through marriage or descent. If there was one principle on which what we would now call 'international affairs' (the phrase did not exist then) was based in 1500, it was that family ties were all-important. The relations of European rulers with one another then were still shaped above all (as they had been for centuries) by the struggles of families to extend, strengthen and safeguard their inheritances. Broadly speaking, most statesmen thought of Europe as a patchwork of personal and family estates; pieces of land belonged to individual rulers and therefore to one or another of the major royal dynasties, just as different farms or houses in different parts of the same country might have the same owner. 'Dynasticism', as it is sometimes called, was the pursuit of the interests of a ruling family, rather than that of the inhabitants of a particular place. It was what Europe's politics were mainly about.

Two or three families stand out. The Welsh Tudors had recently provided a king (Henry VII) for the English throne, but though his son (Henry VIII) made a bid to become Holy Roman Emperor a few years later, English monarchs did not usually count for much in the sixteenth century except when other people were quarrelling among themselves and wanted their help or neutrality. The French Valois were more important. They had ruled France since the fourteenth century and had almost driven the English out after a long struggle; they cut much grander figures than the Tudors and lasted until 1589, when another highly successful line (related by marriage), the Bourbons, took the French throne. In 1500, though, both Valois and Tudor were outshone and were indeed going to be long outlasted by another family, the Austrian Habsburgs.

The ups and downs of the Habsburgs are much of the story of European politics right down to 1918, by which time they had been ruling Austria for about six hundred years. In 1438 one of them became ruler of what became known about that time as the 'Holy Roman Empire of the German Nation'. Usually, it went on being called simply the Holy Roman Empire or even just the 'Empire' and it was the remote but direct descendant of Charlemagne's restoration of an old idea.

Much of the empire still lay outside Germany, but it was German princes who elected the emperor and from the fourteenth century onwards they sometimes chose a Habsburg. From 1438 until the Holy Roman Empire disappeared (in 1806, though an Austrian empire survived, and so the Habsburgs went on using the title of 'emperor'), they did so continuously with only one brief interruption. In 1500 the emperor Maximilian was head of the family. His first wife was the daughter of one of the richest medieval rulers, the duke of Burgundy,

TUTANKHAMON The sheer glamour of ancient Egypt was startingly enhanced in 1922 when the magnificent contents of Tutankhamon's tomb at Thebes were discovered. He reigned only briefly – and died when he was eighteen years old in about 1340 BC – but was buried more sumptuously than any other Egyptian ruler. His stone sarcophagus was set within three shrines, the sarcophagus itself containing three coffins, the last with this mask of pure gold. There was, in addition, a mass of grave-goods, many of precious materials. The imagination of the public was seized by the relics of a king who otherwise left little mark on Egypt's history.

METAL-WORKING AND ART The working of metals brought new possibilities for art as well as new tools to ancient civilizations and barbarian cultures alike. Two thousand years or so separate these works by peoples who left little permanent record other than what they made and embellished. The gold ewer (above) is possibly Hittite work from just before 2000 BC; among its severely geometrical decorations can be distinguished the swastika so often used in Indo-European art. The Celtic shield was found in the Thames at Battersea and is probably from the era of Augustus: its flowing decorative style perhaps indicates it was an offering to a river god.

EASTER ISLAND In 1722 a Dutch expedition in the middle of the Pacific came across a small island two thousand miles or so west of South America whose inhabitants worshipped huge statues of their gods. The Dutch named it Easter Island. Only recently, these great monoliths – some over twenty feet high – have been re-erected as the techniques employed to handle and place them have come to be understood. Yet still little is known of the people who carved them and put them up. One theory is that the skills they needed must have come from Peru, the nearest home of a culture with advanced stone-working techniques. But when? What seems certain is that there were Easter islanders already on the site in the late fourth century AD – while the Roman empire crumbled in Europe.

THE MEDIEVAL WORLD ORDER In 1355, *Andrea Bonaiuti painted this fresco for the church of Santa Maria Novella in Florence. Against the symbolic background of the city's (then uncompleted) cathedral, he set side-by-side the figures of Pope and Emperor. In due order, they are flanked by great officers and magnates of church and state – ecclesiastics to the Pope's right, laymen to the Emperor's left. Christ's flock nestles at the feet of His two Vicars on earth, guarded by dogs who protect them from the wolf of heresy. A pun is intended: these are* Domini canes *– 'the Lord's dogs' – whose coats echo the black and white habits of the Dominicans in whose honour the fresco was commissioned. At the sides of the composition cluster bishops, nuns, doctors on the one side, and, on the other, peasants, nobles and townsfolk (among them, it is believed, representations of Dante, Bocaccio, the painter Cimabue and other notable Florentines).*

THE SAMURAI TRADITION In 1156 the rivalry of two great Japanese clans, the Taira and Minamoto, broke out in open war which culminated in the storming of the capital, Kyoto, (depicted here in this scroll) and Taira victory. The Samurai tradition which wove miltary ideals into so central a role in Japanese culture was shaped in such battles. It had the curious by-product that arms and armour became one of the most important Japanese artforms. The remarkable sword guard from the nineteenth century is an example: the three imbibers sampling the sake jar under a pine tree are three sages, Lao-tze, the Buddha and Confucius, strangely juxtaposed with the opening for the sword's blade.

BENIN The art of the *West African kingdom of* Benin was royal art, focussed upon the glorifying of the Oba or 'divine king', his predecessors, famous ancestors and great events in the kingdom's history. It was as firmly directed to that end as the self-glorifying paintings and sculptures of European princes, but was more limited in form, to bronze, cast brass and ivory. At the end of the nineteenth century, a huge amount of this art became available to European collectors and museums, thanks to a British punitive expedition in 1897 which deposed the Oba, destroyed his palace and carried off thousands of specimens of Benin art. Thriftily, the British government sold them off to pay for the expedition, among them this male head, now in the British Museum.

INDIA On 22 June 1757 a batle took place at Plassey, about a hundred miles from Calcutta, which settled the destinies of India and Great Britain for nearly two centuries. It was won by a small army commanded by Robert Clive, an employee of the East India Company, over the forces of the nawab of Bengal, the Moghul emperor's governor of the province. It was soon followed by the grant to the Company of the right to raise revenue in Bengal, which made it effectively the government, though the Moghul emperor kept his nominal sovereignity. In this picture, Clive points to a group of destitute soldiers while receving from the nawab a sum of money to be used to help disabled veterans and the widows of those who died in the Company's service.

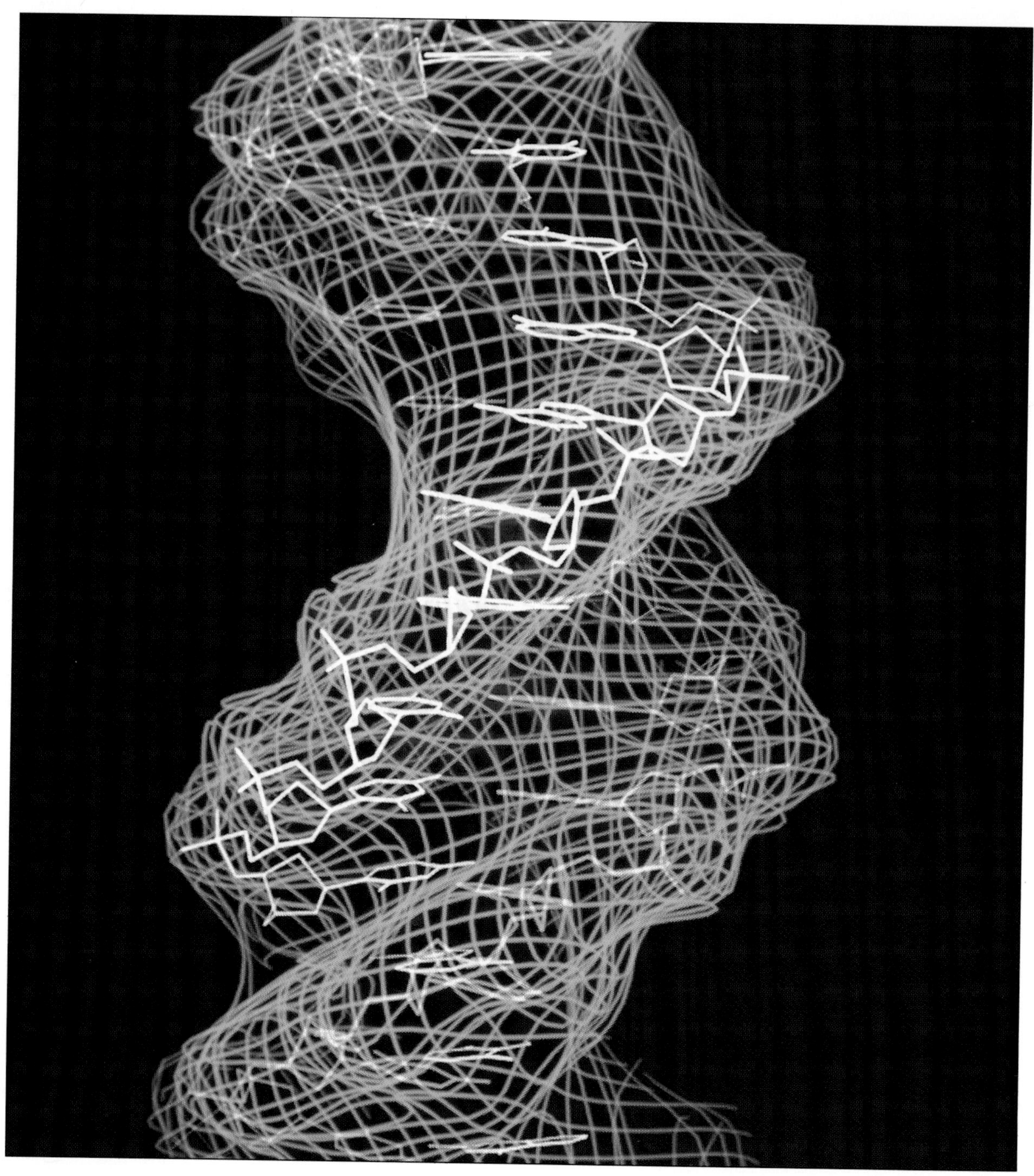

DNA – THE SHAPING OF LIFE New frontiers in science are increasingly reached by developments in technology. Computer graphics have now provided a tool to show structures previously only visible by ultra-microscopical means. This is a representation of a short section of DNA (deoxyribonucleic acid), whose molecule contains all the information from which a particular living organism is built and maintained. It consists of a double helix of strands of sugar phosphate residues linked by pairs of organic, purine and pyrimidine bases arranged as the rungs of a spiral ladder. The precise sequence of the bases determines genetic inheritance by forming genes which may be transcribed to produce RNA (ribonucleic acid) which is translated via specific amino acids into different protein chains. These ultimately produce the molecules which constitute and maintain living organisms. The blue lines in the model show the surface of the structure – the three-dimensional form of the double helix.

who had no son to succeed him. The duke's death therefore caused a lot of trouble, and further complicated the map, for fragments of his inheritance passed into many different hands. But before this there was much quarrelling and trouble. One can view many events of the sixteenth century as a long duel between Valois and Habsburg over the Burgundian inheritance, notably its rich Netherlands provinces (roughly, modern Belgium and Holland). When, in 1519, a Habsburg who was already king of Spain became Holy Roman Emperor and united to the old Habsburg lands the empire of Spain, it seemed that the family might well be on the way to a universal monarchy. This was Charles V, the first man of whom it was said (truthfully) that he ruled an empire on which the sun never set.

Habsburg and Valois joined other families in quarrelling about Italy, seeking allies and satellites there among the dozen or so major states into which the peninsula was divided. Some of these were aristocratic republics (Venice was the most famous and had great overseas possessions in Cyprus, Crete and the islands of the Aegean), some virtually monarchies, whether they admitted it or not (like Florence, nominally a republic but really in the hands of a family of former bankers, the Medici). But these were not the only complications in Italy. Most Italian states fell within the Holy Roman Empire, but some did not, among them three very important ones – Venice, the kingdom of Naples and the Papal States, the lands ruled by the pope as a prince like any other prince.

The last medieval emperor rides out in defence of Christendom much as he must have thought the great paladins who were his predecessors had done; Charles V, painted by Titian in one of his greatest works.

The Empire and eastern Europe

Complicated as this sounds, the map of Italy was simplicity itself by comparison with that of Germany and central Europe. Germany was the heartland of the Holy Roman Empire. The Habsburgs made great efforts to turn the empire into a centralized monarchical state, but they never succeeded. Its constitution was chaos. It was supposed to provide the machinery for harmoniously running the affairs of about four hundred different states, statelets and notables. There were, for example, princes who were the feudal vassals of the emperor (among them the most important were the seven who elected him) but in no other way subordinate to him; dozens of independent imperial cities; the Habsburg family lands in Austria themselves; fifty princes of the Church who ruled in their lands like lay sovereigns; hundreds of minor noblemen – the imperial knights – subject only to the emperor as feudal dependents; the Bohemian and Silesian lands which actually belonged to the crown of Hungary (itself outside the empire) and so on and on. This was a terrible mess, though taken for granted as a proper state of affairs. As Charles V had to govern Spain and its huge possessions outside Europe as well, any chance of exercising real control was virtually non-existent.

Some Germans lived beyond the boundaries of the empire, in royal Prussia, for example. On the Baltic coast Germans were mixed up with Swedes and Poles. Across the sea Sweden was an independent kingdom (including what is now Finland); Denmark and Norway shared another ruler. On the mainland again, the great sprawling kingdom of Lithuania covered much of modern Poland, Galicia and the Ukraine. Russia, farthest east, was expanding, but was then still little more than the northern half of what is now that country west of the Urals, and its Tsar was barely to be regarded as part of the European community of rulers at all. Finally, in central Europe another large and independent Christian kingdom, Hungary, lay between the empire and the Ottomans in the Danube valley, with part of its lands inside the imperial boundary, part outside.

New tendencies in government

The men and women who ruled the political units which made up this varied pattern of realms, estates and nations, did not, in the sixteenth century think of themselves as doing anything very new and often behaved in very 'medieval' ways, or at least in ways which we should find somewhat surprising in modern rulers. French kings still set off to invade Italy very much in the spirit (they thought) of the chivalrous knights of old, while Henry VIII of England turned up in 1520 to a spectacular diplomatic meeting in Flanders – the 'Field of the Cloth of Gold' – which was run much as a traditional medieval social occasion, with tournaments and jousting. Kings still fought, for the most part, for the advancement of their own family's interests rather than those of the people over whom they ruled. As for those at a lower level of public life, noblemen fought back when they thought that kings were encroaching on the independence or dignity they were entitled to by custom. Medieval representative bodies (one of them was the English parliament) also still had a long life ahead of them in some countries.

Yet great political changes were coming. Though these were not always complete even by 1800, the old 'feudal' arrangements which once governed almost the whole of Europe had by then ceased to matter much anywhere west of the Rhine and in some of the lands to the east of it. The process behind this had begun far back in the Middle Ages. Towns had never really fitted into feudal society and they had grown greatly in size and importance since 1100. More and more of the tradesmen and merchants who lived in them were independent men, wealthier than most nobles. Pressures on traditional ways only changed society very slowly and were very complicated, but they already made it impossible by 1500 to think only in 'feudal' terms. Kings, too, for all their conservatism, were likely to be impatient with some of the old ways. They wanted to rule – and that really meant to tax – their subjects without interference from anybody else. They employed lawyers to think up ways of undermining the

INDIA On 22 June 1757 a batle took place at Plassey, about a hundred miles from Calcutta, which settled the destinies of India and Great Britain for nearly two centuries. It was won by a small army commanded by Robert Clive, an employee of the East India Company, over the forces of the nawab of Bengal, the Moghul emperor's governor of the province. It was soon followed by the grant to the Company of the right to raise revenue in Bengal, which made it effectively the government, though the Moghul emperor kept his nominal sovereignity. In this picture, Clive points to a group of destitute soldiers while receving from the nawab a sum of money to be used to help disabled veterans and the widows of those who died in the Company's service.

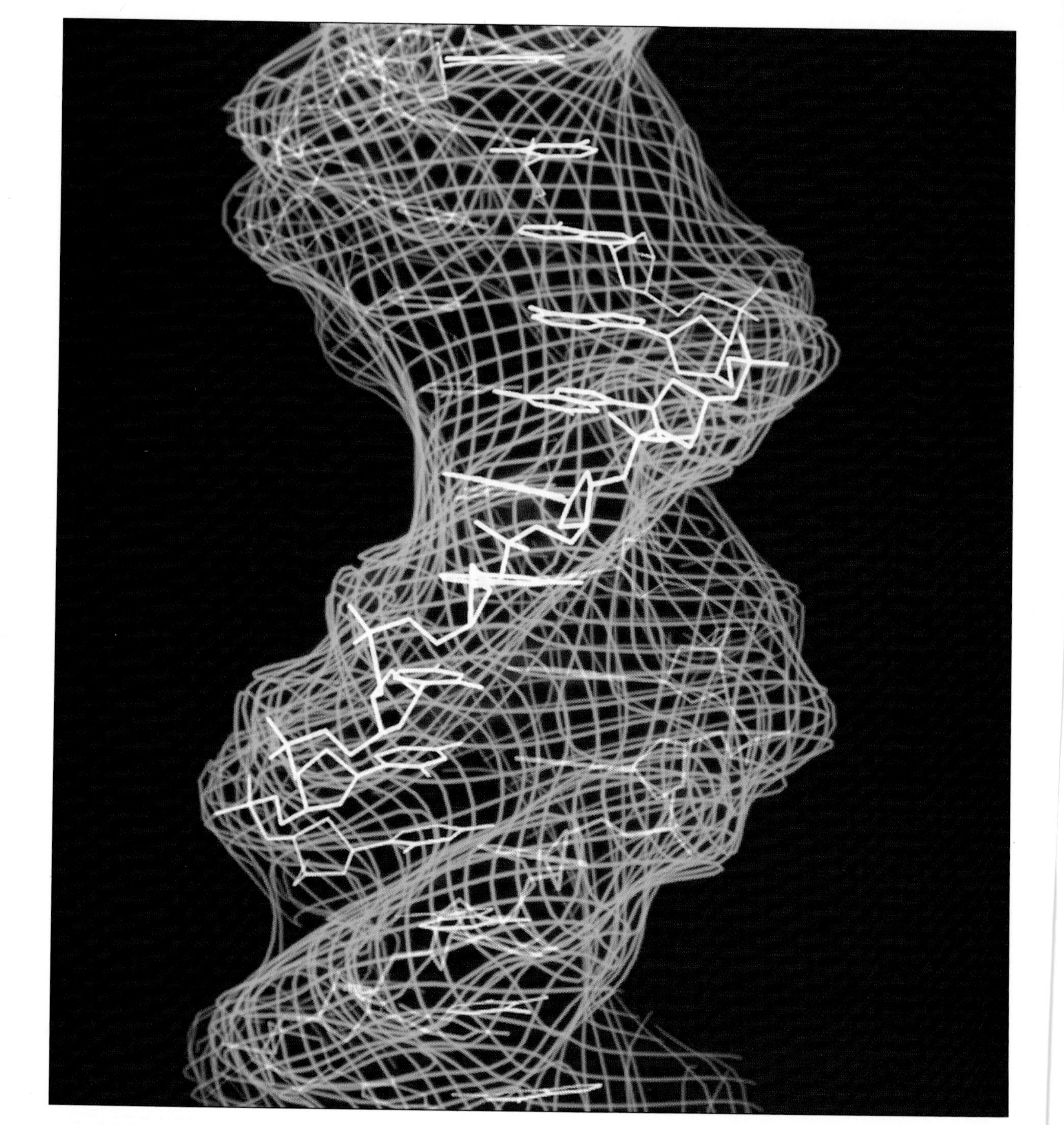

DNA – THE SHAPING OF LIFE   New frontiers in science are increasingly reached by developments in technology. Computer graphics have now provided a tool to show structures previously only visible by ultra-microscopical means. This is a representation of a short section of DNA (deoxyribonucleic acid), whose molecule contains all the information from which a particular living organism is built and maintained. It consists of a double helix of strands of sugar phosphate residues linked by pairs of organic, purine and pyrimidine bases arranged as the rungs of a spiral ladder. The precise sequence of the bases determines genetic inheritance by forming genes which may be transcribed to produce RNA (ribonucleic acid) which is translated via specific amino acids into different protein chains. These ultimately produce the molecules which constitute and maintain living organisms. The blue lines in the model show the surface of the structure – the three-dimensional form of the double helix.

who had no son to succeed him. The duke's death therefore caused a lot of trouble, and further complicated the map, for fragments of his inheritance passed into many different hands. But before this there was much quarrelling and trouble. One can view many events of the sixteenth century as a long duel between Valois and Habsburg over the Burgundian inheritance, notably its rich Netherlands provinces (roughly, modern Belgium and Holland). When, in 1519, a Habsburg who was already king of Spain became Holy Roman Emperor and united to the old Habsburg lands the empire of Spain, it seemed that the family might well be on the way to a universal monarchy. This was Charles V, the first man of whom it was said (truthfully) that he ruled an empire on which the sun never set.

Habsburg and Valois joined other families in quarrelling about Italy, seeking allies and satellites there among the dozen or so major states into which the peninsula was divided. Some of these were aristocratic republics (Venice was the most famous and had great overseas possessions in Cyprus, Crete and the islands of the Aegean), some virtually monarchies, whether they admitted it or not (like Florence, nominally a republic but really in the hands of a family of former bankers, the Medici). But these were not the only complications in Italy. Most Italian states fell within the Holy Roman Empire, but some did not, among them three very important ones – Venice, the kingdom of Naples and the Papal States, the lands ruled by the pope as a prince like any other prince.

The last medieval emperor rides out in defence of Christendom much as he must have thought the great paladins who were his predecessors had done; Charles V, painted by Titian in one of his greatest works.

old arrangements, professional soldiers to make sure that they could squash vassals if they stepped out of line, and civil servants to make sure that government would function even if local bigwigs turned sour. Moreover as time went by they paid less and less attention to their own supposed duties to other magnates; whatever he might be theoretically, they treated the Holy Roman Emperor, for example, like any other prince, intent on promoting the interests of his own dynasty just as they were.

Competition with one another led kings to strive to keep a strong grip on their domains, and made them even more eager to break old-fashioned obstacles to their power. This helped the slow emergence of something people took a long time to take for granted, the idea of 'sovereignty'. In essence this means that (as is now held) within a given area there is only one ultimate law-making authority. This idea began to spread in the sixteenth century. No one, not even emperor or pope, it was argued, had the right to interfere between a sovereign (whether a single prince or a body like the senate of Venice did not matter) and his own subjects. Nor should there be any laws except those made by the sovereign. It was a long time, it is true, before this idea was accepted completely and everywhere, and people went on finding it very difficult to agree with the idea that the sovereign could do anything (even something which cut across the laws of God, for example), but by 1800 the theory that the state or its ruler was the over-riding source of law was accepted throughout most of Europe, even if relics of the older ideas still lingered on.

So kings and princes grew more powerful, and there emerged what has been called 'absolute' monarchy. The first great example was Spain. Under Philip II, the son of Charles V, who had left to Philip the Spanish part of the Habsburg lands when he abdicated in 1556, Spanish government was, at least in theory, centralized to a unique degree. Almost every important decision was in theory settled by the king. In the huge monastery-palace, El Escorial, which he built not far from Madrid, the paper piled up as Philip tried to run a world-wide empire by personally keeping an eye on everything that was going on. It was too much to hope for. Sixteenth-century government did not have the communications system or the responsive power to make it possible for one centre to rule dominions stretching from Peru to the Netherlands. Nevertheless, Habsburg Spain was an outstanding example at least of what absolutism in the sixteenth century could aspire to, even if it did not work in practice. Among its aspirations was a new one: to ensure religious uniformity among the subjects of the empire. This was something which had not much troubled medieval monarchs, but by 1600 religious uniformity was a matter concerning governments everywhere. It was the result of a revolution in the previous century which, more than any other, marks the appearance of modern Europe.

CHURCHES

In 1500, one Church united Europe – and almost defined it. Within fifty years, this was no longer so and that might be taken as the end of the Middle Ages. This was the result of a great upheaval, later called the Protestant Reformation. It marked a new era of European civilization and was to be of outstanding importance in world history. Yet, like so many great changes, few could have foreseen the Reformation and those who launched it would have been horrified had they been able to glimpse the final outcome of what they were doing. They were men with what we should now think of as medieval minds, but they broke a tradition of respect for religious authority going back a thousand years. They ended the unity of Christendom which they deeply believed in. They created new political conflicts though they often thought they were concerned only with unworldly matters. Looking back, we can see too that they were taking the first and most important steps towards greater individual freedom of conduct, more tolerance of different opinions and much more separation between the secular and religious sides of life. All these things would have appalled them. In short, they launched much of modern history.

From the Middle Ages to the present century, pogroms of Jews have been endemic in some parts of Europe. A print of a notorious attack on the Frankfurt ghetto in 1614.

In theory, Europe had been wholly Christian since the Dark Age conversions of the barbarians. Only in Spain did Christian kings in 1500 rule any large number of non-Christian subjects; in other countries a few Jews lived apart from Christians, segregated in their ghettos, taxed and not usually enjoying the same legal protection as Christians. Apart from these special cases, all Europeans were Christians: the words almost mean the same thing in the Middle Ages. Religion was the one Europe-wide tie and Christendom was an undivided whole, held together by a common faith and the work of the Church, Europe's only continent-wide legal institution. Church law operated in every land through courts alongside and separate from the lay system. All universities were governed and directed by churchmen,. Finally, in every country the same sacraments were administered and imposed the same pattern on the great events of people's lives – birth, marriage and death.

Reformers

In spite of its unrivalled position, there had always been plenty of criticism of the Church. There was nothing new about it. Evils which were still worrying critics when the sixteenth century began had been much denounced in the Middle Ages – the ignorance of clergymen, for example, or their misuse of power for personal gain, or their worldly lives. Many such ills – and others – had long been attacked, often by clergymen themselves, and writers had long poked plenty of fun at priests who liked drinking and chasing girls more than attending to their spiritual duties, and had contrasted poor priests devoted to their flocks with their rich, self-indulgent superiors. Yet anti-clericalism – that is, attacking the clergy – did not mean that people wished to forsake the Church itself or doubted the truth of Christianity.

There had long been efforts made by the clergy to put their house in order. As the fifteenth century went on, some critics – many priests among them – began to suggest that it might be necessary to turn back to the Bible for guidance about the way to live a Christian life, since so many of the clergy were obviously not making a very good job of it. They were often labelled heretics and the Church had powerful arms to deal with them. Some of these men, the Oxford scholar Wyclif or the Czech John Hus (who was burnt), for example, had strong popular support, and appealed to the patriotic feelings of fellow-countrymen who felt that the papacy was a foreign and unfriendly institution. Some heretics could draw also on social unrest; no Christian could easily forget what the Bible had to say about the injustices of life.

The followers of Wyclif and Hus, 'Lollards' and 'Hussites' as they were called, were harried and chased by the authorities. It was not they who were to pull down the Church they often criticized. It was still very strong in 1500 and by no means in much worse shape then than at earlier times, even if we seem suddenly to hear more about what is going wrong. Its influence was still taken for granted at every level of society, controlling, moulding, setting in familiar grooves and patterns the accidents of each individual's life, watching over him or her from the cradle to the grave. Religion was so tangled with everyday life that their separation was almost unthinkable. In most villages and little towns, for instance, there was no other public building than the church; it is not surprising that people met in it for community business, and for amusement, at 'church-ales' and on feast days (when even dances were held in it).

Being mixed up in the everyday world was not always good for the Church. Bishops who played a prominent part in the affairs of their rulers had always been in danger of being too busy to be good shepherds of their flocks. The great Cardinal Wolsey, archbishop of York and favourite of the English Henry VIII, never visited his see until sent there in disgrace after falling from favour and power. At the very centre of the Church, the popes themselves often seemed to worry too much about their position as temporal princes. Because the papal throne and the papal bureaucracy had both fallen more or less entirely into Italian hands, foreigners especially felt this. Pluralism – holding many offices and neglecting their duties while drawing the pay for them – was another problem the Church had long faced and did not seem to be able to put right. One reason was that for all the grandeur of the way many bish-

Claims under pressure: Pope (Clement VII) and Emperor (Charles V) are painted as co-equal rulers of a Christendom at a moment when the medieval unity expressed in that idea was about to disintegrate.

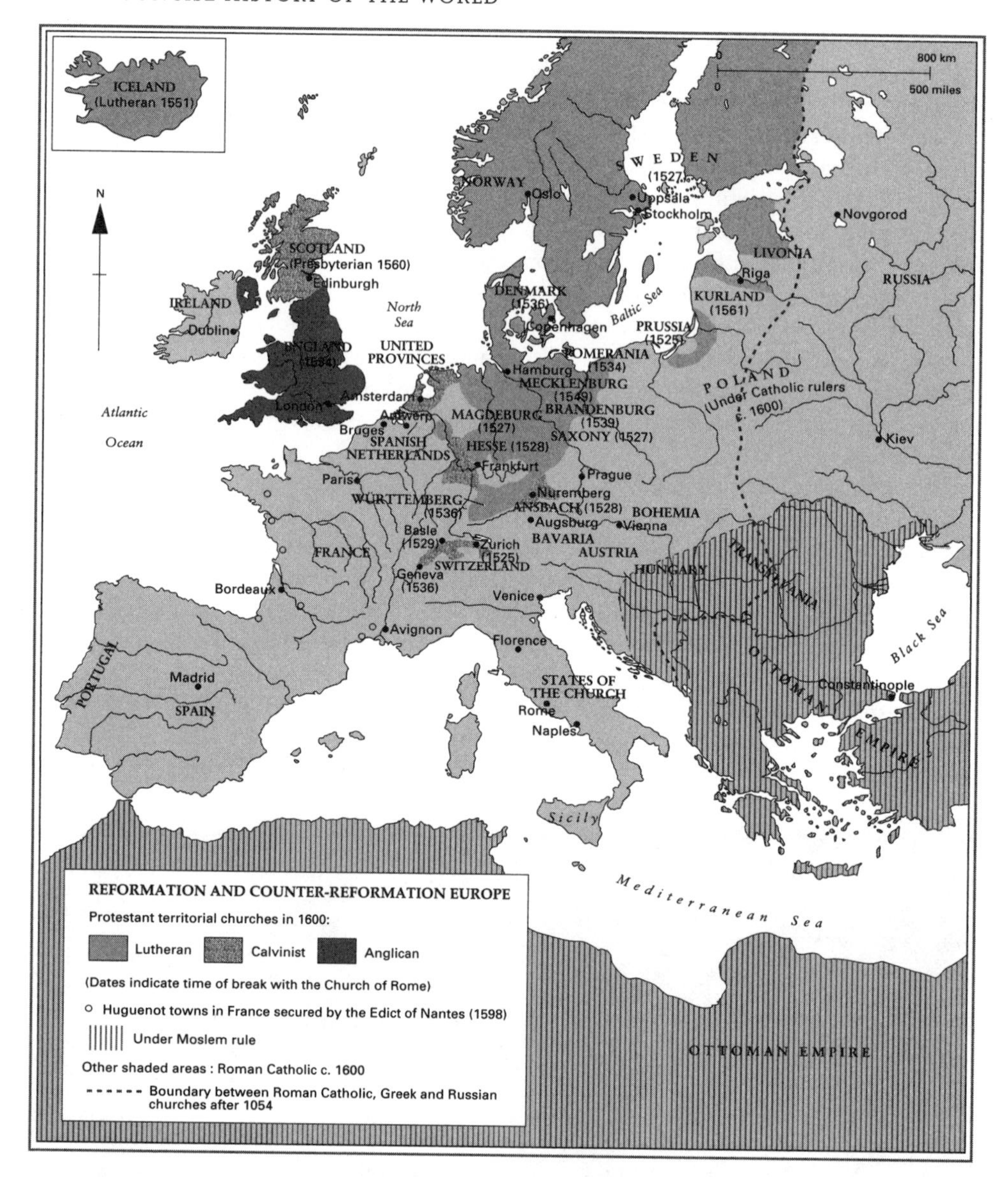

ops and abbots lived, for all the extravagance of the papal court at Rome ('Since God has given us the papacy', one pope is supposed to have said, 'let us enjoy it'), there never seemed to be enough money to go round and, as a result, jobs had to be dished out to reward services. Poverty created other difficulties too. It was unusual for a pope to have to go so far as Sixtus IV, who was finally reduced to pawning the papal tiara, but using juridical and spiritual power to increase papal revenues was an old complaint, and it had its roots in the need to find revenue.

Money was short in the parishes, too. Priests became more rigorous about collecting tithes – the portion of the parishioners' produce (usually a tenth or twelfth) to which they were entitled. This led to resentment and resistance which then tempted churchmen into trying to secure their rights by threatening to refuse people the sacraments – to excommunicate them – if they did not pay up. This was a serious business when men believed they might burn in hell for ever as a result. Finally, poverty was also a cause of clerical ignorance (though not the only one). The standard of education among the clergy had improved since the twelfth century (this owed much to the universities) but many parish priests in 1500 were hardly less ignorant or superstitious than their parishioners.

Against this background, when the papacy began to build a great new cathedral in Rome – the St Peter's which still stands there – it had to find new ways to raise money. One of these ways was licensing more salesmen of 'indulgences'. These were preachers who, in return for a contribution to the funds needed for St Peter's, gave the pope's assurance that subscribers would be let off a certain amount of time in Purgatory, that part of the after-world in which the soul was believed to be purged and cleansed of its worldly wickedness before passing to heaven.

Luther

It was the unexpected spark for a religious revolution. In 1517 a German monk, Martin Luther, decided to protest against indulgences as well as several other papal practices. Like the old-fashioned scholar he was, he followed tradition by posting his arguments in a set of ninety-five 'theses' for debate on the door of the castle church in Wittenberg, where he was a professor at the university. Here began the Protestant Reformation. Soon his arguments were translated from their original Latin into German. They ran through Germany like wild-fire – printing gave them a wider audience than that for earlier criticisms of the papacy.

A portrait of Luther by his friend and contemporary, Lucas Cranach

Unknowingly, Luther was becoming a maker of world history, but he had the temperament for the task. He was a Saxon, the son of peasants, impulsive and passionate, who at the age of twenty-one had become a monk after an emotional upheaval set off by a thunderstorm which broke on him as he was trudging along the highway. Overcome by terror and a feeling of his own sinfulness which made him sure he was fit only to go to hell if he was struck by lightning and killed, Luther suddenly felt the conviction that God cared for him and would save him. It was rather like St Paul's conversion on the road to Damascus in its suddenness and violence. Luther's first celebration of Mass was another overwhelming experience, so convinced was he of his personal unworthiness to be a priest. Later he was to believe Satan appeared to him – and he even threw his ink-pot at him. Luther's nature was such that, when convinced he was right, he was immovable and this explains his impact. Germany may have been ripe for Luther, but the Reformation would not have been what it was without him.

An enormous dislike of the Italian papacy waited to be tapped in Germany. Luther turned to writing and preaching with a will when the primate of Germany, the archbishop of Mainz, tried to silence him. His fellow-monks abandoned him, but his university stood by him and so did the ruler of Saxony, the state he lived in. Eventually his writings divided Germans into those who came to be called 'Lutherans' (though he was first called a 'Hussite') and those who stood by the pope and the emperor. Support came to him not only from clergy who disapproved of the teaching and practice of the Roman clergy, but from humble folk with grievances against tithe-gatherers and church courts, from greedy princes who coveted the wealth of the Church, and from others who simply took his side because their traditional or habitual rivals came out against him.

Luther in the end set out his views in the form of new theological doctrines – that is to say, statements about the beliefs a Christian ought to hold in order to be sure that he really was a Christian and that he would be saved from Hell after death. He said that the Church itself and even attendance at the sacraments was not absolutely necessary to salvation, but that men might be saved if they had faith in Jesus Christ. This was very important. He was teaching that in the last resort it was possible to hope to be saved even without the Church, by simply relying on your own private relationship with God. It has been said that he dethroned the pope and enthroned the Bible, God's Word, which every believer could consult without the Church coming between him and it. A view putting such stress on the individual conscience was revolutionary. Not surprisingly, Luther was excommunicated, but he went on preaching and won wider and wider support.

Protestantism and Counter-Reformation

The political quarrels Luther's teaching aroused between Germany's rulers broke out in wars and revolts. After a long period of turmoil, a general settlement had to be made. By the peace of Augsburg of 1555 (nine years after Luther's death) it was agreed that Germany should be divided between Catholic and Protestant (the word had come into use after the signing of a 'Protestation' against the papacy in 1529). Which religion prevailed in each state was to be decided by its ruler. Thus yet another set of divisions was introduced into that divided land. The emperor Charles V had to accept this; it was the only way of getting peace in Germany, though he had struggled against the Reformers. For the first time Christian princes and churchmen acknowledged that there might be more than one source of religious authority and more than one recognized Church inside western Christendom.

Something else, of which Luther himself disapproved, had already begun to happen by then. Protestantism tended to fragment, as more and more people began to make up their own minds about religious questions. Other Protestants had soon appeared who did not share his views. The most important were to be found in Switzerland, where a Frenchman, John Calvin, who had broken with Catholicism, began to preach in the 1530s. He had great success at Geneva, and set up there a 'theocratic' state – that is, one governed by the godly

The vigour and self-confidence of Counter-Reformation Europe was expressed vividly in its art. Rubens, one of its master-painters, enthusiastically poured out his gifts in a flood of canvases of religious subjects (as in this picture of a miracle attributed to St Ignatius Loyola) as well as in classical and pagan subjects. Catholicism remained a religion of visual imagery while Protestantism turned to the word.

(the Calvinists). Geneva was not a place for the easy-going. Heresy was punished by death; and though that was not surprising in those days, it was distinctly more ferocious to impose the death penalty (as Calvin did) for going off with someone else's wife or husband. Yet Calvinism had great success also in France, the Netherlands and Scotland, whereas, except in Scandinavia, Lutheranism did not for a long time spread much beyond the German lands where it had been born. The result in any case, was further division – there were now three Europes, two Protestant and one Catholic, as well as several minor Protestant sects.

One country where Protestantism was to be particularly important for the future was England. In that country many of the forces operating elsewhere in favour of throwing off allegiance to the papacy were at work, and so was a very personal one, the wish of Henry VIII to get rid of his queen who was not able to give him a son and heir. Yet Henry was a loyal son of the Church; he had actually written a book against Luther which earned him papal approbation as 'Defender of the Faith', a title still borne by his descendant today. It is very likely that he would have been able to get his marriage to his queen 'annulled' – that is, deemed not to have been a valid marriage – by the pope, had she not been aunt to the emperor Charles V, whose support was needed by the Church against the German heretics. So, as the papacy would not help, Henry quarrelled with the pope, England broke away from allegiance to Rome, and the lands of the English monasteries were seized by the Crown. Some Englishmen also hoped to make the English Church Lutheran, but that did not happen.

Once religion muddied the issues of diplomacy and war, it was comforting to find that God was on your side. The English Protestants who commemorated the miscarrying of the Spanish Armada (an invasion fleet) in 1588 gratefully gave credit where they thought it was due: 'God blew, and they were scattered' runs the inscription on this medal.

Protestantism's successes forced change on Rome. Whatever hopes Roman Catholics might have of returning to the former state of affairs, they would have to live for the foreseeable future in a Europe where there were other claimants to the name of Christian. One effect was that Roman Catholicism became more rigid and intransigent – or, to put it in a different way, better disciplined and more orderly. This was the 'Counter-Reformation'. Several forces helped but the most important of them was a general Council of the Church which opened at Trent in north Italy in 1545 and sat, on and off, until 1563. It redefined much of the Church's doctrine, laid down new regulations for the training of priests and asserted papal authority. Putting its decisions into practice was made a little easier by the work of a remarkable Spaniard, Ignatius Loyola, who had founded a new order of clergy to serve the papacy, the Society of Jesus, or 'Jesuits'. Sanctioned in 1540 and bound by a special vow of obedience to the pope himself, the Jesuits were carefully trained as an elite corps of teachers and missionaries (Loyola was especially concerned to evangelize the newly discovered pagan lands). More than any other clergy they embodied the combative, unyielding spirit of the Counter-Reformation. This matched Loyola's heroic temper, for he had been a soldier and always seems to have seen his Society in very military terms; Jesuits were sometimes spoken of as the militia of the Church. Together with the Inquisition, a medieval institution for the pursuit of heresy which became the final court of appeal in heresy trials in 1542, and the 'Index' of prohibited books first issued in 1557, the Jesuits were part of a new armoury of weapons for the papacy.

Religion and war

Reformation and Counter-Reformation divided Europeans bitterly. The Orthodox world of the east was little affected, but everywhere in what had been Catholic Europe there were for more than a century religious struggles political struggles envenomed by religion. Some countries successfully persecuted minorities out of existence: Spain and (in large measure) Italy thus remained strongholds of the Counter-Reformation. Rulers usually made up their minds for themselves and their subjects often fell in with their decisions. Foreigners occasionally tried to intervene; Protestant England had the Channel to protect her, and was in less danger than Germany or France. Yet religion was not the only explanation of the so-called 'religious wars' which devastated so much of Europe between 1550 and 1648. Sometimes, as in France, what was really going on was a struggle for dominance between great aristocratic families who identified themselves with different religious parties. In the end a representative of a Protestant family came out on top there – as the king Henry IV – but did so by changing his religion to Catholicism. So the French monarchy stayed Catholic, though many 'Huguenots' (Protestant Frenchmen) for a long time enjoyed special rights and were allowed to hold fortified towns where they could protect them.

In the Netherlands, which were under Spanish rule, religion gradually transformed a rebellion which had been started by the local nobility who wanted more local self-government. In the end the aristocratic leaders of the southern provinces (modern Belgium) felt they had better stay Catholic and under Spanish rule, while the northern provinces (roughly, the modern kingdom of the Netherlands) identified themselves with Protestantism, even though they contained a large Catholic population. After a long struggle – the Dutch call it the 'Eighty Years' War' – Europe found that a new country had come into existence, the United Provinces, a little federation of tiny republics, led by Holland, within which religious toleration was practised. The worst abuse of religion for political ends was in Germany. The religious quarrels settled at Augsburg broke out again when a seventeenth-century Habsburg emperor, strongly imbued with Counter-Reformation principles, tried again to advance Catholicism. The outcome was the appalling 'Thirty Years' War' – it raged intermittently from 1618 to 1648 – in which religious questions were often lost to sight in the politics and the carnage. At one moment a French cardinal of the Church was allied to a Protestant Swedish king to thwart the interests of the Catholic Habsburgs. Meanwhile armies marched back and forth over Germany leaving misery in their wake, and spreading disease and famine. Some areas were virtually depopulated; once-prosperous towns disappeared.

In the end there had to be another compromise. But the peace of Westphalia, which ended the war in 1648, opened a new era. Although even then many people still thought of religion as well-worth fighting over and certainly as something which justified murdering or torturing your errant neighbours, statesmen for the most part began to take more account of other matters in dealing with one another. The world became a tiny bit more civilized when they turned their attention back to arguments about trade and territory, and away from religion. Europe by then, in the second half of the seventeenth century, was divided into states, most of which did not officially tolerate more than one dominant reli-

A sidelight on the miseries of Germany in the Thirty Years War: mass execution depicted in an etching by the great Lorraine engraver, Jacques Callot, as a part of a series on atrocities.

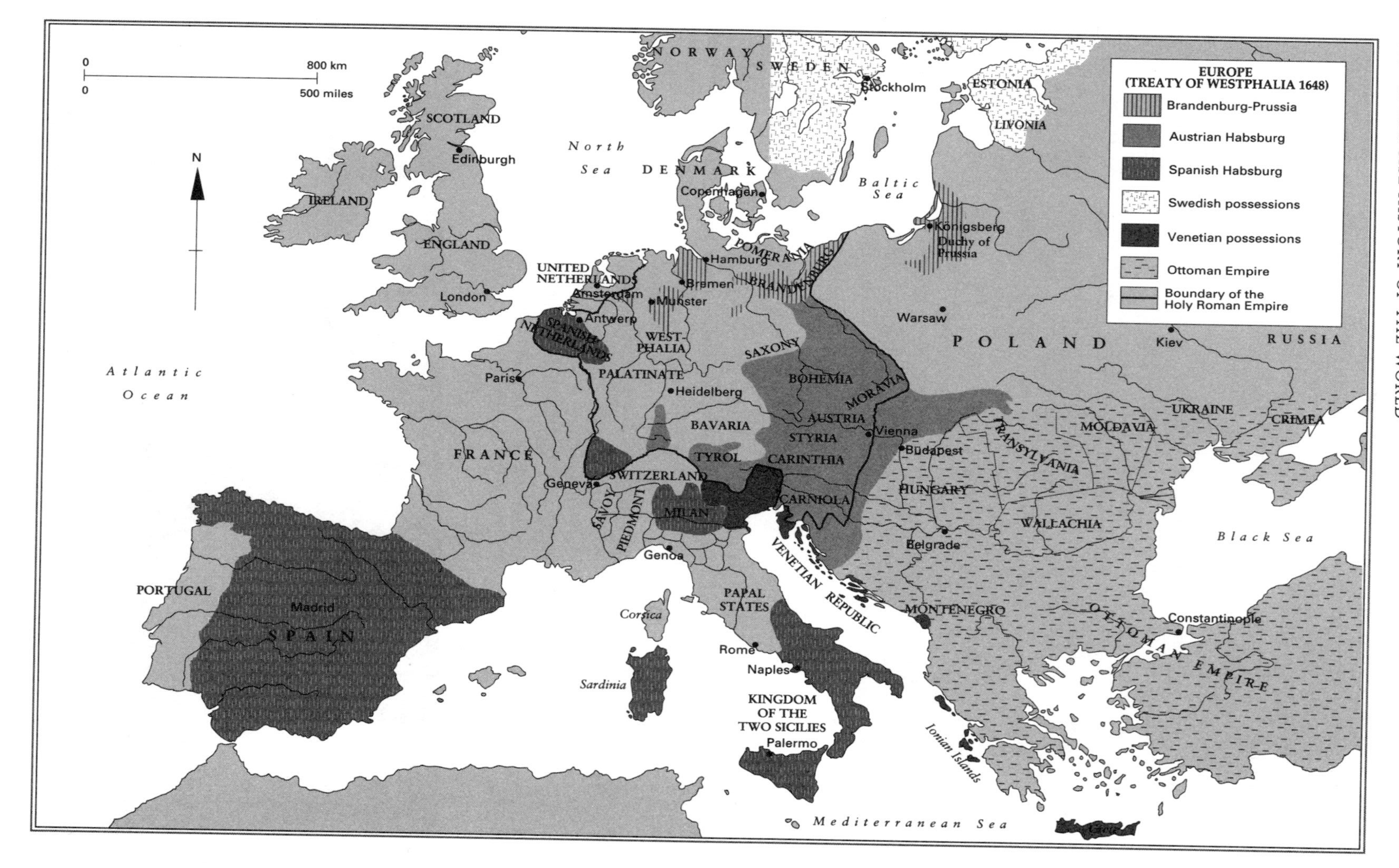
EUROPE
(TREATY OF WESTPHALIA 1648)
Brandenburg-Prussia
Austrian Habsburg
Spanish Habsburg
Swedish possessions
Venetian possessions
Ottoman Empire
Boundary of the Holy Roman Empire
0
800 km
0
500 miles
N
Atlantic Ocean
North Sea
Baltic Sea
Black Sea
Mediterranean Sea
NORWAY
SWEDEN
Stockholm
ESTONIA
LIVONIA
Königsberg
Duchy of Prussia
DENMARK
Copenhagen
POMERANIA
BRANDENBURG
Hamburg
Bremen
Münster
WESTPHALIA
UNITED NETHERLANDS
Amsterdam
Antwerp
SPANISH NETHERLANDS
PALATINATE
Heidelberg
SAXONY
BOHEMIA
MORAVIA
AUSTRIA
Vienna
STYRIA
CARINTHIA
CARNIOLA
BAVARIA
TYROL
SWITZERLAND
MILAN
PIEDMONT
SAVOY
Geneva
Genoa
POLAND
Warsaw
RUSSIA
Kiev
UKRAINE
CRIMEA
MOLDAVIA
TRANSYLVANIA
WALLACHIA
HUNGARY
Budapest
Belgrade
MONTENEGRO
OTTOMAN EMPIRE
Constantinople
Ionian Islands
VENETIAN REPUBLIC
PAPAL STATES
Rome
Naples
KINGDOM OF THE TWO SICILIES
Palermo
Corsica
Sardinia
SCOTLAND
Edinburgh
ENGLAND
London
IRELAND
FRANCE
Paris
SPAIN
Madrid
PORTUGAL

gion, but in some of which – in particular, England and the United Provinces – a fair degree of tolerance was practised.

A NEW WORLD OF GREAT POWERS

Nonetheless, such a shift in the nature of government did not come about swiftly. The most spectacularly successful of all the European rulers who stood for absolute, centralized monarchy, was Louis XIV, ruler of France from 1660 (when he came of age: he had legally been king since he succeeded to the throne at the age of five in 1643) to 1715. Once he had himself taken up the reins of government, he pitched the claims of monarchy higher than any of his contemporaries. Under him, there was no more trouble from the French nobility. The privileges of the Huguenots were taken away. Higher taxation (and France's relatively great pool of manpower) supported a more powerful army than ever before and made possible a successful run of conquests (at least for the first half of his reign).

The Dutch United Provinces and England in fact remained very untypical. In the first case, it sometimes looked as if the Dutch could have done with rather more centralization and strong government, because rivalries between different provinces often made it very difficult for them to co-operate against pressure from the outside. But this was a price paid for having more freedom than was perhaps to be found anywhere else. This freedom was basically a matter of defending the independence of action of the relatively small ruling groups of rich citizens who dominated the government of each state. The most important were the merchant rulers of Amsterdam, capital of the province of Holland and centre of Dutch commercial life. But because their outlook on most matters was similar to that of the majority of those they ruled, and because their economic interests often coincided with those of poorer people – everyone suffered, for example, if business was bad in Amsterdam, not just the rich

Le roi s'amuse. *Louis XIV, retaining his hat while others remove theirs in deference, relaxes from affairs of state over a primitive form of billiards.*

– and because they were very anxious to do nothing which would diminish their freedom to trade and make money, the care of the rich to preserve the freedom of the states assured freedom for the individual citizen. They were remarkably successful during most of the seventeenth century, though they had to fight hard against Louis XIV (who combined his dislike of them as republicans with a liking for their tulips, of which he bought millions each year). By the eighteenth century, partly because of the strains this imposed, the Dutch were entering a period of decline and were never again to be quite such an important world-power as they had been in the previous hundred years.

The English story was very different. It had looked for a time as if the early Tudors might well develop a strong centralizing monarchy like those in Europe. They had the oldest national monarchical tradition in Europe to draw on and a national feeling in many ways more developed than elsewhere behind it. These made it easier for Henry VIII to carry out his nationalization of the Church in England; Protestantism became identified with national feeling in England as it did nowhere else except in Germany. Yet Henry VIII turned to an old institution, parliament, to make the necessary laws. This was very important for the future. There were bodies rather like parliaments in other countries, but almost everywhere in the next couple of centuries they went down before the demands of absolutism; instead, the English parliament grew stronger and stronger.

Ironically, much of this was the work of the Tudors, though they probably would not have wished it so. When Henry asked parliament to pass laws about the fate of the Church, he was conceding that it had a right to legislate on such an important matter; that made it very difficult for later kings to act without parliament's support over questions of such national interest. Another factor was the uncertainty of the succession (none of Henry's children had offspring). The reign of Elizabeth I is rightly seen as a great age, and so it was, yet the queen was for a long time insecure and afraid she might lose her throne (and perhaps her head: hence she cut off that of one possible rival, Mary Queen of Scots). The European situation was against her, and other possible claimants to the throne had a good chance of foreign support. She was therefore careful not to antagonize her subjects. One of the ways in which they made themselves heard was through parliament which voted taxes. It gradually became clear that the monarchy could not levy taxes without parliament's approval of the purposes for which they were raised.

'Good Queen Bess' was so good at handling people that she was able to conceal much of this. Her more unimaginative successors, the first two Stuart kings, were not (possibly in part because James I was a Scotchman who did not much like or understand ways Englishmen had grown used to under the Tudors). Under them, the Crown's relations with parliament broke down. In the middle of the seventeenth century, a great civil war finally settled that England would not develop towards continental absolutism (though, oddly, the country was for a time ruled as a republic by someone with powers very like a dictator's, the 'Lord Protector', Oliver Cromwell). The victory of the cause of a constitutional – that is, limited – monarchy, was confirmed in 1688, when an almost bloodless upheaval, the 'Glorious Revolution', pushed off the throne the last Stuart king, James II, who was believed to be trying to reverse the trend of the last hundred and fifty years in order to establish Catholicism again in England.

After that, England was really ruled by its landowners, who dominated parliament. But, just as the interests of the ruling rich in the Dutch republic often turned out to be the interests of many other people too, so England's rulers looked after national interests pretty well. After all, agriculture was England's main industry; what was good for the landlord and the farmer was likely to be good for the country. Nor were other interests – bankers' and merchants', for example – ignored. They might grumble about policies, but government usually took account of their views. Gradually educated and uneducated Englishmen alike began to feel there was a natural connexion between obvious advantages they enjoyed – personal freedom, equality before the law, Protestantism, safeguards against absolute monarchy – and the

Elizabeth I, greatest of the Tudors; she brought to her office a style and glamour which surpassed even those of her father, Henry VIII; in sagacity and caution she rivalled her grandfather, Henry VII. After her death, she became a folk memory against which were tested the acts of the Stuarts – who were usually found wanting.

growing wealth of the country. After 1660 or thereabouts, though there were many setbacks on the way, most Englishmen found it much easier to stand by the constitution and the idea of limited monarchy.

Already in the eighteenth century many Europeans admired England, not only because she was ruled by elected representatives (even if mainly chosen by landowners) and aristocrats, instead of by a despotic monarch, but because Englishmen were much freer from interference with their private lives than they. It was less usual to keep Englishmen locked up without trial, for example, than elsewhere, and they were not used to having their homes entered and searched without a magistrate's warrant. Rank was very important in English society, but, if a great lord committed a crime, he could be brought before a court like anyone else. All this seemed very strange and admirable to many people who lived on the continent. Yet it too was largely the consequence of England's being ruled by landowners who wanted to protect themselves and thought the best way to do so was by putting behind such privileges the force of laws which only parliament could change. So, 'constitutionalism' came to be associated with one of the major powers as an ideological factof international life .

New issues in international relations

This began to be a force in relations between European states in the eighteenth century, but even in the seventeenth the issues at stake had begun to change somewhat. In the great struggles of Habsburg and, first Valois, and then Bourbon, France, dynastic domination had been the prize at the heart of the struggle. There was Italy to be fought over, then Germany. In the second case, religion had made things worse: Protestant princes had looked for protection to Catholic France against Catholic Habsburg emperors. Across this had already cut other struggles though: the Anglo-Spanish hostility to which special edge was given by religion, but which was fed also by rivalry in the New World and fear over Spanish control of the Low Countries, and the Dutch Revolt.

On 3 September 1650 at Dunbar, Cromwell overthrew the Scottish army supporting Charles II. Another victory at Worcester a year later to the day confirmed the triumph, marking the real end of the Civil War; soon, Scotland was united to England and the Commonwealth was safe – until its supporters fell out among themselves.

Issues which were once dynastic and European had, in fact, already before 1700 begun to be seen in both a wider geographical and a new ideological context. In retrospect, the first was especially important. In their extension, even the world wars of ancient times were small beer by comparison with those fought between 1500 and 1800 on battlefields all over the globe, and often over issues at stake thousands of miles away from the countries engaged. It was, moreover, the beginning of an age (lasting down to 1917 at least) when the way Europeans settled their own quarrels also settled willy-nilly the fate of millions of black, brown and yellow men who had never heard of Paris or London. Some of the reasons for this will already have become clear. The growing European command of the sea, European economic activity all round the globe, the technological advantages Europeans increasingly enjoyed over non-Europeans all contributed to it. They had enabled Europeans to invent the oceanic empire, which depended on sea-communications. It led irresistibly to quarrels between European predator nations right round the globe.

The origins of these quarrels lay for the most part in the growth of trade, which was, as a French minister put it to Louis XIV, 'the cause of a perpetual contest in war and in peace between the nations of Europe'. For roughly two centuries, Spain, Portugal, the United Provinces, England and France sent out ships and built forts in order to try to keep to themselves trade with their own possessions or with local peoples with whom they were the first Europeans to begin to trade. The coastlines of these countries gave them destinies different from those of land-locked central Europe or the Mediterranean basin. Though the New World of the Americas was the main scene of their competition, it was not the only one.

The oceanic empires

The first into the business of overseas empire had been the Portuguese and Spanish, who, without consulting anyone else (though the pope afterwards said they might annex any land

An African view of the European intruder: a Benin bronze of a Portuguese musketeer.

not already belonging to a Christian prince), agreed to divide between them any new lands which might be found anywhere in the world. One treaty, in 1494, said that everything west of a north-south line 370 leagues west of the Azores should be Spanish and everything east of it Portuguese (this is why Brazil, which falls east of the line, became Portuguese and was the only part of South America to do so); and another, in 1529, drew a similar line 297.5 leagues east of the Portuguese Moluccan islands. (A 'league' was usually about five kilometres). This gave everything on the Pacific side of the line to Spain, everything west of it to the Portuguese (except the Philippine islands, kept by the Spanish). Roughly, this meant that the New World of the Americas was the Spanish sphere, and Africa, the Indian ocean and the Spice islands were Portuguese. The respective spheres reflected in a measure the two different kinds of imperial expansion; America, important though trade with it was, and interested in its exotic products though people became, was from the first a continent where Europeans settled; the Portuguese sphere of empire, on the other hand, was for the most part (coastal Brazil is the exception) not one of settlement but of trade.

So things remained for a long time: Europeans went to the Americas in growing numbers for three centuries but few settled in Asia and Indonesia. Those who did were usually planters or long-term residents who nevertheless hoped one day to go home after making their fortune. Because of this difference, what Europeans fought over in the Far East (and on the African routes to the East) was not large tracts of territory but ports and stations – 'factories' was a favourite English word – where native traders could meet Europeans and business could be done. Native rulers who gave permission for their creation had to be respected. European expansion into Asia in its early phases did not usually advance by conquest but by diplomacy and negotiation.

In the East the sixteenth century was dominated by the Portuguese, whose king called himself by the splendid title of 'Lord of the Conquest, Navigation and Commerce of Ethiopia, Arabia, Persia and India'. South of Cape Verde they long had a virtual trading monopoly round to the Indian ocean and across it to the Spice islands. They carried goods between Asian countries – Persian carpets to India, cloves from the Moluccas to China, Indian cloth to Siam – and won dominance over their Arab rivals from bases at the entrances to the Red Sea and the Persian Gulf. All this rested on sea-power and careful diplomacy with native rulers, and it set a pattern for Europeans for the next two hundred years in the Indian ocean and Asia. At the end of the sixteenth century, though, the Portuguese were elbowed aside by the Dutch, who set up an 'East India Company' in 1602 with the aim of replacing them in the spice trade to Europe – a rich prize. With ruthless skill they did so. Having pushed the Portuguese aside, they fought bitterly to keep the English out of trade in the Spice Islands, and in the end were largely successful. By 1700 they had established a general supremacy over what is now Indonesia. Meanwhile a scatter of English 'factories' had appeared round the coasts of India, from Gujarat to Calcutta. The Portuguese also kept some of their older stations in India, and the French and Danish had footholds there too.

Interestingly, not a word about non-European matters appeared in the peace treaties of 1648. Less than twenty years later, though, in 1667, the Treaty of Breda between, England, the Dutch and the French was just as much concerned with affairs outside Europe as with those within. It ended the second of three naval wars between England and the United Provinces over trade. Seventy years further on still, in 1739, two European nations, the United Kingdom and Spain, fought the 'War of Jenkins' Ear' over an entirely non-European matter and this was the first time such a thing had happened. It is a good landmark for the end of a period during which overseas questions had gradually come to loom as large in the eyes of diplomats as the more familiar European ones. For decades English seamen had been trying to break into trade with the Spanish colonies and to take more of it than treaties entitled them to. For the same time the Spanish squadrons had been trying to catch them and had handled them roughly when they succeeded. (Thus, a certain Captain Jenkins lost his ear, he claimed.) What was at stake was a great prize, the right to sell goods to the inhabitants of the Spanish empire. The Spanish, who wanted to keep a monopoly of that trade for themselves, were always under a handicap in defending it; they had to support Habsburg interests in Europe and therefore to divide their forces. Spain could not abandon her colonial empire, because she depended on its revenues, but neither could she stop squandering its wealth on expensive dynastic involvements in Europe.

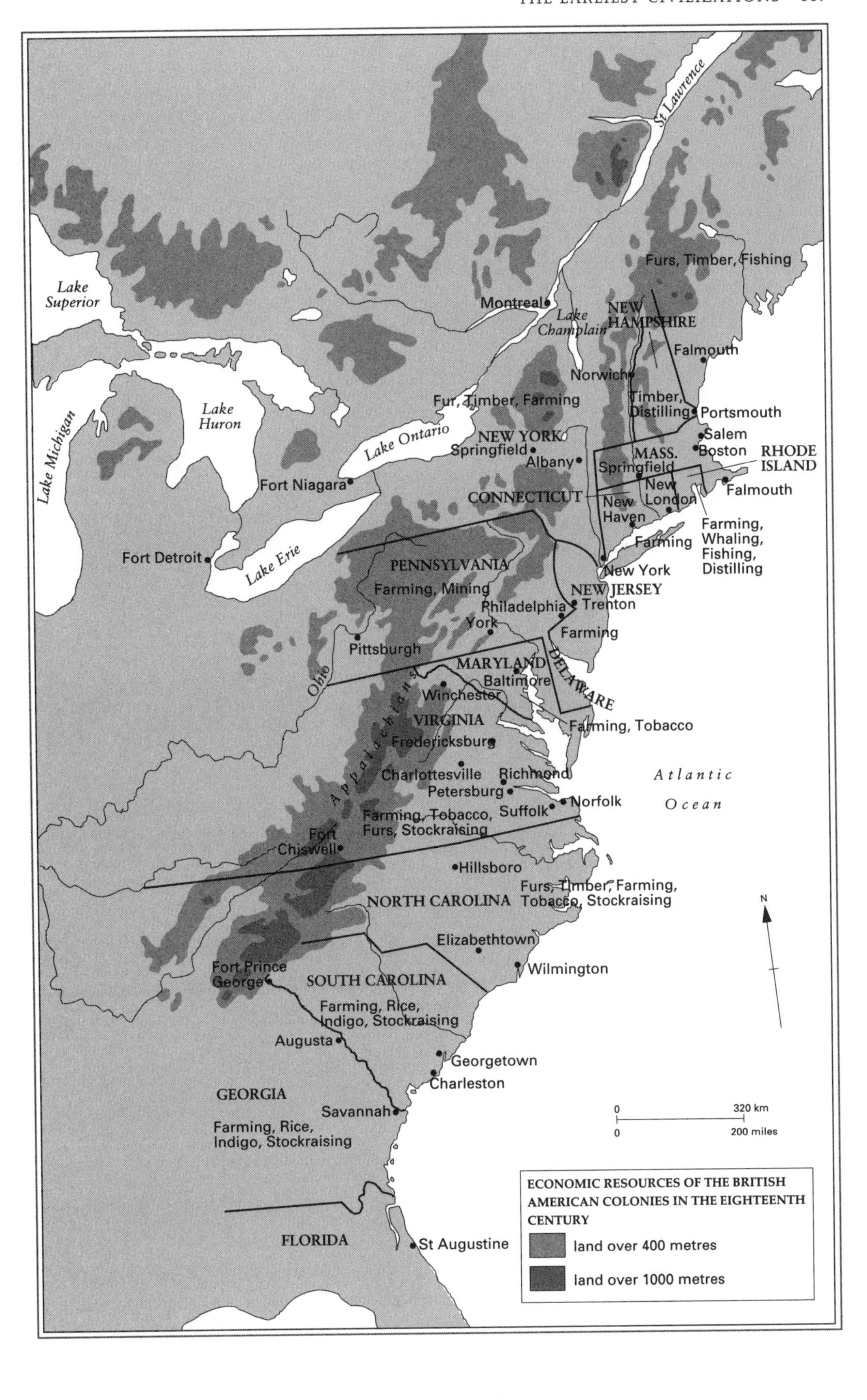

St Lawrence
Lake Superior
Lake Michigan
Lake Huron
Lake Ontario
Lake Erie
Lake Champlain
Montreal
Furs, Timber, Fishing
NEW HAMPSHIRE
Falmouth
Norwich
Fur, Timber, Farming
Timber, Distilling
Portsmouth
Salem
Boston
NEW YORK
Springfield
Albany
MASS.
Springfield
RHODE ISLAND
Fort Niagara
New London
Falmouth
CONNECTICUT
New Haven
Farming, Whaling, Fishing, Distilling
Farming
Fort Detroit
PENNSYLVANIA
New York
Farming, Mining
NEW JERSEY
Philadelphia
Trenton
York
Farming
Pittsburgh
MARYLAND
DELAWARE
Baltimore
Ohio
Winchester
VIRGINIA
Farming, Tobacco
Fredericksburg
Atlantic Ocean
Charlottesville
Richmond
Petersburg
Appalachians
Suffolk
Norfolk
Farming, Tobacco, Furs, Stockraising
Fort Chiswell
Hillsboro
Furs, Timber, Farming, Tobacco, Stockraising
NORTH CAROLINA
N
Elizabethtown
Wilmington
Fort Prince George
SOUTH CAROLINA
Farming, Rice, Indigo, Stockraising
Augusta
Georgetown
Charleston
GEORGIA
Savannah
Farming, Rice, Indigo, Stockraising
0 320 km
0 200 miles
FLORIDA
St Augustine
ECONOMIC RESOURCES OF THE BRITISH AMERICAN COLONIES IN THE EIGHTEENTH CENTURY
land over 400 metres
land over 1000 metres

Anglo-French imperial competition

The English were more favourably placed. They were involved in Europe, of course, but not so completely. And after 1707, when England was united with Scotland, they were truly an island people, safe from foreign invasion across a land frontier. This was why the Act of Union of that year marks an epoch in more than constitutional affairs; it was also a landmark in a long drawn-out contest between England and France which had become entangled with Anglo-Spanish issues. When James II, the last Stuart king of England, was pushed off the throne in 1688 in the 'Glorious Revolution', 'Dutch William' (William of Orange) and his Stuart queen, Mary, then succeeded her father.

This brought England to the support of the Dutch (their bitter enemies only a few years before) against Louis XIV. The most important of the wars which followed was called the 'War of the Spanish Succession' because the prize at stake in it was the Spanish crown, to which France and Austria both had claims when the Habsburg Spanish king died without an heir in 1701. France was, like Spain, handicapped by having to fight in Europe as well as at sea, where Louis XIV was at war with a coalition headed by the Habsburg monarchy. In 1713 the peace of Utrecht which ended the war divided the Spanish inheritance: the Netherlands went to the Austrian Habsburgs, whereas a French prince was allowed to become king of Spain and its empire on condition that the Spanish crown was never united to that of France.

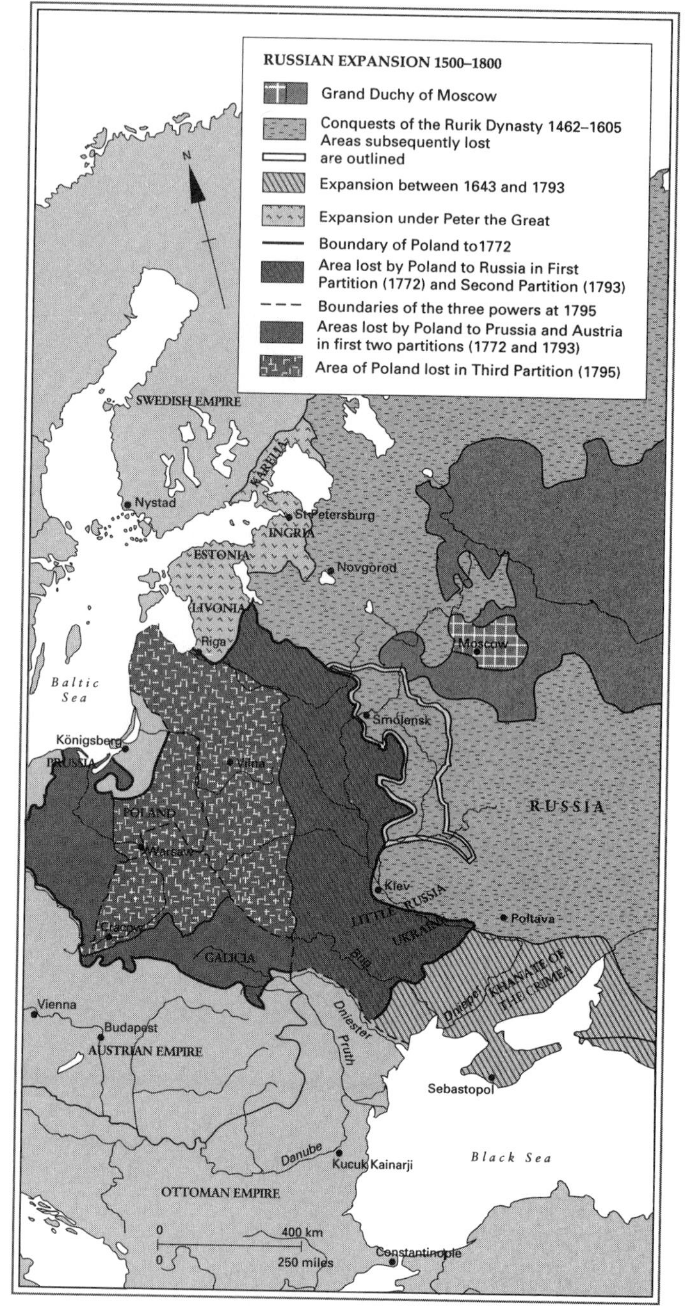

At the same peace, the United Kingdom won many of the French Caribbean islands (she had started collecting from her rivals there as early as the 1650s, when Cromwell's men had seized Jamaica from the Spanish) as well as a new, somewhat bleak but strategically important, part of North America, Acadia, now renamed Nova Scotia (New Scotland). The British also won the right to trade with the Spanish colonies by sending one ship a year to Portobello – a concession to be used as a wedge to prise open the door farther and thus to lead in 1739 to the 'war of Jenkins' ear'. When it came, it soon sucked in France and Prussia, on the one hand, and Austria and Great Britain on the other. The British and French fought in India, where the French East India Company was by the 1740s hard at work dabbling in local politics in order to try and outwit its rivals and gain advantages. The French had also much expanded their activities in North America, where they had set up posts near the mouth of the Mississippi river, the entrance to the huge river system which dominates the centre of the continent. During the early eighteenth century an expedition had pushed up it from the south, while others had made their way down it from the region of the Great Lakes. The result looked to the British colonists of the eastern coast more and more like a huge pincer operation, cutting them off from further expansion inland. In fact the French did not really settle the Mississippi valley and had no solid belt of territory inland. Nonetheless they built forts at strategic points (in this way future cities were founded, St Louis in 1682, Memphis the same year, Detroit in 1701 and New Orleans in 1718) and armed and encouraged the Indians against the British. It was clear that the French were not going to give up their claim to the interior without a struggle.

Fighting never stopped in India and America though peace was again officially made in Europe in 1748, and in 1756 yet another war broke out between France and England. Spain was only of a secondary importance by now; India and Canada were at stake. In this 'Seven Years' War' (peace was made again in 1763) their fate was settled at the same time as that of the lands over which Prussia and Austria (now allies of the British and French respectively) were contending in Germany. The climax of the war came for Great Britain under a government dominated by William Pitt, who has a good claim to be the first British statesman wholly to grasp the possibilities of imperial power. He spoke of winning Canada in Germany, by getting his allies to pin down the French there, and he succeeded. At the peace, though it was less ferocious than some Englishmen hoped, Canada passed to Great Britain and India was made safe for the British East India Company. A string of British islands, added to by fresh acquisitions, all but enclosed the Caribbean, now dotted with British colonies in Jamaica, Honduras and the Belize coast.

TWO EUROPES

While the rivalries of Catholic and Protestant Europe had burgeoned into world-wide struggles in which more than the interests of dynasties were at stake, another new set of elements had entered the calculations of diplomats in eastern Europe. That vast and shapeless region had been for centuries in part a battleground between Teutonic and Slav peoples, in part a zone of confronting cultures, pressed upon from the south by the Ottomans. Swedish kings, too, anxious to extend their lands south of the Baltic, dabbled in its affairs in the seventeenth century. Three things were nonetheless gradually giving a more clear-cut character to eastern Europe. One was the firmer and further extension of serfdom in the northern plains of eastern Germany, Poland and Russia, and in the Danube valley. Another was the obliteration of such old medieval political landmarks as the Teutonic Knights and the kingdoms of Poland and Hungary. The third was the emergence of three great dynastic powers, Hohenzollern Prussia, Habsburg Austria and Romanov Russia as dominant in the region.

Prussia in 1500, was only a small Baltic duchy under the lordship of the kings of Poland. A line of soldier and administrator rulers from Brandenburg, one of the 'electorates' whose rulers voted in the election of a Holy Roman Emperor, took it over in the sixteenth century and thereafter steadily added to their lands. The electorate became famous for having the best army and civil servants in Europe. Seventeenth-century rulers of Prussia turned back the Swedes, and in the eighteenth century one soon known as Frederick the Great, became the first challenger to the supremacy of the Habsburgs among the German princes. He opened a struggle with Austria which, though sometimes interrupted, was to last well into the next century. Austria – or to be exact, the Habsburg monarchy – was challenged in Italy by the French, and then in Germany by France and Prussia in turn, and finally excluded from Spain and its empire at Utrecht. Austria gradually limited her ambitions more and more to central and eastern Europe, winning big gains as Poland decayed and the Ottoman empire declined. But so did Russia, whose emergence was the most important change of all in the east. By 1800 she was to be the strongest military power in Europe, a position hardly conceivable in 1500.

The 'bronze horseman': a nineteenth-century drawing of the monument to Peter the Great in St Petersburg.

The core of the new Russian empire was still the old princely state of Muscovy. The princes of Muscovy were autocrats and it was to go on being very important that Russian government was shaped by their ways and by Tatar rule rather than by, say, the more republican traditions of Novgorod. It was to Moscow, too, that the patriarchate or headship of the Orthodox Church had moved from Vladimir, its old location. The authority of that Church had been thrown behind the princes of Muscovy.

The huge territorial gains of Ivan III and his successors had been further added to in the first half of the seventeenth century, particularly in Siberia. Yet though an enormous transformation of the map, expansion in that direction did not for a long time seem to affect the rest of Europe much: Muscovy was simply too far away and inaccessible and too little known. Moreover, in spite of her autocratic tradition, she was for much of the seventeenth century in a state of anarchy; despotism needs strong rulers. Even the arrival of a new dynasty, the Romanov, on the throne in 1613 only brought slow improvement. Then, in 1682, there came to the throne another exceptional ruler, a man determined to extend his empire still further, and to seek the means to do in western Europe, Peter the Great. His greatest monument remains the new city he founded on the Gulf of Finland, St Petersburg, capital from 1715 to 1918. This was his symbol of 'westernizing', of modernization by borrowing western ideas, and he was the first of many authoritarian reformers to look westward for ways of overcoming backwardness and under-development. He also gave Russia a firm grip on the Baltic coast and eliminated a threat from Sweden which had hung over most of the seventeenth century and seized the formerly Swedish Latvia, Estonia and Karelia. Nevertheless, he had much less success than he hoped, and was not able to retain Azov, Russia's first outlet to the sea in the south, but one taken back by the Ottomans after only a few years.

At home, Russia was – and long remained – very conservative. For all the importance of trade in the great days of Kiev Rus and Novgorod, the merchant class was small. Towns were few. Most artisan trades were practised at a simple level by peasants rather than (as in the West) by specialists. Russia was overwhelmingly a peasant country. Of local trade there was plenty, but it often rested on barter. Even when, as under Peter, deliberate attempts were made to encourage industrial enterprise, they did not change society as the coming of industry was to do in western Europe. Instead of throwing up a new 'middle' class of wealthy traders and manufacturers, pursuing their own interests and standing between noble and peasant, industry was tied to the regime. It was the state which decided to open a mine, or set up a factory, not independent businessmen. This made Russia very different from western Europe. Perhaps more striking still, even in eastern Europe, was Russia's dependence on serfdom. As 1800 approached, not only was the absolute number of serfs steadily increasing, but so was the serf proportion of the Russian population (about two thirds of it by then). The legal powers of serf owners grew too.

The contrast between eastern and western Europe was most vivid in Russia, in spite of the superficially westernized life of the court and aristocracy at the new capital on the Baltic, St Petersburg, which Peter had given his country, a 'window to the west', but no more than that. For all Russia's powers and the attempts of some of his eighteenth-century successors to modernize her, she remained the heartland of a huge region also encompassing much of eastern Germany, central Europe and Poland, where centuries-old layers of historic experience had produced economies, governments and cultures increasingly unlike those further west. Russia itself, with its Byzantine and Tatar tradition, was the extreme example. She had not experienced the Renaissance or Protestant Reformation and was to go on missing great historical experiences which led western Europe to draw away more and more from her as the pace of modernization quickened after 1700. Serfdom was the symbol.

10

WORLD HISTORY IN THE MAKING

NEW VIEWS AND VALUES

Between 1500 and 1800 the way was opened to sweeping, violent, accelerating change and thus to the modern world. Part of the explanation lies in the ideas of early modern Europe. They were, of course, the ideas of very untypical men (and a few women), the leading writers, scholars and scientists of an age. Yet in their own day they may actually have influenced (or even been known to) only very few people and it is somewhat distorting to allow them to dominate the story of what people were 'really' thinking about at a particular time. We now live in an age when science has great prestige and is obviously doing things all around us which demonstrate its power to handle the natural world, yet many of us still believe (or act as if we believe) that crossing fingers and not walking under ladders will ward off bad luck, that astrologers who write in the newspapers can 'predict' the future course of events from the stars, or that an 'auspicious' day should be chosen for a wedding or a journey. When trying to pin down the ways in which the ideas first of Europeans, and then those of other peoples, changed so importantly, shedding one set of assumptions and then taking on others, we have to be careful not to forget the qualifications.

Something that certainly changed by 1800 was the way educated Europeans thought about the past. One effect of the Renaissance had been to interest them in comparisons. In the seventeenth century, there began to be debate about whether mankind had done finer things in ancient times, and, as time passed, about whether other civilizations (the Chinese was one popular candidate) had reached greater heights. In the early nineteenth century, too, people were already beginning to feel that there was more to the Middle Ages than their critics allowed, and that they were in some ways to be admired.

Erasmus of Rotterdam, where he was said to have been born, the son of a priest. Later attacked by Catholics and Lutherans alike, he wrote for a wide public and has been called 'the apostle of common sense.'

From a historian's point of view this is all to the good. More people were coming to look at the past more carefully, however far short of seeing its true nature they still were. But something else had been going on at the same time, and it was one of the most important changes ever to occur in the outlook of Europeans. The gist of it was that large numbers of them became convinced that mankind was moving forward, that history showed a pattern of continuing progress. They came to believe that they were more advanced in civilization, taste, knowledge, science, art than any earlier age and some even that their successors would be more advanced still. The world, in short, was getting better all the time. This was a huge break with medieval views which had often stressed either how much worse things were getting or that, anyway, they could not be changed.

Some of the roots of the new outlook lay in the revival of classical learning already under way before 1400 and reaching its peak in the sixteenth century, when the admirers of classical literature and art made their biggest claims for what they could learn from Greece and Rome. Gradually, some of the 'humanists', as they were called, began to emphasize things about classical antiquity, moreover, which had nothing to do with Christianity and might even be opposed to it. To take a simple example: Christianity said a great deal about the virtues of turning the other cheek, and showing meekness and humility, but such behaviour had not been much admired by the Greeks and Romans. One effect of the revival of classical learning was to suggest to some people that non-Christian standards and values might have much to be said for them and so to contribute to a sense of a break with the past and to the weakening of ideas which had held European culture together for many centuries. Like the Protestant Reformation it made for a more diversified civilization, and for a more secular one.

This must not be overstressed. Humanists who admired pagan virtues and put them before those of Christianity were a minority, and a very small one at that, within the world of educated men – which was itself a very small minority within Europe as a whole. Most humanists found a love of classical learning quite compatible with their Christian beliefs. The most famous of them all, perhaps, was the Dutchman, Erasmus of Rotterdam, and his main purpose in perfecting his learning was to use it to provide accurate texts of the New Testament and the works of the Fathers of the Church.

The coming of print

Humanists and religious controversialists alike had from the fifteenth century a new advantage available to them: print. Movable metal type, oil-based inks and better presses were brought together in Europe for the first time. The real hero of this achievement was the German, Gutenberg, whose enterprise left him financially ruined. But what he did had huge effect. It made it possible, for example, for Erasmus' editions of the New Testament in Greek to reach more people more quickly than the work of earlier scholars. Erasmus offered them a text more accurate than any earlier one and therefore a much better basis for discussion of what the New Testament really meant. It was not, of course, the case that startling or new books were usually those first put into print. The most frequently printed book in the early days of printing was the Bible. People also wanted other familiar works, by great theologians and lawyers, famous texts by ancient authors, not novelties. Nevertheless the existence of the printing-press was to prove of huge importance in circulating new ideas – especially scientific ones – among the small numbers of the specially interested.

Print helped to make Europe a much more literate society. Though most Europeans could not read even in 1800, it was much more common by then for the better-off to be able to do so than three hundred years earlier. Moreover, often those who could not read had books read aloud to them. What they heard read was written in the vernacular. Learned men continued to follow for a long time the practice of writing in Latin, the international language of scholarship. But more and more books were published in English, French, German, Italian, Spanish and other European languages, and just as the invention of writing in early times had helped to 'fix' language in certain ways, printing standardized spelling and vocabulary over wide areas formerly distinguished by dialects and local idioms. Such changes gained momentum as print began to be used for things other than books. Broadsides, printed illustrations with explanations, newsletters, pamphlets and finally the true newspaper or periodical magazine all appeared before 1800. They by no means took the same form everywhere. Englishmen poured out political pamphlets in the seventeenth century (one of the most famous was Milton's *Areopagitica*, a great plea for liberty of the press), while France (because of censorship) for another hundred years or so had far fewer. Newspapers were published in Germany from the seventeenth century onwards. But much more printed material was available everywhere in 1800 than three centuries earlier, and it is probably true that, whatever its quality, public discussion of ideas and events was then going on as never before.

As the end of the eighteenth century approached, more and more demands were heard in countries other than England, the Dutch republic and the English colonies in America, for greater freedom to print and publish. A famous French writer said that although he might disagree violently with what someone said, he would fight vigorously for his right to say it. By then, what such a statement meant was that there ought to be a legal right to print and publish opinions. This was to be something fought for in many countries by liberals in the nineteenth century and then – when the battle might have been thought to be won – again in the twentieth.

SCIENTIFIC REVOLUTIONS

By 1700, printing had already helped to create an international public of learned men. Scientific discoveries and observations had come to be published in the 'proceedings' of the Royal Society in England, or of royal academies elsewhere. This is one good reason for speaking of a 'Scientific Revolution' after 1500, though it might be better to emphasize that there were several distinct big changes, by no means all of which were closely connected. Some had begun simply with observations; Renaissance artists worked out rules of perspective, doctors described in detail human anatomy, map-makers tried to arrange and classify new geographical knowledge won by the voyages of the great discoverers. Others, though, went further.

One of the most important steps forward from medieval science was the invention of intellectual enquiry by systematic experiment. One of its great advocates (though he was not listened to much in his own day) was Lord Bacon, Lord Chancellor of England. He was a man of wide interests; some people believe he really wrote the plays of Shakespeare and, though this seems very unlikely, it tells us something about his reputation that it should be thought possible. Bacon was sure that, if scientific research were done systematically, it would give man enormous power over nature. He was right. He is reported to have died in a highly symbolic way, catching cold one bitter March day while stuffing a fowl with snow to discover how refrigeration affected the flesh.

The feeling that experiments could be made to yield more fruitful results grew stronger as better instruments became available. Telescopes, microscopes, more accurate time-keepers all opened up new areas of investigation. It was inevitable that because some instruments were developed before others science tended to be channelled in specific directions. On the whole it was not until fairly late in these three centuries that chemistry began to advance strongly and only towards the end of the seventeenth century did the biological sciences take their first big steps forward. Before this, physics, astronomy and mathematics were the great fields of scientific achievement; advances in them did more to change the way men looked at the world than any others before the nineteenth century.

The first name which must be recalled is that of Nicholas Copernicus, a Polish priest who in 1543 finished a book (dedicated to the pope) which provided a theoretical demonstration that the planets (including the earth) orbited about the sun. The theories of Ptolemy and commonsense both suggested that this was nonsense: did not the sun rise each day and set

The Royal Observatory was built at Greenwich In 1675 and its primacy is commemorated in the phrase 'Greenwich Mean Time'. A portrait of its royal patron, Charles II, looks down from the wall of the room shown in this engraving.

The title-page of the great work on human anatomy by the Flemish teacher of surgery at Padua, Andreas Vesalius. The word 'physiology' was coined in the year of its publication.

each evening? Clearly, then, the sun went round the earth. No one took much notice of what Copernicus said at first (though, interestingly, Protestant churchmen seem to have been quicker to condemn him than Catholics, who did not officially ban his doctrines until 1616), because it was not possible for a long time to demonstrate that this central idea of his book (which contained many other false ones) was true. Only when the telescope was available in the seventeenth century could Copernican astronomy become visible (its errors, it must be said, as well as its truths). The telescope was used in this way by an Italian professor of physics and military engineering, Galileo Galilei. But he did much more than simply tell people to look through an instrument at what was actually going on in the skies. He also worked out the explanation of the universe this revealed. He produced a new mathematics of the movement of bodies, statics and dynamics, building on the work of fourteenth-century scholars at Oxford who had formulated the first satisfactory law of acceleration.

The book Galileo published in 1632 with the title *A Dialogue on the Two Great Systems of the World* (that is to say, the theories of Copernicus and those of Ptolemy) provoked an uproar. It led in the end to his trial before the Inquisition in Rome where he recanted what he had said (though legend has it that even as he agreed that the sun went round the earth, he muttered 'yet it does move'). But formal suppression of his book hardly mattered, for his views were already public property. The *Dialogue* has since been hailed as the first outright statement of a scientific revolution because, whatever its author might say under pressure, the ideas in that book meant the end of the views of the universe upheld by the Church and traceable back to Aristotle. Even simple men could see the problems this raised – what had happened to heaven? Where was God located? Furthermore Galileo's case advertised the fact that authority – believing something was true just because someone said so – had been worsted in debate by arguments based on observation and logical deduction. Galileo presented a picture of a universe of which the earth – and therefore man – was not the centre but merely one of several similar bodies, and suggested it was possible to define its workings without mystical or religious explanations.

The impact of Newton

In 1642, the year Galileo died, the greatest scientist of the century, Isaac Newton, was born in Lincolnshire. The achievement for which he is most famous is his demonstration that one force – gravity – sustained the physical universe. His theory of gravitation was the core of his most famous book, the *Principia Mathematica*, published in 1687 and said to have been fully understandable to only three or four other men of his time. It brought together the explanation of the heavens and the earth – astronomy and physics – and drew a picture of the universe which was to prove adequate for most purposes for the next two centuries. Newton did much else, too, for he was a man of vast and varied scientific interests and outstanding intellectual powers, so obviously a genius that his own professor at Cambridge resigned his chair when his pupil was only twenty-seven so that Newton could have it. Yet like Galileo, Newton changed the way laymen looked at the world by what he suggested as much as by what he said. At last it began to seem that almost all its secrets might be unlocked by science and if that was so, thought one or two adventurous people (though certainly not Newton, who was a very religious man), what need for churchmen to explain it? What need, even, to talk about God as part of the explanation, since science might explain all of it by the discovery of further great regulating laws?

In the eighteenth century much more was heard of such ideas. Some men even went on to say that the world was a completely self-contained, mechanically determined system, that all that men had to explain and understand in order to live happy lives was the material world. For the first time (though only in the eyes of very few people) 'atheism' – the belief that there is no God – became respectable. It must never be forgotten that only a tiny minority of Europeans – themselves a minority in the world – would have thought like this in 1800. The overwhelming majority, even then, believed still in some sort of invisible world, some kind of God, some form of life after death – and many would have believed in much cruder things, such as magic and witchcraft. Much of the ferocity of the religious wars of the sixteenth and seventeenth centuries had been due to the fact that people believed that great stakes were being played for; divine punishment might fall on the country which allowed God's will to be thwarted by heretics. Witches and wizards had been harried and hunted, as people sought in them the explanation for misfortunes which befell them, and such ways of looking at the world persisted among the population at large. Still, at least educated men knew that some thinkers had pushed a long way down the road along which science (and some other signposts) seemed to point. That is why it is fair to say that the scientific advances of the sixteenth and seventeenth centuries really did amount to a revolution in thought. After it, educated men gradually came to cease to be satisfied with gazing at nature's wonders with bemused awe and the reflection that God had his own though mysterious reasons for creating them. Instead they sought increasingly to find ways of manipulating and exploiting nature. This was an attitude which was to spread much more widely in the next century.

ENLIGHTENMENT

As the eighteenth century went on, European writers made increasing use of a figure of speech based on a particular image: light. The French spoke of an age of *Lumières*, the Germans of *Aufklärung*, the Italians of *Illuminismo*, all of which came to be rendered in English as 'Enlightenment'. It was never the whole story of what was going on even in the minds of educated men and women and it was, of course, organically linked to the past, above all to the shattering by the Protestant Reformation of the old notion of an undivided Christendom. Some Christians, too, favoured the idea that change for the better might be sought within history before it came to an end: human beings could, by their efforts, advance the cause of truth and spiritual improvement. Other landmarks of change could be discerned in the classical humanists' rediscovery of the classical past and the artistic release which followed it. Then there were the voyages of discovery, and their revelations of the inadequacy of long established ideas and of the remarkable achievements of some non-Europeans. It was only with

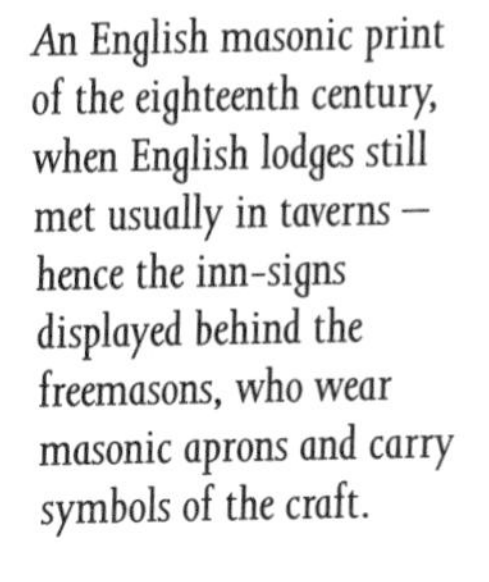

An English masonic print of the eighteenth century, when English lodges still met usually in taverns – hence the inn-signs displayed behind the freemasons, who wear masonic aprons and carry symbols of the craft.

the eighteenth-century Enlightenment, though, that many educated men and women began consciously and openly to reject a great deal of what their predecessors had accepted and did so against a background of spreading literacy and more and more cheaply available printed works. One of the most important cultural changes of the whole of history came about when people ceased to distrust the spread of knowledge. Here lay the greatest success of the Enlightenment. Somehow, by the end of the eighteenth century it was coming to be commonly accepted that more knowledge was good for society. With that, the Enlightenment thinkers had won. The spread of knowledge could be trusted.

New faiths

The Enlightenment may have been the most crucial stage in the emergence of one master-concept of modern European culture, the idea of Progress. While it was obviously rooted ultimately in Judaeo-Christian assumptions about direction and purposefulness in history, in the eighteenth century it attached itself firmly to the notion that the world is increasingly controllable by human will and by reason. This could now begin to be argued from what was going on in some European countries. Though medicine, for example, was at best a rudimentary science (and often not that) and though doctors could do virtually nothing to cure disease, administration and policy were beginning to improve public health, if only marginally and only here and there. Control by quarantine of migration from plague-stricken areas had begun as long ago as the fourteenth century in Italy, and had been generalized by the eighteenth century to the extent of closing frontiers by military means – the longest being the Habsburg Military Border of lookouts within musket shot of one another, running for more than a thousand miles, dotted with quarantine stations where inspections and fumigations could be carried out. Such arrangements were far from perfect and western Europe had another big outbreak of plague in 1720 (though the last of any importance). Yet the practical importance of such successes on the eve of an age of huge growth in European towns and cities is obvious. So was the fact that it came about as a deliberate choice of administrative answers to something once thought of as an unavoidable visitation of God's Wrath.

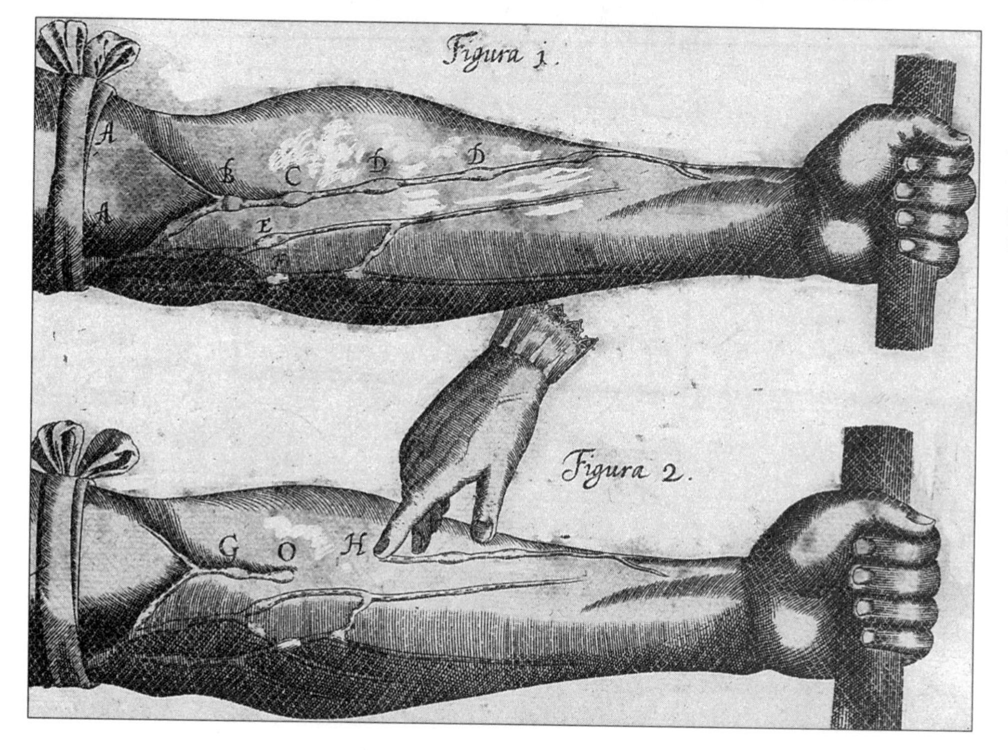

William Harvey, most celebrated for his demonstration in 1628 of the circulation of the blood, was also the author of a treatise on human anatomy from which this illustration is taken.

Perhaps the most important source of the Enlightenment's new confidence in human power was to be found in the new science. Faith in its power was a religious and ideological fact confined at first to a few men, but now shared by millions. It was also, it can be remarked in passing, to give Europeans an enormous and increasing advantage in tapping the world's resources and therefore helps to explain much of their growing dominance of the non-western world. Yet there had been a time when Islamic and Chinese science (to say nothing of Indian mathematics) were well developed while Christendom had no knowledge of what science might mean, beyond a few scraps surviving from antiquity. The Greeks had left behind many ideas which later proved fruitful and had recorded much valuable information, but they had also written down many totally erroneous ideas and had not arrived at the experimental method. Science as we know it today is an artifact of modern Europe. For complicated historical and cultural reasons, it first came into being only after Europe had recovered from Islamic and Byzantine sources all that could be of use to it in the legacy of the ancient world.

For a long time science tended to promote a fundamental optimism about the cosmos. Perhaps because so many scientists had no difficulty in integrating their discoveries with their Christianity, it was long assumed that in some not very clear but certain way, the nature of the universe was ultimately benevolent. God the Creator could not be presumed to have intended evil or suffering; the workings of His marvellous machine, moreover, increasingly exemplified what could easily be interpreted as a wonderful prescience and far-sightedness in the promotion of the well-being of His creatures. The problem of evil was still there, but surely it, too, could be solved? Even men and women could be made much better, given good and rational government, some began to think.

WEALTH AND WELL-BEING

During the eighteenth century English people began to use the world 'Improvement' about many aspects of society. It had first been used to talk about farming, but it began soon to be applied much more widely, and this was partly because there were signs that it was reasonable to think life in some European countries did, in fact, show signs of improvement, and partly because the Enlightenment was suggesting to people that other aspects of life – the treatment of the poor, or the punishment of the criminal, for instance – would follow suit. What fundamentally underlay Improvement, but often escaped notice, was the slow longterm growth of wealth. One of the clearest signs was a remarkable expansion of commerce. Though in 1500 Europe had already swarmed with merchants, on the whole they were then doing only local business. By 1800, their successors were running enterprises of often world-wide scope.

International trade

The earliest great trading cities in the West had been Italian; Venice and Genoa had all but monopolized trade with the Near East, but other towns like Pisa and Florence were trading as far afield as Sicily and with the great north European fairs as early as the twelfth century. In the north the German Hansa towns of the medieval Baltic were already involved in trade with Russia and Scandinavia. In the sixteenth century such pioneering centres were overtaken in prosperity by Antwerp, a great shipping and manufacturing centre, through which wool from England and grain, fish and timber from the Baltic reached the growing populations of the Netherlands, Flanders and Picardy (the last two of which were important textile centres needing imported wool). When, thanks to foreign competition and Spanish rule, Antwerp declined, Amsterdam succeeded to the domination of European trade and finance in the seventeenth century. Finally, after 1688, came the turn of the City of London.

These cities – and many others only slightly less famous brought together the threads of a network of trade steadily growing more complicated and more extensive. Well before 1500

The demand for business manuals of all sorts provided a profitable field of publication to printers. They began to appear in large numbers from the seventeenth century; this example, from France, and dedicated to an outstanding finance minister, deals with the complicated science of the exchange of specie, an important part of the banker's skills in an age which still depended heavily for its business on the physical transfer of coin.

Venice, Genoa and the cities of Catalonia had linked Europe to the sea and caravan trade of Asia, the Indian ocean and the Persian Gulf, most of which flowed to Constantinople in the first place. Some of this business fell off after the disappearance of the Byzantine empire, but soon the North African coast began to provide new products, demands and markets.

Nevertheless the main expansion in trade was for a long time inside Europe. The traditional fairs continued to channel commerce into its well-worn land routes. But sea-borne trade was cheaper than land transport. The first people really to exploit it were the Dutch, partly because of their position, partly because they had to earn money by trade in order to survive, partly because they had large numbers of seamen trained in the North Sea fishing fleets, and partly because they invented a remarkably efficient cargo-carrying ship, the 'flute' or 'fly-boat', which could carry a great deal and be managed by a small crew. Dutch commercial prosperity, which reached its peak in the seventeenth century, was based in the first place on bringing Baltic produce to western Europe, and on the selling of the marvellous salted and pickled herrings, still one of the delights of the Netherlands.

All the earliest developments in the actual machinery by which business was done – the first banks, stock exchanges and devices like 'letters of credit', or 'discounted bills' which made it possible to make payments at a distance without actually carting bags of gold and silver about – were at first confined to exchange inside Europe. Families who rose as moneylenders turned gradually into the first international bankers as kings found it convenient to use them to pay armies operating abroad, or to transfer loans raised in one country for use in another. Paying, supplying and moving Spanish armies about in the sixteenth century, when they were operating over much of Italy, Lorraine and the Netherlands, created business for financiers and merchants and required complicated networks of agents and offices.

In the sixteenth century Spanish America came into the picture. A vast silver mine at Potosi in Peru was discovered and from it can be dated an abundance of bullion (America was to be Europe's main supplier of currency until the nineteenth century) which both encouraged business (there was more money about) and gave people the first explanation to come to hand of an experience forgotten since the later centuries of the Roman empire: inflation. Scholars are cautious about explaining the rise in European prices of something like 400 per cent during the sixteenth century. Given some modern rates of inflation, this does not seem very shocking, but at the time it was very disturbing. Food prices were especially affected and it seems that real wages – that is, the standard of living – of the ordinary working man fell. Yet inflation had important results, one of which was to encourage trade; the commercial atmosphere in the sixteenth century was 'buoyant', even if there were hard times. Profits could be made by shrewd investors.

The slave trade

Some of the biggest profits between 1500 and 1800 were made by selling human beings to other human beings – slaving, as it was called. Slavery had been the basis of economic life in the ancient world and, though the enslaving of co-religionists had more or less disappeared in Europe during the Middle Ages, the Islamic world rested on it. After 1500 Europeans went back into dealing on a large scale in slaves, but in non-Christians; they built up a huge business by tapping a new source of supply, the western coast of Africa. The Portuguese had initiated the new European slave trade there in the preceeding century. They had built forts to serve as collecting points for the slaves rounded up by the native rulers; the steady trickle which soon began to arrive in Europe was to turn into a huge flood.

Nobody planned this. The first black arrived in America in 1502, when the Spanish governor of Haiti was given permission to take with him slaves born in Spain. A few years later a Spanish priest, Bartolomé de las Casas, was so appalled by his countrymen's treatment of the Indians that he suggested that the Spanish settlers in Haiti should each be allowed to import a dozen black slaves. There were not enough Spaniards to do the work and las Casas believed that Africans would stand the labour better than the Indians. As a result, a favourite of the Spanish king (the later emperor Charles V) was allowed to import 4000 Africans a year to the Caribbean islands. This privilege was sold in due course to Genoese merchants

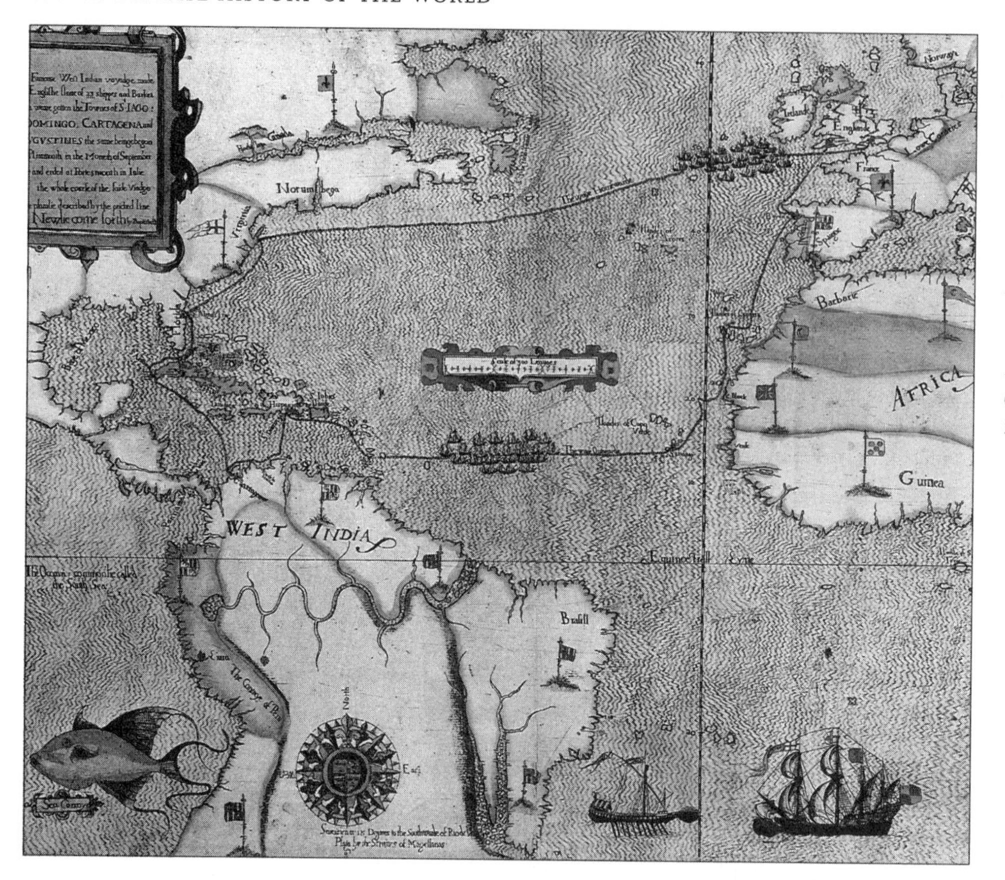

Drake's route to the Caribbean on his voyage of 1584–6 recorded on a contemporary map. This ferocious raid was one of the straws which finally broke the patience of Phillip II of Spain and provoked the launching of the Armada. The story that this was the voyage which introduced both tobacco and the potato to Europe remains, at best, plausible.

and thus the slave trade became international business as a result of trying to give protection to American Indians.

The slave-trade expanded hugely when it was found that many of the Caribbean islands could grow sugar, a crop best handled on a big scale, on large plantations requiring a great deal of labour. Europe was short of manpower to develop the New World; Africa could make up the deficiency. As it grew more profitable to supply slaves to the Americas, so others joined the Portuguese in gathering slaves on the African coast. The trade was soon well worth fighting over, and the Elizabethan 'sea-dogs' sought to break into the slaving monopoly. The Spanish, though, with no West African bases of their own, had to rely on foreign suppliers.

Soon several Caribbean islands had acquired big black populations. The mainland Spanish possessions did not import many slaves though the Portuguese did in their colony in Brazil. A Dutch ship first sold blacks to British colonists as early as 1619 in Virginia, a tobacco-growing area, where slave-labour was useful. Later the cotton and rice plantations of the Carolinas also began to use African slaves. From that time the mainland North Americans, too, built up both a market and a trade to supply it which grew steadily until the late eighteenth century. By then, there were something like forty trading posts on the West African coast concerned in slaving; Dutch, British, Portuguese, French and Danish. It was a huge business. In the eighteenth century, as many as 100,000 blacks were taken across the Atlantic in some years. Though exact numbers will never be known, many more left Africa than arrived in the Americas. Disease, despair and brutality could kill half a ship's cargo before it got to the New World.

Oceanic commerce

For all its grim and repulsive drama, slaving was only one of many new patterns of trade spanning the oceans. The slow building of a new world commercial system much more widespread than any earlier one was irreversible by 1700. Overall expansion went on faster than ever (though there were a few hiccups as things went wrong in particular places), and trade with the non-European world steadily loomed larger and larger in the creation of European wealth. Atlantic trade with European colonies and possessions in America was the most important part of this transoceanic business. Ships would set out from the Atlantic European ports with trading goods to be used in buying slaves on the African coast. After taking blacks to the Caribbean and selling those who had survived the voyage, they would load sugar or coffee and take it back to Europe or the British North American settlements. From the latter, other goods – rum, indigo, rice, corn – would be exported to Europe or to the Caribbean colonies. The Spanish, like the English and French, tried to reserve trade with their own possessions to themselves, though unsuccessfully, because of the huge profit to be made by smugglers and 'interlopers'.

The long-term victors in the struggle for the profits of world trade were the British. One reason was that the government in London was more single-minded in upholding the interests of English (and, after 1707, Scotch) merchants and sea-captains than were the kings of France in looking after those of their subjects. From the French court at Versailles the view towards Europe always seemed more interesting than that out to sea; French kings were more concerned about conquest (or at least holding their own) in Europe than with fishing for cod off Newfoundland, selling slaves to the West Indies, or importing sugar and coffee. The greater British awareness of what these things might mean in terms of profit was one reason why the Royal Navy played so important a part in world politics in the eighteenth century.

The gravestone of a servant in an English churchyard.

Politics and commerce were all the time becoming more and more mixed up. Sea-power did not just guarantee that you could get to other parts of the world and settle colonies there; it could also be used to force open the Spanish colonial markets (illegal entry had already been won by the freebooters and pirates of the previous century, the great age of smuggling and buccaneering). Sea-power was also essential – especially in wartime – if your own traders were to be protected. It could back up diplomacy in negotiating favourable terms over, for example, customs duties levied by a country on imports. Such things mattered more to Great Britain than to any other power, because she gradually emerged as the country which more than any other depended on overseas trade in order to earn money, above all by importing colonial goods for resale in Europe or in the colonies.

In the world picture the trade with Asia, though far from being the most important in terms of bulk or value, remained glamorous and it offered great profits to merchants. Both the Dutch and the English established 'East India' Companies early in the seventeenth century with monopoly rights to trade in the Far East, and the French later followed suit. They became the main ways of contesting for trade in Asia, but they suffered from the disadvantage that, except for a few mechanical novelties, there was very little made by Europeans which the Asians wanted. With India, China and Indonesia the European countries therefore usually had an adverse balance of trade; they could not sell the Asians enough European goods to pay for what they bought, and so had to pay for it in silver. It was another example of how, without anyone really intending it, the world was getting more tied together; the Spanish brought silver from the New World to Europe, where it paid the debts of the Spanish monarchy to bankers who then passed it on to merchants to use to buy goods in Asia. On the miners of Peru rested the financing of trade in Canton. This is, of course, only a tiny section of the whole truth. The main lines of what was happening in these three centuries, though, are fairly clear. World trade was growing; the first part of it to grow really fast was the Atlantic trade; it became more and more tied up with politics and sea-power; and – above all – it was dominated by the Europeans. No Chinese junk or Arab dhow ever docked at a European or American port in these centuries, though thousands of European and American ships went to the Moluccas, India, the Persian Gulf and China.

Increasing knowledge

Trade helped to promote further discovery and geographical knowledge. By 1700 the shape of all the main continents was pretty well known; only the outlines of eastern Australia and northern Siberia, and those of the far American north-west and Bering Strait region had not yet been mapped. Though huge unknown patches remained within Africa and Australia, world maps of a high degree of accuracy were available for the rest of the globe. If he took his chance of ship-wreck, storm, piracy and disease, a traveller might put himself in the hands of a ship's captain with reasonable confidence; as far as navigation and seamanship could assure it, he could be delivered anywhere on the coasts of the world that he wished (and certainly at any of its ports) within three or four months. This was a great advance on the situation two hundred years or so earlier; once launched, change had come very rapidly and at a quickening pace. Geographical and technical knowledge was cumulative: the more there was, the easier the next advance became, even if the maritime technology did not much change.

The great voyages which first mapped the world and brought back reports of new-found lands had been the keys to all that followed. Sebastian Cabot, sailing out of Bristol, made his second voyage to a landfall on the coast of North America in 1498, when Vasco da Gama reached India. Amerigo Vespucci in 1499 began the exploration of the coast of South America and eventually got as far south as the Falkland islands. In 1508 a Portuguese sailed into the Persian Gulf. In 1513 Europeans looked for the first time at the Pacific. Then, in 1519, began what was to prove the greatest achievement of the early navigators: the Portuguese Magellan set out from Seville and in the following year turned the tip of South America by the straits still named after him to sail into the vast unknown of the Pacific. In 1521 he was killed in the Ladrones islands, but one of his ships sailed on, via the Philippines and Timor, crossing the Indian Ocean, rounding Africa and making it back to Seville. Its Spanish commander, del Cano, was the first captain to sail round the world, and to show that all the great oceans were interconnected. Men had known in theory that it could be done; now someone had done it.

After this, knowledge of the Pacific hemisphere slowly began to accumulate. By the early seventeenth century many of its islands, as far south as the New Hebrides had been discovered. In 1616 Dutchmen began to explore the coasts of Australia and in 1642 one of them, Tasman, sailed past the island to which his name was later given, on his way to New Zealand, thus demonstrating that Austrlia was not simply part of an Antarctic continent. By the end of the following century exploration – above all, the voyages of Bougainville and Cook, had incorporated the southern Pacific and Australasia into the known world. The evidence of that was the dumping of the first cargo of convicts in Australia in 1788 and the arrival of the first missionaries in Tahiti in 1797.

Captain Cook's crew improve their acquaintance with a threatened species.

The northerly waters remained longer unexplored. In 1553 an English ship reached the site of what then became the Russian port of Archangel and returned with a letter from the Tsar to Mary Tudor. A series of English voyages beginning with one by Frobisher in 1576 vainly sought to find a 'North-West Passage' round the Americas to Asia. In 1594 a great Dutch navigator, Barents, set off in the opposite direction, following the earlier English explorers of the north-east. On his third attempt to find a way east through the Arctic, three years later, he died in the remote wastes of Novaya Zemlya (no one was in fact to make a successful North-West Passage by ship until 1905, though the first complete north-eastern voyage to Asia was made in 1879).

ISLAM AND THE WESTERN WORLD

Long after the fall of Constantinople in 1453, millions of Europeans lived under Islam, and even more lived under its menace. Paradoxically, just as the long Reconquest of Spain had been completed, Islam was advancing in the east. This was misleading. Islam was divided; Persia was at times at war with both the Turks and the Moghul emperors of India, and Arab states contested Turkish power in the west. Still, the feelings of Europeans are comprehensible, for they were confronted with the sharpest of the Islamic cutting-edges, Ottoman Turkey.

In the fifteenth and sixteenth centuries the Ottomans wrested many of her remaining possessions away from Venice – the Ionian islands at the mouth of the Adriatic in 1479, the islands of the Aegean in the 1550s and 1560s, Cyprus in 1571 – and forced the Spanish monarchy to fight hard to maintain its communications with Italy. Briefly the Turks had a foothold even there, while Spanish gains on the North African coast were quickly lost again to them as they conquered Cyrenaica, Tripoli, Tunisia and Algeria. They had by then over-run Serbia, Bosnia and Herzegovina in Europe itself. In 1526 they shattered the Hungarian army in a defeat so appalling that 'Mohács Field' is remembered still as a black day in the nation's history. Three years later they besieged Vienna for the first time, though unsuccessfully. After a pause this advance was resumed, Hungary being over-run for a second time (this was the last time the Turks overthrew a Christian kingdom), Podolia (the lower Ukraine) being taken from Poland, and Crete from the Venetians. Finally, Vienna was besieged again in 1683. This was the highwater mark of Turkish power.

The Ottoman empire was not built only at the expense of Christians. In North Africa the Turks had established their overlordship over Moslems. By 1520 much of the Hejaz, Syria, upper Mesopotamia and Kurdistan were in Turkish hands. The sultan Suleiman the Magnificent added to these conquests lower Mesopotamia, much of Georgia and Armenia, and pushed farther into the Arabian peninsula too. By 1683 the Ottoman empire ran from the Straits of Gibraltar to the Persian Gulf and the Caspian, and it was to pick up a few more even after that date.

Yet now the tide turned. Until the late seventeenth century, no real danger had threatened the Turks from Europe, but western Europe's territorial questions were broadly settled at Utrecht and though the Habsburgs continued to be preoccupied in Germany the appearance of two major new eastern monarchies – Prussia and Russia – altered very gravely the balance of power which faced Turkey. Ottoman power had begun to roll back before the Austrians and Russians even before 1700. Hungary was recovered by then. Much worse was to follow, especially an important symbolic retreat in 1774, when the Russians won overlordship of the Crimean Tatars, the first surrender by the Turks of power over a Moslem people. By 1800 the Russians had taken much of the northern Black Sea coast and their frontier lay along the Dniester; the Austrians had advanced to the Danube. Yet the final break-up of Ottoman power was to take a long time – until 1918 in fact. In the Middle East the problem of deciding how the formerly Turkish lands should be divided still has to be settled and the wars of the Ottoman Succession continue in our own day.

The explanation of Ottoman decline lies partly in internal weakness. For all its huge extent on the map, Ottoman power varied very much from place to place. In Mesopotamia (almost always disputed with Persia) and Syria the desert Arabs were never really under control. There was no centralized administration worth the name; the Ottoman empire was in most places a matter of arrangements between the 'pasha' – the Sultan's officer – and local bigwigs about the way in which taxes could be raised. This gave the pashas much power and some of them came to resemble dynastic princes as time went by. As a result the empire could never fully mobilize its resources, nor rely on loyalties among its subjects to over-ride the many divisions between provinces, peoples and religions.

The Ottoman 'state' had been put together more or less haphazardly in order to fight the infidels. Such organization as it had was basically military; it was meant to provide recruits

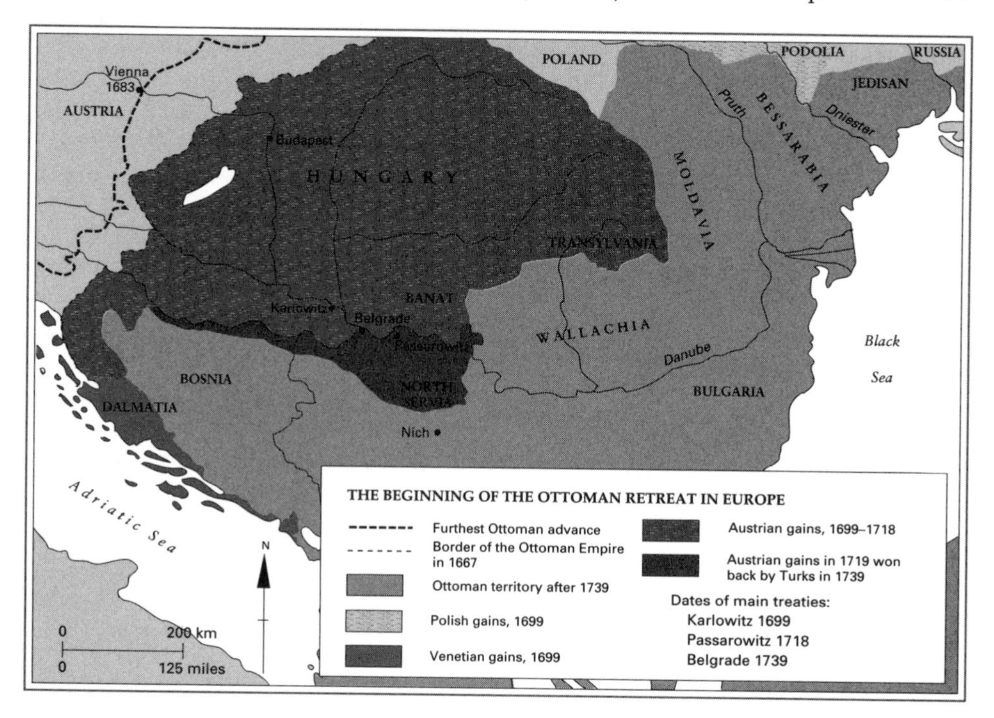

and taxes to pay soldiers and did this by arrangements not unlike the 'feudal' tenures of western Europe. This structure had already become corrupt by the seventeenth century. The sultan's officers padded out their muster rolls in order to draw pay for more men than they could actually produce. Locally they abused their position as recruiters and tax-gatherers and there was no civil service to check them. The sultan himself was the centre of intrigue; favourites, the women of the harem, generals and religious leaders all sought to influence him. The grand vizier, who held the major office of state, had to contend all the time with attempts to undermine his position. The most professional regiments which the Turks possessed were the Janissaries, but they were sadly decayed by 1700 and more of a danger to the sultan than a support; they frequently mutinied or went on strike over pay. Finally, throughout the Moslem community at large, real power was exercised by the religious leaders – the ulema – whose attitude could determine popular support or discontent. There were frequent tumults at Constantinople. After one especially violent upheaval in 1730 the Bosphorus was said to have been covered for days by bobbing corpses.

The Retreat of Ottoman Power in Europe before 1800

1683	Ottoman (second) siege of Vienna fails.
1691	Ottomans lose Transylvania to Austria.
1697	Ottomans lose Hungary.
1699	Ottoman losses to Austria confirmed by Peace of Karlowitz.
1716–17	New Austro-Turkish war: Austrians conquer Belgrade.
1718	Austro-Turkish war concluded with treaty of Passarowitz.
1739	Ottoman recovery in Treaty of Belgrade (Austria gives up Belgrade; Russia agrees not to build a Black Sea fleet).
1768–74	Russian occupation of Moldavia and Wallachia; Ottoman fleet defeated by Russians. Russian conquest of Crimea.
1774	Treaty of Kutchuk Kainarji gives Russia basis for future interference in Ottoman empire.
1787–92	Austria and Russians combine against Ottomans.
1791	Treaty of Sistova restores Belgrade to Ottomans; Austria receives part of Bosnia.
1792	Treaty of Jassy gives Russia frontier on the Dniester.

Of modernization there was little. Almost all that was successfully achieved was the conversion of the navy in the 1690s from the old, oared galleys to sailing-ships of the European kind. But trained sailors were harder to get than galley-slaves had been (one sign of Ottoman decline during this period is the increasing employment of Europeans in the navy and army).

On one other front Ottoman power was slowly undermined. At the beginning of the sixteenth century a new dynasty of rulers had established itself in Persia, the Safavid family. It belonged to the Shi'i sect – a seventh-century form of Islam which had long survived in opposition to the official 'Suni' version. Shi'i doctrines were almost always more widespread in Iraq and Persia than in Syria and there were many ramifications and sub-sects. They all rejected, though, the authority of the caliphs (their rulers). When the Safavid set themselves up in Persia therefore, they were almost bound to quarrel with their Ottoman caliph neighbour, who claimed the headship of Suni Moslems.

In 1514 Persia went to war with the Ottomans, and over the next two centuries the Ottomans had to fight on two fronts while the Safavid rulers – notably their outstanding shah, Abbas the Great – built up a new Persian empire, a centre of high civilization and wealth. Yet the Safavid state was a harsh and intolerant one and it, too, had to fight on two fronts at times, against the Moghul emperors of India as well as the Ottoman sultans. Occasionally the advocates of old-fashioned Shi'i austerity had some success in complaints about declining standards; religious leaders rejoiced under a drink-loving shah at the end of the seventeenth century, when 60,000 bottles of wine from the royal cellars were publicly smashed. But puritan zeal could not reverse the continuing tendency of the Safavids to go downhill. In 1722 the last of their line was overthrown by an Afghan warrior. A few more years of trouble followed until in 1736 a new strong man emerged, Nadir Shah. After expelling the Afghans, he took back provinces seized by the Ottomans and Russians, but this too was to prove only an interlude.

A new eastern Europe

Partly at the expense of the Ottomans, the map of eastern europe was transformed between 1600 and 1800. One of the three monarchies which were the beneficiaries of change was already a great power before the changes began: the Habsburg monarchy. The other two Russia and Prussia, only emerged as great powers during these two centuries.

Russia's transformation was the more striking. Over two hundred years she vastly extended her territories to the west and to the south, became a major military power of great importance in Europe's diplomatic calculations, developed industrial strength which was, for that day, outstanding, and had made a significant break with her traditional and isolated cultural heritage (albeit one which was to prove incomplete). All this was overwhelmingly the result of political action, and the monarchy was its origin and motor. This established a pattern which has endured to the present: modernization in Russia was always to come from government downwards, from the centre to the periphery, to be imposed rather than to grow spontaneously.

The stamp of modernization was first applied by Peter the Great, who came to the throne as a ten-year-old in 1682 and set to work to use (often harshly) the traditional power of Tsarist autocracy to drag Russians into modernity – which meant, for him, the culture of western Europe. His aim was thus to strengthen Russia for international competition, with the aim, first, of securing Russia's Baltic coast. Though also interested in central Asia and Siberian expansion, a great war with Sweden was the heart of his foreign policy achievement and it ended in 1721 with Russia firmly established in Livonia, Estonia and the Karelian isthmus, and her new Baltic capital in process of construction. The physical shift of government from the isolation of old Muscovy to the neighbourhood of the West was highly significant as a symbol of Peter's ambition.

Peter also pressed his ambitions southwards. At one time he had annexed Azov and had a fleet on the Black Sea. But he could not maintain the thrust towards the Ottoman empire: that was left to his successors, who, by 1800 controlled the northern coast of the Black Sea from the Dniester to the Kuban. Meanwhile, the industrialization Peter had encouraged and based on extractive industries (above all, minerals and lumber) had given Russia a favourable balance of trade and a higher pig-iron output than any country in the world. The debit side was that much of this was achieved by the use of serf labour and an alliance of monarchy and gentry which bound Russia slowly into a system of social and political arrangements which inhibited progressive change. For all the glitter of the Russian court under Catherine the Great, Peter's most conspicuously successful successor on the throne, innovation flagged, and for all Russia's power in an age when numbers counted so heavily in military strength, the system of autocracy, serfdom and Orthodoxy which so evidently blocked the way to real modernization was already producing its first critics before she died in 1796.

Prussia and Austria

Catherine the Great was sometimes admired as an 'enlightened' autocrat, in that she patronized men of letters and philosophers from the West who were regarded as the standard bearers of advanced, even radical, ideas. The same was also said of some rulers in other states, among them, the other two 'new' powers of eastern Europe, Prussia and Austria. In each case, though, it seems that royal policy owed more to the pressure to seek change which would generate power for international competition than to advanced ideas.

Prussia, which became a kingdom in 1701, was then a scatter of territories belonging to the former Electors of Brandenburg. Her story, in the eighteenth century, was of the consolidation and extension of these territories by diplomacy and military conquest, the resources for this being found by rigorous management and exploitation of the subjects of the monarchy by a bureaucracy whose efficiency became legendary. Above all, these characteristics

The 'poor girl with three or four dresses', as she called herself, of thirty years earlier: Catherine II, empress of Russia, painted at the end of her life, after she had won her adopted country more new territory than any of its rulers since Peter the Great.

manifested themselves under Frederick the Great, who particularly pressed the interests of Prussia in Germany against those of Austria. He then launched a struggle between the Habsburg dynasty and the Hohenzollern (his own) which was often very bloody and was to be finally settled only in 1866, when the Habsburgs at last conceded Prussian hegemony over the other German states.

The struggle with Prussia animated efforts to reform the sprawling and ramshackle Habsburg dominions of the eighteenth century in order to generate the resources for international competition. While these were sufficient to ensure notable gains from the Ottoman empire (the Habsburg southern frontiers had been carried back to the Save by 1795), they could not overcome the Prussian threat and major concessions of territory had been made in Silesia. In one direction, nonetheless, Habsburg ambition had done well.

Poland

In 1795 there disappeared from the European map the once great state of Poland. Even in the seventeenth century she was still a great military power, decisively engaged against the Ottoman empire. In the eighteenth century, though, her constitution and succession problems fatally weakened her cohesion and gave opportunities to foreign intrigue and intervention. Increasingly, these were the work of the three great powers engaged in competition for territory and predominance in eastern Europe – Russia, Prussia and Habsburg Austria. Attempts to reform the state came to nothing. In 1772, dangerous tension between Russia and Austria over Russian success against the Turks was relieved by an agreement on the first 'Partition' of Poland, which lost to her three neighbours one-third of her territory and half her population. Another followed in 1793, and the last partition in 1795.

The significance of this brutal episode was immense. The three great eastern powers were now brought face-to-face: no further possibility of compensation at someone else's expense now remained except, for Russia and Austria, in the Ottoman Balkans. Secondly, a great conservative interest had now been established between the three powers: each of them now had a large population of Poles with intense national consciousness to hold down.

A NEW AMERICA

In the 1770s there also began a major upheaval in the British North American colonies. In 1760 there were about two million colonists there. Their numbers were still increasing at a rate doubling the population in a generation and besides the English, Irish and Scotch there were Dutch and Germans, too. There were also Indian subjects of the Crown and (mainly in the southern colonies) black slaves numbering about one-sixth of a population which was at most about one-third of that of the mother country. The colonies had grown greatly in area since 1700. Broadly speaking, settlers always tended to press inland from the coast until they came to the mountains which run almost the whole length of the eastern coast, and did so at the expense of the Indians. This had led to fighting and bad blood on the frontiers of some of the colonies – in particular, those of New York and Pennsylvania, whose settlers showed especial keenness in pressing into the river-valleys which led down from the other side of the mountains to the huge Mississippi basin beyond. When the French lost Canada in 1763 the danger that they might block this advance disappeared.

There were finally thirteen colonies in all. People sometimes called their inhabitants 'Americans' – but they thought of themselves much more as New Yorkers or Carolinians or New Englanders. They were often at odds with one another, sometimes quarrelling (and even fighting) over boundaries. The larger colonies were aware of big differences between the frontiersmen, who lived in the western wilderness, and the townsfolk and planters of the coastal plains or 'tidewater'. There was really very little to hold Americans together except that they were all subjects of the English Crown.

In 1763 they still saw themselves as loyal subjects. They were grateful for protection from the French and Indians during the wars, and the government in London had not made enough demands on them to irritate them. Yet Americans were different from Englishmen in their attitude to authority. Socially more easy-going than the British, the colonies, although they had rich and poor, had few titled people and nothing like the English tradition of respect for an aristocracy. Differences between religious sects, too, were accepted more readily in the colonies than at home; many of the early New England settlers had gone there in the first place to get away from the Church of England, and Maryland had been founded to provide a haven for Roman Catholics.

The collapse of the British empire in America which followed within twenty years of the peace of 1763 nevertheless came to most people as a great surprise. The fact that colonists came to feel they no longer much needed the protection of the British was a root cause.

The famous Boston Massacre, a godsend to radical propagandists though the soldiers (who were tried by a local court) were acquitted of the charge of murder.

Another British ban on further settlement in the west was a grievance; the rights of the Indians were to be safeguarded against settlers. This would need soldiers, too, and that would cost money. As these soldiers would be protecting the white frontiers against Indian raids, it seemed to many Englishmen only fair that the Americans should pay for them.

For years one government after another tried to find acceptable and workable ways to tax the colonists with this end in view. One colonial politician coined the slogan 'no taxation without representation'; if Americans did not send members to the parliament at Westminster, it was argued, why should they pay the taxes it imposed? Gradually discontent grew. People began to feel they might like to remain subjects of King George III, but that they would do better if they were not ruled by laws made by parliament, but by laws they made themselves. In practice this was what many of them had really been doing through their own assemblies for years, with very little interference by parliament and the idea caught on. But a desire for complete independence only emerged very slowly. Even when the first shots of the American Revolution were fired (at a British column on its way to seize some illegal arms held in a little town not far from Boston in 1775), many Americans remained loyal.

Rebellion

By 1776 those who wanted a break had grown more numerous. A congress of representatives from each of the colonies met that year in Philadelphia and agreed to a Declaration of Independence which can be taken as in principle the final parting of the ways. From this moment the only chance the British had of keeping their colonies was to crush the rebellion by force. It took seven years for them to admit that they could not do it. In 1783 peace was signed, and the two groups of British subjects went their separate ways. Fighting and diplomacy had decided the matter. Though it may not have seemed so at the outset, the odds had never been clearly in England's favour. Certainly she had a powerful and well-trained army and navy, while the rebels had none and large numbers of Americans were loyal (at the end of the war, thousands went to live in Canada rather than stay in their former homes). The mother country was rich and her colonies were poor. But, on the other side huge distances separated the colonies (where the fighting took place) from the home base of the British army and this led to vast problems of transport and supply. The terrain was often difficult

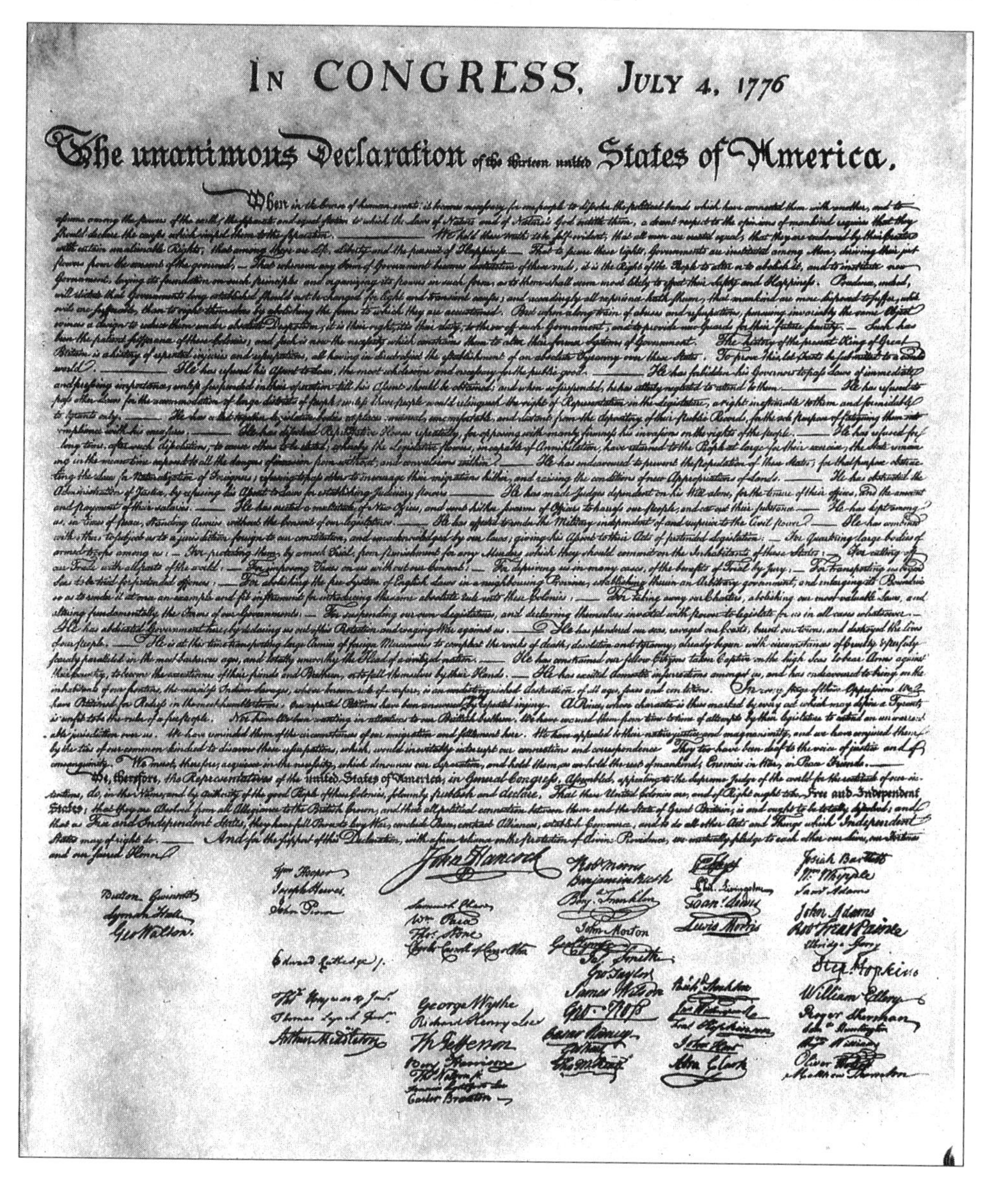
IN CONGRESS, JULY 4, 1776

The unanimous Declaration of the thirteen united States of America.

The Declaration of Independence, the central document of the American Revolution and still a source of inspiration and renewal in the public life of the United States. Perhaps that is because the man who drafted it, Thomas Jefferson, resolutely set out to write a profoundly conservative document: it looked back to accepted truths rather than strove to startle or excite men with new discoveries. 'It was', he said, 'intended to be an expression of the American mind.' His purpose was 'not to find out new principles, ornew arguments never before thought of, not merely to say things which had never been said before, but to place before mankind the common sense of the subject'.

The American Revolution

1763	Peace of Paris ends Seven Years' ('King George's') War.
1764–7	Various revenue measures increase agitation in the colonies.
1773	Boston 'Tea Party' symbolises resistance to taxation by Great Britain.
1775	Skirmish at Lexington and first bloodshed of the war of American Independence.
1776	Declaration of Independence signed at Philadelphia.
1777	British defeat at Saratoga. Articles of Confederation propose structure for United States of America to the individual colonies.
1778	France signs treaty of alliance with USA.
1779	Spain enters war against British.
1781	British surrender at Yorktown.
1783	Peace at Paris between Great Britain and USA and recognition of American independence.

and badly mapped, hard to live in for European soldiers used to campaigning with good communications and supply depots. Nor could the British fight a really savage campaign – burning farms and so on – which would make impossible the survival of the colonists' army, because they could not afford to alienate the friends they had. Finally, foreigners were keen to profit from England's troubles; by the end of the war the British were fighting not only the Americans but the French, the Spanish and the Dutch as well. That tipped the naval odds against the British at a crucial moment, forcing the British army to surrender at Yorktown in 1781. After that disaster it was really only a question of when and how the British would come to terms.

The United States of America

So emerged a new nation and the first decolonialized country: the United States of America. The links between its thirteen states were not very tight, even after they accepted the constitution which brought them together in a federal republic in 1789. But at least some Americans had grasped that if the new states were to survive at all they would have to have some kind of national government. Among them was the former commander of the American army, George Washington, who became the first president of the Union.

The two great changes – separation from Great Britain and the creation of even a weak central government – were of enormous eventual importance for all mankind. The American Revolution can now be seen as the first of a wave of colonial revolts which was to unroll for about fifty years in the Americas and was to take even longer to have its full effects. Another outcome still unguessed at was that North America was to be settled and dominated by people speaking English and sharing much of English culture. They took for granted religious, legal and constitutional traditions established in England, and spread them across a continent. World history would have been very different if the colonists had spread, say, French or Spanish ideas of absolute monarchy. Indeed the founding fathers of the United States pressed some English ideas much farther than they had gone in the mother country. Religious tolerance was taken so far that the government was forbidden by the constitution to support any religion at all.

The United States of America was also the first major nation to be a republic; the general eighteenth-century view was that republics were feeble and suited only to small states. It was a great thing for mankind when the United States proved this wrong (though the proof owed much to good luck, remoteness and natural wealth). Finally, the new republic was a democracy, too – not perhaps in a perfect sense, but more completely than any other. The opening words of the constitution are 'We the People' and democracy was to spread ever deeper and further into American life in the next couple of centuries. With it went distrust of central government, and the slow spread of a more equal degree of political and practical freedom for all Americans in their everyday lives. There is still no other major country of which this is so true.

At first, though, the new state seemed to make very little difference to the world: it was too far away. The British were soon doing more trade with Americans than before the war; political separation seemed not to have much changed business. The French, in spite of their victory, did not get their colonies back, and had been forced to spend a great deal to support the Americans. The war made a difference to the way British governments thought about settler colonies, though. Henceforth, they distrusted them. They spent most of the next century trying to find ways of giving them as much independence as it was safe for them to have as soon as possible so that they would not cost the British tax-payer money or threaten him with a repetition of the American disaster. As for the Americans, they got on with consolidating their new country and enlarging its borders.

THE FRENCH REVOLUTION AND ITS OUTCOME

From the days of Louis XIV until well into the second half of the nineteenth century France was often top dog in Europe. Nonetheless she was showing signs of strain as the second half of the eighteenth century rolled on. The loss of Canada was to some extent offset by British defeat and humiliation, but the French did not get Canada back, and had added hugely to the vast debts of the monarchy. One after another, ministers tried to find a way of reducing her debts and giving France sensible new financial arrangements. They all failed, basically because they could find no way of making the better-off pay their due share of taxes. It turned out that in some ways the impressive French monarchy was not very effective at home. It was certainly not so good at raising revenue as the British parliamentary system, for example. More and more the blame for this state of affairs was laid at the door of the French nobility. When, at last, the king announced that in 1789 he would summon the nearest thing to a parliament that France had ever possessed – the medieval 'Estates General' – there was great rejoicing, for times were hard. Almost everyone seemed for a time to believe that France would be governed better if governed in accordance with the will of the majority.

The process of change

In the end that was more or less the outcome, but only after long and bitter political struggles. When the Estates General met, in May 1789, there began a process by which more and more grievances about many matters other than fair taxation broke out in new demands. So, more and more people turned to politics to put things right. In the course of this the historic constitution of France was swept away, the absolute monarchy turned first into a constitutional one and then into a republic, the king and queen were beheaded, thousands of people died in civil war, the old national religion of Roman Catholicism was given up and the Church's property sold for the good of the state, and a thousand and one other changes, big and small were made. This was the French Revolution.

Many arguments have taken place about when it began and ended; one defensible pair of dates is 1789 to 1799, when a young general, Napoleon Bonaparte, seized power from the politicians and set France again on the road back to monarchy. Nothing like that decade had ever been seen before. Almost all the lasting changes it brought about were made by the end of 1791. The following years to 1795 were the most turbulent of the Revolution. Things settled down somewhat after that. France had by then broken with much of her past, rebuilding her constitution on the basis of equality before the law (nobility had been abolished), religious toleration, and the government of France by an elected representative National Assembly which could make laws on any matter, without regard to rights or tradition.

Yet a great deal remained. Life in much of the countryside can hardly have changed much, given its deeply ingrained ways. The new decimal currency of francs and centimes (still in use today) was not used in country markets for decades; fifty years after 1789, some peasants still counted in the old coins – crowns and sous – and used old measurements rather than the new kilometres and hectares. The revolution nonetheless turned France upside down. Many people never forgave nor forgot this, and for the next century or so the revolution was a touchstone of political opinions. If you were for the revolution, you wanted more people to have the vote, probably wanted a republic and certainly wanted the Church to have less influence than before 1789. You probably believed in free speech and the wickedness of

The ancient Estates General of France meets under the presidency of its king in May 1789 at Versailles.

press censorship too. If you were against the revolution, you looked for a strong government; you sought to restore the influence of the Church in the national life; you believed it was wicked to allow the spread of harmful opinion, and you thought discipline and good order more important than personal freedom. This was, roughly speaking, the division into 'left' and 'right' which spread into the politics of many other continental countries in the next fifty years. It was invented (and the words began to be used) in 1789 when conservatives began to sit together to the right of the president in the National Assembly and liberals began to sit together on his left.

The fact that this division spread later to other countries shows the enormous influence of the Revolution outside France. From the start some of the revolutionaries had said that what they wanted to do in France by way of reform could and should be done in other countries too, and suggested other people should follow the same recipe. Later, when the new France found itself (as it did from 1792 for most of the rest of the decade) at war, they exported revolution by force and propaganda to other countries. French generals set to work to organize revolutions and set up new republics in the lands they invaded.

This was one reason for the many wars which followed 1792. Not only did it seem that France was again setting out on a career of conquest under Napoleon Bonaparte (crowned emperor in 1804) as she had done under Louis XIV, but conquest now seemed likely to be followed by revolution. Great Britain was France's most unremitting enemy. Only once and briefly between 1793 and 1814 did Great Britain make peace. In the end this meant she won the old colonial game hands down, after French sea-power had been broken in 1805 by the great naval victory of Trafalgar. Fighting on land was a different matter. The British long had an expeditionary force in Spain, but the huge numbers which eventually defeated France (first in 1799 and then in 1812-13) came from the peasant masses of Austria, Prussia and, above all, Russia.

For all the aggressiveness of the Revolution, French armies often brought liberation with them. French occupation usually led to the abolition of feudalism and the destruction of old tyrannies, and promoted the equality of men before the law. This was one reason why, from the outset, the French Revolution was (as it has remained) a great inspiration and ideal. All over the word in the next hundred years men would turn against oppressors real or imaginary in the name of ideals summed up in one of the Revolution's slogans: Liberty, Equality,

Bonaparte, 'calm on a fiery horse', as he asked the painter David to depict him, crossing the St Bernard pass in 1800. In 1802 a plebiscite confirmed him as First Consul for life and in 1804 another declared him Emperor of France.

Fraternity. This is why tyrants feared it. And even when men did not look to revolution as a way of getting what they wanted, they would be inspired by the revolutionaries' claim that men had rights as human beings, not simply because they inherited them from any particular system or law or because they had historical traditions supporting them. It is another reason why the French Revolution was an event in world history as well as in the history of France.

The birth of modern politics

After 1815, world politics were increasingly though slowly to take their language and principles from Europe. There, one of the most important trends making itself felt after the French Revolution was that more and more people were getting involved in public life, even if only in a pretty formal way. The main sign of this in most countries was the winning of politically useful rights by more and more people. Some of these rights were negative; a right not to be silenced, for example, without a due legal cause being shown or the right not to be locked up without trial, such as was pretty well guaranteed already to the English by the legal device of the writ of *habeas corpus* (the Latin words with which the text of the writ began).

European government was changed much more during the nineteenth century by such blind forces as those of population growth, urbanizat ion, railways and mass literacy than by revolution. Yet the idea of revolution went on fascinating and terrifying men even when it almost always failed in practice. France, understandably, was the supreme example of a country where the myth of revolution bemused politicians and artists alike. Delacroix's celebrated painting of Liberty leading the people across the barricades, symbolic Phrygian cap on her head and tricolour in her hand, is still one of the most moving evocations of the idealization of what a priest once called la sainte canaille – the mob whose brutality is seen as purged of evil in its commitment to the cause of freedom for all men.

Other rights were positive and allowed you to do something; the most important of those was, undoubtedly, the right to vote and so to have a share in deciding who would govern.

Among major countries, only the United Kingdom and the United States of America could in 1815 show either sort of political rights to be in good fettle or widespread, and there, to them, too, there remained important restrictions (on qualification for the vote in England, for example). But everywhere there was a much greater demand for rights than a few years before largely thanks to the French Revolution. If it had not done much to protect rights, it did much to advertise them. Successive French governments since 1789 had soon shown that in practice they did not like trusting their fellow-citizens with political rights and when they invaded other countries they often behaved in a very high-handed way. Nevertheless, they had cleared a lot of the ground by sweeping away the old absolute monarchy and much of the law that went with it. Often, their armies did the same thing abroad; between 1796 and 1814, for example, much of Italy, Germany, the Netherlands and Switzerland were ruled by republican governments which had laws modelled on those of revolutionary France. More important still, the great 'Declaration of the Rights of Man and the Citzen' to which the National Assembly agreed in 1789 (there were to be two more in the next few years), started a debate which spread across Europe.

The Revolution also launched another provocative idea into Europe's political national sovereignty. However they were chosen, the representatives of the nation were, insisted the French revolutionaries, those with whom the last word rightfully lay in law-making. This was not an idea which would cause much stir in the United Kingdom in 1801, where a parliament part-hereditary and part-elected (on a narrow franchise) enjoyed very great powers, but it was disturbing elsewhere where people thought old institutions and ways should not be interfered with, even by parliaments. It was especially a revolutionary idea in Russia, where the Tsar claimed the divine god-given right of his line to rule as he thought best for

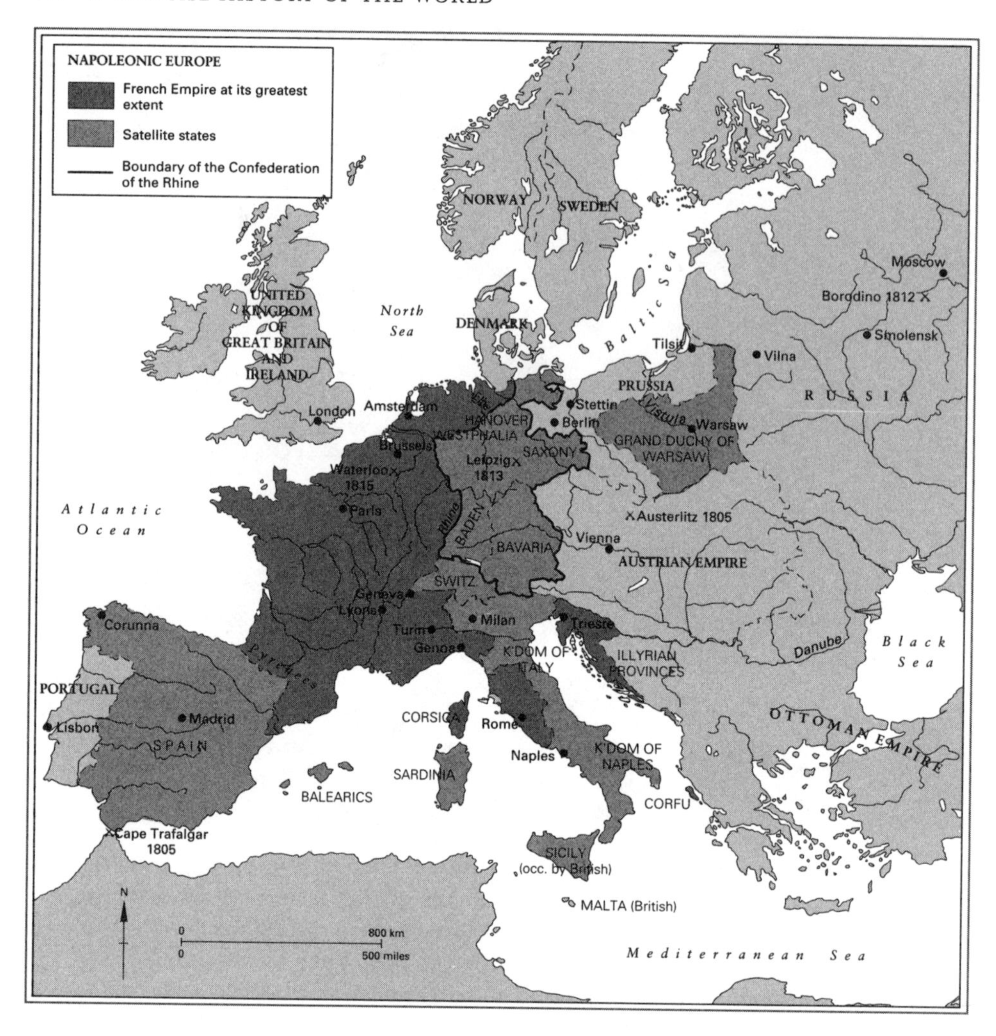

Russia (as his last descendant was still to be doing in the twentieth century) and among peoples ruled by foreigners – Poles, for example.

Finally, the Revolution also brought into question the place of religion in national life. Some thinkers of the Enlightenment had deplored the effects of religious dogma on law and government. Some French revolutionaries came in the end to see the Church as an enemy of the state. They could not tolerate the claim that the Church appealed to a higher authority than the nation itself. Later, the relations of Church and state became an issue in almost every country with a large Roman Catholic population.

Besides raising new issues, the French revolutionaries also changed ways of talking and thinking about politics. By making the degree to which you were for or against the Revolution the touchstone of political views, they helped to spread the notion that everyone could be placed on a spectrum running from extreme republican democracy at one end, to extreme supporters of absolutism at the other. It was assumed that your attitude to the Revolution or the old order as a whole would define where you stood on any particular issue (such as how many people should have the vote, whether you wished to confiscate church property or not, or even whether or not you believed in progress). This simple, two-sided politics of Right and Left fitted much of Europe for the next hundred years or so (though it never fitted British or American politics well – indeed, for most of the period since the French Revolution, it did not fit them at all).

Voltaire, notorious in his day as the enemy of organized religion and a hero of the Enlightenment, gets up, dictating busily to a secretary as he dresses.

Restoration after 1815

After the final defeat of France in 1815 restored much of the old structure, real political life did not exist in Europe except west of the Rhine and in a few small German and Italian states. There, some progress was made towards the winning of 'constitutional' government – the running of public affairs within the limits of constitutional laws which prevent the arbitrary use of power – and often towards some degree of representative government, too. Sometimes this was helped along by revolution (in Spain, parts of Italy and France, for instance), sometimes it progressed peacefully (in Great Britain the constitutional government which already existed was given a broader basis by enlarging the electorate in 1832 and by removing lingering restrictions on certain religious denominations). In all such countries, there was a growing sense that government ought to move in step with public opinion.

In much of Germany and the Habsburg empire (and some Italian states, too) things were very different. In part because of the wishes of their own rulers, in part because of the domination after 1815 of this area by a 'Holy Alliance' of Prussia, Austria and Russia, all fearful of the rebirth of revolution, the control of political behaviour was much stricter, constitutional government was much rarer and even quite elementary freedoms of speech, movement and political activity were hardly to be seen.

Nowhere did republicanism make headway before 1848, and no major European state was a republic when that year began. The old ruling classes, often still led by the great aristocratic families which had so long dominated Europe, still ruled much as before, too, though sometimes (and notably in Great Britain) they had made concessions towards sharing their power with men drawn from the gentry and middling classes. Working-class organizations had appeared, but if they were effective at all it was by winning specific concessions

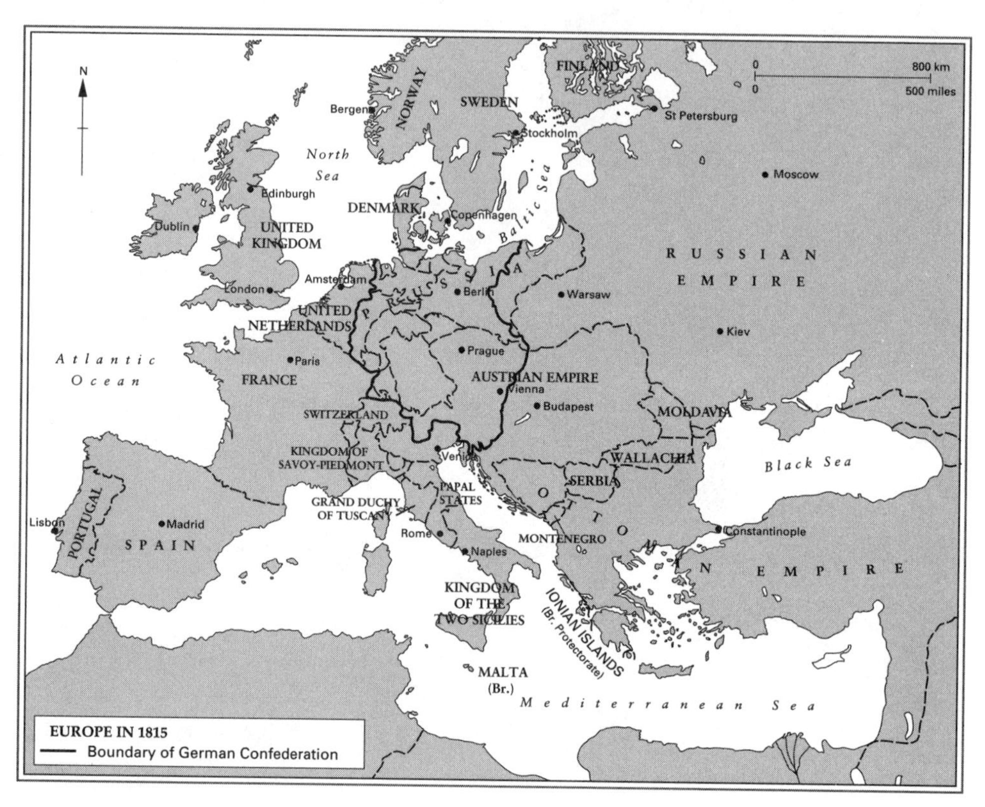

for their members rather than by transforming political arrangements. The biggest threat to the settled order in the 1830s and 1840s long seemed to be the English 'Chartist' movement (so-called because its aims were summed up in a 'People's Charter' to be presented to parliament), but though all its main aims except one were eventually to become law, that was long after the movement had faded away. Nowhere in 1848 was a majority of the population legally entitled to take part in politics by 1848; even the United States only gave the vote to free, adult males). Even before the 'Great Reform Act' of 1832 which increased the British electorate by something like fifty per cent, though, there were more than 400,000 voters in the United Kingdom; this was already more than the expanded electorate of France after a revolution there had installed a more liberal regime in 1830.

Yet for all the qualifications to be made, and although in fits and starts and revolutions and plots in some countries, it is broadly true that after the final overthrow of Napoleon in 1815 there was real progress in Europe towards more liberal and constitutional government, even if it was confined to a few countries. Then, in 1848, those on the side of progress everywhere experienced a tremendous surge of excitement as a wave of revolutions throughout continental Europe suddenly raised their hopes to a pinnacle never before reached and left hardly a government unshaken between the Pyrenees and the Vistula.

1848

Why this happened is still much debated. One or two facts are clear. The 1840s had been, on the whole, bad years for the European economy. A business depression threw many city-dwellers out of work. Harvest failures and bad weather in some countries produced starvation and near-famine from 1846 onwards. But such general explanations do not take us very far. What is much clearer is that once the revolutionary year of 1848 was under way, one revolutionary success made another much easier – there was a sort of chain-reaction.

The first revolution of that year came in Sicily, over a local grievance, the government of the island from Naples. It soon had wide repercussions outside Italy, yet the revolution which really mattered came the following month, February, in Paris. In 1789 revolution in France had eventually engulfed all Europe in war and even the 'July' revolution of 1830 had touched off others in other countries. As someone said, 'when Paris sneezed, Europe caught cold'. So, everywhere east of the Rhine and the south of the Alps, the French revolution was felt as a shock and an inspiration. Revolutions flickered across Germany; ministries fell and constitutions were conceded. In March came the greatest upheavals: revolutions in Vienna and in Berlin, the capitals of Germany's major states. The first drove the Habsburg Chancellor, Count Metternich – a man widely regarded as the very pillar of the conservative system of the Holy Alliance – into exile. Soon other revolutions occurred in other parts of the Habsburg empire – in Italy, Hungary, Croatia and Bohemia. More shocking still, a second great popular revolt took place in Paris in June. Yet it was ruthlessly crushed in a week of street-fighting not by a king but by the new French republic. This was the beginning of the turn of the tide. By the end of 1849, with the exception of France, where the new republic survived, and some Italian states which kept the constitutions their rulers had granted during the troubles, very little seemed to have been achieved. Slowly, conservative forces regained control. With the help of the Russian army (for Russia had remained unshaken and untroubled during the year of revolution) even the rebels of the Habsburg empire were brought to heel. The Pope went back to Rome.

1848–49: Major Events

1848

(Jan) Popular rebellion in Sicily, spreading through kingdom of Naples.

(Feb) Constitution granted in Naples. Revolution in Paris, Louis Philippe abdicates and Second Republic proclaimed. Granting of constitutions in Piedmont and Tuscany.

(Mar) Uprising in Vienna. Demands for Czech and Hungarian autonomy, Venetian and Lombard independence. Uprising in Berlin, King Frederick William IV grants constitution in Prussia. Other German states follow suit.

(Apl) Hungary separates from Austria within the Habsburg Empire. Constitution granted in Austria.

(May) Frankfurt Parliament, dominated by German liberals, opens debate on a new constitution for Germany as a whole.

(Jun) 'Pan-Slav' congress meets in Prague. Uprising in Prague crushed by Habsburg forces: first recovery of the reaction. Radical Parisian revolt suppressed in 'June Days'.

(Sep) Serfs freed in Austria.

(Oct) Insurrection in Vienna suppressed by Windischgrätz.

(Nov) Berlin occupied by troops and Prussian revolution ends.

(Dec) Abdication of Emperor Ferdinand of Austria. Franz Joseph succeeds him. Louis Napoleon elected President of France.

1849

(Feb) Proclamation of the Roman Republic; pope flees.

(Mar) Austrians defeat Sardinian army at Novara, Charles Albert abdicates and Victor Emmanuel succeeds him. Frankfurt Parliament completes constitution making and offers crown of a united Germany to Prussian King Frederick William IV – he refuses it.

(Apl) Hungary claims independence from Austria after centralist constitution is adopted in Vienna.

(Jun) Frankfurt Parliament (German National Assembly) forcibly dispersed by Prussian troops.

(Jul) French troops suppress the Roman Republic.

(Aug) Russian forces crush Hungarian resistance. Venetian Republic surrenders to Austrians.

The outcome of 1848–9

Yet the triumph of reaction was not the whole story. Broadly speaking, the demands of the various revolutions of 1848 had been of three kinds. In eastern Europe peasants rose in revolt to demand the abolition of compulsory labour dues and the feudal rights of landlords. They were after what the French had got in 1789 and had brought to some parts of Germany in the wake of the French revolution. Over the Habs-

Clemens Wenzel Lothar von Metternich, a Rhinelander who had never seen Vienna until he was 22, but who for thirty-three years after 1815 strove from there to conduct a war against revolution. He once said 'error has never approached my mind' and certainly believed it.

burg empire, Germany and much of Poland, 1848 now brought the end of feudalism and serfdom and this was an enormous advance. In what most educated Europeans would have thought of as the 'civilized world', bond labour could from this time only be found in Russia and the Americas.

The second sort of demand made in 1848 was, on the whole, made by middle-class liberals, intellectuals and professional men who wanted more constitutional and representative government and more jobs on the public payroll, at the expense of the old aristocracies. They for the most part did less well than the peasants in getting what they wanted. The reasons for this were complicated and varied from place to place, but one was that when revolution really began to get under way and look as if it was threatening the foundations of society and property (as it seemed to do in a 'socialist' revolt such as that of the 'June days' in Paris) the liberal revolutionaries decided that they had gone too far. They often rallied to the forces of order – which meant the old order of kings and princes, who recovered their nerve and used their armies to re-establish their power. Yet some constitutional improvements survived in Germany after 1848; the full oppressiveness of the Metternich age was not restored.

The liberal and constitutional reformers also failed because they were divided about another issue, the third revolutionary demand of 1848. That year was called 'the springtime of the nations' because many of its revolutions were in the name of peoples seeking to govern themselves instead of being governed by someone else; in particular, Hungarians and Italians tried to shake off rule by Austrian officials. Unfortunately, many of the patriots who fought for their peoples in 1848 were, just because of that, willing to fight against other peoples when they felt threatened by them, and some of their descendants have gone on doing so ever since.

NATIONS AND NATIONALISTS

Since the nineteenth century the idea of the nation has been treated with quite special respect. It was not a new idea. Shakespeare's plays are full of signs of an English sense of nationhood and pride in it, and there is much evidence that people were at an early date pleased to think of themselves as French, or Spanish. Such feelings nevertheless became much more widespread in the last two hundred years or so. More important still, people began to feel that the fact of belonging to a particular nation ought to mean also that they were governed only by men of the same nationality, that the state and nation should be the same thing, or two different sides of the same coin.

This is the political idea called 'nationalism'. It claims that nationhood is the only legitimate foundation for government. Like most other general statements about how governments should be organized, it has led to a great deal of suffering and violence. It is hard to see why unjust or bad government by people of your own nation should be thought morally better than just and good government by outsiders. But nationalism has undoubtedly had more success and has been a more revolutionary force than any other political idea. In the last two centuries it has transformed the map of the world and the lives of hundreds of millions.

Once again, the French Revolution is a landmark. In the first place the French revolutionaries themselves continually harped on the rights of nationality – the nation was sovereign in their eyes, and no higher authority existed. No later French regime ever went back on this principle, and propagandists of the Revolution preached it to sympathizers in other countries. In the second place, the French Revolution led to nearly a quarter-century of almost continuous warfare. The resulting upheavals, as frontiers changed, old rulers were set aside and new ones installed, and ancient institutions were uprooted, gave great opportunities to people to think they might set up new arrangements on the basis of the new principle of nationalism.

In this way, for example, Poles, obliterated as an independent nation in the eighteenth-century Partitions, began to hope that Napoleon might give them back their freedom. He did not (though he set up a feeble imitation of the old Polish state, called the 'Grand Duchy of Warsaw'), but brooding on the possibility was of great importance in keeping Polish patriotic feeling alive. In Italy, where the French armies from 1796 onwards overturned government after government, appearing to some at times as liberators, and to others at other times as oppressors, some began to think of themselves for the first time as Italians – rather than as Romans, Milanese, Venetians or anything else – and to seek to find ways of uniting the old patchwork of Italian states under a national government. Similar things happened elsewhere.

All this much worried European rulers after Napoleon's disappearance from the scene to exile and a lonely death at St Helena. Not all the governments overthrown during the previous twenty years reappeared: the ancient republic of Venice, so important in Europe's history for hundreds of years, survived until 1796, but was not resurrected in 1815. Instead, its former territory then passed under Austrian rule. Many sovereign princes in Germany, too, failed to make their reappearance, their domains being absorbed into those of bigger and

Makers of a new Europe: plenipotentiaries and ministers attending the Congress of Vienna which met in September 1814 and ended by agreeing its Final Act on 9 June 1815, a few days before the battle of Waterloo which ended the Napoleonic threat. The Congress effectively settled the frontiers of a new Europe, and thereby the framework and pattern of relations between the Great Powers until the Crimean War, as well as establishing roles and diplomatic practice on which were based international relations for more than a century.

luckier colleagues. But most of the kings came back and some showed very clearly that they would like to put back the clocks. Others were wiser and made concessions to changing ideas. The returning Bourbon king of France did not try to go back to the old order but accepted a constitution.

Yet to minimize the danger of revolution, the Congress of Vienna, which met to settle peace terms in 1815, ended up by giving chunks of formerly independent states in Italy to the Austrians. The Habsburg rulers were meant to act as policemen and keep the peninsula quiet. This was a blatant conflict with the nationalist principle. Once you admitted that nations ought to govern themselves, then there was no possible justification for Austrians governing Italians – or Poles, Hungarians, Bohemians, Slovaks and Ruthenes, to mention only a few of those ruled from Vienna. Nor was there any for Russians governing Poles, Ukrainians and Finns, or for Prussians governing Poles or Danes. Nationalism was especially likely to be dangerous, that is to say, to the three eastern European states which were the foundation of conservative order after 1815. France had no national problems, and England long pretended she did not (though Ireland really was one).

By and large, cautious police work and diplomatic cooperation, together with a readiness to act ruthlessly if need be, kept Europe at peace between 1815 and 1848, the longest period without war between major powers which Europe had known for centuries. Nationalism had only two real successes in this period, in the 1820s, when a rebellion in the European Ottoman empire led to the appearance of an independent Greece, and in 1830 when the Belgians threw off the rule of the Dutch government under which they had been placed in 1815.

Then came the 1848 revolutions. In them nationalism was inextricably tangled with other causes. In Italy, those who wanted constitutional government knew they could only get it if the Austrians did not interfere. But only by force could the Austrians be kept from interfering. Roman, Tuscan and Milanese liberals, therefore, tended to come together in attempts to organize national resistance whether they sympathized with the radical nationalists or not. This strengthened the claims of some who, like the enthusiastic conspirator Mazzini, saw the point of revolution as making a nation and did not worry much about liberalism or constitutionalism at all. In Germany there was probably more real enthusiasm than in Italy for a unity which would overcome the political divisions still separating Germans under different governments. But the cause of German unity was almost bound to make German liberals turn against the claims of Czech and Polish patriots within German-ruled lands. Fear of self-government in Bohemia and Poznan threw German liberals back in the end to dependence on the armies of kings (notably of Prussia), and as kings disliked constitutions and liberal principles, this led in the end to a sacrifice of liberalism to nationalism.

The monarchy most threatened in 1848 was undoubtedly Austria. She ruled a greater tangle of peoples than any other state. At one moment, the young emperor Franz Joseph who had only come to the throne that year had lost Vienna to revolutionaries, faced armed revolts in Hungary, Bohemia and Slovakia and the almost complete ejection of his armies from Italy; there seemed to be no escape from total collapse for the ancient monarchy. It was avoided only because the revolutionary nationalities fought one another, and Russia came to the emperor's help. Alone among the great conservative powers she was unshaken by revolution. Like London, Madrid and Istanbul, other capitals on the edge of Europe, St Petersburg had no revolution in 1848. So the Russian army could restore the old order again in Central Europe as the revolutionary wave ebbed. That happened in 1849, at the end of which almost all the pre-revolutionary regimes were back in place again. The main exception was France, where a new 'Second' Republic had replaced the constitutional monarchy. Its president had an ominous name: Louis Napoleon Bonaparte.

11
THE GREAT ACCELERATION

AN OPTIMISTIC AGE

Nineteenth-century European and North American conservatives often took a pessimistic view of the future. Nevertheless, since the Enlightenment educated opinion had tended to be tinged by greater optimism (a word which people only began to use in the eighteenth century). It may well be (though it is impossible to measure) that by 1900 most educated Europeans and Americans took it for granted that their civilization had been moving for about three hundred years along a path of progress and growing enlightenment. They saw the Renaissance and Reformation as the first big steps in shaking off the fetters of the past. Thereafter history seemed to them to be going one way; the growing mastery of nature by science, the rise of political institutions which took power away from kings and nobles and gave it to sensible, responsible citizens to control their own lives, the spread of literacy, obvious improvements in life and health for millions and many other changes all persuaded them in a vague but convincing way that the culture they belonged to pointed towards a better future for all mankind. What is more, they tended to think that things would go on like that. They thought, when considering the political world, for example, that there had been a growth in self-government and that this was good: on both sides of the Atlantic it could be seen, they sometimes argued, as one people after another shook off alien rule. What the Americans did in 1776 in rebelling against the British, what the Italians and Germans were thought to have done in the middle of the nineteenth century towards unifying themselves, and what the Balkan nations were doing to undermine Turkish misrule and replace it with their own at the turn of the century could all be seen as part and parcel of the same progressive movement. Some also thought that a struggle for the freedom of the individual conscience which had begun in the Protestant Reformation had opened the way to a general questioning of superstitious ideas, to the triumph of science and the sweeping away of outdated dogmas, though Roman Catholics would not have agreed.

Jeremy Bentham devoted a long life (1748–1832) to the promotion of ideas and schemes for reform reflecting what became known as the philosophy of 'Utilitarianism'. Its aim, in a phrase Bentham borrowed, was the promotion of 'the greatest happiness of the greatest number', and it was widely influential, in spite of much opposition, in the early nineteenth century.

This is a defensible view of what an eighteenth-century English thinker, Jeremy Bentham, called 'the climate of opinion' over much of the nineteenth century. The phrase is a handy way of designating the general intellectual tendency of what was going on – not so much its specific and individual theories, concepts, discoveries, but rather the context in which intellectuals thought and the world did its business, and the fundamental attitudes which underlay that. It draws attention to what is taken for granted as unquestionable. For the nineteenth century, it seems fair to say that the climate was increasingly broadly progressive and welcoming to innovation.

Even in 1900, though, a few people recognized that optimistic conclusions were not the only ones you could draw from history. Others were at least possible. It is now easier to see that they were right to be cautious, in the light of the hindsight which is always so helpful to the historian. Nationalism, for instance, which so many people applauded, depended on your point of view. It was not just a matter of long-established top dogs not wanting to give way. New national states often showed signs of behaving very competitively towards one another, as well as towards old opponents, and that might be dangerous to peace. As soon as one nationalism was satisfied others seemed to turn up; the Hungarians had got what they wanted from the Habsburgs in 1867 when the old monarchy turned itself into a 'Dual Monarchy', but well before 1900 they were being accused by their own Slav and Romanian subjects of oppression in their turn. And if it was alright to sympathize with nations who wanted to throw off the yoke of the wicked Tsar, did it follow that you should also support the efforts of the Catholic Irish to shake off British rule, constitutional and parliamentary as it was? And if you believed in Irish nationalism, should you back that of the Catholic Irish or of the Protestant Ulstermen? Overseas, too, other nationalist clouds were just beginning to gather. What should be the attitude of the European liberals to the national demands of Asians and Africans who might use independence to uphold old and backward social customs? And did not the well-being of European nations in the end rest in some degree on their colonial empires? Perhaps this particular aspect of progress and liberalism needed rather more careful scrutiny before one could be sure that it really did point towards a happier and better future for mankind. Some people, at least, thought so. These were questions which would force themselves forward much more terrifyingly in the twentieth century.

LIVING AND DYING

One source of pessimism at the beginning of the nineteenth century could be found in a book by an English clergyman, Thomas Malthus. Entitled, somewhat cumbrously, *Essay on the Principle of Population as it affects the future Improvement of Society* and published in 1798, it appeared to demonstrate that the world was, demographically speaking, a self-balancing mechanism. Population growth would always take place unless it was deliberately discouraged, and so, since the world's food supply was in the end limited, would always move relentlessly towards disaster: eventually, there would not be enough food; famine and disease (and perhaps wars over resources) would follow. Millions would die, down would come population until there was enough to eat – so that the whole cycle could start all over again.

Numbers

Yet it did not happen. The nineteenth century seemed instead to confirm a long-term story now traceable over thousands of years. It is simple enough: world population has gone up, and in recent times has done so faster and faster. In the nineteenth century it was to more than double: the last doubling had taken about four times as long. Moreover, since 1800 it appears at last to have grown without the occasional setbacks of earlier ages. Some countries, of course, grew faster than others; so did some continents. In Europe in 1800, France had more people under one flag than any other country west of Russia; by 1914 she was running in only fourth place behind Germany, Austria-Hungary and Great Britain. The United States was the fastest growing country of the lot in the same period; at one point in the 1840s its population

The problems of paternity: a Mr Quiverful grapples with the problems of the new census return in an English cartoon of the early nineteenth century.

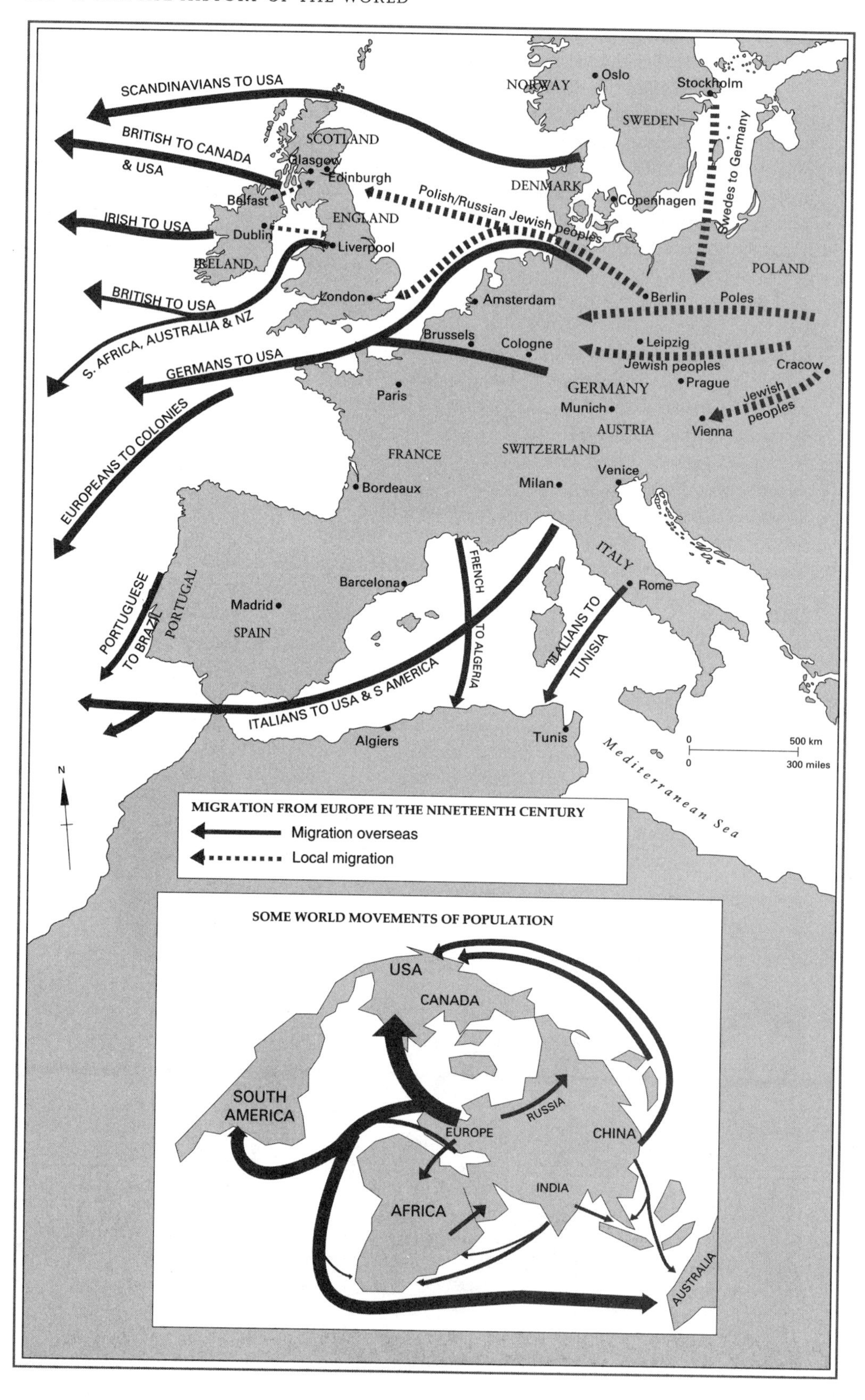
SCANDINAVIANS TO USA
BRITISH TO CANADA & USA
IRISH TO USA
BRITISH TO USA
S. AFRICA, AUSTRALIA & NZ
GERMANS TO USA
EUROPEANS TO COLONIES
PORTUGUESE TO BRAZIL
FRENCH TO ALGERIA
ITALIANS TO USA & S AMERICA
ITALIANS TO TUNISIA
Polish/Russian Jewish peoples
Swedes to Germany
Poles
Jewish peoples
Jewish peoples
NORWAY
Oslo
Stockholm
SWEDEN
SCOTLAND
Glasgow
Edinburgh
DENMARK
Copenhagen
Belfast
ENGLAND
Dublin
Liverpool
IRELAND
POLAND
London
Amsterdam
Berlin
Brussels
Cologne
Leipzig
Cracow
GERMANY
Prague
Paris
Munich
AUSTRIA
Vienna
FRANCE
SWITZERLAND
Venice
Bordeaux
Milan
ITALY
Barcelona
Rome
Madrid
PORTUGAL
SPAIN
Algiers
Tunis
Mediterranean Sea
0 500 km
0 300 miles
N
MIGRATION FROM EUROPE IN THE NINETEENTH CENTURY
Migration overseas
Local migration
SOME WORLD MOVEMENTS OF POPULATION
USA
CANADA
SOUTH AMERICA
EUROPE
RUSSIA
CHINA
INDIA
AFRICA
AUSTRALIA

equalled that of Great Britain but Americans had by 1900 spread all over the continent (much of which had still been unknown and unexplored in 1800), growing in the process to number 76 millions – a rise of over 1000 per cent since the beginning of the century.

There is better information about countries in Europe and America than for those in Asia or Africa, but population was rising everywhere. It looks, for example, as if China rose by over 40 per cent, to about 475 millions, while Japan's population went up from 28 to 45 millions, and that of India from 175 to 300 millions, in the nineteenth century. These were all very large increases.

World population had never risen so fast and uninterruptedly before, and because Europe for a long time grew much more rapidly than other parts of the world, the European share of world population (in 1900 about 24 per cent) went up, too. This high share (higher than ever before or since during most of the nineteenth century) helps to explain why Europe had so great an impact on world history. In this perspective, too, we may count as European not only those who lived in Europe, but those who left it to settle abroad – and their descendants. Without emigration, the population of Europe would have been fifty million higher still in 1914. Overwhelmingly, the population of the United States was by then of European descent, and Canada, South America, Australia, South and North Africa also had large European populations, speaking European languages, living (with suitable adaptations to climate) in European ways and often still thinking as Europeans, too. The late nineteenth century, when emigration was at its peak (about 1 million emigrants a year left Europe for overseas between 1900 and 1914) has sometimes been labelled 'the great Resettlement' of Europeans, and saw far greater migration than the big folk-movements of earlier history. It was also different in another very important way from almost every other such migration for much of it went not to civilized and settled areas, but to almost unoccupied, virtually deserted regions like the American West, the Australian Outback, or Siberia.

Life chances

One clue to the way in which population rose lies in the fact that in some developed countries in 1914 there were many more middle-aged and old people about; people were living longer. Most countries' populations in 1800 could have been represented in a diagram as somewhat flattened and tapering pyramids. Many children died very young in those days; infancy was dangerous and very many babies died within a year of their birth when their chances of survival were lowest. Childhood was a sickly time, too, but if you got through to your 'teens you had a better chance of survival (though still not one good by modern standards), and so on until you began to face the threats of old age. Similar population diagrams for 1914 would look very different. The sides of the pyramid had by then become steeper in most European countries, because people of all ages were living longer. That they did so was not the only reason populations grew, but it was an important one. If they lived longer, then obviously there would be more people to be mothers and fathers and so more children in the next generation. In this way the population rose.

In 1914 the richer, more advanced European countries showed this changing pattern most clearly (the United States was a special case because of its large numbers of young immigrants). The pattern was already tending to spread, too, from north-western Europe where it first appeared, to the south and east, though it had still a long way to go as was shown by differences between the number of years a baby at birth had a statistical chance of living in different European countries. An English baby born in 1914 already had a much better chance of reaching old age than, say, a Romanian baby (to say nothing of one in India or Africa). This would have been much less true a hundred years earlier and was one of the ways in which greater differences than those of earlier times were opening up between human life in different parts of the world; in earlier times people could expect much the same degree of hardship and hunger everywhere. What had happened to Malthus's prediction?

KILLING AND PRESERVING

Paradoxically, Europe and North America were also just the parts of the world where people were beginning to produce much more efficient ways of killing one another. The nineteenth century was a great age of military and naval technology. Gunpowder was supplemented by 'high' explosives: gun-cotton, lyddite, cordite, TNT. Muzzle-loading guns gave way to breech-loaders; rifles replaced smooth-bore cannon, providing faster rates of fire, greater accuracy and longer range. Repeating rifles enormously increased the fire-power of infantry and were themselves duly surpassed by the machine-gun. Naval ships and guns grew to huge sizes; submarines, mines and torpedoes all made their appearance by 1914. Yet warfare seems to have had almost no effect on European population history (the story may not be the same in Asia). In every war fought by Europeans before 1914 where there is reliable data, fewer soldiers were killed by the action of their enemies than by sickness.

Medical advance and public health

While war did not keep population in check, in many countries disease still did for most of the century. Science and technology were to change this, nonetheless, by finding out ways of saving life faster than they invented ways to destroy it and the nineteenth century also began, though only feebly, the era in which real victories over disease were won by the conscious application of scientific knowledge. The process had started tentatively much earlier, it is true, when Europeans had grasped that ships and sailors – though they did not quite know how – somehow carried disease about with them and had begun gradually to improve quarantine arrangements in ports. There had followed the virtual stamping out of plague in western Europe. Alarming outbreaks at Marseilles and Messina as late as the eighteenth century, did not spread like those of the 1660s (when England had its last bad attack). Early in the nineteenth century grain imported from the Black Sea and the Near East again brought plague with it, and it raged again in parts of North Africa and the Balkans still under Turkish rule, but without spreading further into Europe. An outbreak at Glasgow in 1910 produced only 34 cases – and only 15 of them were fatal.

The water-cooled Maxim gun, fully automatic, self-loading, and the model for most machine-guns in service down to the Great War.

Home: a Russian working class family in the corner which they rented in a St Petersburg room in the 1890s.

On the other hand, in this period other diseases were still causing havoc. Outbreaks of typhus, smallpox, dysentery and cholera occurred repeatedly, and over many decades. They may even for a while have got worse, in rapidly-growing new towns and cities. Probably as many people died of disease as of starvation in local famines like the terrible Irish famine of 1846, too. Yet in most west European countries such diseases were on their way to being mastered by 1900. This still left such notable killers of young children as scarlet fever, typhoid or diphtheria very prevalent, though.

Doctors could for a long time do nothing very much except recommend careful nursing in cases of infectious disease, but preventive medicine had already made one great stride forward in the eighteenth century with the discovery that vaccination and inoculation could immunize likely sufferers against some dangers. But prevention was also directed towards the places and conditions in which disease flourished, through an enormous and Europe-wide nineteenth-century effort to make town-life healthier. Huge efforts went into the provision of clean water supplies, the removal of sewage, the proper cleaning of streets, once it was grasped that these had a big effect on death-rates. By 1914 many cities were struggling to let air and light in to the over-crowded rookeries and slums at their hearts – 'slum' was a word invented in the nineteenth century – and began to regulate building so as to ensure that dwellings were at least tolerably well-lit, uncrowded and clean. European and North American cities were far healthier in 1914 – to judge by the length of people's lives and the decline of some diseases common to people living together in large numbers in the past – than the squalid rural life of eastern Europe and the Balkans. The terrible descriptions of the filth and overcrowding of, say, the English industrial cities of the early nineteenth century were written before such changes had really begun to produce an effect.

Much of the change was brought about by law, but some of it just happened: it was part of the 'spin-off' of rising prosperity and advancing technology. Cheaper good-quality building material – brick is the obvious example – made better homes possible, without the old timber, plaster, and thatch which provided nesting for rats, fleas and lice. Cast-iron pipes made it easier to supply running water and decent drains. Cheap transport – trains and trams – allowed people to live further from their work and so reduced the tendency to overcrowd the centres of cities. Many other such changes bore indirectly but importantly on public health. Hospitals changed, too; they ceased to be the terrible dumps for dying and derelict people which many had been in the eighteenth century. A whole new profession came into existence: the nurses, thanks above all to an Englishwoman, Florence Nightingale. Hospitals could not have been staffed without this new professionalization.

Only gradually, though, did medical science do much to increase chances of curing disease and injury. One outstanding contribution was made to it by a Frenchman, Louis Pasteur. He discovered a vaccine for rabies, made major investigations into the diseases of wine and beer, saved the French silk-growing industry from destruction by finding a way of attacking bacilli which infected silk-worms, produced methods of inoculating against diseases of cattle and poultry and, above all, he established the germ theory of the transmission of disease. It was through the study of infection that medicine was most affected by what he did. Pasteur made possible the work of the Englishman Lister, who went on to introduce to surgery the use of antiseptics, having established that the infection of open wounds during surgical operations could be prevented by the use of a carbolic acid spray. This dramatically reduced the death-rate during surgery and opened the way to other uses of antiseptics to reduce infection. Anaesthetics were another great step forward in surgery. They not only won a battle against mankind's old enemy, pain, but made possible prolonged and complicated operations which would have been impossible a few years earlier. Towards 1900, too, chemistry began to provide medicine with new weapons. New drugs made possible the selective treatment of diseases; they could be 'aimed', as it were, at specific targets. Other drugs were invented to control symptoms. It is very difficult nowadays to envisage what the world was like without aspirin, and few inventions have so obviously reduced human suffering.

By 1914, such advances had much changed prospects of life and death in advanced countries. Both sexes were far less likely to die of surgical injury or of infection by one of the diseases of childhood than a hundred years before, and women were less at risk in childbirth. Humans were becoming more likely to live longer and to have more chance to escape from pain. True, they faced new problems: to live long means encountering the special hazards and helplessness of old age. But it would be very difficult to argue that what had happened was not real progress. Even if it was progress confined to a few, comparatively rich societies which could afford such advances, the new methods spread world-wide. Medical knowledge could not be confined. Europeans took their techniques abroad with them, and they were by 1914 being employed in Africa and Asia, though they had not by then shown the impact they were later to produce in dramatic changes of population as they had done at home.

Contraception

Science also, paradoxically, had begun to put a brake on population growth by 1914 by providing new means of contraception. The effects first showed in a trend for the more prosperous classes to have fewer children, a part – though only a part – of the explanation of the narrowing of the base of the demographic pyramid in the most advanced and richer countries noticeable in 1914. The existence of fewer young people in relation to the numbers of middle-aged and elderly reinforced the effect of longer life in making population structures look more like somewhat bulging pillars. The total numbers did not go down as birth-rates decreased, because people were living longer. But the average age of the population went up. Frenchmen may have felt this – and worried about it most – thinking it showed that the French nation was in decline, and would soon not have enough soldiers to defend it. But birth-rates fell in other rich countries, too, often those which had shown the first rapid increases in population a few decades previously. Perhaps it is a demographic law that growing prosperity is first followed by a rise in population and then by a slowing-down of the rate of increase as the birth-rate falls. We cannot be sure, because many different forces – religion, social custom, and economic need are among the most important help to shape the patterns of population growth and population history; it is foolish to be dogmatic. What was clear by 1914 was that the better-off and better-educated were likelier to have small families than the poor, whether because they consciously put off marriage and so lived as married couples for fewer years during which the mother was fertile, or were prudent enough to restrict by one means or other the number of children they had within marriage.

FEEDING MANKIND

Science was nevertheless only one of several reasons why human numbers shot up, and not the main one. The basic explanation was that the world was getting richer. For such a jump to take place, there had to be more food to go round; by 1914 it was available as never before. The output of world agriculture had enormously increased in the previous century, and at a rate surpassing the more gradual growth of earlier times. To that extent it was like the population increase, but just as population did not grow at the same rate all round the world, nor at a smooth, even rate, neither did food production and consumption.

More food was certainly grown in Africa and Asia as time went by, but by 1914 it was not true that everyone got much more to eat than before. The usual diet of the Indian or Chinese peasant can have changed hardly at all for centuries; there was a little more to go round in their countries, but many more mouths among which to share it. Nevertheless, the rise in the output of food was the most fundamental addition to the wealth of mankind during the century before 1914. There were better statistics to measure such changes than ever before. Agricultural production grew at an unprecedented and dramatic rate. Much of it took place in new cultivated lands – Argentina, Canada, the United States – where huge surpluses to local needs were grown for export. But productivity, too, had risen. It has been calculated that it took 373 hours of work to produce 100 bushels of wheat in the USA of 1800; a hundred years later, only 108 hours were needed. Other calculations suggest that between 1840 and 1900 productivity on the land increased by 190% in Germany, 90% in Switzerland and 50% in Italy, to cite European examples. The basic source from which most people have always got food is grain, in whatever local form it was available. Grain production is fundamental to food supplies. Between 1851 and 1913 Germany, for instance, nearly trebled her production and Hungary pushed up hers nearly fivefold.

The mechanical reaper was one of the major developments of nineteenth century agricultural technology. By the end of the century giant machines like this were at work in the vast cornlands of the American and Canadian prairies.

This further huge increase in the available supply of calories was added to by the meat producers. The European cattle herd rose steadily during the nineteenth century to supply this consumption, and so did the numbers of sheep and pigs, but the fastest growth in livestock farming took place overseas, in the Americas, North and South, Australia and New Zealand. For centuries meat had been a luxury: now, in the nineteenth century, it became much more common, not only on the hooks at the local butcher's, but in prepared forms – tinned, or processed into essences and extracts, for example.

Nor was it only the production of traditional food which rose sharply; quite new foods began to be eaten in some countries, while others which had once been luxuries became almost commonplace. Even in the eighteenth century, sugar had replaced honey as a sweetener for the better-off European, but in those days, sugar was still a colonial product and relatively expensive. In the nineteenth century, much more sugar was imported to Europe from overseas while huge quantities of sugar began to be made there from beet. The result was a huge rise in consumption. Prices came down to a level at which sugar could be taken for granted as an everyday commodity, a notable change in European diet. Tea and coffee also became common, to say nothing of the exotic fruits which were available in greater quantities because of technical advances.

Some Innovations in Western Agriculture

17th century	Potato introduced into Europe. Norfolk crop rotation becomes widespread in England.
c. 1701	Jethro Tull developed the seed drill and the horse-drawn hoe.
1747	First sugar extracted from sugar beet in Prussia.
1762	Veterinary school founded in Lyon, France.
1785	Cast-iron ploughshare patented in England.
1793	Invention of the cotton gin by Eli Whitney.
1800	Power-driven threshing machines developed in England.
1820s	First South American nitrates for fertilizer imported to Europe.
1830	Reaping machines developed in Scotland and the US.
1850–1890s	Major developments in transport and refrigeration technology.
1892	First petrol-driven tractor in the USA.
1938	First self-propelled grain combine harvester in the USA.

Agricultural change

Why what people in rich countries ate improved in quantity as well as quality has, once again, no single, simple explanation. It was the outcome of several processes. One was the agricultural 'improvement' whose roots lay back far before 1800 and there is no point in trying to pin it down too closely in time: an 'agricultural revolution' could be said to begin in England somewhere around 1690-1700, in the United States about ninety years later, in Germany a few years later still, and in Russia not really until after 1860. Though it spread eastward through Europe in the nineteenth century, even a century earlier European landowners had customarily come to England to look for tips, to buy livestock and machinery and to seek advice. When they returned to their own estates and tried to put into practice what they had seen they did not always succeed in doing so very quickly. In the middle of the nineteenth century much of (say) Germany and France still looked much as it had done for centuries, though yields on the best farms had slowly begun to rise even before 1800. Altogether then, it is best not to be dogmatic about dates and safest to recognize how patchy and uneven, but also how sweeping in the end, this great change was.

Some contributors to the change were often very noticeable, though, because they did take place fairly suddenly. What the French called 'feudalism' (but was less neatly defined as a whole set of traditional practices and rights which encumbered the free exploitation of land) was legally abolished by them in 1789; it took about the next half-century for similar change to spread through the rest of continental Europe east of Russia. When in 1861 the Russian government at last decided to abolish serfdom, the era of rural European history which had begun with the appearance of the medieval manor was at an end. Henceforth, everywhere in Europe agricultural labourers worked for wages or on their own property, and the full stimulus of self-interest could operate in farming. This encouraged investment, new methods and the bringing together of the little traditional strips and plots into larger, more efficient units.

Another set of reasons for a revolutionary increase in world food supplies lay in the technology which made possible the opening of new lands overseas. It was not just that the world's cultivatable land suddenly and dramatically increased in acreage, though that was true: it became possible to tap the North American plains, the pampas of the South American states and the temperate regions of Australia in the second half of the nineteenth century as never before. Farmers came to them because settlement and earning a livelihood were both made possible by a revolution in transport. As steam-driven railways and ships came into service in growing numbers from the 1860s onwards down came transport costs. Food from these areas became cheaper. As demand for it grew, more people sought to exploit the untapped reservoirs of virgin land. The same thing happened less dramatically in eastern

Europe; as railways were built in Russia and Poland which could bring grain to the central European cities, and as the Black Sea ports began to export more Russian grain in steamships, the effects on the grain-growing regions were dramatic. Those on distant lands became even more so as changes in food-processing (canning, for example, or the invention of the refrigerator ship) made stock-raising more profitable than ever before.

For some European farmers, this could be a disaster. They could not compete with the new cheapness of imports and in the last few decades of the nineteenth century that showed everywhere. In some countries the cultivated acreage dropped dramatically, in others, producers changed to specialized farming. Dairying underwent a huge transformation, especially in Denmark (which also did wonders for and with the pig). Market gardening and specialised fruit-growing provided farmers elsewhere with escape routes from cereal disaster. Almost everywhere in western Europe, though, weathering the depression years of the late nineteenth century required drastic adaptations. A shrinkage of the numbers of those depending on agriculture for a livelihood began.

Ecological change

Agricultural revolution often had biological results going far beyond the life of humans. Plants were moved about to become acclimatized in new parts of the world; mankind was again interfering with natural selection and, of course, that was not new. The introduction of silk to Europe, or the sweet potato to Africa had been earlier examples. But now larger populations and new industrial processes drove forward such changes on a far larger scale. Rubber plantations appeared in Malaya in the early years of the twentieth century after trees were brought from South America; tea began to be grown on a large scale in Ceylon and east Africa; vines were taken from Europe to California and South America. Animals migrated, too. Much earlier, Spaniards had brought the horse to the New World; in the nineteenth century much attention was given to the selective breeding of cattle suitable for non-temperate climates. It helped to provide the huge herds of the Americas and southern Africa, while Merino sheep flourished in Australia. Not all such transplantations were fortunate. The four pairs of rabbits brought to Australia in the 1850s became a plague within a few years. Indigenous animal species suffered from the depredations of those who exploited them for meat or other products. People were slow to wake up to the ecological danger of indiscriminate slaughter. Some appalling results followed, the virtual extinction of the American bison by the hunters who fed the railway construction gangs, for example, or the ravages inflicted on the seal and whale populations of Antarctica and the southern oceans in the early nineteenth century.

Agriculture was only one part of a great revolution in the exploitation of natural resources, but it was the most fundamental. In the early twentieth-century most human beings still got their living by working directly on the land. For a few of them, though, living in the countries of the European world, another change was possible, that to an economic life based on industrial production, a change possibly the most important in the history of mankind since the invention of agriculture itself, or even since the discovery of fire. But it could only come about because there was more to eat than ever before. The agriculture that provided for that was one quite different from that which had held humanity in check for so long by its inability to increase output more than marginally. One of humanity's oldest activities, food-production, had ceased to be a check on history's acceleration and became much more one of its propellants.

THE NEW FACE OF INDUSTRY

Another old phrase (in this case, nineteenth-century) which is still familiar today is 'industrial revolution'. It was coined by a Frenchman to label one of the great social changes he saw going on around him. By it he meant the transformation of society, the production of man-

Coalbrookdale, Shropshire, in 1758. Fifty years after the first smelting of iron with coke instead of charcoal, the appearance of this early industrial centre was still remarkably rural.

ufactured goods in larger quantities and on a larger scale than ever before. This relied on the gathering together of large numbers of workmen to do it, and the use of power-driven machines in ever-growing numbers. Ever since, the word 'industry' has, if not exclusively, usually been used to mean large-scale manufacturing.

Even so, in 1914, not more than a few countries could truly be called 'industrialized' (another nineteenth-century word). The most obvious were Great Britain (the outstanding example, where by 1901 less than 10% of the work force was still engaged in agriculture), the United States (with the biggest overall output) and Germany. Among smaller countries, Belgium stood out sharply, too, while several European countries had biggish industrial sectors – France, Italy, Russia (growing fast) and Sweden, for example – though often such industrialization was localized or specialized. Spain, for instance, had important textile factories in Catalonia, and some mining and steel-making towns in the Asturias and Biscay, but few factories anywhere else.

People quickly noticed that the coming of industry could change the whole pattern of people's lives. In the early stages of the process, it sometimes did so pretty harshly. Artisans working from their homes for small markets were often driven out of business. When and where textile manufacture was reorganized on the basis of factory production weavers and knitters who worked at home could not compete with the cheaper goods made possible by machines and bigger markets, and so they had to give up and find – if they could – work in a factory. Here was another change: the gradual emergence of a society in which most people in work earned wages in manufacturing or in the businesses which sprang up round it. They could not always do this, because there was plenty of cheap labour about. Wages could be kept low and profit high, because there was always a pool of unemployed.

Urbanization

Another change was a transformation of town life. Many more people and a much greater proportion of total populations in many countries lived in towns as the years passed. It is hard to compare countries directly, because definitions varied, but it gives some idea of what was happening to note that while roughly 16% of Englishmen lived in towns in 1801, ninety years later the figure was over 53%. In all the industrializing regions, towns and cities grew in number and size. Between 1800 and 1910 Berlin grew some tenfold in size (to over 2 million inhabitants), Vienna about eightfold (to roughly the same size), and London nearly sevenfold (to a monstrous 7.2 millions). Sometimes such expansion led in the end to old neighbourhoods blurring into one another in that it was impossible to see where one began and another ended. This happened in places which came to have a quite different landscape within a half-century or so – the English 'Black Country' or West Riding of Yorkshire, for example, or the German Ruhr. Old-established ports, like Hamburg, Marseilles or Liverpool, also showed rapid growth as international trade prospered.

In the first half of the nineteenth century, governments tended to think they should not interfere much with a process which was going on unplanned as a result of thousands of individual decisions by individual inventors, manufacturers, builders and businessmen. The results were often grim. Places like Manchester which had been small country towns a few decades earlier faced enormous growth with no public resources. Even in the 1850s, a male child born there only had a life-expectancy of about 25 years. Builders put up rows of back-to-back cottages opening on streets where the only drainage was a stagnant ditch in the middle of it. Not only unpaved, the streets were also unlit and uncleaned. In continental cities big tenement blocks could be found in which people lived 10 and 12 to the room.

An awakening to the needs of health and sanitation was slow in coming not merely because of a lack of knowledge and resources, but because those who influenced affairs and opinions tended to agree that to leave economic life alone and let the market regulate the way people lived really was the best way to ensure growing wealth for everyone. This was the period when what were called the '*laissez-faire*' ideas of the eighteenth century had their greatest effect. It was only in the second half of the century that reforming mayors like Joseph Chamberlain of Birmingham (later prominent as an imperial expansionist) and Karl Lueger of Vienna (who owed much of his support to anti-semitism) began to urge that public authorities (in their cases those of local government) should own and run such services as water-supply or transport.

'*Laissez-faire*' also left working life as hard as ever. Hours were long in factories, discipline strict and wages low. Where possible women and children were employed, because they cost less than men. Employers called on the law to help them maintain this state of affairs – for example, by forbidding workmen to form trades unions in their own defence, and by persuading public authorities that strikes were subversive and threats to social order.

The horrific conditions of some sides of early industrialization – which, it turned out, would be replicated as one country after another became industrialized – have made some people think that its onset was a bad thing for everyone except the few who did well out of it. But many of those in the early generations of town-dwelling factory workers came from

The soup-kitchen of the Victorian slum replaced the monastery's distribution of its broken meats in the Middle Ages. Once destitution was identified as a responsibility for society as a whole, the first relief measures sought to improve, rationalize and enlarge the traditional devices of charity.

A new scale of industrial organization: the big iron looms of an English textile mill in the1840s, centrally powered from one engine-house.

poverty-stricken villages. If they were in work, they might well live better on their wages than as farm labourers. Child labour and women labour were familiar in the countryside (and still are today, as much of Asia, Africa and South America show). In the long run, too, it is obvious that the wealth of the industrial nations must have been of some benefit to their citizens because they lived longer and grew in numbers. However, as someone remarked in another connexion, 'in the long run we are all dead'.

Socialism

Some men were far-sighted enough to see that what was going on in the process of industrialization was so unprecedented that it required quite new ideas to understand it and new programmes of action to remedy some of its consequences. One way to meet the second need, many thought, was by the direct organization of working men into trades unions whose collective strength – resting ultimately on the withdrawal of labour in strikes – would force employers to make concessions of better wages or working conditions. In every European country and in the United States there was prolonged resistance by the employing and wealthier classes to such organization, often with the aid of the law and police. But in every industrial country by 1900 it was also true that hundreds of thousands and in some countries millions of workers had formed trades unions which had already had considerable success in improving their members' lot.

At the same time, intellectual and political criticism of society based on *laissez-faire* principles had also led to the emergence of varied doctrines which were loosely lumped together under the label 'socialism'. The thinkers who advocated such doctrines shared a detestation of what some of them saw as conscious exploitation, and some as the inevitable, unavoidable working of a capitalist, market-dominated system. Some looked forward to struggles to overthrow injustices and inequalities, some, more complacently, to the inevitable march of history which, they believed, was on their side and would eventually see capitalism replaced by more rational and equal ways of sharing the huge potential wealth of the industrial world.

In the 1870s England's grip of the landlord was still very strong. These labourers and their families are standing outside the cottage from which they have just been evicted for membership of a trades union.

Among them, the writer and thinker who was to have the greatest single influence and fame was a German philosopher, Karl Marx.

Although he appears to have come to deplore it, what was called 'Marxism', and purported to be drawn from his writings, came to dominate European socialism by 1900. What he certainly asserted – and what gave his disciples great confidence – was that the irresistible tendency of history was to create, in the industrial proletariat, a class whose exploitation by the property-owners would inescapably lead to social revolution, the overthrow of capitalist oppression and, in the end, to the institution of a rationally ordered society where human beings could at last be truly free.

Marxism came to operate for many people virtually as a religious faith, nerving them to discipline and action such as had been made possible by earlier creeds, while appearing to fit the materialist, 'scientific' ideological current so apparent in an age when religious faith seemed to be on the ebb. Marxists also took up the inspiring myth of the French tradition of popular revolution which was identified by many with struggles of political rights and democracy. By 1900, there even existed an international organization of Marxist political parties and working-class groups, the Second International, which confidently looked to an early transformation of society. Curiously, the evidence was already strongly against such outcome. By the end of the century no even briefly successful popular revolution in a major state had taken place for decades, and growing prosperity, government regulation and the beginnings of welfare provision had already begun to improve the workers' condition in some countries, however depressed it must seem in retrospect to any modern observer.

This was one of the long-term results of industrialization which can be discerned as the twentieth century opens. More than ever before, the economic map of the world was a unity in 1914 although, with a few exceptions – mainly in India, China and Japan – industrial development was then still confined to Europe and North America. It was one more instance of the way in which these parts of the world had a historical destiny not only different from the rest of the globe (as had, in one way or another, long been true), but one different in a

Mr and Mrs Karl Marx.

new way: they were materially much richer. This was the most striking change in Europe since the barbarian invasions (and incidentally affected the actual appearance of the landscape much more).

Society in the industrial age

More important, though, it was a turning-point in the history of mankind. Within 150 years or so it turned societies of peasants and craftsmen working by hand into societies of machinists and book-keepers. It ended for millions of people the sense they shared (like all their ancestors) that life was dominated by agriculture and its rhythm set by the farming calendar, or even by the rising and setting sun. Outside Europe, it transformed many more millions of lives by enormously increasing the demand for the raw materials needed in growing quantities by the industrial nations. This was not by any means entirely to the disadvantage of people in the countries which supplied 'primary' commodities (that is, those needed by the industrial countries), but they usually benefited much, much less than the inhabitants of nations economically more developed.

In many ways, those nations showed the growing difference in wealth separating them from other parts of the world. Everywhere in Europe in 1914, for instance, public transport was better than a hundred years earlier, it was easier to get about, medical attention and education were more plentiful, and shops to distribute goods were more common. Such facts were part and parcel of the same increase in wealth which made it possible for Europeans, and the peoples of European stock overseas, to eat more than the inhabitants of other parts of the globe. Their standards of living rose in many ways; Europeans and their overseas cousins did better than anyone else out of the fact that the world was growing richer faster than ever before. This great expansion in wealth, therefore, increased inequality between different parts of the globe. Yet very few people – if any – could have envisaged bringing such a change about. It was like other great historical upheavals; its effects went far beyond what most of those who brought it about could have dreamed.

The most obvious and most easily measured change is one of sheer industrial output. One or two commodities were especially important, and they provide good indicators of what was going on. Coal supplied most of the non-muscular power needed for manufacturing and transport in this era, either directly through the chemical operation of its heat (as in smelting) or indirectly through steam and, later, through electricity. Coal output in the United Kingdom rose from 11.2 to 275.4 million tons a year between 1800 and 1910. More startlingly still, the United States put up its annual output from 30 to 474 million tons between 1850 and 1910. As for iron, the material basic to any increase in the number of machines – and therefore to all sorts of manufacturing – as well as, increasingly, for building, approximate production figures are given below:

Pig Iron Production – thousands of tons

	UK	Germany	France	USA
1850	2,716	245	561	564
1900	8,778	7,92	2,665	13,79
1914	9,792	14,836	4,664	30,970

Steel was above all the key industrial material for industry because its greater hardness made it superior to iron. For a long time it could only be made in very small quantities, but improvements in method made it possible to make it much more cheaply in the later nineteenth century. This is what happened in consequence to output:

Steel Production – millions of tons

	UK	Germany	France	USA
1890	3.60	2.89	0.77	4.30
1900	5.04	7.71	1.70	10.40
1910	6.93	16.24	4.09	26.50

This emphasizes the sheer scale of the new wealth of some countries and therefore, again, the widening of the gap between rich and poor. It shows also how countries' relations with one another changed dramatically. Great Britain, long in the lead, had been the first country to experience industrialization. In the middle of the nineteenth century she was the 'workshop of the world', as her citizens liked to remind themselves. By 1900, though, she was no longer in front: Germany was in many ways ahead of her. But even then, the overall world manufacturing supremacy of the United States was becoming obvious.

Isambard Kingdom Brunel, cigar in mouth, greatest of Victorian engineers, was a Portsmouth boy. After attending a famous Parisien lyceé, he made his reputation as a bridge-builder and railway engineer. England is still studded with his achievements. Here, he was viewing his last ship, the Great Eastern, whose trial voyage began only a few days before his death at the age of fifty three

Such figures for centrally important commodities also suggest how complicated the structure of industry rapidly became. Coal made steam railways possible. Railways needed iron rails and so encouraged the building of more iron works. These increased the demand for ore (which could be brought much further and more cheaply to the works by rail than by horse and cart). More miners were needed. Their wages made it possible for them to buy more clothes. A bigger European demand for textiles led to a growth in the production of fibre crops (cotton, wool) on other continents. It also led businessmen to put newer machines into their factories. Which required more iron .... And so on, and so on.

GLOBAL COMMERCE

This process came in the end to involve the whole globe. Europe's factories sucked in raw materials – cotton, jute, timber, minerals – from overseas. By 1850 more than half the wool used in English mills came from Australia; France was taking more than half of hers from outside Europe, too, by 1914. Sometimes raw materials which had long existed were suddenly discovered to have individual use – rubber, towards the end of the nineteenth century, was an outstanding instance and this transformed economic life thousands of miles away from where the rubber was used. As more such materials, as well as food and commodities like manufactured goods, were moved about the world, there was a dramatic increase in world trade. The greatest trading nation of all was Great Britain, and the total value of her exports and imports together rose from about £55 million a year in 1800 to over £1400 million in 1913.

For the first time, there was a true world-wide market. People could buy and sell round the world as they already bought and sold inside one country; by 1914 virtually all mankind directly or indirectly formed part of the same great trading community, whether they knew it or not. The price of grain in Chicago, of meat in Buenos Aires, or of steel in Essen could cause changes in other prices right round the world. The first world market with world prices showed that at least in an economic sense 'One World' already existed, its completion having become possible when China, Japan and Africa were at last fully opened up to trade with Europe and America in the nineteenth century. It rested on long-established credit and exchange arrangements of which the most important was a system of paying for goods anywhere with bills 'on' (that is, to be settled by) European bankers and merchants. This system was the descendant of the credit transactions which had begun in the Middle Ages between, at first, a few big European commercial centres. In its fully developed form it was centred, above all, in London, which was by 1914 the centre of the world network of trade, and it concentrated financial institutions as did no other city. Paper – in the forms of bills authorized for settlement, bank notes, cheques kept the whole system going and this paper was always redeemable in credit for other goods or paid off in the last resort in gold. All civilized nations based their currencies on gold. They therefore did not much fluctuate in price. With a bag of gold sovereigns or twenty-dollar or hundred-mark pieces you could go anywhere in the world and pay your way. This made international business very easy. International trade was based on this international 'gold standard'; merchants did not need to make guesses about how much a currency would be worth in a few weeks' or months' time.

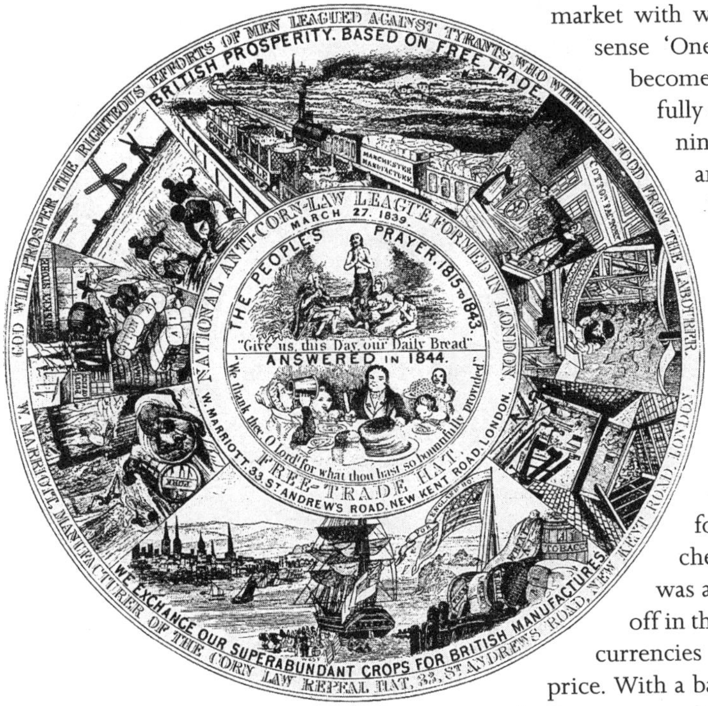

From the inside of 'the FreeTrade hat', a graphic representation of the creed of FreeTrade which for some Victorians came to be revered almost as much as the teachings of Holy Scripture themselves.

It was true that (for what they thought were good reasons) different countries sometimes interfered with the flow of trade across their frontiers. During the 1880s and 1890s, when there was a widespread slump, some governments tried to protect their own manufacturers and farmers by imposing 'tariffs' of duties on foreign goods. Great Britain was virtually the only big nation which refused to do this, and stuck to the 'Free Trade' practices which (it was believed) had done so much to make her a great trading nation and gave her cheap food. But even the tariffs of the 1890s still left plenty of elbow room for merchants to do international business.

Within the world market different regions of the world tended to play different roles; Europe was, on the whole, an importer of raw materials (both food and industrial requirements) from other continents, and to export manufactured goods in exchange. In this sense, Europe was the power-house of world trade. It was the growth of European population, its increasing wealth and its ever more voracious factories which sucked the huge quantities of food, minerals, timber and manufactured goods from one country and one part of the world to another. Right up to the 1860s, Great Britain still produced the majority of the wheat and meat she consumed, but by 1900 80% of her wheat and 40% of her meat was imported. But manufacturing countries were one another's best customers nonetheless, and great quantities of goods were exchanged between European countries and between Europe and the United States (which also, of course, supplied much farm produce). By 1914 Europe took over 60% of the world's imports and supplied about 55% of its exports.

Europe also exported capital to other parts of the world, usually in the form of a loan, or the value of money in goods, which could be used to buy needed materials or to pay labour for agricultural or industrial development in the country importing the capital. In this way, many of the railways of the United States and South America were built, mining was expanded in Africa and tea and rubber plantations were set up in Asia. The interest on the capital was often paid with the profits from the enterprises thus launched. As time passed, this led to irritation. It looked more and more as if Europeans – or, rather, European banks and merchants – owned too much of the business of non-European countries, because they depended on European capital. Local businessmen in particular tended to think that it was unreasonable that the profits of these countries went to the benefit of Europeans.

Income from such overseas investment made up a big part of the earnings of some European countries and to no nation were earnings of this sort more important than to the British. Huge sums from these and other 'invisible exports' – not only dividends from overseas investments, but charges for shipping, insurance, and financial commissions of all kinds – were needed in order to balance the British imports and exports accounts. Only the earnings of invisible exports made it possible for the British people to pay for the imports which enabled them to enjoy a high standard of living. This was one reason why British governments were always keen to preserve international peace and normal business conditions; more than any other nation, the British depended on selling huge quantities of goods on other people's behalf, re-exporting much that they imported, and on the freedom of their merchant ships to plough the seas and that of their bankers and insurance brokers to take calculated risks abroad. The results of this very complex system went far beyond the merely economic. Bringing down prices and stimulating innovation and investment advanced the sort of civilization which Europeans had created. In many ways it made the world a better place. But it was also true that when everyone's affairs were more closely tied up with everyone else's, a glut of grain in America could ruin European farmers and the collapse of a bank or merchant house in London could throw people out of work in Valparaiso or Rangoon.

THE BRITISH LION IN 1850;

OR, THE EFFECTS OF FREE TRADE.

A certain complacency appears in British public life during the prosperous mid-century.

The ups and downs of trade – the patterns of boom and slump called the 'trade cycle' which were first pointed out in Europe in the early nineteenth century – began to have a world impact as time went on. In the 1870s, there began a long slump (sometimes called 'the

Great Depression') which affected most of the richer countries. The search for 'protection' through tariffs was always liable to undermine the great world trading system. Yet it was strong enough to weather the storms until a mortal blow was given to it by war.

Even after that, though, many people assumed that the old international world of trade was a natural state of affairs to which one day they might reasonably expect to return. It had been such a success that it had come to be taken for granted and was not seen as the extraordinary achievement it really was.

A NEW AGE OF MACHINES

In the nineteenth century, new machines began to appear everywhere in large numbers. Though they became more and more complicated than earlier ones, they often seemed easier to use. They were to be seen everywhere in Europe and North America, in the streets of the major cities (cars, trams and bicycles), in factories (looms, lathes, drills), in offices and shops (cash-registers, typewriters). They transformed life in many ways.

First and most obviously, they enormously increased the value of labour: with their aid a worker could produce much more quickly. This was an essential contribution to the huge growth in the output of wealth of the age. The results could even be seen in the countryside. Soon after the nineteenth century began, English agricultural machinery was being shown at European fairs, and by the middle of the century steam was being used to drive machinery and drag ploughs. 20,000 mechanical reapers on German farms in 1880 had become over 300,000 by 1907. Many similar figures could be cited. By 1910 the petrol-driven tractor had made its appearance.

Even more obviously than agricultural improvement, manufacturing rested on the coming of better, cheaper, more ingenious machines. They directly provided the tools needed to make things – lathes and other machine-tools, drop-hammers, blast-furnaces, boot-making machinery, and a thousand and one other examples could be cited – and saved labour in other ways less directly connected with manufacturing itself. The electric trams and underground trains, for example, which were by 1914 taking millions of people in many cities to work, were conserving energies that would have been used fifty years earlier in walking long distances, as well as helping to shorten the time that had to be spent earning a living.

Even in the home the effects of machines were very great. The supply of gas for cooking (from the local gasworks) enormously reduced the trouble and expense involved in getting fuel into and around the house. Piped water (from central waterworks) was available in millions of homes by 1914, and the effect of this can easily be grasped by anyone who has seen the procession of village women in parts of southern Europe to the local stream or well to draw what they needed for their family tasks (in Asia or Africa, of course, the sight is even more common). The sewing-machine actually changed the production of goods in the home, but in the houses of the American and European rich there was also to be found other household machinery – the first 'vacuum' cleaners, electric 'lifts', washing-machines, and at all levels of society that fine old standby which had taken so much effort out of drying and pressing clothes, the mangle.

Just as it is hard to list chronologically the most important mechanical innovations of the century because the interrelation between them all is so complicated and rapid, so it is difficult to trace the overall effects of the coming of machines on the way work was organized and occupations were shaped though it can be done (at some length) for any particular trade. A whole range of new tasks appeared. 'Engineer' was an old word, but its meaning was extended much further in the nineteenth century. Specialization appeared within engineering, in construction, in electrical work, in ships and in chemicals. Institutions of technical education had appeared in many countries to give advanced instruction in engineering and

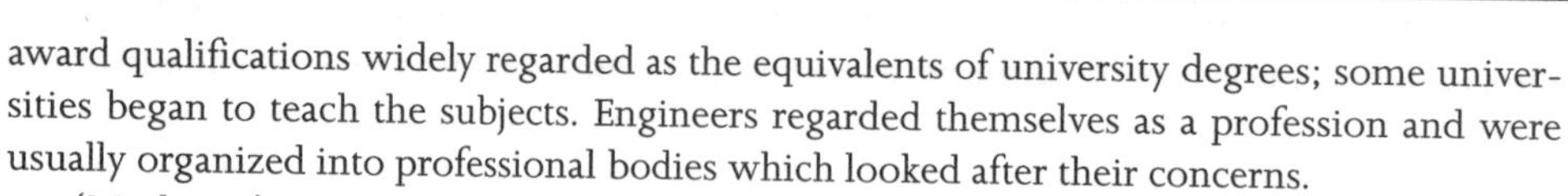

The key to the cotton mill was centrally-supplied power whether from running water or steam, usually as in this example, by driving bands from overhead shafts to the machines.

award qualifications widely regarded as the equivalents of university degrees; some universities began to teach the subjects. Engineers regarded themselves as a profession and were usually organized into professional bodies which looked after their concerns.

'Mechanic' was another word which came to have a new meaning: a craftsman specializing in the handling of machinery. The number of such skilled workers grew at a huge rate in industrialized countries. Mechanics, too, were becoming specialized by the end of this period, and their trade qualifications were increasingly defined by standards set by apprenticeship rules and national certificates. But 'mechanic' is a very vague term – as is 'machinist', another which reflected the technical change of the era – and new applications of the word were emerging all the time. Boiler-makers, artificers, tool-makers, jig-setters, mechanical draughtsmen were all needed in greater and greater numbers and all of them did more and more specialized jobs as time passed. It is not true, then, to say that industrialization and mechanization decreased the variety and individuality of trades; it killed off many handicrafts but increased the number of special skills needed in many ways. On the other hand, it also required large numbers of unskilled workers who were at best machine-tenders and whose work was very dull and unstimulating.

Psychological and intellectual effects of mechanization

Changes on this scale were bound to affect ideas and outlooks. Not only did many more people begin to believe that machines might be made to do almost anything, given sufficient time and trouble (and this meant that the universe looked much less mysterious and unmanageable), but many people saw technical progress as proof that European civilization was on the right lines and moving in the right direction. A few men did not think so and said so loudly and often. But most people before 1914 found the evidence of their own eyes convincing. Visibly, tasks which had once been enormously laborious had come to be carried out much more easily. Goods which had once been luxuries had become commonplace. Outside the rich world, the social benefits of technology appeared as railways pushed into Africa and Asia, taking with them other benefits – better pumps, wells, communications, medicine. Somehow, the world's leading civilization had at last come to take machines for granted and to treat them as an indispensable part of progress. This was an enormous change in humanity's outlook.

Energy

New machines required new power. The consumption of energy rose after 1801 at unprecedented rates. Eighteenth-century improvements to the steam engine made it possible for most of the new power needed in the nineteenth century to be found from steam. Although railways in some parts of the world were wood-fired, they sent up the demand for coal, too. A world output already of 800 million tons a year in 1900 went up to more than 1300 million by 1913 (nine-tenths of this came from Europe and the United States). Other sources of power had also become available. With the discovery that electricity could be generated – the principle of the dynamo was discovered by Faraday in 1831 – the demand for coal (to run generators) was to rise again, but a new way of tapping running water in 'hydro-electric' schemes was also made available, and with it a further extension of energy resources.

Vegetable and animal oils had long been used for lighting when the refining of petroleum (first commercially produced in Pennsylvania in 1859) opened the way to the use of mineral oils for, first, lighting (in the form of paraffin lamps), and then for fuel. Oil and petrol in turn made possible the internal combustion engine – one in which the energy from the fuel is not utilized by turning water into steam to drive a piston, but in which the explosion of the fuel inside a cylinder itself drives a piston. From this was to stem not only the motor-car but small, portable engines for all kinds of work, oil turbines for use in ships at sea and in power generation, and eventually engines powerful and light enough to be used in aircraft. With this increase, world oil production – most of it in the United States – jumped from 5.75 million barrels in 1871 to 407.5 million in 1914. This implied not merely the building of new wells, but a huge refining industry and the necessary transport arrangements for moving oil to where it was needed.

New machines demanded larger supplies of the materials they turned into their products. Railway-building and ship-building absorbed huge quantities of iron. A very important step forward was taken when a new process for making steel (before that time a widely-used but still very expensive metal) was discovered in the 1850s, and it became sensible (and cheap) to use it instead of wrought iron. Then came improvements which brought down the price of steel and so increased demand for it. In the next decade came inventions which made it possible to make aluminium from bauxite (this meant that a hitherto precious metal became a commonplace). As for non-metallic materials, a huge development in the chemi-

The 'machinery corner' inside the Crystal Palace set in Hyde Park, London, the world's largest glass house, built to contain the Great Exhibition of 1851 which was the century's most splendid and confident revelation of the new wealth of industrial society.

cal industry led to the production of celluloid, in 1879, to artificial fibres about twenty years later, and to 'bakelite', one of the first of what we now call 'plastics', in 1909.

By 1914 almost no side of the life of the industrialized countries was untouched by new machines and new materials. Men even wore celluloid collars. Not only the arts of peace but those of war had been transformed, too. The first military use of the railway was in 1854 (in a campaign fought by the British and French against the Russians in the Crimea), and soon generals were planning the use of railways to deploy hundreds of thousands of men. Meanwhile, armies themselves were equipped with more and more machines. It was not just a matter of their better and more powerful weapons; many other military machines were to be seen by 1914 – motor-cycles, trucks and tractors, aeroplanes, signalling apparatus. As for navies, they were revolutionized by the appearance of steam (the first steamship in the Royal Navy was commissioned in 1821) and then again by bigger and better guns and armour-plate. The Leviathans which made up the great fleets of the powers in 1914 would have been unrecognizable to Nelson as ships, except in so far as they floated – whereas he would have not been very much out of place on the deck of a fighting ship a century before his time. All major navies, too, had submarines in service by 1914. In many ways, the machines of war were both the most sinister and also the most impressive signs of the way in which the age of machines had arrived all over the world.

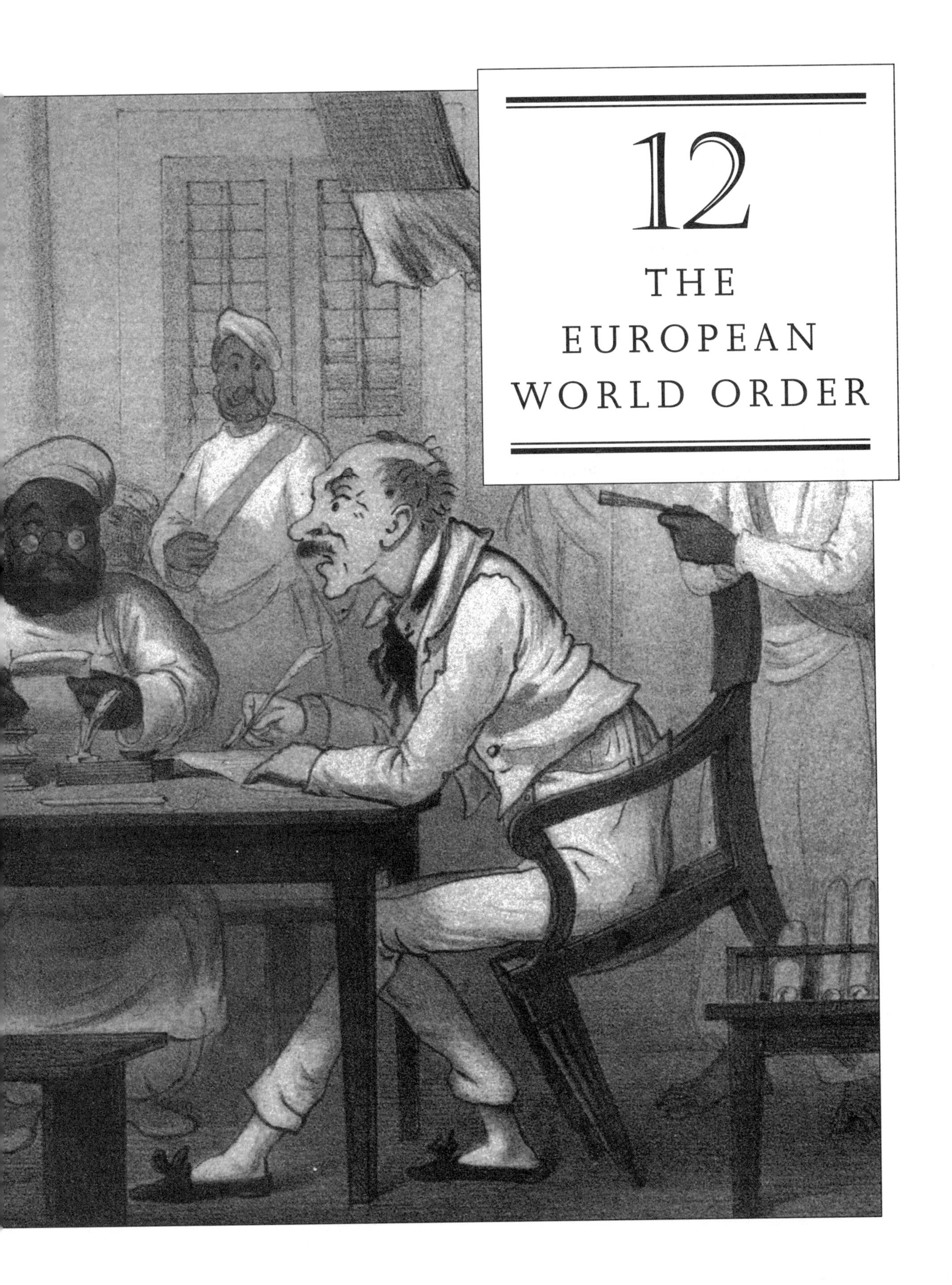

12

THE EUROPEAN WORLD ORDER

THE FORMS OF EUROPEAN ASCENDANCY

Though Europeans in the nineteenth century went in for empire-building even more vigorously than before, European world power in that era was not just a matter of running up the flag over new territories. It presented many different threats to the non-European world, some all the more disturbing because less direct than simple military or political takeover. In many places the arrival of European or American traders, prospectors and financiers led to economic concessions by formally independent local rulers which in fact, though not in name, soon tied their subjects to the wheels of the western chariot, whether this was intended or not. The shape of life in Malaysia was quite altered when Europeans brought the rubber plant there from South America, so creating a new industry on which many of the inhabitants were soon dependent for a living. Mining operations could give a country new political importance; the rulers of Morocco found that their country was being interfered with and squabbled over by Europeans as soon as it was suspected that there were minerals to be found there.

Interference with the internal government of formally independent states could go a long way without getting as far as outright annexation. Since the sixteenth century, when they were first negotiated with the Ottoman Turks, 'Capitulations' had often been made with non-Christian powers in order to ensure security and privileges for resident Europeans. They entitled them to exemption from local lawcourts and to appear instead before special officials or tribunals manned by European judges, and so to escape the operation of the native law. In China in the later nineteenth century Europeans and Americans lived in special areas – 'concessions', as they were called – in cities where they did business, and the governments of these areas was responsible not to the Chinese authorities but to the foreigners. Sometimes they had western garrisons and police, too. Such arrangements weakened the prestige of the local rulers in the eyes of their own people. With some rulers Europeans also negotiated treaties which gave them control over their foreign policy. Altogether, a large and vague zone of practical interference in the affairs of many non-European states stretched a long way beyond the formal boundaries of empire.

At the beginning of this century Christian missionaries were still the pioneers andspearhead of European influence in many parts of the world. When photographed, these indomitable ladies were about to set off on a West African journey.

Finally, there was also another, more subtle form of domination which the European civilization of the nineteenth century began to exercise more and more as time passed. In many places it was to outlive more overt forms of imperial rule. This was the domination of western ideas and ways – of European civilization in the deepest sense. It is very hard to define this precisely except in individual instances. Over huge areas of the world millions of people went on living undisturbed within patterns of behaviour and belief quite alien to the civilization we call western or European. To overlook this would be silly. Nonetheless, nationalism and nationality were western ideas which were to be taken up enthusiastically and to have enormous success in Asia and Africa; so were the ideas of science and technology and the notions of progress attached to them, as well as western notions of law, economics, religion, politics, government and a hundred and one other things. Even if

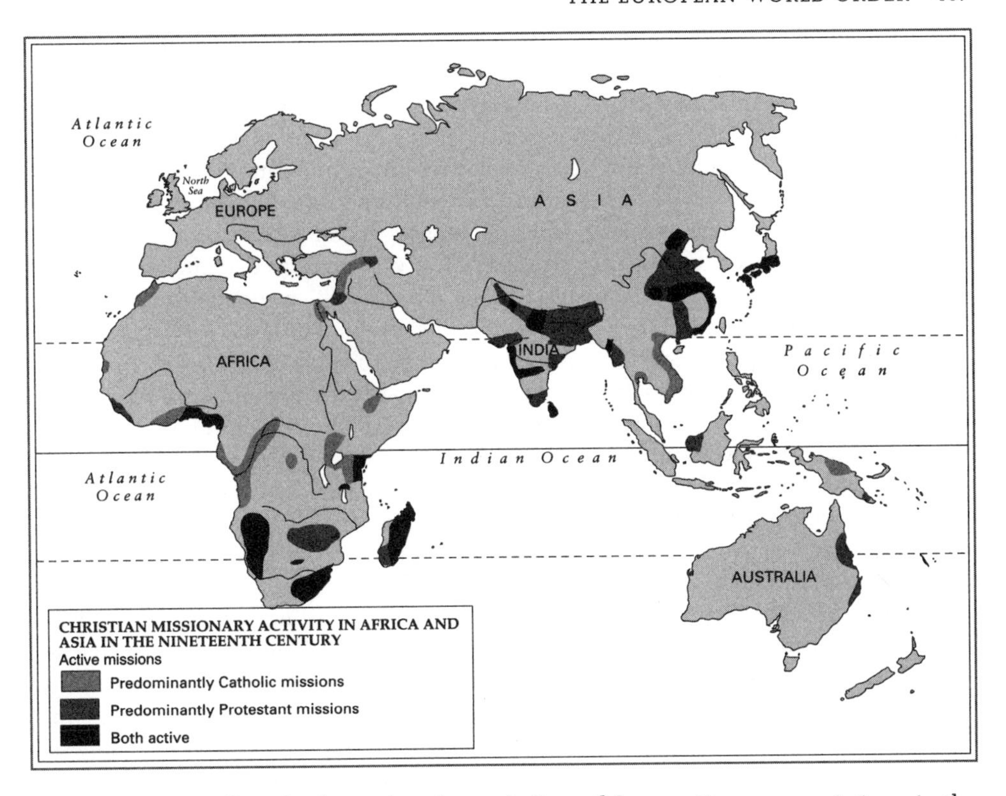

they at first only affected a few – the educated elites of the non-European societies – in the end they bit deeply into the old ways of doing things and had effects far beyond these narrow circles.

All these currents of the age of imperial expansion played upon different parts of the world in different ways. Broadly speaking, the direct acquisition of new territory was most marked in Africa and the Pacific islands, while the more indirect forms of western domination spread in the Asia of the old empires. It is only a rough guide, but a useful one.

Motives and opportunities

The variety of ways in which Europeans came to exercise power over other peoples makes it tempting to over-simplify their motives. Ever since the fifteenth century one of them, clearly, was the hope of economic gain. Men had always eagerly sought new places to trade and make money, new resources in land, minerals or labour, or just opportunities for straightforward theft. In the nineteenth century, many of these possibilities became more attractive still as the European demand for raw materials from other parts of the world grew with industrialization. But it is not necessary to govern another part of the world in order to do business there and many businessmen actually preferred to operate out of reach of European law and order. What is more, even when imperialist nations competed most vigorously to acquire new territory, their officials and politicians were often very disinclined to take on new colonies, knowing that they cost money to govern and protect and that there was no guarantee that they would pay for themselves in the end.

Nor does a search for places in which to invest at good returns explain why people should want new possessions. No country had more money invested abroad in 1900 than Great Britain. Yet though she had the largest empire in the world, British investors had not put money into it on anything like the scale of their investments in the United States and South America, which offered much better returns than most of Africa. Though there was a rough coincidence in time between expansion in the free-enterprise economies of Europe and North America and the building of new empires, and though individual businessmen sometimes tried to interest governments in taking on new colonies in which they had a special interest, the belief that capitalism by itself explains an imperialist 'wave' does not fit all the facts.

In practice, the mixture of motives and purposes in any one part of the world differed strongly from that in another, as different governments listened in different degree to different special interests – those of soldiers, humanitarians, missionaries, crackpots and settlers, as well as those of businessmen. They also listened in varying degree to public opinion. In many countries, the last age of imperialism was also one in which attention was beginning to be paid to what mass electorate wanted for the first time. Those electorates were more likely to read newspapers than a few years earlier, and journalists picked (as they still do) on issues which could be easily dramatized and turned into good 'copy'; imperialism provided many. As a result, statesmen who did not themselves really believe in imperial expansion sometimes went along with the popular tide, or what appeared to be one. Even in that most undemocratic of imperialist countries, Russia, the government seems to have felt that imperial advance would help to rally support behind the regime.

The last complication is that much of the story of imperialism is a story of different degrees of resistance to it, or of difficulty in pursuing it. Imperialism was a matter of opening doors, but sometimes the door was locked, or there was someone on the other side pushing it to keep it closed, while behind other doors there was no resistance at all. When the new empires expanded, they often faced very different opportunities. European settlers overseas soon found this out. Some of them had gone to parts of the world new to Europeans – Australia, New Zealand, Pacific islands, East Africa – and thus played a part in the great imperialist advance. Yet, in the first place, not all nations were equally represented among them; the biggest settler populations were all to be found in the colonies of one imperialist country, Great Britain (emigrants from other European countries tended to go to the United States, or South America). Secondly, although what they found on arrival might vary much from place to place, it was never an old, highly developed civilization, a once-great empire or a major religion such as might be found in India or China; the settlers found little to admire or respect. Nor were there usually big native populations. White settlers therefore could set about making lives for themselves with much greater freedom than the rulers of other British possessions, who had to confront much more complex local conditions. As for non-settler colonies, European states were often tempted to expand because of the difficulty of establishing settled and orderly frontiers if they did not take a hand in the affairs of the peoples who lived on them. Russians in Central Asia and British in India both thought themselves in this position, whether rightly or not.

As for the established great powers of the non-European world, the Ottoman empire in both western Asia and in Europe was already in serious difficulties in 1800. These worsened as the century went on. The Turks seemed no longer able to govern properly their subject peoples, some of whom were likely to ask European states for help. Further east, the once great empire of Persia was under pressure (especially from Russia) and was internally divided and weak. Further east still, the Moghul empire was a shadow of that of the seventeenth century, one Indian state after another seemed unable to provide itself with stable government, and even the once powerful Chinese empire was looking feeble by the early nineteenth century. There was nothing in an Indonesia dominated by the Dutch or in a south-east Asia slipping out of the control of its Chinese overlords to provide firm native resistance to the dominant and aggressive civilization of Europe. Elsewhere in the world, in Africa and the Pacific islands, even more backward communities faced white imperialists. All that had protected many such places for so long from foreign domination were the simple, but formidable obstacles of climate, distance and disease. But the nineteenth century brought ways of overcoming these.

KNOWLEDGE AND TECHNOLOGY

Intellectual advance played a large part in building a new global order. Geographical knowledge gave Europeans great advantages over, say, the Chinese government even at the beginning of the nineteenth century. The South Polar regions excepted, the coasts of the world and the shape of its main land-masses were then pretty well-known. Much of North America was already explored. Spanish and French explorers had opened up the South-West and mapped much of the Great Lakes and Mississippi valley before the end of the seventeenth century. The trans-Mississippi plains and the North-West were left for the nineteenth-century investigators, it is true, and the greatest names in that story are those of Lewis and Clark who in 1804-6 traced the Missouri river to its sources and then crossed the watershed of the Rocky mountains to go down the Snake and Columbia rivers to the Pacific coast of what was called the Oregon territory. That coast, including Vancouver Sound, had already been looked at from the sea, but this was the first land crossing to it. Traders and settlers soon followed them into the north-west. Yet there was still a lot of the world's map still needing to be filled in; islands still waited to be discovered in the huge expanse of the Pacific, while much of the interior of Africa, South America, and even some of Asia remained unexplored. By 1914 this was no longer true and there was very little of the world's land surface then still unmapped.

Africa

At the end of the eighteenth century continuous and well-organised efforts to get into the African interior had begun. The British arrived in the sub-Saharan and Niger regions; a German was probably the first European to cross the Sahara since Roman times (he set out from Cairo and died just before reaching the Niger). Meanwhile, other expeditions were launched from the West Coast. The last, mounted in 1805 by Mungo Park, a great Scotch explorer, makes it clear what dangers lay in the way of such attempts: of its 40 European participants who set out from the coast, only 11 were alive when the upper Niger was reached and by the time the expedition was ready to return only 5 survived (and one of them had gone mad). This handful set off again, only to be all killed or drowned on the way. Nevertheless, explorers still went to Africa. In 1828 a Frenchman reached Tangier from the south, and became the first European to visit Timbuktu and return. A few years later the Niger's mouth was reached for the first time from the interior. Gradually, knowledge of the Sahara desert and the savannahs to the south of it was built up. Meanwhile a series of expeditions from the East Coast was seeking the source of the Nile.

Mungo Park (1771–1806), surgeon and explorer, traced the Niger river in two long journeys 1795–7 and 1805–6 and settled the direction of its flow. He drowned in a fight with natives on the way back from the second journey.

Livingstone

Geographical and scientific enthusiasm inspired most of the explorers, but the most famous of them was a Scotch missionary, David Livingstone. Many Christian missions were already at work in Africa when he landed in South Africa in 1841 but he captured his countrymen's imaginations and linked the ideas of European civilization and Christian evangelism in the 'Dark Continent' as did no one else. He became a true popular hero. He first went north, looking for sites for new mission stations, and only after crossing the Kalahari desert (accompanied by wife and child) to reach the Zambesi did he decide to march westward 1500 miles across unknown territory to the Atlantic (which he reached at Luanda in 1854). He then decided to turn round and go back – and did.

Many more journeys followed. In 1866 Livingstone joined in the search for the origins of the Nile. Increasingly horrified by the signs he found almost everywhere of the tragedies caused by the Arab slavers (slaving had long been abolished on the West Coast by international agreement), he made yet another crossing of the continent on foot, this time following the upper Congo down from the region west of Lake Tanganyika. While he was engaged on this march there occurred one or the most famous incidents in the whole history of exploration, his meeting in 1871 with an American newspaper correspondent, Henry Stanley, who had been sent to find the famous explorer. Stanley's own words tell the well-known story best: 'I would have run to him, only I was a coward in the presence of such a man – would have embraced him, but I did not know how he would receive me. So I did what moral cowardice and false pride suggested was the best thing – walked deliberately up to him, took off my hat and said, "Dr Livingstone, I presume?"'

In 1873 Livingstone died, kneeling in prayer, on his last terrible journey (his servants devotedly buried his heart and then carried his embalmed body for eleven months 1000 miles to the coast). By then, the great age of African exploration was coming to a close. In a few years, the Niger, Zambesi, Nile and Congo were all pretty accurately mapped. Much still had to be known in detail, but the age of railways, roads and telegraphs was dawning and at least the geographical darkness of Africa was shrinking further and more rapidly every year.

The exploration of Africa captured people's imagination in Europe and the Americas in the nineteenth century for many different reasons. There was the impulse to evangelize the continent's native peoples which meant that Livingstone enjoyed an appeal like that of a modern footballer or pop-star. There was the self-interest of individuals (and governments) who backed expeditions looking for the natural wealth the continent was known to hold. Another influence was the anti-slavery movement and guilty feelings about Africa over the damage done by European slave traders in the past. Even national rivalries helped, as governments sought to get information on which to base claims to territory or influence over African rulers. There was sometimes something of a snowball effect in the way this worked, because European governments often acted in Africa from fear of someone else getting in first.

Australian exploration

Very few such motives explain the opening of the other great land-mass waiting for exploration in 1801, the continent of Australia. Containing comparatively few native inhabitants (and those much more backward in civilization than many African peoples), until well into the nineteenth century Australia also seemed to have few natural resources. It was far away

Early Sydney; a scatter of buildings interspersed with farms and still preserving styles familiar in Old Europe.

from both Europe and America, and to all intents and purposes had hardly been touched upon until the end of the eighteenth century, whereas much of the African coasts had been well-known to Europeans long before that. Nor was there the stimulus of rivalry between states to give impetus to exploration.

The Australians themselves, for the most part, provided the explorers of their own continent. Often in the teeth of huge difficulties expeditions were already beginning to push inland in the first half of the nineteenth century. To begin with, it was a little like the penetration of the North American West. Only after the settling of South Australia and Victoria came the first major attack on the deserts. In 1840-41 an Englishman made a terrible march along the desert south coast of the continent to Albany in the west, but the continent was not to be crossed completely by explorers until 1860, when an expedition (using imported camels for transport) went from Melbourne to the Gulf of Carpentaria in the north. This was followed by another crossing from Adelaide to Port Darwin in 1862. After this, the map of Australia slowly filled up. To this process a big, but neglected contribution was made by the real Australians – the aborigines. It was often they who provided the knowledge and skills (where and how to dig for water, which grubs were edible) which made the explorers' survival possible.

The Poles

The Polar regions – north and south – were the scenes of the other big efforts of discovery in this period to capture popular attention. The dream of a passage to Asia round North America or Siberia had died hard. The British government as late as 1818 once more offered a prize – £20,000 – for the first person to make the voyage. Attempts drew attention further north still. One British naval officer who tried to reach the North Pole got as far north as 82° 45' in 1827 (starting from Spitzbergen) and this remained a record for fifty years, though four years later another reached the magnetic North Pole. Meanwhile efforts to seek a north-west passage went on, but only in 1906 did the Norwegian Amundsen for the first time turn down into the Bering Strait after sailing his ship right across the north of Canada and Alaska. He was probably beaten to the North Pole a few years later – though it is not exactly certain – by an American, Peary, but Amundsen was later to be one of the first men to fly over the North Pole in an airship (in 1926). Before that he had achieved an even greater feat in Antarctica.

Roald Amundsen (1872–1928), the greatest of twentieth-century explorers, who diverted his expedition southwards to Antarctica after hearing the Americans had beaten him to the North Pole. He later disappeared when looking by aeroplane for an Italian airship missing in the Arctic.

Cook has been the first man to take a ship across the Antarctic circle; a Russian expedition was the first to sight land within it (in 1821) and a British sailor, Captain Ross, got to within 710 miles of the pole and chartered a thousand miles of the coastline of Antarctica in 1842. This, too, was a record which stood unbroken until the end of the century, when the first party to winter on the continent got further south by sledge. Knowledge was now being accumulated more rapidly; a Swedish expedition (thanks to bad luck, for their ship was crushed and sunk by the ice) was forced to spend two winters in Antarctica before being picked up in 1903. Expeditions were becoming more frequent; one British team turned back only ninety-seven miles from the South Pole in 1909. Finally Amundsen reached it in 1911, on 16 December. That may well stand as a fitting date to symbolize the end of the great age of terrestrial exploration which had begun in the fifteenth century.

WHITE SETTLEMENT

One way in which Europeans changed the course of world history was by planting white populations overseas. In 1800 there were already the United States and large populations based on Spanish and Portuguese stocks in Central and South America. There were British and French settlers in Canada, too, Dutch at the Cape of Good Hope, and a few British – mainly convicts – in New South Wales. By 1914, these populations had grown vastly; they had become new and mature nations.

Outside Central and South America, Great Britain was their main progenitor. There were two main reasons for this. Not only could she provide plenty of emigrants but her rulers long had a deep prejudice against ruling white colonies. They liked them to come quickly to maturity and independence. Memories of the American War of Independence went deep; it was for a long time taken for granted in England that settler colonies were likely to turn out badly and would certainly be costly. When such ideas began to fade away, it was too late to stop the drift towards independent nationhood in British white colonies. For much of the century in which the British flag flew over an empire on which the sun indeed never set, many Englishmen viewed the large pink spaces on the map with complicated feelings, and not necessarily much enthusiasm.

Canada

British Canada lived next door to a republic which had born of rebellion against the British Crown; many American citizens believed Canada should simply be absorbed by the United States. A war between that country and Great Britain from 1812 to the end of 1814 was the only attempt to bring this about; the would-be invaders of Canada were then very unsuccessful. Still, frontier problems nagged on for the next half-century or so. Inside Canada, there was the difficulty of governing two groups of settlers, the French, who had been there first and were mainly settled in Quebec, and the later British arrivals (some from the former American colonies, but many Scotch) who went mainly to the Maritime Provinces and the West. In 1837, some of both races took part in a rebellion helped by Americans. It was suppressed, but the British government began to take steps by which Canadians were first given control of their internal affairs and then entire independence under the Crown. The setting-up of the Dominion of Canada as a federal state in 1867 set the seal on the process by providing a national government. From that moment, Canada can be reckoned as an independent nation in its own right, even though tied by many practical and sentimental links to Great Britain.

Canada was in 1867 poor and underpopulated, and the first transcontinental railway (opened twenty years later) was to be of great importance. As well as pulling the country together as an economic and governmental unit (it was used as early as 1885 to transport troops to put down a rebellion in the North West), it opened up much more of the country to immigration. As in the United States, railways completed the steamship link between the New World and Europe's great centres of population. The 500,000 new Canadians who arrived from Europe between 1815 and 1860 combined with natural increase to push the population up to 3 million by the latter date, though, because of wastage to the United States, this figure only doubled by 1900, when a new burst of settlement and faster growth began.

Australia and New Zealand

The growth of Australian population had at first been more spectacular. The 736 convicts, women and guards who arrived in the original shipment of 'new Australians' in 1788 had multiplied into 100,000 or so settlers in the 1830s and a million by about 1860. Australia presented fewer problems of government than Canada and had no powerful neighbours, but the precarious economy and the continuing transport of convicts (the last were landed only in 1867) gave the British governors on the spot difficulties. The Merino sheep was one answer to the economic problem, followed, later, by the refrigerator ship. In 1850 the colonies into which Australia was divided were each given internal self-government. British troops were withdrawn in 1870, and in due course, the Commonwealth of Australia (like Canada a federal state) came into being on the first day of the twentieth century, 1 January 1901.

Better communications had helped Australia as they had Canada. The first regular steamship run to Sydney from England began in 1856. It made immigration easier, as did railway-building (though chaos ensued when different colonies adopted different gauges). In 1872, a telegraph line was built from Adelaide to Darwin, in the north, from which it was soon possible to reach Indonesia and India (and therefore Europe) by direct line. As the white population (coming mainly from the United Kingdom) grew, so did resistance to settlement by Chinese or Japanese. Individual colonies took up 'White Australia' policies. There were parallels on the west coast of Canada and in the United States' limitation of oriental immigration; the world-wide European folk-movement was, perhaps, more successful than earlier migrations in keeping possible competitors at bay.

By consent of all parties, immigration restriction was built into the agreements of 1901 on which the Commonwealth of Australia was founded. Soon, Japan's defeat of Russia in war renewed alarms about the possible 'Yellow Peril' from the north. Because of this Australia took over British New Guinea for strategic reasons (renaming it Papua). So a former colony itself became an imperial power. The building of an Australian navy was begun a few years later, and compulsory military training was adopted in 1910. Meanwhile, a society basically British, though much more democratic and relaxed in its social attitudes, was taking shape. It startled some Europeans by its democratic style (women had the vote in Australia in the 1890s) and by its generous labour and social service legislation.

In New Zealand too there appeared a new nation predominantly British in culture, but more democratic and (like Australia) inclined to more generous social and welfare provision than the mother-country. The first settlers – whalers, escaped convicts from Australia and traders seeking to make a penny or two out of supplying the native Maori peoples with guns – had been so disreputable a lot that the British government had resisted taking responsibility for the islands at all. The first missionaries were vigorous and hard-working (there was already an Anglican bishop of New Zealand in 1827), but respectable settlement received little encouragement until it looked as if the French might seize the islands. Then, at last, treaties were made in 1840 with the Maori chiefs by which they accepted British sovereignty. New Zealand's short colonial history then began.

Since the settlers were greedy and took away their lands, the Maori were twice driven into rebellion. Unlike the aboriginal inhabitants of Canada and Australia, who had been too weak to resist, the Maoris were numerous and militarily formidable. Nevertheless, especially in the southern island, where there were few Maori, the colony grew fast; there were 300,000 settlers in 1875. Perhaps more important, there were by then over 10,000,000 sheep (in the south island alone). New Zealand found in wool a staple commodity for export. With the introduction of refrigerated cargo ships in 1882 the sheep-farmer could think of raising stock for meat as well as for wool, and cold transport opened the way also to the export of dairy produce. Both islands were under one local government from 1875, London retaining responsibility only for native affairs. The New Zealanders, like the Australians, took steps to keep out Asiatic immigrants, prescribed an eight-hour working day and an old age pension scheme in the 1890s, and also gave women the vote. Finally, in 1907, New Zealand was recognized as an independent 'Dominion' within the British Empire.

South Africa

Australia and New Zealand enjoyed slightly different degrees of legal sovereignty in 1899, but, practically speaking, by then, like Canada, they had freedom from British control. This makes something which happened in that year all the more surprising; the start of a war in South Africa to which Canada, Australia and New Zealand all sent contingents to fight on the side of the mother-country.

The background was a long and unhappy story of Anglo-Dutch conflict. The Dutch had arrived in South Africa in the seventeenth century. They numbered about 25,000 by 1800.

Unlike the settlers of North America or the future Australia, they were established in a country in which there was already a large native population which would not go away or die out, and which grew in numbers as time passed. Even in 1900, when many more white people (mainly British) had gone to South Africa, its white population was only about one-quarter of the black. Different views about the treatment of the native Africans always made it difficult for the British governors and Dutch farmers to come to terms. But there were other difficulties, too. The Dutch were a community closely-knit by custom, language and religion, who did not want their ways to be contaminated from alien sources.

Trouble began soon after 1815, when the British (who had occupied the Cape because of its strategic importance during the war with Napoleon) annexed the area. Soon afterward, British settlers began to arrive. They were accompanied, to the annoyance of the Dutch, by missionaries who almost at once took up the question of the defence of the rights of the native population, whom they sought to convert. English became the official language instead of Dutch, and the old judicial arrangements were replaced by British. When slavery was abolished throughout the British Empire (in 1834), there was much grumbling among the Dutch over the terms of compensation. All in all, it is scarcely surprising that in 1835 there began the Great Trek, a migration to the north of some 10,000 Boers (as the Dutch were often called) with their families, cattle and possessions, across the Vaal river. This was the origin of the later Boer republic of the Transvaal. A few years later, another British author-

Freedom fighters of 1899: three generations of Boer farmers in dress typical of that in which most Boers took the field against the British.

ity was established in Natal, with the aim of protecting the natives against the Boers there – and so more settlers of Dutch stock went north in search of their own kind.

Fifty years of bitterness, occasional fighting and attempts to find solutions to the problem of government in South Africa followed. All the time, the stakes rose higher. More British settlers arrived. Diamonds were found on the Orange river and then gold on the Rand in the Boer Transvaal. Native wars (with, in particular, the Zulu) sent up the cost of government. By the 1890s the Boer leaders were convinced that the British were intent on destroying their republics, while the British thought that the Boers might obtain a seaport on the Indian Ocean and then threaten imperial communications with India. The outcome was the South African (or second Boer) War of 1899-1902.

The Boers had several spectacular successes at the outset, and were able to keep a guerrilla war going a long time after the defeat of their main armies. But in the end they had to give way. The British took over the former republics, but promised that representative institutions would be set up before long. This soon happened; by 1907 elections had given the Boers internal self-government once again in the Transvaal. Almost at once, they began to pass laws against Asian immigrants (mainly Indians). Two years later, a draft constitution for a Union of South Africa was drawn up which allowed each province to regulate its own voting arrangements; unlike the former British colonies, the former Boer territories at once restricted the vote to white men. But it looked as if the quarrels of Dutch and English had at last been settled. On 31 May 1910 the British parliament passed the South Africa Act and a new state, with a vigorous future ahead, came into existence within the British empire.

The former British colonies were the most important European settlement area which actually grew into new nation-states. In other major areas of European settlement this did not happen, though Frenchmen and Italians settled in large numbers in north Africa during the nineteenth century. With the exception of Algeria (which ended up being treated as legally a part of France) these settlement areas either remained formally under the native authorities (as in Tunisia) or were straightforward colonies with no prospect of independence ahead (which was the case in Libya and Tripolitania, seized by the Italians from the Ottomans just before 1914 and hardly settled at all).

Latin America

The only other place where nation-states appeared as off-shoots of European settlement was South America. French occupation of the mother countries had brought about a virtual severing of ties between Spain and Portugal and the Americas during the Napoleonic wars. Some of the creoles (as the American-born of European blood were called) had already noted that the North Americans had shaken off British rule. Now seemed to be the moment to do the same with Spanish. In 1810 a series of risings in widely-separated places began what were to become the 'Wars of Independence'. Two outsiders then entered the story. One was the United States, which declared in 1823 that no European power was to contemplate the Americas as a place for further conquests or colonies. This 'Monroe Doctrine', as it was called after the president who declared it, depended on another outsider, Great Britain, whose government was quite happy for commercial reasons to see South and Central America independent of Spain and Portugal. As the Royal Navy was the only naval force which could thwart any attempt to reconquer them, the survival of the new republics was assured. From the Wars of Independence emerged a collection of new states, often ruled by soldier-dictators (though Brazil for some time had an emperor of the Portuguese royal family). Geography and history made a federation like that of the northern continent impossible. But the new states were not threatened from the outside; nor would many internal weaknesses have been removed by merging in a United States of South America. Quarrels and wars finally resulted in the appearance by 1900 of four mainland central American republics (of which Mexico was the biggest), two island states in the Caribbean (soon to be joined by a

third, Cuba) and ten republics in South America. Their politicians, at least in public stance and language, seemed remarkably like their European equivalents; a French emperor, Napoleon III, coined the term 'Latin America' for the continent in the middle of the nineteenth century.

South America attracted European emigrants much more than did Central America, though it was notably less of a draw than North America; only 6 of the 41 million Europeans who crossed the Atlantic between 1845 and 1914 went south of the Rio Grande. Still, this helped to confirm the European flavour of societies many of whose inhabitants were Amerindian or (in the case of Brazil and some Caribbean islands) African in descent. Population in Central and South America did not, however, grow so rapidly as that in the United States overall; there were about 80 million in the whole area in 1914.

THE HIGH TIDE OF EMPIRE

Although the presence overseas of new nations of European stocks and much of European culture was crucial for much of the future development of the world, European ascendancy was of course most obvious in the form of colonial empires and the direct rule of non-Europeans. Two such, the British and Russian, shared about a third of the land surface of the globe in 1914. The British empire contained some 400 million people – about a fifth of humanity at that time – of whom about 35 million lived in the United Kingdom. The French, with over 50 million overseas colonial subjects, also ruled more people of other nations than their own, and millions more – and vast areas of land – were ruled by other European powers. Outright control of so much of the earth and its inhabitants was one of the clearest signs at the beginning of the twentieth century that the Europeans really were the world's top dogs.

The contrast with 1800 was enormous. A century later, the only places outside the Americas still not under direct white rule were China, the (much-shrunken) Ottoman and the Persian empires, Japan and a handful of smaller countries. The change to this state of affairs had come about very rapidly, and to a great extent in the last thirty years of the nineteenth century. What happened was so striking that in 1900 people were talking much more about empires and 'imperialism', a word which only seems to have begun to be used in English in the 1850s, than fifty years earlier. Not everybody liked empire-building, but everybody agreed that it was an outstanding fact of the age. There had never been so imperialist a century or such seemingly successful empires.

Empires have existed almost from the beginnings of civilization, but different empires at different times have meant very different things in practice. Chinese imperial officials, for instance, seem to have been satisfied that subject peoples acknowledged the empire's superiority by providing periodic tribute and showing respectful behaviour and that was all; yet the

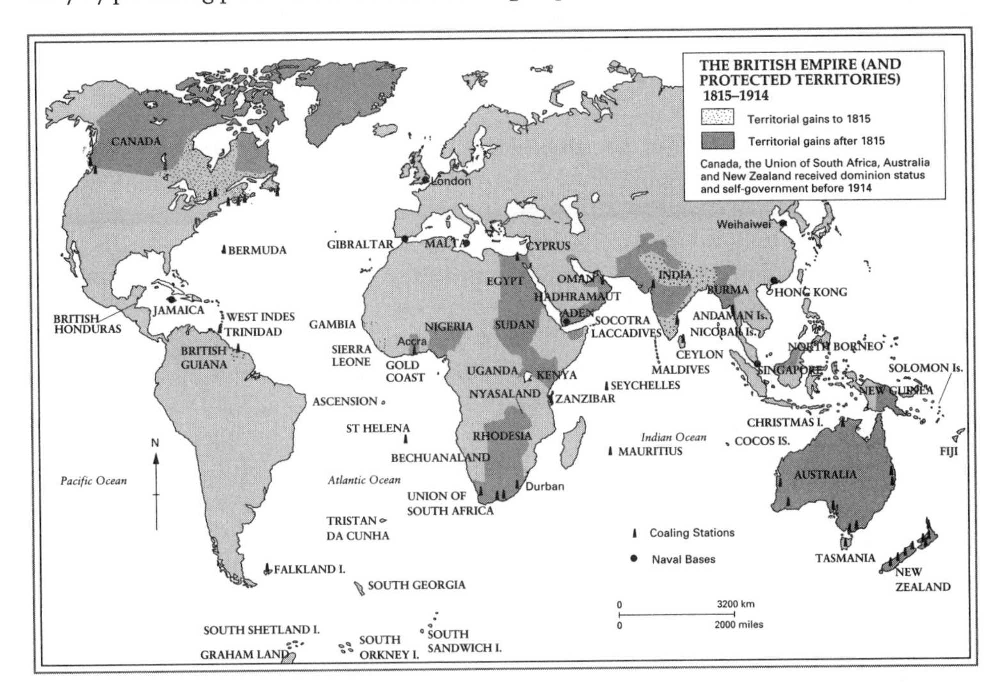

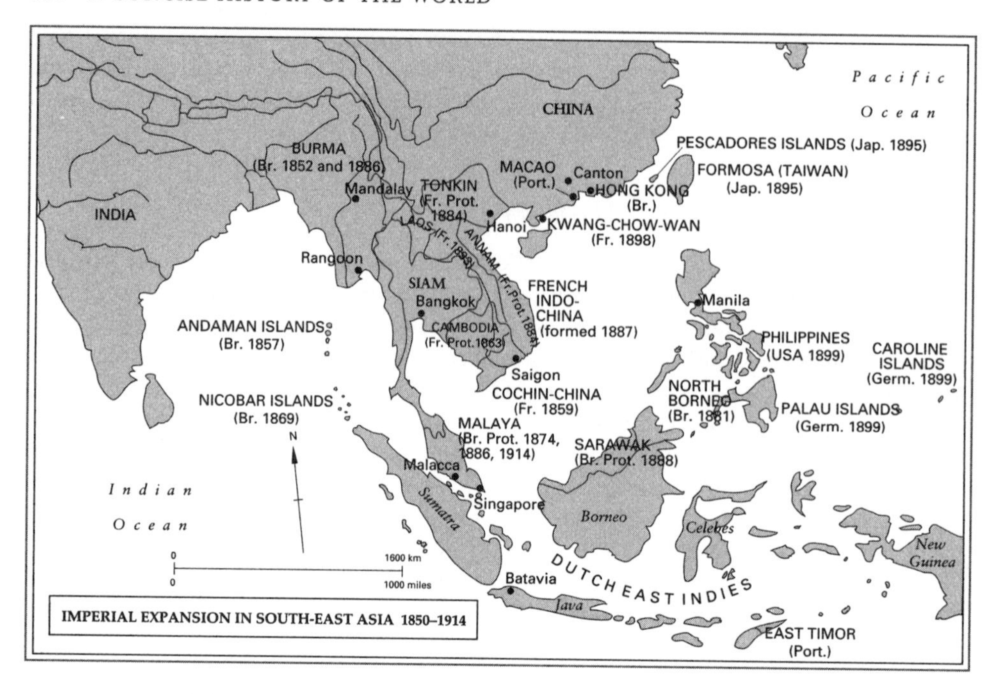

IMPERIAL EXPANSION IN SOUTH-EAST ASIA 1850–1914

theory on which this was based was that all mankind was subject to their emperor. Unsurprisingly, the new, mainly European, empires of the nineteenth century turned out to have some special features of their own.

The most obvious was their geographical reach; in the end, some of them were claiming rights over the frozen wastes of Antarctica while others disputed the waterless deserts of the Sahara (both, at first sight, unattractive regions). There seemed to be no part of the world in which empire-builders were not interested or where they were not prepared to operate. This was partly because of the new accessibility of many parts of the world thanks to exploration, technology and science, and the new military irresistibility of the empire-builders. Almost the only non-European nations who halted them before 1914 and thus kept their independence were the Ethiopians, who managed to destroy an Italian army which invaded their country in 1896, and the Japanese, who went a long way towards adopting European ways in order to survive.

Both these facts emphasize that nineteenth-century imperialism was overwhelmingly European, even if only in origin. Only one Asian country – Japan – took part in the empire-building of the age; the older Chinese, Ottoman and Persian empires, all of which had made great conquests in the past, were losers in the nineteenth-century, contracting, not expanding. The nations acquiring new possessions were more numerous than ever, but except for the United States and Japan, they were all European. Some of them – Russia, Great Britain, France – had long been in the business of empire. One of the old European imperial powers, Spain, lost more than she gained during the century and was really no longer in the race by the end of it, though she picked up a little new territory. Two other peoples prominent in an earlier phase of empire-building, the Portuguese and Dutch, were in much the same position. But two European states which had not existed even in 1850, Germany and Italy, were also acquiring new territory overseas, and so was Belgium, which had only appeared in 1830.

Overwhelmingly, therefore, this was an age of European imperialism, though the United States eventually joined in. Hers, though, was in more than one way a special case. Additions to her territory on the American mainland (like those of Russia in Asia), do not usually spring to mind as an instance of empire-building although they continued for much of the nineteenth century. They too, though, fit the broad pattern: an imperialism which (the Japanese example excepted) was the work of the 'white' peoples – those, that is to say, of European stocks and cultures. But the process is part of another major theme in the new ordering of the world which was going on in the nineteenth century, the growth of a new world power.

A new world power

Between independence and 1850, Americans took control of half of a continent and 6 million of them in 1800 became 23.5 million in fifty years. They were by then already the products of what a nineteenth-century writer called a 'melting pot': they were moulded into a new nation by the experience and setting of a new continent and by the institutions of the republic. The fact that so many Americans had chosen voluntarily to cross the Atlantic to their new country, or had accompanied parents and relatives who had done so, or (even if they had been born in America) had still been brought up in families which had members who had made that choice, long helped to foster a strong patriotic feeling. Unlike every other great power the United States was one to which people had consciously chosen to belong. With a frontier region still offering abundant land and resources for exploitation and an expanding eastern commercial and industrial economy presenting opportunities of a different kind, Americans were very much aware that they were better-off than other European peoples.

Over the nineteenth century, the United States took as many European emigrants as the rest of the world combined. Many of them arrived unable to speak English. Yet English remained the language of the country, and for a long time leading Americans still looked to England for their cultural inheritance and many of their ideas. Only in 1837 was an American president elected who did not have an English, Scotch or Irish name (there was not to be another until 1901). Many of the basic institutions, too, were English; English legal ideas, an English emphasis on Protestant Christianity, English beliefs in the sanctity of property, were the foundation of the Republic itself.

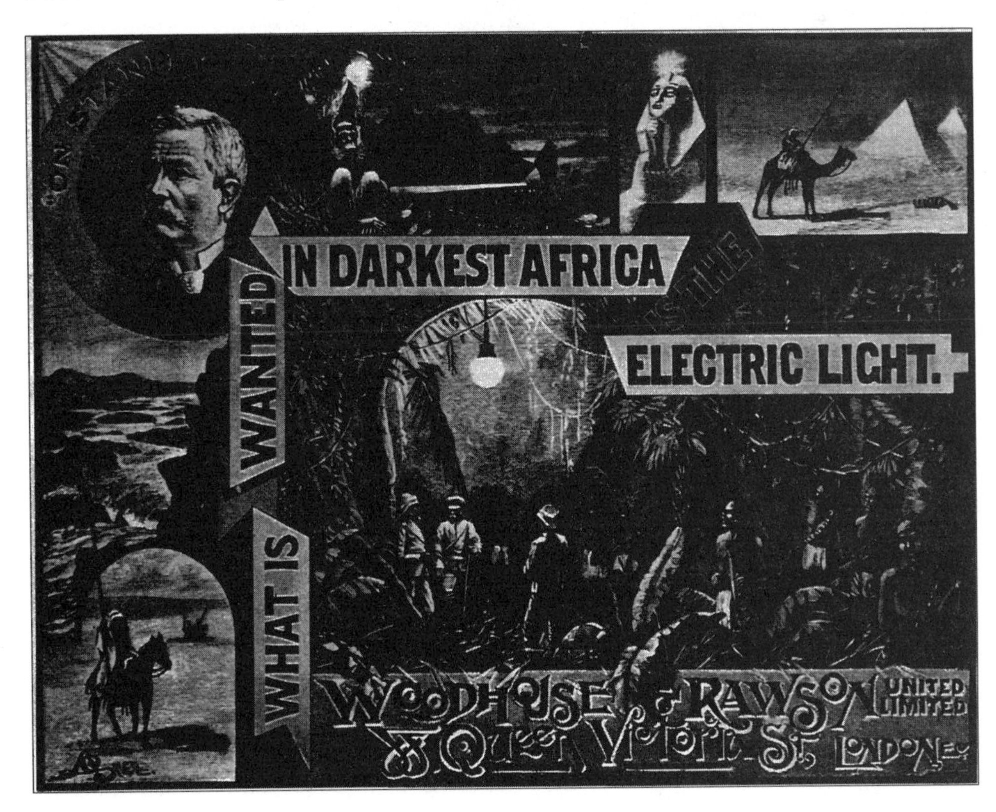

An advertisement of the 1890s and a true sign of the times, tapping, as it does, the glamour of exploration, technological pride and a conviction of racial superiority.

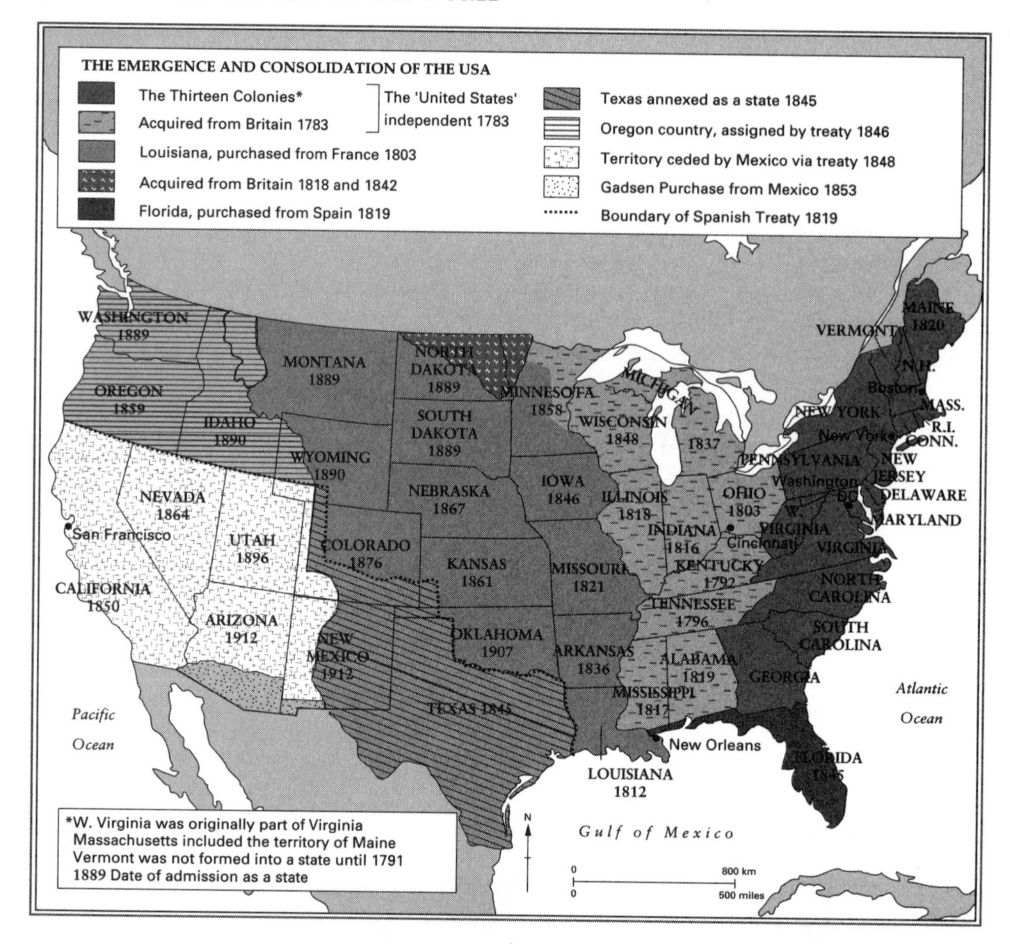

By the standards of continental Europe, both countries were 'free', though that meant different things in each of them. England was not a democracy while the United States was. For a time in the early nineteenth century, this did not much challenge the old leading classes in American politics, but in the 1830s Andrew Jackson (who has been seen as the first truly democratically supported president, speaking for large numbers of Americans on the basis of a national programme) took office. From his day, it began to seem likely that one of the big issues of American politics would be, increasingly, the question whether the will of the nation as a whole, expressed in a democratic vote, should prevail over the protection which the Constitution gave to minority interests, in particular, those of individual states.

The first expansionists

So far as the outside world was concerned, what went on inside the United States was for a long time not of great interest except to those monarchies which in 1783 still had North American territories of their own: Great Britain, France, Spain and Russia. When George Washington said farewell to his fellow-countrymen on laying down office in 1796 he told them to keep clear of political entanglements with Europe. In what he said, there was no hint that his country would one day have a world role to play. Though the United States went briefly to war with Great Britain in 1812, she took little part in the international relations of the French Revolution and the Napoleonic era. Outsiders (except for the British) seemed not to bother much about the United States, because Americans did not much bother about them. As settlers from the old Atlantic colonies had by 1800 still got no further west than the Ohio valley, this detachment is easy to understand. The United States had a small population (about six times that of London at the same date), many of whom had deliberately turned their backs on the Old World. They had enough to do in this new country. From the start, the outlook of Americans was deeply marked by a tendency towards what would later be called 'isolation' and this was, if anything, confirmed by the most important single act of American statesmanship in the first half of the nineteenth century, the 'Louisiana Purchase'. For $11,250,000 the United States in 1803 bought from France territory bigger than the whole area of the republic at that moment. It gave the young nation not only the future states of Louisiana, Arkansas, Iowa, Nebraska, North and South Dakota, much of Minnesota, Kansas, Oklahoma, Wyoming, Montana and a sizeable piece of Colorado, but also access to the trans-Mississippi western half of the continent, previously cut off by first Spanish and then French possessions. As emigrants entered these lands, the whole balance of the United States was to begin to change.

The war of 1812, though hardly defensible and wholly unsuccessful, was another milestone in the story of expansion, but also one in the history of American nationalist feeling (a caricaturist then invented 'Uncle Sam' as a symbol of the nation, and the 'Star-spangled Banner', now the national anthem of the United States, was composed) and policy. It made both sides anxious to settle their differences: after this the danger of another Anglo-American war over Canada was never very grave. Future boundary disputes would be settled by peaceful negotiation. Before the middle of the century, the outstanding border questions were settled; no English statesman dreamt of taking more territory south of the 49th parallel. But after the Treaty of Ghent which ended the war it was already clear that the United States was going to be the most important state in the hemisphere.

There was now a lot of United States territory to fill up, and other boundaries to be drawn, though. As the settled area extended first west of the Alleghanies and then west of the Mississippi, many Americans came to feel that they had a special destiny – and therefore a right, to dominate the whole breadth of the continent. The words 'Manifest Destiny' began

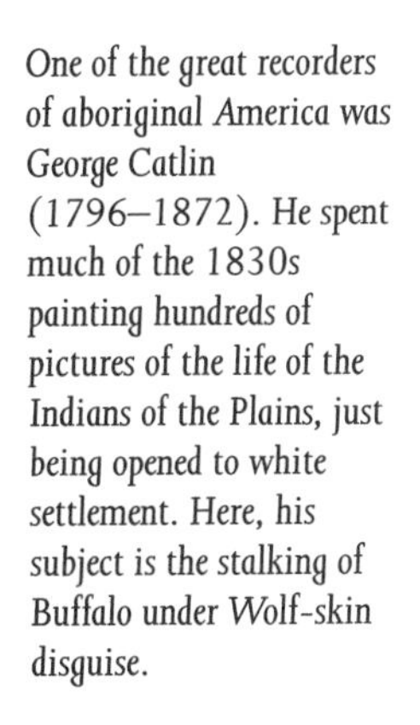

One of the great recorders of aboriginal America was George Catlin (1796–1872). He spent much of the 1830s painting hundreds of pictures of the life of the Indians of the Plains, just being opened to white settlement. Here, his subject is the stalking of Buffalo under Wolf-skin disguise.

to be heard. This boded ill for other Americans. Canada might be safe, because she was the colony of a great power, but the American Indians were not. Broadly speaking, they were simply swept aside, dispossessed of their hunting-lands and homes, killed if they resisted. They were seen as savages with no right to resist the onrush of a higher civilization. This was one of the darker sides of America's expansion.

Another was the treatment of Mexico. The Spanish neighbours of the United States in the south had been succeeded after the Wars of Independence by a Mexican republic which was to be the main victim of 'Manifest Destiny'. When American settlers in Mexico rebelled against Mexican rule and set up an independent Texan republic, it was soon annexed to the United States. Mexico lost the war which followed. She had to make a peace in 1848 which gave up not only Texas but all the land which would one day be the states of Utah, Nevada, California and most of Arizona. With a little more Mexican land bought by the United States in 1853, this rounded off the core mainland territory of the republic in the form which has remained to this day. In 1867, Alaska was bought from the Russians (who had already long before abandoned their claims to the posts they had once established in California).

Slavery and secession

Although Americans could never see the success story of overland expansion in the moral terms they applied to European imperialism it nonetheless posed moral questions for them. This was because it raised constitutional and political issues along the line of the old potential clash of the democratic majority and the interests of individual states of the Union. These became confused with the destinies of the biggest single group of people under American law who did not benefit from its democratic protection – the American blacks – and so the stage was set for a great and tragic conflict.

When George Washington became president there had been about 700,000 blacks in the United States. The overwhelming majority of them were slaves. They belonged absolutely to their owners, who could require them to work as they wanted, could discipline them if they would not to the extent of flogging or other physical punishment, and could sell them or leave them by will to new masters. They lived for the most part in the southern states, where they were largely employed in field-labour or as household servants. Some were treated well, some badly. Some masters were deliberately cruel, but others were affectionate and paternal. Whether slaves were happy or unhappy, though, they were not free, like white Americans; they belonged to white men and women.

Few people questioned this state of affairs. Washington himself was a slave-owner, and so were almost all the 'Founding Fathers'. Yet by 1850 the blacks of America had become an appalling political problem. There were many more slaves by then (4 million in 1860), and they were spread over more states than in Washington's day. As the importation of slaves from Africa had been made illegal, most of them were American-born. Their numbers had grown because more slaves were needed as cotton-planting spread to new areas. 'King Cotton' had an assured market in the mills of England to which America was the main supplier; the total crop doubled between the beginning of the century and the 1820s and then doubled again in the next ten years. By 1860, two-thirds of the value of the total exports of the United States was provided by cotton.

This enormous change transformed the southern United States. Cotton-planting and slavery spread together across the south, away from the old Atlantic coast states where slavery had first been established and into Alabama, Mississippi, Tennessee and Arkansas. Those states became more and more dependent on slavery, which most southerners gradually came to see as essential to everything that made them different from northerners. By the middle of the century some of them were beginning to think of themselves almost as a separate nation within the United States, and that the things that made them distinctive were threatened from the outside by the government in Washington.

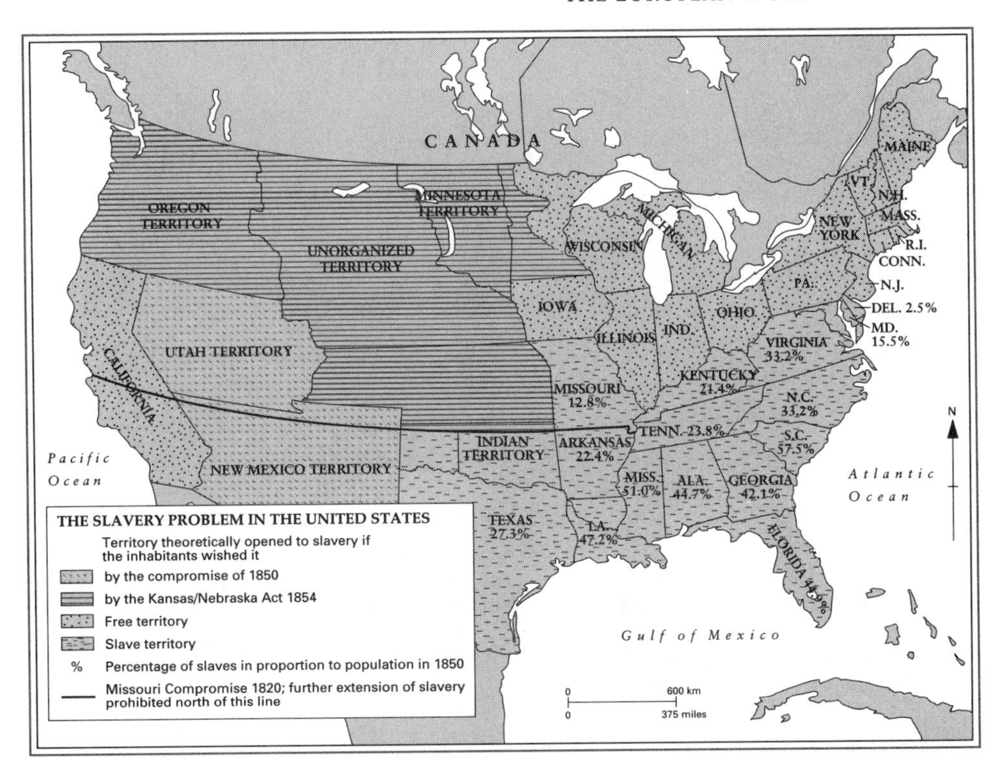

This was because slavery had become mixed up with territorial expansion. As the West was opened up after the Louisiana Purchase, and new states came into existence there, great questions increasingly hung over them: should slavery be allowed in the new states since it already existed in the older ones? Or could it be kept out of them by laws made by the Congress? Southerners said it could not be kept out, or if it could, it could only be kept out by the inhabitants of the new states themselves, since the Constitution left slavery to be regulated by state authorities. Opponents of slavery denied this; an Act of Congress (they said) could keep slavery out of any new United States territory. So what the constitution meant became a matter of dispute; did it set up a national law-making body which could always override the states in the end, or did the states have certain rights which could not be taken from them even by Act of Congress?

It became harder and harder to deal with these questions peacefully, largely thanks to the activities from the 1830s onwards of those who were against slavery and came to be called 'abolitionists'. Whereas some of those against slavery were simply worried about preventing it from spreading into new states, the abolitionists wanted to abolish it even where no-one had hitherto questioned its legal right to exist. They had the advantage that since the eighteenth century, opinion in all civilized countries (few of which had substantial slave populations, even abroad) had been turning against slavery. It had been abolished (temporarily) in French colonies in 1794, in the British (permanently) in 1834. But in the United States slavery was actually growing – and growing fast – at a time when it was in retreat elsewhere. This made many Americans uneasy. Above all, though, the abolitionists had on their side democracy itself – they argued that the majority of the people of the United States should decide, and should, if necessary, change what the Constitution had said five or six decades earlier about the rights of individual states.

Abolitionists steadily raised the temperature of the debate by provocative acts. They helped slaves to escape from the South, resisting their return by the courts in the North, and circulated propaganda for their cause. Meanwhile, the politicians did their best to arrange

Those for and about whom the United States divided: black field hands on a Tennessee plantation after the War between the States.

compromises. For a long time this was enough; the South did not need to feel that it had its back to the wall. Only in the 1850s did the spirit of compromise at last break down. A new territory, Kansas, had then to be organized as a state. Abolitionists and anti-abolitionists fought to decide whether slavery should be allowed in the territory. Men were killed and people began to talk of 'bleeding Kansas'. A new party, the Republican, emerged over this issue, saying that Congress should decide what should happen in Kansas and as a result, it was at once identified by the South as an enemy. In the 1860 presidential election, the Republicans only went so far as to say that slavery should be kept out of any future territories to be added to the Union (the party was not abolitionist), but even that was too much for many southern politicians. When the Republican candidate won, the state of South Carolina announced in December 1860 that she would secede from the Union in protest. Within a month or so, a further six states joined her. They set up a new union, the Confederate States of America, with its own constitution, government and president.

Civil War

The greatest tragedy of American history had begun, thanks to good, even invincible, arguments on both sides. In the North – where most of the states staying with the Union and the strongest abolitionist feeling were both to be found – the government claimed that Congress was able to make laws binding the whole Union, because it represented the majority. This did not mean, said the Republicans, that slavery was to be abolished in the South, only that it was not to be introduced into new states. But, replied those in the South, if this is so, then those who do not agree are entitled to withdraw from a Union set up on a different understanding. Why, they asked, in effect, should not the inhabitants of South Carolina and rest of the southern states be as free to run their own internal affairs as the Hungarians or Italians who were claiming national freedom in Europe? What is more, southerners feared that, whatever might be said, if they conceded that Congress could make laws about slavery elsewhere in the Union, it would not be long before it began to make laws on its working inside southern states. Like all those about great and tragic questions, such arguments divided friends, neighbours and even families. They brought upon the United States a huge and bloody struggle called, according to your point of view at the time, the 'Rebellion' or the 'War between the States'. Most historians still call it simply the Civil War.

The new president of the United States was a lawyer from Illinois, Abraham Lincoln, who is to this day probably the greatest man ever to have filled his office. He was determined to do all he could to make the return of the southern states to the Union possible, but he was even more determined to save the Union. To begin with, he mobilized federal forces to restore the ordinary operation of the government in the southern states (this did not please the abolitionists, who wanted more). He once said that 'if I could save the Union without freeing any slave, I would do it; and if I could save it by freeing all the slaves, I would do it'. But he went on to proclaim the freeing of all the slaves in the rebel states on New Year's Day 1863, feeling that this was by then needed to win the war. This made the South even more determined to resist, though, and it took two and a half more years to defeat the Confederacy. In 1865, after that defeat and after Lincoln's death at the hands of an assassin, the final step was taken and the Constitution was changed to prohibit slavery anywhere in the United States.

On 7 November 1861 a British steamer homeward bound from Havana was stopped by a United States sloop, boarded and forced to surrender two passengers, representativesof the rebellious Confederate government of the South. The result was the last crisis in which there seemed a real danger of war between the two great Anglo Saxon powers. Popular feeling was greatly excited on both sides of the Atlantic. During the weeks of uncertainty Punch published these two cartoons. In the end, the United States government released its prisoners. There was never again to be so great a danger of collision, and even at that date no British government could contemplate a war with theUnited States with anything but dismay.

WAITING FOR AN ANSWER.

COLUMBIA'S FIX.

COLUMBIA. "WHICH ANSWER SHALL I SEND?"

The war was appalling; it killed over 600,000 Americans (out of a population of about 30 million at the outset), more than any war the United States has ever fought with foreigners. Most of them died of disease, but breech-loading rifles and cannon, together with the numbers and material which could be mobilized thanks to the railway made the battlefields scenes of dreadful carnage. The South started with many disadvantages. She was inferior in numbers to the North (the odds were about 2 to 1), had little industrial backbone, and only her cotton crop with which to buy goods abroad. But she had some fine soldiers and her people were fighting for what they believed to be survival. Nor did the slaves prove disloyal. So, in the end, it took four years of savage fighting, the South being gradually hemmed in to the old seaboard states and suffering terribly in the devastation of her territory and loss of life in the process.

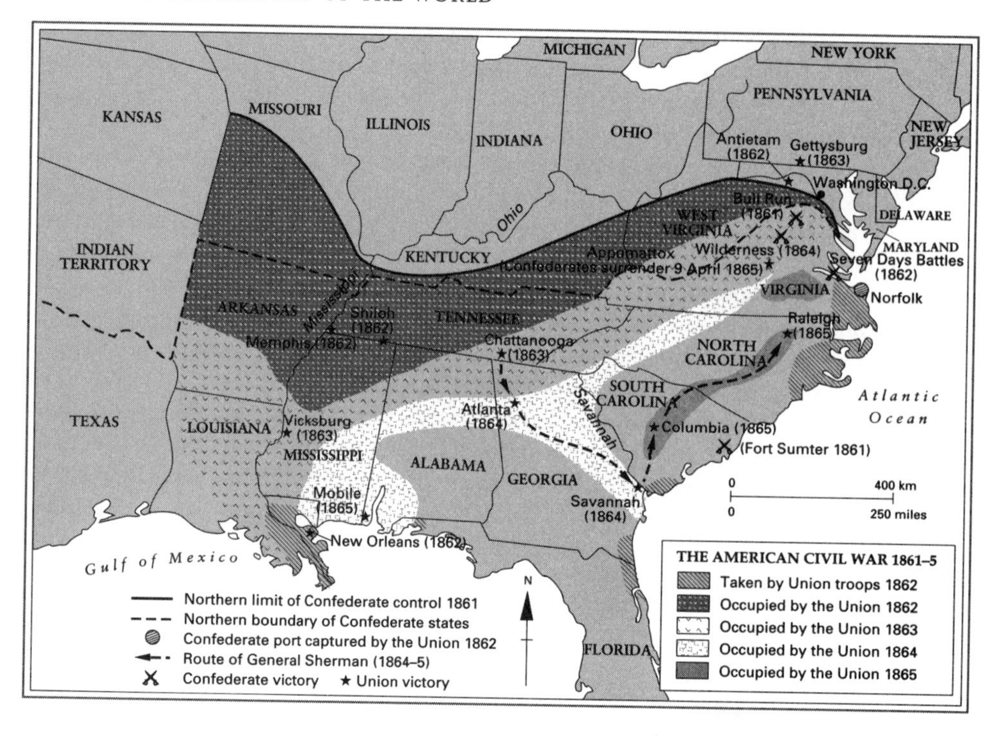

Unlike many wars, though, the Civil War was decisive. It settled some important matters for ever, not only for America, but for mankind. In the first place, it confirmed that the Americas would continue to be dominated by one great power. The threat that the United States would break up disappeared. That one great power would, therefore, exploit the riches of this great area was to decide in the next century the outcome of two world wars. The war also decided that this huge area would be under a democratic government; it was a triumph for democracy. Lincoln once gave a famous definition of it as a 'government of the people, by the people, for the people'. In its full meaning, that is still not, perhaps, to be seen anywhere on earth; nevertheless, the Civil War decided that the majority would in future have the last word through the national government of the United States, and not the individual states.

The outcome for those over whom the war had, in the end, come to be fought, the blacks, was clear in legal and constitutional terms: slavery came to an end and blacks became Americans, with the same constitutional and legal rights as other Americans. But this was not the whole story. Though millions of former slaves in the South (where most of them continued to live) were suddenly free, they were also uneducated, pretty much unskilled except for field labour, and had few leaders among their own people. Northern armies occupied parts of the South for a few years, and while they were there they protected the newly-freed in the use of their new rights. But when they went away, the black Americans had to live among white people who bitterly resented the changes brought to their ways of living by the new laws and hated them as a symbol of the South's defeat. They turned to bullying and economic pressure to hold them down; race relations were far worse twenty years after the war in the South than they had been before and the blacks' position grew worse, not better. The race relations question had been born as slavery died.

The war also led American politics to settle down in a two-party system which has lasted to this day. The Republican and Democratic parties which had been the main contenders in the election of 1860 have divided the presidency between them ever since. For decades, the Democrats' cause was to be associated with the South, and Republicanism with the North,

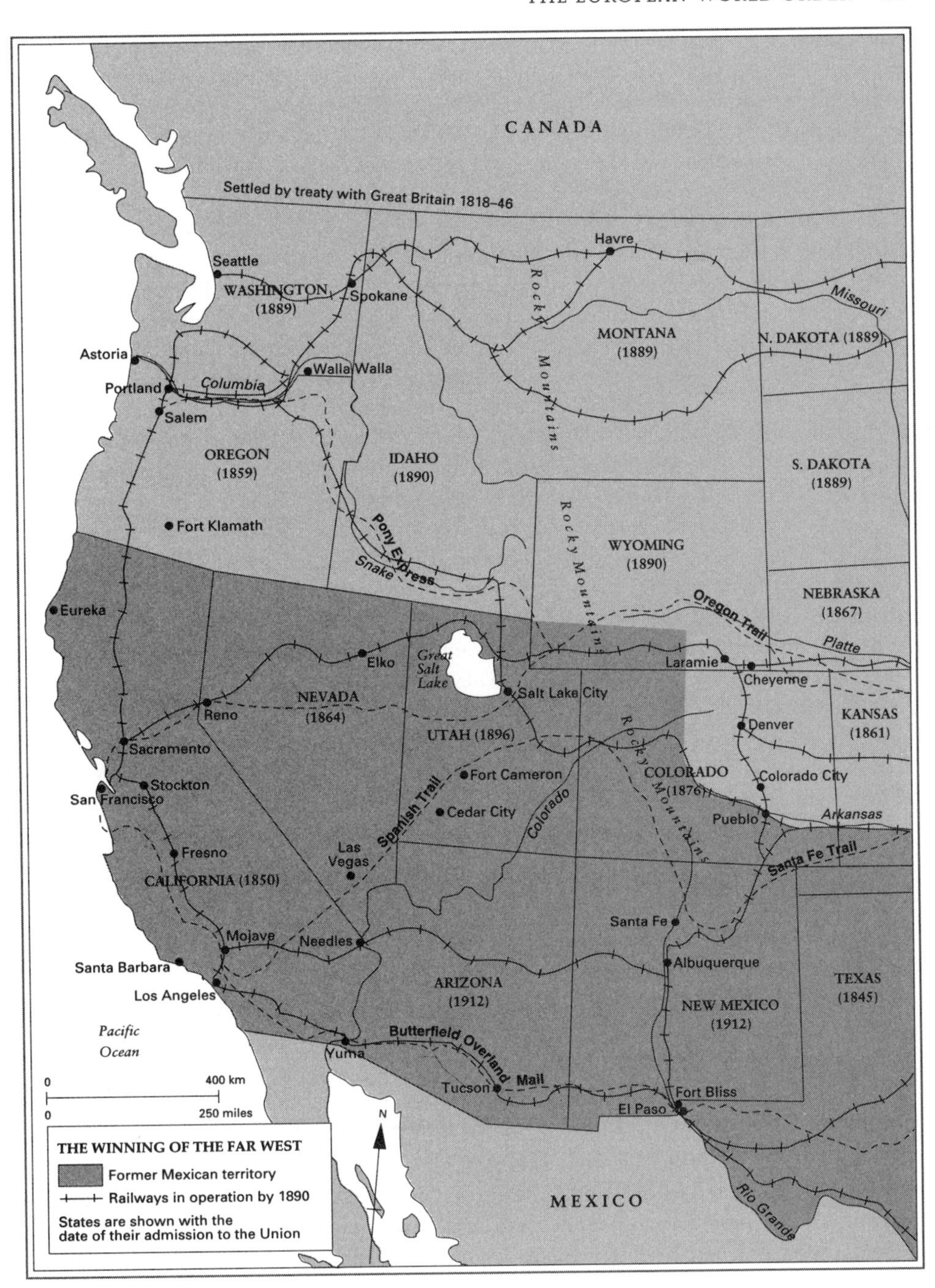

as the Union emerged from the nightmare of the war to resume the march of expansion interrupted in 1861.

The American economic surge

Soon the economic tide was with the Republicans as the great expansion interrupted briefly by the war was resumed. Its most striking manifestation had hitherto been territorial; it was about to become economic. The phase of America's advance to the point at which her citizens would have the highest per capita income in the world was just opening in the 1870s. In the euphoria of this huge blossoming of confidence and expectation, all political problems seemed for a while to have been solved. Under Republican administrations America turned, not for the last time, to the assurance that the business of America was not political debate but economic advance. True, the South remained largely untouched by the new prosperity and slipped even further behind the North, but Americans in the North and West could look forward with confidence to even better times ahead. Foreigners felt it too; that is why they were coming to the United States in growing numbers - two and a half million in the 1850s alone. They fed a population which had grown from just over five and a quarter millions in 1800 to nearly forty millions in 1870. About half of them by then lived west of the Alleghenies and the vast majority of them in rural areas. The building of railroads was opening the Great Plains to settlement and exploitation which had not yet really begun. In 1869 the golden spike was driven which marked the completion of the first trans-continental railroad link. In the new West the United States would find its greatest agricultural expansion; already, thanks to the shortage of labour experienced in the war years, machines were being used in numbers which pointed to a quite new scale of farming, the way to a new phase of the world's agricultural revolution which would make North America one granary of Europe. There were a quarter of a million mechanical reapers alone at work by the end of the war. Industrially, too, great years lay ahead; though the United States was not yet an industrial power to compare with Great Britain (in 1870 there were still less than two million Americans employed in manufacturing), the groundwork was done. With a large, increasingly well-off domestic market the prospects for American industry were bright.

Poised on the brink of their most confident and successful era, Americans were not being hypocritical in forgetting the losers. They understandably found it easy to do so in the general sense that the American system worked well. The blacks and the poor whites of the South had now joined the Indian, who had been a loser steadily for two centuries and a half, as the forgotten failures. The new poor of the growing northern cities should probably not be regarded, comparatively, as losers; they were at least as well off, and probably better, than the poor of Manchester or Naples. Their willingness to come to the United States showed that she was already a magnet of great power. Nor was that power only material. Besides the 'wretched refuse', there were the 'huddled masses yearning to breathe free'. The United States was in 1870 still a political inspiration to European radicals.

American overseas imperialism

Not until the 1890s was the 'frontier of settlement' declared no longer to exist and so the peopling of the West accomplished. Nevertheless, once the Civil War was over and the linking of the two coasts by rail and telegraph had taken place, more had already begun to be heard about United States interests abroad and the need to look after them. At the end of the century this bore fruit in an American decision to join in the imperialist movement like everyone else – except that, like every other example of imperialism, it was not exactly like everyone else's and had its own special features. One was that many Americans felt uneasy about it. They argued that the republic had itself come into existence in a revolt against a colonial power and were sure it should not go into the business of empire for itself. The constitution made no provision for ruling colonies, only for territories which might eventually become full states within the Union: how could this be expected to follow for places hundreds, even thousands, of miles away? (This overlooked the dubious circumstances in which much American territory had already been acquired; even the agreed Alaska purchase from the Russians was certainly an extension of American government over foreign territory that was not contiguous with that of the United States.) Nevertheless, American imperialism went ahead.

Geography led beyond American shores in two directions. One was west, across the Pacific; the other was south, towards the Caribbean and South America. American trade and whaling were well established in the Far East at an early date; in the 1820s the US Navy already had a Far Eastern squadron. The first American arrivals in Hawaii occurred at about the same time. Soon after this, the spectacle of other powers getting concessions from the Chinese empire led the American government to make similar treaties and then to send Commodore Perry to make the Japanese open their ports to foreign trade.

An explosion which detonated a war. The USS Maine blows up.

In the second half of the century the Americans came to share in the administration of Samoa, to acquire Hawaii, and then took from Spain the Philippines and Guam. The motives behind these steps were complex; some Americans were anxious to look after America's interests when other powers seemed to be acquiring territory, so that they should not go by default, some talked economic nationalism and about the need of markets for exports (though as the United States was a huge domestic market for its own manufactures, this did not have much substance to it). Still other people argued that you could deduce from what Darwin said (or what they believed him to have said) that the struggle for survival between nations was like the struggle between natural species for survival; those which were obviously vigorous would come out on top, showing that they were meant to do so by ruling other peoples.

Yet, so far as acquiring new territory went, American overseas imperialism did not in fact go on for very long. The final annexation of Hawaii in July 1898 came in a burst of aggressiveness and expansion whose main victim was the old imperial power of Spain. In February 1898, an American cruiser, the USS *Maine*, mysteriously blew up in harbour at Havana, Cuba – which was then still a Spanish possession. American business interests had long been prominent in the island, and American sympathy had long been given to a Cuban rebellion which the Spanish, in spite of great efforts and much brutality, had been unable to master. Without very good reason – no-one has found out why the *Maine* blew up – the United States went to war with Spain. It was, said a future American president, a 'splendid little war'. American marines and soldiers defeated the Spanish in Cuba, and the entire Spanish Atlantic squadron was sunk in a battle in which Americans suffered barely a scratch. Across the Pacific, the Spanish Pacific fleet was destroyed in Manila Bay and a rebel movement to overthrow Spanish rule in the Philippines was given American support. At the peace, Guam, the Philippines and Puerto Rico passed to the United States and Cuba gained its independence, though only on terms which allowed the United States to reoccupy the island under certain circumstances (as happened from 1906 to 1908, for example).

The Caribbean

When enthusiasm for imperial conquest died down after the Spanish War (as it did, and pretty rapidly) the southern front of American power still preoccupied the United States in a special way. It was the old interpretation of the Monroe Doctrine that the hemisphere was of special concern to the United States, which therefore had special rights to act in it in defence of its interests. Now a new aspect of those interests became apparent as modern technology

Two ocean power. A United States battleship passes through the Panama Canal soon after its opening.

made possible a canal across the isthmus between North and South America, connecting the Pacific to the Atlantic through the Caribbean. American strategists were especially concerned by its possibilities. The rise of Japanese naval power made the maintenance of a strong fleet in the Pacific more desirable than ever; its reinforcement could be made much easier and more rapid if it were done via Panama rather than round Cape Horn.

In 1903 a treaty for the acquiring of land from Colombia through which a canal was to be built was rejected by the Colombian government. A revolution in Panama, where the planned route ran, was, therefore, engineered with American support. The United States prevented the suppression of the revolution and a new Panamanian republic which emerged from it then duly handed over to American jurisdiction and occupation a strip of territory which was to be the Panama Canal Zone. It also conceded to the United States the right to intervene in its affairs if necessary, in order to maintain order. Work then began. Outstanding engineering (unlike Suez the Panama Canal had to be given locks) enabled the canal to be opened in 1914.

The Panama Canal changed American strategy. It gave a new twist to American policy in the whole Caribbean area. Since the Canal was the key to America's naval defences, it had to be specially protected. There followed much more American interference, sometimes with armed forces, in the affairs of the Central American and Caribbean republics; if they fell into disorder, it was argued, a situation might arise which a power unfriendly to the United States could exploit. Americans who did not like this policy were quick to attack it as imperialism under a new guise.

In the Far East it soon began to seem as if the forebodings of anti-imperialist Americans might be justified. Not long after the Philippines had been taken over, a revolt which had begun there against Spain turned against the Americans. There began a lengthy and costly guerrilla war. By 1902, when it was brought under control, American opinion was only too anxious to hand over government to the Philippinos if there was a way to do so safely. Unfortunately, this was difficult (it took, in the end, until the 1930s), and there was always the danger that if the United States abandoned the islands, then another imperialist power would step in and take them – Japan, perhaps. That might imperil other American interests in the area, notably those of trade with China. Alarmed at what would happen if China collapsed, the Americans supported what they called an 'Open Door' policy there; foreign powers should keep their hands off China, they said, preserving the treaties which gave them trading rights and competing with one another peacefully through economic means. As this was already British policy, the United States was not likely to be challenged if it took this line.

President Theodore Roosevelt, who had engineered the Panama revolution which made possible the building of the Canal, was also the first president to assert the right of intervention in Caribbean states. (This was called the 'Roosevelt corollary' to the Monroe Doctrine.) He sent marines to Santo Domingo to make sure its debts to foreign investors could be paid and so deprive foreign powers of an excuse to intervene there. Under his successors, such intervention among her neighbours came to be called 'dollar diplomacy'. Marines were in due time sent to Nicaragua by President Taft. President Woodrow Wilson, who took office in 1912, went on a great deal about renouncing imperialist ways, but in practice did much the same as his predecessors. From 1914 to 1916 American marines occupied Santo Domingo, the government eventually being suppressed so that a new constitution could be imposed by the Americans. Haiti was for a time occupied in 1915. The most flagrant example of Wilsonian intervention, though, was in Mexico. When a military dictator took over that country Wilson withheld recognition of him on the grounds that his regime did not measure up to the moral standards of the United States. Marines were landed at Vera Cruz in 1914, only to be withdrawn after the dictator was forced from office. But American punitive expeditions were back in Mexico a few years later, though, it is true, in response to a Mexican general's raid into New Mexico.

By then, though, many Americans were heartily sick of external adventure. It had cost a lot of money and brought little profit. Nor was there any chance of further expansion anywhere except in the Caribbean area, for the rest of the globe was by then almost completely divided among other powers. When, in 1914 a great war broke out in Europe, foreign entanglements once more seemed unattractive to Americans.

ASIA IN THE EUROPEAN AGE

China

In 1800, two great empires, in very different stages of decay, unknowingly faced a century of increasing humiliation at the hands of white men. Legendarily, it is said that Napoleon had once described one of them, China, as 'a sleeping giant'. 'Don't wake him', was his advice. Yet twenty years after Napoleon died, that was exactly what the Europeans had begun to do, without, as first, it seeming that there was much danger in ignoring his advice. Since its great days in the early eighteenth century, Manchu power had been undermined at home and had weakened abroad. Yet its officials still sent a British emissary away in 1793 with a dismissive and patronizing message to his ruler – George III – in what they called 'the lonely remoteness of your island, cut off from the world by intervening wastes of sea' which was quite in line with their long-standing views of the outside world. For them China was the centre of civilization, the 'Middle Kingdom' surrounded by vassal peoples under the influence of her ways (Tibetan, Vietnamese, Koreans, for example) or, beyond that, by inferior barbarians whose doings were of no account.

Yet the Manchu had passed their peak. Great revolts had already begun to break the long internal peace, a traditional sign of decay in the imperial authority. A surge of population since the middle of the seventeenth century much more than doubled the population in the next century and a half; by 1800 it stood at about 330 million. This was too much for Chinese agriculture; almost all cultivatable land was being used and the most arduous efforts could not raise much larger crops with existing knowledge and technology. The upheavals

caused by rebellions expressed the sufferings of the empire's subjects. Secret societies and religious cults could exploit them to keep alike old dislike of the dynasty (the Manchu, after all, were foreigners) and popular grievance showed every sign of growing fiercer after 1800 as inflation began to send up prices.

Though former dynasties had sometimes shown that they could survive bad times for centuries if need be, the Ch'ing (who were, in the end, to hold out until 1911) also faced an unprecedented challenge from the outside. The problems posed by 'barbarians', usually from Central Asia were not new. Sometimes the Ch'ing's predecessors had been overthrown by them. But always this had been followed by cultural digestion of the barbarians: after each conquest, the imperial administration had remained in the hands of the scholar-gentry trained in the Confucian tradition, and the people had often been virtually unaffected by the change of rulers. The barbarians were then 'sinicized' by the higher civilization they took over. In the nineteenth century, though, China faced for the first time barbarians (the Chinese did not make distinctions between whites for a long time and called them all 'feringhi', which was their version of the old word 'Franks') who were not going to be impressed by Chinese ways but would view them with contempt. Instead, the Europeans would try to inject Chinese life and China's rulers with their own ideas, often by military and political means.

Manchu China

1644	Manchu seize Peking. Beginning of Ch`ing dynasty
1662–1722	The K`ang-Hsi emperor
1683	Conquest of Formosa
1696	Control established over Mongolia
1724	Control established over Tibet
1736–96	Ch`ien Lung emperor. Empire reaches its maximum extent with victories in the Ili region (1729–34, 1754–61), Burma (1767–9) and Tibet (1791–2)
1839–42	Opium War between Britain and China; Hong Kong ceded to Britain
1851–64	Taiping rebellion
1856–58	Second Anglo-Chinese War ends with Treaty of Tientsin
1859–60	Third Anglo-Chinese War culminates in French and British troops capturing Peking
1894–95	Sino-Japanese War ended by Treaty of Shimonoseki
1898	Hundred Days reform movement. After *coup d'état* dowager empress Tzu Hsi becomes sole ruler
1900	Boxer rebellion
1905	Founding of the Kuomintang (National People's Party or KMT)
1911	Overthrow of the Manchu dynasty

Lord Macartney's reception by the emperor – as seen by an English cartoonist who reveals more of his countrymen's attitude than those of the Chinese.

The opening of China to the West

The new challenge came upon China much more rapidly than might have been expected. Ever since the sixteenth century, trade between China and Europe had shown what economists now call an 'unfavourable balance' from the European point of view. Put simply, Europe had very few goods to offer which the Chinese wanted. European traders in China had, therefore, to pay for what they bought in cash – silver, since that was the basis of the Chinese currency. They had no goods to hand over in exchange. The British East India Company, for example, had to ship silver bullion out to the East to pay for the tea and other goods its ships picked up in eighteenth-century Canton. In the first years of the nineteenth century, though, this situation changed, and did so very swiftly.

Opium is a drug made from poppies with a long history as a useful pain-killer. But it has also been very popular for other reasons; it seems to make life pleasanter by taking away troubles and worries. For this purpose, some people have always used it rather as others have used alcohol, another popular drug with some similar uses. The parallel is not exact, though. Alcohol can make people show increasing excitement and obstreperousness, but opium produces a drowsy, foggy contentment which ends in sleep and happy dreams. Opium (or its derivatives) can be taken in many forms, but one of the most common is by inhaling its smoke from a pipe like that of tobacco, and this was how the Chinese of the south, who rapidly developed a great taste for opium, took it. The British at last found in opium a commodity which the Chinese wanted and it could be grown in India.

Unfortunately, like many another drug, opium is addictive; people become dependent on it, and will then break many of the ordinary rules of social life to satisfy their craving for it. What is more, the particular effects of opium (in making people lethargic, careless of the future, irresponsible) all seemed very undesirable to Chinese officials. The Manchu officials therefore forbade the drug's import (which had the additional disadvantage in their eyes that because it came from abroad, it might make China dependent on foreigners). When the ban was imposed, consignments of opium already landed were seized and destroyed.

So began the awakening of China. A great outcry by British merchants followed the burning of a large quantity of opium in 1839 at Canton. Reasonably enough, Lord Palmerston, in charge of foreign affairs in London, told them that his government could not intervene to help its subjects break the laws of the country where they were asking to do business. But British officials on the spot thought otherwise and opened hostilities. Attempts to patch up the quarrel on the spot failed; much bigger naval operations followed and in what became known as the 'Opium War' British troops were landed to occupy not only several southern ports but other places. Outright bullying forced on China a peace treaty in 1842 which required the opening of five of her ports to foreign trade, the levying of a single rate of fixed duty on imports, and the cession of Hong Kong to Great Britain, all of which were interferences with her internal sovereignty. It is not an episode on which many Englishmen now look back with pride but to many at the time civilization meant not only lining one's pockets but overcoming backwardness. Free trade was not only expected to create a prospering economy of benefit to both sides, but to enable Christianity and humanitarianism to be brought to bear on what they saw as the barbarities of a pagan society – the Chinese subjugation of women, for example, or the persistence of judicial torture.

Within ten years, the Americans and French too had signed 'unequal treaties' (as they were later called) with the empire which gave them rights of trade and diplomatic representation, won special legal protection for their citizens and, in the end, admitted missionaries and permitted the toleration of Christianity. The 1840s thus began the visible undermining of the authority and prestige of the empire, though that had not been the aim of European governments. The treaties had forced the Manchu to acknowledge the ending of the age-old principle of Chinese foreign relations that all foreign peoples were really tributaries; now Chinese diplomacy had to accept western ideas about the sovereignty of individual states. Worse still, though, the arrival of foreign merchants and Christian missionaries in increasing numbers and the fact that they could not be brought before Chinese courts showed that the imperial government could not resist the will of barbarians it officially despised.

The missionaries were particularly insidious. Not only did they preach and teach in ways undermining the Confucian tradition and social system (the idea that all human beings were equal in the sight of God was a revolutionary one in China), but their converts began to claim the protection of the European consuls and courts. They tried to live in European areas where they could not be harassed by Chinese officials. Even when the missionaries met popular hostility, too, as they often did, this did not help the officials much. If they protected the missionaries against a riot, then they would become unpopular, if they did not, the missionaries were likely to get killed, the local European consul would send for a gunboat or soldiers to seize the murderers, and the imperial administration would be shown up as unable to protect Chinese against the foreigners.

Such strains were felt against the background of deepening social distress and a growing danger of rebellion. But the Manchu and their officials took a long time to recognize that the empire was approaching a crisis which might pose a mortal danger. Though a few of them saw that some concessions might have to be made to foreigners, almost all the officials felt that China had often been in difficulties before and that she had always come through them. The last significant intrusion of ideas from the outside into Chinese culture had been Buddhism and that had been absorbed successfully in the end. The superiority of her culture was such (they were sure) that China must, in due course, resume her rightful position in the world, however bad things might look for a time. A few thought it might be wise to find out from the barbarians some of the secrets of their steamships and cannon so that they could be put to use by the empire but even educated Chinese could not believe that traditional ways might have to be changed or even abandoned and that quite new ideas might be needed if the empire was to survive at all.

Concessions and decline

Such attitudes made it very difficult for China to react effectively to European civilization. A half-hearted borrowing of machines and the employment of European generals (rather as barbarian generals from the Central Asian deserts had been employed by earlier dynasties) was one response. In the 1860s Europeans were employed as military help in mastering the greatest of all the revolts of the century, the Taiping rebellion which raged from 1850 to 1864. It had begun as a local outbreak under a leader who nevertheless soon showed he might win a wider influence and who owed his ideas in part to American missionaries, and talked a sort of Christian communism. In the end the rebellion was crushed, but not before the Ch'ing had been forced to make yet more diplomatic and commercial concessions to the foreigners in order to buy time and support against the rebellion. Nor had its suppression even with foreign aid been easy; it seems to have cost about twenty million lives.

Amid the disorders of the Taiping rebellion an Anglo-French invasion between 1857 and 1860 led to the occupation of Peking and the sacking and burning of the Summer Palace before further treaties extracted more humiliating concessions from China. In 1858 her lands north of the river Amur were given to Russia, and the Ussuri peninsula (on which the Russians were to build Vladivostok) was handed over two years later. Big surrenders of Chinese territory were also made to Russia in Central Asia, beyond the province of Sinkiang. Russia's appetite was not surprising; she was the power with the longest land frontier with China and had been pressing forward in Central Asia for decades before this – and on the Amur since the days of Peter the Great. But other European states also nibbled away at territories over which the Chinese claimed overlordship, even if they did not rule them directly: the British took Burma, the French much of Indo-China. Before the end of the century, perhaps encour-

Soldiers of the 'ever victorious' army raised and trained to fight the Taipings by the British general, Gordon, on behalf of the imperial government. Later, in 1883 at Khartoum in the Sudan, Gordon was to die what many of his countrymen regarded as a martyr's death.

aged by a Japanese seizure of Formosa (Taiwan) and spurred on somewhat by the fear that rivals might get ahead of them in the race if China collapsed altogether, the Europeans were again at work in the land-grabbing business in China itself. The Russians installed themselves in Port Arthur; England, France and Germany took new seaports on long leases, and even the Portuguese (who had been in Macao longer than any other Europeans had been in China) turned their old lease there into outright ownership (they said). In the background, moreover, was a continuing stream of concessions, loan agreements and interference in the Chinese administration all of which made China look a country really under foreign control, even if still legally independent.

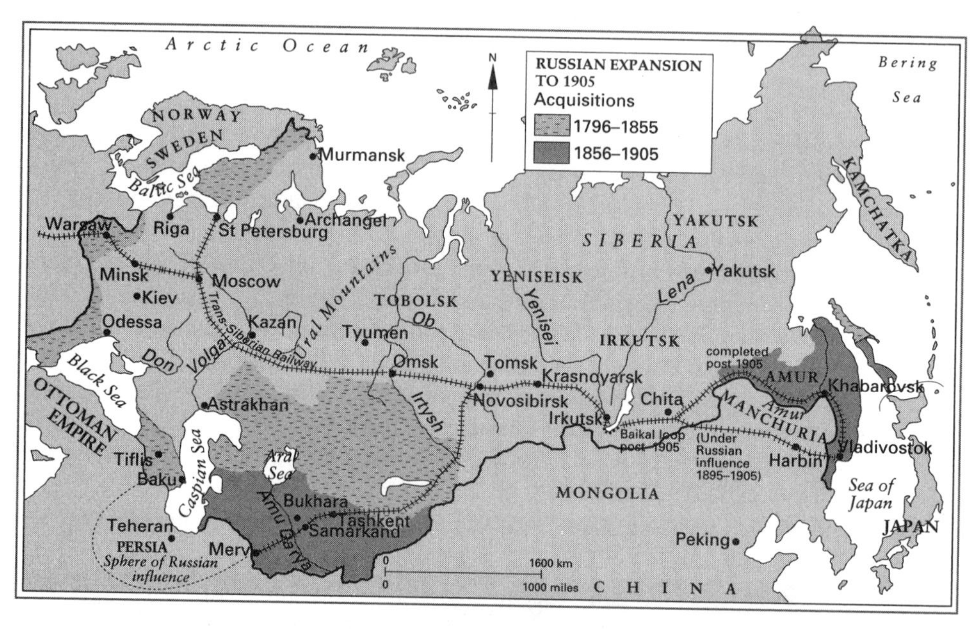

By 1900 many Europeans were expecting China to break up or crumble away as the Ottoman empire seemed to be doing. Far from the sleeping giant awaking, it seemed to be likely to undergo the death of a thousand cuts – a legendary Chinese torture – being sliced up bit by bit by the predators from the west. Some Chinese, though, were determined this should not happen. As early as the 1870s, a 'society for self-strengthening' was founded to consider western ideas and inventions which might be useful. Its members drew attention to the efforts of Peter the Great and to contemporary efforts to modernize another Confucian society, Japan. Students were for the first time officially sent abroad to study in Europe and the United States. Yet even those who sought reform still found it hard to shake off the idea that they should root it in the Confucian tradition.

Reform and revolution

It made things worse that reform became an issue tangled up in court politics. The emperor (who had come to the throne in 1875 as a child) was at loggerheads with the dowager empress soon after his personal rule began in 1889. When, in 1898, the reforming party seemed at last to be getting somewhere and a stream of reforming edicts and laws was issued in what was known as 'The Hundred Days of Reform', the empress enlisted the support of Manchu officials and soldiers whose sinecures and privileges were threatened, seized the emperor, locked him up and swept the reformers aside. At about the same time, signs of popular support for sticking to old ways could be seen in an outbreak of troubles in the provinces where certain militia units had come under the influence of a widespread and secret society called (somewhat oddly to western ears) the 'Society of Harmonious Fists'. Its members

Every intervention by a foreign power in China had implications for other powers who might find the balance of their own interests threatened. The anti-Boxer campaign attracted wide support for just this reason, as Punch noted.

were usually called 'Boxers', for short. They were violently anti-foreign. They attacked Christian Chinese converts and, soon, foreign missionaries.

The Boxers were secretly favoured by the Manchu officials and the court, which hoped to use them against the foreigners. When there were diplomatic protests and demands that they be suppressed by the government, a full-scale rebellion broke out, egged on by the dowager empress and her agents. European troops seized Chinese forts in order to secure the route to Peking, where there was a large foreign community to be safeguarded. The empress declared war on all foreign powers, the German minister at Peking was murdered and the legations there were then besieged for several weeks; elsewhere, more than two hundred foreigners – mainly missionaries – were killed.

Retribution was swift and disastrous. An international expedition fought its way to Peking and relieved the legations. The Russians occupied southern Manchuria. The court fled from the capital, but after a few months had to accept terms: the punishment of the responsible officials, the payment of a huge indemnity, the razing of forts, foreign garrisons on the railway to Peking, and the fortification of an enlarged legation quarter. The Boxer rising had not only failed, but had done further damage to the already shaky Manchu regime. The internal outlook was now more uncertain than ever and there were now Chinese who had begun to think about revolution.

THE BRITISH RAJ IN INDIA

By that time, there was also rising opposition to imperial rule in the sub-continent of India, though the imperial power there was no longer native but European. India had become enormously important to the British. Their imperial history indeed makes no sense without India, and even its shape was in large measure given it by India. A great many other parts of the empire were only there because they were important either to the defence of the sub-

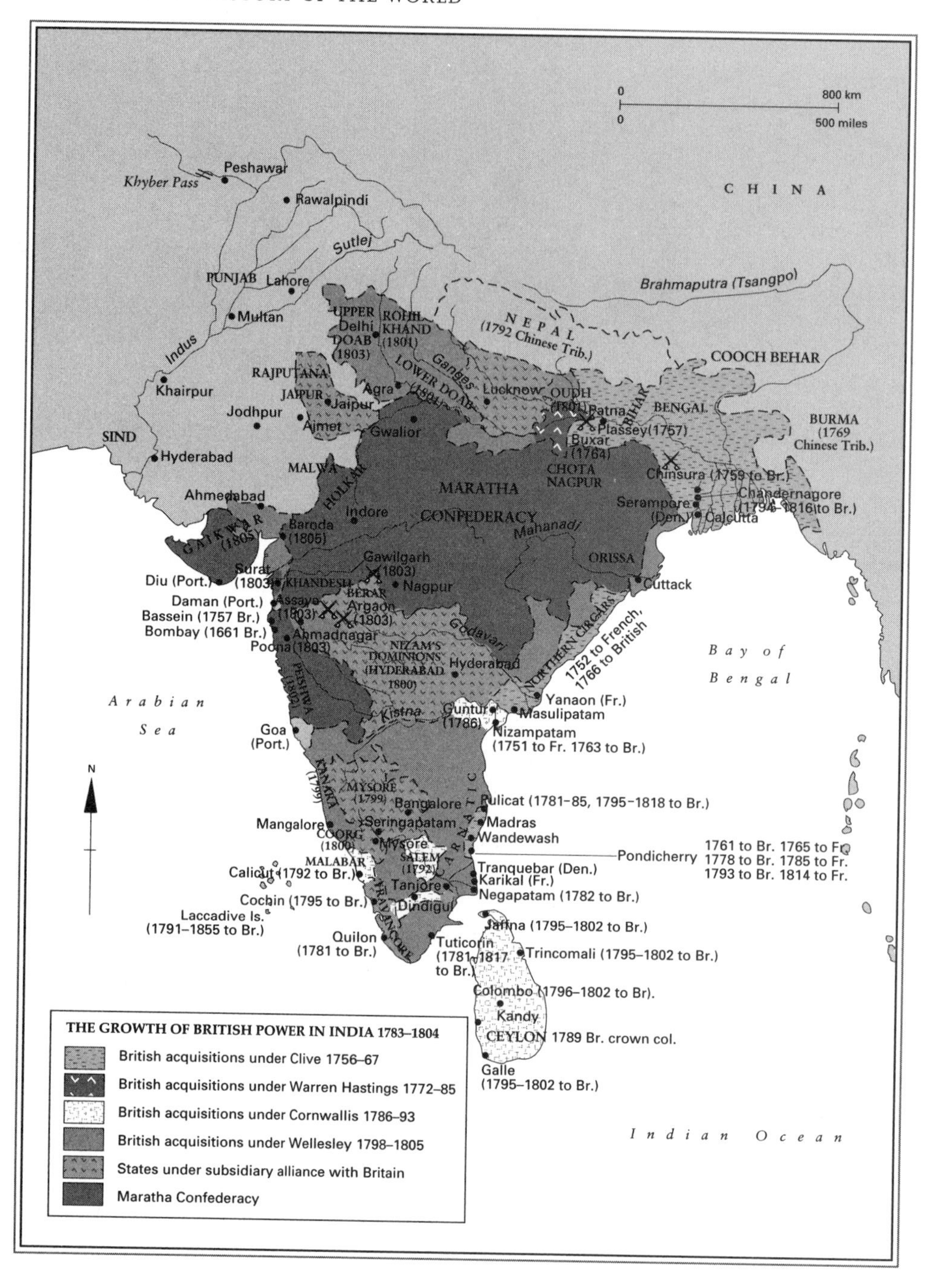

continent, or of the sea-route to it from England. There had been even in 1800 more people under British rule in India than in any other imperial possession – and by 1900 there were more than all the rest of the empire put together. As time went by, great numbers of Indians migrated to other parts of it, and Indian communities appeared as far away as Fiji, East Africa and the British West Indies. Indian trade, too, was always important; the sub-continent took large quantities of British manufactures. Indian soldiers helped defend many other parts of the empire, and at one time or another fought on Great Britain's behalf in every continent except the Americas. Finally the inhabitants of the sub-continent and the British have always had an important and continuing influence on one another's cultures, with results still visible today.

When the 'Raj' (as it came to be called, by people who saw in British rule the successor to that of the Moghuls) was in the making, very few people envisaged such an outcome. The ruler of British India in 1800 was still in name the East India Company, though its governor-general had since 1784 been appointed by the British government. It had been created to trade, and for a long time its agents went on seeing India in this light, as a place where all they wanted from government was to make sure that they could get on with business. Even in the eighteenth century, though, the acquisition of taxing rights in Bengal from the native ruler had entangled the Company in Indian and English politics, and the British government's share in running India grew steadily. Just as steadily the trading privileges of the Company declined. It lost its monopoly of trading in India (in 1813) and that with China twenty years later. So it came to rely on taxes for its revenue; it behaved more and more like an ordinary colonial government, in fact.

A young Indian prince and his advisers. A British political officer was staioned at each of the courts of the native Indian states, whose internal administration was left in their own hands – unless intervention became necessary because of intolerable inadequacy or potential danger to the Raj.

This system was called 'Dual Control', and it lasted in name until 1857, the British government steadily taking more and more of a hand as time went on. Meanwhile, more and more native Indian states were either annexed or brought under British control by treaty. The Moghul emperor was powerless to stop this, though he remained the nominal ruler of much of the sub-continent. Persian ceased to be the language of law and administration and was replaced by English. Missionaries were allowed to work in India after 1813 and began to attract more converts (there had always been some Indian Christians in the Portuguese and French settlements). Colleges and schools were founded, and the first Indian railway was built in 1853. The British governors-general by and large favoured such changes. They saw them as enlightening improvements, just as they introduced new legal codes as an advance on Hindu and Muslim practice. The population grew and in 1850 numbered about 200 million. About 70% of these were Hindu, about 20% Moslem.

The Mutiny and its results

More and more Englishmen (and, after the opening of steamship lines to Europe, more Englishwomen) came out to India to make their careers there, but their numbers remained tiny by comparison with this huge mass. In their daily lives, most Indians remained largely untouched by British rule, living in villages where tradition regulated their lives. It long looked as if the future was one of continuing enlightened despotism – no-one envisaged that the Indians might one day rule themselves. Then, suddenly in 1857, a terrible shock was given to British confidence. A series of uprisings then broke out after mutinies by native soldiers in Bengal who believed that a new kind of cartridge issued to them was greased with the fat of animals which their religion taught them to regard as unclean and so would pollute them. Other outbreaks followed. Soon, over most of northern India British rule seemed to be in danger. The mutineers attracted support from other Indians, Moslems and Hindus alike, who feared the modernization brought by the British because of the threat it presented to their traditional customs and ways. Some native rulers, both Hindu and Moslem, jumped at the chance, as they thought, of recovering their independence. Most Indians, though, and most of India, took no part in what was called 'the Indian Mutiny'.

Few though the British were, they struck back ruthlessly with the aid of loyal Indian soldiers. In a few months, the danger was over. Violent punishments followed. The old Moghul emperor whom the mutineers had claimed as their leader was deposed. The rule of the Company came to an end and the governor-general became a 'viceroy' who reported directly to the government in London. From now on until its end ninety years later, the British government of India was to be that of the crown, whether directly or indirectly.

This might have happened anyway. What the Mutiny brought about in other ways is less certain. Of the conservative and reactionary aims of the rebels, none was achieved. In one respect, though, it was decisive, for the shock it gave to the British (and especially to those who lived in India) was never to be forgotten. In India, from this time onward, British and Indians lived increasingly separate lives hardly associating with one another except on business. British residents tended to feel that India was a strange, incomprehensible place, where an untrustworthy and childlike population had to be kept in hand, if necessary by force. Such attitudes hid the fact, nonetheless, that there were hundreds of Englishmen in India, many of them in government, who were devoted students of India's languages, culture and civilization (the launching of serious study of classical India was the work of British scholars). It was also true that Indians and British would go on influencing one another willy-nilly. The trading connexion with Great Britain and the rest of the Empire was bound slowly to change Indian economic life. The ideas and principles taught in Indian schools and colleges and practised by the administration helped to shape many young Indians' ideas of the future their country should have; it was one often conceived on European lines, with as its goals democratic and representative political institutions, and a state based on the western ideal of nationalism.

On the other side, Great Britain's role as a great power was increasingly shaped by both the power India provided and the new necessities it imposed. 'As long as we rule India', said one viceroy, 'we are the greatest power in the world. If we lose it, we shall drop straightaway to a third-rate power.' In order to keep India safe the British were to be involved in almost continuous fighting with the tribes of the North-West frontier, the conquest of Baluchistan and Kashmir, and diplomatic struggles with Russia over the question of influence in Afghanistan which on at least one occasion looked like starting a war. In the 1880s, Burma was annexed to protect India from a possible French advance from Indo-China, and a few years later the Malay states were taken over with the same idea. An expedition was sent to Lhasa in Tibet in 1907 in order to ensure that it remained free of foreign influence. And, of course, India influenced British strategical thinking about Africa, which lay on the long steamer routes via Suez or the Cape.

Medieval English kings held 'crown wearings' which asserted their title to their office and their dignity. In 1911, in India, a similar ceremony, but blended with the monarchic rituals of Asia, was held. It was the only occasion in the whole history of British India when the reigning monarch went there, to attend a great Coronation Durbar at Delhi. George V, the King Emperor, and his consort, Queen Mary, wore the crowns and robes of English majesty, and were accompanied by heralds in the tabards of the Middle Ages, but were paged by the sons of Indian princes who had come to do homage as they would have done to a Moghul emperor.

Meanwhile, the government of India's 300 million people over the whole sub-continent except a few tiny Portuguese and French enclaves had to be carried on. In 1892 there were only 918 white civil servants to do the job. Unusually there was one British soldier in India to about 4000 Indians. Clearly, the 'Raj' did not rely on European numbers. It rested instead, on two foundations: the participation, help and goodwill of Indians, both on the civil and military sides, and on not trying to do too much. After the Mutiny, too, the British were cautious for reasons of public order about needlessly antagonizing Indians by interference with tradition. They drew the line at allowing the killing of girl babies (whose future dowries were seen by parents as an unnecessary expense) but did not interfere to stop child marriage. They formally regulated the rights of Indian princes and buttressed their rule.

All the time, though, the economic and cultural consequences of the British power were changing India in ways which would in the end make the preservation of the British Raj more difficult. When British industrialists and trades unionists thousands of miles away used parliament to put obstacles in the way of Indian entrepreneurs anxious to profit from the most settled period of government India had ever known, Indian merchants and manufacturers were annoyed. Young Indians of the Hindu elite studied at British universities or read for the English bar – and were then irked, on returning to India, to be looked down on as 'natives' by Englishmen in India. They wondered why words like 'fair play' and 'democracy' could not be applied in India, too, and so demonstrated the powerful contagion of British culture. Slowly, fed by these and other influences, Indian nationalism began to crystallize and take political forms. A few villages had encouraged this by advocating more Indian local self-government. National self-consciousness was handicapped, though, by divisions between Hindu and Moslem elements within it. Partly because of this, the Raj still had not by 1914 greatly relaxed its grip. The forces at work undermining it nevertheless continued to accumulate.

A NEW ASIAN POWER

Nineteenth-century Japanese were among the keenest observers of events in China and India. Though, in 1800, the country was still largely unknown to Europeans (other than a few Dutchmen), there were beginning to be signs that things could not go on like this for much longer. The Europeans were relatively much stronger than two hundred years earlier, and the Japanese much weaker. They would find it harder to keep out foreigners if they really wanted to break into Japan's isolation and, if they did, there were the different fates of China and India to consider as possible outcomes. The social strains produced by a long peace and the growth of new economic interests, together with her military obsolescence, meant that Japan faced the likelihood of European and North American pressure in a disadvantageous position. Some Japanese knew this. They had already begun to get round the regulations designed to keep out foreign ideas by importing books of what was called 'Dutch learning'. Even the shogunate had begun to authorize the translation of some European books on technical subjects. An ingenious people, the Japanese showed a willingness to copy and borrow which was quite unlike the superiority with which Chinese reacted to western influences. Without help except from illustrations in Dutch books, for example, a group of Japanese physicians carried out in 1771 their first human dissection (of the body of a criminal). Their readiness to learn

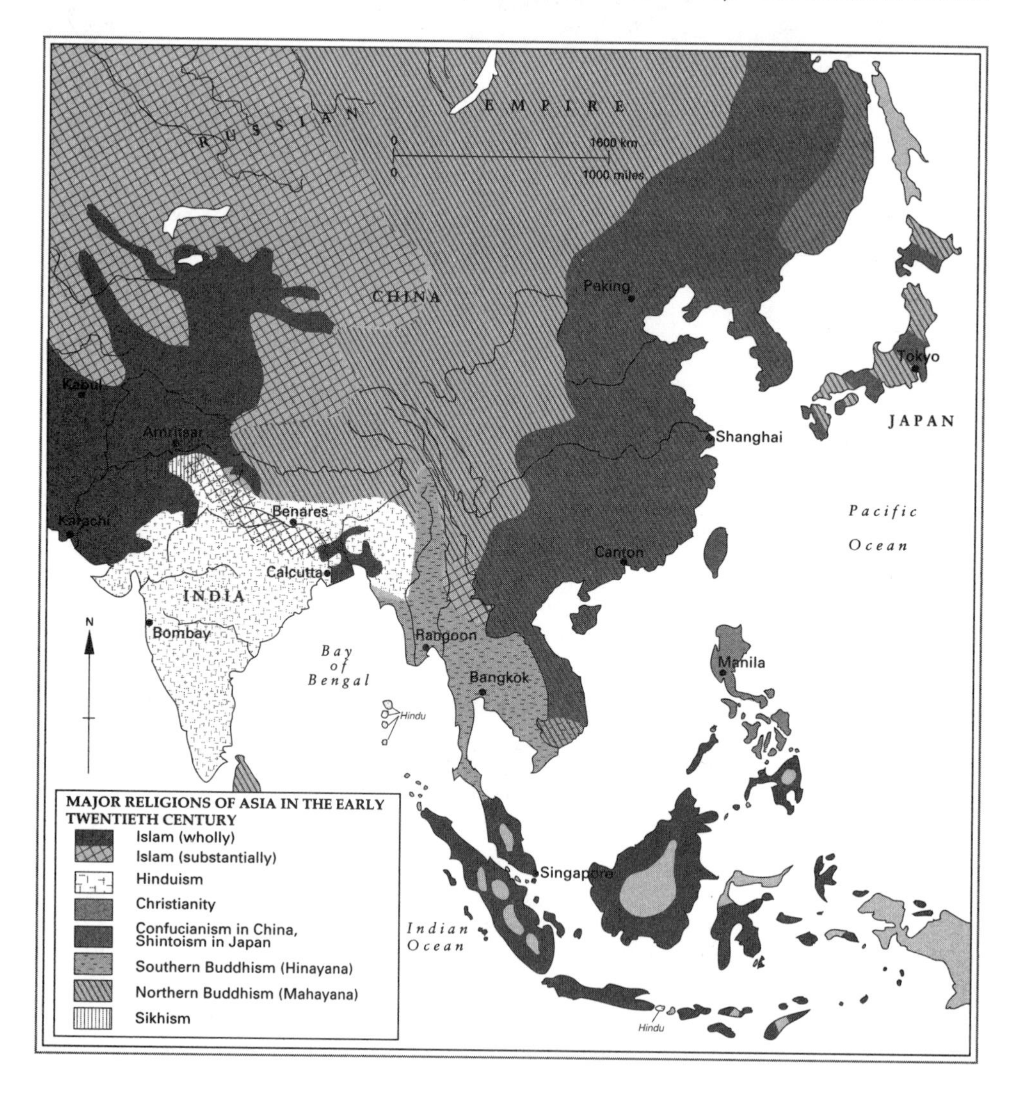

and to change their ways for more effective ones was to the advantage of the Japanese in facing the challenge of the foreigner, though there was little agreement about what was actually to be done. Some talked of 'expelling the barbarians'; others of 'opening the country', but both courses had risks.

The bullying of China by European states, Great Britain in the lead, which forced the once-great empire to accept humiliating treaties, was in the end decisive. In 1842, it was conceded that foreign ships should be provisioned in case of need, though individuals would still not be allowed to land. Then, in 1851, the President of the United States took up the question of the treatment of American sailors shipwrecked on the Japanese coasts and of the shelter and supply needs of American whalers and ships engaged in the Far East trade. It was decided to send a naval squadron to Japan to assure the opening of Japanese ports to foreigners. In 1853 Commodore Perry sailed into Yedo Bay. Yedo depended on supply from the sea. The Japanese were impressed: the Americans had better guns than theirs and steam-driven ships. They accepted an American consul and opened two ports to American trade. There followed treaties with European powers admitting other foreign merchants to Japan and authorizing diplomatic missions.

The Meiji restoration

To many Japanese it now seemed that their country might, if they did not take care, be driven along by the foreigners as helplessly as was China. Clearly, the Tokugawa regime was not able to deal with the crisis. Leaders of two major clans began to take up European military methods and to send emissaries abroad to learn from the barbarians. They had been impressed by Perry, even, perhaps, by the little steam-train he brought with him and showed off on specially constructed track at the big party which marked the signing of the first treaty, and by the vast quantities of whisky and champagne then consumed. Soon after his arrival some of the clan estates set up the first manufacturing establishments on western lines shipyards, arms factories and cotton mills. The next step was the organization of the military opposition to the Tokugawa. It seemed for a little while as if there might be another collapse into division and anarchy. Instead, the nobles who opposed the shogunate called a new centralizing force into play – or, rather, recalled an old one. In a *coup d'état* at Kyoto on 3 January 1868 they seized control of the Imperial Court. The hereditary office of Shogun was abolished and the emperor was brought back from the wings of Japanese government to the centre of the stage, reasserting his direct responsibility for governing the country. This, symbolized by the movement of the court to Yedo, was the beginning of the 'Meiji restoration', and it was really a revolution. It launched deliberate modernization.

Japan's new leaders sought to develop further the earlier initiative of the major clans. Their aim was to learn what the western countries could teach them and to use it to modernize without becoming westernized and losing their own heritage in the process. In this, they were to be remarkably successful. Only a few years earlier, in January 1860, a symbolic achievement had showed the native resources which could be drawn upon. The *Kanrin-Maru*, a sailing-ship with a steam-engine producing only a hundred horse-power, which could only be used to manoeuvre in port, had then left Yedo for San Francisco, where she put in just over five weeks later. She had been navigated under sail across the Pacific by her Japanese crew, the first to take a ship so far, and they had done it only seven years after Perry had first brought steam-ships into Yedo bay. Even more recently the first Japanese to do so had gone to the Dutch for instruction in navigation. As a young member of the crew later wrote, in a magnificent comparison, 'Even Peter the Great of Russia, who went to Holland to study navigation could not with all his attainments have equalled this feat of the Japanese'.

Modernization and its limits

Patriotic pride was a large part of the mood in which many Japanese approached the task of modernization. Patriotism fed a passionate anxiety to escape the fate of the Chinese and Indi-

A parliament house of what had become the standard European layout was built for the new Japanese parliament, here being opened in 1891 by the emperor, seated at the right. Not only have the members been made to look very Europeanized in this print, but the ladies in the gallery, too, wear European dress.

ans. It reinforced a cultural willingness to learn and to borrow knowledge and technology. Thus stimulated, Japan was to change fast. The abolition of the old semi-feudal system of clan rule in the emperor's name was the first step towards creating a national state. Clan rivalries had played a big part in bringing to an end the Tokugawa supremacy. Now the major clans took the lead in surrendering their lands to the emperor 'so that', they said, 'a uniform rule may prevail throughout the empire'. There followed the adoption of many European institutions of government. The country was divided for administration into 'prefectures'. In 1889 a bicameral parliament was instituted. Japan had by then already adopted military conscrip-

The first railway in Japan ran from Tokyo to Yokohama. One of its very European looking stations appears in this print.

tion so that she could have an army on the European model; she had also set up her first postal system, first railway and a daily newspaper. The European calendar was adopted.

Much of the past still remained, especially in the patriotic cult and the respect paid to imperial authority. In 1890 a declaration on education was drawn up, to be read on great days to generations of Japanese school children for more than a half-century to come, which urged them to keep up the traditional virtues of respect for their parents, obedience, and, if necessary, self-sacrifice in the cause of the nation. The tradition of the samurai was kept alive, too. In the first ten years or so of the Restoration some of them had followed their discontented lords into rebellion, to be defeated by the new conscript army. Most of them now eagerly accepted service in the civil service, army or navy of the new order, while their lords, who had been compensated for the loss of their land with incomes guaranteed by the government, still had much influence; some of them were soon sitting in the new house of peers. Even though an outsider might be struck at the rate at which she was modernizing, Japan remained in many ways unchanged.

Yet some change was very visible. Power-driven machinery began to be introduced into the silk-spinning industry in the 1870s and soon became wide-spread (though twenty years later more than half the silk reeled in Japan was still processed by hand). By the early 1890s, too, Japan had a new cotton industry (though only a hundredth of the number of British spindles). It was only with the second half of the 1890s that really rapid industrial growth began. Japan's annual 5 million tons of coal production of 1895 had nearly quadrupled by 1913, and over the same period the output of raw silk trebled, while the output of cotton yarn went up more than sixfold. By 1914, Japan was industrialized as was no other Asian country.

In this great economic surge, the part of agriculture was less obvious, but was in fact more important still. Agricultural output per head much more than doubled between 1868 and 1914. Yet this did not make much difference to the lives of the great majority of Japanese who remained peasants. Farming had to provide the taxes which paid for the capital investment needed by industry, the new services and administration and education; under their heavy burdens the peasants stayed poor. Traditional ways in the villages were barely disturbed. Women, notably, remained downtrodden, oppressed and cramped by the old-fashioned ways. Yet Japan had joined the modern world.

GATHERING CLOUDS

It is a curious paradox that the approach to the climax of European power in other parts of the world was accompanied by growing signs of instability in Europe itself, where the revolutionary emergence of a new international order was as marked as elsewhere in the second half of the nineteenth century. The best point of departure is 1848, not because of its significance as an episode of social revolution, but because it punctuates the story of European nationalism, revealing the power of that force and, though it was hard to see at the time, dividing a period of prolonged international peace from one of war. In the next quarter-century there were wars between Great Britain, France, Turkey and Sardinia on the one side, and Russia on the other (the 'Crimean' war, 1854-6), between France allied with Sardinia against Austria (1859), and three others fought by Prussia against Denmark (1864), Austria (1866, Italy joining tin on the Prussian side) and France (1870). The first of these wars was really about an old question: should Russia dominate and perhaps overthrow Turkey? All the others were about nation-building.

New nations

The defeat of Austria in Germany (where the Habsburgs had to acknowledge the supremacy of Prussia) and Italy, where very little territory remained to her after 1866, meant that she

The actual appearance of the battlefield began to be photographed in the Crimean War. A Russian battery after its capture by the French.

had also to make concessions to other nationalities within her borders; the monarchy was no longer strong enough to resist their claims. So a 'Compromise' was arranged in 1867 with one of the subject peoples, the Magyars. They were given considerable independence in what was henceforth called 'the Dual Monarchy' (because it was really a combination of two independent units under one ruler) or 'Austria-Hungary'. Franz Joseph was now emperor in one half of his lands and king in the other. The other subject nations of the empire remained dis-

In 1848 monarchs everywhere trembled. In this Punch cartoon, the uncrowned figure at the stern of the boat is Louis Philippe, already turned off the French throne when it was published.

appointed, though. Hungary had been bought off by allowing the Magyars who ruled it to join the Austrians in oppressing Serbs, Slovenes, Romanians, Slovaks and others.

This was not all the nation-building of these years. One delayed outcome of the Crimean war was the emergence as an independent nation-state of Romania – though the name was not used until the 1860s. And the unification of Italy and Germany (together with concessions to the Magyars) made it all the more likely that other peoples in Central Europe and in the Balkans (particularly under Turkish rule) would clamour for political independence, too. Thus the outcome of these years was very complicated, but very important. A glance at the map before and after shows they were of revolutionary effect. Yet the statesmen who did most to bring about these changes (the Prussian minister Bismarck and the Italian Cavour) transformed the map and conditions of European diplomacy as the men of 1848 had hoped to do, but in the interests of conservatism and in order to thwart a revolutionary nationalism which they feared.

Europe's structure was thus by 1871 predominantly one of nation-states. But there were two defects to it. One was that there were still places where trouble was being stored up for the future. Ireland was one example: the British came near, it seemed, to giving it 'Home Rule' (that is, self-government under the Crown) in the later nineteenth century, but party politics thwarted that in the end. Norway and Sweden were still left in one state until they separated peacefully in 1905. Russia was left (like Prussia and Austria) ruling much of Poland and had its discontented non-Russian Baltic peoples and the Finns. Inside the Hungarian half of the Dual Monarchy, Croats, Romanians, Slovaks, Slovenes, Serbs all felt oppressed. Above all, Turks still ruled Bulgars, Macedonians, Albanians, Bosnians (until 1878, when the actual government of Bosnia was given to the Austrians, though the Sultan kept his nominal authority). The Balkans were, indeed, a nationalist's nightmare, so tangled up were its peoples, languages and religions.

Meanwhile, the balance of power in Europe had wholly changed. The old Holy Alliance of conservative states had come to an end in the Crimea and a new German empire (formally set up in 1871) replaced France as the dominating power of Europe. This was the political side of a change already appearing in population and economic trends. Until 1945 German predominance was to be a major problem facing European statesmen.

For forty years, Giuseppe Mazzini was the advocate of Italian unification and revolution. More generally, he is rightly remembered as one of the purest embodiments of the belief in revolutionary nationalism. Yet he died in exile, deeply disappointed and embittered by the unity Italy achieved through foreign alliances and war.

The German ascendancy

Nonetheless, for more than forty years after 1871, the great powers managed successfully to live side by side in peace. This was a great achievement, for during these forty years there were many and increasing signs of dangers lurking just beneath the surface of international life. Germany had forced France to make peace on humiliating terms in 1871, giving up two of her provinces, Alsace and Lorraine, and paying a huge indemnity. From that moment, the new Germany had clearly succeeded to France's long ascendancy in Europe. With a bigger population, fast-growing, and with a greatly stronger economy (also growing at a rate which challenged even Great Britain), she was by 1900 the strongest military power on the continent. But France was not reconciled to the loss of her provinces.

Italy, an only slightly less new nation than Germany, had taken the city of Rome from the Pope to give herself the historic capital many Italians had wanted for a long time in 1870. New nations are often touchy and difficult in foreign affairs. Their rulers are often very much aware of division and weaknesses at home and of the temptation to overcome them by blustering policies abroad so as to appeal to patriotic sentiment. Italy's leaders went in for colonial adventures (culminating in going to war with Turkey in 1911 to seize chunks of North Africa) while other Italians kept reminding their fellow-countrymen of Italian communities living under Austrian rule who were (as they said) 'unredeemed'; they thought these Italians and the territories they lived in should be 'liberated'. Here was another disturbing force.

For a long time, though, Germany seemed to pose no further danger. No-one in Germany who mattered wanted to unite all Germans under German rule, and for nearly twenty years her foreign affairs were directed by one man, a highly intelligent, quick-tempered and intolerant Prussian squire, whose main aim was that life in Germany should go on being run with the Prussian ruling class in the saddle, Count Otto von Bismarck. He had engineered the wars of German unification in the 1860s. With them successfully concluded, he feared social upheaval and even revolution at home if there should be another war, and made every effort to avoid one. His management of the affairs of the strongest European powers was crucial in keeping the peace. Nonetheless, Germany was changing in spite of Bismarck's wish that she should not. Her growing population and industrial strength threw up new ideas, attitudes and demands. After his dismissal in 1890, German foreign policy was more and more shaped by these forces. Some Germans in position of influence sought for Germany greater international respect and prestige – a 'place in the sun' – and at the same time began to feel more envious or fearful of other states.

It is true, also, that at the level of diplomacy, a breakdown of the European balance was already beginning to appear possible even in Bismarck's day. National minorities, for example, were growing more vociferous in the Ottoman and Habsburg empires. More important still, rulers and people alike in many countries gradually lost the sense that peace would suit them better than war for obtaining the ends they sought. At times, it even seemed that some people would positively welcome a war; memories of the last European wars faded, of course, as time passed.

Bismarck had tried to guarantee peace and German security by making alliances with Russia, Austria-Hungary and Italy. That meant that France could not seek revenge after 1871, since she could find no ally to help her and could not defeat Germany single-handed. Bismarck worked hard to ensure the friendship of his allies with one another and that Great Britain would stick to her own fairly isolated line about European affairs in which she did not feel herself directly involved. Yet the old rivalry between Russia and the Habsburg empire in south-east Europe was a constant danger to his policy. It went back to the eighteenth-century question, posed at the very beginning of the long decline of the Ottoman empire: what was to take its place? The Austrians did not want it to be the Russians, who would bar their route to the south down the Danube. The Russians did not want it to be the Austrians, who would thus block their way to seizing the mouth of the Black Sea. When the Russians fought the Turks from 1876 to 1878 it looked for a moment as if both the Austrians and the British might join in to help the Ottoman empire as the latter had done in 1856. Bismarck successfully averted the danger at a great congress in Berlin which somehow rewarded or silenced everybody and thus set Russo-Austrian relations once more on a fairly easy course until the early years of the twentieth century.

Even Bismarck, though, had felt that if it came to a showdown between the Habsburg Monarchy and Russia he would have to side with the first. This led his successors to allow Germany's alliance with Russia to lapse. In 1892 Russia made an alliance with France; it was a perfectly natural thing for the two colonial rivals of Great Britain to do this, and it was to bring pressure to bear on Great Britain that they made it. Nonetheless, it meant that France was no longer isolated, and might, therefore, be one day able to stand up to Germany. Europe was beginning to divide, as yet almost unnoticed, into two camps.

Tsarist Russia

One source of uncertainty and, possibly, instability was nonetheless apparent. That Russia was to be numbered among the great powers in 1900 was unquestioned, but beyond that point it was hard to go. Her vast manpower and huge natural resources suggested she was bound to dominate east European affairs, and perhaps those of much of Asia, too. Yet various weaknesses were also obvious; in many ways she lagged behind western Europe. Relatively speaking, she was weaker than in 1800. At that time, although her size, history and geographical position made her unique, much of the rest of Europe (largely unindustrialized and a continent of country-dwellers and small towns) still resembled her. It was different a hundred years later.

The path towards the modernization of Russian society had been strewn with obstacles. To begin with, there was the tradition of autocratic rule. The Tsar's authority had never been held in check as had absolutism elsewhere by obstreperous vested interests. If reform was to come to Russia at all, it had to come from above, for there were few channels for demands to come up from below. For a long time, that meant it did not come at all. Alexander I may (as some people believed) have wished to introduce changes, but in the end he disappointed those who looked to him for reform. His successor, Nicholas I, a cold, brutal man, imbued with a barrack-square outlook, never even thought of permitting any degree of liberalization. So, for the first half of the nineteenth century, Russian autocracy grew more and more congealed than ever, and the country more isolated than ever from what went on elsewhere.

Alexander II, the 'Tsar Liberator' reigned from 1855 to 1881, presiding over the emancipation of the serfs and many other reforms. But his regime bore down severely on dissent and revolutionary movements, and he died the victim of an assassin's bomb.

The failure to get on with solving Russia's problems which followed led in the end to weakness. They hindered economic growth, for example. Although in the eighteenth century Russia had important mining and metallurgical industries, these were quickly outstripped by those of other countries as the nineteenth century went on. Russian agriculture, too, could not achieve the increases in production seen in other countries; while the population continued to grow, the condition of most Russians grew worse. During the nineteenth century it seems unlikely that rises in grain production ever matched the rate at which the population grew. One outstanding reason for this was the survival in Russia of an archaic institution: serfdom.

While serfdom declined and disappeared elsewhere, it spread in Russia and worked increasingly harshly. Serf revolts and attacks on overseers became commonplace (and at time seemed to threaten wide-spread rebellion). But this was not all that was wrong; serfdom also took away any incentive for better cultivation by the peasant, and prevented the free movement of labour which would have made it possible to find workers for new factories. Poverty, too, kept down the peasant's demand for goods. On the other side, it had to be conceded that serfdom was embedded so deeply in Russian society that sudden abolition might mean the breakdown of much of government itself, since the autocracy relied upon the estate-owner to carry much of the burden of what would have been called local government elsewhere.

Defeat in the Crimean war (in the last year of which Nicholas I died) forced the government to turn to reform. The crucial measure underlying all the rest was the freeing of the serfs in 1861 (four years before the ending of slavery in the United States). This enormous undertaking was greatly to the credit of the regime. Much thought had been given to it. The crux of the reform was that serfs ceased to be the private property of estate-owners and became legally free individuals. This did not in practice at once mean complete freedom. In various ways, it was still made hard for peasants to get permission to leave their native villages. Nonetheless, though such restrictions slowed the effects of the change, in the end it opened the way to the modernization of Russian agriculture and industry.

Alexander II, under whom these reforms were carried out, is remembered as the 'Tsar liberator' for the ending of serfdom, but his reign produced other reforms too. But they never touched the central principle of autocracy: they were all handed down like gifts from the Tsar. They were not recognized as the rights of the Russian people and could be withdrawn. This is one reason why some enemies of the regime refused to compromise with it or to try to make the reforms work successfully. They continued to plot and struggle to overthrow the state. Often they assassinated officials; once they assassinated a Tsar. This, of course, strengthened the fears of conservatives who thought that no concessions should be made and that such concessions as had been made should be withdrawn.

Given the hardship under which most peasants continued to live (they suffered heavy tax burdens because of the need to find money to pay for railways and other investment) and the irritation of the growing numbers of liberal-minded businessmen and farmers thrown up as economic development speeded up, it is not surprising that under Nicholas II,

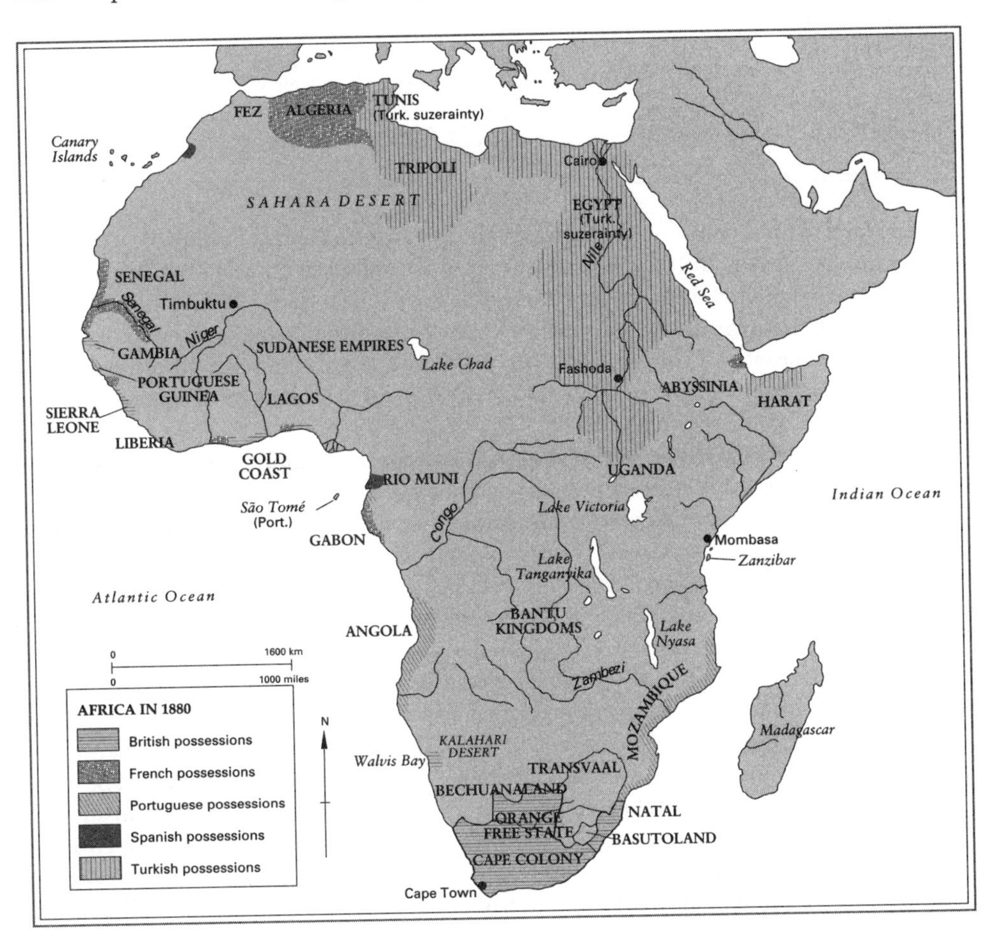

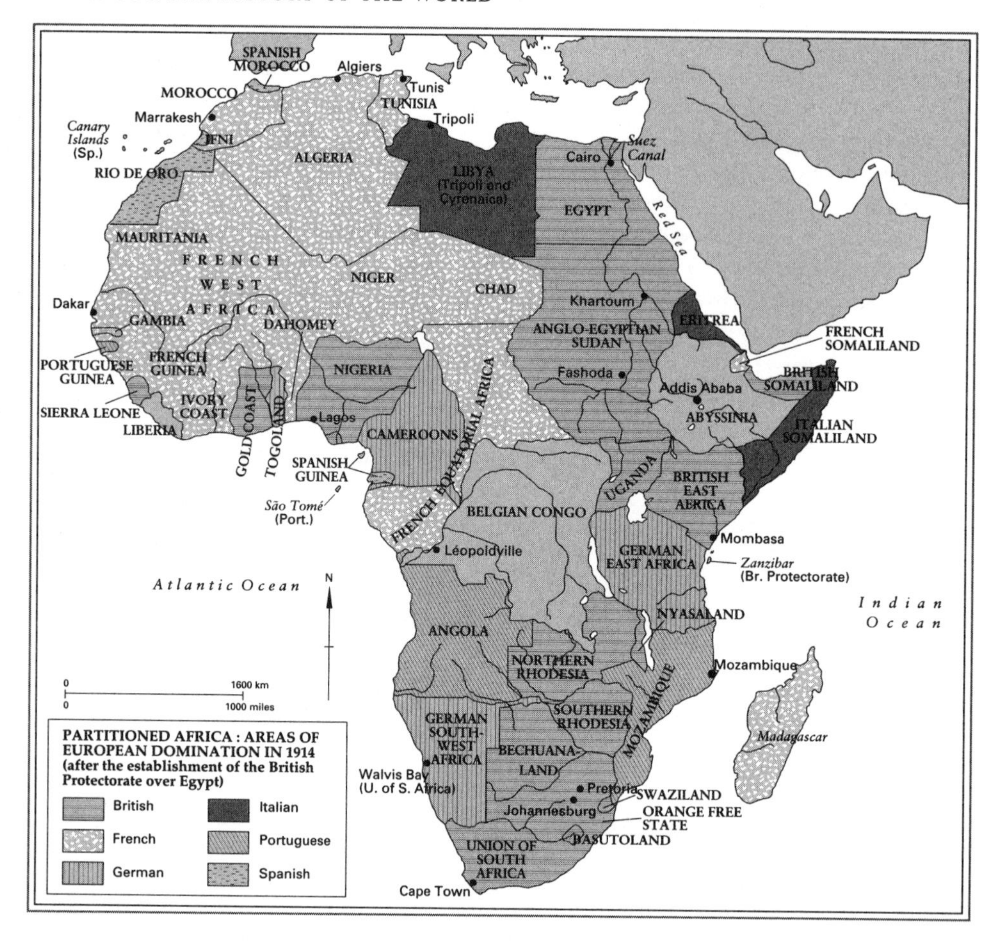

the last (and in some ways one of the most unimaginative and narrow-minded) of the Tsars, there was a revolution. In 1904, Russia suffered grave defeats in a war with Japan. In the following year, revolution broke out and the regime for a time tottered. More concessions were made. A sort of parliament or consultative council called the Duma was set up. It was not much, but it meant that the slow business of training Russians in self-government could begin at last. Unhappily, it was only to have a few years of life before involvement in another war led to a restriction of its powers.

Yet by 1914 Russia's great power status was to look much more certain once again. She was once more on the road to industrial strength; even if well behind that of Germany or England, her industrial output was growing at a faster rate than that of either. A great industrial future lay ahead, it was clear. The agricultural problem was at last beginning to come right. New legislation accelerated the emergence of a new class of substantial peasant-farmers called Kulaks, yeoman-farmers (as they might be called), who were interested in efficiency and profits. Their efforts began to raise productivity at last.

As Russian confidence grew, its rulers became sure they could stand up for her interests and confident that the Russian army, supported by a growing railway network and expanding industrial base, gave them the means to do so. On the other hand, though Russia was a European country in name, she was still one where poverty of Asiatic horror could be found. She was still a country where the Orthodox Church was mixed up in government and society in a way unknown for a century or so past in most of Europe. She had one or two good universities and schools and some distinguished scientists and scholars, but the overwhelming majority of the population were illiterate peasants. Above all, she was still a country where government rested in the last resort on what was regarded as the God-given power of the autocrat. And it was as a consequence of all these things that she was also the only important country in which there existed a seriously menacing revolutionary movement quite happy to overthrow the regime by force.

Bismarck's successors at the head of affairs in Germany were less able and less wise than he. They had, too, more complicated political situations to deal with at home, where new interests were clamouring for attention, some of them implying changes in foreign policy. They sometimes sought the support of the young, excitable (and easily irritable) German

'One empire, one people, one God' says the inscription on this stamp. Germany had been united less than forty years when it was issued.

emperor, Wilhelm II, and this was crucial because the monarch's powers were considerable. Under his rule, Germany was to become a much more unpredictable element in the diplomatic chemistry of the next century, and even in that of the 1890s.

One further achievement of European diplomacy before the long peace broke down was the settlement of a whole range of imperial and colonial questions without war. When, in the end, war came, it was over European issues and not, as so often was predicted, over European empires overseas, although it had at times seemed possible that Great Britain would go to war with either Russia or France. At the heart of this achievement lay the peaceful partition of almost all Africa between Europeans by the end of the century, for the most part after 1881. It was managed by a long series of deals between individual powers which by 1914 left Great Britain with a protectorate over Egypt, Ottoman Libya in the hands of the Italians, the French dominating Algeria and sharing with the Spanish effective control in Morocco, and then, southwards, a west African coast divided between European powers except for tiny and backward Liberia. The Sahara, Senegal basin and much of the Congo were French; the Belgians had the rest of it. British territory ran up from the Cape to the Congo border, but were cut off from the coast by the Germans in Tanganyika and Portuguese Mozambique. From Kenya, though, British territory stretched inland to the borders of the Sudan. Ethiopia remained the only other independent African country.

Elsewhere, too, there were great agreed settlements. The Pacific was partitioned and the British, French and Russians all extended their possessions in Asia. By the end of the century there was even talk of a peaceful partition of China. Few doubted that Europeans still determined the ordering of the world outside the Americas.

13

THE LATEST AGE: THE LONG RUN

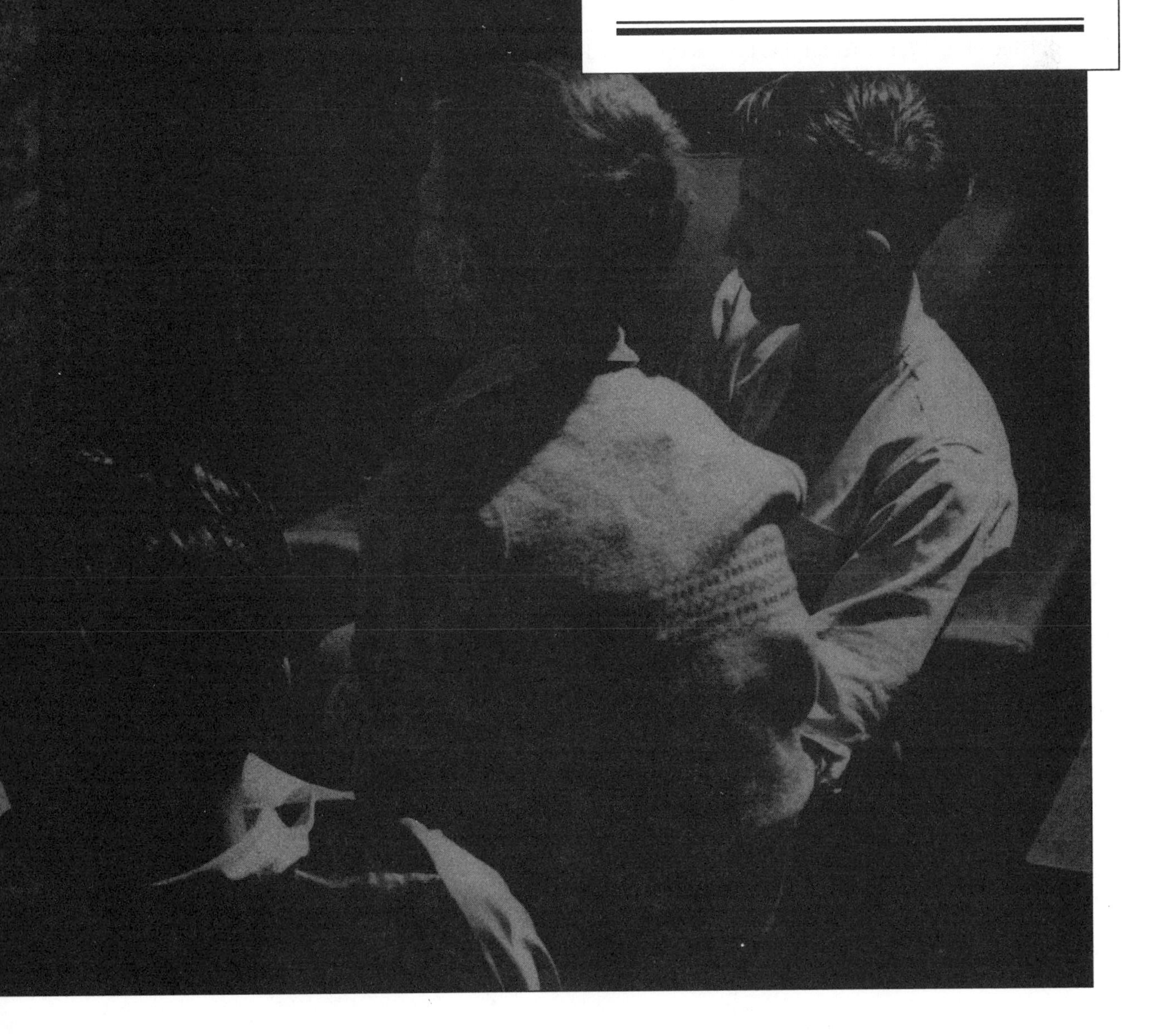

RECENT HISTORY

Historical change seems to move on an 'exponential' curve, one that gets steeper and steeper as time passes. Faster growth in the mastery of nature is only one indicator. Politics have been just as swiftly transformed. Of the European 'Great Powers' of 1900, none now remains a world power. Among them, only Great Britain and France, one a constitutional monarchy and one a republic, are still governed even in form as they were at the start of the century. Outside Europe, the colonial empires which looked so solid a hundred years ago vanished almost overnight in the 1950s and 1960s. It is very difficult to keep our bearings in such an ever-changing historical landscape, or even to grasp the main facts that have made the twentieth century so revolutionary. Only a few stand out clearly.

One is the completion of something begun a few hundred years ago: mankind truly lives in one world at last. Technology, politics, economics and, more and more, culture, have now made the world one, even if only a few recognize it or behave accordingly. A second great fact is that while this owes much to the achievement of supremacy by the peoples of European origin over the whole globe, so far as it was political and military, that supremacy is over: the empires of yesterday have crumbled and they are (to quote a late-Victorian Englishman) 'one with Nineveh and Tyre'. Yet this has been accompanied by a unique cultural success; much of European civilization has been taken up universally. Whether non-Europeans know where it came from or not, its influence is more widespread and more visible than ever before. It explains much of that 'one world' already touched on. The third great obvious fact is science. It has almost become the religion of the age; everyone now expects it to do marvels and is only surprised when miracles fail to appear. It has transformed life and has done as much as anything to make the history of this century dynamic and accelerating. Some people have found that not exciting, but appalling; they fear that change is too rapid for mankind's habits and standards of behaviour to grapple with it without disaster. Fortunately, historians do not have to predict – and, indeed, they should not, for all they know about is the past, which is full of examples of predictions which turned out badly. They will do better to stick to saying what happened. The best place to begin is with those developments and trends which run through the last century or so with little or only temporary interruption.

POPULATION

In 1900 there were about 1600 million human beings alive, by 1950, 2500 million or so. As these words are being written (in 1993) there are well over 5000 million of us. The last 1000 million or so have been added in only fifteen years or so and we may number just under 6000 millions or so before the end of this century. This is one of the best examples of accelerating change. Population growth is now much, much faster than ever before. One result has been widespread alarm. As at the beginning of the nineteenth century, some again fear Malthusian disaster ahead. Further misuse of the environment, over-crowding and competition for resources reflect growth which has been very uneven between countries and peoples and looks as if it will go on being so.

Some societies now seek to control their shape and size. But to do so is an uncertain business, and at the very least, many poor countries cannot for a long time hope to achieve significantly slower growth. Natality only began to drop in the last century in a few countries after prosperity made it attractive to have smaller families. Further medical, nutritional and sanitary advances must for a time make things worse; they preserve the infants, sick and aged who would in earlier times have died, but only to share (in growing numbers) a slower-growing pool of resources. Furthermore, in most of the world falling mortality has yet to make the impact it did in Europe between 1800 and 1900. When and where it does so, population is likely to rise faster still.

The consequences – or at least some of them – are already visible. Instead of resembling pyramids, developed societies are now even more like slowly tapering columns; the proportion of older people is much bigger than a century ago. The reverse is true in poorer countries, which often have huge preponderances of young people. Two-thirds of China's population is under 33. There are threatening growth rates in many countries: Mexico's population quadrupled between 1900 and 1975 and Brazil's increased sixfold. Only a few developing countries show any signs of success in slowing down, if not stemming, population growth. It is hard to shake off pasts and traditions, above all in a matter so important to the individual as sexual activity.

Population and power have usually been related, with whatever qualifications, and a comparison of the ten most populous independent states in 1900 and 1990 is interesting, even if the figures are only approximate.

Population comparisons 1900 and 1990 – in millions

	1900		1990
China	475 m.	China	1200 m.
Russia	133 m.	India	800 m.
USA	76 m.	USSR	290 m.
Austria-Hungary	46 m.	USA	248 m.
Japan	45 m.	Indonesia	180 m.
Germany	43 m.	Brazil	150 m.
UK	42 m.	Japan	125 m.
France	41 m.	Pakistan	108 m.
Italy	34 m.	Nigeria	105 m.
Ottoman empire	25 m.	Bangladesh	105 m.

This shows some startling changes in relativities. On any reckoning each list contains the three most powerful countries in the world of the day. Yet sheer population now counts for much less than it did in 1900. China is certainly a great power and would probably be one on grounds of population alone, for in a sense it makes her militarily invulnerable. Her social revolution has begun to increase her wealth too. In other well-populated countries, though, poverty is an obstacle to power which looks insurmountable, whether absolute, in the sense that natural resources are poor (Bangladesh), or relative, in that they are swallowed by population growth which is too fast (until recently, Indonesia). It is not easy to generalize. In the early 1970s India was thought to be about to become self-sufficient in food; her agricultural output doubled between 1948 and 1973. Yet this only just succeeded in holding the line for a population growing by a million a month.

GROWING WEALTH

Though many have starved, though, more have lived and this means the world has produced more – has, in fact, grown still richer. Whether this can continue is not a question for historians. They can only point out that, viewed most generally, the world economic trend is in the long run upward. A long climb to a peak of activity and wealth was interrupted in 1914; many of the conditions which had made it possible were then shattered in the First World War; a partial resumption of wealth-generation in the 1920s was followed by world slump and a fragmentation of the international economy in the 1930s; the war of 1939-45, bringing further distortions but an enormous recovery of production, was then followed by resumed growth world-wide after 1950 and greater economic interdependence in spite of new political divisions. Though with hiccups in the 1970s and later 1980s, this has persisted.

In 1900 the assumption that continuing economic growth was assured was well-established in a few countries. By the 1980s many more people shared that idea. Many, indeed, actually felt aggrieved when it was not confirmed by daily experience. This was a huge change in human thinking. Yet though growth has been widely shared, the outcome has been uneven. Gross Domestic Product (GDP) has risen almost everywhere since 1900, and one set of calculations has provided the following per capita estimates (in 1988 dollars):

Per Capita Gross Domestic Product (GDP)
1900 and 1988

	1900	1988
Brazil	436	2451
Italy	1343	14432
Sweden	1482	21155
France	1600	17004
Japan	677	23323
UK	2798	14477
USA	2911	19815

Such selected (and questionable) figures require cautious interpretation, but they drive home the fact that although the world has become much richer, some countries remain woefully poor. In 1988 Afghanistan, Madagascar, Laos, Tanzania, Ethiopia, Cambodia and Mozambique all *officially* had a GDP per capita of less than $150.

Like population, wealth has tended to grow faster as the century has gone on. Since 1945 there has been peace between major powers; for all the warlike operations often going on, they rarely fought openly with one another. Their rivalries, indeed, often helped to promote transfers of resources and knowledge which increased real wealth.

The first such transfers came about in the later 1940s, when American aid made possible the recovery of Europe as a major world centre of industrial production. The enormous wartime expansion of the American economy which had brought it out of the pre-war depression (together with the immunity of the American home base from physical damage by war) had not only won a great victory and rebuilt American economic strength, but sustained a huge expansion of world trade for about thirty years. International circumstances helped: for a long time there was no alternative source of capital on the scale needed, and nations were willing as never before to set in place institutions for co-operation in regulating the international economy. In their determination to avoid a return to the near-fatal economic anarchy of the 1930s, they produced the International Monetary Fund, the World Bank and the General Agreement on Tariffs and Trade (GATT). The economic stability thus provided to the non-communist world underpinned after 1950 two decades of growth in world trade at nearly 7 per cent per annum in real terms.

Less formally, and often less visibly, scientists and engineers also did much for long-term economic growth through technology, and the improvement and rationalization of processes and systems. That this was another rising, exponential curve then became clear, especially in the second half of this century. In agriculture, this contributed to huge growth in food production. Effective herbicides and insecticides were not available commercially until the 1940s and 1950s, but by then the mechanization of agriculture was already commonplace in developed countries. The tractor was its symbol. Now, the fields are not the only part of the farm to be mechanized. Electricity has made possible automatic milking, grain-drying, threshing, the heating of animal sheds in winter and, finally, the computer and automation. Human labour matters less; in both the United States and western Europe the agricultural

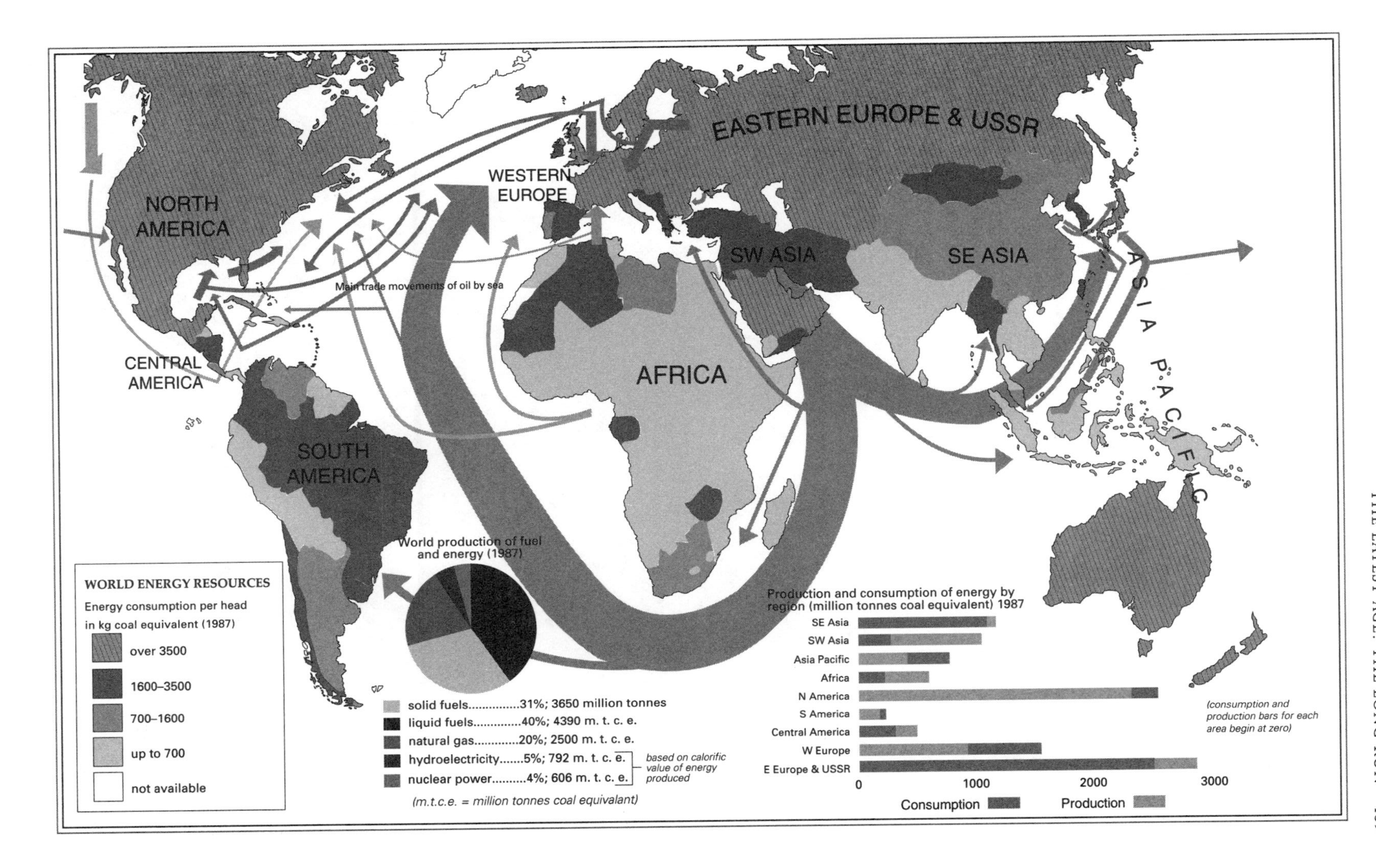
NORTH AMERICA
CENTRAL AMERICA
SOUTH AMERICA
WESTERN EUROPE
EASTERN EUROPE & USSR
SW ASIA
SE ASIA
AFRICA
ASIA PACIFIC
Main trade movements of oil by sea
World production of fuel and energy (1987)
WORLD ENERGY RESOURCES
Energy consumption per head in kg coal equivalent (1987)
over 3500
1600–3500
700–1600
up to 700
not available
solid fuels..............31%; 3650 million tonnes
liquid fuels.............40%; 4390 m. t. c. e.
natural gas.............20%; 2500 m. t. c. e.
hydroelectricity.......5%; 792 m. t. c. e.
nuclear power..........4%; 606 m. t. c. e.
based on calorific value of energy produced
(m.t.c.e. = million tonnes coal equivalant)
Production and consumption of energy by region (million tonnes coal equivalent) 1987
SE Asia
SW Asia
Asia Pacific
Africa
N America
S America
Central America
W Europe
E Europe & USSR
0
1000
2000
3000
Consumption
Production
(consumption and production bars for each area begin at zero)

workforce has continued to shrink while productivity per given area has risen. Yet there are probably more subsistence farmers in the world today than in 1900, simply because there are more people. The subsistence farmers' relative share of the world's cultivated land and of the value of what is produced has fallen, though.

Rich and poor

Agricultural plenty is unevenly spread and has been easily disrupted. Russia's once fed the cities of central and western Europe with grain, but as recently as 1947 the USSR suffered famine so severe as once more to provide reports of cannibalism. Improvements in productivity which had been achieved over the previous hundred years virtually came to a stop in some east European countries after 1945 and in a few underwent recession in the next three decades. In countries with large and rapidly-growing populations subsistence agriculture is still common and productivity remains low. Just before the First World War, the British wheat yield per acre was already more than two and a half times that of India; by 1968 it was roughly five times. Over the same period the Americans raised their rice yield from 4.25 to nearly 12 tons an acre, while that of Burma, the 'rice bowl of Asia', rose only from 3.8 to 4.2. Advanced agriculture is usually found in countries advanced in other ways. Unless they have a particular agricultural speciality, countries which most need to grow more food have found

Following upon war, revolution and civil war, famine brought immeasurable suffering to millions of Russians and Ukrainians in 1921.

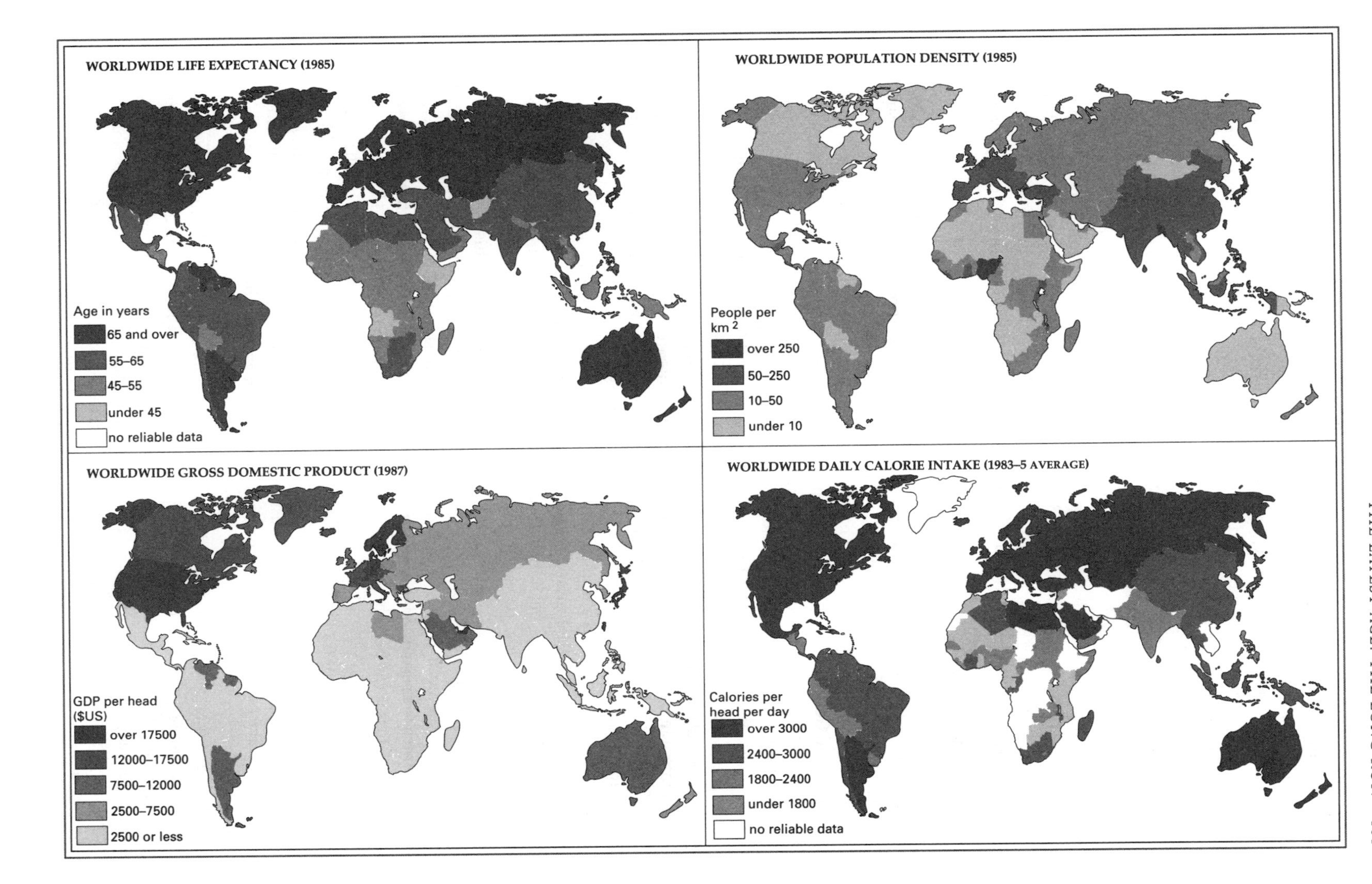
WORLDWIDE LIFE EXPECTANCY (1985)
Age in years
65 and over
55–65
45–55
under 45
no reliable data
WORLDWIDE POPULATION DENSITY (1985)
People per km 2
over 250
50–250
10–50
under 10
WORLDWIDE GROSS DOMESTIC PRODUCT (1987)
GDP per head ($US)
over 17500
12000–17500
7500–12000
2500–7500
2500 or less
WORLDWIDE DAILY CALORIE INTAKE (1983–5 AVERAGE)
Calories per head per day
over 3000
2400–3000
1800–2400
under 1800
no reliable data

it very difficult to produce crops more cheaply than can the developed world. Russians, Indians and Chinese, big grain and rice producers, now find themselves buying American and Canadian wheat.

One simple measure of disparity in wealth is consumption. Roughly half mankind consumes about six-sevenths of the world's production; the other half shares the rest. Electricity is a good test; relatively little electrical power is traded internationally and most of it is used in the country where it is generated. At the end of the 1980s, the United States produced nearly 40 times as much electricity per capita as India, 23 times as much as China, but only 1.3 times as much as Switzerland. The poor have not usually grown poorer (though it has sometimes happened), but the rich have grown much richer. Even spectacular improvements in production have often failed to change the position of poor countries in relation to the rich, because of rising populations – and rich countries, in any case, began at a higher level. Most of the countries which enjoyed the highest standards of living in 1900 still, by and large, enjoy them today – and they are the major industrial countries of the developed world.

The industrial world

The landscape of world industry, both in its distribution and nature, has undergone vast changes since the beginning of the century. Even in 1970 three of the great industrial agglomerations of the world were still, as they had been in 1939, the United States, western Europe and the USSR. By 1990, Japan had become the third greatest while the USSR had even fallen behind Germany. Old heavy and manufacturing industries, long the backbone of economic strength, are no longer decisive indicators. Of the three greatest steel-making countries of 1900, the first two (the USA and Germany) were still among the first five eighty years later (though this was a smaller Germany than that of 1900), but in third and fifth places respectively; the United Kingdom (third in 1900) came tenth in the world table – with Spain, Romania and Brazil close on her heels. Newer industries have often found a better environment in some developing countries than in the mature economies. By 1988 the peoples of Taiwan and South Korea had per capita GDP, in the first instance, nearly eighteen times that of India, and, in the second, more than fifteen times.

Industries – electronics and plastics are examples – have appeared which barely existed even in 1945. Coal replaced running water and wood in the nineteenth century as the major source of industrial energy, but long before 1939 was itself joined by hydroelectricity, oil and natural gas; very recently, power generated by nuclear fission has been added to these. Nonetheless one trend shows a deep continuity with earlier times. This is the enormous growth in the production of commodities directly for the use and pleasure of the individual consumer. Of the immense ramifications, one example must suffice. A four-wheeled contraption which is a recognizable ancestor of the modern motor-car, the French Panhard, was only created in the 1890s. When the first motor show was held in London in 1896 cars were still few, and rich men's toys, but in 1907 Henry Ford set up a production line deliberately planned to supply a mass market at an unprecedently low price. By 1915 a million Ford cars were made each year; eleven years later the Model T sold for less than $300. Ford had provided for the masses something previously regarded as a luxury, and changed the world as much as had the coming of the railways a century before because others copied and followed him. He helped to spread amenity and a new pollution round the globe.

By the 1980s, a world-wide and internationally integrated car manufacturing industry had come into being. Nearly three out of four of the world's cars are now made by eight large producers. By 1990, Japan, whose post-1960 economic growth owed much to its car industry was already consciously running it down in anticipation of new challengers abroad. Other more positive changes, have flowed from the car; today half the robots employed in the world's industry are welders in car factories (another quarter do the painting in them). Over a much longer term, a huge demand for oil was created (though this was already in

Old and new: to the assembly line, long the typical form and symbol of industrial production, is now joined the robot.

sight even before 1914). Many people in many economic roles came to be employed in jobs dependent on the car. Ford also helped to transform industrial production. Like many great revolutionaries he brought other men's ideas to bear on his own. He did not invent the assembly line, the characteristic modern way of making consumer goods by moving the article under manufacture from worker to worker (and now from robot to robot), but gave it wider application than ever before. The psychological effect on the worker was often deplored, but the technique was fundamental to a wider sharing of wealth. Ford saw that such work would be very boring and paid high wages to compensate for it (so contributing to the fuelling of economic prosperity by increasing purchasing power and, therefore, demand).

Communications

Since 1945, industry has been revolutionized by information technology, the complex science of devising, handling, and managing (mainly electronic) machines to process information. Few innovatory waves have rolled in so fast. Much of the fundamental inventive and development work was done during the Second World War. In a couple of decades it was diffused over a huge range of services and industrial processes. Rapid increases in the power and speed, combined with reduction in size and improvement in visual display capacity, of computers meant that much more information could be more rapidly ordered and processed than hitherto. Quantitative change brought qualitative transformation. Calculations which until recently would have required the lifetimes of many mathematicians to complete can now be run through in a few minutes. Intellectual advance has never been so suddenly accelerated. At the same time the capacity and power of computers grew with astonishing speed, so it became easier and easier to pack their potential into smaller and smaller machines. Within thirty years, a 'microchip' the size of a credit card was doing the job which had first required a machine the size of the average British living-room. The transforming effects have been felt in virtually every human activity, from money-making to war-making.

Computers are nevertheless only the latest link in a long chain of communication innovations. The nineteenth century had brought the application of steam to land and sea trans-

Gugielmo Marconi seated at the receiving 'set' at which on 12 December 1901 he received in Newfoundland morse signals transmitted from Cornwall, over two thousand miles away on the other side of the Atlantic.

port, and then the production of the petrol or oil-driven internal-combustion engine and electric tram. Balloons were eighteenth-century inventions and the first 'dirigibles' existed before 1900, but the first flight by a man-carrying, powered, heavier-than-air machine was only made in 1903. Eighty years later, the value of goods imported and exported through Heathrow, London's biggest airport, was greater than that of any British seaport, and aeroplanes are now the normal form of long-distance travel, offering a service hardly imaginable at the start of this century. By then, information transmission had been revolutionized for half-a-century or so; poles carrying lines for the electric telegraph alongside railway tracks were already a familiar sight. Once Marconi had exploited electro-magnetic theory to make possible the sending of the first 'wireless' messages, transmitters and receivers no longer needed physical connexion. The first radio message to cross the Atlantic did so, appropriately, in 1901, the first year of a century to be deeply marked by his invention. By 1930 most people who owned 'wireless' receivers (and there were millions) had ceased to believe that windows had to be kept open to allow the broadcast 'waves' to reach them. Large-scale radio broadcasting was by then going on in all major countries.

Pictures were soon communicated as easily as sound. In 1896 the first cinema show in London took place at the Regent Street Polytechnic. By 1914 there were 3500 cinemas in Great Britain and many more in other countries. Film-making industries sprang up, above all in the USA (though India was in the end to be the most prolific film-producing country in the world). Already by 1939 film and radio had already begun to change habits, tastes and ideas; both had been taken up by politicians, governments and businessmen anxious to promote their wares. Even the twentieth century's vast and world-wide expansion in primary schooling, literacy and the newspaper press may not have done so much to spread knowledge of what life had to offer in material terms as did radio and the cinema. And, although Soviet Russia, India and Japan all made very distinctive films for local consumption, the cinema often spread ideas and standards based on the life of North America and Europe.

The impact of television was greater still. The first crude transmission of pictures had been achieved by a German in 1911. In 1936 the BBC opened the first regular television broadcasting service. Only after 1945 did it really take hold though, first in the United States, but twenty years later the medium had become commonplace in the leading industrial countries. It is now the primary source of mass entertainment and information throughout the world. About its effects, argument continues, but it has cut into the appeal of newspapers, radio and the cinema. It may have opened a new visual age of communication in which images, rather than literacy, will play the main part. It may be the biggest force for cultural and social change since the coming of print, taking mankind away from words to pictures, and from thought towards the ebb and flow of emotion, impression and imprecision.

NEW WAYS OF SEEING THE WORLD

Although it is tempting to look back at the world before 1914 and see it as almost unimaginably different from our own, many of the ideas and attitudes in the twentieth century cannot be understood unless due weight is given to the depths of their roots in the nineteenth. Confident, optimistic and liberal though nineteenth-century culture was, much in it also anticipated a more pessimistic, anguished age. Liberty of opinion and debate struck some as double-edged benefits. Whatever might be said among educated people, for millions of ill-educated and superstitious Europeans old-fashioned religion was the main regulator of their lives. Was it really a good idea to weaken something on which they relied for their ideas of what was permissible? More deeply still, if you allowed that everything in the end could be questioned and that no standards were to be taken for granted, were you not destroying the basis of any society at all? Society needs some unquestioned assumptions.

Some doubts went back to an unease produced by the Enlightenment itself. Others arose over new problems. There was a self-destructiveness implicit in the critical vigour of western civilization and it can be seen in one of its greatest and most celebrated nineteenth-century scientific achievements, the work of Charles Darwin. Though often misunderstood and over-simplified, what he said (or was believed to have said) shaped new ways of thinking about much besides biology. Some people believed that what they understood him to say in his book, *The Origin of Species* (1859), about the natural selection of those fittest to survive in the biological world meant that the human world functioned similarly. Some of them went on to justify unrestrained economic competition on these grounds. They said that it ensured that those with the best qualities – of courage, intelligence, determination and practical acumen – would come to the top. This was a very comforting idea for those who did not know what to do about those who obviously lost in the competition of life. It implied that no-one was to blame; their plight was the result of a natural process.

Determinism

Such ideas are of a kind sometimes called 'determinist'; at bottom they suggest or assert that certain given facts, often material, determine what will happen in the long run, and that individual efforts will not be able to alter this, or at least not in any important degree. Men and women who could not accept, like their forebears, that God ruled the world now seemed ready to accept that mindless physical processes did. In the Darwinian example, the 'determining' facts were presumably the genetic inheritances which led to some men being successful and others not. But biological theories were not the only source of determinist ideas. Some intellectuals stressed the importance of geography or climate, others that of economics. The official 'Marxist' doctrines which were upheld by many socialists of the Second International were such; they often seemed to boil down to saying that the world was the way it was because of economic forces, that it was moving steadily and irresistibly towards the triumph of the proletariat over its oppressors, and that nothing could prevent this – a very comforting or very depressing view according to where you sat to watch the drama of history.

Determinist theories of one sort or another were certainly more widespread and acceptable at the end of the nineteenth century than when it began. They all had one important point in common: they tended to weaken the feeling that men have responsibility for their own lives and can freely take the decisions which shape them. They were therefore very different from the Christian ideas which lay at the very roots of European civilization, from the ideals of the free, enquiring individual dreamed of by the Renaissance and Enlightenment thinkers, and even from the confidence of the men who made the breakthrough into industrial society, all of whom had believed that what individuals decided to do mattered a great deal and could ultimately change the world. The new ideas were symptoms that some people were beginning to feel doubts at a very deep level indeed about something very basic to their culture.

The idea of biological evolution by natural selection was the most influential scientific idea of the nineteenth century. Although at least one other scientist arrived at the idea independently of him, and although he himself attributed crucial importance to the influence of Malthus, it is associated above all with the name of Darwin.

Yet in the intellectual ferment of the nineteenth century almost every new idea was soon challenged. It is hard, therefore, to say what 'society' on the whole 'thought', and perhaps to try to do so is not sensible. People pointed out, for example, that the things which often seemed most to dwarf the individual and remove his power to run his life – the growth of huge, anonymous cities, the building-up of industrial empires in which people were only little cogs in big machines, the increase in the power of government – were all likely to leave them passive, apathetic and feeling helpless. But it could equally well be argued that in much of their day-to-day life millions of people actually had more freedom of choice than in the past because science and technology gave them greater control than ever before over their environment. Electricity gave men and women the chance of a better use of their time by providing cheaper, cleaner and simpler methods of lighting houses and workshops. Once the bicycle had been invented, millions of people had a new freedom of movement which they used both in leisure and in work. As the idea of contraception spread, men and women could more easily shape their family lives as they pleased and not leave as much to chance. Yet such freedom in very down-to-earth, practical matters must itself have told towards a general change in outlook in advanced societies which has been labelled 'materialism'. It was not just a matter of a growing taste for material comforts and pleasures, there were also ideas at work which hark back to the empiricism of some Enlightenment thinkers. One more important sign was a slow ebbing of belief in a supernatural world; more people came to believe that life could be explained in wholly material terms, and that the world could be manipulated so as to provide better and better material conditions for human life. From one point of view this was another a very optimistic outlook, but suggest that human beings were themselves simply to be explained as a result of material forces: how, after all, could they be the only natural objects in the world not subject to the material laws which governed the rest of it? And if that were true, how could they be said to have any special or ultimate value which entitled them to be treated in a special way?

Racialism

Sinister implications of one kind of determinist materialist thinking appeared in the rise of theories about race. A number of writers and thinkers, taking up ideas vaguely related to those of Darwin, claimed that different races were distinguishable not only by their physical characteristics (skin colour, shape of features, type of hair and so on), nor just by their culture (language, institutions) but also by different innate qualities of mind and ability. Some said that these differences arose because some races stood higher on the evolutionary scale, or fulfilled more important natural purposes than others. Almost always, such theorists claimed that the white 'race' was the best of the lot. Some Europeans and North Americans even made distinctions within it and asserted that 'Teutonic' or 'Anglo-Saxon' whites were superior to the 'Mediterranean' or 'Latin'. Nowadays, members of other 'races' have begun to follow the same arrogant trail and claim innate superiority in their turn.

Wrong-headed though such ideas are, it is easy to see that they fitted well into the current tendency to look for big determining and explanatory facts which could clarify the overall picture. Unfortunately, though, people acted on them, Politicians and propagandists, too, used them to work up excitement about the 'Yellow Peril' of expansion supposedly threatening Europe from the Mongoloid peoples. Racialist ideas were used also to justify appeals to patriotism, or claims to rule peoples who were said to be 'naturally' inferior because they were backward. Nevertheless, before 1914 such ideas mattered less than later, when they had really horrific consequences.

Anti-semitism

Jews had been persecuted throughout the Middle Ages. The excuse then found most satisfactory for this was that they deserved it: the Jews had crucified Jesus Christ, the founder of the Christian religion (those who said this did not dwell on the fact that the first Christians and, of course, Christ himself, were all devout Jews). This accusation was an effective rabble-rouser in medieval society, always ready to put devotion to bad use, and which had, in any case, other reasons to dislike Jews. Because they were for a long time the only money-lenders and figured prominently in commerce, Christians often tended to be in debt to them. They had no self-evident and necessary place in the predominantly rural society of medieval Europe; even today people do not tend to think bankers indispensable. Jews congregated together in towns, where they stood out conspicuously, often even in their dress, though they were comparatively few in number.

Though the days of persecution, mobbing and massacre slowly died away in western Europe, more and more Jews moved eastwards during the Middle Ages to the kingdom of Poland-Lithuania, to the area later called under the Tsarist empire 'the Pale'. Gradually, the smaller numbers of Jews in the United Provinces, England, France, Italy and Germany came to be treated in a more tolerant way, especially after the French Revolution. In the early nineteenth century there was a great liberation of Jews in every western country from many civic and social injustices previously inflicted on them. At least in the middling and higher levels of society, by 1900 they normally led untroubled lives, though often unassimilated, a community set apart by religion, education and language. Hebrew was the language of the Jewish religion while Yiddish (a mixture of German dialect and Hebrew) was spoken by most Jews in eastern Europe. Social prejudice still survived, though, and even Jews who won great eminence in the arts, sciences, commerce and finance, tended to remain on the fringes of Europe's ruling circles (though not those of the United States or South Africa).

This was the background against which pseudo-Darwinian ideas about race began to be spread in the second half of the nineteenth century. Anti-semitism had never died out. The Russian Orthodox church encouraged it, as did the tsarist regime. Some Roman Catholics blamed the Jews (along with others – freemasons, for example) for the French Revolution. When a big commercial and financial crisis occurred in Germany and Austria in the 1870s, Jewish banks and financiers were blamed by many for the loss of their savings. The migration of Jews from the east (often from the very conservative and traditionally-minded Jewish communities of the Pale, who stood out in clothing and appearance) into the major cities of central Europe – especially Vienna – also led to friction over jobs with the natives. A series of financial scandals in France in the 1880s, too, led to the sale of over a hundred thousand copies of a book attacking Jews – and this in a country which had the most successful record of toleration in continental Europe and a Jewish population which was probably smaller than the public which bought the book.

Yet in no western European country before 1914 did Jews fear any return to their former position of legal inferiority. Instead, leading Jews became more and more fully accepted into society: they entered the professions in increasing numbers, went into politics and rose to high office, continued to prosper in business, found it easy to acquire higher education and generally looked forward to more and more successful assimilation in societies of which they were citizens equal with all others. To the life of the United States in particular, they made a great contribution and in it they were especially influential. Only a few Jews before 1914

The Black Hundreds. A demonstration in Odessa by the reactionary, anti-semitic movement used by the Tsarist authorities against revolutionary forces. They carry a picture of the Tsar.

thought that their people might seek another aim than that of assimilation and that they should constitute themselves a nation like any other with a territorial basis in an independent Jewish state: these were the Zionists.

This picture was only badly marred in tsarist Russia. About five million Jews lived there at the end of the nineteenth century – about a fifth of all the Jews in the world – and most of them were confined to the Pale. Tsarist government deliberately drew on the old superstitious hatreds fanned by the Orthodox Church, in order to divert disaffection and divide its subjects against one another. From the 1880s onwards there were frequent 'pogroms'; Jewish houses and shops would be sacked and looted, and Jewish ghettoes were invaded by thugs who beat up and sometimes killed the inhabitants, or raped the girls. Such harassment was sometimes organized by the police; even when it was not, the authorities usually turned a blind eye to mobs doing their work for them. The fact that Jews achieved eminence in Russian scholarship, art and business did not deter the regime; indeed, it took away some legal rights Jews already possessed and made it harder still for them to enter schools and universities. It is perhaps not surprising that Jews became very prominent – and much more so than their total numbers would have suggested – in Russian revolutionary groups.

Apart from Russia, the only other European country with legalized anti-semitism at the beginning of this century was Romania. In the days of Turkish rule, Romanian Jews had been reasonably well treated, but with political independence came anti-semitism. The fight for national freedom had been seen very much as a Christian crusade against Islam, and Jewish communities which had been settled for centuries in the Danubian provinces were treated as aliens by the new Romania until 1919. Yet by then no cultivated European would have said that eastern Europe should be taken as the standard of the civilization to which he belonged.

THE MANAGEMENT OF NATURE

The twentieth century has brought unprecedented prestige to natural science. No intellectual achievement in any other field can match what it has done to advance understanding of the natural world. Yet technical advances are perhaps still the only way in which this comes home to most people. In the nineteenth century most of the practical results of science were still by-products of scientific curiosity, some of them accidental. By 1900 scientists had seen that consciously directed and focused research was sensible; fifty years later modern industry was already dependent on science, directly or indirectly, whether obviously so or not, and the link is now taken for granted. Nowadays, the ordinary citizen of a developed country cannot lead a life which does not rely on applied science.

This all-pervasiveness, coupled with spectacular achievements, was one of the great reasons for the ever-growing overt recognition given to it. Money is one yardstick; the attention of governments another. When, during the war of 1939-45, the British and Americans mounted a major effort to produce nuclear weapons, the resulting 'Manhattan Project' (as it was called) was estimated to have cost as much as all the scientific research previously conducted by mankind from the beginnings of recorded time. A search for better weapons explains much of the huge scientific investment of the United States and the Soviet Union after 1945. Curiously, this has not meant that science has grown more nationalistic; indeed, the reverse is true. There is not only a splendid, centuries-old, tradition of international communication among scientists but they have good theoretical and technical reasons to ignore national frontiers.

The new physics

So far as theoretical progress is concerned, the story can be picked up in the 1870s, when James Clerk Maxwell, the first professor of experimental physics at Cambridge, published work in electro-magnetism which broke effectively into problems left untouched by seventeenth-century science. Henceforth, the Newtonian view that the universe obeyed natural, regular and discoverable laws of a somewhat mechanical kind and that it essentially consisted of indestructible matter in various combinations and arrangements, could no longer be regarded adequate. Electro-magnetic fields now had to be fitted into it. The foundation of modern physical theory by experimental work followed. By 1914, Röntgen discovered X-rays, Becquerel radioactivity, Thomson identified the electron, the Curies isolated radium, and Rutherford carried out the investigation of the atom's structure. The result was a new picture of the universe. It began to look less like an aggregation of lumps of matter and more like tiny solar systems of particles in particular arrangements. Their behaviour blurred the distinction between lumps of matter and electro-magnetic fields. Moreover, the arrangements of particles were not fixed; in nature one might give way to another. Thus chemical elements could change into other elements. When Rutherford showed that atoms could be

Albert Einstein
Old Grove Rd.
Nassau Point
Peconic, Long Island

August 2nd, 1939

F.D. Roosevelt,
President of the United States,
White House
Washington, D.C.

Sir:

Some recent work by E.Fermi and L. Szilard, which has been communicated to me in manuscript, leads me to expect that the element uranium may be turned into a new and important source of energy in the immediate future. Certain aspects of the situation which has arisen seem to call for watchfulness and, if necessary, quick action on the part of the Administration. I believe therefore that it is my duty to bring to your attention the following facts and recommendations:

In the course of the last four months it has been made probable - through the work of Joliot in France as well as Fermi and Szilard in America - that it may become possible to set up a nuclear chain reaction in a large mass of uranium,by which vast amounts of power and large quantities of new radium-like elements would be generated. Now it appears almost certain that this could be achieved in the immediate future.

This new phenomenon would also lead to the construction of bombs, and it is conceivable - though much less certain - that extremely powerful bombs of a new type may thus be constructed. A single bomb of this type, carried by boat and exploded in a port, might very well destroy the whole port together with some of the surrounding territory. However, such bombs might very well prove to be too heavy for transportation by air.

This letter, drafted by a Hungarian scientist, Leo Szilard, and signed by Einstein, was sent to President Roosevelt less than a month before war began in Europe. Six years and four days later, an atomic bomb was dropped at Hiroshima. The technology of a new order of weapon had been created with amazing swiftness, and the post-war era was dominated by the awareness that in consequence quite unprecedented dangers faced mankind.

'split' because of their structure as a system of particles, it meant that matter at this fundamental level could be manipulated (though as late as 1935 he said that nuclear physics would have no practical implications – and no one rushed to contradict him). Two particles were soon identified, and the discovery of others has gone on ever since.

A new theoretical framework to replace the Newtonian was beginning to appear by 1930. By 1905 Max Planck and Albert Einstein had showed experimentally and mathematically that the Newtonian laws of motion could not explain why energy transactions in the material world took place not in an even flow but in discrete jumps – quanta, as they came to be termed. Planck showed that radiant heat (from, for example, the sun) was not, as Newtonians physics required, emitted continuously; he argued that this was true of all energy transactions. Einstein argued that light was propagated not continuously but in particles. These contributions were unsettling, but though Newton's views had been found wanting, there seemed to be nothing of equivalent generality yet to put in their place.

Meanwhile, after his work on quanta, Einstein had published in 1905 his statement of the special theory of relativity. This (together with later work, confirmed experimentally in 1919) showed that the traditional distinctions of space and time, and mass and energy, could not be consistently maintained. Einstein directed his colleagues' attention to a 'space-time continuum' in which the interplay of space, time and motion could be understood, and astronomical observation in due time confirmed that this made sense of facts for which Newtonian cosmology could not properly account. A major theoretical advance by Schrödinger and Heisenberg, two mathematicians, finally provided a mathematical framework for Planck's observations and, indeed, for nuclear physics. Quantum mechanics seemed to have inaugurated a new age of physics (though presenting difficulties for the theory of Relativity). Further development led to predictions of the existence of new nuclear particles duly verified by observations and in the end made possible the huge achievement of actually first tapping and then harnessing the energy of the nucleus through weapons research in the 1940s. When that happened, it became clear that Einstein had also formulated a mathematical statement of the relations of mass and energy which experiment confirmed.

By 1950 much more had disappeared in science than just a once-accepted set of general laws (and in any case, for most everyday purposes, Newtonian physics was still all that was needed). For all its mathematical sophistication, Newton's universe had been essentially and structurally simple, based on fundamental laws which could be grasped by laymen. The picture provided by the new physics was emphatically not one readily understandable, even in outline. The whole notion of a general law was being replaced by the concept of statistical probability as the best that could be hoped for, and this tendency spread from physics to other sciences. The idea as well as the content, of science was changing. Furthermore, the boundaries between sciences collapsed under the onrush of new knowledge made accessible by new theories and new instrumentation. The conflations involved in importing physical theory into neurology, or mathematics into biology made still more unattainable the synthesis of knowledge which had still been the hope of the nineteenth century. The rate of acquisition of new knowledge (some in such quantities that it could only be handled by the newly-available computers), faster than ever, was another difficulty. There was no clear advance towards an overarching theory intelligible to lay understanding as Newton's had been.

Biological sciences

There was a sense in which, somewhere in the middle of the 1950s, the baton passed (or appeared to pass) from the physical to the biological sciences. Their advance had begun with the invention of the microscope early in the seventeenth century. This revealed the organization of tissue in discrete units later called cells. In the nineteenth century it was grasped that cells could divide and that they developed individually. By 1900 it was widely accepted that individual cells provided a promising and fundamental approach to the study of life, and the application of chemistry to this became one of the main avenues of biological research. Meanwhile, nineteenth-century biological science had also inaugurated a new discipline, genetics, the study of the inheritance of characteristics by offspring from parents. Darwin had invoked the principle of inheritance as the means of propagation of traits favoured by natural selection. The first steps towards understanding the mechanism which made this possible were those of an Austrian monk, Gregor Mendel. From a meticulous series of experiments breeding pea plants, Mendel concluded that there existed hereditary units controlling the expression of traits passed from parents to offspring. Their physical reality came to be accepted. In 1909 a Dane gave them the name 'gene'.

Gradually the chemistry of cells was deciphered. The presence in the cell nucleus of a substance which might embody the most fundamental determinant of all living matter had already been established in 1873. Experiments then revealed a visible location for genes in chromosomes, and in the 1940s it was shown that genes controlled the chemical structure of protein, the most important constituent of cells. In 1944 the first step was taken towards identifying the specific effective agent bringing about changes in certain bacteria, and therefore controlling protein structure. In the 1950s it was at last identified as 'DNA' (deoxyribonucleic acid), whose physical structure (the double helix) was established in 1953. The crucial importance of this substance is that it is the carrier of the genetic information which determines the synthesis of protein molecules at the basis of life. The chemical mechanisms underlying the diversity of biological phenomena were at last accessible. Physiologically, this implied a transformation of man's self-perception unprecedented since the general acceptance of Darwinian ideas in the last century, and its implications as yet are far from being wholly discerned.

The identification and analysis of the structure of DNA was perhaps the most conspicuous step towards the manipulation of nature. It appeared to point the way to the conscious shaping of life-forms. Once again, not only more knowledge but new definitions of fields of study and new applications followed. 'Molecular biology', 'biotechnology' and 'genetic engineering' quickly became familiar terms. The genes of some organisms could, it was soon shown, be altered so as to give those organisms new and desirable characteristics. By manipulating their growth processes, yeast and other microorganisms could be made to produce novel substances, enzymes or other chemicals. The empirical technology accumulated informally and unconsciously by thousands of years' experience in making bread, wine and cheese was at last to be surpassed. Genetic modification of bacteria can now be used to grow chemicals or hormones. By the end of the 1980s there was launched a world-wide collaborative investigation, the Human Genome Project, whose almost unimaginable aim was the mapping of the human genetic apparatus so as to identify the position, structure and function of every human gene (there are from 50,000 to 100,000 in each cell, each of them having up to 30,000 pairs of the four basic chemical units that form the genetic code). Screening for the presence of certain defective genes, and even the replacement of some of them, is already achieved; the medical, social, and moral implications are enormous. In day-to-day police work, what is called DNA 'fingerprinting' is already used to identify an individual from a blood or semen sample, though argument continues about its limitations.

Space

To establish the level at which ideas have cultural, social, or political effect is a long-standing problem for historians. For all the revolutionary work in physics or biology, it is unlikely that even their approximate scientific importance can be sensed by most of us, and the same may well also be true of the recent vast extension of our physical environment by cosmonauts and satellites. Visions of space exploration and what it might mean were already appearing in fiction in the last years of the nineteenth century, and the technology which made it possible goes back almost as far. Before 1914, a Russian scientist, K. E. Tsolikovsky, had designed multi-staged rockets and devised many of the basic principles of space travel. The first Soviet liquid-fuelled rocket went up (three miles) in 1933, and a two-stage rocket six years later. The Second World War stimulated a major German rocket programme, which the United States drew on to begin its own in 1955. For most people, though, the space age began in October 1957, when an unmanned Soviet satellite called Sputnik I was launched by rocket and was soon in orbit around the earth emitting radio signals. That ended the era when the possibility of human travel in space could still be doubted.

Sputnik I entangled space exploration with superpower rivalries. The Americans started with more modest hardware than the Russians (who already had a good lead) and the first American satellite weighed only three pounds; Sputnik I weighed 184 pounds and its success shattered American confidence that their technology was bound to be superior to that of the USSR. A much-publicized first launch attempt by the Americans failed, while within a month of Sputnik I the Russians had already put up Sputnik II, an astonishingly successful machine, weighing half a ton and carrying the first passenger in space, a black-and-white mongrel called Laika; to the outrage of thousands of dog-lovers, she was not to return after the satellite's six-month orbiting of the earth. The Russian and American space programmes then somewhat diverged. The Russians emphasized power, size, and the lifting of big loads by their rockets. This had military implications more obvious than those (equally profound but

A twentieth century Diaz or Columbus sets off: Yuri Gagarin preparing to leave on his globe-orbiting space mission, 12 april 1961. He reached a maximum altitude of 188 miles above the surface of the earth, which he circled in one hour and 48 minutes.

less spectacular) flowing from American concentration on data-gathering and on instrumentation. Although people soon spoke of a 'space race' the contestants were really running towards different goals.

When Vanguard, the American satellite which failed in December 1957, was successfully launched again the following March, it went much deeper into space than any predecessor. Tiny as it was, it has provided more valuable scientific information in proportion to its size than any other satellite. At the end of 1958 the first satellite for communications purposes was also successfully launched by the Americans, who soon scored another 'first' – the recovery of a capsule after re-entry. The Russians then orbited and successfully retrieved Sputnik V, a four-and-half-ton satellite, carrying two dogs, who returned to earth safely. In the spring of the following year, on 12 April 1961, a Russian rocket took off carrying a man, Yuri Gagarin, who landed again 108 minutes later after one orbit around the earth. The invasion of space by the greatest of earthly predators, *homo sapiens*, had begun.

In May 1961 the American president proposed that the United States should try to land a man on the moon and return him safely to earth before the end of the decade. He argued that such a project provided a good national goal, that it would be 'impressive to mankind', that it was of great importance for the exploration of space, and (somewhat oddly) that it was of unparalleled difficulty and expense. Money was soon found for it. Though the Russians continued to make spectacular progress, after 1967 the glamour passed to the Americans. In 1968, they sent a three-man vehicle around the moon and transmitted television pictures of its surface. In May 1969 a vehicle put into orbit by the tenth rocket of the project approached to within six miles of the moon to assess the techniques of the final stage of landing. A few weeks later, on 16 July, a three-man crew was launched in 'Apollo XI'. Its lunar module landed on the moon's surface four days later. On the following morning, 21 July, the first human being to set foot on the moon was Neil Armstrong, the commander of the mission. The goal had been achieved with time in hand. But it was more than a triumphant new assertion of what America could do. It was also a sign of the latest and greatest extension of mankind's environment, the beginning of human life on other celestial bodies.

Ten years before the Americans planted the American flag there, a Soviet mission had dropped a pennant on the moon. This seemed ominous; nationalism might, it seemed, provoke squabbles in space. Yet though the rivalry of the United States and the USSR

The Exploration and Use of Space: Major Steps Down to 1969

1903 Konstantin Tsolikovsky (1857–1935) publishes paper on rocket space travel using liquid propellants.

1933 **1 May:** Tsolikovsky predicts that many Soviet citizens will live to see the first space flights.

1944 German V.2 rockets used to bombard London and Antwerp.

1954 President Eisenhower announces a small scientific satellite, Vanguard, will be launched 1957-8.

1957 **1 July:** Launch of Sputnik 1 (USSR), weight 184lbs.
3 November: Launch of Sputnik 2 (USSR), weight 1120lbs., with the dog Laika as passenger.

1958 **31 January:** Launch of Explorer (USA) and discovery of Van Allen radiation belts.
17 March: Launch of Vanguard 1 (USA), weight 3.25lbs. The first satellite with solar batteries.

1959 **13 September:** Luna 2 (USSR) crashes on Moon, the first man-made object to arrive there.
10 October: Luna 3 (USSR) photographs far side of Moon.

1960 **11 August:** Discoverer 13 (USA) recovered after first successful re-entry to atmosphere.
19 August: Sputnik 5 (USSR) orbits earth with two dogs which return unharmed.

1961 **12 April:** Major Yuri Gagarin (USSR) orbits earth.
25 May: President Kennedy commits USA to landing man on the moon by 1970.
6 August: Vostok 2 (USSR) makes seventeen orbits of earth.

1962 **10 July:** Launch of Telstar (USA) and first television pictures across Atlantic.
20 February: First manned orbited space flight.

1965 **18 March:** On Voskhod 2 mission (USSR) Alexey Leonov makes ten-minute 'walk in space'.
2 May: Early Bird commercial communication satellite (USA) first used by television.
15 December: Launch of Gemini 6 (USA) which makes rendezvous with Gemini 7.

1966 **July–November:** Gemini mission 10, 11, 12 (USA) all achieve 'docking'.

1967 **27 January:** first deaths in US programme.

1968 **21–27 December:** Apollo 8 (USA) makes first manned voyage round moon.

1969 **14–17 January:** Soyuz 4 and 5 (USSR) dock in space and exchange passengers.
16–24 July: Apollo 11 (USA) lands two men on the Moon.

Men first descend onto the moon, brought by television to approximately 500,000,000 people.

undoubtedly led to wasteful duplication of effort and inefficiencies, space exploration became more internationally co-operative as it went on. European and Asian countries joined in. Mercifully, it was soon agreed that celestial objects were not subject to appropriation by any one state; there was to be no interstellar repeat of Tordesillas. In July 1975, some hundred and fifty miles above the earth, cooperation became a startling reality when Soviet and American machines were connected so that their crews could move from one to the other. In spite of doubts, exploration continued in a relatively benign international setting. The visual exploration by unmanned satellite of space beyond Jupiter, the first landing of an unmanned exploration vehicle on the surface of the planet Mars, and the maiden voyage of the American Space Shuttle, the first reusable space vehicle, in 1977 were great achievements. So rapidly did there grow up a new familiarity with the idea of space travel that in the 1980s it seemed only mildly risible that people should begin to make commercial bookings for it – and even for space burials (if that is the word), too. As the decade drew to a close what was to prove the last big enterprise of the Soviet space effort was launched in 1988, a satellite to prepare the way for a future manned voyage to Mars. In the 1990s, though, it began to be clear that scepticism about cost-effectiveness was likely to mean that such ambitions were not soon to be realized, even in the USA.

WOMEN

Among the most visible and dramatic changes of this century have been some primarily affecting women, though they have, of course, both consequentially and indirectly deeply affected the other half of the human race too. Such changes are far from completed, but they truly mark a great historical divide. It is now hard to believe they could ever be reversed. Yet although they have recently become so notable they too have deep historical roots and can only be usefully assessed over the long term.

Virtually all the positive impulse behind the liberation of women into new possibilities of choice has come from within the western European cultural tradition and a full account would have to start by deciding what it was in the legacy of the classical and Judaeo-Christian traditions that provided the potential for what followed. Here there is no space to trace the chronological story back so far, but the background should be kept in mind. Many ideas now taken for granted about what women might do with their lives and how better education could help them to do it can first be clearly discerned in the eighteenth century. There then surfaced the first clear demands for more equal treatment for women. The French Revolution played an important part, both positively and negatively, in their emergence. Much talk about universal 'Rights of Man' stimulated some women into agitating for greater rights for themselves, as well as tarring them with the brush of association with subversion. In France itself, they did not get very far. However keen the revolutionaries were on the liberation of French men, they thought that the French woman's place was in the home. Women who tried to take part in revolutionary politics were ignored. One of their leaders was actually guillotined. In England – where women enjoyed a greater degree of freedom in day-to-day life than in much of Europe at that time, a remarkable woman, Mary Wollestonecraft, took up the cause of women and in 1792 published a book called *A Vindication of the Rights of Women*. She awoke violent dislike (one politician called her 'a hyena in petticoats'), no doubt because any suggestion of changes in the roles of the sexes was liable to arouse alarm among males. Her book is as good a foundation-stone as any for the modern women's movement and drew more attention to her cause than any earlier document.

Political rights

During the nineteenth century, pressure for the enlarging and strengthening of women's rights grew. The cause prospered, though slowly and unevenly, and had notable achieve-

The campaign for women's political rights in the United Kingdom provoked extraordinary outbursts of feeling on both sides – perhaps because the contenders were not themselves conscious of the depth and complexity of the emotions awoken by feminist demands. This lady attempted to chain herself to the railings of Buckingham Palace in 1914.

ments to its credit by 1914. By then, women in some countries had won what some of them thought was the key to political power – votes in national elections. In 1890 the state of Wyoming gave women the right to vote for members of Congress and the President of the United States (in several other states they already had a vote in local government by then and a woman had already stood as a candidate for the presidency). In the next ten years, three more American states followed suit, while New Zealand, and Western and South Australia also gave women the national franchise. Finnish women got the vote in 1907, and six more American states joined the movement by 1914. By then, political movements advocating voting rights for women were to be found in many countries, even India.

Some countries whose male legislators were unresponsive to argument faced vociferous demands on this issue. In Great Britain some women took to violence – smashing windows, pouring acid into letter-boxes and physically assaulting politicians in order to draw attention to their claims. The attention thus won by the 'suffragette' movement as it was called (because those who took part in it wanted the 'suffrage', or right to vote) may not have been always to its benefit. It violently antagonized many (both men and women) because it awoke fears of very deep changes in the relations of the sexes. Sudden steps in the direction of sexual equality have continued to do so.

Women and the professions

Yet the forces bearing women forward to greater equality and freedom drove on whether their opponents liked it or not. The spread of the idea that women might properly be educated in the same way as men had transformed girls' lives in many families already between 1800 and 1914. By the latter date, women were being admitted to universities in the United States and all the major European countries and a huge structure of girls' schools had grown up to provide the teaching which, if available at all to girls, had in 1800 been obtainable only from private tutors or in a few convents. Women had begun to make their mark in the sciences; the first one famous for doing so was Marie Curie, the Polish-born scientist who shared a Nobel prize for physics in 1903, succeeded her husband as professor at the Sorbonne in 1906 and won a second Nobel prize outright in chemistry five years later for her work on radium. (Women were not to get the vote in France, her adopted country, until 1946.)

By 1914 there were women doctors, lawyers, university teachers and social workers. Education – though, of course, available at the highest level only to a small minority – had helped to transform the choice of careers open to women. Access to the professions had to be won by overcoming not only the barriers imposed by the lack of educational facilities for much of the nineteenth century, but also others imposed by prudery and fears about women's modesty. In this respect, one outstanding servant of her sex was Florence Nightingale, an Englishwoman who came to the public notice thanks to her almost single-handed creation of decent hospital services for the British army of the Crimean war and her subsequent unflagging (and largely successful) efforts to improve the lot of the ordinary soldier. Not only did the soldier benefit; so, in the long run, did women. One of Miss Nightingale's many contributions to the betterment of humanity created another career opportunity for her sex; she made the profession of nursing respectable. Until her day, virtually the only women of standing and respectability to nurse the sick had been members of religious communities, both Roman Catholic and Protestant (it was to German Protestants that Florence Nightingale went to seek training). Apart from these religious women, nursing had usually been left not only to the untrained and ignorant, but to the unscrupulous and sometimes the downright criminal as well. Florence Nightingale insisted not only on a high level of cleanliness and discipline (to say nothing of respectability) in her nurses, but trained them in a new way, so that they could make a systematic and serious contribution to the process of healing the sick – also a major contribution to medical progress.

Politics and the professions offered already by 1914 unmistakable signs that a crucial point had been passed, however much remained to be played for. The struggle was eagerly pressed forwards on many fronts, even if many women remained ignorant of what was being asked for on their behalf. An American lady, Mrs Amelia Bloomer, had thought women should not be obliged to wear skirts (which were at this time inconveniently long and likely to collect dirt). When she invented a kind of trousers suitable (she thought) for women and she aroused much ridicule by wearing them. Still, her name has passed into the English language in the term 'bloomers', and it is an immortality of sorts. But in some ways the most important changes in women's lives – important because they affected the largest numbers of women – were coming about without anyone thinking much about what they might mean for reasons quite other than the heroic efforts of the feminist politicians.

Women's employment

It began to be apparent in the eighteenth century in a few places that women were going to be offered many more ways of earning a living by industrialization. They had always – per-

In Great Britain, two world wars were great disturbers of settled ideas about a woman's role. A propaganda picture of 1939 showing a member of the Auxiliary Territorial Service (ATS) in which, at its height, over 200,00 women served.

haps since the invention of agriculture – toiled in the fields (as, today, they still do in many countries). They had long been given a living as household slaves, and where slavery disappeared they became paid domestic servants. They had also always spun at home (weaving, because the loom made it heavy work, was usually a man's task); hence the word 'spinster', because it was a way of earning your daily bread if you were unlucky enough not to get married. Here industrialization made a difference. As the demand for thread rose it became easier for women to obtain more work at home. The next step was for them to go to the town where the first factories were appearing, to work in cotton-spinning. This was not always healthy or very stimulating, but neither was the life of the peasant woman. It was a real extension of choice.

The greater acceptance of education for women and the provision of more industrial jobs for them went much further in the nineteenth century. In developed societies there appeared scores of new occupations and millions of new jobs for women. Sometimes a single invention made the difference: the typewriter was one which was crucial. Sometimes it was a new way of doing things – the development of big stores for retail trading, for instance. More and more, as typists, secretaries, telephonists, shop assistants and factory hands, some women, earning their own livings, could enjoy greater freedom than in the old, male-dominated world only a few decades before. By 1900 jobs in industry or commerce were giving millions of women the first chances anyone in their families had ever enjoyed of escaping either from a parental tyranny which often lasted well into adult life, or from married drudgery. More and more women would grasp this opportunity in more and more countries as the twentieth century went on – and, of course, men would resist, as they felt their own jobs and roles threatened.

Technology also offered women other freedoms. A huge number of separate inventions and innovations, in all sides of life, have reduced drudgery and made work in the home easier. Some were as simple as the coming of piped water to the house which meant an end of long, often weary trips to a neighbourhood pump – or the arrival of gas for lighting and then cooking – which reduced the dirt and trouble inseparable from oil-lamps and open ranges. Outside the house, better shops, carrying larger stocks of mass-produced goods, increased the choice open to housewives and, therefore, the ease with which they could meet their families' needs. Imported foods (made available by steamships and railways, canning or processing) slowly made family catering different and easier than that once based – as it still often is in many parts of Asia or Africa – on visits to the market twice a day. Cheaper soap and washing soda were the products of nineteenth-century chemical industry, and even the first domestic machines – vacuum cleaners and washing-machines for the rich, hand-turned mangles for the poor – were in use by 1914. Historians too often overlook such humble innovations.

The last force already beginning to affect women's (and men's) lives before 1914, though not outside the most advanced countries (and even there, it was hardly ever talked about openly), was contraception, the conscious control by physical or chemical means of the number of children in the family. Societies in the past had relied upon infanticide or delay in marrying to achieve this. By 1914 the application of technology and spreading knowledge to family limitation had already begun to produce measurable effects in the more advanced countries of Europe and North America. In the early years of this century it was a trend most noticeable among the better-off; the more educated and the wealthy were keenest to limit their family size by artificial means. Still, the mere notion that this possibility existed spread rapidly except where it ran up against strong religious or popular objection. And although important to both sexes, it affected women most. They had for the first time the possibility of reducing the demands made on them by childbearing and raising a family which for the whole of earlier human history had dominated the lives of the overwhelming majority of their sex.

All the forces already at work to change women's lives before 1914 operated more widely and more forcefully as the twentieth century unrolled, most evidently in the most developed countries. But two great wars had their transforming effects in every country, bringing about a questioning and rejection of much in traditional ways, and producing a forced draught of economic, military and even intellectual mobilization which thrust millions of women into new roles – not always to their regret. It was in those countries, too, that there was felt most strongly the effect of developments in communication. It was not only the formal communication of feminist propaganda which promoted women's independence. There was also the generation of new notions of the possible, new alternative models of behaviour by, first, the cinema and then television, which entered the home itself. Above all, there was advertising, from which not only knowledge of new facts – especially technological facts – reached the home, but values and attitudes as well.

The non-western world

One of the most striking developments was the spread to non-western societies of what may be broadly called 'western trends' in the way women were regarded. Different societies have always treated women somewhat differently. European women have long lived less cramped lives than their Asian or African sisters. Recently, though, the gap between the way women are treated in countries of European origin and in more traditional and old-fashioned societies has widened dramatically, and this has produced demands for change in the latter. Nowadays, even very backward societies are, so far as the freedom of women is concerned, having to make concessions. Their representatives on international bodies or at the United Nations pay lip-service to steps to improve the lot of women, though this does not always mean much in practice. About half the agricultural labour force of the world is female and it is not to be found in the developed nations. In India or Africa, a woman can still work as a field labourer on the family plot under the direction of the men of the family. Often she has still to rely on marriage or family charity as her only insurance against starvation. In some countries, the urge to produce children is still a powerful check on the liberation of women along 'western' lines. Nonetheless, even very traditional societies can change. It looks as if once again the example set by the 'western' civilization which has its roots in Europe will, in the end, alter the habits of the rest of the world through the same agencies which have operated in western societies – economic and educational opportunity, technology, contraception (now greatly simplified through the availability of contraceptive pills), as well as through conscious agitation and campaigning by feminists. What is new, nonetheless, is that these forces will have to operate in societies without the background of Christian and liberal culture which so much of Europe and North America could take for granted and will also have to face strong and sometimes violent resistance from traditional authority.

14
THE LATEST AGE: UPHEAVAL

THE APPROACH TO DISASTER

In the first half of the twentieth century two great European wars destroyed the old European power system and with it the nineteenth-century international order. At a deeper level a shared consensus which had still underpinned the political and economic structures of the civilized world at the beginning of the century was also dissolved. Meanwhile, a fatal undermining of the colonial empires, which had shaped so much of the previous two or three centuries also took place. These themes underlie the tumultuous story of 'what happened' and the chronological story makes more sense if they are kept in mind.

The first of the two great wars in which Europe all but destroyed itself began in 1914. There then came to an end a long peace between the European great powers which had lasted since 1871. A deep-seated quarrel between Austria-Hungary and Russia finally boiled over and did so in such a way that Germany, France and Great Britain were all drawn in.

In 1908 the Dual Monarchy had much irritated the Russians by annexing Bosnia, a province still legally Turkish though occupied by Austria-Hungary. The Austrians were beginning to feel (like some of the smaller Balkan nations) that reformers might rebuild the strength of the Ottoman empire – where a 'Young Turk' party had emerged, ambitious to do just that. A firm Habsburg grip on Bosnia was desirable, it seemed, if it were not to be taken back by the Turks or, worse still, seized by the Serbs. In Vienna Serbia was suspected of wanting to unify all the South Slav peoples. There were so many Slavs inside the Dual Monarchy (and especially in its Hungarian half) that Serbian ambition was seen as a grave threat.

The Austrians under-estimated the outrage felt by the Russians. There was no compensating advantage for them and they had not been squared in advance. Russia now ceased to hope to work with the Dual Monarchy to manage change in the Balkans (which was bound to continue) by agreement. Serbians, too, were much annoyed, but Serbia was not powerful enough to resist. Russia was the greatest Slav power, though, and if trouble boiled up again the Serbians might well find the Russians readier to support them. Though the crisis ended without war the sky had darkened. Russia had for nearly a century confidently awaited the collapse of the Ottoman empire: was it now to be saved at the eleventh hour by the Young Turks? The Straits of Constantinople now mattered more than ever to Russia since she had begun to export huge quantities of grain from the Black Sea provinces.

In many ways, indeed, Russia had good cause to assert her great power status once again. She was well on the way to being an industrial power; even if well behind Germany or England, her industrial output was growing at a faster rate than that of either. The agricultural problem was at last beginning to come right. New legislation accelerated the emergence of a new class of Kulaks (yeoman-farmers as they might be called) who were interested in efficiency and profits. Their efforts began to raise productivity at last.

As Russian confidence grew, its rulers were sure they could stand up for her interests and confident that the Russian army, supported by a growing railway network and expanding industrial base gave them the means to do so. On the other hand, though Russia was a European country in name, she was still one where poverty of Asiatic horror could be found. She was still backward, a country where religion was mixed up in government and society in a way unknown for a century or so past in most of Europe. She had one or two good universities and schools and some distinguished scientists and scholars, but the overwhelming majority of her people were illiterate peasants. Above all, in spite of the 1905 revolution she was still a country where government rested in the last resort on what was regarded as the god-given power of the autocrat.

Germany and France, separated by the question of Alsace and Lorraine, taken from France in 1871, were bound to be involved in any Austro-Russian quarrel; France and Russia's ally. The German generals aimed to avoid a war on two fronts by first defeating France and then switching their forces; though Russia had the greater numbers, they would take longer to mobilize. So the German military planned to defeat France first in a quick campaign (as in 1870) by passing through the neutral state of Belgium. It was hoped that the powers which had guaranteed Belgium's neutrality by a solemn treaty as long ago as 1839 would not mind, or would at least look the other way until the week or so needed was over. This was at best a gamble.

One power with whom Germany had no grounds to quarrel was Great Britain, though pressure groups interested in colonial affairs and in naval expansion had long been trying to stir up anti-British feeling in Germany in support of their aims. This led to British anxiety about German policy. When Germany bullied France about influence in Morocco, some British statesmen began to think that a halt would have to be called if Wilhelm II was not to end up lording it over the continent of Europe as Napoleon and Louis XIV had done in their day. Military talks with the French were begun to explore what might be done if the two nations found themselves on the same side – a great change for two traditional rivals – and the British army was reorganized so as to make it possible to send an expeditionary force to France. It was not clear in what conditions it would be sent, only that if it was sent it would be to help confront a German invasion.

Although many Englishmen sought good relations with Germany, the atmosphere was made worse by German efforts to build a great navy. This hardly seemed to make sense unless a contest with the Royal Navy was envisaged, thought England's rulers. Alarmed, they first carried out certain reforms and reorganization to make the Royal Navy stronger in home waters and then launched a technical revolution by building a battleship of wholly novel design. This was HMS *Dreadnought*, more powerful, bigger, faster than any capital ship afloat and carrying more than twice as many heavy guns in her main armament. All older battleships at once became obsolete; soon, everybody began to build 'dreadnoughts' as the new type was soon known (the older kind of ship now became referred to as the 'pre-dread-

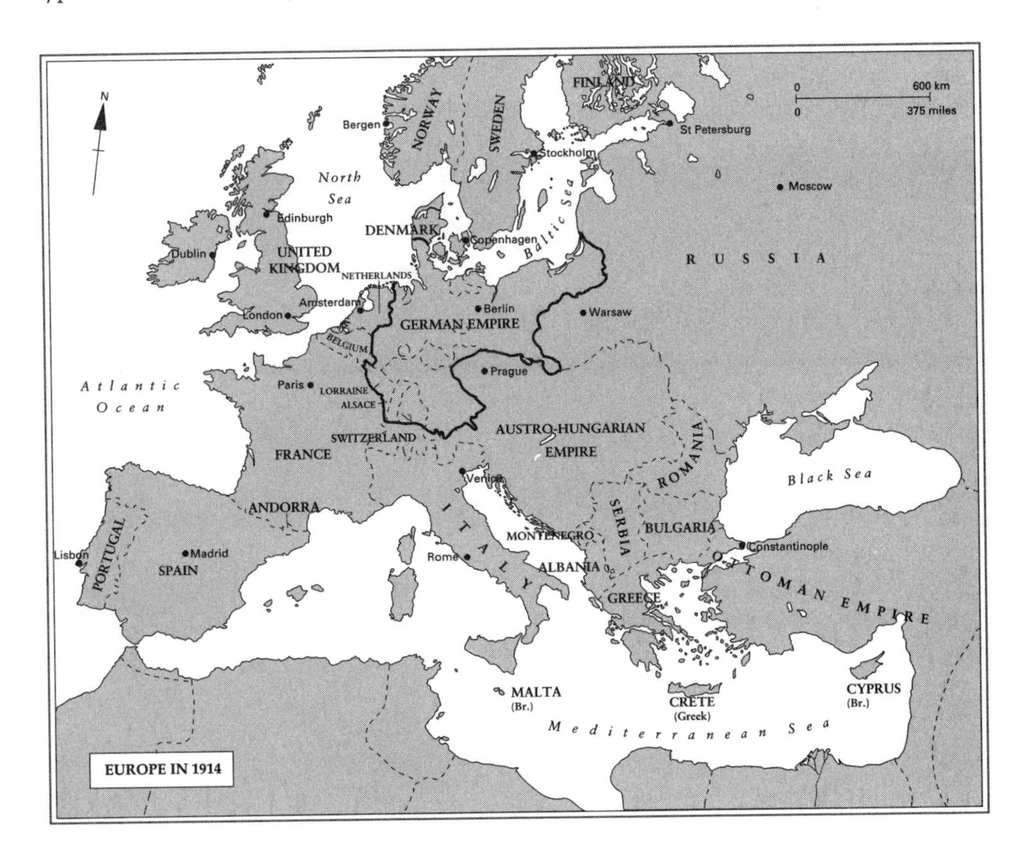

EUROPE IN 1914

nought'). But Germany persisted in challenging British naval supremacy and a naval 'race' to build dreadnoughts began. After a slow start, the British determined that even if they could not stop the race, they would win it; soon they were out-building Germany and by 1914 were again well ahead of her. Nothing had been gained by Germany's naval programme; instead, a lot of money had been spent, much damage had been done to British trustfulness of German intentions, and British public opinion had been alienated.

Sarajevo

The powder-keg, though, was still the Danube valley. The Austro-Hungarian government still distrusted Serbia's intentions, Russia was getting stronger; in a few years her armies, reorganized and re-equipped, would be much more powerful, and her network of strategic railways would be complete. If the Serbs were to be taught a lesson it would have to be before the Russians were strong enough to back them up with the threat of war. This was why the assassination in June 1914 of an Austrian archduke in the little Bosnian town of Sarajevo launched a world war. The murder plans were very amateurish, but so were those made to protect the victim. Franz Ferdinand, the archduke, had been warned of the dangers of visiting Bosnia. It was full of Slavs who resented the Austrian occupation and 28 June, when the visit to Sarajevo was to take place, seemed chosen as if to infuriate them, for it was the day of the greatest of the Serbian patriotic festivals. There had been several attempts to assassinate Habsburg dignitaries in recent years. Yet almost no special precautions were taken. Except for a few extra detectives from Budapest and Trieste the local police force – only 120 men – had to guard by themselves the archduke while he travelled in an open car through the streets for about four miles.

The last photograph of the archduke alive shows him and his wife leaving the town hall to enter their car. As they did so one of the conspirators standing not far away asked a policeman which car was the archduke's. The detective told him and he promptly threw a bomb at it. The archduke was not hurt, though many people were wounded, some gravely. Not unreasonably, the archduke (who was no coward) asked to change the planned route, and the cars moved off. But no-one had told the drivers of the change of plan. When the military gover-

The last photograph of the Archduke Franz-Ferdinand and his wife, the Duchess of Hohenberg, before they were assassinated at Sarajevo. Moslem Bosnians present then wore the fez.

nor shouted that the archduke's car was going the wrong way the driver (a Czech) braked hard and stopped dead. Among those at that spot was a young man called Gavrilo Princip, one of the would-be assassins. At point-blank range, Princip drew a pistol and fired. The archduke and archduchess both died almost at once. Princip was seized and the whole incident was over. Also, Europe was on the road to war.

The terrorists had been armed by a Serbian patriotic secret society, but the murder was taken as a wonderful opportunity to teach the Serbian government a lesson. The Monarchy could now force on her such a humiliation that its rebellious Slavs would never again look to her for support. The Germans agreed that Austria-Hungary should act, if necessary by force. Nearly four weeks after the murder, therefore, on 23 July, the Serbians were given an Austro-Hungarian ultimatum imposing heavy demands. With Russian advice, they accepted almost all of it. But this was not enough, and on 28 July, a month after the assassination, the Austrians declared war on Serbia.

Events now rolled forward almost automatically towards disaster. When Russia began to mobilize in order to bring pressure to bear on Austria-Hungary, the Germans without waiting declared war on her. As they had long before decided that they would have to do, they also declared war on France and invaded Belgium. The violation of Belgium's neutrality gave the British government the excuse it needed and united public opinion. On 4 August it declared war on Germany. By an odd paradox, the last two great powers formally to go to war with one another were Austria-Hungary and Russia, whose mutual fears and rivalries lay at the root of the struggle.

There is no single or simple explanation. If the Austrians had not humiliated the Russians in 1909, if the archduke had been less courageous, if Princip had been sitting at another cafe, if the Germans had not built a fleet, and if any of a hundred things other than these had gone otherwise, then the outcome would have been different. But there would still have been deep-lying problems to be solved. What was to be the final outcome of the collapse of Turkish power in the Balkans? Was the imperial German government to run Europe? Were Alsace and Lorraine ever to be returned to France? Was the Dual Monarchy capable of governing its Slav subjects and contenting them with Habsburg rule? Attempts to solve any of these were bound in the end at least to run the risk of general war.

THE GREAT WAR 1914–18

One of the paradoxes of 1914 is that in every country huge numbers of people, of all parties, creeds and blood, seem, surprisingly, to have gone willingly and happily to war. Many of them saw it as an opportunity, not a disaster. Reality turned out to be very different from what had been expected. It was far more appalling than anything any of those who had helped to bring the war about had imagined. It was long to be known as 'the Great War', for it was different in scale from any earlier conflict, and it led to operations all over the globe. It lasted over four years. This was unusual; earlier wars had not brought such continuous fighting. Only the American Civil War had anticipated the slogging match of 1914–18, when millions of men faced one another for month after month, year after year, separated by only a few hundred yards of battlefield, trying to grind their enemies into submission. From the start, too, the sea war was fierce and it became worse as each side tried to starve the other by blockade. Even the air, finally, became a place of battle. Military aeroplanes had been first used in war in 1911, when the Italians attacked the Ottoman empire in North Africa, and balloons had been used over a century before, by the French in the Revolutionary Wars, but now for the first time the skies became a zone of combat extending far beyond the lines of battle.

War was mechanized as never before. By the end of it lorries had become as important as horses in supplying the soldiers in the field. The railway had transformed military mobility in the previous century; now petrol-driven transport joined it. And, of course, weapons

DH4 day bombers of the Royal Flying Corps, France, early 1918. The design was one of the most successful of the Great War, in spite of the near impossibility of pilot and observer communicating in flight, given the distance between them, and a tendency to burst into flames in any crash landing.

improved (if that is the word) terrifyingly. By 1914 every army had breach-loading rifles, machine-guns and field-guns. Their power and precision led to vast slaughter. The regular British infantryman who went to France in 1914 had a rifle which in his hands could hit a man-sized target at ranges of up to half a mile. He – like his opponents and allies – was supported by machine guns (firing 600 rounds a minute), field guns firing three or four times a minute at ranges up to 10,000 yards or so, heavier guns which could bombard targets six or seven miles away, and some monster cannon with even longer range. The slaughter brought about by these weapons was relentlessly continuous. For four years about 5000 men were killed somewhere every day. Among the great powers, the French and Germans suffered most in proportion to population, and the Americans (who entered the war in 1917) least. In an appalling five-month battle before the French fortress of Verdun in 1916, the French and German armies between them suffered over 600,000 casualties (dead, wounded and missing), and on the first day of the battle of the Somme that year the British army (at that time entirely composed of volunteers) lost 20,000 dead and almost another 40,000 wounded. On the great monument at Thiepval to British soldiers who died over a year or so on the Somme there are over 70,000 names – and they are the names only of those whose bodies were never found.

In all previous wars the worst killer had been sickness. Large numbers of men, cramped together in makeshift conditions, with temporary sanitation, possibly with polluted water supplies and no fresh food provided ideal conditions for epidemics of dysentry, cholera, smallpox, typhus. Sickness killed three times as many British soldiers as the Boers in the South African War of 1899-1902. By the Great War, more was known about treatment and prevention, and industrial societies could supply and maintain huge armies in the field with food, proper clothing and medical supplies. For the first time since records were available most of the military casualties of a war were caused by direct enemy action.

Civilians, too, suffered more heavily as the war went on. Food shortages and disease tended to kill children and old people first; they were less able to stand up to privation and

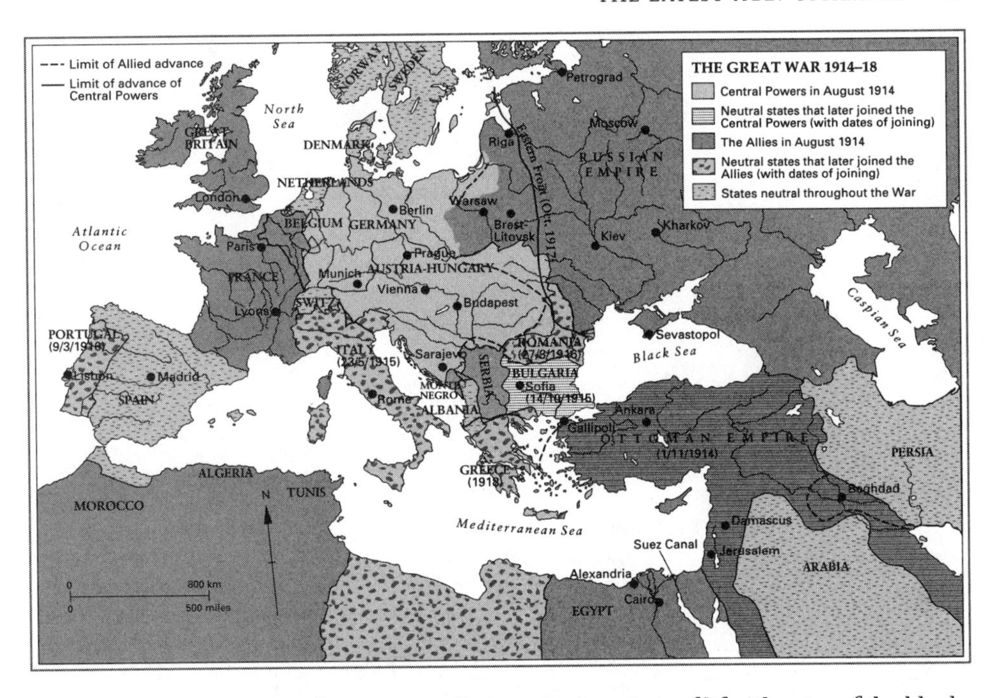

sickness than the soldiers, who were usually men in the prime of life. The aim of the blockades each side sought to impose was nevertheless not only to starve people but factories – of minerals, chemicals, fuel, imported machinery. Military needs were enormous: boots, uniforms, barbed wire, timber for building, tools for digging – all were needed on a scale unimaginable a few years earlier. As for weapons and ammunition, there seemed never to be enough. A British infantry battalion of 1914 was allocated two machine guns; a few years later it had over fifty, which meant a huge increase in the use of ammunition. Artillerymen fired shells at a rate which quickly exhausted supplies in the first year of war. Later, huge bombardments took place; that before the Battle of the Somme, delivered by nearly 2000 guns on a ten-mile front, was heard on Hampstead Heath, three hundred miles away.

By 1918, the war was world-wide, too. The 'Central Powers' (Austria-Hungary and Germany) had been lined up at the outset against the 'Allied' or 'Entente' powers (Great Britain, France and Russia). Within a few months Japan joined the Entente, and Turkey the other side. Italy came in against Austria-Hungary in 1915. In 1917, the United States entered the war on the allied side; in Europe a year and a half later, when the war ended, only Spain, Switzerland, Holland and the Scandinavian countries were still neutral. Even China had formally joined the Allied cause.

Stalemate in Europe had spread the war thanks to the stopping-power of up-to-date weapons. Even the most pulverising bombardments left defenders armed with machine-guns still able to halt an attack within a few thousand yards, and sometimes within hundreds. The Germans held Belgium and a big area of northern France into which they had advanced swiftly in the first weeks of the war and the western front settled down to a kind of siege warfare, with millions of men living in trenches and underground. Meanwhile, in the east, steady fighting slowly sapped the strength of the Russian army and undermined the logistical base on which it depended.

One attempted solution to deadlock was to invent new weapons – poison gas and the tank, for instance. Another was to look for allies and superior numbers. Another was blockade. At the end of 1916, after failing to win the summer battles in France and with Russia still on her feet, the German High Command concluded that Germany was going to lose the war. The British naval blockade would strangle Germany unless something could be done quickly.

'Tank' was a word adopted by the British to keep secret a new invention they were preparing – a self-propelled fortress which could cross all but very large obstacles, impervious to the barbed-wire and machine-gun fire which brought infantry to a bloody halt so many times. By the end of the war, tanks had been developed for specialized roles. This French example was a light, two-man machine, and comparatively fast.

The Western Front, 1916. A British battalion moves to the rear with its wounded past a supply column moving up to the front.

It was decided to blockade Great Britain in its turn with submarines – 'U-boats', or 'undersea-boats'. They began to sink without warning any ships going to a British port, neutral or belligerent, unarmed or armed, carrying war material or not. This at last brought the United States into the war. After that, the Allies had only to win the battle against the U-boat; time would then steadily tell on the Allied side, as America's huge potential armies were brought to bear on the battlefield. Germany had one last stroke of good fortune: Russia collapsed in revolution in 1917. This released German forces for the western front, and with them the German generals launched in 1918 their last great attack. It only just failed. Allied counterattack followed, and by late summer the Germans and their allies were everywhere (except in Russia) in retreat. In October, Germany asked for a suspension of fighting. A severe armistice was granted, and at eleven o'clock in the morning, on 11 November 1918, the Western Front at last fell silent.

THE POST-WAR WORLD

When the fighting stopped, many people thought that things could simply go back to 'normal'. This was impossible. The world of 1914 was gone beyond recall, at least in Europe. In eastern Europe and the Near East, four empires had collapsed. The Russian army, ill-fed, badly-equipped and armed, had fought with magnificent courage, even winning a great victory against the Austrians in 1916. But by 1917 it had shot its bolt, and Russian industry alone could not meet her soldiers' needs. Much of it had been in Poland, one of the main battle zones, and the Russian railways had by 1916 all but collapsed. The country was paying the price for belated industrialization. The allies could only supply Russia through northern ports frozen for much of the year or through Vladivostok, six thousand miles from the front line.

In 1917 a 'March revolution' (which the Russians call the 'February revolution', because of the different calendar Russia then followed) began with food riots in the capital. The disobedience and mutiny of the soldiers who should have suppressed them followed. The appearance of a new 'provisional' government then led to the abdication of the Tsar. At first, Russia's allies welcomed the change: the new government said it would go on fighting the Central Powers. It was a democratic government and looked a much more respectable ally than the old tsarist regime. But the Russian people wanted peace. Many also wanted to use the revolution to overturn what they saw as old injustices; peasants coveted the lands of the nobility, suppressed nationalities wanted independence, some industrial workers wanted to end private ownership of factories.

In the cities the influence of the extremist Marxist majority of the Russian socialist party – the Bolsheviks – was strong. In November (October in the old calendar) they ousted the provisional government from power. It took two or three years for the Bolsheviks to establish themselves securely in the teeth of foreign invasion, civil war and opposition from other revolutionary groups, but in the end they were successful. So Russia became the first state in the world with a Marxist government, dedicated formally to the advance of the cause of the world's workers, at least as the Bolsheviks saw it.

Meanwhile, in September 1918, Austria-Hungary had begun to break up in revolution. A few weeks later, revolution in Germany led William II to abdicate. In the Ottoman empire, long before this, revolt had broken out in its Arab territories, and when the war ended there was not much left except Turkey itself. From the former Ottoman territories in the Near East and Arabia were to emerge in due course a series of new Arab states, as well as a new Turkey. Out of former German, Austro-Hungarian and Russian territory there appeared three new Baltic states (Latvia, Lithuania and Estonia), a new country called Czechoslovakia, a new Austrian republic, a much reduced Hungary, a reborn Poland and a new South Slav state (later to be called Yugoslavia) which included the former kingdoms of Serbia and Montenegro. The details took years to settle, but that eastern Europe would be organized round new units was settled while the war was still going on.

The peace settlements

Of the treaties of 1919 the most important was that with Germany. The whole settlement was very much the work of the leaders of the victorious powers, Great Britain, France and, above all, the United States. President Woodrow Wilson was idealized by many Europeans. He had proclaimed his support for the principles of nationalism and democracy. The French, though, wanted above all guarantees against any future German revival and another invasion, and the British were anxious to revive a realistic balance of power in Europe. The outcome was a series of punitive impositions on Germany (which also had to give back Alsace and Lorraine and lost much territory in the east), and an unsystematic patchwork of attempts to settle frontiers so as to respect the demands of nations which had emerged *de facto* from the former Russian and Austro-Hungarian empires. Ominously, the United States did not, in the end, ratify the Versailles treaty with Germany, and Russia had not been represented in any of the peace negotiations.

It is scarcely surprising that the new Europe did not please everyone nor that some bitterly resented it. Nevertheless, it seemed to settle a lot of questions which had been troubling people for much of the previous century. It was at least possible to think that there would now be no more oppressed nations in Europe ruled by foreigners, the bugbear of nineteenth-century nationalists.

Unhappily, satisfying some nationalists almost always meant irritating others. Poland might be revived but many of her citizens were not Poles. Czechs and Slovaks might agree to live together in their new democratic republic, but the Germans in the Czech lands would have preferred to go on under the Habsburgs. South Slavs and Romanians might be pleased to have done with Magyar rule, but the Magyars felt bitter over their lost territory. Soon, Croatians were complaining about their treatment by the Serbs in the new state of Yugoslavia.

The End of the Great War and the Peace Settlements

1918	
March 3	German-Soviet treaty of Brest-Litovsk
Apr 10	Congress of Austrian subject peoples in Rome
May 7	German-Romanian treaty of Bucharest
June–Sept	Allies recognize independence of Czechoslovakia
Sept 30	Allies grant armistice to Bulgaria
Oct 29	Yugoslav independence proclaimed
Oct 30	Allies grant armistice to Ottoman empire
Nov 3	Armistice between Allies and Austria-Hungary
Nov 9	German republic proclaimed
Nov 10	Romania re-enters war on Allied side
Nov 11	30-day Armistice ends fighting on the Western front
Nov 13	Austrian republic proclaimed
Nov 16	Hungarian republic proclaimed
1919	
Jan 18	Peace conference opens at Paris
June 28	Signature of treaty of Versailles with Germany
Sept 10	Treaty of St Germain with Austrian republic
Nov 27	Treaty of Neuilly with Bulgaria
1920	
June 4	Treaty of Trianon with Hungary
Aug 10	Treaty of Sèvres with Ottoman monarchy
1921	
Mar 16	Kemalist government of Turkey makes treaty with USSR
1923	
July 24	Treaty of Lausanne and final peace terms between new Turkish government and allied powers

The League of Nations

One possible ground for optimism existed in an attempt to organize international life as never before. As a first step towards regulating the behaviour of independent sovereign states towards one another, a 'League of Nations' was set up, with its headquarters at Geneva. This owed much to Woodrow Wilson, whose enthusiasm pushed the idea past his allies (though he subsequently failed to persuade his fellow country-men to join the League). Almost at once, the League began to intervene – with some success – in disputes between states which might in the past have led to armed conflict. It also took up economic problems and the tragedies of the 'refugees', millions of whom were confronting central and eastern Europe and the Near East with grave demands on their limited and over-strained resources.

The shattering of empires, the triumph of long-suppressed national demands and the creation of the League were the most obvious features of a new international order. It did not spring to the eye so readily that for the first time since the Turks had menaced it in the sixteenth century, Europe's future had been settled by an outside power. Germany's generals had cracked a year earlier than their opponents had expected, because they knew the war was lost once America's full weight could be brought to bear. The days of the European political supremacy in world affairs was over. Most of the signatories of the Treaty of Versailles were non-European. The surviving colonial empires were now to be threatened by new nationalist movements. Japan was a victorious great power, too, and much more was to be heard of her demands in the next few years. Finally, European economic strength had been savagely wounded by the war. It was against this unpromising background that the continent appeared to face a grave new threat.

INSTITUTIONALIZED REVOLUTION

Ever since 1789, some Europeans had hoped for or dreaded popular revolution. The nineteenth century's history could be argued to support both stances, at least at first sight. Between 1821 and 1914 many risings had taken place, assassinations been staged, strikes held, bombs let off: it was a violent age. New political forces – above all, Marxist socialism – were often to be found using revolutionary slogans and sometimes what looked like revolutionary methods. Yet for all this excitement, there had never been a successful popular revolution in a major country. Outbreaks of violence and popular unrest were usually handled without much difficulty by confident regimes. In some countries, a growing liberalism in political arrangements had provided safety-valves for the expression of anger as well as ways of meeting social grievance. Though some of Europe's rulers feared revolution in 1914, the way their peoples responded to the demands war made on them showed that they need not have done so.

After 1918, things were different. The war had not only badly damaged traditional authority, economic well-being, and shattered the political structures of the old order. For the first time there had come into being a state, a great power, whose rulers, sincerely or not, claimed to seek the overthrow of all existing society and its replacement by a different model. This was the new Russia, the Union of Soviet Socialist Republics (USSR).

The USSR

The two men who had done most to win power for the Russian Bolsheviks were Vladimir Ilych Lenin and Leon Trotsky. Before 1914 Lenin taught his party to be a small, highly-disciplined revolutionary elite, ruthlessly purging their ranks of anyone disposed to argue with the decisions of the party's leadership or not accepting its interpretations of the teachings of Karl Marx as truth. The onset of war had sent him into exile, but with the help of the Germans (who were only too anxious to do anything they could to hasten Russia's collapse) he had returned in 1917 after the February revolution. The collapse of the tsarist state owed nothing to the Bolsheviks, but was the work of the German army, which had broken the will of the Russian people to fight. Once returned to Russia, though, Lenin strove to clear the ground for the successful establishment of a Bolshevik-led state on socialist lines.

Lenin conducted a brilliant political campaign to undermine the authority of the new government. The moment to remove it from power came in October. Almost without bloodshed (and largely thanks to the tactics and planning of Trotsky), the Bolsheviks occupied the Winter Palace (the seat of government) and other crucial points in the capital. Though the Congress of 'Soviets', the workers' and soldiers' councils which had sprung up during the

May Day 1919 in Moscow. Lenin in Red Square. Admirers still wear middle-class clothes and there is already a hint of the 'cult of personality' in the background.

summer and were largely dominated by Bolshevik sympathizers, was under their control, the Bolsheviks had to fight hard to survive in the next few months. Trotsky now made his other great contribution to the revolution by organizing and commanding the new 'Red Army' which, though the USSR was forced by the Germans to make a humiliating peace at Brest Litovsk, nonetheless crushed those who wanted to restore the old regime and fought off the Poles. The use of terror to wipe out or terrify internal opponents was, of course, traditional in Russia; her new rulers did not break with the methods of autocracy at the revolution.

More importantly, the new regime gave the poor of the towns and the peasants what they wanted: peace and land. Its first decree proposed that all warring governments should at once discuss peace terms, without claiming annexations. No other government responded, but this mattered little; it was a signal to Russians as much as to foreigners. The second decree passed by the Congress of Soviets the day after the seizure of the Winter Palace declared all land to be

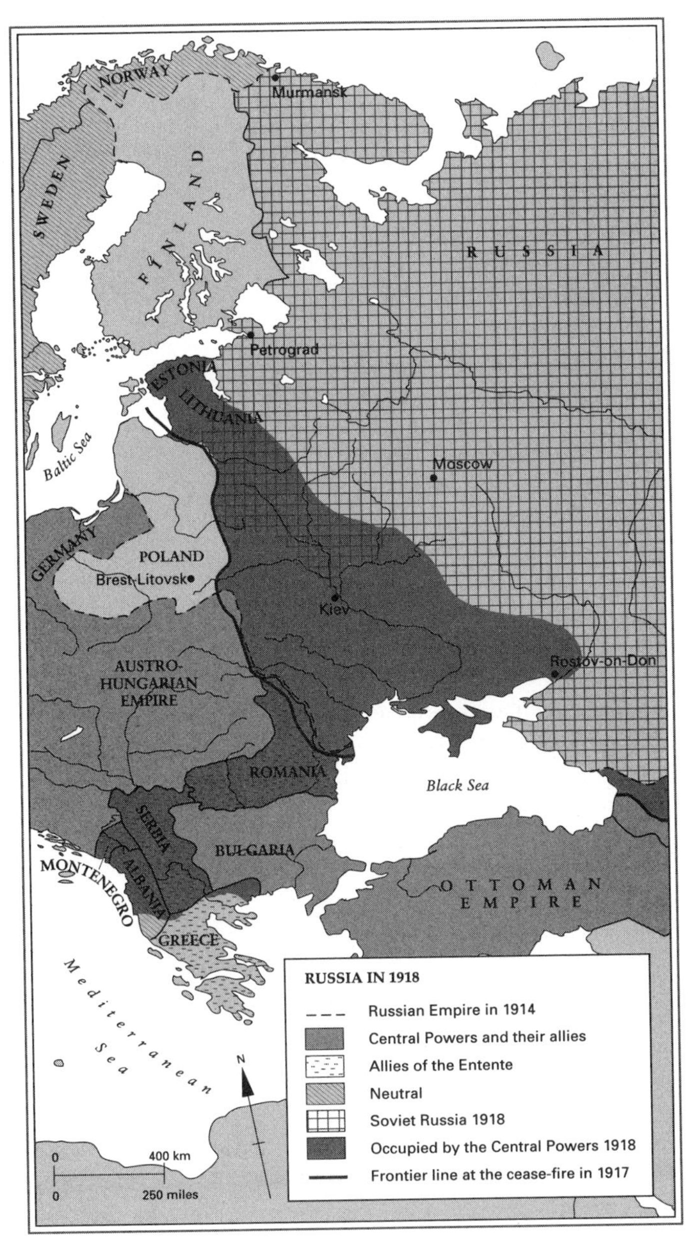

the property of the people; within a few years over 500 million acres were transferred to the poorer peasants, and the estates of the old landowning class, the Church and the royal family were wiped out. A huge majority of Russians were thus given a stake in keeping the new regime going.

SURVIVAL

Life in the young USSR was very harsh. The Germans had exacted savage peace terms. Fighting in the civil war was fierce; atrocities were committed and more of the country's feeble economic resources were destroyed. Some parts of the old tsarist empire tried to break away from the new regime; a few (Finland, the Baltic provinces) were successful, others (the Ukraine) were not. Confiscating food from the peasants to feed the towns led to more resistance to the regime and, therefore, more brutal repression. Even some of those who had first supported the Bolsheviks turned against them. At the great naval base of Kronstadt in 1921 a revolt of sailors demanding democratic elections, freedom of speech and the press, and the release of all political prisoners was mercilessly repressed. But it was a risky moment. By 1921, nearly half Russia's whole grain-growing area was out of production. An appalling famine swept through much of southern Russia when the year brought a drought; millions died, and survivors fell back on eating straw from roofs, leather harness, and even on cannibalism.

Lenin decided that concessions had to be made. Producers were given greater freedom to take their goods to market and get market prices for them. Communist diehards disliked it, but it worked. Slowly, the country began to pull round, though not until 1928 did industrial and agricultural production climb back to the level of 1913. Even then, the new Russia was relatively much less powerful than its predecessor had been in 1914 and the economic and technical base of the military power which numbers gave her was still very fragile. Yet a vast change had in fact begun. Russia was again back on the road to modernization she had entered under the tsars.

The revolution had given Russia rulers who, however barbaric in western eyes, were dogmatically sure that history was on their side and that the socialist cause of which they were the vanguard was destined to triumph world-wide. This doctrine, presented as the true interpretation of Karl Marx' teaching, provided an encouraging and powerful myth. Meanwhile, Russians who were not communists could also feel that what they were doing was in the best interests of the fatherland whose potential was indeed vast. The revolution had triumphed in a backward and poverty-stricken land (a fact that did not fit Marxist predictions) but it could become the base of one of the strongest powers on earth.

The ebbing of revolution

The Russian revolution and the Bolshevik seizure of power were events in world history. In 1919 there was set up in Moscow a Third Socialist International, soon known as the 'Comintern'. Its purpose was to organize internationally the 'communist' parties which had appeared in virtually every country where the old socialist parties were blamed first for failing to avert the coming of war in 1914 and then for not promoting revolution. The test of true socialism for Lenin was adherence to the Comintern; in every country, therefore, Marxist socialists soon divided into two camps. One, comprising those now usually called communist parties, looked to Moscow and Russian direction; these became to all intents and purposes the instruments of Soviet international policy. They denounced bitterly and fought with those (many equally sincere in claiming to be Marxist) who remained in the old socialist parties. So, the European 'left' was for decades condemned to division.

The new revolutionary threat looked very frightening to some non-Marxists. Yet it soon faded. A Bolshevik government appeared briefly in Hungary, and there were a few coups by communists in Germany, some briefly successful. Yet though the government of the new republic which had emerged there was dominated politically by socialists, it looked to conservative forces – notably the professional soldiers of the old army – in order to prevent revolution. Communist policy in fact made united resistance to conservatism more difficult, frightening moderates and alienating potential allies on the Left. In eastern and central Europe, the social threat was often clearly a national threat and the war did not end there until in March 1921 a peace treaty between Russia and the new Polish Republic provided frontiers which would last until 1939. Poland was the most anti-Russian by tradition, the most anti-Bolshevik by religion, as well as the largest and most ambitious of the new nations. But all of them felt threatened by a recovery of Russian power. This connexion helped to turn many of these states before 1939 to dictatorial or military governments.

DEMOCRATIC DIFFICULTIES

If the world of conservatives and revolutionaries had been transformed by the Great War, so had that of liberals and democrats. On the one hand, high hopes were awoken by the appearance of democratic constitutions in many places where they had never been seen before. On the other were alarming social and economic realities. By 1918 conditions in many big European cities were harrowing as a result of blockade. A large tract of France had been blighted and ravaged by fiercer fighting than any seen before; whole towns had been reduced to rubble and villages expunged from the map. In eastern Europe physical devastation was less intense, but there had been less to destroy in the first place. Everywhere sowing and harvesting had been interrupted again and again. Europe's grain-growers, though, could not have fed the starving cities even if they had the seed-corn and even if the labour to sow it and reap the first post-war harvest had been available, because often there were no railways.

All European countries, too, had spent savings and money which ought to have been ploughed back into investment; they had produced less during the war because labour had been taken from farm and factory to serve in the armies. Between 1913 and 1920 Europe's manufacturing output went down by almost a quarter. Germany had been Europe's greatest prewar industrial power, but after Versailles the need to make what were called 'reparations' payments to the Allies stood in the way of her recovery. Russia, now Bolshevik, neither could nor would play her pre-war part in the smooth working of the European economy as a customer for manufactures and capital, and as a supplier of grain. The economic unity which the Habsburg monarchy had given to much of the Danube valley had gone. New political frontiers cut old economic links. Some of the new states were economically so crippled that they dared not even allow their rolling-stock to cross frontiers in case it should not come back. Much of eastern Europe starved through the first winter after the war. Soldiers returned to find no jobs and children and old folk likely to die from disease or malnutrition. As if this were not enough, in 1919 one of the last great world epidemics reached its peak. A wave of influenza killed more people than the Great War itself (somewhere between five and ten million in Europe alone).

Many 'new' states (including Germany) had to make their first experiments with democracy in these appalling conditions. Two existing constitutional monarchies, Great Britain and Italy, widened their electorates to include all adult males and some British women were given the vote in 1918 (and all of them in 1929). The League tried to assist civilized politics by taking up the rights of minorities, guaranteed in some of the peace treaties (for example, that with Poland) and a number of questions left over from the peace negotiations at Versailles were settled by 'plebiscites' – that is, by direct votes of the people in the area concerned. All this added up to an expansion of formal democracy. Yet this was by no means the whole story. Democratic government in Russia had been swiftly swept aside by the Bolsheviks, who dissolved the only freely elected Constituent Assembly Russia had ever had shortly after seizing power. In eastern and central Europe old-fashioned conservatives who hated republicans, democrats and socialists alike, and regretted the passing of the old empires were helped by fears of Bolshevik-inspired revolution to put dictators and 'strong men' in power. Another threat to democracy came from those who lost by it. Many did not like plebiscites which left them living under alien rule. Some, in the defeated states (above all, Germany) complained that the Allies talked a lot about democracy, but would not let their former enemies run their own affairs without interference, and crippled their economies by demands for reparations.

Benito Mussolini, 'Duce' of Fascist Italy, at the height of his prestige and seeming success, a photograph taken in 1935.

Fascism

In Italy an anti-democratic movement took power in the 1920s which gave politics a new term – 'fascism'. It was supported and promoted by Italians who sought to win support by terrorizing their opponents, advertising their strength and ruthlessness, and recommending the adoption of tough, authoritarian measures to solve Italy's problems. Although Italy had been on the winning side, many Italians felt bitter that she had not got more out of the peace settlement. Their patriotic feelings were exploited by the Fascists, who blamed Italy's constitutional government and democratic allies for the betrayal of the country. In proportion to Italy's numbers and wealth her casualties had been heavy. Grave damage had been done to her economy, which had never been strong. After the war, inflation had ruined people in all ranks of society, and the plight of the poor grew worse. Prices soared and they could not buy food, while unemployment plagued the cities. Though some Italians turned to socialism and communism, fear of revolution drove many others into the arms of the Fascists.

By 1922, there were several Fascist members of parliament and in many Italian towns the Fascists had used violence to turn out communist local authorities, and had broken up the offices of trades unions and socialist newspapers. The existing government could not (or would not) maintain law and order, and in many places a majority of Italians seemed willing to let the Fascists do what they liked. Their undisputed leader was a former socialist journalist called Benito Mussolini. He had a bombastic, bullying style and was something of a virtuoso as an orator and public relations expert, though it is now hard to see why he was so successful. Mussolini succeeded in bluffing the king into dismissing the government of the day and letting him form a new one with members of other parties in it. Once installed at the levers of government, he then used them to bring about fundamental changes, step by step. Dictatorship was only gradually imposed, but in 1925 the old liberal constitution of 1861 itself was set aside and democratic parliamentarianism came to an end. Soon, opponents of the regime were being rounded up; a few were murdered. Mussolini's regime was not as brutal as the Bolshevik one he admired, but it was bad enough. And for all its claims to solve Italy's problems by dynamic and vigorous action, it did nothing of the kind.

A drift to authoritarianism

Not only Soviet Russia and Fascist Italy had turned their backs on democracy by 1930. Lithuania and Yugoslavia had both by then become dictatorships, and Czechoslovakia was the only one of the 'new' states of 1918 to retain its democratic constitution twenty years later. Bulgaria, Romania and Greece (of the pre-1914 constitutional states) had all by 1938 passed into the hands of generals or authoritarian monarchs, and, at the other end of Europe, Portugal was ruled by a very authoritarian regime, while the democratic republic of Spain was being throttled by one of its generals, Francisco Franco. There is no single explanation of democracy's sickliness. Economic hardship, fears of communism, and violent nationalism all helped undermine it, as well as the minorities and the frontier grievances left over from 1919. Only in a few western and Scandinavian countries were enough people familiar with the traditions needed to make democracy work. In one or two countries where there were old clerical and secular power rivalries, Catholics, too, often saw democracy and liberalism as enemies of the Roman Church. It is not really surprising, therefore, that over much of Europe democracy soon showed that its potential had been exaggerated in the euphoric days of Wilsonian rhetoric and aspiration.

Rosa Luxemburg, the Polish-born leader of the German socialist Left and, briefly, of the communist movement in 1918 – for which she was murdered by right-wing forces in 1919.

Weimar Germany

Yet ten years after the war some liberals still felt optimistic. A recovery of prosperity had helped, above all in Germany. The 'Weimar republic' (Weimar was the town where its constitution was drawn up) had started under big handicaps. To many German patriots, the republic was an affront from the start: it only existed at all because Germany had been defeated. It had signed the peace terms (and would always be blamed for them); it had been born in revolution. It then faced awful practical problems.

The socialist politicians of the new government set out to give the new Germany a democratic and liberal constitution but were at once deserted by those who should have been their allies but who wanted, instead, a revolutionary republic based on workers' and soldiers' councils (rather like the Russian Soviets). For a few months the issue hung in the balance; in the end the army put down the radicals. By then the German Communist Party (KPD), looking to Moscow for leadership, had emerged as a rival of the old Social Democratic Party (SPD). The Weimar republic had now to fight off old-fashioned monarchists on the right, and communists on the left. The Allies meanwhile made things worse by their harsh peace terms.

Though they brought the end of the blockade in July 1919, they were bitterly resented. 'Versailles' was soon blamed for an appalling inflation. Money lost its value at a staggering rate:

prices rose by a multiple of about 1,000,000,000 between 1918 and 1923. This helped to turn the saving classes against the republic, which they believed to be dominated by Marxists.

A turning-point came in 1924 when a big international loan then opened the way to stabilizing the currency. Spectacular economic recovery followed in the next few years. Statesmen and economists elsewhere had come to see that Germany, with her large population, huge reserves of skill, ingenuity, organizing capacity, natural resources, industrial plant and high level of culture, simply could not but play a major part in Europe's life. A question followed from this which still remained to be answered: was it also not true that with so many strengths to build on, together with a strategic position at the heart of Europe, an unrivalled military tradition and strong national feeling, Germany was bound to play a big, and perhaps a dominating political role as a European great power? This was the German problem which dominated European diplomacy between 1918 and 1939.

With prosperity, the republic looked safer. The dangers of revolution and violence faded away – or seemed to do so. Weimar Germany flourished, a free, democratic society, admired abroad because of the vigorous artistic, scientific and scholarly life which went on there. Its constitution guaranteed fundamental rights; its supreme court upheld them. Its elections gave support to coalition governments anxious to maintain the constitution. Yet many Germans remained unreconciled to it. The KPD steadily and bitterly attacked the SPD which upheld it. Nationalist groups and conservatives looked back with regret to the great days of Bismarck when Germany dominated Europe (or seemed to). They also appealed to a new, mass nationalism which wanted to submerge internal differences in a more-or-less tribal belief in the German 'Folk' (*Volk*). True, the treaty of Versailles was being whittled away as the 1920s went on – by, for instance, an easing of the reparations. A new treaty at Locarno between the main European states in 1925 to which Germany freely acceded seemed to mean an end to quarrels in the West. But still the lands lost by Germany in the East and the fate of people of German blood in the new states of central Europe excited nationalist anger.

Adolf Hitler

Such ideas were to be exploited by one of the few men who unquestionably shaped the course of modern history, largely for ill: Adolf Hitler. He was an Austrian. After an unhappy early life he found psychological release and contentment in the Great War, and was a good soldier, twice decorated. Defeat was a bitter experience. He was to devote the rest of his life to changing the verdict of 1918. In the early 1920s he became a nationalist agitator, denouncing Versailles. He took part in an attempt to overthrow the local Bavarian government in 1923 as a prelude to marching on Berlin This failed and he was locked up for a time. But he continued to orate and to write. His rambling political tract, *Mein Kampf* (My Struggle), was written in prison; it contains a mish-mash of Darwinian notions of natural selection by struggle, anti-semitism, admiration for a German medieval empire which had never really existed, and so on. Hitler soon had a small following, the National Socialist German Workers' Party, whose members were called 'Nazis' for short.

Prosperity in the later 1920s helped to hold Nazis and other extremists in check. They could do little except preach their vague but violent views and brawl with their opponents, denouncing the Versailles peace and claiming that all Germans should be united in one nation-state which would acquire new lands for the *Volk* in the East. They called for a crusade against Germany's enemies, in particular, against Marxists and Jews. Some of these ideas had roots deep in German culture and proved very appealing. But the Nazis also had a modern look: they talked social revolution and totally and completely rejected liberal democracy. For a long time many people did not take them seriously. As the end of the 1920s approached, the Nazis had yet to gather momentum and it was still possible to feel optimistic about German democracy.

THE ECONOMY 1919–39

During the war, Japanese and Indian manufacturers had received an enormous boost, farming countries beyond the oceans had a boom and so did suppliers of primary products needed by manufacturers: tin and rubber, lumber, iron ore, bauxite, and nitrates. Above all, the United States had prospered. Already the greatest manufacturing economy in 1914, she now became a great exporter of manufactures. But Britannia ruled the waves; there was no chance of the Central Powers getting important quantities of material through the British naval blockade and this meant that during the Great War the Allies were the main customers of American industry and American farmers. Allied money fuelled an American wartime economic boom.

It also changed the whole structure of world trade. Before 1914, Great Britain, Germany and France had been exporters of capital, the United States had been a capital importer. The war reversed the position. The Allies had to pay for what they bought. In theory they could do so by exporting their own goods, but in fact the Americans did not want many of them and, in any case, British industry had enough to do to meet orders from its own government. The bills had to be settled, therefore, in dollars or some other acceptable currency (which in the end meant gold, the international currency of the day). To raise dollars, the Allies first sold off their investments in the United States to Americans and then began to borrow money there. So, the United States ceased to be a 'debtor' nation, paying interest on capital it borrowed from abroad, and became a 'creditor' nation, with capital of its own for export. After the war, this gave the United States a new weight in the world economy.

The world economy between 1919 and 1939 was troubled by very large swings and fluctuations. Very roughly speaking, down to 1924 Europe was busy repairing the damage of the war. Then came five years or so of prosperity and optimism, when all seemed well. In 1929 there began a phase of collapse which spread right across the world, reaching its full intensity in the early 1930s. Recovery only came at the end of the decade. One result of the 1930s slump was that, by 1939, governments in every industrial country were tending to interfere more and more in the economy; the old pre-1914 world of *laissez-faire* had vanished. This change was so unplanned, piecemeal and gradual, and varied so much from country to country that it is easy to overlook it, important though it was. Most people, of course, noticed little more than the triumphs and disasters of their own lives – the switch-back changes in their personal fortunes as currencies gained and lost value, the overnight drop from security to helplessness because of the loss of a job.

The worst physical damage of the war was repaired by 1925; harvests were back to normal, Europe's total food and raw material production had passed pre-war levels and currencies had been stabilized after bad bouts of inflation. For most countries, economic well-being seemed at least to be again in sight, though Great Britain, Germany and Russia were still not producing as much as in 1913. During the next four years things continued to go well (1929 was the best year for European trade until 1954). World output of manufactures rose by over a quarter, world trade by nearly a fifth. Major currencies were steady again: you could exchange them for gold at fixed prices. Even the laggards caught up; Great Britain's manufacturing production was back to its 1913 level by 1929, and Germany's was by then well ahead of it. The main explanations were the change in the political climate in Europe as Germany came to be treated as an equal again, the repair of war damage and, above all, a long boom in the United States, the biggest national economy in the world.

She had paid off her foreign debts, had a big domestic market to supply and was able to help in the physical re-equipping of other countries. There was an immediate post-war slump – especially on the farms – as demand fell off, but soon the world's first mass consumer market for manufactured goods began to gather momentum. Wealth accumulated. Americans had money to spare and much of it was invested in Europe, above all in Germany.

This underpinned economic recovery in the mid-1920s. Between 1925 and 1929 Europeans borrowed about $2,900 million from Americans – and this was on top of the debts left over from the war. The happy effects of this on European prosperity then rolled outwards to the rest of the world as Europe's appetite for the products of Africa, South America, Australasia and Asia again grew larger.

American slump and world depression

In 1928, the American boom began to draw to a close. Americans who had lent money to European countries began to call in their loans. This placed the borrowers in difficulties. At the very least they had to tighten their belts and consume less in order to repay – and some could not repay at once. Meanwhile, businesses in the United States began to fail; confidence ebbed as more and more people wanted their money in hand. One disastrous symptom was the collapse of the New York stock market in October 1929, remembered as 'the Wall Street crash'. It shattered what remained of American confidence. By 1930 American money for investment abroad had dried up. At the same time Americans could afford fewer imports. A world slump was on the way.

Taking 1929 levels of industrial production as 100, by 1932 that of the United States had fallen to 52.7, that of Germany to 53.3, and that of the United Kingdom to 83.5. These figures are shorthand for appalling disaster. As manufacturing countries cut back production workers lost their jobs; the demand for imports fell; overseas buyers in their turn could not buy manufactured exports. As world trade fell off and so less business was done by shipping firms, insurance and banking, money ceased to be available to lend to people who want to start up new businesses or improve their existing ones, and so on, and on, and on. The national income of the richest nation in the world, the United States, fell by 38 per cent in these years: in other words, if the burden could have been equally shared, everyone in the United States would have had less than two-thirds as much coming in during the year 1932 as they had three years earlier.

Debtor nations tried to cut imports in order to save foreign currency and protect home markets. As a result, world prices fell still faster – and so blighted primary producers in other continents. On top of this came a European financial crisis when an Austrian bank failed in 1931, which led to the collapse of the gold standard. By then, factories were closing everywhere.

Although industrial countries were the most visibly afflicted (there were over 40,000,000 unemployed at the worst moment) disaster was unequally shared. Russia, poor as she was, was protected from the start by her political and economic system; she had never been dependent on world trade. In Europe, Sweden suffered least, Great Britain less than many other countries, France later, and Germany most. Probably the worst-affected industrial countries were the United States and Japan. The non-industrial world suffered even more. East European farmers were badly hit by the collapse of agricultural prices as producers strove to undercut one another on a falling market. The South American or African peasant, often tied to the market for one product, like wheat, sugar or cocoa, was in the worst plight of all.

World prices for agricultural produce remained low right through the 1930s, so that the later part of the decade was a good one if you had a job and lived in an industrial country; the cost of living was low, and in real terms had gone down since 1929. International trade in 1939 was still running at less than half the level of 1929. One reason for the slowness of recovery was that countries sought 'protection' behind high tariffs to keep out foreign competition. It was natural to adopt what looked like a short-term solution, but it handicapped manufacturing countries relying on exports. In some other ways governmental intervention – which increased enormously as people clamoured for something to be done about the slump – was more helpful. In the United States and Great Britain, for example, it encouraged certain kinds of investment, in particular in public works. Hard times also increased the demand for the provision of relief by governments; countries already well advanced down the road to what was to be called the 'welfare state' (such as the Scandinavian nations and Great Britain) now went further. The unemployed British worker had a higher real income on his 'dole' of unemployment relief in the 1930s than the average British wage-earner in employment at the beginning of the century. But that was not to say much. Resentment of an economic system which could induce such upheaval in people's lives everywhere stimulated new and violent political demands.

A TROUBLED ASIA

Beyond Europe, too, the world had changed. Though (except for Germany's) the old colonial empires survived, Europe's overseas power was seeping away, almost unnoticed. This was most true in Asia. 1911 had been a landmark there: the founding of a Chinese republic then brought two thousand years of empire to an end. Japan was already modernized to the point at which she was a major power. Then came the Great War. Both Japan and the new Chinese republic joined the Allies, though the Japanese prudently side-stepped requests to send an army to France and the Chinese sent only a labour force there. But their manufactures were important. Japanese industry and the new Chinese industrial districts of the big coastal cities both prospered. India, too, was of huge economic importance in the war; her leaders (including many of the nationalists to whom concessions had begun to be made before 1914) rallied loyally behind the imperial war effort. The Indian army recruited millions of men without having to turn to conscription.

In other ways, too, the war stirred things up. Some Asians picked up new ideas from their wartime travels; something like a hundred thousand Indo-Chinese served in the French army in France, and many of them must have seen a very different face of the French empire from that presented in Saigon or Hanoi. In India, China and Japan, nationalists hoped peace would bring further advances (though of different kinds) and, indeed, all three countries were independently represented at the peace conference. Long before it met, though, the Bolshevik revolution had directly changed the destiny of millions of Asians and of a huge part of the continent – Asiatic Russia is about four times as large as the India, and nearly twice as large as China. Indirectly, though, the revolution was to be even more important to Asia. Almost at once, the Russians began to build up political influence there by propaganda. Asian communist parties appeared, both in the colonial possessions of other powers and in independent countries. Suspicion of Russia by the imperial powers deepened, particularly in Great Britain, always uneasy about India and the Arab world.

Revolution in China

Communism had much to exploit in China. After a shaky start to the new republic, civil war led to division and disorder. Nothing was done to meet the pressing social and economic needs of China's still-growing peasant population; the numbers of the landless and indebted went up steadily, and so did poverty, starvation and misery. The Japanese took advantage of Chinese weakness to make territorial and other demands on the new republic. But in 1919 the first widespread and mass movement in support of Chinese independence from outside interference took place. It was called (after the day it began) the 'May 4th movement' and led to a boycott of Japanese goods, and an angry denunciation of China's treatment in the peace treaties (which had given former German possessions in Shantung to Japan).

China's leaders were divided. Some of them looked to Marxism and Moscow for inspiration and help. A Chinese Communist Party (CCP) was formed in 1921, But attacking 'capitalism' raised difficulties: many Chinese capitalists and landlords supported the reforming and modernizing nationalist party, called the Kuomintang (KMT). While the first president

PuiYi, last emperor of China. In 1912, a year after this picture was taken, he abdicated and twenty-five centuries of imperial rule came to an end. PuYi was once more restored to the Chinese throne for a few days in 1917 and then again reigned – nominally – as a puppet emperor of Manchukuo on behalf of the Japanese from 1934 to 1945. Five years in Russian hands were then followed by political 're-education' by the Chinese communists, from which the former emperor emerged to take a job as an archivist in 1959. In 1967 he died aged sixty one.

of the republic, Sun Yat-Sen, lived, CCP and KMT cooperated and were able to win further concessions from foreigners – notably the British. The Russians looked on approvingly. But after his death in 1925, the KMT decided to end the challenge from their rivals. After much bloodshed, the communists in the cities were virtually wiped out. They remained strongly entrenched in much of the countryside where they had the support of the peasants. By 1930, in Kiangsi, one of the southern provinces, they had organized an army and claimed to be ruling fifty million people. The KMT government determined to destroy this sanctuary. In 1934 its attacks forced the communist army to begin in October a 'Long March' to preserve itself. About 100,000 soldiers, many with their families, broke out of Kiangsi through mountainous country. Other scattered communist units joined them. In 1936 they reached northern Shensi, an area hard to blockade, where the peasants were cruelly oppressed and even more likely than in the south to support the communists. The KMT had not won the

The Japanese capture of Tientsin in the Sino-Japanese war of 1894 celebrated in a contemporary Japanese print.

Chian K'ai-shek in the 1920s. The strong man of a new China.

civil war, therefore; even though the communists had been dislodged from the south, their army survived, though terribly reduced by the end of the Long March, henceforth the legendary epic of the Chinese revolution.

The Japanese, too, had felt aggrieved by the peace. Even though they had done well out of it territorially, they did not get a declaration in favour of racial equality in the Covenant of the League of Nations, as they had hoped. They felt they had been treated as inferiors. Japan's successes against Russia in the war of 1904–5, and her strength (she had the third largest navy in the world in 1918) inspired her leaders. In 1929 they could look back on twenty years in which steel production had grown tenfold, textile production had been tripled and coal output doubled. Yet Japan more than ever needed foreign markets if she was to feed a population which had risen from forty-five to sixty million since 1900. These markets were mainly in Asia and collapsed in the world slump. By 1931 half Japan's factories were at a standstill and millions were destitute. Some extremists saw that the other great powers were in disarray themselves and would not be likely to resist decisive Japanese action to secure markets in China. But, if Japan was to be the dominant power in Asia it looked as if she would have to act soon – the KMT might, otherwise, rebuild China's independence first.

In 1931 the Chinese government seemed to be about to reassert old claims in Manchuria where there had been heavy Japanese investment since the Russians had been driven out in 1905. Japanese officers on the spot organized a clash with Chinese soldiers and used it as an excuse to take over the whole province. A new puppet state, Manchukuo, was set up under Japanese control. Further quarrelling followed, and in 1933 Japanese troops crossed the Great Wall, occupying for the first time part of historic China. They resumed their attack in

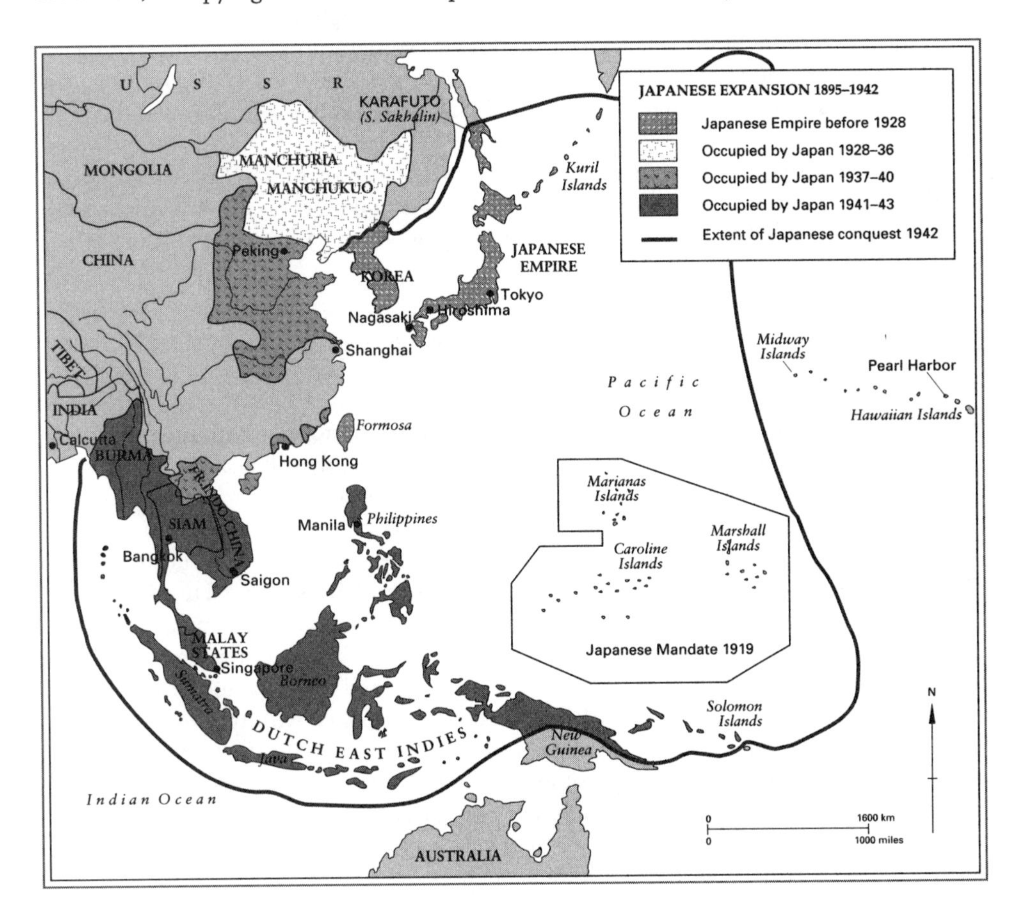

China 1932. Under the gaze of the languid sirens of westernized film posters, Japanese sailors and marines man a strongpoint. Identification markers are laid out on the ground for the benefit of the Japanese air force.

1937 and so began what they called 'the China incident' and eight years of struggle; this was, in a way, the opening of the Second World War. The Western powers had other things on their minds, and could not intervene. By 1941, China was all but cut off from the outside world, though the attack had brought KMT and CCP back together again in alliance against the Japanese.

In the race between China and Japan for modernization and power in East Asia, Japan had won. Yet Japan's efforts to win the war demanded bigger and bigger economic resources. In the long run, that meant enlarging the conflict, if Japan was to be sure of the oil and minerals she required.

REVOLUTION IN THE USSR

Circumstances were just as important as Marxism in shaping a new Russian empire. Its rulers could not wipe the slate clean and start afresh; they had to start with the ruins of the most backward of European countries, illiterate, overwhelmingly still peasant, and in many ways barbaric. They had to govern peoples of many different stocks and tongues who might try to break away. The former subjects of the tsar were used to the idea that government was a brutal business and took police bullying for granted. Their new rulers, still unsure of their grip on the country, did not break with that style. As the Bolsheviks believed that history was on their side and justified their use of power to crush opposition to the Party, which they saw as the vanguard of the proletariat, they only showed such respect for democratic government or individual rights as tactics demanded. Famine and civil war, too, made them all the more ruthless. They had soon replaced the old secret police with their own. By 1922 anarchist and other left-wing politicians were being locked up and the Communist Party itself had been purged of about a fifth of its members. Lenin's concessions brought a relaxation in political as well as economic life, but only briefly. There followed terror and economic centralization as never before – a real revolution, which transformed Russia as 1917 had not done.

Stalin

Lenin dominated the first years of the USSR. He was a powerful speaker and debater and even those who disagreed with his policies admired his personal devotion to the Party. From 1921 onwards, though, he was often ill, and rivalries and personal difficulties grew among his colleagues. When he died (in 1924) there followed a complicated struggle inside the Party from which there emerged a new leader whose power was to become far greater than Lenin's had ever been. Josef Stalin is the most important figure in Russian history since Peter the Great and for much the same reason: both changed that history. Both were ruthless and cruel, in the tradition of the great autocrats. Stalin, a Georgian in whom some people have seen as something of an oriental despot, out-manoeuvred and drove into exile the more brilliant but conceited Trotsky, the one major figure who might have displaced him. Yet Stalin took over a policy which Trotsky had advocated: Russia was to be industrialized as quickly as possible.

This revolution can be dated from 1928, when the first of two 'Five-Year Plans' for the economy was launched. Official Marxism had always taught that the economy settled the shape of politics and government; Stalin's revolution, carried out in the name of Marxism and with much Marxist theory as window-dressing, in fact showed precisely the opposite – that with government, police and army firmly in control you could transform the economy by force. At the cost of huge suffering and great crimes, Russia was, by 1941, strong enough to meet the test of war again.

In 1928 industrial and agricultural production again more or less reached pre-war levels. A 'New Economic Policy' adopted by Lenin had led to the growth of a number of private concerns and also to the prosperity of the peasant farmer, who at last got a good price for his grain. Yet within ten years, by 1937, private enterprise had been more or less completely blotted out, and a spectacular increase in industrial production was claimed. Pig-iron output quadrupled in ten years and electricity generation increased sevenfold. Capital investment was high and 80 per cent of Russian industrial production was coming from plant built in the previous ten years.

The cost to the people was heavy. The regime held down consumption; real wages fell so that the state could save more for investment. But the fall in living standards was not equally shared; the peasants lost most. To make them give up their grain (which they would otherwise have eaten or held back for higher prices) the land in the main grain-growing regions had been brought by Stalin into huge 'collective' farms. There was fierce resistance. The Party had always been weak in the countryside and the crushing of opposition was undertaken by the secret police and army. Millions of poorer peasants – as well as the better-off small-holders, or Kulaks – were killed in what was virtually a second civil war; their grain was carried off to feed the workers of the industrial cities. The worst came in 1933, when famine followed massacres and mass deportations. Even official figures admitted that in every year down to 1935 the grain harvest was lower than in 1928. Angry peasants slaughtered their animals, rather than give them up; seventy million head of cattle in 1928 were reduced to forty-nine million by 1935. In seven years, five million families in European Russia disappeared. Stalin said later that collectivisation had been as severe a trial as the Second World War. But Russia became a great industrial power, and that was the aim of the whole process.

Silence about the true facts and relentless propaganda helps to explain why there was no opposition among the urban masses to Stalin's rural brutalities. There were misgivings among the Party leaders, but Stalin steadily tightened his grip. A great series of purges and trials took place between 1934 and 1938. The world looked on puzzled as old Bolsheviks confessed in court to improbable crimes, and were then shot or lost to sight in the prisons and labour camps of the secret police. But trials of the well-known were only the tip of the iceberg. Hundreds of thousands of civil servants and Party officials disappeared; half the officers of the army were removed and nine out of ten of its generals were shot. By 1939, more than half of the delegates who had attended the Party Congress of 1934 had been arrested.

Thus Russia passed into the hands of Stalin's men. By 1939, 70 per cent of the Party's members had joined since 1929. A new generation had grown up who took Stalin's regime for granted and admired him. They learnt about the past only through the official history, and doubts were set aside by the USSR's enormous and often visible gains since 1917. Over one-sixth of the land area of the world it had enormously reduced illiteracy, set up a basic network of welfare services, tapped new sources of intelligence, talent and skill, emancipated women and had created a huge educational and scientific system to supply the technicians and teachers a new society needed. It had also built up huge armed forces to defend these gains: defence took just over 3 per cent of Russian budget expenditure in 1933 and 32.6 per cent in 1940.

Whether this could not have been achieved by other means, without such brutality and at less cost in suffering, remains an open question. Stalin put Russia back on the highroad towards modernization first embarked on by Peter the Great, but she might have got there earlier but for the Great War. Market economies had transformed other countries just as dramatically in the previous two centuries, and Russia, given her huge resources, was almost certain to become a world power sooner or later. Whether terror and a command economy were necessary we shall never know.

THE AMERICAN ALTERNATIVE

In 1918 the United States was the richest and strongest of the victors. Only fourteen years later it had a quarter of its workforce unemployed and its industrial output almost halved. Some thought America might be heading for revolution. A surge of post-war prosperity and isolation had left it unprepared for such a crisis. Republican administrations in the 1920s had presided over Americans troubled by little except 'prohibition', a problem created by a constitutional amendment banning the making and sale of alcoholic drink. It had dramatic effects, many of them unhappy. It encouraged organized crime to go in for illicit distillation, brewing and smuggling and so delivered a blow to civic life and public morality whose effects, some believe, have never been erased. It also mattered that the issue split the Democratic Party and so ensured a long Republican ascendancy.

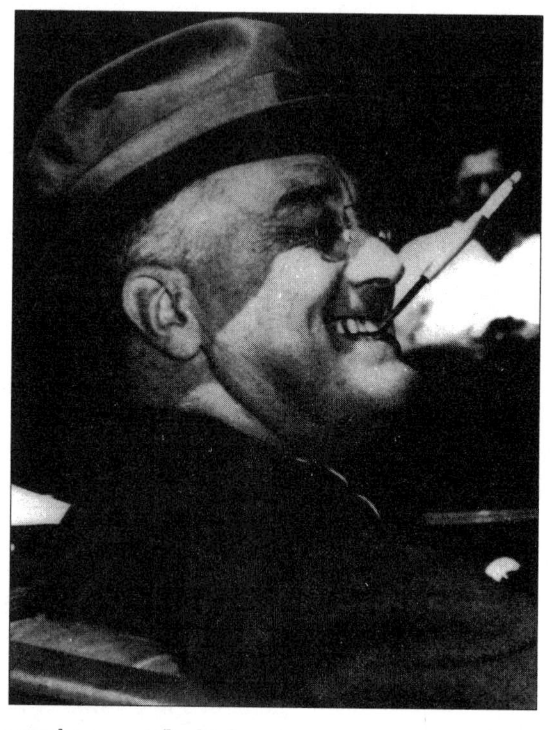

Franklin D. Roosevelt, a cripple of magnetic personal quality whose greatest gift when he became President in the depths of the economic slump may have been the contagion of his jaunty confidence that America's problems were manageable if only tackled with courage.

A third new Republican president in a row took office in 1928, just as the economic boom was showing signs of flagging. In October the following year the Wall Street crash cut at the root of the confidence which had sustained decade-long investment. Suddenly, wage-income shrank, service and retail trading crumbled. Mortgages were foreclosed, bankruptcies multiplied, banks could not collect debt and closed their doors leaving depositors penniless. The Great Depression began and in the 1932 presidential election, the Republicans were punished for it.

The New Deal

The new president was the first Democrat since Wilson: Franklin Roosevelt. He may well have been the saviour of democracy in the United States. A skilled politician, he won widespread support (he had carried 42 out of 48 states). He created and held together a new electoral coalition of those who had suffered most – farmers, east-coast Catholics of immigrant stock, industrial workers and their unions, blacks and the white liberal middle class. It was to give his party supremacy in

Washington until 1952. Yet his greatest triumphs were psychological. Millions of Americans believed he cared about them and that he had the will to tackle the country's problems. In his inaugural address he had told them that 'the only thing we have to fear is fear itself'; within a week a half-million letters poured in to thank him for a message of hope. Congress was persuaded to pass legislation to deal with the most urgent problems (and, incidentally, ended prohibition) and this formed a basis for a 'New Deal' which transformed American history. Although not so sweeping as its critics feared and its supporters hoped, it brought welfare benefits on a new scale, reformed the banking system, rescued agriculture, and poured federal money into poor states and regions. It could not get unemployment down below 10 per cent until 1941, but it showed that American democracy could react effectively in an emergency. It was a tonic to democracy not only in the United States, but abroad. It preserved the American constitutional system for what were by 1939 already appearing to be even greater challenges ahead.

REVOLUTION IN GERMANY

In 1933 the once-tiny Nazi movement took control of Germany and it was the most important political change in Europe since 1918. In 1929, as the economic weather began to roughen, the Weimar republic had run into choppy water. Soon the storm reached hurricane force; at its worst, in 1932, Germany had six million unemployed, and people feared inflation as bad as that which had wiped out savings ten years earlier. Hitler and the Nazis reaped the political benefits. They appealed to the wish of many Germans for strong measures and national unity. They exploited impatience with parliamentary politicians who had failed to prevent economic disaster, desire for scapegoats and resentment against the Versailles settlement, believed by many Germans to be the root of their troubles and morally unjust to boot. The semi-military 'Storm Troops' of the Nazi movement (the SA organization) grew rapidly in numbers as the crisis got worse. They had been formed to protect Nazi meetings from interruption but evolved into bands of thugs brawling on the streets with their Communist equivalents; soon, like the early Italian Fascists, they began terrorizing political opponents and Jews without interference by the police. In 1930 the Nazis won 107 seats in parliament – just under a fifth of the total. They had become a major political force. When in July 1932, new elections made the Nazis the biggest party in parliament the President of the Republic, Field-Marshal Hindenburg, decided that its leader must be given a chance to show whether he could deal with Germany's problems. He asked Hitler to be Chancellor (head of government). On 30 January 1933 Hitler took office, asked for new elections to be held (as he was entitled to do) and promised a coalition with conservative groups. Germany's greatest soldier of the First World War, Ludendorff, wrote to Hindenburg condemning his action and prophesying national disaster.

The Nazi revolution

What Hindenburg had done was perfectly legal and the Nazis' take-over continued along constitutional lines. Their party newspaper warned Germany that if they had their way the forthcoming elections would be Germany's last, but they set to work to win them. As the government, they controlled the radio and used it to push their own campaign. The police looked the other way while the Nazis' opponents were terrorized and beaten up. In a new style of political campaigning Hitler was flown round Germany in an aeroplane, so getting the greatest advantage from his magnetic personality. Yet, though seventeen million people voted for them – nearly 44 per cent of the electorate – the Nazis did not get a majority of seats or votes. Nevertheless, Hitler asked parliament, where he had secure alliances with other groups, for extraordinary powers to rule by decree and got them in March 1933. Armed with them the Nazi revolution could begin.

With one hand on a standard of the SA – the street-fighting units which carried Nazis to power – Hitler renews the ties of loyalty of the party rank-and-file at the Nuremburg rally of 1934.

The Nazis soon thrust their temporary allies out of government. The communist deputies had by then been imprisoned and the KPD and SPD were dissolved. The Nazi party was soon the only one permitted. Strikes were forbidden, trades unions were dissolved; there were thousands of arrests and hundreds of political murders. German life was shaken-up from top to bottom; the churches were harried, the professions dragooned, the universities purged – even the Boy Scouts were banned (as in Russia). The one conservative force which Hitler feared, the German army, which alone could stand up to the Storm Troops, soon caved in; it transferred its loyalty to Hitler, taking a special oath to him after Hindenburg died in 1934.

Propaganda depicted the new Germany as a transformed, dynamic, united people. Yet little was actually achieved in home affairs, and almost nothing that was done lasted long. A public works programme had already been launched before Hitler took power: the Nazis continued and intensified it so that unemployment figures dropped by 40 per cent within a year, but economic recovery had come to a peak by 1936 and after that real earnings did not rise. Rearmament, a high priority for Hitler, soaked up benefits that might have gone to the consumer. Yet the regime stayed in power, still governing under the emergency measures of 1933.

The reasons were complex. One was psychological. Hitler made Germans feel pride in themselves again. Soon, too, he provided an indisputable run of successes in foreign affairs. As for his followers, the movement gave them a sense of status, though its leaders were, for the most part, a second-rate lot. The Nazis also drew strength from an abominable race policy. At first directed against Jews, it was to be extended later to others. Hitler detested Jews, whom he accused of tainting German racial purity. From 1935 onwards, new laws began to deprive them of the legal and civic rights they had enjoyed since the beginning of the nineteenth century; Jews were brutally terrorized, stripped of their possessions, and their houses, shops and synagogues were desecrated and sacked. They were the most obvious victim of a regime in which terror, blatantly enforced by the security police (the *Gestapo*), played a bigger and bigger part. By 1939 Germany had no free press, no free speech, no free parliament, and an economy controlled by force, which kept workers in line and held down wages.

TOWARDS A SECOND WORLD WAR

Hitler's lasting achievements were all to prove to be destructive. He led Germany on a course which in the end not only shattered the national unity won by Bismarck, but also handed over nearly all eastern Europe to the Russians, whom he despised, under Bolsheviks whom he hated. In the process, millions of people were killed in another world war, in whose slaughter the Germans were not spared. Yet his foreign policy had at first brought enormous successes.

Hitler sought to undo the treaty of Versailles and to win Germany territory in the East, at the expense of the Slav peoples whom he regarded as inferiors. Circumstances helped him. The last Allied occupying forces had left Germany in 1930. Reparations had finally crumbled away in the economic collapse. More important still, other countries were for a long time unwilling to resist his demands. Many people, notably in England, had thought the peace too harsh and had bad consciences about it. More important still, the memory of the Great War was so dreadful that people were for a long time willing to make almost any concession rather than risk another such conflict. Some saw in a strong Nazi Germany a barrier against communism. As for the two great outsider nations, the Americans were after 1929 wrapped up in their own troubles while between Hitler and the Russians lay other, new states, whose help they would need before they could act against Germany. Mussolini, though at first wary about Hitler's ambitions, in the end became his ally in what was called the 'Axis'. This left France and Great Britain with the unpleasant recollection that it had taken four years of bloody fighting and blockade (three of them with Russia on their side), and finally the help of the United States to beat Germany in the Great War.

Hitler at once took up the demand (already made by his predecessors) that Germany should have the same right to armed forces as the victors of 1918. He withdrew from the League (which was trying to promote disarmament) and in 1935 reintroduced conscription – a clear breach of the Versailles treaty – and announced he had an air force. The Versailles territorial settlement was next to go. In March 1936 German soldiers moved into the Rhineland – German territory where Germany had been forbidden to keep soldiers or build fortifications. France and Great Britain did not retaliate. At the same time, Hitler said he would no longer stick to the guaranteed frontiers in the west, accepted by previous German governments under Weimar. The League did nothing to restrain these aggressive acts, but the British and French began to rearm faster.

The outbreak of civil war in Spain in 1936 posed another problem. The Germans and Italians supported one side, and the Russians the other. For many people this simplified the issues to a simple ideological conflict, but supporters for both sides could be found in the democracies. A badly divided public opinion hampered the French and British governments in dealing with the Axis powers. They had other problems, too. One was the USSR; Stalin might perhaps help against Hitler, but could only do so by crossing the territory of France's ally, Poland. In any case, some asked, what confidence could be placed in a Russia engaged at that moment in shooting half her general staff? Further afield there were other troubles, notably the ominous advance of the Japanese in the Far East.

1938 brought Hitler further success. In March came the unification of Austria and Germany. A plebiscite took the wind out of the sails of critics of this breach of Versailles, and there was little protest in France or Great Britain, though German territory now outflanked Czechoslovakia, a state with three million German inhabitants. Hitler determined to use the old cry of national rights of which so much had been heard at Versailles, nineteen years earlier; he demanded self-determination for the Sudeten Germans (as they were called). Weeks of negotiation gave him more; appalled by the danger of having to fight for Czechoslovakia and believing that the last defects of the Versailles settlement could be remedied by accepting Hitler's demands, the British and French agreed at a meeting in Munich to transfer large areas of Czechoslovakia to Germany. This crippled the one democratic state in central Europe and their one sound ally there. It offended the Russians, who had not been consulted, and convinced Germans that Hitler was a wonder-worker who could be followed with blind confidence. As for Hitler, he concluded that the democracies would always give way to a threat of war.

Munich was the high-water mark of the policy called 'appeasement' – that of meeting what were said to be reasonable German grievances. When Hitler seized what remained of the former Czechoslovakia in the following March (on the pretext that the republic had broken down), there was a great revulsion of feeling in Great Britain. The government introduced conscription in peacetime for the first time in British history and gave guarantees to several eastern European countries that they would be protected against aggression, among them the Poles. Many Germans wanted to get back from Poland its former German lands, above all the corridor which connected the republic with the sea and cut off Germany from East Prussia and the historic German city of Danzig (Gdansk), since 1919 a 'free city' under the League of Nations. The Poles welcomed the British guarantees but were adamant that they would not allow Soviet forces on their soil. This made military co-operation between the British and French and the USSR impossible. Stalin saw he could make a better deal with Hitler. Sharing Germany's dislike of Poland's frontiers, in August 1939 he made a treaty with the Nazi regime. War began on 1 September, when the Germany army invaded Poland.

THE SECOND WORLD WAR 1939–45

On 3 September the British and French governments reluctantly declared war on Germany. So began another struggle to settle Germany's place in Europe. Poland, isolated and soon invaded by Soviet forces from the east, collapsed and was divided between her conquerors in a new partition. When the USSR a few months later also swallowed Lithuania, Latvia and Estonia, a Germany bigger than that of Bismarck's and something like a revived tsarist empire were left facing one another in eastern Europe.

Hitler's victories

The British and French remained on the defensive in the west, haunted by the bloodshed of 1914–18 and hoping that blockade would bring them victory. The threat it presented to Germany's ore supplies from Scandinavia led to a brisk German campaign in April 1940 to overrun Norway and Denmark, but before this was complete the Germans attacked in the west, too. Holland and Belgium soon collapsed, while the French and British armies were hurled back, the British saving itself (at the cost of its equipment) only by a remarkable seaborne evacuation from Dunkirk. A little later, the French signed an armistice ceding occupation of about three-fifths of the country, including the entire northern coasts, to the Germans. By then, Mussolini had joined in on the victor's side, so that by the end of June Germany was left without a single opponent still in the field against her on the European mainland.

As the USSR was collaborating with Germany by supplying her with raw material and the handful of remaining neutral countries had to watch their step, Great Britain was left virtually alone. She had allies in a gallant handful of European governments in exile and the Dominions of the Commonwealth, but that was all. On the other hand, she had a secure base so long as she controlled the seas, and she had recently changed her political leadership. A new coalition government was led by Winston Churchill, something of a political outsider who now proved himself the greatest Englishman of his age. As had no earlier British war leader, he animated his people to effort and sacrifice; he was the man the hour demanded.

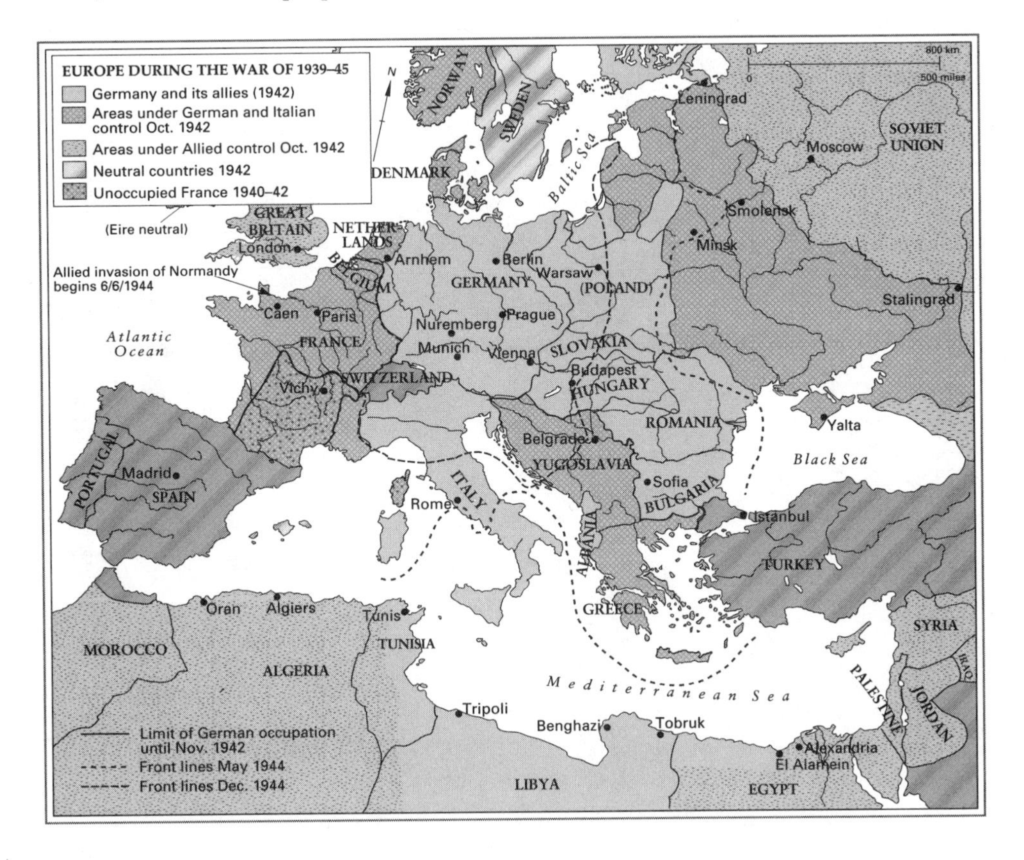

Almost at once victory followed in a great air battle over southern England in August and September 1940. After it, Hitler could at best hope only for stalemate in the west, since Great Britain could not be invaded without German control of the skies over the Channel.

Yet for a long time Hitler's opponents had little about which to be cheerful. The British had to put up with a winter of bombing by night which, though nothing like so appalling as what was in due time to be visited on the German cities, was hard to endure. In the spring of 1941 Hitler added Yugoslavia and Greece to his conquests and was badly damaging British shipping by submarine warfare. But he was returning to an old dream, for in December 1940 he had told his generals to prepare the invasion of the Soviet Union.

1941: the crucial year

It began on 22 June 1941, huge losses being quickly inflicted on the Russian armies and vast tracts of territory occupied. Some Soviet citizens, particularly in the Ukraine, eagerly welcomed the invaders. But though they came within sight of Moscow, the Germans had not broken Soviet resistance before the winter set in. The Germany army was poorly prepared for it and the first Soviet counter-attacks were successful. Hitler was now bogged down with a huge land power as an undefeated opponent. While these great events unfolded, sympathy for Great Britain and hostility to Nazi Germany had been growing in the United States. American opinion remained firmly against direct involvement in the war and only slowly could Roosevelt edge official policy forward. In 1941 he was given power to lend or lease defence materials to any country whose security appeared to involve that of the United States. This was of crucial importance: access to the 'arsenal of democracy' as Roosevelt called it, was essential to the maintenance first of the British and then of the Soviet war-effort. As the 'Battle of the Atlantic' between British convoys and German submarines, surface raiders and aircraft grew fiercer, Roosevelt was also able to win support for steps to protect American shipping likely to be under German attack.

Kerch, 1944. Bodies of those murdered by the retreating German army discovered by survivors.

7 December 1941: an American battleship stricken by the Japanese air attack on Pearl Harbor.

Meanwhile, in the Far East Japan had in 1940 used the embarrassments of Great Britain and France to occupy Indo-China and obtain a closure of the Burma road along which supplies had been sent to the Chinese. There was much sympathy for China in the USA, and soon American citizens were forbidden to supply goods of strategic importance – oil, above all – to Japan. The American Pacific fleet was moved from its California base to Pearl Harbor, Hawaii. Opinion in Tokyo was long divided about what to do. Finally, in the autumn of 1941, the Japanese government decided to go to war with the United States. Intelligence intercepts meant that by the end of November Roosevelt knew that war was coming.

On Sunday, 7 December, early in the morning, waves of Japanese aircraft swept in to attack Pearl Harbor. By 9.30 a.m. they had all but wiped out the American air units there and had sunk three battleships and several other vessels. But by firing the first shot the Japanese had ensured the rallying of Americans behind a declaration of war on 8 December. Foolishly, Germany declared war on the United States on 11 December (Hitler's second great strategical error) and so wars worldwide were finally brought together in one great struggle.

The Japanese quickly won a sensational series of victories which gave them Dutch Indonesia, the Philippines, Malaya and much of Burma. They threatened India with invasion. Across the Pacific their naval and air forces established a huge maritime shield of island bases from New Guinea and the Solomon Islands north to the Marshalls and Wake Island. They threatened at one end the Aleutians and at the other Australia. Yet the enjoyment of victory was brief. In May and June 1942, in the battles of the Coral Sea and Midway, Japanese naval air power was blunted in two great long-range actions fought by aircraft from carriers which hardly saw enemy ships. From that moment the American counter-offensive in the Pacific slowly but inexorably began to unroll.

Burma, 1944. Indian soldiers armed with American weapons fight other Asians on a battlefield dominated by the temples of Burmese Buddhism – which are themselves built from that most characteristic product of European industrialism, corrugated iron.

In the same decisive year the worst phase of the Battle of the Atlantic was survived and, after, allied shipping losses steadily fell and those of German submarines mounted. On the eastern front, the German army achieved its deepest penetration of the Soviet Union, reaching the Caucasus and almost the Caspian, but succumbed to a Russian winter offensive which deprived it of a quarter of a million troops encircled in Stalingrad by the end of the year. On the home front, bombing at last began to inflict real hardship on the German cities. Finally, in North Africa, a British offensive in Egypt and Anglo-American landings in North Africa resulted in the elimination of Axis power in 1943 and of the brief-lived Italian empire in that continent (the east African Italian possessions had been seized during 1942). Henceforth, there could be no doubt of the eventual outcome.

For a long time, the Red Army continued to bear the main brunt of fighting Germany. Anglo-American landings in Italy in 1943 increased the pressure but only in June 1944 did the western allies make a successful re-entry into northern France and establish another major front. By then, Mussolini had been overthrown and the Germans were in all but continuous retreat everywhere. At the end of 1944 the territory of the USSR was clear of German forces and the Red Army was deep into Poland, Romania and Bulgaria. In April 1945 it entered Berlin at last, a brilliant ending to a heroic achievement, while the western allies had

The Second World War in the Far East and Pacific

1940	
Aug	British withdraw from Shanghai and northern Chinese garrisons
Sept	Japanese forces begin to arrive in French Indo-China Japan joins in triple pact with Germany and Italy
Oct	US places embargo on export of iron and steel scrap to Japan
1941	
Apr	Japan and USSR join in neutrality ageement
Dec 7	Japanese attack on Pearl Harbour, Hong Kong, Malaya followed by US and British declaration of war on Japan
Dec 11	Germany and Italy declare war on US
1942	
Jan–June	Japanese occupation of Malaya, Burma, Indonesia, Solomon islands Philippines, Timor, part of Aleutian islands, part of New Guinea
Feb 15	Fall of Singapore
May 7 and June 4–7	Battles of Coral Sea and Midway
Aug 7	US landings in Solomon islands
1943	
July	Opening of US South Pacific offensive
1944	
Feb	US forces take Marshall islands
June	Systematic bombing of Japan begins
Oct	US landings in Philippines
1945	
Feb–Mar	US capture of Iwojima
Apr	Completion of British destruction of Japanese forces in Burma
Apr–June	US capture of Okinawa
Mar–Aug	Major phase of air offensive against Japan
Aug 6	Atomic bomb dropped on Hiroshima
Aug 8	USSR declares war on Japan and invades Manchuria
Aug 9	Atomic bomb dropped on Nagasaki
Aug 10	Japanese cabinet decides to surrender
Sept 12	Final signature of surrender

reached the Baltic coast and had overrun most of southern Germany and Austria. The 'Third Reich' Hitler had proclaimed came to an end on 8 May after his own suicide and the unconditional surrender of what was left of Germany's forces.

Japan was not far from defeat. Her air power had virtually ceased to exist, most of her fleet was sunk, her protective glacis of islands had been lost, and from its bases American bomber fleets were devastating her cities one by one. Then, in August, two weapons of a quite new kind were used against her: atomic bombs, the first to use not conventional explosive but the immense energy contained in the atomic nucleus. The weight of the huge advantages of the revolution in physics had been thrown into the scale. One bomb fell on

The reward of the Red *Army*: the Russian flag over the Reichstag building in Berlin, May 1945.

The European War 1939–45

1939	
Sept 1	German invasion of Poland
Sept 17	Soviet invasion of Poland
Sept 27	Polish resistance comes to an end
Nov 30	Soviet attack on Finland
1940	
Mar 12	Finns make peace with USSR
Apr	British and French mine Norwegian waters to hinder German shipping
Apr 9	Germans invade Norway and Denmark
May 3	Allied forces withdraw from Norway
May 10	German invasion of Netherlands, Belgium, Luxembourg
May 14	Dutch army lays down arms
May 26	Belgian forces ordered to capitulate
May 28–June 4	Evacuation of bulk of British forces and 140,000 French from Dunkirk
June 10	Italy declares war on France and Britain
June 22	French armistice with Germany (and June 24 with Italy)
July 9	End of French Third Republic and formal initiation of new regime at Vichy
Aug 8–Oct 10	Battle of Britain
Oct 8	German troops enter Romania
Oct 28	Italian attack on Greece from Albania
Nov	Hungary and Romania join German-Italian-Japanese pact
1941	
March	Bulgaria joins Axis
Apr 6	German invasion of Yugoslavia and Greece
Apr 17	Yugoslav capitulation
Apr 23	Greek armistice with Germans; British forces withdrawn
May 20	Successful German airborne attack on Crete begins
June 22	German invasion of USSR. By the end of October German forces have occupied Odessa and Kharkov, entered the Crimea, and are on the outskirts of Moscow
1942	
July 2	Opening of German summer offensive – capture of Sevastopol and entry of Northern Caucasus
Nov 8	Anglo-American landings in North Africa provoke occupation of Vichy by Germans and scuttling of French fleet at Toulon
Nov 19	At furthest extent of German success, Russian counter-offensive begins.
1943	
Jan	Russian raise siege of Leningrad
Feb 2	German surrender at Stalingrad
March	German spring offensive begins
July	Soviet summer offensive opens
July 10	Allied landings in Sicily
Sept 3	Allied invasion of Italy and armistice with new Italian government
Nov 6	Russian recapture of Kiev
Dec 31	Russian recapture of Zhitomir
1944	
Feb	Soviet forces enter former Estonia
Mar	Crimea retaken by Soviet forces
June 4	Anglo-American forces enter Rome
June 6	Anglo-American landings in Normany open the invasion of northern Europe
August	Soviet forces enter Poland, Romania and East Prussia
Aug 15	Allied landings in south of France
Aug 24	Surrender of Romanian government
Sept 2	Liberation of Brussels
Sept 12	American forces enter German territory near Eupen
Sept 25	USSR declares war on Bulgaria: surrender three days later
Oct 20	Russians enter Belgrade
Dec 16–25	German counter-offensive in France defeated
1945	
Jan–Apr	Battle of Germany
Jan 17	Soviet forces take Warsaw
Feb 7	Yalta Conference
Feb 13	Final Russian mastery of Budapest
Feb 20	Russians near Berlin
Mar 7	Allied forces cross the Rhine
Apr 20	Russians enter Berlin
Apr 25	US/USSR forces meet on the Elbe
Apr 28	German forces in Italy surrender
May 1	Death of Hitler announced
May 7	German surrender
May 8	VE day – the end of the war in Europe
June 5	Allied Control Commission takes control of German territory as of 31 December 1937

Hiroshima, the other on Nagasaki, where Europeans and Japanese had first established real contact four centuries before. The results were dreadful and the emperor determined to save his country further disaster by surrender. With that, the Second World War at last came to an end.

The balance sheet

It may well be literally true that no member of the human race was unaffected by the Second World War. It had exceeded any earlier conflict in horror and destructiveness. In it had been deployed unparalleled resources and power. Immense slaughter and physical destruction was only a tithe of what it had cost. Yet it had eliminated what had been certainly the worst threat ever posed to civilization and humanity.

It would take many years for the whole story of the moral cost of the war to appear, but one vivid sign of it – and of what had been overcome – became immediately and horrifyingly visible as the allied armied advanced into Germany and central Europe. They found themselves over-running camps where sadistic brutality and callous neglect had gone further than anyone had yet conceived. The prisoners in them had for years suffered torture, starvation and crushing labour. They had done so sometimes because they were political opponents of the Nazis, sometimes because they were hostages or enslaved labour, sometimes simply as prisoners of war. But this was not the worst. The majority of those who had suffered were Jews condemned to inhuman treatment and death simply because of their race.

On 17 July 1945, at Potsdam, the heart of the Prussian military tradition, the Allied war leaders met for the first time after the defeat of Germany. President Truman had taken up his office automatically on the death of Roosevelt; Winston Churchill was soon afterwards to lose his after defeat in a general election. Only Stalin straddled war and peace.

The Nazis had made special efforts to wipe out those they deemed genetically undesirable. In the case of the Jews they had spoken glibly of a 'Final Solution' to a Jewish 'problem'. Rightly, the word 'Holocaust' has been given to what they did. Full figures may never accurately be known, but five, perhaps six million Jews perished, whether in the gas chambers of extermination camps, or in factories and quarries where they died of exhaustion and starvation, or in the field where they were rounded up and shot by special extermination detachments. To overthrow the system that made this happen, was a great and noble achievement, a victory for civilization and decency. Ironically, none of the allied powers had gone to war consciously to bring about so moral an end. The only ideological warrior from beginning to end of the struggle had been Hitler and the goals he had sought were morally loathsome.

15

THE LATEST AGE: AN ERA OF UNREST

THE WORLD OF 1945

The United Nations Organization

One of the most important decisions taken during the Second World War was to set up a new international organization. The United Nations Organization (UNO) came into being at San Francisco in 1945. Its structure resembled that of the League. Its two essential organs were a small Council and a large General Assembly where, at first, fifty-one nations were represented by permanent representatives. The Security Council had as permanent members only the representatives of the United States, the USSR, Great Britain, France and China; its other members were drawn in rotation from the other members of UNO. It had greater power than the old League Council largely thanks to the Soviet fear of being outvoted in the General Assembly. As a result, permanent members were given a veto power sufficient to defend their essential interests. Not all the smaller nations liked this; but it had to be adopted if the organization was to work at all. Soon, though, the influence of the General Assembly as an area of debate became clear. For the first time, a world public linked as never before by radio and film – and before long, by television – could expect a case to be made at the General Assembly for what sovereign states did. It took much longer to provide effective action over many of the world's problems.

The General Assembly met for the first time in London in 1946. Quarrels broke out at once; when complaints were made about the continued presence of Soviet soldiers in Iranian Azerbaijan, occupied during the war, the Russians promptly replied by attacking Great Britain for keeping forces in Greece. Within a few days the first (Soviet) veto was cast; there were to be many more. The instrument which other powers envisaged as an extraordinary measure for the protection of special interests soon became a familiar piece of Soviet diplomatic technique. Already in 1946 the USSR appeared to be contending with a still inchoate western bloc in the UNO – and this did much to make appearance reality.

Cold War at the UN, 28 October 1948: the USSR and (nominally independent) Ukrainian delegates raise their hands to cast another veto in the Security Council. Andrei Vyshinsky, the Soviet representative, was foreign minister at the time, and had been in the 1930s, the brutal prosecutor in many of the treason trials then held in Moscow.

The superpowers

Both officials and members of the public in the United States in 1945 were much less suspicious of the world than they later became. The Soviet Union appeared much more distrustful and wary. There were really no great powers left in 1945 except these two. For all the legal fictions expressed in the composition of the Security Council, Great Britain was gravely overstrained, France, barely risen from the living death of occupation, was stricken by internal divisions, and China in modern times had never yet been a great power. Germany was in ruins and under occupation. Japan, too, was occupied and only slightly less damaged physically. The Americans and Russians enjoyed an immense superiority over all possible rivals; they were the only real victors. They alone had made positive gains from the war.

Though at huge cost, the USSR had won a stronger position than tsarist Russia had ever had. She had a vast European glacis, much of it new Soviet territory and the rest organized in friendly, weak states. Her forces garrisoned eastern Germany, a major industrial entity. Beyond the glacis lay Yugoslavia and Albania, the only communist states to emerge since the war without the help of Soviet occupation; in 1945 both were Moscow's allies. Above all, the Soviet strategical preponderance in central Europe faced none of the old barriers to Russian power. An exhausted Great Britain and slowly-reviving France could not be expected to stand up to the Red Army, and no conceivable counterweight on land existed if the Americans went home (and in 1945 they had already begun to do so).

Russian armies stood also on the borders of Turkey and Greece – where a communist rising was under way – and occupied northern Iran. In the Far East they held much former Chinese territory in Sinkiang, as well as Mongolia, northern Korea and the naval base of Port Arthur. From Japan they had taken the southern half of the island of Sakhalin and the Kuriles. In China a strong communist movement already held much of the country. By 1948, you could walk from Erfurt in Eastern Germany to Shanghai without setting foot outside communist territory.

The new world power of the United States rested much less on territorial occupation. She, too, had at the end of the war a garrison in Europe, but American electors wanted it brought home as soon as possible. American naval and air bases round much of the Eurasian land mass were another matter. The elimination of Japanese naval power, the acquisition of island airfields and the building of huge fleets had turned the Pacific Ocean into an American lake. Above all, the United States alone possessed the atomic bomb. But the deepest roots of American empire lay in economic strength; overwhelming American industrial power had been decisive in bringing Allied victory. The United States had been unharmed by enemy attack; the home base and its fixed capital were intact. Americans had actually seen their standard of living rise during the war, which had ended the depression. Finally, old commercial and political rivals were staggering under the burdens of recovery while the United States was a great creditor country, with capital to invest in a world where no one else could supply it. Their economies drifted into her ambit because of their own lack of resources, while hers were greater than ever. The result was a world-wide surge of indirect American power, visible even before the war ended.

Even before the fighting stopped in Europe, it had been clear that the Russians would not be allowed to join in occupying Italy or dismantling her colonial empire, and that the British and Americans would have to accept the Polish settlement Stalin wanted. Outside their own hemisphere, the Americans were not happy about explicit spheres of influence; the Russians, on the other hand, liked them. But neither power seems to have sought confrontation. The main concern of the American military after victory was to demobilize. Lend-Lease arrangements were cut off even before the Japanese surrender, thus weakening friends who could not provide from their own strength a new security system. As for the Soviet Union, over twenty million of its citizens had died and a quarter of its gross capital had been destroyed. Stalin may well have been less aware of Soviet strength than of Soviet weakness in 1945.

Europe in 1945

Yet within a few years, relations between the two world powers deteriorated very badly, mainly because of quarrels over Europe, a continent desperately in need of coordinated reconstruction. The cost of the war's destruction there has never been accurately measured, and it was moral as well as material. Civilized society had given way over much of it to the atrocities of deportation and mass slaughter, and struggles against German occupying forces had bred new divisions; as countries were liberated by the advancing Allied armies, the firing squads got to work in their wake and old scores were wiped out. In France more perished in the 'purification' of liberation than in the great Terror of 1793. But, above all, Europe's economic life had disintegrated. Leaving out Russians, nearly 15 million Europeans had died. Millions of dwellings had been destroyed in Germany and the USSR. Factories and communications were in ruins, currencies had collapsed. Though the flywheel of European economic life had been industrial Germany, the Allies' first thoughts were to prevent her recovery. The Russians carried off capital equipment from the east as 'reparations' to repair their own ravaged lands.

The Soviet economy was itself burdened by Stalin's decision to develop nuclear weapons, and to keep up huge armed forces. For the Soviet citizen, the years immediately after the war were as grim as had been those of the industrialization race of the 1930s. Yet in September 1949 an atomic explosion was achieved and in the following March it was officially announced that the USSR had an atomic weapon. By then, the international scene had already changed beyond recognition.

COLD WAR

Like all helpful slogans or labels, the phrase 'Cold War' runs a risk of over-simplifying things. There was always much more to world history between 1945 and 1990 than just the hostility of two great powers and their allies. Nonetheless, it is true that a global contest between the United States and the USSR which never boiled over into outright war, but which was stubbornly fought in ideological, political, economic terms, dominated international affairs for over thirty years and tended to infest and complicate every other issue facing mankind.

In 1917 there had appeared a state formally committed to changing the world by revolution. When Russia, after temporary eclipse, re-emerged as a major power in the 1920s she did so under new management which soon made it clear that there was a new sort of performer and a new Russian style on the international stage; principles of international life hitherto taken for granted would be accepted by the Soviet leaders only as and when they thought fit. Even in 1945, it is true, some people found this hard to credit. Important though it was, though, five years later it had become a somewhat abstract point. By then, 'Cold War' had begun, and it meant primarily a bitter and growing antagonism between the United States and the USSR.

After the armies of the western powers met those of their Soviet allies in 1945 in central Europe, they withdrew, as agreed, from those parts of Germany already assigned to Soviet occupation. Austria they shared with the USSR, while the rest of eastern Europe north of Greece was left under Soviet occupation or sway. Only Yugoslavia and Albania then had communist governments. Before the end of 1945, another had been set up in Bulgaria and in 1947 non-communists left the hitherto nominally coalition governments of Hungary, Romania and Poland. When, in February the following year, Czechoslovakia followed suit after a coup under Soviet patronage, Europe was effectively divided into two camps. In western Europe, communists had taken up a blatantly pro-Soviet and revolutionary stance.

Roosevelt had been sure that, broadly speaking, the United States could get on with the Soviet Union. His successor, President Truman (Roosevelt died in April 1945) and his advisers came to different ideas, though not at once, and largely as a result of their experience in

Like fascism and Nazism before it in Italy and Germany, communism came to power in Czechoslovakia by management from above; the armed police and militia, the keys to power, march past the new government after the coup.

Germany, whose government the four occupying powers (France was the fourth) had envisaged as a unit. They from the start shared in the administration and occupation of Berlin. Yet Soviet efforts to prevent German recovery led to increasing practical separation of the Soviet zone of occupation from those of the three other powers. Stalin obviously feared any reunification of Germany unless under a government he could control; perhaps there were too many memories of attacks from the west for Russians, whatever the ideological character of their government, to trust a united Germany. This led in the end to a solution by partition to the German problem which no one had envisaged. It started when the western zones of occupation were economically integrated and the entrenchment of communism by favouritism and intimidation began in the Soviet zone. In 1946 the outline of an all-communist eastern Europe was already appearing.

Yet when Winston Churchill drew attention in 1946 to the increasing division of Europe by an 'Iron Curtain' many Britons and Americans condemned him. Opinion began to change only as the Soviet veto was more and more employed to frustrate her former allies and the communists of western Europe were clearly seen to be manipulated in Russian interests. Perhaps Stalin was expecting economic collapse in the capitalist world.

The Truman doctrine and Marshall Plan

British and American policy began to converge as it became clear that the British intervention in Greece had made possible free elections there; Soviet intervention had done the reverse in Poland. President Truman had no prejudices in favour of Russia to shed, and in February 1947 he took a momentous step. It was prompted by a message from the British government which, more than any other, revealed that Great Britain was no longer a world power: the British balance of payments was forcing the withdrawal of British forces from Greece. The British economy had been gravely damaged by the war; there was urgent need for investment at home and the first stages of decolonialization were expensive and already under way. The financial burden was too great. Truman at once decided that the United States must fill the gap and much more than propping up two countries against Soviet bullying was implied. Although only Turkey and Greece were to receive help (and then only in the form of finance), he deliberately offered the 'free peoples' of the world American leadership to resist, with American support, 'attempted subjugation by armed minorities or by outside pressures'. This reversed the return to isolation many Americans craved in 1945. The decision to 'contain' Soviet power, as it was called, was possibly the most important in American diplomacy since the Louisiana Purchase.

A few months later, the 'Truman Doctrine' was completed by the Marshall Plan (named after the American Secretary of State). It offered economic aid to European nations which would come together to plan jointly their economic recovery. Its aim was a non-military, unaggressive form of containment by removing the dangers of economic collapse. The British Foreign Secretary, Ernest Bevin, was the first European statesman to grasp its implications. With the French, he pressed for acceptance of the offer by western Europe. The Russians would not participate, nor allow their satellites to do so, though the Czechoslovakian coalition government declined with obvious regret. The USSR violently attacked the Plan and set up a new instrument of ideological warfare, the 'Cominform', in September 1947. It at once began to denounce what it called 'American imperialism'. When western Europe set up an Organization for European Economic Cooperation (OEEC) to handle the Marshall Plan, the Soviet half of Europe was contrarily organized in a Council for Mutual Economic Assistance (Comecon) which provided window-dressing for the Soviet integration of the command economies of the east.

BERLIN AND KOREA

The Cold War (as it came to be called) had now clearly begun and it was to run on into the 1980s, as the 'superpowers' (as they came to be called) strove to assure their security by all means short of war. The first crisis came over Berlin.

The western powers, remembering the years after 1918, sought German economic recovery as a step to that of western Europe as a whole. In 1948, without Russian agreement, they introduced in their zones a much-needed currency reform. With Marshall Aid, available (thanks to Soviet decisions) only to the western-occupied zones, this further cut Germany in two. Eastern Germany was thenceforth decisively on the other side of the Iron Curtain while a distinct western Germany could now emerge. Currency reform divided Berlin too, and the Soviet response was to disrupt communication between the city, isolated well within the Soviet zone, and western Europe. The dispute escalated. Without interfering with the access of the western allies to their own forces in their sectors of Berlin, the Soviet authorities stopped the traffic which ensured supply to the Berliners who lived in them. The aim was to show them that the western powers could not protect them. So, a trial of strength was soon under way. The western powers, at enormous cost, organized an airlift to keep up a flow of food, fuel and medicine to West Berlin. Its only airfield handled over a thousand aircraft a day for most of the time, with an average daily delivery of 5000 tons of coal alone. The implication was that this could be stopped only by force. For the first time since the war American strategic bombers moved back to bases in England.

The blockade lasted over a year. It never came to shooting, but it was decisive. It established one point at which the United States was prepared to fight. Supply was not interrupted nor were the West Berliners intimidated, though the city was now split. Meanwhile, the western powers signed a treaty setting up the North Atlantic Treaty Organization (NATO) in April 1949, a few weeks before the blockade was ended by agreement. It was the first Cold War creation to transcend Europe. The United States and Canada and most western European states joined (only Sweden, Switzerland and Spain did not). It provided for mutual assistance to any member attacked and was yet another step away from the almost vanished Washingtonian tradition in American foreign policy. In May a new German state, the Federal Republic, emerged from the three western zones of occupation. In October, a German Democratic Republic (the GDR) was set up in the Soviet zone. Henceforth, there were to be two Germanies, with a frontier of barbed wire and landmines between them.

Cold War then re-erupted in east Asia. In 1945 Korea had been divided; its industrial north was occupied by Russians and the agricultural south by the Americans. After both had withdrawn and vain efforts to obtain nationwide elections, the United Nations recognized a government set up in the south as the only lawful government of the Republic of Korea. A separate government in the north also claimed sovereignty over the whole country. North Korean forces invaded the south in June 1950 and within two days President Truman sent American forces to fight them, acting in the name of the United Nations. The Security Council voted to resist aggression, and as the Russians were at that moment boycotting it, they could not use their veto.

After a few months it seemed likely that the North Koreans would be overthrown. When fighting drew near the Manchurian border, though, Chinese forces intervened and drove back the UN (but mainly American) army. This raised the possibility of direct action, possibly with nuclear weapons, by the United States against China. Prudently, Truman refused to countenance involvement in a greater war on the Asian mainland. When further fighting showed that the Chinese, though able to keep the North Koreans in the field, could not overturn South Korea against American wishes, armistice talks were started. A new Republican and unequivocally anti-communist American administration which came into office in 1953 knew its predecessor had sufficiently demonstrated American will and capacity to uphold an independent South Korea and an armistice was signed in July 1953. In the Far East as well as in Europe the Americans had thus won the first battles of the Cold War.

American bombers attacking a chemical works in North Korea. Until the coming of the intercontinental ballistic missile, American air-power was the first-line deterrent to what many people in the West regarded as an aggressive and expanding Communism; no comparable offsetting force existed in the East.

Shortly before the Korean armistice, Stalin died. There appeared to be no break in the continuity of Soviet policy, and his successors soon revealed that they too had the improved nuclear weapon known as the hydrogen bomb. It was in a way Stalin's final memorial, guaranteeing (if it had been in doubt) the USSR's status in the post-war world. He had carried to their logical conclusion the repressive policies of Lenin, using them to rebuild most of the tsarist empire after giving his countrymen the strength to survive (just, with the help of powerful allies) their gravest hour of trial. Yet it is hard to believe that Russia would not have become a great power again without communism, and her people had been rewarded for their sufferings with precious little except survival and a sense of international standing. A political culture based on isolation continued to be a handicap to modernization and humanization.

By 1953 western Europe was substantially rebuilt thanks to American economic support. NATO protected its member states. The German Federal Republic and the GDR moved further and further apart and on two successive days in March 1954 the Russians announced the eastern republic's full sovereignty and the West German president signed a constitutional amendment permitting the rearmament of his country. In 1955 the Federal Republic entered NATO; the Russian riposte was the Warsaw Pact, an alliance of its satellites. The GDR agreed to settle with old enemies: the line of the Oder-Neisse was to be the frontier with Poland. The greater Germany dreamt of by the nineteenth-century nationalists and Hitler had thus ended in the obliteration of Bismarckian Germany. The new West Germany was federal in structure, non-militarist in sentiment and dominated by Catholic and Social Democratic politicians whom Bismarck would have seen as enemies of the state, while historic Prussia was now ruled by revolutionary communists. There had been no peace treaty, but the problem of containing German power was tacitly settled for thirty-five years. Also in 1955, Austria re-emerged as an independent state and the last American and British troops were withdrawn from Trieste.

A world-wide division by then existed between what we may call capitalist and command (or would-be command) economies. After 1945 all earlier divisions of the world market were transcended; two methods of distributing resources were increasingly to divide first the developed world and then other regions (the most important being East Asia). The essential determinant of one system, the capitalist, was the market – though a market very different from that envisaged by nineteenth-century Free Trade ideologies and one in many ways very imperfect – and that of the other, the communist-controlled group of nations (and some others), was political authority. Or, at least, so it was intended. This distinction between two systems remained a fundamental of world economic life from 1945 to the 1980s. Neither system remained wholly unchanged and some trade went on between them. One came to be much less completely dominated by the United States and the other less com-

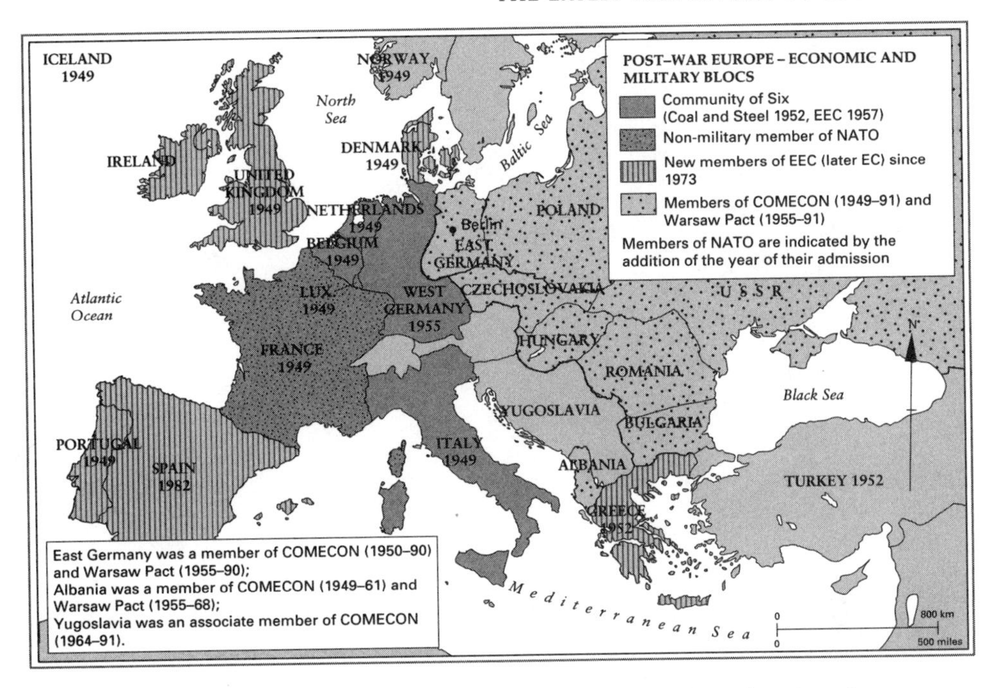

pletely dominated by the Soviet Union than in 1950. Nonetheless, they were long seen as alternative and exclusive models for economic growth. Their competition was inflamed by the Cold War and helped to spread its antagonisms.

THE END OF THE COLONIAL EMPIRES

The most revolutionary change in world politics after 1945 was the end of European empires. At the end of the war the British, French, Dutch, Portuguese and Belgian empires were still there (the Italian disappeared between 1941 and 1943). Thirty years later, Europeans ruled less of the world than they had done even four centuries earlier. The confusion and tensions of the process of dismantling empire were bound to make heavy demands on those who had to manage it, and presented huge potential dangers. It is one of the most remarkable achievements of our century that the era of decolonization should have been navigated without world war or huge regional conflicts.

The new Middle East

The last extensions of old empires had come between the wars, when 'mandates' intended to evolve towards statehood over much of the near and middle East were given to the French and British by the League of Nations. The future of the region seemed to lie in structures based on the European idea of nationalism; it might answer one question of the Ottoman Succession – how was the Arab world to be organized? In fact, nationalism was to prove a mirage. During the Great War the British and French had agreed that Turkey should be cut down to more or less its present size and that the French were to install themselves in what had been Ottoman Syria and the British in Iraq. But this made it more difficult to decide what should be given to the Arab rulers. Another complication had been the British government's announcement, in 1917, that it favoured establishing a 'national home' for Jews in Palestine. This pleased Zionists. It was not clear, though, whether the British meant by this a Jewish national state, and it appeared to cut across what had been promised to the Arabs.

Between 1918 and 1939 nationalist feeling on European lines was most advanced in Egypt, where the intellectuals and mobs of the cities clamoured against western imperial

power. Slowly, the British relaxed their control. In 1936 they agreed to garrison the Suez Canal Zone only for a definite number of years. Elsewhere in the Arab world, trouble came from a few (often rival) ruling families of a more traditional kind, though the French had to deal with the urban nationalists of Syria and the Lebanon. The British did their best to get rid of their Mandates quickly; the one they had been given for Palestine gave them most difficulty.

The establishment of Israel

In 1914 about 90,000 Jews lived in Palestine. From the Mandate's beginning, the Arab population showed alarm over Jewish immigration. The British were attacked by Arabs if they allowed in more Jews, and by Jews if they talked of restricting their numbers. Soon, the new Arab governments nearby began to interest themselves in the question. There were riots, murders, and acts of terrorism against Jewish settlements. In 1936 the British proposed to partition Palestine, but the Arabs rejected this. By then Hitler had come to power in Germany. Fearing persecution, more German and Central European Jews now wanted to come to Palestine – or, as some already called it, Israel. As the war of 1939 approached, the British had to suppress a full-scale Arab rising, but in dealing with it they put an absolute limit on the number of Jews to be allowed into Palestine in future. British relations with Arab states were by then more complicated than ever. The approach of war in Europe had made the Suez Canal more and more important to the British who now also had a new interest: the flow of oil

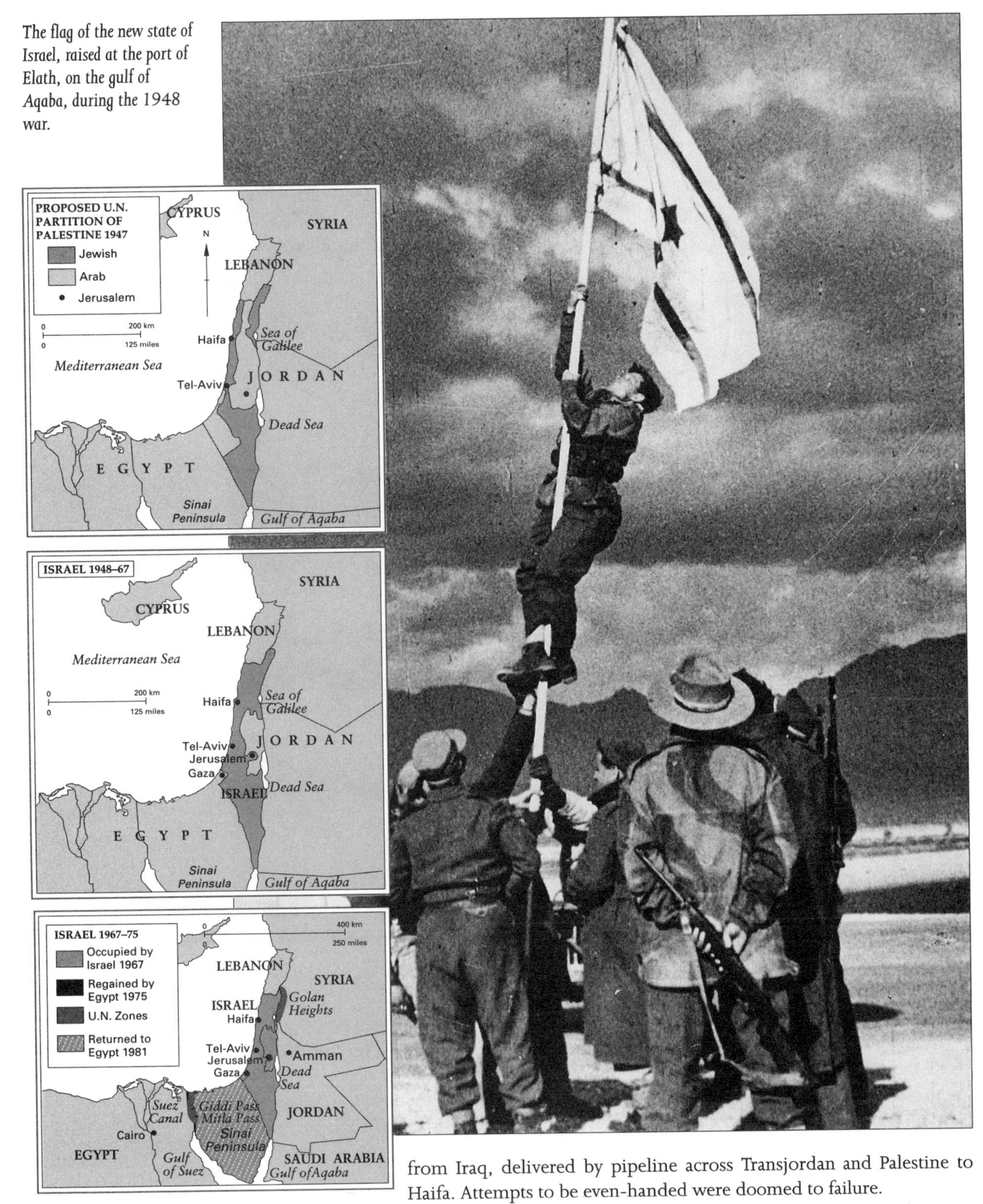

The flag of the new state of Israel, raised at the port of Elath, on the gulf of Aqaba, during the 1948 war.

from Iraq, delivered by pipeline across Transjordan and Palestine to Haifa. Attempts to be even-handed were doomed to failure.

War brought a crisis. The British government was not the only one restricting the entry of Jews who fled from the German extermination programme but it was the government ruling Palestine, the place where many of them wanted to go, and to which it was easy for

In 1954 an attempt was made on the life of Gamal Nasser, already the hero of Egyptian nationalists and a rising star in the political firmament of the whole Arab world. The photograph was taken as he was on his way back to Cairo after the incident.

other countries to say they should go. Zionist terrorism was now added to Arab violence. American pressure mounted, because Jewish votes were important to American politicians. Victory made things worse. The Russians took up the Zionist cause, seeing in it a way of making trouble for a Cold War opponent, and of extending their influence in the area.

The British took the matter to the United Nations, which agreed on a plan of partition (both the United States and Russia voted for it); the Arabs would not accept. Violence mounted. Finally, the British announced that they would give up Palestine. They left on 15 May 1948, the day after a new nation state, Israel, had been proclaimed by the Jews. Almost at once, it was attacked by Egypt and its Arab neighbours, who claimed to protect the Palestinian Arabs. Israel survived and won. Yet victory left her surrounded by defeated but revengeful enemies, who now had the additional grievance – of which they made much useful propaganda – of an exodus of 700,000 Palestinian Arabs who fled from Israel rather than live under Jewish rule. The Jews appeared to be persecutors in their turn, though many Palestinian Arabs returned to Israel in 1949.

In the next few years, three big upheavals changed the Arab position in almost every respect except its continuing hostility to Israel. One was the worsening of the Cold War. Russia had long been dabbling in the politics of the area; now the United States joined in. The second was a dramatic rise in the consumption of oil by the major industrial nations and in its production (largely from vast new oil-fields discovered in the Persian Gulf, Arabia and Libya). The third great change was a more complicated and drawn-out political upheaval which removed colonial domination from the Islamic world and replaced many traditionalist Islamic monarchs with more radical and revolutionary regimes.

The 1950s were a revolutionary decade. The French had to give up Lebanon and Syria after a struggle. They recognized the full independence of Morocco and Tunisia. The Egyptian king was overthrown in 1952. In 1954 a full-scale rebellion began against the European settlers of French Algeria – of whom there were over a million. The British had already withdrawn their garrison from Suez when a 'strong man' who had emerged in Egypt, Gamel Abdel Nasser, seemed to be the anti-imperialist, reforming leader for whom Arabs waited. The British, French and Israelis conspired to overthrow him in the 1956 Suez adventure, the last fling of old-style imperialism. It failed (though the Egyptians suffered a disastrous defeat), and his prestige rose even higher.

Algeria finally became independent in 1962, after dreadful suffering; Libya (independent already and rid of its king) began to jockey with Syria, Egypt and Saudi Arabia for leadership of the Arab world. By way of redressing the balance, the Egyptian and Jordanian governments got ready to attack Israel. But in 1967 the Israelis struck first, inflicting a third major defeat on Arab armies. Victory gave them more easily defended frontiers, which they announced they would keep and so made a fourth war almost inevitable. It came in 1973. This time Israel's enemies struck at a moment disadvantageous to Israel; the industrial world was having to take more account of the most powerful of the Arab states, the small but immensely rich oil-producers, whose decision to raise the prices of their oil created almost overnight an international economic crisis. As the world began to fear an economic slump like that of the 1930s, it looked as if Israel might not be able to rely for ever on the guilt felt in Europe and the United States about Hitler's treatment of European Jews.

Decolonization in Asia

Indian sub-continent

1942 British offer of autonomy after the war refused by Indian independence movement

1946 **Mar 14** British offer of full independence to India

1947 **Feb 20** British set terminal date for withdrawal
June Endorsement of partition by Hindu and Moslem leaders
Aug 15 Inauguration of Dominions of India and Pakistan

1948 Ceylon (Sri Lanka)) becomes a Dominion

1950 Inauguration of Republic of India within Commonwealth

1956 Inauguration of Islamic Republic of Pakistan within Commonwealth

South-East Asia

1945 Independence of Vietnam, Cambodia, Laos all proclaimed, but French slowly regain control of each country

1948 Burma becomes independent republic

1949 Independence of Vietnam within Frech Union conceded by French

1954 France concedes complete independence of South Vietnam, Laos, Cambodia. Vietnam divided by internal agreement

1957 Federation of Malayan states, independent within Commonwealth

1959 Self-government for Singapore

Indonesia

1945 Independence of new Republic of Indonesia proclaimed, followed by re-establishment of Dutch authority

1949 Dutch transfer sovereignty to new United States of Indonesia

1954 French dissolution of all links with Netherlands

East Asia

1943 UK and US relinquish extra-territorial rights in China

1945 Division of Korea between USSR and USA occupation

1946 Inauguration of Republic of Philippines

1948 Republic of Korea (South) and Korean Peoples' Democratic Republic (North) set up; UN recognizes only the former

Imperial withdrawal in Asia

In Asia a decisive blow had been struck against white hegemony by the Japanese victories of 1941–2, above all when 70,000 British, Australian and Indian troops surrendered to tyhem at Singapore. Not only colonial territories, but the impalpable asset of white prestige had gone.

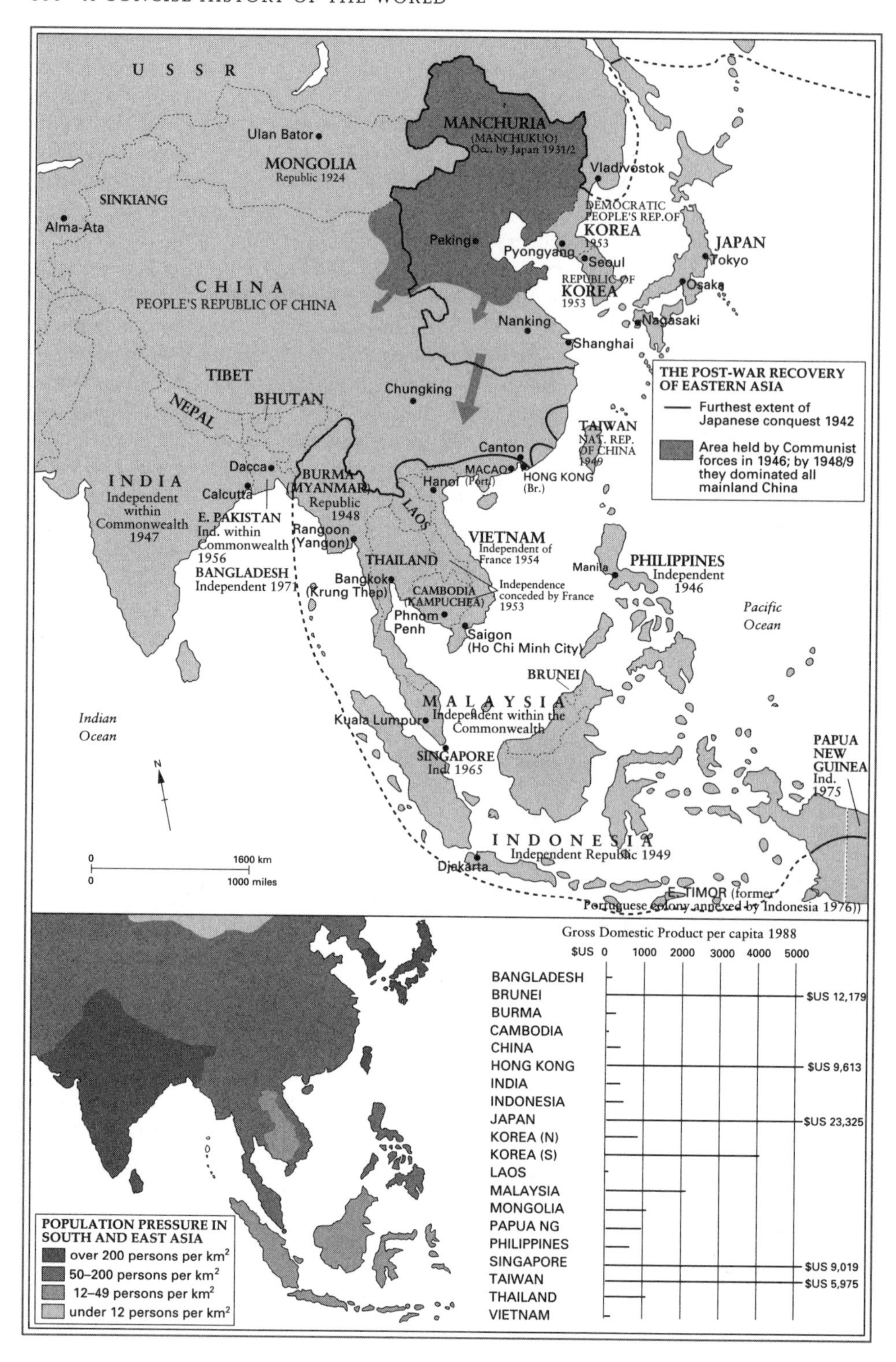

Japanese patronage of anti-white feeling – 'Asia for the Asians' was an evocative slogan – and, paradoxically, resistance to the Japanese by some native leaders both strengthened forces opposed to the return of a colonialism which had failed. Only the speedy arrival of British forces restored French authority in Indo-China, and Dutch in Indonesia in 1945. Within a few years, each was to lose control again, and an era of colonial rebellions had opened. Like those in the Americas a century and a half before, the colonial rebellions would profit from the rivalries of great powers, this time, those of the Cold War, but unlike them they were rebellions of indigenous stocks against intruders, not of white settlers and their descendants.

Not that rebellion was always needed. The most important single landmark was the departure of the British from India in 1947. This gave self-government to a sub-continent of 400 million people, at once destroying the only political unity its peoples had ever enjoyed. A predominantly Hindu India and an Islamic Pakistan (which later divided again, when in 1971 Bangladesh broke away to form a separate state) emerged, amid much bloodshed. Burma became an independent republic in 1948 and in the following year fighting which had been going on in Indonesia since the Dutch return came to an end with the emergence of a new Indonesian republic. The former British Malay territories were given freedom as an independent federation of Malaya in 1957.

Indo-China

By then the French had left Indo-China, divided into Cambodia, Laos and two Vietnamese republics. This was the area in which Cold War thinking and great power interest did much to prolong the birth-pangs of a new order. Communists had always played an important part in the Indo-Chinese independence movements though less in Cambodia and Laos than in Vietnam – the southern and coastal portion of Indo-China – where they were bitterly resisted by the French until 1954. In that year, a French garrison at Dien Bien Phu was defeated in pitched battle, a moral and propaganda defeat as important as that at Singapore twelve years previously. It broke the will of the French to fight for empire, and they now withdrew leaving behind a communist north Vietnam working hard to undermine non-communist south Vietnam. The United States began to give aid to the south, and Russia and China to its rival.

DECOLONIZING IN AFRICA

In 1900 there were about 110 million Africans; at present rates of growth by the year 2000 there may be 700 million. This is the most important change in Africa's history. It has tipped the continent's population balance towards black Africa and south of the Sahara. The demands and pressures which population growth has produced run through all recent African political and economic history. Yet decolonization has undoubtedly been an even more dramatic change. In 1918, almost the entire continent was ruled or bossed from Europe. The only two truly independent native African states were Liberia and Ethiopia. Now, including off-shore islands and Madagascar, there are forty-eight. At the beginning of the century, little modern African industry existed except on the South African Rand. Since then, South Africa has become a major industrial power while Rhodesia, Zaire and Zambia are important mining countries (producing copper, coal, manganese, iron, uranium). Algeria, Nigeria and Libya are major oil-producers. Though still the object of interference and exploitation from outside, African states have been courted by non-Africans for their resources and strategic position.

Black Africa's story has its own unity in this era of change. The absence of a dominating, highly developed and literate indigenous civilization (as in India or China) meant that white skills and direction in producing change counted for much in them. Where climate encouraged European settlement – particularly in South Africa – there were white communities which complicated the politics of modernization. Geography was important, too. Though the continent was of great interest to outsiders, not all of them could get at it easily. Whereas even in the nineteenth century Russia was a major imperial player in Asia, she was not involved in Africa until the movement towards independence was well developed. Communism mattered little outside the white industrial cities of South Africa until the arrival of communist advisers (from the USSR and China) in the 1960s, and soldiers (from Cuba) in the 1970s.

The two world wars had stirred things up, but in the rise of the first black African political leaders they were probably less important than the effects of education (often by white missionaries) and the acts, for good or ill, of the colonial governments. Between 1918 and 1939, these did not change much in the areas they controlled, although the former German colonies had been replaced by Mandates. In the war of 1939-45 Ethiopia, overthrown by the Italians in 1936, was restored and Italian Somalia and Eritrea disappeared from the map. Yet south of the Sudan and Sahara that map still looked in 1945 much as it had in 1939. The area shaded pink because it belonged to the British Commonwealth (a term the British had begun to use as they lost the taste and the nerve to rule an empire) included two areas which were only formally a part of it. Southern Rhodesia had, since the 1920s, got used to governing itself as a virtually independent dominion whose white settlers did not expect interference from London, and the Union of South Africa had steadily been pushed by its Boer voters towards an almost complete independence. There did not seem in 1945 to be much chance that the African map would change for a long time – if ever.

Facing page: In 1966 the British colony Beuchuanaland became independent. The Union Jack was hauled down in a ceremony at which British administrators were by then already practiced hands. Its successor in this case was the flag of Botswana, Africa's fortieth independent state at that time.

New African nations

Yet it did, and with revolutionary speed. In 1957 there appeared the first black 'new nation' in sub-Sahara Africa: Ghana. Less than twenty years later not one European colony remained, except for a couple of tiny Spanish enclaves. Also, South Africa had become a fully independent republic in 1961, and the Rhodesians had seceded from the Commonwealth and were engaged in bitter internal struggle over who their future rulers should be. Though bloodshed tended to follow the end of colonial rule, it did not usually play much part in its downfall (except in Algeria, where a rebellion was needed to turn out the French). Recurrent strife showed the dangers facing the new Africa. In the former Belgian Congo (now Zaire), the mineral-rich region of Katanga erupted into civil war in 1960. Before long, the Russians and Americans were indirectly involved and the Cold War had come to Africa. Later, revolutionary movements in the Portuguese colonies of Angola and Mozambique sought support from communist countries and that led to civil war after the independence Portugal had conceded (following revolution at home in 1974). Appropriately, the first colonial power to establish itself in Africa was the last to leave.

At the end of 1960 civil war also broke out in Nigeria. Many wars, rebellions, coups were to follow elsewhere, and they still continue. Politicians had to struggle to make democracy work amid peoples with no experience of it, but with traditional loyalties and animosities still to fight about. Some African states resembled the new nations of the nineteenth-century Balkans or South America and turned to military leaders. Often, there were not enough Africans to provide administrators and technicians for the new regimes, so that many of them still had to rely for a time upon key white personnel, while the supporting structures of colonial rule – education, communications, armed forces – were often far less soundly based than those in, say, India. Literacy rates were low and new African nations were much more dependent on foreign help than newly independent Asian countries. Such handicaps drove their leaders to seek popularity and success by whipping up hatred of external enemies. African states have often found it very difficult to cooperate except in verbal and diplomatic attacks on white racism. Political turmoil was likely also because huge economic and social difficulties faced them. Many of them had no real social or geographical unity and existed only because European diplomats had long ago drawn boundaries between their colonies. Some suffered the handicap of specialized economies. During the war of 1939–45, many African agriculturists had shifted towards the growing of cash crops on a large scale for export and the consequences were sometimes profound. An inflow of cash in payment for exports

looked at first attractive, but the spread of a cash economy could generate disruptive change, often in the form of unanticipated urban and regional development. Meanwhile, the tying of some Africans to particular patterns of development led later to economic vulnerability and rigidity. Even the benevolent intentions of outsiders, whether through colonial development programmes or, later, international aid, could sometimes end by shackling African producers more firmly still to world markets sensitive to any fall in international demand.

As populations rose more and more rapidly after 1960 and as a disappointment with 'freedom' set in, discontent was virtually inevitable, and this meant instability. The especially bad outbreaks of strife in Katanga and Nigeria (the biggest of the new nations and the one which had looked one of the most stable and promising and had enjoyed the advantage of large oil reserves) were only the worst examples. In other countries, too, fierce struggles between factions, regions and tribes soon led small westernized élites of politicians to abandon the democratic and liberal principles much talked of in the heady days when colonialism was in retreat. Between 1957 and 1985, independent Africa saw thirteen heads of state assassinated and twelve major wars. The need, real or imaginary, to prevent disintegration and strengthen central authority strengthened the trend towards one-party, authoritarian governments and the exercise of political authority by soldiers. Tyrant-like figures appeared, some simply bandits, though seen by some Africanists as the inheritors of the mantle of pre-colonial African kingship. However they should be understood, coups and revolutions were frequent and even the oldest of African states succumbed; in 1974 a revolution in Ethiopia brought to an end the oldest surviving Christian monarchy in the world and a line of kings supposed to run back to the son of Solomon and the Queen of Sheba. A year later, the soldiers who had taken power seemed just as discredited as their predecessors.

Their troubles did not stop African politicians blaming the outside world and, indeed, encouraged them to do so. Much was made of the mythological drama of the old European slave trade as a supreme example of racial exploitation (African participation in it being denied or overlooked and the Arab slave trade ignored). More immediate sources of exasperation were to be found in the recalcitrance of economic and social problems and a sense of political inferiority always near the surface in a continent of relatively powerless states (some with populations of less than a million). An abortive attempt in 1958 to overcome such weakness as arose from division sought to found a United States of Africa. There followed alliances, partial unions, and essays in federation which at last produced in 1963 the Organization for African Unity (OAU), largely thanks to the efforts of the Ethiopian emperor, Haile Selassie.

Meanwhile, the economic record of black Africa got worse. Africa has been the only continent since 1960 where annual per capita growth in GDP has turned downward, by -0.1% in the late 1970s, and -1.7% from 1980 to 1985. Agricultural decline was widespread in the 1970s. Political concern with urban voters, corruption and prestigious investment played havoc with commercial and industrial policy. Meanwhile, population inexorably rose and famine inexorably recurred. The onset of world economic recession after the 1973 oil revolution had a shattering effect, worsened within a few years by the impact of recurrent drought. Against this background, political cynicism flourished and heroes of the independence era lost their way. Lack of self-criticism (or at least of its expression) led to new resentments often exacerbated by Cold War. Yet, Marxist revolution had little success. Paradoxically, only in Ethiopia, most feudally backward of African states, and the former Portuguese colonies, the least-developed former colonial territories, did Marxist régimes take root; former French and British colonies proved less vulnerable.

The Lion of Judah, descendant of Solomon, Haile Selassie, emperor of Ethiopia. Driven from his country by Italian invasion in 1936, he was restored to the throne in 1941, to rule until 1974, when he was deposed by a coup. He died in prison the following year.

Apartheid

Increasingly, and explicably, scapegoats for disaster tended to be found at hand; resentments gradually focussed less on former imperial rulers than on the racial division of black and white in Africa itself, flagrant in the greatest African state, the Union of South Africa. The Afrikaans-speaking Boers who politically dominated it cherished their own ancient grievances against the British, intensified by defeat in the Boer War. Ties with the British Commonwealth had been steadily weakened. Though South Africa entered the war of 1939 against Germany and supplied important forces to fight in it, intransigent 'Afrikaners', as they increasingly called themselves, supported a movement favouring cooperation with the Nazis. Its leader became Prime Minister in 1948. In 1961 South Africa left the British Commonwealth and became a republic. Afrikaners had by then built up their economic position in the industrial and financial sectors, as well as in their traditional rural strongholds, and imposed, as they did so, a structure of separation of the races – apartheid – which systematically sought the reduction of the black African to the inferior status demanded by Boer ideology.

Boer beliefs were rooted in a belief that God had set his face against miscegenation, and in a crude Darwinian notion of genetic inferiority. They had some appeal to whites elsewhere in Africa. When Southern Rhodesia seceded from the Commonwealth in 1965, it was feared that its rulers wanted a society more like the South African. The British government dithered; there was nothing that the African states could immediately do. Nor could the United Nations do much, though 'sanctions' forbidding trade with the former colony were imposed on its members. Many black African states ignored them and the British government winked at breaches of them. It did not feel able to intervene militarily to suppress a colonial rebellion as flagrant as that of 1776 – a depressing precedent, of course.

South Africa, the richest and strongest state in the area, together with Rhodesia and the Portuguese colonies, was the object of mounting black African anger as the 1970s began. The drawing of the racial battlelines was not offset by minor concessions to South African blacks even if some black states depended on their economic links with the Republic. After the Portuguese withdrawal from Angola, though, a Marxist régime took power there. This led the South African government to take thought. Further, Rhodesia's prospects had sharply worsened; a guerrilla campaign could now be launched from Mozambique where Portuguese rule had ended. The American government contemplated with dismay the Cold War implications if Rhodesia collapsed at the hands of black nationalists depending on communist support. It applied pressure to the South Africans, who, in turn, applied it to the Rhodesians. In September 1976 the Rhodesian prime minister sadly told his countrymen that they had to accept the principle of black majority rule. The last attempt to found an African country dominated by whites had failed. African nationalists still sought to achieve unconditional surrender, but in 1980 Rhodesia briefly returned to British rule before re-emerging into independence, this time as Zimbabwe. This left South Africa alone as the sole white-dominated state in the continent.

A NEW CHINA

At the end of the Second World War, China was still engaged in struggles over the way her own modernization should proceed. The defeat of Japan had not owed much to her, except by the drain the occupation imposed on Japanese strength. The KMT, nominally in alliance with the communists, reserved its own forces for the day of reckoning with them, while the communist leader, Mao-Tse-Tung, urged his comrades to sink their roots in the countryside by winning over the peasants. In contrast, the KMT usually brought fresh oppression to the areas it controlled; peasants were terrorized into paying higher taxes and the rents demanded by the landlords who supported the KMT. When the Japanese collapsed more suddenly than expected, the communists were therefore in many areas well-placed to take over their arms and assume control. In the north and Manchuria they were helped by the Soviet forces; conversely, the Americans landed at some of the main seaports as soon as they could, and held them until KMT forces could get there.

Three years of civil war then began. The Kuomintang government rapidly lost grip. Its own soldiers and bureaucrats began to show increasing signs of thinking that communism might be the best way ahead for China. In 1947, fed up with the inefficiency and corruption of the regime, the Americans withdrew their own forces and began to cut down their aid to it. In 1948 the government had to withdraw to Taiwan (where its successor remains today) and the communists could consolidate their hold on the mainland. On 1 October 1949 the Communist Peoples Republic of China was formally inaugurated at Peking, once more the country's capital; China was reunited at last under an unequivocally revolutionary regime.

Communist China

The USSR had been the first state to recognize the new regime in China, closely followed by the United Kingdom, India and Burma. China's leaders faced no real danger from the outside and could concentrate on the long overdue and immensely difficult task of modernization. Poverty was universal, disease and malnutrition widespread. Material and physical construction and reconstruction were overdue, population pressure on land was as serious as ever, and the moral and ideological void presented by the collapse of the ancien régime over the preceding century had to be filled. Rural China was the starting-point. The overthrow of local village leaders and landlords was often violent (800,000 Chinese were later reported by Mao

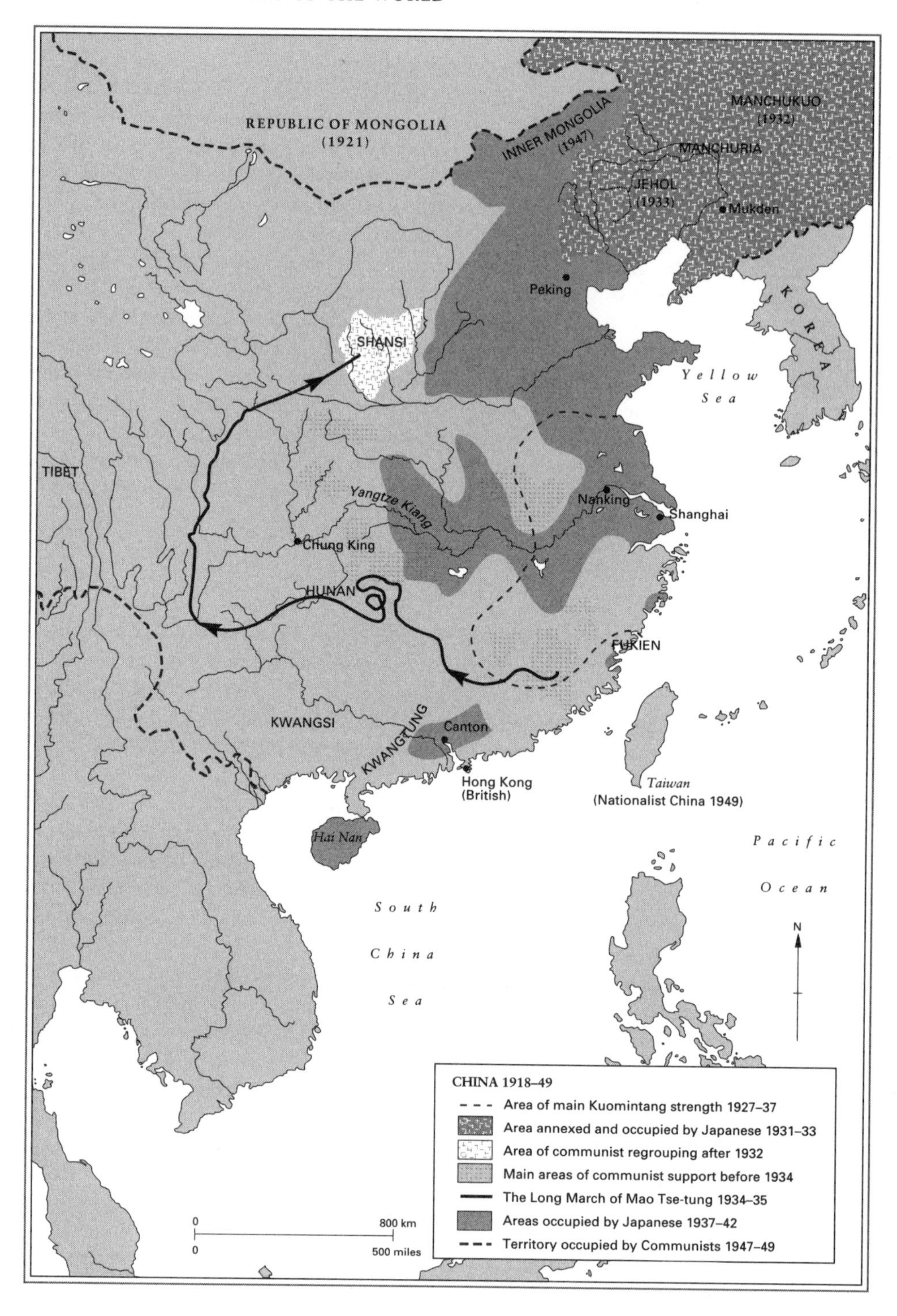

himself to have been 'liquidated' in the first five years of the People's Republic; it is certainly an underestimate). Meanwhile industrialization was pressed forward with Soviet help.

The superficial unity of the communist bloc and a Sino-Soviet treaty of 1950 were meanwhile interpreted – especially in the United States which insisted on her exclusion from the United Nations Organization – as evidence that the new China was entering the Cold War.

Her rulers certainly talked revolution and anti-colonialism and their choices were bound to be confined by the parameters of the international situation. They nonetheless showed from the start many of the traditional concerns of Chinese policy, notably in seeking to re-establish the historic sphere of Chinese power. A Chinese occupation of Tibet in 1951 recalled the fact that it had for centuries been under imperial suzerainty, while Korea was another old zone of Chinese interest and the reaction on the Yalu to the American threat of 1950 hardly surprising. But from the start the most vociferous concern over the Chinese periphery was shown about Taiwan, lost by China in 1895 to the Japanese and only briefly restored in 1945 to mainland control. The United States government, deeply committed to the KMT régime, announced in 1955 that it would protect its satellite. About this issue American policy seemed to crystallize obsessively for over a decade; the KMT tail seemed at times to wag the American dog. Conversely, during the 1950s, both India and Russia supported Peking over Taiwan, insisting that the matter was one of Chinese internal affairs; it cost them nothing to do so.

International complications

Yet the next decade brought tension between China and these seeming friends. When the Chinese further tightened their grasp on Tibet in 1959, territorial disputes with India had already begun. The Chinese would not recognize a border drawn by a British-Tibetan negotiation and never formally accepted by any Chinese government; forty-odd years' usage was hardly significant against China's millenial historical memory. Fighting took place on the border in the autumn of 1962, and the Indians did badly; the Chinese did not withdraw. Then, early in 1963, China suddenly startled the world by denouncing the USSR for helping India and cutting off economic and military aid to China. There was a long history behind this, too. Chinese communists could remember tension between Soviet and native influence in the leadership of the Chinese party as far back as the 1920s. Mao himself represented the latter. But, puzzlingly, the issue had to be presented to the rest of the world in Marxist jargon. When a new leadership in Russia began tentatively to dismantle the Stalin myth, the Chinese sometimes sounded Stalinist when they were pursuing anti-Soviet policies.

Sino-Soviet quarrels in fact had very deep roots. Long before the foundation of the CCP the Chinese revolution had been driven by resentment of the foreigners. Russians had always been pre-eminent among them. Peter the Great had begun encroachment upon the Chinese sphere and his nineteenth-century successors seized more Chinese territory than any other foreign aggressors. The historical clock did not stop with the fall of tsardom. A Russian protectorate over Tannu Tuva had been claimed in 1914; in 1944 the area was annexed by the USSR. In 1945 Soviet armies entered Manchuria and north China, thus reconstituting the tsarist Far East of 1900; they remained in Sinkiang until 1949, in Port Arthur until 1955. In Mongolia they left behind a satellite Mongolian People's Republic. With something like 4500 miles of shared frontier (if Mongolia is included), the potential for dispute between the two communist states was immense. Skirmishing and quarrelling began soon after the proclamation of the Peoples' Democratic Republic.

Although Mao Tse-tung could be as brutal as Bolshevik theory demanded, he had a firm belief in pragmatism and the lessons of experience, and advocated a sinicized Marxism. This was not likely to make him over-respectful of Soviet views in the 1960s. Mao's attitude to knowledge and ideas was predominantly utilitarian and pragmatic, and so very much in the Chinese tradition. In part because of this, indeed, his relationship with the CCP had not always been smooth. Only when disaster had overtaken urban communism, while he had identified the peasantry as the way ahead, could he come to the top. The notion of a protracted revolutionary war, waged from the countryside and carried into the towns, came to look promising in other parts of the world, too, where the orthodox Marxist belief that industrial development was needed to create a revolutionary proletariat was not persuasive.

Mao's ascendancy

Soviet withdrawal of economic and technical help in 1960 came just as natural disasters – floods are said by official Chinese sources to have drowned 150 million acres of agricultural land – had followed the collapse of the 'Great Leap Forward', an economic offensive launched by Mao with the aim of decentralizing the economy and repudiating centralized planning on the Russian model with its bureaucratic dangers. The 'Great Leap' turned rural life upside down, driving huge furrows not only through traditional agricultural practice, but the social life of the family and village. It was a disaster. Mao's standing suffered and his rivals put the economy back again on the road to modernization. (One striking symbol was the explosion of a Chinese nuclear weapon in 1964, an expensive admission card to a very exclusive club.) They managed to avoid crippling famine and kept the loyalty of the people. As China's population continued inexorably to rise and her share of world population may have sometimes been higher in the past, her leaders even talked unmoved of the possibility of nuclear war; Chinese would survive in greater numbers than the inhabitants of other countries. Yet she still lagged in her modernization.

What was happening in the USSR continued to shape Chinese policy. After Stalin's death, corruption and conservatism was apparent in the Soviet bureaucracy. The fear of something similar in China was evident in Mao's last attempt to give his ideas dominance over a new generation, the 'Cultural Revolution' of 1966-69. He felt that the revolution might congeal and lose the moral élan which had carried it so far; to protect it, old ideas had to go. In an attempt to offset the emergence of a new entrenched ruling class, universities were closed, physical labour was demanded of all citizens in order to change traditional attitudes towards intellectuals and there was a renewed emphasis on self-sacrifice and the thought of Chairman Mao. By 1968 the country had been shaken from top to bottom. At first a supporter of the 'Red Guards' who led the Cultural Revolution, Mao in the end had himself to concede that things had gone too far. Finally, after three years of uproar, the army intervened to re-establish order and institute new cadres. Though a party congress reconfirmed his leadership, Mao had again failed (and possibly another half-million Chinese had been murdered or driven to suicide).

The Cultural Revolution was an aberration, but of the great revolutions of world history, the Chinese had to be one of the most far-reaching. Society, government and economy were always enmeshed and integrated in China as nowhere else. The traditional prestige of intellectuals and scholars embodied the old order; deliberate attacks on the family's authority were attacks on the most conservative of all Chinese institutions. The advancement of women and propaganda to discourage early marriage were assaults on the past such as no other revolution had ever made, for in China the past meant a role for women far inferior to anything to be found in pre-revolutionary America, France or even Russia. Attacks on party leaders which accused them of flirtation with Confucian ideas were much more than jibes. China had an enormous history to overcome.

The Chinese revolution may be seen as one of the great surges of history, comparable to such processes as the spread of Islam, or Europe's assault on the world in early modern times. Yet, paradoxically, it is unimaginable without conscious direction; government had the mysterious prestige of the Mandate of Heaven. Chinese tradition still gives authority a moral endorsement long vanished in the West; no great society for so long drove home to its peoples the lessons that the individual mattered less than the collective whole, that authority could rightfully command the services of millions at any cost to themselves in order to carry out great works for the good of the state, that authority was unquestionable if exercised for the common good. The notion of opposition is distasteful in China because it suggests social disruption. Mao fitted this tradition and benefited from the Chinese past as well as destroying it. He was the dictator of a moral doctrine which was presented as the core of society just as Confucianism had been.

'Bombard the Headquarters!' A poster from 1966, launching the Great Proletarian Cultural Revolution.

A NEW EAST ASIA

Even by the time of Stalin's death, a prophecy made by the South African statesman Smuts more than a quarter-century before that 'the scene had shifted away from Europe to the Far East and the Pacific' seemed more and more to the point. After Korea further evidence of it had come in Indo-China. Traditionally tributary to China, it was a country where Soviet and Chinese policy diverged, blurring Cold War divisions. After Dien Bien Phu a conference at Geneva agreed to partition Vietnam pending elections which might reunite the country but they never took place. Instead, there gradually unrolled in Indo-China what was to become the fiercest phase since 1945 of the Asian war against the West begun in 1941. The Western contenders were no longer the former colonial rulers but the Americans. On the other side was a mixture of Indo-Chinese communists, nationalists and reformers with Chinese and Soviet support. Anti-communism and belief in indigenous governments led the United States to back the South Vietnamese as it backed South Koreans and Filipinos. In South Vietnam, though, there did not emerge régimes of unquestioned legitimacy in the eyes of those they ruled. Instead, they came to be identified with the western enemy so disliked in east Asia. An apparently corrupt ruling class seemed able to survive government after government. Meanwhile, the communists sought reunification by supporting from the north an underground movement in the south, the Viet Cong. In 1962 the American president, John Kennedy, decided to send 4000 American 'advisers' to help the South Vietnam government. It was a clear step towards American involvement in a major war on the mainland of Asia such as Truman had feared.

The Third World idea

By then, a new complexity in international life was apparent. One symptom was the appearance of professedly neutralist or 'non-aligned' nations. Representatives of twenty-nine African and Asian states had met at Bandung in Indonesia in 1955, most of them (except the Chinese) from former parts of the old European empires (the Yugoslavs were soon to join them, though they had not been imperial subjects since 1918). Their countries were poor and needy, more suspicious of the United States than of Russia and more attracted to China than to either. They came to be called the 'Third World' nations, a term apparently coined by a French journalist in a conscious reminiscence of the legally underprivileged French 'Third

Estate' of 1789 which had provided so much of the driving force of the French Revolution. They felt disregarded by the great powers, excluded from the economic privileges of the developed countries and worthy of a greater say in running the world. Much was said to that effect in the United Nations. Yet the expression 'Third World' masked important differences within it. More people have been killed since 1955 in Third World wars and civil wars than in conflicts external to it.

Russians and Chinese each sought the leadership of the underdeveloped and uncommitted. At first this emerged obliquely. They differed over the Yugoslavs and as time passed Pakistan drew closer to China (in spite of a treaty with the United States) and Russia closer to India (which down to 1960 had received more economic aid from the USA than any other country). When the United States declined to supply arms to her in 1965, Pakistan asked for Chinese help. This made evident a new fluidity in international affairs.

Indonesia, too, made difficulties for great powers. Its vast sprawl encompassed many peoples, often with widely diverging interests, for whom the departure of the Dutch had meant release from the discipline an alien ruler had imposed just as familiar post-colonial problems – over-population, poverty, inflation – began to be felt. In the 1950s its new central government was increasingly resented; by 1957 it had faced armed rebellion in Sumatra and unrest elsewhere. The time-honoured device of distracting opposition with nationalist excitement did not work for long. Sukarno, the president, had already moved away from the liberal forms adopted at the birth of the new state and in 1960 parliament was dismissed. In 1963 he was named president for life. Both the USSR and the United States, fearing he might turn to China, long stood by him, which enabled him to throw his weight about and to turn on the new federation of Malaysia, put together that year from fragments of British empire in South-East Asia. Malaysia mastered the Indonesian attacks with British help and this setback seems to have been the turning-point for Sukarno. Food

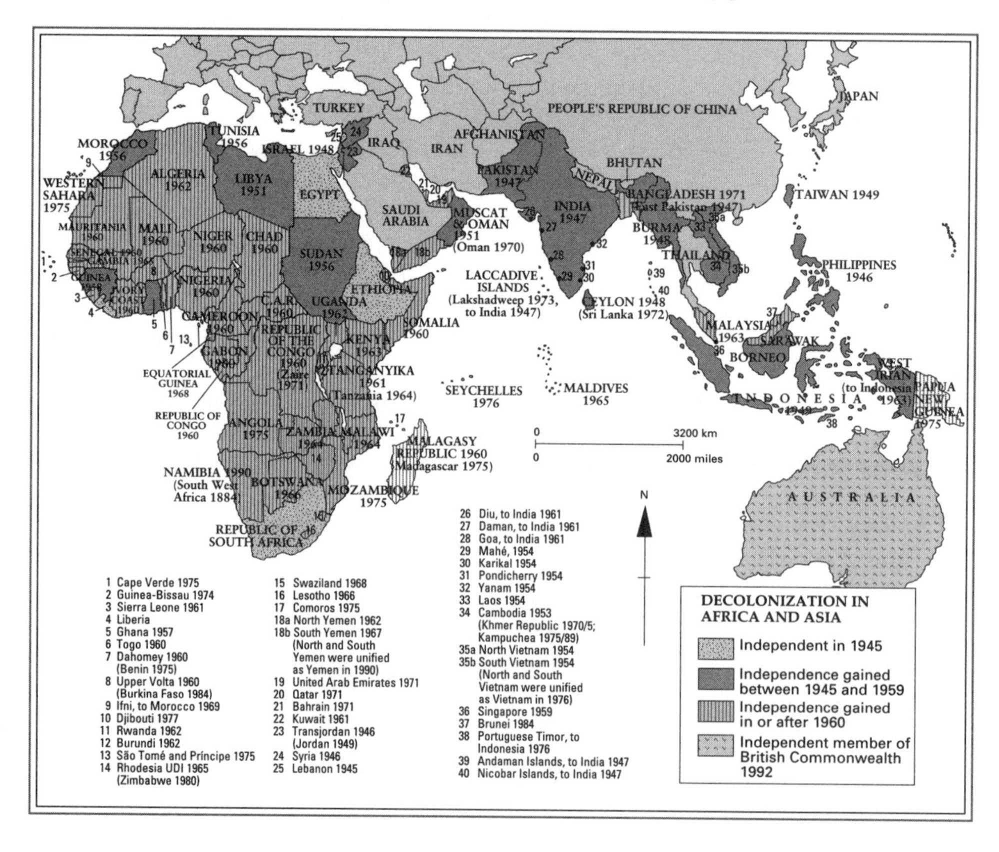

President Sukarno of Indonesia addressing the opening session of the Bandung Conference, 18 April 1955.

shortages and inflation led to an attempted coup by the communists (or so the military said), and in 1965, the army stood back ostentatiously while popular massacre removed the communists to whom Sukarno might have turned. He himself was duly set aside the following year and a solidly anti-communist regime took power. It broke off diplomatic relations with China.

Nonetheless, even by 1960 the recreation of Chinese power was already the dominant strategical fact east of Singapore. Even South Korea and Japan benefited from the Chinese revolution; it gave them, too, leverage in dealing with the West. Yet East Asians buttressed their independence in both communist and non-communist forms and rarely succumbed to direct Chinese manipulation. It is difficult not to link this, ironically, to the deep and many-faceted conservatism of their societies. In their discipline, capacity for constructive social effort, disregard for the individual , respect for authority and hierarchy and deep self-awareness as members of civilizations proudly distinct from the West, East Asians had many sources to draw on in protecting their new independence.

The recovery of Japan

Japan's surrender had taken Stalin by surprise. The Americans firmly rejected his requests for a share in an occupation Soviet power had done nothing to bring about and embarked alone upon the last great episode of western paternalism in Asia. Once again, though, the Japanese showed their astonishing gift for learning from others only what they wished to learn. Yet 1945 was a dividing line for them. Defeat forced them spiritually into a twentieth century they had previously entered only technologically and confronted the Japanese with deep and troubling problems of national identity and purpose. The dream of 'Asia for the Asians' was blown away and the rolling back of colonialism left Japan with no obvious Asian role. Moreover, the war's demonstration of her vulnerability had been a great shock. The economy was wrecked: over eighty per cent of Japanese shipping alone had been destroyed. Finally, it had been followed by the loss of territory and occupation.

Yet positive elements could also be discerned. The monarchy had made the surrender possible; many Japanese saw the emperor as their saviour from annihilation. The American commander in the Pacific, General MacArthur, took care to have a new monarchical constitution adopted by the Japanese before republican enthusiasts in the United States could interfere. Japanese social cohesiveness and discipline was another asset, though for a time the Americans seemed to threaten it by their determination to democratize the country. A major land reform, democratic education and careful demilitarization was by 1951 deemed adequate for a peace treaty between Japan and most of her former opponents (except the nationalist Chinese and the USSR, who followed suit within a few years) and Japan regained her full sovereignty and control of her own affairs, though American forces remained on her soil. Though facing a China stronger and much better consolidated than for a century, Japan's position was not disadvantageous.

The implications of the American occupation had soon begun to change. Japan was separated from China by 500 miles of water, but Korea, the old area of imperial rivalry, was only 150 miles away (and Soviet territory only ten). The coming of the Cold War to Asia brought Japan real benefits. The Korean war made her important as a base, and stimulated the economy. Industrial production quickly climbed back to pre-war levels. United States diplomacy promoted Japanese interests abroad. As, too, Japan was until 1951 forbidden to have any armed forces, she long had no defence costs, but enjoyed the protection of the American nuclear umbrella.

A twentieth-century Shogun: General Douglas MacArthur, the architect of the Japanese monarchy's preservation and dictator of Japan 1945–1950.

Japan soon emerged as a key part of the American security system in Asia and the Pacific. This system also rested on treaties with Australia, New Zealand and the Philippines (which had become independent in 1946). Others followed with Pakistan and Thailand, the Americans' only Asian allies other than Taiwan. Indonesia and (much more important) India remained aloof. These alliances reflected, in part, the new conditions of Pacific and Asian international relations. For a little while there were still British forces east of Suez, but during the war Australia and New Zealand had discovered that the British could not defend them and that the Americans could. Though they could sustain Malaysia against the Indonesians, the British knew that Hong Kong survived only because it suited the Chinese that it should. Yet the new Pacific could not simply be organized in terms of the Cold War. Americans saw Japan as a potential anti-communist force, but Australian and New Zealanders remembered 1941 and feared a revival of Japanese power. American policy was not, then, made solely by ideology, though long obsessed by communist success in China and by Chinese patronage of revolutionaries as far away as Africa and South America. In reality, too, China's re-emergence made nonsense of the dualist, Cold War system. Russia had become one corner of a triangle, as well as losing its unchallenged pre-eminence in the world revolutionary movement. Never a simple business, the Cold War was now more complex than ever. Not only in the crudities of the propaganda it generated, it now looks somewhat like the complex struggles of religion in sixteenth- and seventeenth-century Europe, when ideology could provoke violence, passion, and even, at times, conviction, but could never exactly contain complexities and cross-currents arising from interests (above all, those of nationality). But rhetoric and mythology roll on long after they cease to reflect reality, and in this, too, Cold War was somewhat like the religious struggles of former days.

COLD WAR COMES TO THE WESTERN HEMISPHERE

After the Berlin blockade both sides in Europe were more anxious not to run risks or rock the boat. A revolution in Hungary against its communist regime in 1956 was crushed by Warsaw Pact forces. In 1962, there was a new flurry of diplomatic excitement when the GDR, with Soviet support, suddenly and rapidly sealed off west from east Berlin by building a wall which stemmed the drain of valuable man-power to western Europe. In the end this reduced tension, for it removed a thorn in the GDR's side. When a crisis, threatening nuclear war, arose in the same year, it was not in Europe but on the doorstep of the United States.

Latin America had not been troubled much by Cold War politics. Its twentieth century had moved to different rhythms from that of Europe or Asia. In 1900 much of Latin America was already stable and prosperous. The modernity of many of its great cities showed this, and so did their attractiveness to European immigrants. Most Latin American countries exported agricultural or mineral products; their manufacturing sectors were inconsiderable and for a long time seemingly untroubled by the social and political problems of Europe, though in rural areas there was class conflict a-plenty.

The First World War then brought important changes. Before it, though the predominant political influence in the Caribbean, the United States had not exercised much economic weight in South America's affairs. The liquidation of British investments in the war changed this; by 1929 the United States provided about forty per cent of South America's foreign capital. Then came the world depression. Many South American countries defaulted on their payments to foreign investors. It became almost impossible for them to borrow abroad. The collapse of prosperity led to growing nationalist assertiveness, sometimes against neighbours, sometimes against the North Americans and Europeans; foreign oil companies were expropriated in Mexico and Bolivia. Traditional ruling oligarchies were compromised by their failure to meet the problems posed by falling national incomes. From 1930 onwards there were military coups in every country except Mexico. Yet 1939 again brought prosperity as commodity prices rose, and the trend was prolonged by the Korean war. Argentina's rulers flirted with Nazi Germany, but most of the republics were sympathetic to the allies, who courted them; most joined the United Nations side before the war ended, and one, Brazil, sent a small expeditionary force to Europe, a striking gesture. The most important effects of the war on Latin America, though, were economic. An intensive drive to industrialize gathered speed in several countries. On the urban work forces which industrialization built up was founded a new form of political power to compete with the military and the tra-

The briefly successful Hungarian revolution of 1956 released the frustration and bitterness of years, often with brutal results for members of the regime's former security police, many of whom were lynched.

ditional élites in the post-war era. Authoritarian, semi-fascist, popular mass movements appeared in several countries.

A significant change had also come about (though not as a result of war) in the use by the United States of its preponderant power in the Caribbean. Twenty times in the first twenty years of the century American armed forces had intervened directly there, twice going so far as to establish protectorates. Between 1920 and 1939 there were only two such interventions and indirect pressure also declined. In the 1930s President Roosevelt proclaimed a 'Good Neighbour' policy which stressed non-intervention. Yet while American policy was dominated by European concerns in the early post-war years it began after Korea slowly to look southwards again. Washington had not been unduly alarmed by manifestations of Latin American nationalism which tended to find a scapegoat in American policy, but became increasingly concerned lest the hemisphere provide a lodgement for Russian influence. The Cold War had come to the continent. In 1954 a government in Guatemala which had communist support was overthrown with American help.

Anxious that the footholds provided for communism by poverty and discontent should be removed, the United States provided economic aid and applauded governments which said they sought social reform. Unfortunately, whenever the programmes of such governments moved towards the eradication of American control of capital by nationalization, American policy tended to veer away again. On the whole, therefore, while it might deplore the excesses of an individual authoritarian régime the American government tended to find itself in Latin America, as in Asia, supporting conservative interests.

Cuba

The only victorious Latin American revolution was in Cuba, an island within a relatively short distance of the United States. Cuba had been especially badly hit in the depression; it was virtually dependent on one crop, sugar, which had only one outlet, the United States. This economic tie was only one of several which gave Cuba a closer and more irksome 'special relationship' with the United States than had any other Latin American state. Until 1934 the Cuban constitution had included special provisions restricting Cuba's diplomatic freedom and the Americans kept a naval base on the island (as they still do). There was heavy American investment in urban property and utilities, and Cuba's poverty and low prices made it an attractive holiday resort for Americans.

The United States was seen as the real power behind Cuba's conservative post-war governments. In fact, this ceased to be so; the State Department disapproved of the island's dictator, Batista, and cut off help to him in 1957. By then a young nationalist doctor, Fidel Castro, had already begun a guerrilla campaign against the régime. In two years he was successful and became something of a hero in the United States. In 1959, as prime minister of a new and revolutionary Cuba, he described his régime as 'humanistic' and, specifically, non-communist. He worked with a wide spectrum of people who wanted to overthrow Batista, from liberals to Marxists and the United States patronized him as a Caribbean Sukarno. Yet the relationship quickly soured once Castro turned to agrarian reform and the nationalization of sugar concerns and to denunciation of those Americanized elements in Cuban society which had supported the old régime. Anti-Americanism was a logical means – perhaps the only one – for Castro to unite Cubans behind the revolution. Soon the United States broke off diplomatic relations with Cuba and began to apply economic pressure as well. Before long, it decided to promote Castro's overthrow by force. Exiles were already training with American support in Guatemala when President Kennedy took office in 1961. He was neither cautious nor thoughtful enough to hold back an expedition which failed miserably. Castro now turned in earnest towards Russia, and at the end of the year declared himself a Marxist-Lenninist. Henceforth, Cuba was a revolutionary magnet in Latin America. Castro's torturers replaced Batista's and his government pressed forward with policies which badly damaged

The young Dr Castro just after his successful seizure of Havana, 1959.

the economy, but sought to promote egalitarianism and social reform (in the 1970s, Cuba claimed to have the lowest child mortality rates in Latin America). Russian economic aid kept the country going.

The crisis

There soon followed the most serious confrontation of the whole Cold War and probably its turning point. The Soviet government decided to install in Cuba rocket missiles capable of reaching anywhere in the United States. American photographic reconnaissance confirmed in October 1962 that the Russians were building sites for them. When this could be shown to be incontrovertible President Kennedy announced that the United States Navy would stop any ship delivering further missiles to Cuba and that those already there would have to be withdrawn. One Lebanese ship was boarded and searched in the days that followed; Russian ships were only observed. The American nuclear striking force was prepared for war. After a few days and some exchanges of personal letters between Kennedy and Khrushchev, the Soviet leader, the latter agreed that the missiles should be removed.

The effect on the relations of the superpowers and on their assessments of one another was profound. Soviet space technology had already alarmed Americans in the late 1950s, but it now seemed that the Unites States still had, after all, too great a preponderance of strength to be challenged. While the USSR made huge (and successful) efforts to reduce the American lead in the next few years, the Cold War had passed its most dangerous point; though it went on, there lay ahead a period of often interrupted but closer contact and negotiation between the two superpowers on all sorts of questions. By the middle of the 1970s, there was a growing sense that the days of the great simplicities were over. Though two giants still dominated the world as they had done since 1945, and still at times talked as if they divided it, too, into adherents or enemies, the prospect of nuclear war as the ultimate price of geographical extension of the Cold War had been faced and found unacceptable.

The superpowers' new relationship

The arms race, it is true, was far from over. A huge Soviet effort was made in the later 1960s (with some success) to achieve superior power over the United States. But a compelling tie now existed between the two nuclear giants. Superiority in nuclear killing-power, it came to be seen, is a rather limited affair, a truth summed up in the convenient acronym 'MAD'; that is to say, both countries had the capacity to produce 'Mutually Assured Destruction'. Each knew the other had enough striking power to ensure that even after a surprise attack had deprived it of the cream of its weapons, what remained would be sufficient to ensure a retaliation so appalling as to leave the successful attacker's cities smoking wildernesses and to deprive victory of any meaning.

In 1973 talks began on arms limitation and the possibility of a comprehensive security arrangement in Europe. In return for the formal recognition of Europe's post-war frontiers (above all, that between the two Germanies), the Soviet negotiators finally agreed in 1975 at Helsinki to increased economic intercourse between eastern and western Europe and a paper guarantee of human rights and political freedom. Though the guarantee was unenforceable,

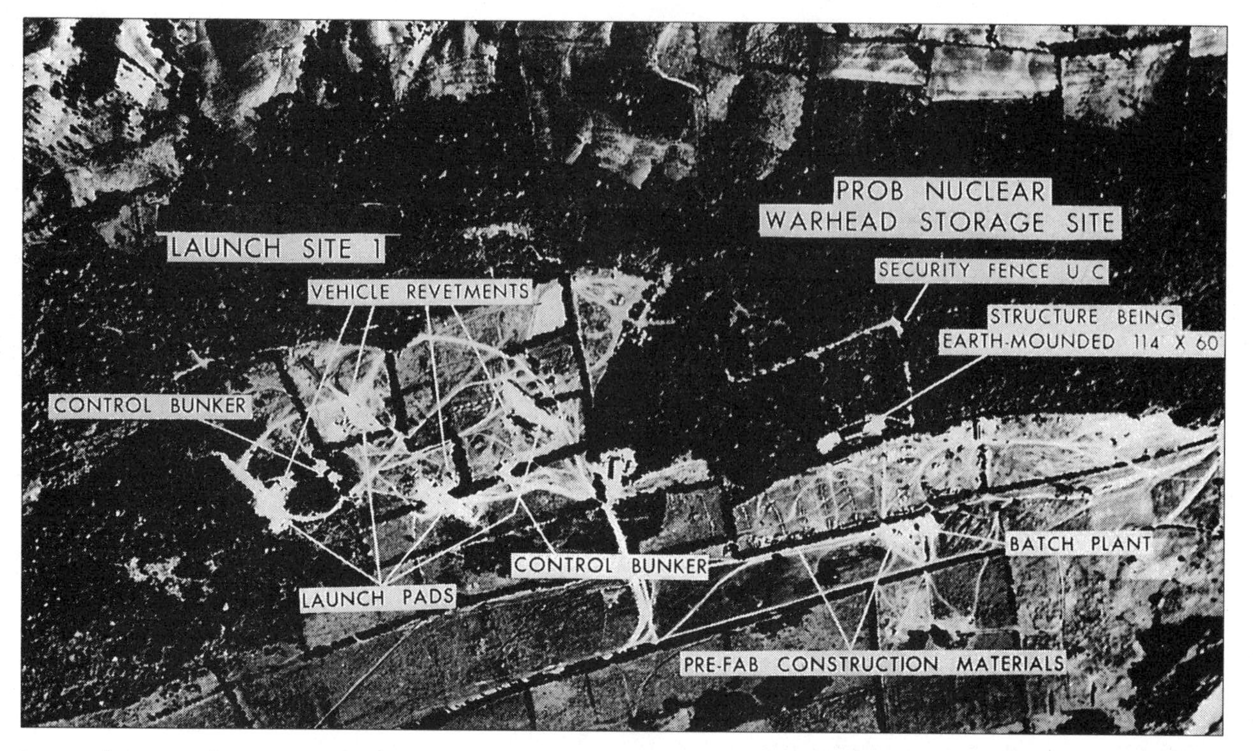

The importance to modern military intelligence of photographic and television observation from aircraft and satellites was dramatically illustrated when the UnitedStates government produced its photographs (of which this is one) of missile sites under construction in Cuba in 1962.

it proved immensely important both as an inspiration to dissidents in communist Europe and Russia, and because the flow of trade and investment between the two Europes slowly fed other contacts. It was virtually the long-awaited peace treaty ending the Second World War, giving the Soviet Union what it most wanted, recognition of its right to the territorial spoils of victory.

Change in the USSR

Possibly because of Cuba, Nikita Khruschev the dominant figure in Soviet government since 1958, had been removed from office in 1964. His personal contribution to Soviet development had been evident in a great shaking-up of the Party, a qualified 'destalinization' of Soviet life, a huge failure over agriculture, and a new emphasis in the armed services towards the strategic rocket services which became their élite arm. His fall made clear that the Soviet Union was getting better at bloodless political change; he was not killed, sent to prison or even to run a power-station in Mongolia. Crucially, too, a speech he had made at the Twentieth Party Congress in 1956 denouncing Stalin's cruelty and errors could not be unsaid. Symbolically, Stalin's body had been removed from Lenin's tomb, the national shrine. In the next few years there was what some called a 'thaw', when marginally greater freedom of expression was allowed to writers and artists. Yet the authoritarian nature of Soviet government was unchanged in principle, though, optimistically and exaggeratedly, some argued in the 1960s and 1970s that the USA and USSR were in fact growing more and more alike.

This theory of 'convergence' seized on the fact that the Soviet Union was a developed economy. It overlooked its inefficiencies and distortions. Russian agriculture had once fed the cities of Central Europe and paid for tsarist industrialization but under communism it was a continuing failure; paradoxically, the USSR often had to buy American grain. Soviet per capita national income still lagged in the 1970s far behind that of the United States. When Soviet citizens were given old age pensions in 1956 it was nearly half a century after the British. Though they had union-wide public health services, they fell further and further in quality behind those available in the West.

Yet to the Third World the USA and USSR were both rich. Millions of Soviet citizens, too, were more aware of the contrast with the stricken and impoverished Russia of the 1940s than of lagging behind the United States. There had been a long legacy of disruption to master; only in 1952 did real wages get back to their 1928 level. Soviet citizens were not inclined to feel sorry for themselves. Moreover, by 1970 a Soviet scientific base existed which at its best rivalled that of the United States. Soviet space technology justified the revolution; it showed the USSR could do anything another nation could do, and much that only one other could do.

Whether this meant that the Soviet Union was in some sense becoming a satisfied nation, with leaders more confident and less suspicious of the outside world, is doubtful. It remained a police state where the limits of freedom and the basic privileges of the individual were defined in practice by an apparatus backed up by administrative fiat and political prisons. Its bureaucracy showed increasing tendencies to conservatism and corruption in the interest of the ruling class. Criticism, particularly of restraints upon intellectual freedom, began to be heard clearly in the 1960s. Symptoms of antisocial behaviour, such as hooliganism, black-marketeering and alcoholism were reported, as in other large countries. It was probably more important that in the 1970s native Russian–speakers for the first time became a minority in the Soviet Union.

Change in the USA

Changes in the United States were easier to assess. Of the continued growth of American power and wealth there could be no doubt. Already in the middle of the 1950s the United States produced more than half the world's manufactured goods. Its population passed 200 million in 1968; only one in twenty Americans were then not native-born (yet within ten years there would be worries about a huge Spanish-speaking immigration from Mexico and the Caribbean). More Americans than ever lived in cities or their suburbs, and they were living longer. The likelihood that they would die of some form of malignancy had trebled since 1900 – paradoxically, a sure sign of improvement in public health, for it showed that other diseases had been mastered. In 1970, few doubts existed about the republic's ability to sustain the huge military potential upon which American global power rested – even if there were already many about the way that power was used.

The smiling face of the post-Stalinist USSR: Khruschev, secretary-general of the Communist party 1953–64, flanked by (on his right) Andrei Gromyko, foreign minister 1957–85 and Nikolai Bulganin, prime minister 1955–58.

As presidents came and went government continued to hold and indeed increase its importance as the first customer of the American economy. Hopes of balanced budgets and cheap, business-like administration always ran aground upon the fact that government spending was a primary economic stimulant. The United States was a democracy. The Welfare State slowly advanced because voters wanted it that way. This helped to prolong the Democratic coalition. Thanks to war-weariness, Republican presidents were elected in 1952 and 1968, but neither was able to persuade Americans that they should also elect Republican congresses. On the other hand, signs of strain were already apparent in the Democratic bloc before 1960 – Eisenhower appealed to many southern voters – and by 1970 something a little more like a national conservative party had appeared under the Republican banner. It was the start of the disappearance to the 'Solid' (Democratic-voting) South as a political constant.

The American race problem

President Kennedy's election by a disputable margin of the popular vote in 1960 had brought at first a striking sense of innovation. Both in foreign and domestic affairs, the eight years of renewed Democratic rule after 1961 were indeed to bring great changes to the United States, but they were not those which Kennedy or his vice-president, Lyndon Johnson, foresaw when they took office. One was in the position of black Americans. In 1960 a century after emancipation, the black American was (as he or she is still today) likely to be poorer, more often on relief, more often unemployed, less well housed and less healthy than the white American. This had been a local and southern problem, but had been turned into a national one by migration. Between 1940 and 1960 the black population of northern states almost trebled. New York became the state with the biggest black population in the Union. It also came to be clear that the problem was not just one of legal and constitutional rights, but one of economic and cultural deprivation. Meanwhile, the world outside had changed; many of the new nations which were becoming a majority at the UN were nations of coloured peoples. Communist propaganda always made good use of the plight of American blacks.

Storage of new cars from General Motors in the 1950s, a symbol of huge productive power and problems of traffic management and pollution to come.

A demonstration by black campaigners for civil rights. Washington, 1963. The memorial to Lincoln – the president who issued the Emancipation Proclamation a hundred years earlier – was a frequent focus of such rallies.

Incontrovertibly, the long-debated legal and political position of black Americans was altered radically for the better. A struggle for 'Civil Rights', of which the most important was the unhindered exercise of the franchise (always formally, but actually not, available in some southern states) had begun in the 1950s. Decisions of the Supreme Court that the segregation of different races within the public school system was unconstitutional and should be brought to an end where it existed within a reasonable time broadened the issue and threatened social practice in many southern states, but by 1963 black and white children were attending some public schools together in every state of the Union (though integration is still far from complete).

Kennedy also initiated a programme of measures (to be brought to maturity by his successor) which went beyond the voting rights issue to attack unfavourable discrimination and deprivation of many kinds. Yet legislation seemed powerless to reduce a hard core of black poverty and resentment. It broke out in riot and arson in what came to be called 'ghetto' areas of the great American cities in the late 1960s. Poverty, poor housing and bad schools in run-down urban areas were symptoms of deep dislocations inside American society and produced inequalities made more irksome by the increasing affluence in which they were set. Even greater emphasis was given to their removal by Lyndon Johnson, who succeeded to the presidency when Kennedy was murdered in 1963. He was a convinced and convincing exponent of the 'Great Society' in which he discerned America's future. Potentially one of America's great reforming presidents, Johnson nevertheless experienced tragic failure because his presidency came to be overshadowed by a disastrous Asian war.

American policy in Asia

American policy in South-East Asia assumed that Indo-China was essential to Cold War security. South Vietnam had to be kept in the western camp if other nations – perhaps as far away

War in Vietnam: a Vietnamese woman pleads with an American soldier as her home blazes behind her.

as India and Australia – were not to be subverted. President Kennedy had begun to back up American military aid with 'advisers'; there were 23,000 in South Vietnam when he died, many of them in action in the field. President Johnson followed suit, believing that American pledges had to be shown to be sound currency. But government after government in

Saigon turned out to be broken reeds. At the beginning of 1965 Johnson was advised that South Vietnam might collapse without additional American help; soon after, the first American combat units were officially sent there. American participation now soared out of control. By Christmas 1968 a heavier tonnage of bombs had been dropped on North Vietnam than had fallen on Germany and Japan together in the entire Second World War. Over 500,000 American servicemen were then serving in the South.

The outcome was calamitous. The American balance of payments was wrecked by the war's huge costs, which also took money from badly-needed reform projects at home. A bitter domestic outcry arose as casualties mounted and attempts to negotiate seemed to get nowhere. Rancour grew, and with it the alarm of moderate America. The outraged were not just the young people rioting in protest and distrust of their government, or the angry conservatives appalled by ritual desecrations of the symbols of patriotism and evasions of military service. Vietnam changed the way many other Americans looked at the outside world. It was at last borne in on the thoughtful among them that even the United States could not obtain every result she wanted, far less obtain it at any reasonable cost. It was the sunset of the illusion that American power was limitless and irresistible. In March 1968 Johnson, who had already drawn the conclusion that the United States could not win, restricted the bombing campaign and asked the North to open negotiations. He also dramatically announced that he would not stand for re-election.

Only four years after a huge Democratic majority had won Johnson re-election as president, a Republican president, Richard Nixon, was elected. In 1969 he soon began to withdraw American ground forces from Vietnam. In 1970 he opened secret negotiations with North Vietnam, though renewing and intensifying bombing of the north. The United States could not admit it was abandoning its ally, but had in fact to do so. After difficult negotiations, a cease-fire was signed in Paris in January 1973.

Vietnam had cost the United States vast sums and 57,000 dead, gravely damaged American prestige, eroded American diplomatic influence, ravaged domestic politics and had frustrated reform. It had perverted the economy. It had only briefly preserved a shaky South Vietnam, in spite of inflicting terrible suffering on the peoples of Indo-China. President Nixon reaped the benefit of relief at home. His recognition of how much the world had already changed since Cuba was shown by unprecedented efforts to establish normal relations with China. In February 1972 he visited that country to try to bridge what he described as '16,000 miles and twenty-two years of hostility' (he might have added 'and 2500 years of history') and so became the first American president to visit mainland Asia. A few months later he was the first to visit Moscow, a step followed by the first agreement on arms limitation. The stark polarized simplicities of the Cold War were clearly gone. The Vietnam settlement came next and, almost at once, the disappearance of the South in civil war. There was too much relief in the United States over getting out of the morass, though, for nice scruples over the observation of the peace terms by the North Vietnamese.

When a political scandal forced Nixon's resignation, his successor faced a Congress suspicious of any further foreign adventures and determined to thwart them. No attempt was made to uphold the guarantees given to the South Vietnamese régime and by the spring of 1975 all American aid to Saigon had come to an end. As in China in 1947, the United States cut its losses at the expense of those who had relied on her (though 117,000 Vietnamese left with the Americans). Perhaps the hardliners on Asian policy had been right all along – only the knowledge that the United States was in the last resort prepared to fight for them could guarantee the post-colonial régimes' resistance to communism. Yet better relations with China mattered more than the loss of Vietnam.

By the end of the 1970s America and her allies were confused and worried. The situation was not easy to read. Many Americans were troubled by what they saw as their country's military weakness (essentially, in missiles). The traditional leadership of the president in foreign

The opening of a new chapter of history, President and Mrs Nixon visit the Great Wall near Peking in 1972.

affairs had been undermined by a new distrust of the executive power. When Cambodia collapsed, and South Vietnam quickly followed, questions began to be asked about how far a retreat of American power might go. If the United States would no longer fight over Indo-China, would she, then, do so over Thailand? Or over Israel – or even Berlin?

NEW DANGERS

In 1979, the Shah of Iran, long a trusted ally of the United States, was driven from his throne – an almost wholly unforeseen event and a damaging blow to American policy. It also threatened, paradoxically, the stability of the volatile Islamic world. His supplanters were a coalition of outraged liberals and Islamic conservatives, and the second soon swept aside the first. Iranian traditions and society had been dislocated by the Shah's modernization in which he had followed his (more cautious) father. A speedy reversion to archaic tradition (strikingly, in the treatment of women) quickly showed that Iran had repudiated more than a ruler. The new regime emerged as a Shi'ite Islamic Republic, led by an elderly and fanatical cleric. He and his followers abhorred the Americans as patrons of the former Shah, myrmidons of capitalist materialism, but Soviet communism was soon undergoing similar vilification, as a second 'Satan' threatening the purity of Islam.

Islamic resurgence

The new regime voiced a rage shared by many Moslems world-wide. It arose from fears of secular westernization and disappointment over the failed promises of modernization. In the Middle East, as nowhere else, nationalism, socialism and capitalism had all failed to solve the region's problems – or at least to satisfy passions and appetites they had aroused. Instead, they had inflamed them. Many Moslems thought that the modernizers – even Nasser – had led their peoples down the wrong road. Millions felt the contagion of the West to be threatening.

The roots of such feelings were varied and deep, and had been nourished by centuries of conflict with Christianity. They had been refreshed from the 1960s onwards by the growing difficulties of western powers (and, for that matter, the USSR) in the Middle East and Persian Gulf, thanks to the Cold War. A favourable conjunction of superpower embarrassments seemed promising, given the oil factor. But there was also mounting evidence that western commerce, communications and the simple temptations of the West offered to those rich with oil, might be more dangerous to Islam than earlier, more political and military, threats had been. When Moslem Arabs sought to learn from western technology or to acquire academic instruction, for example, they were open to the danger of seduction by western values. For this reason, the Ba'ath socialist movement which attracted many radically-inclined Arabs (and was entrenched in Iraq and Syria by 1970) was anathema to the Moslem Brotherhood, which deplored its 'godlessness' even in the Palestinian quarrel. Islamic fundamentalists

Rioting in Teheran and many other places in Iran in 1978 led to the overthrow of a government, the imposition of martial law in all major cities, and promises to end corruption – all to no avail. The religious leaders of the mobs remained determined to overthrow the Shah.

rejected popular sovereignty, too; they sought Islamic control of society in all its aspects. Before long the world began to hear that Pakistan forbade mixed hockey, that Saudi Arabia punished crime by stoning to death and amputation of limbs, that Oman was building a university in which men and women students were segregated during lectures – and much more. By 1978 even in comparatively 'westernized' Egypt students were voting for fundamentalists in their own elections, while in medical school girls were refusing to dissect male corpses and demanding a segregated, dual system of instruction.

It was (and remains) very difficult to assess this phenomenon. Because it provided a focus for widely-shared emotions, it looked in 1980 as if the Iranian revolution had transformed the rules of the game played hitherto in the Middle East. Yet Islamic resurgence was in some degree only one of those recurrent waves of puritanism which have from time to time re-enthused the Faithful down the centuries. It owed much to circumstance; Israeli occupation of Jerusalem, for instance, site of the third of Islam's Holy Places, greatly enhanced the sense of Islamic solidarity. Yet, seizing on what seemed to be the weakness of a country in revolution, Sunnite (and nominally Ba'athist) Iraq launched an attack on Shi'ite Iran in 1980 which was to lead to eight years of bloody war, the loss of a million lives and, once more, the division of Islamic peoples on ancient lines. What is more, although her revolution could irritate and alarm the superpowers, Iran could not thwart them. At the end of 1979, she had been forced to watch helplessly while a Russian army went to Afghanistan to support a puppet communist regime there. Nor, in spite of seizing American hostages (and exacting a ransom for them after the failure of an American attempt at a rescue by *coup de main*), were the Iranians able to bring about the return of the former Shah to Islamic justice.

A declaration by President Carter in 1980 that the United States regarded the Persian Gulf as an area of vital interest was an important sign. A superpower could not ignore the threat the region's instability posed to international order. In the unhappy Lebanon, ordered government virtually disappeared in the 1980s and the country collapsed into anarchy. This at

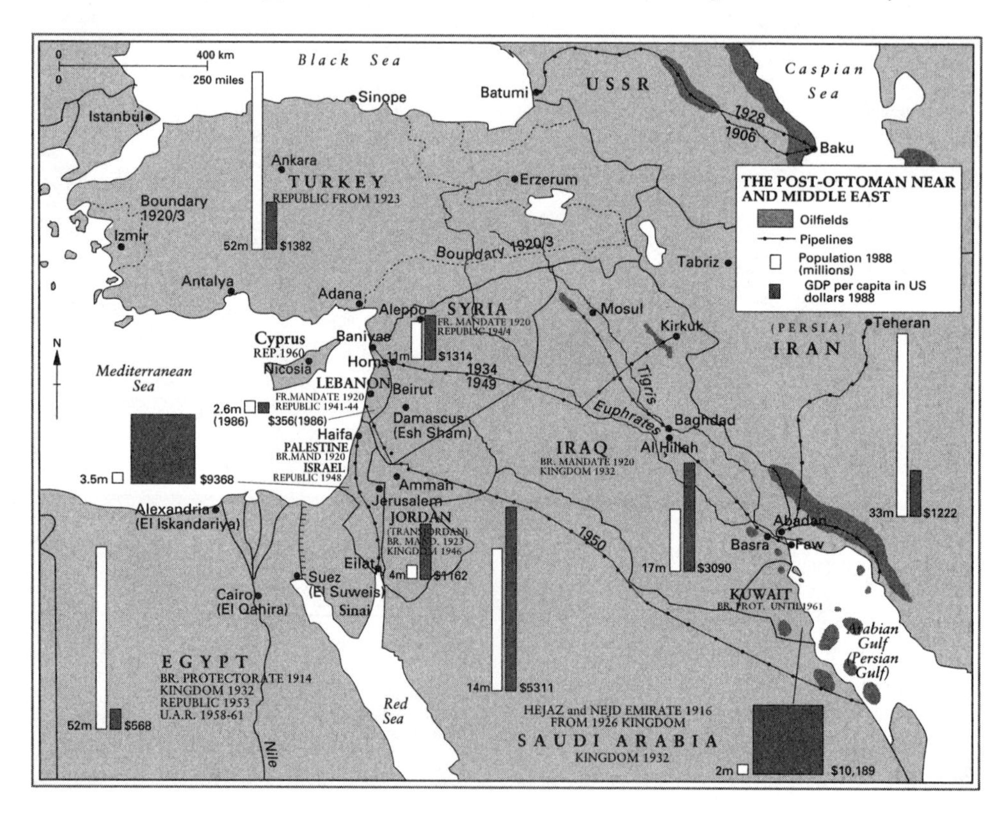

first gave the PLO an even more promising base for use against Israel than in the past. Israel, therefore, took to operating with increasing violence on and beyond her northern borders. In turn, there followed a heightening of tension within Israel where the decade brought increasing Jewish-Palestinian conflict, violence and, in the end, insurrection in the areas of predominantly Palestinian settlement.

The United States was not alone in being troubled by such turbulence. When Soviet soldiers were sent to Afghanistan (where they were to stay for nearly ten years) the anger of many Moslems was bound to affect events inside the USSR, with its huge Islamic population. Some thought Islamic extremism might induce caution on the part of the two superpowers. Fundamentalists murdered the president of Egypt in 1981 because he had made peace with Israel two years before. The government of Pakistan continued to impose Islamic orthodoxy, and winked at assistance to the anti-communist Moslem rebels in Afghanistan. In North Africa evidence of radical Islamic aspiration could be seen in the frequent bizarre sallies and pronouncements of the dictator of Libya (he called upon other oil-producing states to stop supplying the United States while one third of Libyan production continued to find a market there, and in 1980 briefly united his country to Ba'athist Syria) as well as further west. Algeria's promising start to independence faltered as emigration to Europe seemed the only economic outlet available to many of her young men. For the first time in any Arab country, in 1990 the Algerian Islamic fundamentalist party won a majority of votes in elections. In the previous year, a coup had brought a military and militant Islamic regime to power in the Sudan and it at once suppressed the few remaining civic freedoms there. Yet by then there was also plentiful evidence that the tide was not running only one way. Because of preoccupations and changing circumstances elsewhere, the old Soviet-American rivalry was decreasingly available for exploitation. Worse still, two potentially rich Islamic countries, Iraq and Iran, were for most of the 1980s fatally entangled in a costly struggle with one another.

Iraq

The ruler of Iraq, Saddam Hussein, was a Moslem by upbringing, but led a formally secular (Ba'athist) regime actually based on patronage, family and the self-interest of soldiers. He sought power, and technological modernization as a way to it. When he went to war with Iran, the prolongation of the struggle and evidence of its costs were greeted with relief by traditional Arab rulers; it appeared to them to pin down both a dangerous bandit and the Iranian revolutionaries whom they feared at the same time. They were less pleased, though, that the war distracted attention from the Palestinian question.

The alarms and excursions of the 1980s in the Gulf raised from time to time the spectre of interference with oil supplies, and produced incidents which at times threatened open conflict between Iran and the United States. Meanwhile, the Levant festered. Israel's annexation of the Golan Heights, her vigorous operations in Lebanon against Palestinian guerilla bands and their patrons, and her government's encouragement of further Jewish immigration (notably from the USSR) had all helped to strengthen her against the day when she might once again have to confront the threat of united Arab armies. At the end of 1987, though, there came the first outbreaks among Palestinians in the Israeli-occupied territories which persisted and grew into insurrection, the intifada. Although the PLO won further international sympathy by officially recognizing Israel's right to exist, it was not well placed in 1989, when the Iraq-Iran war finally ended. In the following year the ruling Ayatollah in Iran died and there were signs that his successor wished to pursue a less adventurous and violent policy, though supporting both the Palestinian and the Islamic causes.

During the Iraq-Iran war, the United States had seen Iran as the greater danger. Yet when Americans found themselves at last face-to-face at war in the Gulf with a declared enemy, it was with Iraq. Following the peace with Iran, Saddam Hussein took up an old border dispute with the sheikdom of Kuwait, with whose ruler he quarrelled also over oil quotas and prices. Whatever the importance of these issues to him, what seems to have mattered most was his determination to seize the immense oil wealth of Kuwait. His threats increased until, on 2 August 1990, the armies of Iraq invaded Kuwait, and within six hours had subdued it.

There followed a remarkable mobilization of world opinion through the UN. By confusing the pursuit of his own predatory ambitions with Arab hatred for Israel, Saddam Hussein sought to play both the Islamic and the Pan-Arab cards, but they proved to be of very low

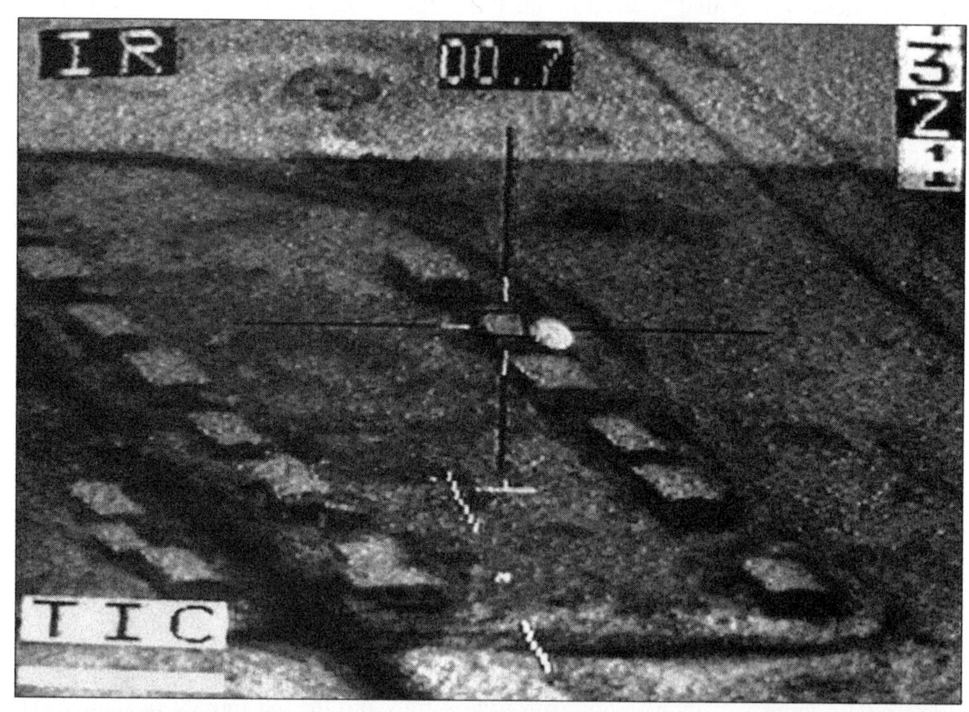

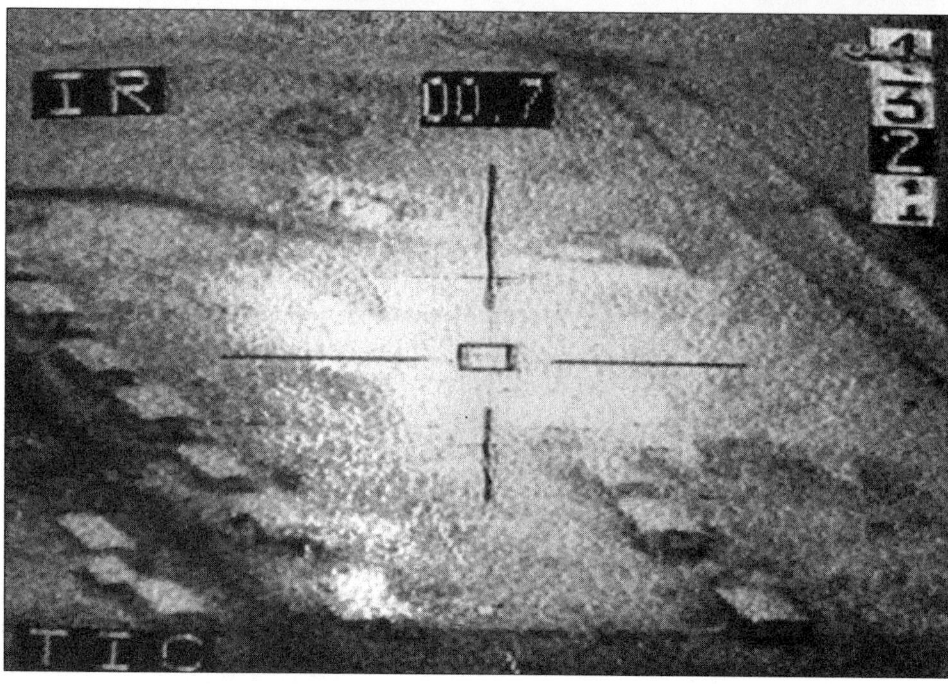

A new kind of war. Photographs of a laser guided missile fired at a munitions dump in Kuwait, January 1991.

value. Only the PLO and Jordan spoke up for him, and, to his no doubt pained surprise, Saudi Arabia, Syria and Egypt were improbable partners in the alliance which rapidly formed against him. Almost equally surprising to him must have been the acquiescence of the USSR in what followed. Finally, and most startlingly of all, the Security Council produced (with overwhelming majorities) a series of resolutions which condemned Iraq's actions and, finally, authorized the use of force to ensure the liberation of Kuwait. On 16 January 1991 American, British, French, Saudi and Egyptian forces went into action and within a month Iraq gave in.

Once again, though, a turning-point in the Middle East had not been reached. What had ended was one more Ottoman succession war. Yet some things had changed. In spite of Hussein's attempts to inspire an anti-Israel Islamic crusade, he had not found takers. The greatest loser had been the PLO; Israel was the real gainer. An Arab military success at her expense became inconceivable for the near future. Syria and Iran had shown signs before the Kuwait crisis that, for their own reasons, they intended to make attempts to get a negotiated settlement of the Israel problem. For the United States, it was clearly more of a priority than ever to do so and that encouraged hopes that Israel might, at last, show less intransigence. In 1991 talks began between the Israeli government and Arab states at which PLO representatives were present. Though these almost at once came to a halt, they were renewed after a change of government in Israel in 1992 and even survived a burst of Israeli heavy-handedness towards Palestinians whom it expelled from its territories.

By then, undoubtedly, the alarming spectre of radical and fundamentalist world Pan-Islamic movement had faded somewhat. For all the unrest and discontent inside Islamic countries, and continuing provocation from Iraq, there was little hope that these could be coordinated effectively against the West. There was, moreover, less sign than ever that the Islamic countries were willing to do without the subtly corrosive technical apparatus of modernization which the West offered. The crisis in the Gulf had also appeared to show (almost incidentally) that the oil weapon had lost much of its power to frighten, let alone damage, the developed world. Symbolically, within a year, all the Kuwait oil wells ignited by Hussein's retreating forces had been quenched. But the powder-keg remained in place. The prospects for the Middle East remain unpredictable as ever.

Latin America after the Cuban crisis

Once the missile crisis was over, the United States had promised not to invade Cuba, but tried to isolate it as much as possible from the rest of the hemisphere, fearing that its revolution would appeal to the young of other Latin American countries. Yet, though Castro strove to present Cuba as a revolutionary centre for the rest of the continent, Latin American revolution did not follow. Cuban circumstances had been very untypical. The hopes of peasant rebellions elsewhere proved illusory and such potential as there was turned out to be urban rather than rural. The brutalities practised in dealing with terrorists, it is true, alienated middle-class support from authoritarian governments in some countries and anti-Americanism continued to run high. A new American initiative, based on social reform, an 'Alliance for Progress', made no headway. Worse still, it was overtaken in 1965 by fresh evidence of the old Adam of intervention, this time in the Dominican Republic, where, four years before, American help had assisted the overthrow of a dictatorial régime. When its successor was pushed aside by soldiers acting in defence of the privileged who felt threatened by reform, the Americans cut off aid; it looked as if, after all, the Alliance for Progress might be used discriminately. But aid was restored – as it was to other rightwing régimes – when President Johnson took office. A rebellion against this military régime in 1965 resulted in the arrival of 20,000 American troops to put it down.

By 1970 the Alliance had virtually been forgotten and it looked as if Latin American nationalism was entering a new and vigorous period. If Cuba-inspired guerrillas had ever

1973 in Chile. An early episode in the coup overthrowing the left-wing government of President Allende (second from right).

presented a danger, they appeared to do so no longer. Once the spur of an internal fear was gone there was little reason for governments not to try to capitalize on anti-American feeling. Chile nationalized the largest American copper company, the Bolivians took over oil concerns and the Peruvians American-owned plantations. A tour by a representative of the President of the United States that year led to protest, riots, the blowing up of American property and requests to stay away from some countries.

Meanwhile, Latin America's real problems were still not being met. The 1970s and 1980s revealed chronic economic troubles and, by 1985, it was reasonable to speak of an apparently insoluble crisis. There were several sources for this. For all its rapid industrialization, the continent faced appalling population growth. The hundred or so million Latin Americans and Caribbean islanders of 1950 are expected to become 500 millions by the year 2000. The problem began to be obvious just as the difficulties of the Latin American economies were again beginning to show their intractability. The aid programme of the Alliance for Progress patently failed to cope with them and failure spawned quarrels over the use of American funds. Social divisions remained menacing. Even the most advanced Latin American countries displayed vast discrepancies of wealth and education. Constitutional and democratic process, where they existed, seemed increasingly impotent to confront such problems. In the 1960s and 1970s, Peru, Bolivia, Brazil, Argentina and Paraguay all underwent prolonged authoritarian rule by military regimes, some of which undoubtedly believed quite sincerely that only authoritarianism could bring about needed changes of which civilian government had proved incapable.

The consequences were vividly brought to the notice of the world not only by reports of torture and violent repression from countries like Argentina, Brazil and Uruguay, all once regarded as civilized and constitutional states, but in Chile, a country with a more continuous history of constitutional democracy than other Latin American states, where a military coup in 1973 overthrew a government many Chileans believed to be under communist control. The counter-revolutionary movement had support from the United States, and many Chileans were willing to endorse it, frightened as they were by the revolutionary tendencies of the elected régime it displaced. The new order in the end rebuilt the economy and, in the late 1980s, even began to look as if it might be able to liberalize itself.

Meanwhile, the oil crisis of the 1970s had finally sent the foreign debt problems of the oil-importing countries of Latin America out of control. By 1990, most orthodox economic remedies had been tried, in one country or another, but had usually proved unworkable or unenforceable in dealing with runaway inflation, interest charges on extended debt, the distortion in resource allocation arising from bad government in the past, and simple administration and cultural inadequacy for the support of good fiscal policies. It remains impossible to guess how the complex and consequent economic crisis can be surmounted. While it is not, Latin America remains an explosive, disturbed continent of nations growing less and less like one another except in their distress. Most Latin Americans are now poorer, if per capita income is the measure, than ten years ago.

Africa

In much of Africa, as elsewhere, stabilization was still unachieved in 1992. In 1974 the General Assembly of the UN had forbidden South Africa to attend its sessions because of apartheid, and in 1977 the UN Commission of Human Rights deftly sidestepped demands for the investigation of horrors perpetrated by black against blacks in Uganda, while castigating South Africa (along with Israel and Chile) for its misdeeds. From Pretoria, the view northwards was coming to look more and more menacing. The arrival of Cuban troops in Angola showed a new power of strategic action against South Africa by the USSR through its clients and satellites. Both that former Portuguese colony and Mozambique also provided bases for South African dissidents who fanned unrest in the black townships and sustained urban terrorism in the 1980s.

Under pressure the position of the South African government changed. To the dismay of many Afrikaners, when a new prime minister took office in 1978, he began slowly to unroll a policy of concession; the question marks over South Africa's future seemed at last to move away from the possibility of abolishing apartheid to the question of terms on which black majority rule might be conceded. This initiative slowed before long, and growing distrust among his Afrikaner supporters led Mr P.W. Botha back towards repression, though he inaugurated a new constitution in 1983; while outraging black political leaders by its inadequacy, it disgusted white conservatives by conceding the principle of non-white representation. Meanwhile, the pressure of economic sanctions against South Africa by other countries was growing. In 1985 even the United States imposed them to a limited extent; as international confidence in the South African economy fell the effects showed at home. Straws before the wind of change in domestic opinion could be discerned; the Dutch Reformed Church conceded that apartheid was at least a 'mistake' and could not (as had been claimed) be justified by Scripture. Divisions grew among Afrikaner politicians. It probably helped, too, that in spite of its deepening isolation, South African military action successfully mastered the border threats. In 1988 peace was made with Angola.

Against this background Mr Botha (president of the republic since 1984) reluctantly and grumpily stepped down in 1989 and Mr F.W. De Klerk succeeded him. He made it clear that the movement towards liberalization was to continue and would go much further than many then thought possible, even if this did not mean the end of apartheid in all respects. Political protest and opposition were allowed much more freedom, imprisoned black Nationalist leaders were released. In 1990 the symbolic figure of Mr Nelson Mandela, leader of the African National Congress, the main vehicle of black opposition, emerged at last from jail and was soon engaged in discussion with the government about what lay ahead. For all the intransigence of his language, there were hopeful signs of a new realism that the task of reassuring the white minority about a future under a black majority must be attempted. Just such signs, of course, prompted other politicians to greater impatience. By the end of 1990 Mr de

1990: a scene startling to many South Africans, black and white alike – Mr. Mandela meets the South African Prime Minister, Mr. F. W. de Klerk, and the Foreign Minister, Mr. R. ('Pik') Botha.

Klerk had even said he would rescind the land legislation which was the keystone of apartheid. It was an interesting indicator of the pace with which events had moved in South Africa that the interest of the worlds was less focused by then on the sincerity or insincerity of white South African leaders, than on the realism (or lack of it) of their black counterparts and their ability (or inability) to control their followers. The hopes surrounding Mr Mandela at the time of his release had given way to misgivings. There were plentiful signs of the division between his followers. It was clear that a stony path still lay ahead of South Africa.

THE EMERGENCE OF A NEW WORLD ORDER

China changes course

In China, although there was fluctuation in policy, a fairly consistent trend towards a relaxation in some sectors of the economy was evident from Mao's death in 1976. Within a few years qualification began to be noted in official statements about his achievements. The dominating figure in the ruling gerontocracy was now Deng Xiaoping, whose name was associated with economic liberalisation. Modernization, it can now be seen, was being slowly given precedence over socialism, though much of the Marxist rhetoric remained intact and there was no likelihood of concessions of political power by the CCP. In the 1980s there was evidence that Deng's policies were at last delivering a remarkable change in China's economic performance.

It was not achieved by reckless abandonment of control. China's leaders were determined to remain firmly in charge. Helped by the persistence of the old Chinese social disciplines, by the relief felt by millions that the Cultural Revolution had been left behind, and by the policy (contrary to that of Marxism as still expounded in Moscow until 1980) that economic rewards should flow through the system to the peasant, they won support. There was a major swing of power away from the rural communes set up in the 1950s, which in many places practically ceased to be relevant, and by 1985 the family farm was restored as the dominant form of rural production over much of China. At the same time, it was apparent to many Chinese that China now enjoyed new respect and status abroad; one striking, if paradoxical, sign was an official state visit by Queen Elizabeth II in 1985, which followed successful negotiations with the United Kingdom and Portugal for the eventual resumption of Chinese sovereignty over Hong Kong and Macao.

Within a few years, nevertheless, there were signs of difficulties. Foreign debt shot up and inflation, by the end of the decade, was running at an annual rate of about 30%. There was anger, too, over evidence of corruption, and divisions in the leadership were widely known to exist. Those believing that a re-assertion of political control was needed began to gain ground, and there were signs that they were manoeuvring to win over Deng Xiaoping. Western observers had been led by the policy of economic liberalization to take unrealistic and over-optimistic views about the possibility of economic relaxation being followed by political, and contemporary changes in eastern Europe (and even some signs in China itself) nourished such hopes. But the illusion suddenly crumbled.

In the early months of 1989, city-dwellers were feeling the pressures both of acute inflation and of austerity measures imposed to deal with it. This was the background to a new wave of student demands for political reform. Encouraged by the presence in the governing oligarchy of sympathizers with liberalization, they demanded that the Party and government should open a dialogue with a newly formed and unofficial Student Union about corruption and reform. Posters and rallies began to call for greater 'democracy'. The leadership was alarmed, refusing to recognize the Union which, it was feared, might be the harbinger of a new Red Guards movement. There were demonstrations and as the seventieth anniversary of the May 4th Movement approached the student leaders invoked its memory so as to give a broad patriotic colour to their campaign. They were not able to arouse much support in the countryside, or in the southern cities. Encouraged nevertheless by obvious sympathy high in the Party, they began a mass hunger strike which won widespread popular sympathy and support in Peking.

The most senior members of the government, including Deng Xiaoping, seem to have been thoroughly alarmed: they believed China faced a major crisis. Some feared a new Cultural Revolution (Deng Xiaoping's own son, they could have remarked, was a cripple as a

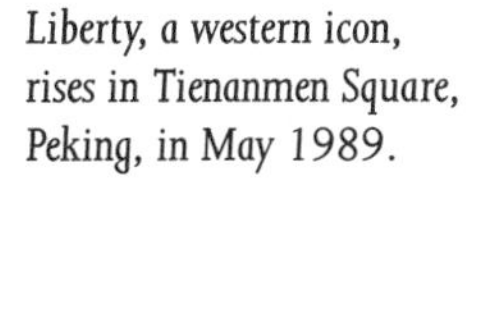

Liberty, a western icon, rises in Tienanmen Square, Peking, in May 1989.

result of the injuries inflicted on him by Red Guards during that outburst). Others were alarmed by contemporary events in the USSR. On 20 May martial law was declared. After a moment's hesitation repression was ruthless. The Student Union leaders were encamped in Peking in Tiananmen Square, where, thirty years before, Mao had proclaimed the foundation of the People's Republic. From one of the gates of the old Forbidden City a huge portrait of him looked down on the symbol of the protesters: a towering plaster figure of a 'Goddess of Democracy', deliberately evocative of the New York Statue of Liberty. On 2 June the first military units entered the suburbs of Peking, brushing aside resistance and barricades. Two days later, they dispersed the students and a few sympathizers with rifle-fire, teargas, and a brutal crushing of the encampment under the treads of tanks which swept into the square. Killing and mass arrests (perhaps as many as ten thousand in all) went on for some days. Much of what happened took place before the eyes of the world, thanks to the presence of film-crews in Peking which had for days familiarized TV audiences with the demonstrators' encampment. Foreign disapproval was almost universal.

As so often in China, the true meaning of this episode is still not clear. Obviously her rulers had felt they faced a grave threat. Yet the rural masses did not sympathize with the protesters. Changes in the ruling hierarchy and vigorous attempts to impose political orthodoxy followed. Economic liberalization was reined in. Neo-Marxist slogans were heard again. China, it was clear, was at least not going to go the way of eastern Europe or the USSR, whose demise was the greatest single indicator that an era was ending.

The end of the Cold War

In 1980, the year of a presidential election, Mr Reagan, the Republican candidate, played on Americans' fears of the Soviet Union. When he took office as president he inherited an enormous budgetary deficit (which he was to increase), disappointment over what looked like recent successful initiatives by the Russians in Africa and Afghanistan, and dismay over what was believed to be the reversal of the nuclear arms imbalance of the 1960s. In the next five years he restored the morale of his countrymen by remarkable (and often cosmetic) feats of leadership. Symbolically, on the day of his inauguration, the Iranians released their American hostages, the close of a humiliating and frustrating episode. He also ended the Cold War.

Mr Reagan's election had re-awoken animosity and suspicion among the conservative leaders of the Soviet Union. It seemed likely that promising steps towards disarmament might be swept aside – or even worse. In the event, the American administration showed a remarkable pragmatism in foreign affairs, while, on the Soviet side, internal change opened the way to greater flexibility. In November 1982 Leonid Brezhnev, for eighteen years general secretary of the Party, died, to be followed by two brief-lived successors before there came to the office in 1985 the fifty-four-year-old youngest member of the Politburo, Mr Mikhail Gorbachev. Virtually the whole of Mr Gorbachev's political experience had been of the post-Stalin era. His impact upon his country's, and the world's, history has still to be properly assessed (as has his personal motivation and the conjunction of forces which propelled him to the succession) but it was immense. His acts and speeches soon showed a new approach. For want of a better term, his course was seen as liberalization, which was an inadequate western attempt to sum up two Russian words he used a great deal: *glasnost* (openness) and *perestroika* (restructuring). The implications of this were profound and dramatic. A Soviet leader was at last recognizing that the economy could no longer sustain its former military might and its commitments to allies abroad, improve (however slowly) living standards at home, and assure self-generated technological innovation. What was more, Americans had been promised wonders by their government in the shape of a new scheme of defensive measures in outer space. Though thousands of scientists said the project was unrealistic, the Soviet government could not face the costs of competing with that. The USSR had again to be modernized – and the implications were vast.

In 1986 President Reagan and Chairman Mikhail Gorbachev held a 'summit' meeting in Iceland. There were no formal results but the meeting was important in establishing personal relations between the two men.

Mr Gorbachev's new course soon surfaced in meetings with Mr Reagan of which the most important was in Iceland in 1986. Discussion of arms reduction was renewed and agreements were reached on other issues – notably, for the withdrawal of Soviet forces from Afghanistan, where they were bogged down in a guerrilla war (they left in 1989). In America's domestic politics, a huge budgetary deficit, and a flagging economy were virtually lost to sight in the euphoria produced by an increasingly rapid transformation of the international scene. Optimism grew as the USSR began to display signs of growing division and difficulty in reforming its affairs. Though Mr Reagan could not convince many of his countrymen that more enthusiastic assertions of America's concern in Central America were truly in their interests, he ended his presidency as a remarkably popular figure; it was only after he had left office that it began to dawn that for most Americans the decade had been one in which they had got poorer. Mr Reagan had only allowed rich Americans to get richer.

In 1987, the fruits of negotiation on arms control were gathered in an agreement over intermediate range missiles. Others were to follow. In spite of so many shocks and its erosion by the emergence of new foci of power, the nuclear balance had held after all. The superpowers had shown they could manage their conflicts and the world's crises without all-out war. They at least, if not other countries seeking to acquire nuclear weapons, appeared to have recognized that nuclear war, if it came, held out the prospect of virtual extinction for mankind. In 1991 both agreed to further huge reductions in existing weapons stocks.

Revolution in eastern Europe

This huge change in the international scene had vast consequences. At the end of 1980 Soviet power seemed in 1980 to hold eastern Europe as firmly as ever in its grip, though this concealed social and political change which had long been going on within the Warsaw pact countries. At first sight, they showed a remarkable uniformity. In each of them, the Party was supreme; careerists built their lives round it as, in earlier centuries, men on the make clustered about courts and patrons. In each (and above all in the USSR itself), there was also an unspeakable and unexaminable past which could not be mourned or deplored, whose weight overhung and corrupted intellectual life. So far as the east European economies were concerned, investment in heavy industrial and capital goods had provided first a surge of early growth (in some of them more vigorous than in others) and then an international strait-jacket of trading arrangements with other communist countries, dominated by the USSR and rigidified by aspirations to central planning. They were increasingly and obviously unable to meet their people's desire for commodities taken for granted in western Europe. Agricultural output everywhere remained low; in most east European countries agricultural yields remained only from half to three-quarters those in the west. By the 1980s all of them, in varying degree, were in a state of economic crisis with the possible exception only of the GDR (whose per capita GDP stood at $9300 a year in 1988, against $19500 in the Federal Republic).

Brezhnev had insisted that developments within eastern bloc countries might require – as in Czechoslovakia in 1968 – direct intervention to safeguard Soviet interests. Perhaps this was realism, a recognition of the possible dangers presented to international stability by disturbance and dissent in communist Europe. Certainly there was likely to be little danger to peace from internal change in western countries, steadily growing more prosperous, and with memories of the late 1940s and the seeming possibility of subversion far behind them. By 1980, after revolutionary changes in Spain and Portugal, not a dictatorship survived west of the Trieste-Stettin line and democracy was everywhere triumphant. For thirty years, the only risings by industrial workers against their political masters had been in East Germany, Hungary, Poland and Czechoslovakia – all communist countries (conspicuously, when Paris was in uproar in 1968, and student riots destroyed the prestige of the government, the Parisian working class had done nothing). As awareness of the contrast with the West grew in the eastern block, dissident groups emerged, survived and even strengthened their posi-

A dramatic confrontation in Prague during the Russian invasion in 1968.

tions, in spite of severe repression. The Helsinki agreement of 1975 helped, and so did radio and television broadcasting from western Germany. Gradually, too, a few officials or economic specialists, and even some party-members, began to show signs of scepticism about centralized planning. The key to stability in the east, nevertheless, remained the Soviet Army. There was no reason to believe that fundamental change was possible in any of the Warsaw pact countries so long as it was there to uphold governments subservient to the USSR.

The Polish lead

The first clear sign that this might not always be the case came in the early 1980s, in Poland. Poles often looked to the Roman Catholic church to speak for them – all the more confidently after a Polish pope was enthroned in 1978. It had done so on behalf of workers who protested in the 1970s against economic policy, and condemned their treatment by the authorities. Then, in 1980, a year of crisis for Poland, a series of strikes came to a head in an epic struggle in the Gdansk shipyard from which emerged a new and spontaneously organized trade union, 'Solidarity'. It added political demands to the economic goals of the strikers, among them one for free and independent trades unions. Solidarity's leader was a remarkable, often-imprisoned, electrical union leader, Lech Walesa, a devout Catholic, closely in touch with the Polish hierarchy. The shipyard gates were decorated with a picture of the Pope and open-air masses were held by the strikers.

The shaken Polish government made an historic concession, by recognizing Solidarity as an independent, self-governing trade union. But disorder did not cease, and with the winter, the atmosphere of crisis deepened. Threats were heard of possible intervention; but the Soviet army did not move. It was one of the first signs of change in Moscow which was the premise of everything which was to follow but tension continued to rise. On five occasions the Russian commander of the Warsaw Pact forces came to Warsaw. On the last, the Solidarity radicals broke away from Walesa's control and called for a general strike. On 13 December, martial law was declared. The fierce repression which followed may have helped to show Soviet invasion was not needed. Solidarity went underground. There began seven years of further economic deterioration, clandestine organization and publication, strikes and demonstrations, and continuing ecclesiastical condemnation of the regime. After 1985, though, the change in Moscow began to produce effects in other Warsaw pact countries. The climax came in 1989.

That year opened with the Polish government's acceptance that other political parties and organizations, including Solidarity, had to share in the political process. Elections were held in June in which some seats were freely contested; in them Solidarity swept the board. Soon the new parliament denounced the German-Russian agreement of August 1939, condemned the 1968 invasion of Czechoslovakia, and set up investigations into political murders committed since 1981. In August Walesa announced that Solidarity would support a coalition government. Mr Gorbachev made a crucial statement that this would be justifiable. By then some Soviet military units had already left the country. In September a coalition dominated by Solidarity and led by the first non-communist Prime Minister since 1945 took office as the government of Poland. Western economic aid was soon promised. By Christmas the Polish People's Republic had passed from history and, once again, the historic Republic of Poland had risen from its grave.

Poland led eastern Europe to freedom. In other communist countries leaders were much alarmed by events there. All eastern Europe, too, had been exposed in varying degrees to a steadily increasing flow of information about non-communist countries, above all, through television (it was especially marked in the GDR), greater freedom of movement, and more access to foreign books and newspapers after Helsinki. A change in consciousness had thus already been under way when Mr Gorbachev had come to power. It was soon to become clear that he had released revolutionary institutional change in the Soviet Union. First, power was

1980, Lech Walesa addresses his supporters in the Lenin Shipyard, Gdansk.

taken from the party. The opportunities so provided were then seized by newly emerging opposition forces, above all in republics of the Union which began to claim greater or lesser degrees of autonomy. But, the economic out-turn was appalling. It became clear that a transition to a market economy, whether slow or rapid, was likely to impose far greater hardship on many – perhaps most – Soviet citizens than had been envisaged. By 1989 the Soviet economy was clearly out of control and running down. Always before in Russian history, modernization had been launched from the centre and flowed out to the centre through authoritarian structures. But that was precisely what could not now happen, because of the resistance of the bureaucrats of the command economy.

The end of the communist order

As more information became available to the rest of the world about the true state of an increasingly accessible Soviet Union; another, rough-and-ready judgements could be made. One was that the discrediting of the party and ruling class was profound and that economic failure hung like a cloud over the liberalizing of political processes. Soviet citizens began to talk of the possibility of civil war. The thawing of the iron grip of the past had revealed the power of nationalist and regional sentiment when excited by economic collapse and opportunity. After seventy years of efforts to make Soviet citizens, the USSR was suddenly exposed as a collection of peoples as distinct as ever. Some of them (above all in the three Baltic republics of Latvia, Estonia and Lithuania) were quick to show dissatisfaction with their lot. Azerbaijan and Soviet Armenia posed problems complicated by the shadow of Islamic unrest which hung over the whole Union. To make matters worse, some believed there was a danger of a military coup.

Signs of disintegration multiplied, as Mr Gorbachev just succeeded in holding his place and, indeed, in obtaining formal enhancements of his nominal power (though this had the disadvantage of leaving responsibility for failure in his hands, too). In March 1990, the

Lithuanian parliament re-asserted Lithuania's independence. Complicated negotiations avoided the armed suppression of the revived republic by Soviet forces and Latvia and Estonia followed suit on slightly different terms. Mr Gorbachev accepted undertakings that the three republics should guarantee the continued existence of certain practical services to the USSR, but by the end of the year even this was already looking out of date and unworkable. Parliaments in nine other republics had already by then either declared themselves sovereign or had asserted a substantial degree of independence of the USSR. Some of them had made local languages official; and some had transferred Soviet ministries and economic agencies to local control. The Russian republic – the most important – set out to run its own economy separately from that of the Union. The Ukrainian republic proposed to set up its own army; in 1991 it claimed control of all Soviet forces on its territory – and of their nuclear arms. The world looked on, bemused and uneasy.

The realization that an increasingly divided and paralysed USSR would not (perhaps could not) intervene to uphold its creatures in the communist bureaucracies had quickly been grasped in the other Warsaw Pact countries. The Hungarians had moved almost as rapidly in economic liberalization as the Poles, even before overt political change, but their most important contribution to the dissolution of communist Europe came in August 1989, when they allowed Germans from the GDR to enter Hungary freely as a way to the west. A complete opening of Hungary's frontiers came in September (Czechoslovakia followed suit) and a flow became a flood. The Russians remarked that this was 'unusual'; for the GDR it was the beginning of the end. On the eve of the carefully-planned and much-vaunted celebration of forty years' 'success' as a socialist country, and during a visit by Mr Gorbachev (who, to the dismay of German communists, appeared to urge the east Germans to seize their chance), riot police battled with anti-government demonstrators on the streets of Berlin. November opened with more demonstrations in many cities against a regime whose corruption was becoming evident and on November 9 came the greatest symbolic act, the breaching of the Berlin Wall. The East German Politburo caved in and the demolition of the rest of the Wall followed.

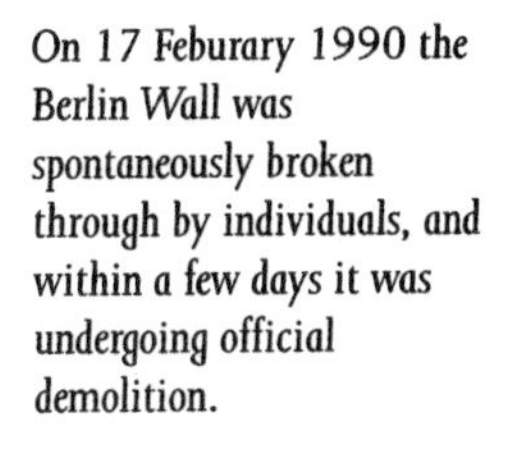

On 17 Feburary 1990 the Berlin Wall was spontaneously broken through by individuals, and within a few days it was undergoing official demolition.

THE SOVIET UNION AND ITS SUCCESSORS
Territory acquired 1939–45
Border disputes with China and Mongolia after 1945
Independent republics emerging 1990–91
N
Arctic Ocean
0
1000 km
0
600 miles
Bering Sea
Sea of Okhotsk
Sea of Japan
Khabarovsk
Khabarovsk
RUSSIAN FEDERATION
MONGOLIA
CHINA
PART OF RUSSIAN FEDERATION
Baltic Sea
Tallinn
ESTONIA
Riga
LATVIA
Vilnius
LITHUANIA
Minsk
BELARUS
Moscow
Kiev
Kishinev
MOLDOVA
UKRAINE
Black Sea
GEORGIA
Tbilisi
ARMENIA
Yerevan
Baku
AZERBAIJAN
Caspian Sea
Aral Sea
KAZAKHSTAN
UZBEKISTAN
Ashkhabad
Tashkent
Alma-Ata
Bishkek
KYRGYZSTAN
TURKMENISTAN
Dushanbe
TAJIKISTAN

All over eastern Europe, communist governments were now shown to have no legitimacy in the eyes of their subjects. The outcome was everywhere a demand for free elections. The Poles gave themselves a new constitution: in 1990, Lech Walesa became President. Hungary had a little before this elected a parliament from which emerged a non-communist government. Soviet soldiers began to withdraw from that country. In June 1990, Czechoslovakian elections produced a free government and it was soon agreed that Soviet forces were to leave by May 1991. Bulgaria was less decisive: there, communist party members turned reformers won a majority. Romania underwent a violent revolution (ending in the killing of its former communist dictator) after a rising in December 1989 which revealed internal divisions foreshadowing further strife.

Change in Germany was the most important of all. The breaching of the Wall revealed not only no political will to support communism, but no will to support the state either. Unification suddenly came in sight. A general election there in March 1990 gave a majority to a coalition dominated by Christian Democrats – the ruling party of the western German Federal Republic. Unity was no longer in doubt; only the procedure and timetable remained to be settled. In July the two Germanies joined in a monetary, economic and social union. In October the former GDR became part of the Federal Republic. Surprisingly, no serious alarm was openly expressed over this momentous change, even in Moscow, yet alarm there must have been. The new Germany would be the greatest European power west of Russia, whose power was now in eclipse as it had not been since 1918. Mr Gorbachev's reward was a treaty with the new Germany promising economic help with Soviet modernization. It might also be said, by way of reassurance to those who remembered 1941–5, that the new German state was not just an older Germany revived, but one shorn of the old east German lands, which it had formally renounced and one not dominated by Prussia as both Bismarck's Reich and the Weimar republic had been. More reassuring still (and of importance to west Europeans who felt misgivings) the Federal Republic was a federal and constitutional state, economically successful, with nearly forty year's experience of successful democratic politics to build on, and embedded in the structures of a democratic western Europe.

Elsewhere, the scene was less reassuring. Some observers soon noted the virulent emergence of new – or re-emergence of old – national and communal division to plague the new eastern Europe. Czechs and Slovaks drew apart; Bulgarians brooded over their Turkish fellow-citizens; Hungarians and Romanians over Transylvania. Above all, over the whole area there lay thickening storm-clouds of economic failure. Liberation might have come, but it was soon clear that it had come to peoples and societies of very different levels of sophistication and development, and with very different historical origins. In 1991 a terrible jolt was given to optimism over the prospects of peaceful change when two of the constituent republics of Yugoslavia announced their decision to separate from the federal state. Their decision was shaped by ancient national animosities held in check in the communist republic. In August, sporadic fighting began between Serbs and Croats. Precedents for intervention by outsiders in civil struggles did not seem promising – different views were held by different countries. This was more evident still as Bosnian Moslems became the target of Serb attacks. In 1992 the United Nations agreed to send forces from member states to ensure the delivery of relief supplies and the savagery of ancient hatreds became more and more evident on the spot. At what point military force might be employed to control them remained uncertain.

THE END OF THE USSR

Meanwhile, in August 1991 an attempt had been made to set aside the Gorbachev regime by force. Though it failed, Soviet politics had taken a lurch forward toward disintegration. The circumstances of the attempted coup had given Mr Boris Yeltsin the leader of the Russian republic, the largest in the Union, an opportunity to appear as a strong man of the Soviet scene without whose concurrence nothing could now be done. The army, the only conceivable threat to his supporters, did not move against him. While the world waited for clarification, the purging of those who had supported or acquiesced in the coup was developed into a determined replacement of Union officialdom at all levels, the redefinition of roles for the KGB and a redistribution of control over it between the Union and the republics. The demolition of the Communist Party of the Soviet Union also began almost at once. Relatively without bloodshed, at least to begin with, the huge creation which had grown out of the Bolshevik victory of 1917 was coming to an end. There were good grounds for rejoicing over that. But it was far from clear that nothing but good would follow. When, on 31 December 1991 the Union flag over the Kremlin was hauled down, to be replaced by the Russian, and the USSR disappeared from history to be replaced by a new Commonwealth of Independent States, little could be discerned except the deepening and complicating economic and political problems faced in major degree by almost all of its republics.

WESTERN EUROPE

As the unity of structure given to eastern Europe by Soviet domination was dissolving, it seemed that western Europe might be approaching a climax in a process of integration which could be traced back to idealists who in 1945 had been convinced that only political unity could preserve the continent from future disaster. In many ways, Cold War favoured their disciples. The Marshall Plan and NATO were in retrospect to turn out to have been among the first practical encouragements of a new European unity.

The Marshall Plan had been followed by the setting-up of an Organization (at first of sixteen countries, but later expanded) of European Economic Co-operation in 1948. In the following year, a month after the signing of the NATO treaty, representatives of ten different European states met as the first 'Council of Europe'. But economics were more important. Customs Unions had already been created in 1948 between the 'Benelux' countries (Belgium, the Netherlands, and Luxembourg), and (in a different form) between France and Italy. There later emerged in 1951 from a French proposal a Coal and Steel Community which embraced France, Italy, the Benelux countries and, most significantly, West Germany. It was the first great step towards taking her into a new international structure.

Meanwhile, in both France and Italy the political weakness symptomized by their domestic communist parties subsided with economic recovery. By 1950, the danger that French and Italian democracy might suffer a fate like Czechoslovakia's had disappeared. Anti-communist opinion tended to coalesce in western continental Europe about parties whose integrating forces were either Roman Catholic politicians or social democrats well aware of the fate of their comrades in eastern Europe. Broadly speaking, these changes meant that (outside Spain and Portugal) continental western European governments of a moderate right-wing complexion pursued similar aims of economic recovery, welfare service provision, and western European integration in practical matters during the 1950s.

In that decade the major thrust towards European unity continued to be economic. The crucial step came in 1957: the European Economic Community (EEC) or 'Common Market', then came into being when France, Germany, Belgium, the Netherlands, Luxembourg and Italy joined in signing the Treaty of Rome. Some enthusiasts spoke of the reconsitution of Charlemagne's heritage. Countries which did not join the EEC set up their own, looser and more limited, European Free Trade Association (EFTA) two-and-a-half years later. By 1986, the six countries of the original EEC (by then it was usually called the EC – the word 'Economic', significantly, being dropped) had grown to twelve; while EFTA had lost all but four of its members to it. Five years later, what was left of EFTA was envisaging merging with the EC.

Western Europe's slow but accelerating movement towards political unity seemed to show at least that the era of western European international war, rooted in the beginnings of the national state system, was over. Tragically, though recognizing that fact, Great Britain's rulers had not seized at the outset the chance to join in giving it institutional expression; their successors were twice to be refused admission to the EEC before finally joining it. Meanwhile, the Community's interests had been steadily cemented together by a Common Agricultural Policy – to all intents and purposes, a huge bribe to the farmers and peasants who were so important a part of the German and French electorates, and, later, to those of poorer countries as they became members.

Though new institutions continued to emerge and, in 1991, a British government went further than ever before in identifying the country with the cause of European integration, the economic climate had by then turned bleak. Germany, the richest community member, was by 1992 finding the economic burdens of reunification far heavier than expected. To make matters worse, the addition of new and poorer countries to the EC in the 1980s and the awareness of other applications for membership on the way from eastern Europe caused alarm. Worst of all, it became clear, suddenly, that the Community was virtually an unreality so far as hopes of common foreign policy action were envisaged. This, ironically, was because of the events in the former Yugoslavia, where the interplay of national hatreds after the collapse of the former Communist federal state let loose bloodshed on a scale which threatened to spill across boundaries and, some thought, threaten international peace. As in 1914, though for very different reasons, the name of Sarajevo became ominously well-known. Like

History re-asserts itself as a Moslem Bosnian accused of sniping at Serb refugees is shot down by a Serb policeman in a north Bosnian village, 7 May 1992.

the barbaric atrocities of which Bosnia was the scene, the political questions posed there provided a bitter reminder of the deep historical problems still left unsolved as the euphoria of 'liberation' faded in the east, and the end to history which some had hailed revealed itself for the illusion it had always been.

CONCLUSION

In the middle of 1993 the world's future could hardly be said to look assured. Nor is any end to human suffering now in sight, or any ground for believing it should be. Historians, fortunate that they are not in the prediction business, are no doubt unsurprised by such facts. They are familiar with the truth that however startling may be some of the ways in which history unrolls, surprises always turn out to have histories. Their roots are worth study; they help to make apparent discontinuities explicable, logically related to what came before them. Perspective and the passing of time help them to become more clearly a part of history.

This needs to be remembered more than ever in so rapidly a changing world as we now live in. It is not just the disappearance of an effective dictator, nor the actions of partisans and occupation forces in the Second World War, nor (to go back a little further) Habsburg policy towards subject peoples, nor even the burgeoning of nationalist ideology and mythology since the French Revolution which explain the present torments of the peoples of a short-lived Yugoslavia. Centuries of Ottoman rule, the medieval rivalry of Catholic and Orthodox, even, perhaps, the antagonism of Frank and Slav are also part of the story. On the other side of the globe, Asia is only faintly beginning to experience the impact of the emergence of China from a temporary eclipse to resume a cultural and political importance shaped by millennia of historical experience. As for global phenomena such as population pressure, environmental strain, the upheavals of traditional societies having to adapt to an ever-accelerating technological disruption, there is not one of them for whose understanding history is not the beginning of wisdom.

That, though, does not mean arrival at solutions. What we can do to confront such huge challenges is also shaped by history and there is no limitless room for manoeuvre. History, too, can be misunderstood – even misrepresented – and disaster can follow from that. Bad history is a dangerous master. But we have never needed sound history more.

ILLUSTRATION SOURCES

CHAPTER 1

2-3 Prehistoric/Acheulean flint hand axes, Swanscombe, England, c 1,500,000 BC. © Michael Holford, Essex

6 From top to bottom: *Australopithecus africanus* from Transvaal, Africa, c 1,500,000 to 60,000 BC; reconstructed skull of *Homo erectus* from China, 1,500,000 to 60,000 BC; *Homo neanderthalensis* from La Chapelle-aux-Saints, France, 60,000 - 35,000 BC; *Cro-magnon*. All photographs C M Dixon, Kent

7 Olduvai Gorge, Tanzania, Africa. © Mary Jelliffe, Ronald Sheridan, (Harrow) London

9 Jawbone of Homo habilis, Tanzania, Africa, 1,000,000 - 2,000,000 million years old. © Ronald Sheridan, (Harrow), London

11 Scrapers and hand-axes. © Cambridge University, Museum of Archaeology and Anthropology, Cambridge

13 Harpoon points, Magdalenian period. © Ronald Sheridan, (Harrow), London

15 Pre-historic flint tools, Mousterian period. © Michael Holford, Essex

19 Bronze age chisel from Hollingbury, England. © Ronald Sheridan, (Harrow), London

20 Wall painting from a cave in Lascaux, Dordogne. © Arch.Phot. Paris/ S.P.A.D.E.M.

21 Cave at Lascaux, Dordogne. © Arch. Phot. Paris/S.P.A.D.E.M.

25 The Round Tower at Jericho, Israel. © Ronald Sheridan, (Harrow), London

28 Prehistoric bone needs, Avebury, England. © Ronald Sheridan, (Harrow), London

30 Copper and bronze treasure (part of). © Ronald Sheridan, (Harrow), London

32 Reconstruction of a mammoth hunter's house made of mammoth bones, c 20,000 BC at Mezhirich, Ukraine. Photograph C M Dixon, Kent

CHAPTER 2

34-35 Early Babylonian, c 2650 BC, steatite carved bowl, possibly Elamite origin. Photograph C M Dixon, Kent

39 Scene at Sumerian dairy. © The British Museum, London

41 Mesopotamian cylinder seal, c 2300 BC. © Michael Holford, Essex

42 Limestone tablet found in Mesopotamia, dated 3500 BC. © Ashomolean Museum, Oxford

44 Mesopotamian statues found at Tell Asmar, Iraq. The Oriental Institute, University of Chicago, Chicago

45 The great ziggurat at Ur, Iraq. © Roger-Viollet, Paris

46 Statue of Gudea, carved out of diorite, third millenium BC. © Michael Holford, Essex

49 Statue of Ibihil, a steward, found at Mari, Syria. © R M N, Louvre, Paris

51 Tomb painting from Thebes, Egypt. © The British Museum, London

54 The pyramid at Saqqara, Egypt. © Barnaby's Picture Libary, London. Photograph Gerald Clyde

56 Painting on the tomb of Pharaoh Rekhmaza, Eighteenth Dynasty. Photograph Andre Held, Paris

58 Musicians and dancers, Egypt. © The British Museum, London

59 The colossi of Memnon at Thebes, Egypt. © Ronald Sheridan, (Harrow), London

61 Akhnaton or Amenhotep IV, New Kingdom. © Ronald Sheridan, (Harrow), London

62 Mortuary temple of Ramses III, Exterior detail of the sea battle with the 'sea-peoples', New Kingdom, Dynasty XX. The Oriental Institute, University of Chicago, Chicago

64 Ruins of Mohenjo-Daro, 3000 BC, Pakistan. Photograph J Allan Cash Ltd, London

66 Soapstone figure from Mohenjo-Daro, Pakistan. The Government of Pakistan, Department of Archaeology

70 Lid of clay pot from pre-Shang Honan, China. © Museum of Far Eastern Antiquities, Stockholm

73 Oracle on tortoise shell, China. Institute of History and Philology, Academia Sinica

74 Fragments of a stone mould for casting, 300 BC, China. © Ronald Sheridan, (Harrow), London

77 Pottery model of a watchtower, Han period, 100-200 AD. Photograph C M Dixon, Kent

CHAPTER 3

80-81 The Dolphin fresco, decoration on the wall of the Queen's room at the palace of Knossos, Crete. A typical Minoan decoration around the door are the rosettes, late Minoan, c 1450-1400 BC. © Michael Holford, Essex

83 Incense burner stand, Phoenician, Episkopi, Cyprus, c 12th century BC. © Michael Holford, Essex

85 Reconstructed fresco at Knossos showing the Minoan distinctive 'bull-leaping'. Photograph C M Dixon, Kent

86 Theseus and the Minotaur, fourth century BC, Greece. © Ronald Sheridan, (Harrow), London

88 Painted vase from Gournia, Minoan neo-palatial period, 1700-1450 BC. © Ronald Sheridan, (Harrow), London

90 Statue of Cretan woman. © Ronald Sheridan, (Harrow), London

91 The Treasury of Atreus, Mycaenae, Greece, 14th century BC. Bernard Cox, The Bridgeman Library, London

92 Silver bowl from Cyprus. © The British Museum, London

94 Hittite cunieform, Turkey. © Ronald Sheridan, (Harrow), London

95 Fresco at Collegiata in north aisle, Duomo, San Gimignano, Italy by Bartolo di Fredi, 1356. Photograph C M Dixon, Kent

97 Jewish king bowing to Assyrian ruler. © The British Museum, London

99 David spies on Bathsheba from a Life of the Holy Virgin Mary in The British Library, London. Photograph the Bridgeman Library, London

100 Soliders with slings from an Assyrian relief from the Palace at Niniveh, Iraq, 650 BC. Photograph e.t. archive, London

101 Felt wall-hanging, applique of a rider and seated person from Altai,Siberia, 5th–4th c BC in the Hermitage, St Petersburg, Russia. Photograph C M Dixon, Kent

102 Tomb of Cyrus the Great, Pasargadae, Iran. © Ronald Sheridan, (Harrow), London

104 Persepolis, Iran. © Ronald Sheridan,(Harrow), London

106 Cup showing merchant ship and warship, Athens, Greece, 540 BC. Photograph Michael Holford, Essex

107 Pot by Exekias of Athens. Photograph Staatliche Antikensammlungen und Glypothek, Munich

110 Statue from the Acropolis. © V and N Tombazi. Courtesy the National Tourist Organization of Greece, Athens

111 Hector and Achilles, fifth century BC vase, Greece. © The British Museum, London

112 Dionysus on a pot of about 500 BC. Staatliche Antikensammlungen und Glypothek, Munich. Photograph Christa Koppermann

113 Statue of youth, found in Milos, about 550 BC. Courtesy the National Tourist Organization of Greece

116 Etruscan couple on sarcophagus, about 500 BC. The Mansell Collection, London

117 Greek warrior (hoplite), 5th c BC, Greece. © Ronald Sheridan, (Harrow), London

118 The Parthenon, Athens. Photogrpah Werner Forman Archive, London

121 Girl reading a scroll, 4th–5th c BC, Greece. © Ronald Sheridan, (Harrow), London

122 Bust of Thales. The Mansell Collection, London

123 Bust of Socrates (469–399 BC), after Lysippus, Ephesus Museum, Turkey. Photograph The Bridgeman Library, London

125 The Erechtheum, Athens. The Mansell Collection, London

127 Spartan pot, Greece. Bibliotheque Nationale, France

128 Strygil held by young Greek athlete. Photograph C M Dixon, Kent

CHAPTER 4

130–131 Colosseum, Rome, 80 AD. © Ronald Sheridan, (Harrow), London

133 Part of a mosaic from Pompeii depicting Alexander the Great in the National Museum of Naples. Photograph Andre Held, Paris

136 The Venus de Milo, Louvre. Photograph Roger-Viollet, Paris

138 Bronze vessel used for cremated remains, about 500 BC. Photograph Museo Civico Archeologico Bologna

140 Bronze head, Gallic chief. Berne Historical Museum

143 Julius Caesar, Fotomas Index, Kent. Propaganda coin. © The British Museum, London

144 Onyx gem, Italy. Aufnahme des Kunsthistorischen Museums, Vienna

146 The Pontius Pilate Stone, found at Caesarea. © Ronald Sheridan, (Harrow), London

151 Roman relief showing a sepulchre being built. Photograph C M Dixon, Kent

152 Roman aqueduct in Segovia. Photograph J Allan Cash Ltd, London

153 Reconstructed theatre built 2nd c BC, Cyprus. © M & J Lynch, Ronald Sheridan, (Harrow), London

154 The Arch of Titus in the Forum, Rome. The Mansell Collection Ltd, London

155 Masada, Herod's The Great's fortress above the Dead Sea. Photograph Weidenfeld Publishers Ltd, London

157 Sassanian silver disk showing King Shapur II (241–272 AD) hunting stags. Photograph C M Dixon, Kent

162 The head of the Emperor Constantine, part of a huge seated statue. Photograph Hirmer Fotoarchiv, Munich

163 Stilicho, the son of a Vandal and last general of the Western Roman Empire. Photograph Hirmer Fotoarchiv, Munich

165 Leo the Great meets Attila the Hun in a tapestry. © Ronald Sheridan, (Harrow), London

CHAPTER 5

168–69 Courtyard of the Ibn Tulin mosque, Cairo, Egypt, 9th c. © Ronald Sheridan, (Harrow), London

171 Mosaics from the Church of San Vitale, Ravenna. Courtesy the Italian State Tourist Office, London

173 Hagia Sophia, looking towards the east end from the West. © A F Kersting, London

174 St Simeon Stylites sitting on his column. Photograph Werner Forman Archive, Fruhchristlich-Byzantinische Sammlung, Berlin, Germany

175 Barkeinni psalter. Weidenfeld (Publishers) Ltd, London

178 Koran, by Muhammad Husayn, two circular medallions with dedication to a Sultan of Bokhara. Photograph e.t. archive, London

181 Caravan of pilgrims on the road to Mecca, British Library. Photograph The Bridgeman Library, London

183 Leaf from manuscript of *Automata* (treatist on mechanical devises) by Abu 'l'Izz Isma'il al Jazari, colours and gilt on paper, probably Syrian, 1315, Mamluk School. The Metropolitan Museum of Art, Bequest of Cora Timken Burnett, 1957, New York

184 The Mosque of Cordoba, Spain. Instituto Amatller de Arte Hispanico. MAS, Barcelona

185 Illumination from a book on chess, Spain. MAS, Barcelona

188 Psalter for Basil II, eleventh century. Hirmer Fotoarchiv, Munich

191 Detail of Scythian gold spherical cup, Kul Oba, 300–500 AD. Hermitage Museum, St Petersburg. © Werner Forman, London

193 Mongol warrior, Persian painting. Topkapi Palace Museum, Istanbul

194 Timur Lang's tomb, Samarkand, Uzbekistan. Robert Harding Associates, London

197 Mehmet II, by Gentile Bellini, 1480. National Gallery, London

200 Suleiman the Magnificent, Turkish painting. Topkapi Palace Museum, Istanbul

CHAPTER 6

202-203 Scene from the Kabuki play "Chushingura", the 47 heroes attack Moronao's castle, woodblock print in Maidstone Museum and Art Galley, Kent, England. Photograph The Bridgeman Library, London

205 Bronze Buddha, Gupta period, 6th c. AD, National Museum of Delhi, India. Photograph The Bridgeman Library, London

209 Descent of the Ganges, rock carvings at Mahabalipuran, 7th c, India. © John P. Stevens (Harrow), London

211 Illustration to Muhammad Juki's Shahnameh, battle scene between the forces of Persia and Turan and Moghul rulers, mid seventeenth century. Photograph The Bridgeman Library, London

213 The Great Wall of China. © Barnaby's Picture Library, London. Photograph Hubertus Kanus

216 Bronze figure of horse, Eastern Han Dynasty, second century, China. Robert Harding Associates, London

217 Earthenware fishmonger, tomb artefact, Eastern Han dynasty, 25–220 AD, China. Photograph The Bridgeman Library, London

219 Portrait of Confucius, anonymous, 17th c, China in the Bibliotheque Nationale, Paris. Photograph The Bridgeman Library, London

220 Painting of an examination, eighteenth century, China. Bibliotheque Nationale

222 Temple of Heaven, Beijing, China. Sally and Richard Greenhill, London

224 Armoured man on a horse, T'ang figure, 618–906 AD, China, in the British Museum. © Michael Holford, Essex

226 Scholars on a scroll, China. Collection of the National Palace Museum, Taiwan, Republic of China

228 Mongol archer on horseback, seals of the emperor Ch'ien Lung and others, Ming dynasty, 1368–1644 in the Victoria and Albert Museum. Photograph The Bridgeman Library, London

230 Minamoto Yoritomo, Japan. Sakamoto Photo Research Laboratory, Tokyo

232 Japanese swordmanship, Japan. © The Board of the Trustees of the Victoria & Albert Museum, London

CHAPTER 7

234–35 Bayeux Tapestry. English spearmen defend a hill, Battle of Hastings, 1066. © Ronald Sheridan, (Harrow), London

238 Ivory of St Gregory, tenth century. Aufnahme des Kunsthistorischen Museums, Vienna

239 The Skellig, Kerry. Bord Faithes, Dublin, Ireland

241 Codex Vigilanus, El Escorial, Madrid. MAS, Barcelona

242 Saxon barbarian warrior, 700 AD. Archiv fur Kunst und Geschichte, Berlin

243 Portrait of Charlemagne from the codex de Archivo Capitolase zi Modena. Photograph e. t. archive, London

246 Psalter, 9th c, made in France, Wurrtemberg Landesbibliotek. Bildarchiv Foto Marburg

249 Psalter, Emperor Otto III, Staatsbibliothek, Munich. Hirmer Fotoarchiv, Munich

251 Viking ship, Norway. Fotograph Mittet, Oslo

252 The Witanagemot and king, Anglo-Saxon, England. The British Library, London

255 St Etienne, Nevers, France. Photo Jean Roubier

256 Miniature of King Edgar, tenth century, England. The British Library, London

259 Psalter map, 13th c. Photograph e.t. archive, London

260 Caerphilly castle, Wales, 13th c. © Ronald Sheridan, (Harrow), London

261 Ivory of the martyrdom of St Thomas á Becket, English, circa 1400. The Metropolitan Museum of Art, The Cloisters Collection, Purchase, 1970, New York

263 Pope Honorius III. © Ronald Sheridan, (Harrow), London

265 Mozarabic art, eleventh century illuminated manuscript. The British Library, London

268 Moscow, the Kremlin: belfry of Ivan the Gret (Ivan III), built 1505–1600. Photograph C M Dixon, Kent

269 Engraving of Ivan the Terrible (IV), 1530–84. Photograph e. t. archive, London

272 Manuscript of warfare, fifteenth century, English. The British Library, London

274 The Nave of Ely Cathedral, built by Norman monk Simeon after the invasion, completed in 1189. © Michael Holford, Essex

276 December in the fifteenth century, English. The British Library, London

CHAPTER 8

278–79 Map of Columbus' discoveries. New York Public Library, New York

282 Vasco da Gama. The British Library, London

283 Magellan's discovery of the straits. Fotomas Index, Kent

286 African, Nok head, northern Nigeria, terracotta, 5th–1st c BC in the British Museum. © Michael Holford, Essex

287 Inscribed votive tablet of red slate from temple at Meroe, second century BC, Africa. © Werner Forman, London

289 Salt cellar in Ivory from Benin, Africa. Werner Forman, London

292 Mayan relief. © The British Museum, London

294 Macchu Pichu, Peru. © Barnaby's Picture Library, London. Photograph Peter Larsen

295 Mixtec manuscript. By permission of the trustees of the British Museum, London

297 Chichen Itzá, Mexico, the pyramid of Kukulcan crowned by the temple dedicated to Quetzalcoatl, Toltec/Maya, 10th–12th c AD. Photograph Werner Forman Archive, London

300 The Portuguese arrival in Japan, 16th c. © Ronald Sheridan, (Harrow), London

302 John Smith captures the King of Pamaunkee. The Mansell Collection, London

303 Painting of William Penn in North America by Benjamin West. The Mansell Collection, London

306 Himeji castle, Kyoto, Japan, 1609. Photograph The Bridgeman Library, London

308 Taj Mahal, Agra, India. © Barnaby's Picture Library, London. Photograph Hubertus Kanus

309 Painted cotton hanging of Indians' vision of Europeans, Shahibag, Ahmedabad, India. Calico Museum of Textiles, Sarabhad Foundation, India

CHAPTER 9

310-311 British defeat Spanish and Bavarian allies at Ramillies, Spanish Succession War, 23 May 1706. Photgraph Mary Evans Picture Library

312 The Plague at Tournai, 1349. Photograph The Bridgeman Library, London

315 Sir Charles Morgan presents William IV (1765–1837) with a shorthorn bull at Tredegar Castle, Monmouth, Wales in a lithograph by J H Carter, 1836. Collection Dr C I Davenport Jones, London. Photograph The Bridgeman Library, London

316 'Haymaking -July' by Pieter Brueghel the Elder, (1515–1569). Photograph the Bridgeman Library, London

319 Titian painting of Charles V, El Prado, Madrid. The Mansell Collection, London

322 An engraving of the Frankfurt ghetto, Germany, seventeenth century. The Mansell Collection, London

323 Charles V and Pope Clement VII, Rome. The Mansell Collection, London

325 Luther by Cranach, circa 1525. The Mansell Collection, London

327 Rubens painting of Ignatius Loyola, founder of the Jesuits. The Mansell Collection, London

328 Medal struck to celebrate the defeat of the Armada, 1588. The Mansell Collection, London

329 The horrors, cruelties and sufferings of the Thirty Years War from "Les Miseres et les Malheurs de le Guerre" by Jacques Callot (1529–1635), in the British Museum. Photograph e. t. archive, London

331 Louis XIV playing billiards with the Duc de Chartres, the Count of Toulouse and the Duc de Vendome. Photograph The Mansell Collection, London

333 Elizabeth I of England by M Gheeraerts. The National Portrait Gallery, London

334 Medallion of Oliver Cromwell by Thomas Simon, 1650. National Portrait Gallery, London

335 Portuguese soldier carrying a matchlock gun, wearing 16th c dress in the British Musuem. © Michael Holford, Essex

340 Monument of Peter the Great at St Petersburg,Russia. © The Hulton Picture Company, London

CHAPTER 10

342-43 Joseph Wright's painting "The Air Pump". The National Gallery, London

344 Erasmus of Rotterdam. The Mansell Collection, London

346 Engraving of The Royal Observatory, Greenwich, London in 1675. The Science Museum, London

347 Title-page of Vesalius' work on human anatomy. The Mansell Collection, London

349 English Freemasons, eighteenth century. The Mansell Collection, London

350 Illustration from Harvey's treatise on the circulation of blood. The Mansell Collection, London

352 Title-page of French business manual, seventeenth century. The Mansell Collection, London

354 Drake's route to the Caribbean. Fotomas Index, Kent

355 Gravestone of black servant, England. Christine Vincent, Hertfordshire

357 Captain Cook's men shooting sea horses. Christine Vincent, Hertfordshire

361 Catherine the Great, painting at the Winter Palace, St Petersburg, Russia. The Mansell Collection, London

363 Engraving of the Boston Massacre, Paul Revere, 1770, Boston, USA. The Metropolitan Museum of Art, Gift of Mrs Russell Sage, 1909, New York

364 The Declaration of Independence, USA. The Mansell Collection, London

367 The Estates General at Versailles, France, 5 May 1789. The Mansell Collection, London

368 David's painting of Napoleon, France. The Mansell Collection, London

369 Delacroix's painting of Liberty, Paris, France. Louvre, Paris

371 Voltaire by Hubert Carnavalet. The Mansell Collection, London

374 Prince Metternich. The Mansell Collection, London

376 The Congress of Vienna, 1815. The Mansell Collection, London

CHAPTER 11

378–79 The opening of the Stockton-Darlington railway, 27 September 1825. Photograph Mary Evans Picture Library, London

380 Jeremy Bentham. Photograph Helicon Publishing Ltd, Oxford

381 George Cruikshank drawing, 'Taking the Census', 1851. Mary Evans Picture Library, London

384 The Maxim gun, England. National Army Museum, London

385 Russian workers in St Petersburg, Russia, 1890s. © Novosti Press Agency, London

387 The mechanical reaper, North America, nineteenth century. Science Museum, London

390 Coalbrookdale, Shropshire, 1758. © The Hulton Picture Company, London

391 Print of a soup-kitchen, England, nineteenth century. Christine Vincent, Hertfordshire

392 Mill in 1840, Silesia, Germany. Mary Evans Picture Library, London

393 Evicted British labourers, 1870s. Photograph Institute of Agricultural History and Museum of English Rural Life, University of Reading

394 Mr and Mrs Karl Marx, London. The Mansell Collection, London

396 Isambard Kingdom Brunel viewing his last ship, *The Great Eastern*. Photograph Institute of Agricultural History and Museum of English Rural Life, University of Reading

397 The 'Free Trade' hat, England. The Mansell Collection, London

398 Engraving of well-fed British lion of free trade, London. Fotomas Index, Kent

400 Cotton Mill, c 1830. Mary Evans Picture Library, London

402 Great Exhibition, Crystal Palace, London, England, 1851. Mary Evans Picture Library, London

CHAPTER 12

404–405 The British in India, the Judge, c 1860. Photograph Mary Evans Picture Library, London

406 Christian missionaries in Africa. Bodleian Library, Oxford

409 Mungo Park. Photograph The National Portrait Gallery, London

410 Early Sydney, Australia. © The Hulton Picture Company, London

411 Captain Roald Amundsen (1872–1928). Popperfoto, Northampton

414 Boer fighters, 1899, South Africa. © The Hulton Picture Company, London

419 Europe's cultural superiority as depicted in an 1890s advertisement. Christine Vincent, Hertfordshire

421 The Buffalo Hunt under the wolf skin mask, 1832–33 by George Catlin in National Museum of American Art, Smithsonian. Photogaph The Bridgeman Library, London

424 Cotton pickers, USA, nineteenth century. Culver Pictures Inc, New York

425 *Punch* cartoons. Punch Publications Ltd, London

429 Destruction of the US Battleship *Maine* in Havana Harbor, 1898. Chicago Historical Society, Chicago

430 The *Missouri* is the first battleship to pass through the Panama Canal. © The Hulton Picture Company, London

433 English cartoon of the reception in at the court in China of Lord Macartney's deputation, 1792. Fotomas Index, Kent

435 General Gordon's Bodyguard trained on half of the Chinese government to resist the Taipings, China, 1860. John Hillelson Agency, London. Photograph Felice Beato

437 *Punch* cartoon on an anti-Boxer rebellion theme, 1900. The Mansell Collection, London

439 A young Indian prince and his advisers. Oriental and India Office Collections, The British Library, London

441 King George V and Queen Mary show themselves to the people, Delhi, Durbar, India, 1911. © Popperfoto, Northampton

444 Japanese parliament house, 1891. © The Hulton Picture Company, London

444 The first railway station in Japan, wood block print, 1872. Christine Vincent, Hertfordshire

446 War in the Crimea, attack on Malakoff and Little Redan, Russia. John Hillelson Agency, London

447 *Punch* cartoon showing the frightened rulers in 1848. Hulton Deutsch Collection.

448 Giuseppe Mazzini. The Mansell Collection, London

450 Alexander II of Russia (1818–1881). © Popperfoto, Northampton

453 German postage stamp of the coronation of Kaiser William II. Edimedia, Paris

CHAPTER 13

454–55 The Wooley family of Bolton looking at new TV set, 1950. © Popperfoto, Northampton

460 Famine in Russia, post-Revolution. © Roger-Viollet, Paris

463 The assembly line of a Nissan car factory showing robots at work. © Picturepoint, London

464 Marconi in Newfoundland listening to transatlantic signals on 12 December 1901. © Popperfoto, Northampton

466 Charles Darwin, a photograph by Captain Darwin. The Mansell Collection, London

468 Reactionaries marching in Odessa, Ukraine, 1905. © The Hulton Picture Company, London

470 Letter to President Franklin Roosevelt from Albert Einstein, 1939. Photograph courtesy of Franklin D Roosevelt Library, New York

473 'Goodbye comrades' – Yuri Gagarin waves goodbye before entering space capsule which took him around the world, 15 April 1961. © Topham Picture Source, Kent

475 Man on the moon, 1969. Photograph NASA, Houston, Texas, USA

477 Woman chaining herself to the railings of Buckingham Palace, 1914. © Popperfoto, Northampton

479 A typical ATS girl, 10 October 1939. © Popperfoto, Northampton

CHAPTER 14

482–83 Germany's first 35,000 ton battleship, the 'Bismarck' takes the water watched by Adolf Hitler, 14 February 1939. © Popperfoto, Northampton

486 The last photograph of Archduke Franz-Ferdinand and his wife, taken as they were leaving the Town Hall, Sarajevo, 28 June 1914. © Popperfoto, Northampton

488 A group of De Havilland long-distance bombing machines, Serny Aerodrome, France, 1918. Photograph the Imperial War Museum, London

490 French tank, First World War. The Imperial War Museum, London

490 Middlesex battalion, 1916, France. © Popperfoto, Northampton

494 Lenin, Krupskaya, Zagorsky, Litvinov, Ludvinskaya in Red Square, May Day, 1919. © Novosti Press Agency, London

497 Benito Mussolini, 1935. © Popperfoto, Northampton

498 Rosa Luxemburg. © Helicon Publishing Ltd, Oxford

503 Pu-Yi, the last emperor of China, 1911. The Mansell Collection, London

504 The Japanese capture of Tientsin in China, 1894, from a contemporary print. © The Board of the Trustees of the Victoria & Albert Museum, London

504 Chian K'ai-shek on Black Dragon, 1929. © Barnaby's Picture Library, London

506 Japanese marines behind sandbags in China, 1932. © Popperfoto, Northampton

508 Franklin Delano Roosevelt, Warm Springs, Georgia, USA, 1939. © Associated Press, London

510 Hitler and the Party – massing of the standards at Nuremberg, 1934. © Barnaby's Picture Library, London

514 Nazi atrocities in Russia, 1942. © Novosti Press Agency, London

515 The attack on Pearl Harbour, 1941. © Popperfoto, Northampton

516 The Battle for Fort Dufferin, Burma, Second World War. © Popperfoto, Northampton

518 The Red Banner of victory over the Reichstag, 1945. © Novosti Press Agency, London

520 Churchill, Stalin and Truman at the Potsdam Conference in 1945. © Popperfoto, Northampton

CHAPTER 15

522–23 Student demonstration, Paris May 1968. Photograph: Sipahioglu, © Sipa-Press/Rex Features, London

524 Soviet vetoes Berlin peace proposal, 26 October 1948. Photo by Rene Helier, © Popperfoto, Northampton

527 Aftermath of Czech crisis-police and militia parade in Prague, 1948. © Hulton Picture Company, London

529 USAF bombers over Korea. Robert Hunt Library, London

533 Raising the flat in 1948 in Israel. © Israel Defense Department, Tel Aviv

534 Nasser greets the masses, Cairo, 1954. © Popperfoto, Northampton

539 The Union Jack makes way for the new country of Botswana's flag, 1966. © Camera Press, London

540 Haile Selassie, 1974. © Popperfoto, Northampton

545 Poster which introduced the Cultural Revolution in 1966, it shows Mao inspiring the people. © Camera Press, London

547 President Sukarno of Indonesia addressing the opening session of the Bandung Conference, 18 April 1955. ©Popperfoto, Northampton

548 Gen Douglas MacArthur. © Helicon Publishing Ltd, Oxford

549 Prague, Czechoslovakia, 1968. © Magnum Photos, London

551 Fidel Castro in 1959 in Havana, Cuba. © Popperfoto, Northampton

552 Missile site in Cuba, 1962. © Popperfoto, London

553 Russian leaders on official visit to Finland, 10 June 1957, Bulganin and Khrushev and Gromyko. © Popperfoto, Northampton

554 Hundreds of new cars await delivery at General Motor's storage area in New Jersey, USA. © Camera Press, London

555 Black demonstrators in Washington D.C. 1963. © Topham Picture Source, Kent

556 US soldier in Vietnam with woman and child. © Camera Press, London. Photograph James Pickerell

558 Mrs and Mrs Richard Nixon in China, 1972. © Topham Picture Source, Kent

559 An effigy of the Shah is burned in Tehran, 1978. © Camera Press, London. Photographer Michel Giannoulatos

562 French Jaguar fighter during bombing of a munitions dump in Kuwait, January, 1991. © Popperfoto, Northampton

564 Salvador Allende, 1973 in Chile. © Topham Picture Source, Kent

566 F W de Klerk and Nelson Mandela confer in Davos, Switzerland, 1992. © Popperfoto, Northampton

567 A statue of Liberty is raised in Tienamen Square, Beijing, China, May, 1989. © Popperfoto, Northampton

569 Gorbachev and Reagan share a laugh at the Geneva Summit concluding ceremony, November 1985. © Popperfoto, Northampton

570 Prague, The Czech Republic, 1968. © Magnum Photos, London

572 Lech Walesa addressing the workers in Gdansk shipyard, Poland, 1980. © Popperfoto/UPI, Northampton

573 The Berlin Wall comes down near the Brandenberg Gate in East Berlin, Germany, February, 1990. © Camera Press, London

577 Moslem sniper accused of firing on Serb convoy in northern Bosnia is executed by local policeman, 7 May 1992. © Popperfoto, Northampton

COLOUR PLATES

Tutankhamon (Ronald Sheridan (Harrow), London); Metal-Working and Art, Hittite ewer (Michael Holford, Essex); Battersea shield (C M Dixon, Kent); Easter Island (Werner Forman Archive, London); The Medieval World Order (SCALA, Florence); the Samurai tradition and Benin (Michael Holford, Essex); India (E T Archive, London); DNA (Science Photo Library, London).

INDEX